Subcellular Biochemistry

Volume 111

The book series SUBCELLULAR BIOCHEMISTRY is a renowned and well recognized forum for disseminating advances of emerging topics in Cell Biology and related subjects. All volumes are edited by established scientists and the individual chapters are written by experts on the relevant topic. The individual chapters of each volume are fully citable and indexed in Medline/Pubmed to ensure maximum visibility of the work.

Akihiko Nakano • Jaakko Saraste
Editors

The Golgi Network, Volume II

Golgi Functions

Springer

Editors
Akihiko Nakano
RIKEN Center for Advanced Photonics
Wako, Saitama, Japan

Jaakko Saraste
Department of Biomedicine
University of Bergen
Bergen, Norway

ISSN 0306-0225 ISSN 2542-8810 (electronic)
Subcellular Biochemistry
ISBN 978-3-032-16832-0 ISBN 978-3-032-16833-7 (eBook)
https://doi.org/10.1007/978-3-032-16833-7

This Springer imprint is published by the registered company Springer Nature Switzerland AG
The registered company address is: Gewerbestrasse 11, 6330 Cham, Switzerland

Preface

In 1898, Camillo Golgi discovered a novel structure in the cerebellar Purkinje cell, which had the appearance of an extensive reticulum around the nucleus. Now, after 127 years have passed, it is again timely to ask what we know about this still mysterious "Golgi apparatus." The title of this book "The Golgi Network" is symbolic. It could refer to the structure of this organelle, which often looks reticular and network-like. The word network itself is indeed used in the case of the *cis*- and *trans*-Golgi networks. The two important main functions of the Golgi apparatus are: (1) glycosylation, which involves an intricate network of enzymes involved in glycan synthesis and modification, and (2) protein and lipid sorting, which critically regulates the complex cellular network of trafficking. Research in these fields is making tremendous progress thanks to the state-of-the-art technologies developed in the past few decades. Human networking, interaction among researchers, has increasingly helped in bringing together such novel techniques and collaboration to better tackle the many questions on this organelle that remain unanswered. The book is divided in two volumes, I. Golgi Organization and Dynamics and II. Golgi Functions. A variety of topics have been chosen to introduce the readers to the frontiers of advanced Golgi research, hoping that this collection will give an update on current ideas regarding many aspects on the Golgi apparatus and the organelles it closely interacts with, thereby providing a valuable resource for researchers and students.

Wako, Saitama, Japan — Akihiko Nakano
Bergen, Norway — Jaakko Saraste

Overview for Volume II

Volume II of the "The Golgi Network" comprises a collection of chapters that address various aspects of Golgi function. To elucidate the fundamental molecular mechanisms of protein trafficking, Chaps. 1, 2, 3, 4, 5, 6, and 7 focus on the transport machinery, including coat/adaptor proteins (Chaps. 1, 2, 6, and 7), the golgin family of coiled-coil proteins (Chap. 4), and Rab and Arf GTPases (Chaps. 3 and 5). Lipid trafficking is also discussed (Chap. 8), along with the introduction of a concept of sorting zones (Chap. 9). When it comes to research on the process of glycosylation, including glycan synthesis and modification, the structural and reactional diversity presents significant complexity. Comprehensive omics projects addressing these issues are currently underway worldwide. We have opted to exclude detailed descriptions of such large-scale glycome analyses from this volume and instead focus on selected topics closely related to protein trafficking (Chaps. 10 and 11). The remainder of Volume II explores the relationship between microtubules and the Golgi (Chap. 12) and role of the Golgi in innate immunity (Chap. 13). Additionally, several examples of the Golgi's emerging involvement in various diseases are presented (Chaps. 14, 15, 16, 17, 18, and 19). We hope that a deeper understanding of these connections will contribute to the development of effective therapies and enhance our future well-being.

Contents

Part I
Golgi Functions in Traffic

Chapter 1
Beyond Vesicular Traffic: The ERES as the Gateway for Biosynthetic Secretion

Meir Aridor

Abstract Biosynthetic secretion mobilizes about a third of the human proteome from the Endoplasmic Reticulum (ER) in route to cellular and extracellular destinations. Traffic from the ER to the Golgi Complex begins at distinct sites on the ER, the ER exit sites (ERES), which are functionally and physically connected with the Golgi. ERES are the first sorting sites in the secretory pathway. This chapter explores key molecular functionalities of the ERES.

Keywords ERES · COPII · TANGO1 · ERGIC · Golgi · COPI

Introduction

The point of exit from the endoplasmic reticulum (ER exit sites, ERES) is a key cellular sorting step that governs biosynthetic secretion or degradation of a third of the proteome. The successful reconstitution of vesicular traffic using cell free assays defined vesiculation components including small GTPases and vesicle forming coats. The vesicular traffic model depicting physical separation between organelles allows for compartment identify preservation. However, accumulating evidence suggest that cytosolic coating machinery involved in vesiculation is integrated within molecular scaffolds that define the fate of cargoes concentrating in the ER and direct vesicular and/or non-vesicular direct traffic between compartments. The ERES gateway also controls selective and global shuttling of cargoes for lysosomal degradation via autophagic pathways. This chapter briefly discusses the makeup of ERES and mechanisms by which ERES-scaffolds maintain compartment identities while supporting direct connectivity between compartments to facilitate secretion or degradation.

M. Aridor (✉)
Department of Cell Biology, University of Pittsburgh School of Medicine, Pittsburgh, PA, USA
e-mail: aridor@pitt.edu

A. Nakano, J. Saraste (eds.), *The Golgi Network, Volume II*, Subcellular Biochemistry 111, https://doi.org/10.1007/978-3-032-16833-7_1

The ER Exit Sites

Camilo Golgi defined the themed cellular organelle morphology, and begun an era that led to the meticulous documentation of the biosynthetic traffic pathway its morphology, membrane organization and function. The introduction of electron microscopy early on uncovered the remarkable organization of the Golgi in flat polarized cisternae [139]. Detailed characterization of traffic moving from cis to trans and the TGN faces of the compartment was parallelled with the documentation of vesicular profiles and buds on cisternae rims. The analysis was correlated with cell free reconstitution studies that resulted in the isolation and characterization of vesicle intermediates [118]. However, the analysis also uncovered tunnels connecting cisternae that support biosynthetic traffic and a role for vesicular and tubular traffic in anterograde or retrograde directions is still debated as is the nature and directionality of traffic for each mode. Models of cisterna progression accompanied by retrograde traffic are considered [64, 142]. Regardless, the molecular principles for vesicle formation have been deduced. Similarly, the mode of traffic and role for vesicles budding of the Endoplasmic Reticulum (ER) has been a major focus in recent years [5, 6]. The feed of biosynthetic traffic to the Golgi begins with the exit of cargo proteins from the Endoplasmic Reticulum (ER) where cargo concentrates at specialized sites or organelles, the ER exit sites (ERES) [16, 17, 68, 75, 200, 207]. The number of these ER portals in cells varies from 200 to 500 [68] (Fig. 1.1a). ERES tend to nucleate at highly curved ER membranes of the reticulum [134, 211]. The ER itself extends throughout the cell and forms multiple membrane contacts with cellular organelles and the plasma membrane [191]. Accordingly, ERES are spread all over the cell, but tend to concentrate near the juxtanuclear region where the microtubular organizing center resides and the Golgi is positioned to assist in nucleating microtubule tracks (Fig. 1.1a). In most cells, ERES partitions with 50%–60% of ERES localizing at the Golgi region [179]. This biased distribution is actively maintained and regulated. Using optogenetics to impose a traffic arrest induces ERES movement from the cell periphery and clustering near the Golgi the, linking ERES distribution with cargo loading and traffic [169]. This re distribution likely functions in optimizing cargo delivery to the Golgi complex. Under unperturbed conditions, ERES are relatively static only showing short range mobility [68]. The sites undergo fusion and fission events that are indicative of liquid-liquid phase separation (LLPS) directed by resident proteins that make up the ERES [23]. ERES redistribute to the cell periphery when microtubules (MT) are depolymerized. Both plus end and minus end directed MT motors associate with ERES suggesting that the quasi-stable localization of ERES is a result of an active tag of war between microtubule motors and it is their regulated association with the sites that define cellular location [11, 31, 67, 198].

Individual ERES are about 400 nm in size and are characterized by concaved membrane [75]. At higher resolution, each ERES is composed of a collection of budding profiles or buds that are clustered to build the ERES. The density of buds varies between an average of 4.4 bud sites at ERES of the cell periphery and 6.1 buds

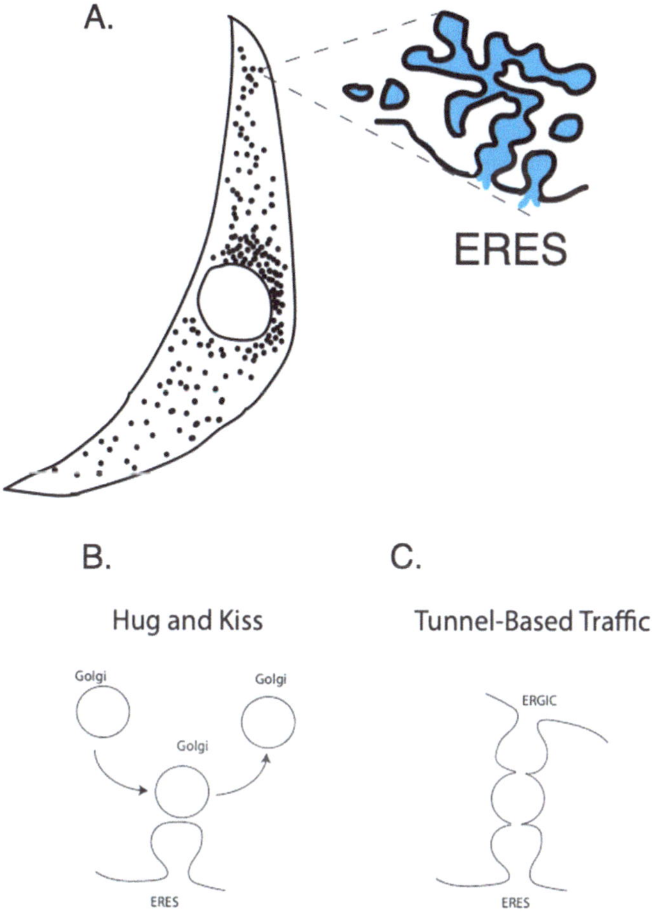

Fig. 1.1 ERES and modes of traffic. (**a**) A model showing a putative fibroblast where the cellular distribution of ERES is shown (dots). ERES are distributed throughout the cell and cluster at the perinuclear region next to the Golgi and MTOC. An ERES is enlarged showing continuities between ERES and tubulation that makes up the ERGIC compartment as well as vesicle budding at this site. (**b**) A model depicting ERES in budding yeast cells, where Golgi vesicles transiently interact with buds on ERES to drive "hug and kiss" based traffic. (**c**) A model depicting non-vesicular traffic at ERES where 3 vesicle profiles are fused to generate a traffic tunnel

at ERES within the Golgi region, supporting a role of ERES localization in directing select traffic events. ERES emanating from ER cisternae with buds extending into central vesicular tubular clusters (VTCs) that make up the ER to Golgi intermediate compartment (ERGIC) of the ER export complex [17]. The introduction of Focused Ion Beam (FIB) milling techniques combined with high resolution SEM or CryoET at nanometer resolution provided a closer look at connectivity and membrane continuities within ERES and united the originally defined membranes of the ER export complex into a single unit of ERES-ERGIC. ERES buds that concentrate a variety of cargoes form tubular connections or tunnels that directly link the ER with the ERGIC-VTCs that make the ER export complex [148, 200]. These tunnels or tubules have a diameter of 40–60 nm and are extended within ERGIC to form tubular assemblies. Yet the utilization of high-pressure freezing and sectioning or CLEM cryo-FIB milling and tomography in a variety of cells demonstrated that free COPII coated vesicles ranging from 53.3 to 91 nm (average size of 70 nm) in size as well as smaller COPI coated vesicles (mean diameter of 55 nm) retrieved from ERGIC and Golgi membranes that also occupy the ERES space [32, 44, 127, 212], suggesting that a dynamic process of vesiculation, membrane contacts and tunnel formation takes place at ERES to accommodate physiological secretion needs [103]. Free COPII coated and uncoated vesicles have also been documented in mammalian cells and nanoscale analysis of ERES in drosophila performed either in imaginary disc or fat cells present similar organization suggesting that the type of cargo mobilized through the sites does not define ERES morphology, which is a constant feature of an ERES organelle. Each drosophila ERES has an average of 3 connecting tubules or tunnels emanating from the ER and linking to a pre-cis--Golgi compartment (the drosophila equivalent of ERGIC, [207]). Tunnel morphology provides information on the mechanism of assembly, with 1–3 connected vesicles make up a single tunnel (Fig. 1.1a). This morphology suggests that a coordinated process of vesicle formation, fission and fusion establish direct connectivity. The proximity of donor acceptor membranes supports a role for direct molecular tethering that holds the membranes together and argues against a role for LLPS in organization, yet lateral clustering may require phase separation of ERES proteins. Furthermore, drosophila ERES contain multiple free vesicles and as such, LLPS may maintain vesicle clustering as observed with synaptic vesicles [207]. Vesicles at ERES in drosophila fall into two distinct size categories likely functioning in anterograde and retrograde traffic within the hub as documented in mammalian cells [44, 127]. Overall ERES appear to support vesicular and direct non-vesicular traffic from the ER and both are driven by a vesicle forming process [103, 173]. The elaborate tubular morphology of ERES supports their role as a major cellular traffic hub, where extended tubulation allows for effective sorting in anterograde, retrograde and degradative directions. The utilization of tubular morphology in what can be defined as "geometrical sorting", where surface to volume ratios and membrane thickness are modified to selectively accommodate transmembrane domains is a common feature of endocytic and secretory sorting sites. The morphology of ERES and cellular distribution define it as the second organelle of the secretory pathway. As with other organelles ERES present a unique and defined molecular composition that is distinct from that of the ER.

The Molecular Composition of an ERES Organelle

If we define ERES as an organelle, it should present unique lipid, protein and chemical composition, ones that differ from the bulk ER. The preservation of defined composition within connected ER and ERGIC membranes collectively defined as ERES, requires specialized mechanisms of sorting that do not rely on spatial and temporal separation between compartments offered by vesicular traffic.

ERES Present Selective Composition of Lipids

Phosphatidyl inositol 4-phosphate (PI4P) dynamically accumulates at ERES-ERGIC and assists in the assembly of ERES and their adaptation to acute cargo load [27, 49, 50]. It is enriched on connected ERGIC membranes [47], supports ER exit and cargo traffic in high eukaryotes and is also required for ER to Golgi traffic in yeast. Budding yeast present undeveloped ERES and lack the stable membrane connectivity that is observed in high eukaryotes yet utilize transient hug and kiss contacts to support ER exit and traffic [27, 30, 49, 95, 104, 190]. PI4P is actively dephosphorylated on ER membranes in a reaction utilized to drive the exchange of incoming PI4P with lipids such as sterols delivered from the ER, providing a driving energy source for the delivery of cholesterol to target membranes against a concentration gradient [119]. Thus PI4P, which is enriched on ERGIC, provides directionality for anterograde traffic moving from the ER. In a similar manner, cholesterol is present at very low levels in ER membranes, yet it is selectively enriched at ERES [200]. Low levels of cholesterol at the ER, its site of synthesis, are critical for physiological sterol sensing that monitors relatively small fluctuations in cholesterol concentration to regulate synthesis and uptake [29]. Cholesterol enrichment at ERES is likely required to provide the bilayer with elastic properties required for membrane remodeling events that support traffic. Cholesterol is required for ER exit [159].

The Protein Composition of ERES

The protein composition of ERES is distinctive and resident proteins are utilized to mark the sites [6]. The repertoire of ERES residents evolved with increased physiological demands. In lower unicellular eukaryotes primitive ERES are marked by cytosolic proteins including coat proteins (COPII) and a scaffold protein Sec16 that localizes to membranes by interactions with a membrane protein Sed4 [211]. The interactions of Sec16 with an oligomerized Sed 4, an inactive homolog of the COPII activating protein Sec12, targets ERES to areas of high membrane curvature on the ER membranes.

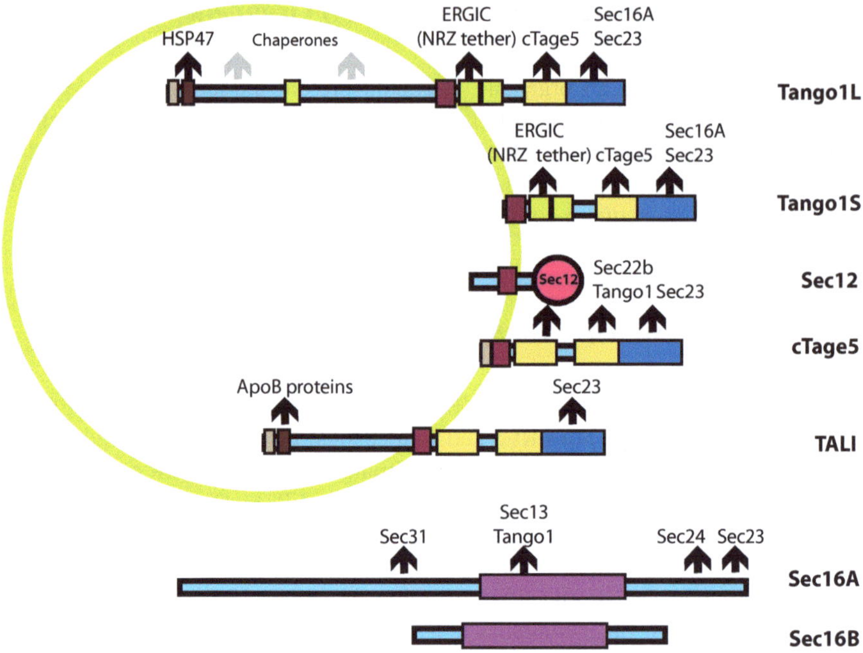

Fig. 1.2 Resident Proteins of ERES. A fence (green) composed of TANGO1 like proteins, recruits Sec12 and Sec16 proteins to assemble the ERES. These proteins interact with cargo receptors and the cytosolic COPII which dynamically regulate the assembly of the fence and drives the exit of proteins from ERES

The biosynthetic machinery of advanced eukaryotes is required to support multicellular organization and intercellular communication. Major classes of clients including nutrient containing apolipoproteins or Extra Cellular Matrix (ECM) components such as collagens, all are too large to be fit and be mobilized in small size vesicles. Fibrillar collagens are integral part of the ECM that anchor cells and organize tissues [80]. Collagens are assemble from three polypeptide chains forming triple helical rod-like structures in the ER with sizes that exceed 300 nm. Similarly, cholesterol obtained by dietary uptake or synthesis in packaged in the ER in apolipoprotein particles that can be 500 nm or larger in diameter [98]. The spatial requirements for mobilization of bulky cargoes, which are major clients of the biosynthetic pathway in mammals, are met by the scaffolding and assembly of elaborated ERES membranes that connect the ER to target ERGIC membranes. The need to build such ERES is met with the evolutionary addition of TANGO1 like proteins including TANGO1L, TANGO1S, cTAGE5 or the apolipoprotein selective TALI [5, 6] (Fig. 1.2). The task of organizing the complex membrane system of the mammalian ERES has therefore expanded beyond Sec16 providing a set of membrane receptors that recruit Sec16 and localize coat assembly at an expanded ERES [6, 154].

TANGO1-like proteins are resident ERES proteins that mark and define ERES and are required for their formation. TANGO1 (Transport ANd Golgi Organization) [160] was discovered by an RNAi-based screen conducted in drosophila S2 cells monitoring Golgi organization and the secretion of a small bulk flow type cargo (i.e. a small sized soluble secreted protein that is less likely to carry a concentrative ER exit motif and is mobilized into budded vesicles by diffusion according to its concentration within the ER) [18].

TANGO1 presents a large cytosolic domain composed of a proline rich disordered domain (PRD) and two coiled-coil domains that are followed by a single transmembrane domain that is extended by a short hydrophobic segment likely embedded selectively in the inner leaflet of the bilayer. The protein extends into the ER lumen with a large intrinsically disordered region (IDR) that is terminated by a structurally distinct SH3 domain [13, 107, 160]. Drosophila express a single TANGO1 isoform whereas in mammals, additional family members were added. TANGO1 has a short isoform (TANGO1S) that lacks the luminal domain. Another family member, cTAGE5, has similar molecular architecture with a PRD and 2 coiled domains in the cytosol and a trans membrane domain, but it lacks luminal parts [161]. In specialized cells in the intestine or liver, an added member dedicated to apolipoprotein secretion is a fusion protein between cTAGE5 and luminal MIA2 [164]. TANGO1L/S and cTAGE5 are complexed together using the second CC domains [107, 108] (Fig. 1.2). Inspection of ERES using super resolution microscopy in drosophila or mammalian cells shows that the proteins form a ring that likely fences ERES [101, 150, 151]. This functional arrangement serves to differentiate the ERES as an organelle from the ER. Importantly, the assembly of ERES, either evolutionary primitive ones as in S. Cerevisiae or advanced metazoans ones are controlled by the recruitment of the cytosolic coat protein complex II (COPII).

The COPII Coat

Morphological characterization of the secretion pathway combined with cell free reconstitution studies led to the conceptualization of the vesicular traffic model [5]. The model addressed two pillars of cellular organelles: First, secretory organelles exhibit distinct morphological features and composition that allow for simple classification. Second, the robust flux of biosynthetic cargoes through these quasi-stable organelles, (about a third of the proteome, accompanied by carrier membranes and machinery proteins) does not perturb organelle identify. Newly synthesized proteins are trafficked from the ER through the Golgi to their cellular and extracellular destinations in vesicles/carriers that bud of donor membranes, tether and fuse with acceptor membranes to deliver cargo while preserving organelle identity. Because of this temporal and spatial separation, it became apparent that vesicular traffic can be reconstituted in the absence of cellular organization. Cells can be broken and the budding of vesicles from donor membranes can be effectively reconstituted with minimally required purified components. Following this logic, all eukaryotic cells including ones that do not present distinct morphological organelle organization would function similarly, allowing for the utilization of genetic screens to identify traffic components [5]. By combining biochemical reconstitution with purified components and organelles with genetic screens, coat protein complexes (COP) that form vesicles were defined, including COPI for ERGIC and Golgi compartments and COPII for initiating exit from the ER [118]. The two identified coats represent two distinct classes of cytosolic coats that govern intracellular traffic. COPI is structurally related to the AP-clathrin configuration of cytosolic coats with structurally homolog adaptor layer that presents a closed (cytosolic) and open (membrane bound) configurations [83, 91, 189]. In contrast, COPII presents mostly rigid adaptor layer that binds membranes [24, 125]. Both COPII and COPI have an outer layer composed of an ACE1 complex, namely alpha solenoids crowned with a WD40 beta propeller domain [51, 202]. In the case of COPII, this layer polymerizes to provide a cage whereas in COPI it coexists as a single complex with the adaptor layer and its beta propellers reach the donor membrane to bind cargo [71, 84, 177, 189]. Both COPII and COPI are utilized to transport newly synthesized proteins from the ER to the Golgi [8]. The activities of the two vesicle forming coats are coupled yet it is COPII that mediates the exit step from the ER.

COPII is composed of five proteins, assembled as three separate cytosolic components (1) a small GTPase, Sar1 (2) an inner layer or adaptor complex composed of Sec23 and Sec24 proteins and (3) An outer layer composed of Sec13 and Sec31 [20, 114].

The small GTPase Sar1

Sar1 is the small GTPase of the ARF family that initiates COPII assembly [129]. It is activated by Sec12, a membrane protein that localizes to ER exit sites [19, 162, 201]. Active Sar1 ejects an elongated amphipathic N-terminus that is embedded selectively in the outer leaflet of the bilayer, initiating membrane curvature and buckling [11, 26, 76, 97, 102]. Molecular dynamic modeling demonstrates that Sar1 assembles minimally in dimers, where spacing of these by the interacting inner layer of COPII defines membrane curvature [145, 146]. The self-assembly of Sar1 in scaffolds that constrict the carrier neck regulates the final fission step as demonstrated for both Sar1 and related Arf family members [102, 140]. The distribution of active Sar1 proteins on ERES membranes depends on cargo size and expands from the ERES collar onto ERGIC membranes potentially facilitating membrane remodeling and tubulation events [109].

The two isoforms of Sar1 in mammals Sar1a and Sar1b are largely interchangeable, yet Sar1b has a selective role in supporting the secretion of apolipoproteins [174, 187]. Sar1b is deficient in assembly with Sec23A mutants that lead to the development of craniofacial disease with collagen secretion defects, perturbing the interface that supports outer layer recruitment [54]. Biosynthetic secretion may be considered a house keeping basic function, but Sar1 levels are tightly controlled. Sar1 proteins are limiting components in cell free assays reconstituting COPII vesicle formation [26]. Sar1a expression is selectively induced by TFG-β1 in hypertrophic scarring fibroblasts modeling fibrosis, where its depletion inhibits collagen 1 secretion [2]. Precise Sar1a levels are critical for collagen secretion and ECM deposition in c. elegance and humans, where the E3 ubiquitin ligases KHLH40a/b target Sar1a selectively for ubiquitination and degradation to control Sar1a levels [113].

The COPII inner layer (Sec23/24)

The cargo selection adaptor complex of COPII, Sec23/24, is recruited to membranes by active Sar1, which forms an extended interface with Sec23 proteins (there are two isoforms in mammals) to assemble the membrane bound pre-budding complex [9, 24, 77]. The complex is a relatively rigid structure only showing minor intermolecular structural changes in Sec23 upon clustering with neighboring Sar1-Sec23 units when densely packed [77]. Sec23 functions as a GAP for Sar1, providing an arginine residue (finger) at the GTP binding site to stabilize a transition state during GTP hydrolysis [3, 24]. Sec24 proteins, of which there are four isoforms in mammals, provide extended cargo binding capabilities using a small number of binding pockets that accommodate a relatively diverse range of signals [106, 111, 112, 120, 121, 125]. The relatively rigid assembly of the Sec23/24 proteins on ER membrane dictates that cargo binding pockets are placed at defined distance from the bilayer. This simple configuration enables selective control of cargo export, where the positioning of ER-exit motives with respect to the membrane and hence presentation for Sec24 binding is regulated. The cholesterol regulated exit of SCAP-SREBP from the ER is an example of how such mechanism is used to provide physiological control over cholesterol synthesis and uptake [29].

However, Sar1, Sec23 and Sec24 all interact with cargoes contributing to the diversity of cargo recognition by the COPII-inner layer [63, 66, 125].

The COPII outer layer (Sec13/31)

The outer layer of COPII is composed of Sec13/31 proteins that polymerizes on top of the coat inner layer to assemble a hedral cage. A platform formed by the membrane bound Sar1-Sec23 complex serves as the binding site for a short non-structured segment within a C-terminal proline-rich domain (PRD) of Sec31, leading to a sequential assembly of the coat [25, 51]. The linkage between the coat layers is aided by multiple charged based interactions between Sec31-PRD and Sec23 and an added binding between Sec23 and triple proline motifs located on the PRD of Sec31 [77, 105, 178]. The interaction of Sec31 with the Sar1-Sec23 platform orients the catalytic histidine of Sar1 to facilitate the Sec23 induced GTP hydrolysis [3, 25]. This two-gear mechanism culminates in destabilization of coat-membrane binding and importantly provides a regulatory step to control coat activities. GTP hydrolysis markedly facilitates the process of membrane fission, when measured on isolated ER membranes or model liposomes [14, 26, 69] yet it is not an absolute requirement for this process [72, 147]. Overall the coat is functional in vesicle biogenesis on ER or synthetic membranes. However, efficiency is relatively low in mammals and effective budding requires additional cytosolic factors [90, 170].

Importantly, COPII assembly regulates the assembly of the TANGO1-cTAGE5 fence [151]. Chemical inhibition of Sar1 activation inhibits overall ERES assembly, with the remaining sites presenting undeveloped and less constricted membranes [200]. Reduced expression of coat components affects the organization of ERES resident membrane proteins, while inhibition of GTP hydrolysis leads to robust clustering of the sites at MTOC [151, 197]. COPI functions in traffic from the ER to the Golgi functioning sequentially following COPII mediated ER exit [8]. COPII assemble on ER buds connecting to the ERES whereas COPI is slightly displaced (300 nm) from the ER and assembles on the distal tubular regions of ERGIC, connected to ERES [122, 173, 179, 200]. It is COPI that coats tubular traffic intermediates that detach from the ER and mobilized to the Golgi. COPI is required for anterograde traffic from the ER to the Golgi but it is not required for the assembly of ERES. Inhibition of COPI activities results in enlarged ERES that concentrate cargo proteins [46, 173]. COPI-driven sorting in the retrograde direction at ERES-ERGIC initiates retrieval at the ER exit step [8]. Thus linked ERES-ERGIC are the first multidirectional sorting station in biosynthetic secretion. In morphological terms, COPI retrograde and ER entry sites surround the site of COPII assembly for exit and inhibiting COPII activities negates this organization [158, 207]. In drosophila, COPI vesicles populate the outer "ring" of ERES while COPII vesicles are found within the ERES perimeter [207].

COPII is an integral part of the ER exit machinery that scaffolds ERES membranes to facilitate biosynthetic traffic.

The Functions of ERES

COPII can form vesicles, select cargo, incorporate or recruit tethering and fusion machinery. As such COPII can function as a minimal sufficient machinery for ER exit. However, its activities are coupled with physiological demands, and this coupling is directed by ERES proteins. (1) ERES proteins control the assembly of the COPII coat allowing for the selection of broad physiological cargoes. (2) ERES proteins organize the space between donor and acceptor membranes, building a traffic hub for effective traffic. (3) The cellular distribution of ERES supports local biosynthetic secretion. (4) ERES control bilayer composition, enhancing local enzymatic remolding that modify the physical properties of membranes to facilitate membrane deformation and fission. (5) ERES establish a sorting platform that refines COPII-mediated ER sorting and expand its repertoire of cargoes. Given the above functions, which will be discussed below, it is perhaps not surprising that traffic from the ER continues even when COPII activities are compromised [40].

ERES Directing COPII Assembly

ERES proteins control the assembly of COPII by regulating the activity of Sar1, either enhancing GTP binding (activation) or delaying inactivation through GTP hydrolysis. In simple eukaryotes such as the budding yeast Saccharomyces cerevisiae, Sec12 is distributed throughout the ER and is not concentrated at ERES [19, 132]. This simplicity likely reflects limited physiological requirements for intercellular communication, biosynthetic material exchange or multicellular organization. Increased biosynthetic load is likely exerted in Pichia pastoris, a more developed fungi, where the ERES resident protein Sec16 contacts and localizes Sec12 to ERES, maximizing Sar1-COPII activation [124]. In Metazoan, it is cTAGE5 that interacts with Sec12, localizing and concentrating Sar1 activation at ERES thus facilitating the initiation of coat assembly [162]. The distribution of Sar1, which is typically broader than that of other COPII components [109], likely reflects this robust, localized activation of the GTPase at ERES.

ERES proteins also control the inactivation rate of Sar1 by directing the coupling between the two COPII coat layers. In budding yeast, the soluble yet peripherally associated large membrane protein, Sec16 provides a scaffold for ERES. Sec16 interacts with all COPII subunits and self-polymerizes using its Ancestor Coat Element 1 (ACE1) central domain to organizes an ERES [202, 210]. The interactions of Sec16 with Sec23 compete with Sec23-Sec31 binding thus inhibiting the linkage between the two coat layers and as a result inhibiting GTPase activation. This delay attenuates coat assembly and budding, optimizing cargo soring [94, 210]. The strategy of regulating coat inner and outer layer coupling is conserved throughout evolution providing a checkpoint for ER-exit at ERES. In metazoans TANGO1-cTAGE5 complexes contact the Sec23 subunits of COPII using multiple triple proline repeats (PPP) within the PRD, five on cTAGE5 and seven on TANGO1 [105]. PPP ligands located on a disordered domain are likely equally accessible for COPII binding, competing for a single binding site on Sec23. These dynamic interactions function as an allovalent trap, concentrating COPII at ERES and controlling its dynamics. The PRDs of TANGO1-cTAGE5 compete with Sec31 for Sec23 binding. Mammalian Sec31A presents two PPP signals for Sec23 binding. By delaying the linkage between the two coat layers, GTPase activity is delayed not only optimizing cargo sorting but also regulating the timing of membrane fission [5]. It is important to note that this putative regulation is deduced from protein binding assays, yet overall contributions to inner-outer layer linkage and regulation of carrier fission require formal demonstration in physiological settings. TANGO1 serves as the membrane receptor for soluble Sec16 proteins [108], which dynamically cycles between the cytosol and ERES. It interacts with the PRD of TANGO1 and assembles within the ring that fences ERES buds [108]. Like TANGO1-cTAGE5, metazoan Sec16 may also compete with COPII outer layer recruitment to inhibit GTPase activation, providing another checkpoint that controls coat functions. Sec16 activities, regulated by post translation modifications, control the residency and activities of COPII at ERES [34, 74, 195].

Overall, control over coat assembly at ERES is critical for sorting efficiency and for the coordination between budding tethering and fusion required for the formation of traffic tunnels [149].

ERES Building Traffic Hubs

While in a simplistic vesicular traffic model, vesicles bud from donor membranes, un-coat, travel in space, tether and fuse with acceptor membranes, ERES optimize COPII mediated traffic by defining the space to be traveled. Traffic at the primitive ERES of budding yeast is mediated by "hug and kiss" type of reaction where acceptor ERGIC and cis Golgi vesicles transiently contact donor ERES for cargo delivery [95, 190] (Fig. 1.1b). Yeast acceptor ERGIC and Golgi compartments undergo molecular transformation characteristic of cisternal maturation, a process that transforms Golgi composition to establish cis to trans polarity [64, 190]. The hug and kiss strategy negates the need for expansive travel and targeting reactions and is likely dependent on tethering activity between buds and acceptor compartment, which are transient in nature. In contrast, metazoan ERES use stabilized contacts (Fig. 1.1c). TANGO1 interacts with the NBAS-RINT1-ZW10 (NRZ) tether complex that links ERGIC with ER membranes and cTAGE5 interacts with Sec22b, a SNARE protein involved in bi-directional fusion between ERGIC and ERES membranes [48, 133, 151, 163]. This molecular scaffolding tethers and connects donor with acceptor membranes supporting effective traffic and enabling the formation of tunnels for direct non-vesicular traffic. Tunnel-based traffic is essentially required for the exit of bulky cargoes such as fibrillar procollagens from the ER. Depletion of NRZ components, regulatory and SNARE proteins including Sly1, Syntaxin18 or YKT6, which function in retrograde traffic, Syntaxin5 that functions in anterograde traffic and components of the TRAPP1 complex involved in Rab1-driven tether and fusion all inhibit collagen exit from the ER. Mutations in the TRAPP component TRAPPC2 (sedlin) lead to the development of Spondyloepiphyseal Dysplasia Tarda (SEDT), a disease characterized by selective defects in collagen secretion, supporting a role for fusion in tunnel formation for anterograde traffic [192]. TRAPP1 interacts with TANGO1 and with Sar1 to control the GTPase activity of Sar1, adding another layer of COPII regulation at ERES.

The space between donor and acceptor membranes is organized by tunable, regulated phase separation reactions using multiple components of the ERES. The scaffolding activity of Sec16A at ERES is augmented by dynamic phase separation that is controlled by cycles of phosphorylation and dephosphorylation. The kinase DYRK3 interacts with ERES proteins and together with a phosphatase activity control the liquid state of Sec16A to manipulate coat availability for traffic [57].

The COPII coat itself contributes to phase separation at ERES that differentiate those from the surrounding cytosol. Sec23 and Sec24 are structural homologues, yet the larger Sec24 proteins have a unique N-terminal IDR that phase separates and regulates COPII activities [194]. Phase separation of Sec24 leads to enlargement of

ERES and enhanced coat recruitment by increasing local retention and concentration of COPII components. LLPS by Sec24 is controlled by manganese released from mitochondrial stores, tunning the transport and secretion of apolipoprotein particles to support organism physiology [194].

COPII interacting adaptors self-organize themselves to contribute another layer of organization at ERES. Track Fused Gene (TFG) was identified as an interactor of Sec16, where its depletion leads to disruption of ERES organization [203]. TFG binds Sec23 and is involved in regulating uncoating within the ERES scaffold [70]. Depletion of deletion leads to impairment of collagen secretion [115] and general traffic to the Golgi following ER exit [87] controlling the availability of Sec23 and thus the GTPase activity of Sar1 [88]. TFG contains a structured N-terminus that assembles in octamers [87] and a C- terminal IDR that phase separates. Recent work shows TFG forming hollow and porous condensates at ERES that potentially filter traffic components, allowing for the diffusion of the smaller COPII into bud sites while preventing the larger COPI complex from entry thus defining the composition of ERES and the spatial distribution of COPI-retrieval activities [199]. As with TFG, Sec23ip contains an N-terminus IDR that phase separates [92, 123]. Sec23ip/p125A was identified as a Sec23-interacting protein [188], yet it is a stable stoichiometric component of the Sec13/31 complex in cytosol that binds the C-terminal α-solenoid of Sec31A [103, 137]. Upon membrane binding, Sec23ip/p125A links the two coat layers and this activity requires membrane recognition, mediated by a PI4P recognition module [78, 92]. p125A serves as a spatial sensor that detects membrane apposition by binding the coat inner layer on the ER and PI4P on acceptor ERGIC and cis Golgi membranes thus coupling coat assembly with membrane organization at ERES. This activity is required to selectively support tunnel-based traffic of fibrillar collagens, providing a mechanism to coordinate the assembly of the coat with the formation of these traffic conduits [103]. While Sec23ip/p125A populates all ERES, recent work has shown that at select Golgi-associated ERES, Sec23ip interacts with VPS13B linking ERES to a lipid transfer protein and the ERES-Golgi interface [45]. Since Sec23ip/p125A harbors an intact PA-PLA1 homologous domain, it may support lipid flow through the remodeling of phosphatidic acid (PA), an activity that may be required to balance the composition of ERES membranes and control their tension. However, PLA1 activity could not be measured with isolated Sec23ip/p125A [128]. Both TFG and Sec23ip participate in regulating COPII assembly, contributing to the scaffolding of ERES.

The assembly of the ERES traffic hub utilizes multiple layers of molecular organization, including tethering between donor and acceptor membranes, and protein assemblies in the cytosol that separate the ERES from is surroundings, constructing a robust traffic hub. The hierarchy and rules that direct scaffold assemblies at ERES remain to be defined.

ERES Facilitate Local Secretion in Cells

The cellular distribution of ERES is regulated and likely supports local biosynthetic needs. In polarized cells such as primary neurons, select ERES and associated Golgi units are localized in dendrites supporting localized traffic [12]. Similarly, ERES are mobilized early on to processes that will become the axon, where they support growth. Axonal localization of ERES correlates with microtubule organization. ERES are depleted from the axon once it is specified and the microtubules polarize [7]. However, transmembrane proteins locally translated in axons of specified neurons and are trafficked to the plasma membrane from axonal ERES. Axonal ERES utilize the NRZ tether complex and Sec22b to support Golgi bypass traffic to the plasma membrane and this local secretion is essential for axon development [130].

ERES positioning is mediated by the interactions of ERES with microtubule motors. Sec23-TANGO1-cTAGE5 interactions control the cellular positioning of ERES, where stabilized Sec23-TANGO1 binding clusters ERES near the cis-Golgi [169]. Golgi associated ERES selectively accumulate bulky cargoes such as collagen 1 and collagen 7, whereas small size cargoes appear uniformly distributed between all ERES. Similarly, clustering of mRNAs encoding basement membrane components at the Golgi region seen in mouse epithelial cells supporting localized, coupled translation and ER exit of bulky cargoes [196]. ERES clustering may optimize bulky cargo traffic by minimizing the distance between ERES and Golgi. It is likely that stabilization of Sec23-TANGO1 binding with optogenetic arrest stabilizes the binding between $p150^{glued}$-dynactin and Sec23 linking COPII and the MT motor dynein for mobilization within the membrane in the minus direction toward the MToC resulting in clustering at the Golgi region [198]. At steady state, ERES, which interact with both minus (dynein) and plus (kinesin KIF5B) directed motors show limited mobility derived from a tug of war between the two motors [11, 67, 68, 198] and regulating Sec23-residency at ERES through interactions with TANGO1 shifts this balance. It is tempting to suggest that TANGO1-like proteins control ERES localization using inside-out and outside-in signaling. In an inside-in mechanism, the engagement of TANGO1 with HSP47-collagens in the ER lumen is transmitted to the cytosol to control TANGO1-PRD-Sec23 binding and dynein residency at ERES. Conversely, the interactions of TANGO1-like proteins with tether and fusion proteins that recruit acceptor ERGIC and cis Golgi membranes may be transmitted from the cytosol to the ER lumen directing HSP47-TANGO1 binding for selective capture of bulky cargo at Golgi localized ERES. The ERES machinery is tuned to cargo traffic and Sec12 is recruited by cargo receptors to maximize COPII initiation as seen with TANGO-1-cTAGE and similarly with wntless (wls), which escorts wnt molecules for secretion [181]. Cargo engagement with Sec24 leads to activation of a heterotrimer G protein, inducing a signaling cascade that culminates in adjustments to ER export and protein synthesis [180]. In addition, interactions of COPII adaptors with membranes as seen as seen for p125A and the utilization of allovalent entrapment and separation from cytosol afforded at ERES preserve local concentration of traffic components to adhere with physiological demands.

Once cargo concentrates in ERGIC tubules, it is trafficked from peripheral ERES to the Golgi using the Rab-GTPase controlled FTS-Hook-FHIP complex (FTS, FHIP2A, and Hook2) that utilize the microtubule tracks to deliver Rab1A and COPI-coated ERGIC tubules to the Golgi [36].

ERES Maintain Unique Lipid Composition

The core function of the ERES is generating carriers to mobilize cargo from the ER. The ERES are composed of deformed, constricted tubulated and vesiculated membranes. These outcomes rely on the activity of membrane-shaping proteins such as Sar1 and the composition of the bilayer. The rapid generation and enlargement of carriers is aided by changes to membrane tension [152]. Membrane shaping proteins such as Sar1 induce localized changes in membrane curvature through amphipathic insertion and the resulting area difference elastic stress can most effectively be generated in confined lipid environment and so are the local changes to membrane tension that enlarge carriers [26, 102]. Fission derived from the buildup of elastic stresses may depend on elastic properties of the bilayer where bilayer organizing lipids such as cholesterol play key roles. The generation of curvature and its reversal in processes of tubulation is aided by enzymatic remodeling of the bilayer with the transient addition and removal of cone-shaped phospholipids. Similarly, lipid signals that define the directionality of traffic at ERES, including PI4P and PI3P need to be restricted to acceptor membranes to perform such function. Yet tunnels that form membrane connectivity between ER, ERES, ERGIC and cis Golgi may interfere with these processes, allowing for the rapid diffusion and exchange of lipids between compartments. To overcome these issues, barriers are placed to allow for localized and thus effective membrane remodeling. The unique trans membrane configuration of TANGO1 provides one mechanism for lipid gating, as it can form a diffusion barrier for phospholipids. Such activity has been demonstrated when reconstituted in model giant unilamellar vesicles (GUVs) and may function similarly at ERES [27, 153, 200]. The interactions of other ERES components with the membranes likely contributes to the unique lipid environment yet their role is yet to be defined. Lipid modifying enzymes including PLD [144], lysoPA re-acylation enzymes [28] and potentially PA-PLA1 activities likely participate in local tubulation, fission and vesiculation and these are likely regulated by biosynthetic demand. One documented example for such adjustments is the response of ERES to acute increase in cargo load that leads to enlargement of ERES, where this expansion depends on the localized formation of PI4P—tunning local the bilayer composition at ERES to biosynthetic demands [49].

ERES Facilitate COPII Mediated Sorting for ER Exit

A core function of ERES is protein sorting. The role requires differentiating cargoes destained for ER exit from resident ER machinery, cargoes that are yet on a productive folding pathway and those which will not fold and are destined for degradation. A third of the proteome is sorted for traffic in the ER and while ER chaperones, ER retention and disposal using the ER Associated Degradation (ERAD) machinery facilitates retro translocation of misfolded or unassembled proteins from the ER to the cytosol for degradation by the proteasome providing a key quality control step [93], ERES are coupled to these activities and provide another layer of sorting, serving as an integral part of the cellular proteostasis network. ERES function in selection of proteins for biosynthetic secretion yet also collect cargo for lysosomal degradation.

The basic sorting activity of COPII utilizes interactions between ER exit motifs on cargo and the coat inner layer that collect and concentrate cargo for ER-exit. These sorting activities are augmented by the ERES resident proteins [6]. This ERES scaffold interacts with cargo receptors that connect luminal cargoes in the ER with the cytosolic COPII for traffic. TANGO1 itself functions as a cargo receptor, where its unique luminal SH3 domain interacts with collagen bound HSP47 to facilitate secretion [81]. Sorting goes beyond the binary interactions between coat and cargo or cargo receptors with added combinatorial recognition that is utilized for effective cargo sorting, allowing the COPII sorting machinery to monitor the thermodynamic and kinetic stability of cargo proteins. In fact, membrane cargoes that directly bind COPII also use cargo receptors to monitor and support their concentration and exit from the ER [33, 73, 138, 168]. The ERES scaffold and its associated cargo receptors also participates in the retention of ER resident proteins, collectively forming a distillation platform at the point of ER exit.

Cargo Receptors as Integral Components of the ERES Scaffold

The traditional view of luminal cargo selection involves interactions of cargo receptors with cargo in the ER lumen and binding of ER exit motives presented by the receptors in the cytosol with the inner layer of COPII. However, the role of cargo receptors goes beyond this simple bridge-like function. Cargo receptors couple protein processing folding and assembly with ER exit, serving as quality-control elements at ERES. The receptors also prevent the entry of ER resident proteins into budded carriers, providing a filtration mechanism that combines selection and enrichment with exclusion and retention. The net result enhances concentrative traffic while reducing the burden on retrograde traffic from later compartments [5, 6]. These roles are exemplified by studies on the p24 family of cargo receptors (also referred to as p23/24 or TMED) that are prominent components of both COPII and COPI vesicles [143]. The assignment of a role for p23/24 as cargo receptors is strait forward. These type 1 membrane proteins present a short cytosolic

C-terminus that carries canonical COPII and COPI binding motifs that directs their cycling between the ER and post ER compartments. The receptors present coiled-coil and GOLD (Golgi dynamics) domains in the ER lumen that interact with and are required for the traffic of a variety of cargoes from the ER. But p24 receptors not only select cargo, they also monitor processing of cargoes, thus ensuring exit of mature cargo as documented for the selection of glycosylphosphatidylinositol (GPI)-anchored proteins for ER exit [55, 126, 157]. The interactions of p24 with these cargoes follows the sequential enzymatic remodeling of both the glycan moieties and the lipid anchors of the GPI reporting on maturation. The variety of cargoes that utilize the TMED receptor system may be explained by a diversity of the p24/TMED family, which includes ten members in mammalian cells (nine in humans) [1]. These are divided between four sub-groups, where heterodimerization and oligomerization between family members generate versatile sorting capabilities [143]. Dimerization and oligomerization also directs the interactions of the receptors with the coats, defining preference for either COPII or COPI binding, in a regulated-manner [22, 41, 204]. Furthermore, the transmembrane domains of p24 receptors interact with selective lipids, including sphingomyelin and potentially cholesterol and such interactions within the bilayer govern the temporal and spatial dimerization of the receptors to direct activity [38, 141]. p24/TMED also regulate the assembly state of cargoes, where cargo assembly governs traffic from the ER. The stabilization of the dimeric state of STING, an activator of innate immunity by the p24 receptor TMED2 directs its traffic to the Golgi and activation regulating the IFN response critical for controlling infection [182].

Yet the activity of the TMED family at ERES goes beyond cargo selection. Deletion of a TMED family members is often accompanied by co-depletion of other TMED members. In mammals or flies such deletions are embryonically lethal likely reporting on traffic defects of proteins that play key roles in development [143]. However, in the single cell system of budding yeast, the deletion of all eight members is tolerable yet leads to secretion of ER resident proteins and activation of the unfolded protein response [21]. Similarly in mammalian cells, disabling COPII-p24 interactions either by mutagenesis of Sec24 subunits that bind p24/TMED or with a drug (4-Phenyl Butyrate) that blocks the binding site for TMED on Sec24 lead to breakdown of retention, entry of resident ER proteins into COPII carriers and secretion [106]. Thus p24/TMED provide filtration at ERES that assists in protein retention. TMED receptors are an integral part of the ERES scaffold [208]. Studies in drosophila demonstrated that the assembly of TMED at ERES is hierarchical with recruitment to ERES of α family members leading the collection of β and δ followed by γ. Importantly, this recruitment requires interactions of the GOLD domain of the receptors with the luminal SH3 domain of TANGO1 (the single member of the TANGO1 family in drosophila). Thus TANGO1, which links donor-acceptor membranes in the cytosol, interacts with p24/TMED cargo receptors within the ER and both interact with COPII in the cytosol assembling a dynamic part of the ERES scaffold that extends from the lumen of the ER to the cytosol. Deletion of TMED or TANGO1 leads to enhanced COPII vesiculation and inhibition of general traffic [208]. Similar excessive vesiculation is documented in mammalian cells deleted of

TANGO1, where cTAGE5 is co-depleted [116]. Thus, the building of an ERES scaffold is required for the selection of cargoes and for the exclusion of ER resident proteins at ERES. The scaffold restricts COPII-mediated vesiculation likely supporting tunnel formation between ERES and ERGIC/cis Golgi. These coupled activities facilitate direct cargo traffic through tunnels while preventing the mixing of donor and acceptor proteins.

Evidence has linked p24s/TMEDs to other cellular process that go beyond the expected functions of vesicular and non-vesicular cargo receptors and filters. For example, TMED10 is involved in the unconventional secretion of cargoes that do not present a signal sequence. TMED10 interacts with these cargoes, oligomerizes and potentially translocate these peptides into ERGIC. Other members are involved in the formation of membrane contacts that govern lipid composition [184, 214]. TMED2 and TMED10 participates in ER-Golgi interactions controlling the formation of lipid domains at the plasma membrane [4]. The C-terminus of TMED9 interacts with the cytosolic domain of Sec12 forming tight membrane contacts between the ER and ERGIC that facilitate local SAR1 activation for the formation of COPII vesicles required to support LC3 lipidation and autophagosome formation from ERGIC [99]. The unifying mechanism behind these diverse functions is yet to be defined. These added capabilities are potentially utilized at ERES.

The roles of other cargo receptors as building blocks of the ERES scaffold is less defined. However, it is important to note that prevalent cargo receptors recognize general determinants that are common to diverse cargoes. ERGIC53, a lectin that selectively binds high mannose moieties on N-glycosylated cargoes provides little specificity that is augmented by combinatorial interactions with other sorting factors [82, 171, 213, 217]. In the case of the highly N-glycosylated coagulation factors F5 and F8, an added factor is MCFD2, a luminal dual EF-hand domain protein that forms a tertiary complex with cargo and the lectin domain, where the ERGIC53-lectin selects the glycan moieties (High mannose N-glycosylation) and MCFD2 binding co-select a linear exit motif to facilitate ERES concentration and traffic [167, 205]. Somewhat similarly, SURF4, another cargo receptor, recognizes a Cardin-Weintraub motif (BBBXXBX, where B is a basic amino acid) and a broad range of N-terminal motifs uncovered by the proteolytic removal of the signal sequence [186]. These later broad signals present a hierarchy of recognition that prefers a Φ-P-Φ (hydrophobic-proline-hydrophobic)signals yet interacts with a range of sequences that conform broadly with this triple amino acid sequence termed collectively ER-ESCAPE motif. The interaction with diverse permutations of the motif provides a range of binding affinities for effective ER export that accelerates the exit for aggregation prone proteins from the ER and this binding hierarchy is reflected in the steady state levels of such proteins in the ER [209]. Similarly, the strength of ER-ESCPE interactions reflect the timing required for cargo folding assuring proper quality control. SURF4 interacts with uncovered ER-ESCAPE motifs co-translationally, coupling protein folding with ER exit [110]. Over expression and thus over representation of ER-ESCAPE motif containing cargo leads to the formation of Rab1 dependent elongated tubular ERGIC (2–20 μm long with a diameter of <30 nm) that are depleted of ERGIC53, demonstrating an adaptation of the ERES scaffold to cargo representation [206]. As with ERGIC53, additional signals direct SURF4 mediated sorting and the receptor plays key physiological roles in lipid homeostasis, employing combinatorial recognition together with a member of the TMED cargo receptors, TMED10, to export PCSK9, a secreted regulator of the surface expression of LDL receptors [110] and also driving the exit of apolipoprotein B from the ER [193].

Cargo receptors integrate with the ERES scaffold and on occasions provide. A "two gear" sorting mechanism, where cargo is first collected by ERES proteins like TANGO1 and subsequently escorted from ERES into carriers by added receptors whereas TANGO1 is left behind. TMEM39A, TMEM131 facilitate collagen exit from the ER likely following recognition of HSP47-Collagens by TANGO1. TMEM131 and TMEM39A function as canonical cargo receptors that interact with collagens in the ER lumen while binding the cytosolic traffic machinery in the cytosol with TMEM131 binding the tether complex TRAPP8 subunit, and TMEM39A, which binds Sec23A, linking cargo with the budding and fusion machinery for traffic [185, 215, 216]. In a similar manner, Ceroid-Lipofuscinosis Neuronal (CLN) CLN6 and CLN8 complex interact with lysosomal acid hydrolyzes in the ER. CLN8 escorts those hydrolyses to the Golgi for delivery to the lysosome, functioning as a canonical cargo receptor that binds COPII to promote ER-exit, while leaving CLN6 behind [15, 43].

Overall, cargo receptors and cargo are integral part of the ERES scaffold that couple cargo maturation and availability with scaffold assembly and traffic. The recognition of abundant and general exit signals on polypeptides such as high mannose N-glycosylation moieties or ER-ESCAPE motifs uncovered co-translationally with the removal of signal peptides, suggest that a combinatorial sorting platform is embedded in the ERES scaffold to enrich for nascent cargoes and direct their exit from the ER. This sorting platform also prevents resident proteins from entry into ERES, providing filtration within the ERES scaffold [106, 208]. Cargo sorting activities are dynamically regulated by assembly and disassembly of the COPII coat. The scaffold's filtration machinery controlling retention of ER residents while enriching biosynthetic cargoes at ERES, provides a gating function at the entrance to the secretory pathway.

ERES Facilitate Quality Control, Shuttling Proteins for Lysosomal Degradation

It has long been established that COPII is utilized at ERES-ERGIC to contribute membranes for the formation of autophagosomes. For this task, ERES undergo remodeling that enhances the connectivity between ERES and ERGIC membranes. Contacts are established between ERGIC localized TMED9 and ER localized Sec12 [99]. cTAGE5 facilitates the mobilization of Sec12 from ERES into ERGIC where FIP200 associates with Sec12 and Phosphatidyl Inositol 3-phospate (PI3P) signaling is activated [42, 58, 99]. Phosphorylation of COPII inner layer proteins directs their interactions with autophagy regulators together driving the formation of COPII

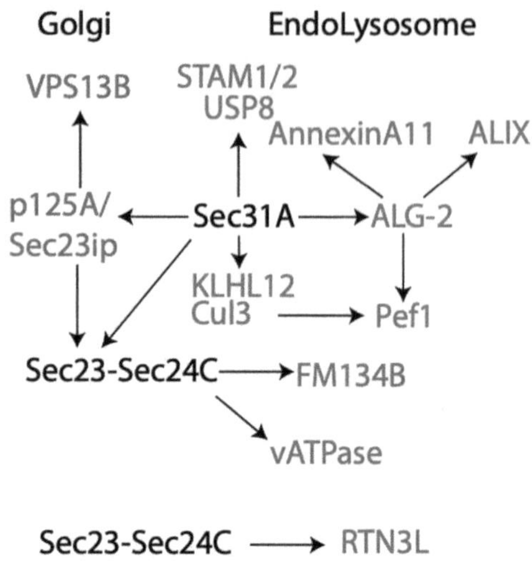

Fig. 1.3 Regulatory interactions of select adaptors with COPII direct biosynthetic traffic or lysosomal delivery for degradation of cargoes at or independently of ERES. Select interactions with COPII inner (Sec23 and Sec24) or outer layer (Sec31) that connect to Golgi or the Endolysosomal System are shown

vesicles budding from ERGIC, providing membranes for LC3 conjugation. Notably, the phosphorylation of Sec23B by ULK1 prevents its degradation and enhances its selective association with Sec24A/B, promoting autophagy [85], suggesting a switch mechanism that governs biosynthetic or degradative utilization of COPII. The documented interactions of COPII inner (yeast Sec24 [42]), outer (mammalian Sec31 [131]) layers with vesicles harboring complementary autophagic components including the lipid scramblase ATG9 [62, 135], and the E3 ligase component ATG16L, as well as the interactions of ERGIC – associated Rab1-ARFGAP1 with the lipid transfer protein ATG2 [56] all suggest that COPII vesicles derived from ERGIC are targeted to support the formation and elongation of autophagosomes [59, 60, 65, 79].

COPII components also participate in the clearance of protein aggregates from the ER (Fig. 1.3). The COPII inner layer Sec24C interacts with the long form of the autophagy receptor, the reticulon Rtn3L, to facilitate ER-phagy. Protein aggregates are sorted and collect at 3-way junctions on peripheral tubular elements, and these ER-tubules are fragmented by the activity of the autophagy receptor, which is ubiquitinated by the E3 ligase complex $CUL3^{KLHL12}$. ER fragments and taken by autophagosomes for delivery to lysosomal degradation. Sec24C driven ER-phagy does not required ERES nor COPII-mediated traffic, as blocking COPII activities does not affect the clearance pathway. [35, 39, 52].

In contrast, the sorting activity of COPII is utilized at ERES to target selective proteins for lysosomal degradation through a process of vesicle-mediated micro-autophagy [53]. Cargo proteins interact with the COPII machinery early in the folding cascade. As a result, interactions with misfolded cargoes become inhibitory to budding of COPII from ERES as observed with Z-AAT, a mutant of alpha antitrypsin. Z-AAT aggregates in the ER and is cleared by lysosomal degradation [10]. The coupling between Z-AAT aggregation and COPII is initiated when PDI interacts with Z-AAT leading to interactions between Z-AAT the cargo receptors including ERGIC 53 and p24 family members TMP21 and TMED9. Cargo receptors link the aggregated protein with COPII inhibiting ER exit [156, 183]. The resulting attenuated COPII inner layer subunit Sec24C interacts with the vATPase, concentrating the proton pump at ERES. vATPase recruits the E3 ligase ATG16L1 leading to LC3C lipidation of ERES membranes [183]. ERES membranes decorated with LC3 recruit the ER-phagy receptor FAM134B-II. leading to membrane fragmentation and fission and the released vesicles fuse with lysosomes using syntaxin17 and VAPM8 delivering Z-AAT for degradation.

How can assembled COPII at ERES differentiate between misfolded cargoes and ones that are on a productive folding pathway? The inhibition of ER exit by COPII-Z-AAT interactions at ERES can be mimicked by the expression of a GTPase-deficient Sar1 prolonging the residency of COPII at ERES and providing a kinetically-driven quality control that directs the modified ERES to lysosomal degradation [183]. Autophagic pathways, induced by inhibition of secretion from ERES, can also bypass canonical ER-Golgi mediated biosynthetic secretion, leading to some level of Golgi bypass non-conventional secretion of misfolded cargoes. Partially unstable cargoes such as the ΔF508 variant of the Cystic Fibrosis Transmembrane Regulator (ΔF508-CFTR) utilizes such pathways to reach the plasma membrane under conditions of ER exit inhibition [61, 176].

ERES play a global role as a hub shuttling biosynthetic proteins for degradation during starvation by using a selective ERES-micro autophagy pathway [100]. This pathway is directed by interactions of Sec31A with endo-lysosomal proteins, many of such interactions were charactered in the past 20 years (Fig. 1.3). As such, Sec31A interacts with STAM1/2 and USP8, a myriad of ALG-2 interacting partners that are recruited to Sec31A upon elevation of cytosolic Ca^{2+} including Pef1-Cul3^{KLHL12} that ubiquitinylate Sec31A, Annexin A11 and ALIX (Fig. 1.3) [86, 89, 96, 117, 155, 166, 172]. In many instances, depletion of such proteins, leads to conflicting secretory outcomes. The versatile Ca^{2+} regulated adaptor ALG-2 can homo or hetero dimerize, forming platforms for the recruitment of effectors that can on one hand adjust sorting and traffic activities with physiological cues [165], yet on the other hand direct ERES for lysosomal degradation [100]. The general inhibition of mTOR in starvation induces micro ERES-phagy. Ca^{2+} release from lysosomes in response to mTOR inhibition leads ALG-2, functioning as a ubiquitin ligase adaptor, to associate with Sec31 leading to its ubiquitination. ERES coated with ubiquitinated Sec31 are targeted for uptake by endo-lysosomes in an ESCRT-dependent process that requires ALIX [100, 175]. The pathway may also be activated when misfolded collagens populate ERES, where induction of ALG-2-Pef1-Cul3^{KLHL12} mediated ubiquitination of Sec31A leads to the degradation of misassembled procollagens in lysosomes [136]. Degradation of misassembled procollagen is an integral part of biosynthesis of collagens where lysosomal degradation is coupled with physiological secretion [37].

The uncovered versatile traffic pathways mobilizing membranes and cargoes to endo lysosomes or Golgi (Fig. 1.3), highlight an underexplored functioning of ERES in non-canonical secretion or lysosomal degradative pathways and the roles of ERES as a central element of cellular proteostasis.

Concluding Remarks

In recent years we begun to develop a functional, unified molecular-morphological view of the ERES organelle. Basic elements in the cytosol that participate in cargo sorting, membrane deformation and carrier release, including coats, coat-adaptors and tethers are integrated with membrane components that build up a scaffold and physically define ERES to mediate and orchestrate sorting while facilitating membrane apposition for traffic.

The rules of scaffold assembly remain largely undefined. The relative contributions, organization and regulation of LLPS of the various components that make up the ERES and support optimal concentrations of COPII traffic machineries for secretion or degradation should be defined. The hierarchy of interactions and signals that drive those and lead to LLPS and the signaling that adjusts LLPS to couple ERES with cellular physiology needs better definition.

The dynamics of the lipid membranes that make up the ERES, balancing bilayer composition, regulating membrane tension, remodeling and sorting lipids within the membranes that facilitate protein sorting, membrane deformation, carrier release and directional traffic require better definition.

Coat-driven sorting platform composed of cargo receptors displaying broad specificities is dynamically assembled at ERES. Studies should map the combinatorial recognition of cargoes at ERES and the mechanisms that allow cargo receptors to actively support retention of ER proteins. A better understanding of mechanisms by which cargo itself controls scaffold assembly, ERES organization and the cellular positioning of ERES for traffic should be defined. A coupling between local translation, protein folding and traffic is emerging. The potential for this coupling in defining ERES composition and function, pertaining to diverse groups of cargoes and/or distribution of ERES within cells needs better definition. Such coupling may play key roles in a potential polarization of ERES in cells like neurons where satellite ERES are positioned in dendrites or axons at distance away from the cell body to support local secretion. The recent realization that ERES not only provide membranes for autophagosome formation but actively participates in sorting for lysosomal degradation via cargo-selective or global ERES-phagy pathways is yet to be defined, mapping lipid signals and post translation modifications on cargo and machinery that drives cargoes for autophagy and degradation.

With newly afforded resolution, cellular and molecular tools, it is likely that many of these questions will be addressed, providing new insights on the cell biology of biosynthetic secretion. The ERES mobilizes a third of the proteome and defects in traffic lead to the development of a plethora of human diseases. The ERES controls cellular signaling that govern broad physiology including lipid homeostasis, innate immunity and the integrated stress response. ERES are commandeered by pathogens to assist in replication and assembly. Better understanding of the ERES organelle will further provide critical insights into human physiology in health and disease states.

References

1. Aber R, Chan W, Mugisha S, Jerome-Majewska LA (2019) Transmembrane emp24 domain proteins in development and disease. Genet Res 101:e14
2. Ahn KJ, Kim JS (2022) TGF-beta1 upregulates Sar1a expression and induces procollagen-I secretion in hypertrophic scarring fibroblasts. Open Med (Wars) 17:1473–1482
3. Antonny B, Madden D, Hamamoto S, Orci L, Schekman R (2001) Dynamics of the COPII coat with GTP and stable analogues. Nat Cell Biol 3:531–537
4. Anwar MU, Sergeeva OA, Abrami L, Mesquita FS, Lukonin I, Amen T, Chuat A, Capolupo L, Liberali P, D'Angelo G, van der Goot FG (2022) ER-Golgi-localized proteins TMED2 and TMED10 control the formation of plasma membrane lipid nanodomains. Dev Cell 57:2334–2346
5. Aridor M (2018) COPII gets in shape: lessons derived from morphological aspects of early secretion. Traffic 19:823–839
6. Aridor M (2022) A tango for coats and membranes: new insights into ER-to-Golgi traffic. Cell Rep 38:110258
7. Aridor M, Fish KN (2009) Selective targeting of ER exit sites supports axon development. Traffic 10:1669–1684
8. Aridor M, Bannykh SI, Rowe T, Balch WE (1995) Sequential coupling between COPII and COPI vesicle coats in endoplasmic reticulum to Golgi transport. J Cell Biol 131:875–893
9. Aridor M, Weissman J, Bannykh S, Nuoffer C, Balch WE (1998) Cargo selection by the COPII budding machinery during export from the ER. J Cell Biol 141:61–70
10. Aridor M, Bannykh SI, Rowe T, Balch WE (1999) Cargo can modulate COPII vesicle formation from the endoplasmic reticulum. J Biol Chem 274:4389–4399
11. Aridor M, Fish KN, Bannykh S, Weissman J, Roberts TH, Lippincott-Schwartz J, Balch WE (2001) The Sar1 GTPase coordinates biosynthetic cargo selection with endoplasmic reticulum export site assembly. J Cell Biol 152:213–229
12. Aridor M, Guzik AK, Bielli A, Fish KN (2004) Endoplasmic reticulum export site formation and function in dendrites. J Neurosci 24:3770–3776
13. Arnolds O, Stoll R (2023) Characterization of a fold in TANGO1 evolved from SH3 domains for the export of bulky cargos. Nat Commun 14:2273
14. Bacia K, Futai E, Prinz S, Meister A, Daum S, Glatte D, Briggs JA, Schekman R (2011) Multibudded tubules formed by COPII on artificial liposomes. Sci Rep 1:17
15. Bajaj L, Sharma J, di Ronza A, Zhang P, Eblimit A, Pal R, Roman D, Collette JR, Booth C, Chang KT, Sifers RN, Jung SY, Weimer JM, Chen R, Schekman RW, Sardiello M (2020) A CLN6-CLN8 complex recruits lysosomal enzymes at the ER for Golgi transfer. J Clin Invest 130:4118–4132
16. Balch WE, McCaffery JM, Plutner H, Farquhar MG (1994) Vesicular stomatitis virus glycoprotein is sorted and concentrated during export from the endoplasmic reticulum. Cell 76:841–852
17. Bannykh SI, Rowe T, Balch WE (1996) The organization of endoplasmic reticulum export complexes. J Cell Biol 135:19–35
18. Bard F, Casano L, Mallabiabarrena A, Wallace E, Saito K, Kitayama H, Guizzunti G, Hu Y, Wendler F, Dasgupta R, Perrimon N, Malhotra V (2006) Functional genomics reveals genes involved in protein secretion and Golgi organization. Nature 439:604–607

19. Barlowe C, Schekman R (1993) SEC12 encodes a guanine-nucleotide-exchange factor essential for transport vesicle budding from the ER. Nature 365:347–349
20. Barlowe C, Orci L, Yeung T, Hosobuchi M, Hamamoto S, Salama N, Rexach MF, Ravazzola M, Amherdt M, Schekman R (1994) COPII: a membrane coat formed by sec proteins that drive vesicle budding from the endoplasmic reticulum. Cell 77:895–907
21. Belden WJ, Barlowe C (2001) Deletion of yeast p24 genes activates the unfolded protein response. Mol Biol Cell 12:957–969
22. Bethune J, Kol M, Hoffmann J, Reckmann I, Brugger B, Wieland F (2006) Coatomer, the coat protein of COPI transport vesicles, discriminates endoplasmic reticulum residents from p24 proteins. Mol Cell Biol 26:8011–8021
23. Bevis BJ, Hammond AT, Reinke CA, Glick BS (2002) De novo formation of transitional ER sites and Golgi structures in Pichia pastoris. Nat Cell Biol 4:750–756
24. Bi X, Corpina RA, Goldberg J (2002) Structure of the Sec23/24-Sar1 pre-budding complex of the COPII vesicle coat. Nature 419:271–277
25. Bi X, Mancias JD, Goldberg J (2007) Insights into COPII coat nucleation from the structure of Sec23.Sar1 complexed with the active fragment of Sec31. Dev Cell 13:635–645
26. Bielli A, Haney CJ, Gabreski G, Watkins SC, Bannykh SI, Aridor M (2005) Regulation of Sar1 NH2 terminus by GTP binding and hydrolysis promotes membrane deformation to control COPII vesicle fission. J Cell Biol 171:919–924
27. Blumental-Perry A, Haney CJ, Weixel KM, Watkins SC, Weisz OA, Aridor M (2006) Phosphatidylinositol 4-phosphate formation at ER exit sites regulates ER export. Dev Cell 11:671–682
28. Brown WJ, Plutner H, Drecktrah D, Judson BL, Balch WE (2008) The lysophospholipid acyltransferase antagonist CI-976 inhibits a late step in COPII vesicle budding. Traffic 9:786–797
29. Brown MS, Radhakrishnan A, Goldstein JL (2018) Retrospective on cholesterol homeostasis: the central role of Scap. Annu Rev Biochem 87:783–807
30. Bryant KL, Baird B, Holowka D (2015) A novel fluorescence-based biosynthetic trafficking method provides pharmacologic evidence that PI4-kinase IIIalpha is important for protein trafficking from the endoplasmic reticulum to the plasma membrane. BMC Cell Biol 16:5
31. Budnik A, Stephens DJ (2009) ER exit sites—localization and control of COPII vesicle formation. FEBS Lett 583:3796–3803
32. Carter SD, Hampton CM, Langlois R, Melero R, Farino ZJ, Calderon MJ, Li W, Wallace CT, Tran NH, Grassucci RA, Siegmund SE, Pemberton J, Morgenstern TJ, Eisenman L, Aguilar JI, Greenberg NL, Levy ES, Yi E, Mitchell WG, Rice WJ, Wigge C, Pilli J, George EW, Aslanoglou D, Courel M, Freyberg RJ, Javitch JA, Wills ZP, Area-Gomez E, Shiva S, Bartolini F, Volchuk A, Murray SA, Aridor M, Fish KN, Walter P, Balla T, Fass D, Wolf SG, Watkins SC, Carazo JM, Jensen GJ, Frank J, Freyberg Z (2020) Ribosome-associated vesicles: a dynamic subcompartment of the endoplasmic reticulum in secretory cells. Sci Adv 6:eaay9572
33. Castillon GA, Watanabe R, Taylor M, Schwabe TM, Riezman H (2009) Concentration of GPI-anchored proteins upon ER exit in yeast. Traffic 10:186–200
34. Centonze FG, Farhan H (2019) Crosstalk of endoplasmic reticulum exit sites and cellular signaling. FEBS Lett 593:2280–2288
35. Chidambaram R, Kumar K, Parashar S, Ramachandran G, Chen S, Ferro-Novick S (2024) PINK1 controls RTN3L-mediated ER autophagy by regulating peripheral tubule junctions. J Cell Biol 223:e202407193
36. Christensen JR, Kendrick AA, Truong JB, Aguilar-Maldonado A, Adani V, Dzieciatkowska M, Reck-Peterson SL (2021) Cytoplasmic dynein-1 cargo diversity is mediated by the combinatorial assembly of FTS-Hook-FHIP complexes. Elife 10:e74538
37. Cinque L, Forrester A, Bartolomeo R, Svelto M, Venditti R, Montefusco S, Polishchuk E, Nusco E, Rossi A, Medina DL, Polishchuk R, De Matteis MA, Settembre C (2015) FGF signalling regulates bone growth through autophagy. Nature 528:272–275

38. Contreras FX, Ernst AM, Haberkant P, Bjorkholm P, Lindahl E, Gonen B, Tischer C, Elofsson A, von Heijne G, Thiele C, Pepperkok R, Wieland F, Brugger B (2012) Molecular recognition of a single sphingolipid species by a protein's transmembrane domain. Nature 481: 525–529
39. Cui Y, Parashar S, Zahoor M, Needham PG, Mari M, Zhu M, Chen S, Ho HC, Reggiori F, Farhan H, Brodsky JL, Ferro-Novick S (2019) A COPII subunit acts with an autophagy receptor to target endoplasmic reticulum for degradation. Science 365:53–60
40. Cutrona MB, Beznoussenko GV, Fusella A, Martella O, Moral P, Mironov AA (2013) Silencing of mammalian Sar1 isoforms reveals COPII-independent protein sorting and transport. Traffic 14:691–708
41. D'Arcangelo JG, Crissman J, Pagant S, Copic A, Latham CF, Snapp EL, Miller EA (2015) Traffic of p24 proteins and COPII coat composition mutually influence membrane scaffolding. Curr Biol 25:1296–1305
42. Davis S, Wang J, Zhu M, Stahmer K, Lakshminarayan R, Ghassemian M, Jiang Y, Miller EA, Ferro-Novick S (2016) Sec24 phosphorylation regulates autophagosome abundance during nutrient deprivation. Elife 5:e21167
43. di Ronza A, Bajaj L, Sharma J, Sanagasetti D, Lotfi P, Adamski CJ, Collette J, Palmieri M, Amawi A, Popp L, Chang KT, Meschini MC, Leung HE, Segatori L, Simonati A, Sifers RN, Santorelli FM, Sardiello M (2018) CLN8 is an endoplasmic reticulum cargo receptor that regulates lysosome biogenesis. Nat Cell Biol 20:1370–1377
44. Downes K, Flood J, Nans A, Van der Verren S, Audhya A, Zanetti G (2025) Multi-scale molecular imaging of human cells reveals COPI and COPII vesicles at ER exit sites. bioRxiv 2025.2007.2029.667472
45. Du Y, Fan X, Song C, Chang W, Xiong J, Deng L, Ji WK (2024) Sec23IP recruits VPS13B/COH1 to ER exit site-Golgi interface for tubular ERGIC formation. J Cell Biol 223: e202402083
46. Dukhovny A, Papadopulos A, Hirschberg K (2008) Quantitative live-cell analysis of microtubule-uncoupled cargo-protein sorting in the ER. J Cell Sci 121:865–876
47. Dumaresq-Doiron K, Savard MF, Akam S, Costantino S, Lefrancois S (2010) The phosphatidylinositol 4-kinase PI4KIIIalpha is required for the recruitment of GBF1 to Golgi membranes. J Cell Sci 123:2273–2280
48. Fan J, Wang Y, Liu L, Zhang H, Zhang F, Shi L, Yu M, Gao F, Xu Z (2017) cTAGE5 deletion in pancreatic beta cells impairs proinsulin trafficking and insulin biogenesis in mice. J Cell Biol 216:4153–4164
49. Farhan H, Weiss M, Tani K, Kaufman RJ, Hauri HP (2008) Adaptation of endoplasmic reticulum exit sites to acute and chronic increases in cargo load. EMBO J 27:2043–2054
50. Farhan H, Wendeler MW, Mitrovic S, Fava E, Silberberg Y, Sharan R, Zerial M, Hauri HP (2010) MAPK signaling to the early secretory pathway revealed by kinase/phosphatase functional screening. J Cell Biol 189:997–1011
51. Fath S, Mancias JD, Bi X, Goldberg J (2007) Structure and organization of coat proteins in the COPII cage. Cell 129:1325–1336
52. Ferro-Novick S, Reggiori F, Brodsky JL (2021) ER-Phagy, ER homeostasis, and ER quality control: implications for disease. Trends Biochem Sci 46:630–639
53. Fregno I, Fasana E, Bergmann TJ, Raimondi A, Loi M, Solda T, Galli C, D'Antuono R, Morone D, Danieli A, Paganetti P, van Anken E, Molinari M (2018) ER-to-lysosome-associated degradation of proteasome-resistant ATZ polymers occurs via receptor-mediated vesicular transport. EMBO J 37:e99259
54. Fromme JC, Ravazzola M, Hamamoto S, Al-Balwi M, Eyaid W, Boyadjiev SA, Cosson P, Schekman R, Orci L (2007) The genetic basis of a craniofacial disease provides insight into COPII coat assembly. Dev Cell 13:623–634
55. Fujita M, Watanabe R, Jaensch N, Romanova-Michaelides M, Satoh T, Kato M, Riezman H, Yamaguchi Y, Maeda Y, Kinoshita T (2011) Sorting of GPI-anchored proteins into ER exit sites by p24 proteins is dependent on remodeled GPI. J Cell Biol 194:61–75

56. Fuller DM, Wu Y, Schueder F, Rasool B, Nag S, Korfhage JL, Garcia-Milian R, Melnyk KD, Bewersdorf J, De Camilli P, Melia TJ (2025) ATG2A engages Rab1a and ARFGAP1 positive membranes during autophagosome biogenesis. bioRxiv 2025-03
57. Gallo R, Rai AK, McIntyre ABR, Meyer K, Pelkmans L (2023) DYRK3 enables secretory trafficking by maintaining the liquid-like state of ER exit sites. Dev Cell 58:1880–1897
58. Gan W, Zhang C, Siu KY, Satoh A, Tanner JA, Yu S (2017) ULK1 phosphorylates Sec23A and mediates autophagy-induced inhibition of ER-to-Golgi traffic. BMC Cell Biol 18:22
59. Ge L, Zhang M, Schekman R (2014) Phosphatidylinositol 3-kinase and COPII generate LC3 lipidation vesicles from the ER-Golgi intermediate compartment. Elife 3:e04135
60. Ge L, Zhang M, Kenny SJ, Liu D, Maeda M, Saito K, Mathur A, Xu K, Schekman R (2017) Remodeling of ER-exit sites initiates a membrane supply pathway for autophagosome biogenesis. EMBO Rep 18:1586–1603
61. Gee HY, Noh SH, Tang BL, Kim KH, Lee MG (2011) Rescue of DeltaF508-CFTR trafficking via a GRASP-dependent unconventional secretion pathway. Cell 146:746–760
62. Ghanbarpour A, Valverde DP, Melia TJ, Reinisch KM (2021) A model for a partnership of lipid transfer proteins and scramblases in membrane expansion and organelle biogenesis. Proc Natl Acad Sci USA 118:e2101562118
63. Giraudo CG, Maccioni HJ (2003) Endoplasmic reticulum export of glycosyltransferases depends on interaction of a cytoplasmic dibasic motif with Sar1. Mol Biol Cell 14:3753–3766
64. Glick BS, Luini A (2011) Models for Golgi traffic: a critical assessment. Cold Spring Harb Perspect Biol 3:a005215
65. Graef M, Friedman JR, Graham C, Babu M, Nunnari J (2013) ER exit sites are physical and functional core autophagosome biogenesis components. Mol Biol Cell 24:2918–2931
66. Guo Y, Linstedt AD (2006) COPII-Golgi protein interactions regulate COPII coat assembly and Golgi size. J Cell Biol 174:53–63
67. Gupta V, Palmer KJ, Spence P, Hudson A, Stephens DJ (2008) Kinesin-1 (uKHC/KIF5B) is required for bidirectional motility of ER exit sites and efficient ER-to-Golgi transport. Traffic 9:1850–1866
68. Hammond AT, Glick BS (2000) Dynamics of transitional endoplasmic reticulum sites in vertebrate cells. Mol Biol Cell 11:3013–3030
69. Hanna MGT, Mela I, Wang L, Henderson RM, Chapman ER, Edwardson JM, Audhya A (2016) Sar1 GTPase activity is regulated by membrane curvature. J Biol Chem 291:1014–1027
70. Hanna MGT, Block S, Frankel EB, Hou F, Johnson A, Yuan L, Knight G, Moresco JJ, Yates JR 3rd, Ashton R, Schekman R, Tong Y, Audhya A (2017) TFG facilitates outer coat disassembly on COPII transport carriers to promote tethering and fusion with ER-Golgi intermediate compartments. Proc Natl Acad Sci U S A 114:E7707–E7716
71. Hara-Kuge S, Kuge O, Orci L, Amherdt M, Ravazzola M, Wieland FT, Rothman JE (1994) En bloc incorporation of coatomer subunits during the assembly of COP-coated vesicles. J Cell Biol 124:883–892
72. Hariri H, Bhattacharya N, Johnson K, Noble AJ, Stagg SM (2014) Insights into the mechanisms of membrane curvature and vesicle scission by the small GTPase Sar1 in the early secretory pathway. J Mol Biol 426:3811–3826
73. Herzig Y, Sharpe HJ, Elbaz Y, Munro S, Schuldiner M (2012) A systematic approach to pair secretory cargo receptors with their cargo suggests a mechanism for cargo selection by Erv14. PLoS Biol 10:e1001329
74. Hu Y, Zeng Z, Ming X, Yin S, Guan Y, Gao L, Li D (2025) Construction of pathogenic Sec16a mutation mouse model using CRISPR/Cas9. Anim Model Exp Med. https://doi.org/10.1002/ame2.70082
75. Hughes H, Budnik A, Schmidt K, Palmer KJ, Mantell J, Noakes C, Johnson A, Carter DA, Verkade P, Watson P, Stephens DJ (2009) Organisation of human ER-exit sites: requirements for the localisation of Sec16 to transitional ER. J Cell Sci 122:2924–2934

76. Hutchings J, Stancheva V, Miller EA, Zanetti G (2018) Subtomogram averaging of COPII assemblies reveals how coat organization dictates membrane shape. Nat Commun 9:4154
77. Hutchings J, Stancheva VG, Brown NR, Cheung ACM, Miller EA, Zanetti G (2021) Structure of the complete, membrane-assembled COPII coat reveals a complex interaction network. Nat Commun 12:2034
78. Inoue H, Baba T, Sato S, Ohtsuki R, Takemori A, Watanabe T, Tagaya M, Tani K (2012) Roles of SAM and DDHD domains in mammalian intracellular phospholipase A1 KIAA0725p. Biochim Biophys Acta 1823:930–939
79. Ishihara N, Hamasaki M, Yokota S, Suzuki K, Kamada Y, Kihara A, Yoshimori T, Noda T, Ohsumi Y (2001) Autophagosome requires specific early sec proteins for its formation and NSF/SNARE for vacuolar fusion. Mol Biol Cell 12:3690–3702
80. Ishikawa Y, Bachinger HP (2013) A molecular ensemble in the rER for procollagen maturation. Biochim Biophys Acta 1833:2479–2491
81. Ishikawa Y, Ito S, Nagata K, Sakai LY, Bachinger HP (2016) Intracellular mechanisms of molecular recognition and sorting for transport of large extracellular matrix molecules. Proc Natl Acad Sci USA 113:E6036–E6044
82. Itin C, Roche AC, Monsigny M, Hauri HP (1996) ERGIC-53 is a functional mannose-selective and calcium-dependent human homologue of leguminous lectins. Mol Biol Cell 7:483–493
83. Jackson LP (2014) Structure and mechanism of COPI vesicle biogenesis. Curr Opin Cell Biol 29:67–73
84. Jackson LP, Lewis M, Kent HM, Edeling MA, Evans PR, Duden R, Owen DJ (2012) Molecular basis for recognition of dilysine trafficking motifs by COPI. Dev Cell 23:1255–1262
85. Jeong YT, Simoneschi D, Keegan S, Melville D, Adler NS, Saraf A, Florens L, Washburn MP, Cavasotto CN, Fenyo D, Cuervo AM, Rossi M, Pagano M (2018) The ULK1-FBXW5-SEC23B nexus controls autophagy. Elife 7:e42253
86. Jin L, Pahuja KB, Wickliffe KE, Gorur A, Baumgartel C, Schekman R, Rape M (2012) Ubiquitin-dependent regulation of COPII coat size and function. Nature 482:495–500
87. Johnson A, Bhattacharya N, Hanna M, Pennington JG, Schuh AL, Wang L, Otegui MS, Stagg SM, Audhya A (2015) TFG clusters COPII-coated transport carriers and promotes early secretory pathway organization. EMBO J 34:811–827
88. Kasberg W, Luong P, Minushkin K, Pustova I, Swift KA, Zhao M, Audhya A (2024) TFG regulates inner COPII coat recruitment to facilitate anterograde secretory protein transport. Mol Biol Cell 35:ar113
89. Kawaguchi K, Endo A, Fukushima T, Madoka Y, Tanaka T, Komada M (2018) Ubiquitin-specific protease 8 deubiquitinates Sec31A and decreases large COPII carriers and collagen IV secretion. Biochem Biophys Res Commun 499:635–641
90. Kim J, Kleizen B, Choy R, Thinakaran G, Sisodia SS, Schekman RW (2007) Biogenesis of gamma-secretase early in the secretory pathway. J Cell Biol 179:951–963
91. Kirchhausen T, Owen D, Harrison SC (2014) Molecular structure, function, and dynamics of clathrin-mediated membrane traffic. Cold Spring Harb Perspect Biol 6:a016725
92. Klinkenberg D, Long KR, Shome K, Watkins SC, Aridor M (2014) A cascade of ER exit site assembly that is regulated by p125A and lipid signals. J Cell Sci 127:1765–1778
93. Kumari D, Brodsky JL (2021) The targeting of native proteins to the endoplasmic reticulum-associated degradation (ERAD) pathway: an expanding repertoire of regulated substrates. Biomolecules 11:1185
94. Kung LF, Pagant S, Futai E, D'Arcangelo JG, Buchanan R, Dittmar JC, Reid RJ, Rothstein R, Hamamoto S, Snapp EL, Schekman R, Miller EA (2012) Sec24p and Sec16p cooperate to regulate the GTP cycle of the COPII coat. EMBO J 31:1014–1027
95. Kurokawa K, Okamoto M, Nakano A (2014) Contact of cis-Golgi with ER exit sites executes cargo capture and delivery from the ER. Nat Commun 5:3653
96. la Cour JM, Mollerup J, Berchtold MW (2007) ALG-2 oscillates in subcellular localization, unitemporally with calcium oscillations. Biochem Biophys Res Commun 353:1063–1067

97. Lee MC, Orci L, Hamamoto S, Futai E, Ravazzola M, Schekman R (2005) Sar1p N-terminal helix initiates membrane curvature and completes the fission of a COPII vesicle. Cell 122:605–617
98. Levy E (2015) Insights from human congenital disorders of intestinal lipid metabolism. J Lipid Res 56:945–962
99. Li S, Yan R, Xu J, Zhao S, Ma X, Sun Q, Zhang M, Li Y, Liu JG, Chen L, Li S, Xu K, Ge L (2021) A new type of ERGIC-ERES membrane contact mediated by TMED9 and SEC12 is required for autophagosome biogenesis. Cell Res 32:119–138
100. Liao YC, Pang S, Li WP, Shtengel G, Choi H, Schaefer K, Xu CS, Lippincott-Schwartz J (2024) COPII with ALG2 and ESCRTs control lysosome-dependent microautophagy of ER exit sites. Dev Cell 59:1410–1424
101. Liu M, Feng Z, Ke H, Liu Y, Sun T, Dai J, Cui W, Pastor-Pareja JC (2017) Tango1 spatially organizes ER exit sites to control ER export. J Cell Biol 216:1035–1049
102. Long KR, Yamamoto Y, Baker AL, Watkins SC, Coyne CB, Conway JF, Aridor M (2010) Sar1 assembly regulates membrane constriction and ER export. J Cell Biol 190:115–128
103. Long KR, Singh G, Villemeur M, Brouwers N, Malhotra V, Raote I, Aridor M (2025) p125A (Sec23ip) couples COPII coat assembly with donor-acceptor membrane organization to facilitate tunnel-based traffic. bioRxiv 2025-05 07.652703
104. Lorente-Rodriguez A, Barlowe C (2011) Requirement for Golgi-localized PI(4)P in fusion of COPII vesicles with Golgi compartments. Mol Biol Cell 22:216–229
105. Ma W, Goldberg J (2016) TANGO1/cTAGE5 receptor as a polyvalent template for assembly of large COPII coats. Proc Natl Acad Sci USA 113:10061–10066
106. Ma W, Goldberg E, Goldberg J (2017) ER retention is imposed by COPII protein sorting and attenuated by 4-phenylbutyrate. Elife 6:e26624
107. Maeda M, Saito K, Katada T (2016) Distinct isoform-specific complexes of TANGO1 cooperatively facilitate collagen secretion from the endoplasmic reticulum. Mol Biol Cell 27:2688–2696
108. Maeda M, Katada T, Saito K (2017) TANGO1 recruits Sec16 to coordinately organize ER exit sites for efficient secretion. J Cell Biol 216:1731–1743
109. Maeda M, Arakawa M, Komatsu Y, Saito K (2024) Small GTPase ActIvitY ANalyzing (SAIYAN) system: a method to detect GTPase activation in living cells. J Cell Biol 223: e202403179
110. Maldutyte J, Li XH, Gomez-Navarro N, Robertson EG, Miller EA (2025) ER export via SURF4 uses diverse mechanisms of both client and coat engagement. J Cell Biol 224: e202406103
111. Mancias JD, Goldberg J (2007) The transport signal on Sec22 for packaging into COPII-coated vesicles is a conformational epitope. Mol Cell 26:403–414
112. Mancias JD, Goldberg J (2008) Structural basis of cargo membrane protein discrimination by the human COPII coat machinery. EMBO J 27:2918–2928
113. Mansur A, Joseph R, Kim ES, Jean-Beltran PM, Udeshi ND, Pearce C, Jiang H, Iwase R, Milev MP, Almousa HA, McNamara E, Widrick J, Perez C, Ravenscroft G, Sacher M, Cole PA, Carr SA, Gupta VA (2023) Dynamic regulation of inter-organelle communication by ubiquitylation controls skeletal muscle development and disease onset. Elife 12:e81966
114. Matsuoka K, Orci L, Amherdt M, Bednarek SY, Hamamoto S, Schekman R, Yeung T (1998) COPII-coated vesicle formation reconstituted with purified coat proteins and chemically defined liposomes. Cell 93:263–275
115. McCaughey J, Miller VJ, Stevenson NL, Brown AK, Budnik A, Heesom KJ, Alibhai D, Stephens DJ (2016) TFG promotes organization of transitional ER and efficient collagen secretion. Cell Rep 15:1648–1659
116. McCaughey J, Stevenson NL, Mantell JM, Neal CR, Paterson A, Heesom K, Stephens DJ (2021) A general role for TANGO1, encoded by MIA3, in secretory pathway organization and function. J Cell Sci 134:jcs259075

117. McGourty CA, Akopian D, Walsh C, Gorur A, Werner A, Schekman R, Bautista D, Rape M (2016) Regulation of the CUL3 ubiquitin ligase by a calcium-dependent co-adaptor. Cell 167: 525–538
118. Mellman I, Emr SD (2013) A Nobel Prize for membrane traffic: vesicles find their journey's end. J Cell Biol 203:559–561
119. Mesmin B, Antonny B (2016) The counterflow transport of sterols and PI4P. Biochim Biophys Acta 1861:940–951
120. Miller EA, Schekman R (2013) COPII—a flexible vesicle formation system. Curr Opin Cell Biol 25:420–427
121. Miller EA, Beilharz TH, Malkus PN, Lee MC, Hamamoto S, Orci L, Schekman R (2003) Multiple cargo binding sites on the COPII subunit Sec24p ensure capture of diverse membrane proteins into transport vesicles. Cell 114:497–509
122. Mironov AA, Mironov AA Jr, Beznoussenko GV, Trucco A, Lupetti P, Smith JD, Geerts WJ, Koster AJ, Burger KN, Martone ME, Deerinck TJ, Ellisman MH, Luini A (2003) ER-to-Golgi carriers arise through direct en bloc protrusion and multistage maturation of specialized ER exit domains. Dev Cell 5:583–594
123. Mizoguchi T, Nakajima K, Hatsuzawa K, Nagahama M, Hauri HP, Tagaya M, Tani K (2000) Determination of functional regions of p125, a novel mammalian Sec23p-interacting protein. Biochem Biophys Res Commun 279:144–149
124. Montegna EA, Bhave M, Liu Y, Bhattacharyya D, Glick BS (2012) Sec12 binds to Sec16 at transitional ER sites. PLoS One 7:e31156
125. Mossessova E, Bickford LC, Goldberg J (2003) SNARE selectivity of the COPII coat. Cell 114:483–495
126. Muniz M, Riezman H (2016) Trafficking of glycosylphosphatidylinositol anchored proteins from the endoplasmic reticulum to the cell surface. J Lipid Res 57:352–360
127. Nair A, Nair A, Stock P, Jain A, Somerville E, Sanyal A, Kirchhausen T (2025) Close-up of vesicular ER exit sites by volume electron imaging using FIB-SEM. bioRxiv 2025.2004.2023.650353
128. Nakajima K, Sonoda H, Mizoguchi T, Aoki J, Arai H, Nagahama M, Tagaya M, Tani K (2002) A novel phospholipase A1 with sequence homology to a mammalian Sec23p-interacting protein, p125. J Biol Chem 277:11329–11335
129. Nakano A, Muramatsu M (1989) A novel GTP-binding protein, Sar1p, is involved in transport from the endoplasmic reticulum to the Golgi apparatus. J Cell Biol 109:2677–2691
130. Nguyen HH, Kersten N, Li CH, de Jong H, Arora T, Liolios T, Nguyen DTM, Bebelman MP, Altelaar M, Koppers M, Farías GG (2025) The axonal ER couples translation and secretion machineries for local delivery of axonal transmembrane proteins to promote axonal development. bioRxiv 2025.2009.2009.674816
131. Nie J, Ma S, Wu L, Li Y, Cao J, Li M, Mei P, Cooper PR, Li A, Pei D (2024) SEC31a-ATG9a interaction mediates the recruitment of COPII vesicles for autophagosome formation. Adv Sci 11:e2405127
132. Nishikawa S, Nakano A (1993) Identification of a gene required for membrane protein retention in the early secretory pathway. Proc Natl Acad Sci USA 90:8179–8183
133. Nogueira C, Erlmann P, Villeneuve J, Santos AJ, Martinez-Alonso E, Martinez-Menarguez JA, Malhotra V (2014) SLY1 and Syntaxin 18 specify a distinct pathway for procollagen VII export from the endoplasmic reticulum. Elife 3:e02784
134. Okamoto M, Kurokawa K, Matsuura-Tokita K, Saito C, Hirata R, Nakano A (2012) High-curvature domains of the ER are important for the organization of ER exit sites in Saccharomyces cerevisiae. J Cell Sci 125:3412–3420
135. Olivas TJ, Wu Y, Yu S, Luan L, Choi P, Guinn ED, Nag S, De Camilli PV, Gupta K, Melia TJ (2023) ATG9 vesicles comprise the seed membrane of mammalian autophagosomes. J Cell Biol 222:e202208088

136. Omari S, Makareeva E, Roberts-Pilgrim A, Mirigian L, Jarnik M, Ott C, Lippincott-Schwartz J, Leikin S (2018) ER exit sites drive procollagen quality control through non-canonical autophagy. Proc Natl Acad Sci USA 115:E10099
137. Ong YS, Tang BL, Loo LS, Hong W (2010) p125A exists as part of the mammalian Sec13/Sec31 COPII subcomplex to facilitate ER-Golgi transport. J Cell Biol 190:331–345
138. Pagant S, Wu A, Edwards S, Diehl F, Miller EA (2015) Sec24 is a coincidence detector that simultaneously binds two signals to drive ER export. Curr Biol 25:403–412
139. Palade G (1975) Intracellular aspects of the process of protein synthesis. Science 189:867
140. Pang X, Zhang Y, Park K, Liao Z, Li J, Xu J, Hong MT, Yin G, Zhang T, Wang Y, Egelman EH, Fan J, Hsu VW, Park SY, Sun F (2025) Structural elucidation of how ARF small GTPases induce membrane tubulation for vesicle fission. Proc Natl Acad Sci USA 122:e2417820122
141. Pannwitt S, Stangl M, Schneider D (2019) Lipid binding controls dimerization of the coat protein p24 transmembrane helix. Biophys J 117:1554–1562
142. Pantazopoulou A, Glick BS (2019) A kinetic view of membrane traffic pathways can transcend the classical view of Golgi compartments. Front Cell Dev Biol 7:153
143. Pastor-Cantizano N, Montesinos JC, Bernat-Silvestre C, Marcote MJ, Aniento F (2016) p24 family proteins: key players in the regulation of trafficking along the secretory pathway. Protoplasma 253:967–985
144. Pathre P, Shome K, Blumental-Perry A, Bielli A, Haney JH, Alber S, Watkins SC, Romero G, Aridor M (2003) Activation of phospholipase D by the small GTPase Sar1 is required to support COPII assembly and ER export. EMBO J 22:4059–4068
145. Paul S, Audhya A, Cui Q (2023) Molecular mechanism of GTP binding- and dimerization-induced enhancement of Sar1-mediated membrane remodeling. Proc Natl Acad Sci USA 120:e2212513120
146. Paul S, Audhya A, Cui Q (2024) Delineating the shape of COat protein complex-II coated membrane bud. PNAS Nexus 3:pgae305
147. Pyle E, Zanetti G (2024) Cryo-electron tomography reveals how COPII assembles on cargo-containing membranes. bioRxiv 2024.2001.2017.576008
148. Raote I, Malhotra V (2019) Protein transport by vesicles and tunnels. J Cell Biol 218:737–739
149. Raote I, Malhotra V (2021) Tunnels for protein export from the endoplasmic reticulum. Annu Rev Biochem 90:605–630
150. Raote I, Ortega Bellido M, Pirozzi M, Zhang C, Melville D, Parashuraman S, Zimmermann T, Malhotra V (2017) TANGO1 assembles into rings around COPII coats at ER exit sites. J Cell Biol 216:901–909
151. Raote I, Ortega-Bellido M, Santos AJ, Foresti O, Zhang C, Garcia-Parajo MF, Campelo F, Malhotra V (2018) TANGO1 builds a machine for collagen export by recruiting and spatially organizing COPII, tethers and membranes. Elife 7:e32723
152. Raote I, Chabanon M, Walani N, Arroyo M, Garcia-Parajo MF, Malhotra V, Campelo F (2020) A physical mechanism of TANGO1-mediated bulky cargo export. Elife 9:e59426
153. Raote I, Ernst AM, Campelo F, Rothman JE, Pincet F, Malhotra V (2020) TANGO1 membrane helices create a lipid diffusion barrier at curved membranes. Elife 9:e57822
154. Raote I, Saxena S, Malhotra V (2022) Sorting and export of proteins at the endoplasmic reticulum. Cold Spring Harb Perspect Biol 15:a041258
155. Rismanchi N, Puertollano R, Blackstone C (2009) STAM adaptor proteins interact with COPII complexes and function in ER-to-Golgi trafficking. Traffic 10:201–217
156. Roberts BS, Mitra D, Abishek S, Beher R, Satpute-Krishnan P (2024) The p24-family and COPII subunit SEC24C facilitate the clearance of alpha1-antitrypsin Z from the endoplasmic reticulum to lysosomes. Mol Biol Cell 35:ar45
157. Rodriguez-Gallardo S, Sabido-Bozo S, Ikeda A, Araki M, Okazaki K, Nakano M, Aguilera-Romero A, Cortes-Gomez A, Lopez S, Waga M, Nakano A, Kurokawa K, Muniz M, Funato K (2022) Quality-controlled ceramide-based GPI-anchored protein sorting into selective ER exit sites. Cell Rep 39:110768

158. Roy Chowdhury S, Bhattacharjee C, Casler JC, Jain BK, Glick BS, Bhattacharyya D (2020) ER arrival sites associate with ER exit sites to create bidirectional transport portals. J Cell Biol 219:e201902114
159. Runz H, Miura K, Weiss M, Pepperkok R (2006) Sterols regulate ER-export dynamics of secretory cargo protein ts-O45-G. EMBO J 25:2953–2965
160. Saito K, Chen M, Bard F, Chen S, Zhou H, Woodley D, Polischuk R, Schekman R, Malhotra V (2009) TANGO1 facilitates cargo loading at endoplasmic reticulum exit sites. Cell 136:891–902
161. Saito K, Yamashiro K, Ichikawa Y, Erlmann P, Kontani K, Malhotra V, Katada T (2011) cTAGE5 mediates collagen secretion through interaction with TANGO1 at endoplasmic reticulum exit sites. Mol Biol Cell 22:2301–2308
162. Saito K, Yamashiro K, Shimazu N, Tanabe T, Kontani K, Katada T (2014) Concentration of Sec12 at ER exit sites via interaction with cTAGE5 is required for collagen export. J Cell Biol 206:751–762
163. Santos AJ, Raote I, Scarpa M, Brouwers N, Malhotra V (2015) TANGO1 recruits ERGIC membranes to the endoplasmic reticulum for procollagen export. Elife 4:e10982
164. Santos AJ, Nogueira C, Ortega-Bellido M, Malhotra V (2016) TANGO1 and Mia2/cTAGE5 (TALI) cooperate to export bulky pre-chylomicrons/VLDLs from the endoplasmic reticulum. J Cell Biol 213:343–354
165. Sargeant J, Seiler DK, Costain T, Madreiter-Sokolowski CT, Gordon DE, Peden AA, Malli R, Graier WF, Hay JC (2021) ALG-2 and peflin regulate COPII targeting and secretion in response to calcium signaling. J Biol Chem 297:101393
166. Satoh H, Shibata H, Nakano Y, Kitaura Y, Maki M (2002) ALG-2 interacts with the amino-terminal domain of annexin XI in a Ca(2+)-dependent manner. Biochem Biophys Res Commun 291:1166–1172
167. Satoh T, Nishio M, Suzuki K, Yagi-Utsumi M, Kamiya Y, Mizushima T, Kato K (2020) Crystallographic snapshots of the EF-hand protein MCFD2 complexed with the intracellular lectin ERGIC-53 involved in glycoprotein transport. Acta Crystallogr F Struct Biol Commun 76:216–221
168. Sauvageau E, Rochdi MD, Oueslati M, Hamdan FF, Percherancier Y, Simpson JC, Pepperkok R, Bouvier M (2014) CNIH4 interacts with newly synthesized GPCR and controls their export from the endoplasmic reticulum. Traffic 15:383–400
169. Saxena S, Foresti O, Liu A, Androulaki S, Pena Rodriguez M, Raote I, Aridor M, Cui B, Malhotra V (2024) Endoplasmic reticulum exit sites are segregated for secretion based on cargo size. Dev Cell 59:2593–2608
170. Schindler AJ, Schekman R (2009) In vitro reconstitution of ER stress induced ATF6 transport in COPII vesicles. Proc Natl Acad Sci USA 106:17775–17780
171. Segal A, Zivelin A, Rosenberg N, Ginsburg D, Shpilberg O, Seligsohn U (2004) A mutation in LMAN1 (ERGIC-53) causing combined factor V and factor VIII deficiency is prevalent in Jews originating from the Island of Djerba in Tunisia. Blood Coagul Fibrinolysis 15:99–102
172. Shibata H, Kanadome T, Sugiura H, Yokoyama T, Yamamuro M, Moss SE, Maki M (2015) A new role for annexin A11 in the early secretory pathway via stabilizing Sec31A protein at the endoplasmic reticulum exit sites (ERES). J Biol Chem 290:4981–4993
173. Shomron O, Nevo-Yassaf I, Aviad T, Yaffe Y, Zahavi EE, Dukhovny A, Perlson E, Brodsky I, Yeheskel A, Pasmanik-Chor M, Mironov A, Beznoussenko GV, Mironov AA, Sklan EH, Patterson GH, Yonemura Y, Sannai M, Kaether C, Hirschberg K (2021) COPII collar defines the boundary between ER and ER exit site and does not coat cargo containers. J Cell Biol 220: e201907224
174. Shoulders CC, Stephens DJ, Jones B (2004) The intracellular transport of chylomicrons requires the small GTPase, Sar1b. Curr Opin Lipidol 15:191–197
175. Shukla S, Chen W, Rao S, Yang S, Ou C, Larsen KP, Hummer G, Hanson PI, Hurley JH (2024) Mechanism and cellular function of direct membrane binding by the ESCRT and ERES-associated Ca(2+)-sensor ALG-2. Proc Natl Acad Sci USA 121:e2318046121

176. Song MS, Sim HJ, Eun SH, Jung MK, Hwang SJ, Ham MH, Kwak K, Lee HJ, Kim JY, Jang DG, Chung HC, Shin DH, Kim YJ, Noh SH, Mun JY, Lee JM, Lee MG (2025) Tubular ER structures shaped by ER-phagy receptors engage in stress-induced Golgi bypass. Dev Cell 60: 1568.e11–1585.e11
177. Stagg SM, LaPointe P, Razvi A, Gurkan C, Potter CS, Carragher B, Balch WE (2008) Structural basis for cargo regulation of COPII coat assembly. Cell 134:474–484
178. Stancheva VG, Li XH, Hutchings J, Gomez-Navarro N, Santhanam B, Babu MM, Zanetti G, Miller EA (2020) Combinatorial multivalent interactions drive cooperative assembly of the COPII coat. J Cell Biol 219:e202007135
179. Stephens DJ, Lin-Marq N, Pagano A, Pepperkok R, Paccaud JP (2000) COPI-coated ER-to-Golgi transport complexes segregate from COPII in close proximity to ER exit sites. J Cell Sci 113(Pt 12):2177–2185
180. Subramanian A, Capalbo A, Iyengar NR, Rizzo R, di Campli A, Di Martino R, Lo Monte M, Beccari AR, Yerudkar A, Del Vecchio C, Glielmo L, Turacchio G, Pirozzi M, Kim SG, Henklein P, Cancino J, Parashuraman S, Diviani D, Fanelli F, Sallese M, Luini A (2019) Auto-regulation of secretory flux by sensing and responding to the folded cargo protein load in the endoplasmic reticulum. Cell 176:1461–1476
181. Sun J, Yu S, Zhang X, Capac C, Aligbe O, Daudelin T, Bonder EM, Gao N (2017) A Wntless-SEC12 complex on the ER membrane regulates early Wnt secretory vesicle assembly and mature ligand export. J Cell Sci 130:2159–2171
182. Sun MS, Zhang J, Jiang LQ, Pan YX, Tan JY, Yu F, Guo L, Yin L, Shen C, Shu HB, Liu Y (2018) TMED2 potentiates cellular IFN responses to DNA viruses by reinforcing MITA dimerization and facilitating its trafficking. Cell Rep 25:3086–3098
183. Sun Y, Wang X, Yang X, Wang L, Ding J, Wang CC, Zhang H, Wang X (2023) V-ATPase recruitment to ER exit sites switches COPII-mediated transport to lysosomal degradation. Dev Cell 58:2761–2775
184. Sun Y, Tao X, Han Y, Lin X, Tian R, Wang H, Chang P, Sun Q, Ge L, Zhang M (2024) A dual role of ERGIC-localized Rabs in TMED10-mediated unconventional protein secretion. Nat Cell Biol 26:1077–1092
185. Sung JY, Lim GE, Goo J, Jung KJ, Chung JM, Jung HS, Kim YN, Shim J (2025) TMEM39A and TMEM131 facilitate bulk transport of ECM proteins through large COPII vesicle formation. J Genet Genomics 52:189–203
186. Tang X, Wang T, Guo Y (2022) Export of polybasic motif-containing secretory proteins BMP8A and SFRP1 from the endoplasmic reticulum is regulated by surfeit locus protein 4. J Biol Chem 298:102687
187. Tang VT, Xiang J, Chen Z, McCormick J, Abbineni PS, Chen XW, Hoenerhoff M, Emmer BT, Khoriaty R, Lin JD, Ginsburg D (2024) Functional overlap between the mammalian Sar1a and Sar1b paralogs in vivo. Proc Natl Acad Sci USA 121:e2322164121
188. Tani K, Mizoguchi T, Iwamatsu A, Hatsuzawa K, Tagaya M (1999) p125 is a novel mammalian Sec23p-interacting protein with structural similarity to phospholipid-modifying proteins. J Biol Chem 274:20505–20512
189. Taylor RJ, Tagiltsev G, Briggs JAG (2023) The structure of COPI vesicles and regulation of vesicle turnover. FEBS Lett 597:819–835
190. Tojima T, Suda Y, Jin N, Kurokawa K, Nakano A (2024) Spatiotemporal dissection of the Golgi apparatus and the ER-Golgi intermediate compartment in budding yeast. Elife 13: e92900
191. Valm AM, Cohen S, Legant WR, Melunis J, Hershberg U, Wait E, Cohen AR, Davidson MW, Betzig E, Lippincott-Schwartz J (2017) Applying systems-level spectral imaging and analysis to reveal the organelle interactome. Nature 546:162–167
192. Venditti R, Scanu T, Santoro M, Di Tullio G, Spaar A, Gaibisso R, Beznoussenko GV, Mironov AA, Mironov A Jr, Zelante L, Piemontese MR, Notarangelo A, Malhotra V, Vertel BM, Wilson C, De Matteis MA (2012) Sedlin controls the ER export of procollagen by regulating the Sar1 cycle. Science 337:1668–1672

193. Wang X, Wang H, Xu B, Huang D, Nie C, Pu L, Zajac GJM, Yan H, Zhao J, Shi F, Emmer BT, Lu J, Wang R, Dong X, Dai J, Zhou W, Wang C, Gao G, Wang Y, Willer C, Lu X, Zhu Y, Chen XW (2021) Receptor-mediated ER export of lipoproteins controls lipid homeostasis in mice and humans. Cell Metab 33:350–366
194. Wang X, Huang R, Wang Y, Zhou W, Hu Y, Yao Y, Cheng K, Li X, Xu B, Zhang J, Xu Y, Zeng F, Zhu Y, Chen XW (2023) Manganese regulation of COPII condensation controls circulating lipid homeostasis. Nat Cell Biol 25:1650–1663
195. Wang MJ, Wang YC, Masson E, Wang YH, Yu D, Qian YY, Tang XY, Deng SJ, Hu LH, Wang L, Wang LJ, Rebours V, Cooper DN, Ferec C, Li ZS, Chen JM, Zou WB, Liao Z (2024) SEC16A variants predispose to chronic pancreatitis by impairing ER-to-Golgi transport and inducing ER stress. Adv Sci (Weinh) 11:e2402550
196. Wang S, Matsumoto K, Mehlferber MM, Zhang G, Aronova MA, Yamada KM (2024) Microtubule-dependent apical polarization of basement membrane matrix mRNAs in mouse epithelial cells. Cells Dev 177:203898
197. Ward TH, Polishchuk RS, Caplan S, Hirschberg K, Lippincott-Schwartz J (2001) Maintenance of Golgi structure and function depends on the integrity of ER export. J Cell Biol 155:557–570
198. Watson P, Forster R, Palmer KJ, Pepperkok R, Stephens DJ (2005) Coupling of ER exit to microtubules through direct interaction of COPII with dynactin. Nat Cell Biol 7:48–55
199. Wegeng WR, Bogus SM, Ruiz M, Chavez SR, Noori KSM, Niesman IR, Ernst AM (2024) A hollow TFG condensate spatially compartmentalizes the early secretory pathway. bioRxiv 2024.03.26.586876
200. Weigel AV, Chang CL, Shtengel G, Xu CS, Hoffman DP, Freeman M, Iyer N, Aaron J, Khuon S, Bogovic J, Qiu W, Hess HF, Lippincott-Schwartz J (2021) ER-to-Golgi protein delivery through an interwoven, tubular network extending from ER. Cell 184:2412–2429
201. Weissman JT, Plutner H, Balch WE (2001) The mammalian guanine nucleotide exchange factor mSec12 is essential for activation of the Sar1 GTPase directing endoplasmic reticulum export. Traffic 2:465–475
202. Whittle JR, Schwartz TU (2010) Structure of the Sec13-Sec16 edge element, a template for assembly of the COPII vesicle coat. J Cell Biol 190:347–361
203. Witte K, Schuh AL, Hegermann J, Sarkeshik A, Mayers JR, Schwarze K, Yates JR 3rd, Eimer S, Audhya A (2011) TFG-1 function in protein secretion and oncogenesis. Nat Cell Biol 13:550–558
204. Xiao L, Pi X, Goss AC, El-Baba T, Ehrmann JF, Grinkevich E, Bazua-Valenti S, Padovano V, Alper SL, Carey D, Udeshi ND, Carr SA, Pablo JL, Robinson CV, Greka A, Wu H (2024) Molecular basis of TMED9 oligomerization and entrapment of misfolded protein cargo in the early secretory pathway. Sci Adv 10:eadp2221
205. Yagi H, Yagi-Utsumi M, Honda R, Ohta Y, Saito T, Nishio M, Ninagawa S, Suzuki K, Anzai T, Kamiya Y, Aoki K, Nakanishi M, Satoh T, Kato K (2020) Improved secretion of glycoproteins using an N-glycan-restricted passport sequence tag recognized by cargo receptor. Nat Commun 11:1368
206. Yan R, Chen K, Wang B, Xu K (2022) SURF4-induced tubular ERGIC selectively expedites ER-to-Golgi transport. Dev Cell 57:512–525
207. Yang K, Liu M, Feng Z, Rojas M, Zhou L, Ke H, Pastor-Pareja JC (2021) ER exit sites in drosophila display abundant ER-Golgi vesicles and pearled tubes but no megacarriers. Cell Rep 36:109707
208. Yang K, Feng Z, Pastor-Pareja JC (2024) p24-Tango1 Interactions ensure ER-Golgi interface stability and efficient transport. J Cell Biol 223:e202309045
209. Yin Y, Garcia MR, Novak AJ, Saunders AM, Ank RS, Nam AS, Fisher LW (2018) Surf4 (Erv29p) binds amino-terminal tripeptide motifs of soluble cargo proteins with different affinities, enabling prioritization of their exit from the endoplasmic reticulum. PLoS Biol 16: e2005140
210. Yorimitsu T, Sato K (2012) Insights into structural and regulatory roles of Sec16 in COPII vesicle formation at ER exit sites. Mol Biol Cell 23:2930–2942

211. Yorimitsu T, Sato K (2023) Sec16 and Sed4 interdependently function as interaction and localization partners at ER exit sites. J Cell Sci 136:jcs261094
212. Zeuschner D, Geerts WJ, van Donselaar E, Humbel BM, Slot JW, Koster AJ, Klumperman J (2006) Immuno-electron tomography of ER exit sites reveals the existence of free COPII-coated transport carriers. Nat Cell Biol 8:377–383
213. Zhang B, Kaufman RJ, Ginsburg D (2005) LMAN1 and MCFD2 form a cargo receptor complex and interact with coagulation factor VIII in the early secretory pathway. J Biol Chem 280:25881–25886
214. Zhang M, Liu L, Lin X, Wang Y, Li Y, Guo Q, Li S, Sun Y, Tao X, Zhang D, Lv X, Zheng L, Ge L (2020) A translocation pathway for vesicle-mediated unconventional protein secretion. Cell 181:637–652
215. Zhang Z, Bai M, Barbosa GO, Chen A, Wei Y, Luo S, Wang X, Wang B, Tsukui T, Li H, Sheppard D, Kornberg TB, Ma DK (2020) Broadly conserved roles of TMEM131 family proteins in intracellular collagen assembly and secretory cargo trafficking. Sci Adv 6:eaay7667
216. Zhang Z, Luo S, Barbosa GO, Bai M, Kornberg TB, Ma DK (2021) The conserved transmembrane protein TMEM-39 coordinates with COPII to promote collagen secretion and regulate ER stress response. PLoS Genet 17:e1009317
217. Zheng C, Page RC, Das V, Nix JC, Wigren E, Misra S, Zhang B (2013) Structural characterization of carbohydrate binding by LMAN1 protein provides new insight into the endoplasmic reticulum export of factors V (FV) and VIII (FVIII). J Biol Chem 288:20499–20509

Chapter 2
COPI Vesicle and Tubule Formation at the Golgi Complex

Kunyou Park, Jia-Shu Yang, Victor W. Hsu, and Seung-Yeol Park

Abstract The Coat Protein I (COPI) complex forms both vesicles and tubules for transport at the Golgi complex. Anterograde Golgi transport occurs with the Golgi cisternae acting as carriers, a process known as cisternal maturation, while COPI vesicles transport in the retrograde direction. The cargoes of retrograde transport include Golgi glycosylation enzymes, enabling these enzymes to maintain their distributions as residents of the cis, medial, and trans cisternae of the Golgi. COPI also forms tubules that connect the Golgi cisternae, which enables a faster mode of anterograde Golgi transport than that mediated by cisternal maturation. Recent advancements in our understanding of COPI vesicle formation have significantly contributed to the current view of Golgi transport. Particularly, the discovery of COPI tubules has provided new insights into how Golgi transport occurs.

Keywords COPI · Golgi · Vesicles · Tubules · ARF1 · BARS · LPAAT · PLA2 · PLD

K. Park
Department of Life Sciences, Pohang University of Science and Technology (POSTECH), Pohang, Gyeongbuk, Republic of Korea

Division of Rheumatology, Inflammation and Immunity, Brigham and Women's Hospital, and Department of Medicine, Harvard Medical School, Boston, MA, USA

J.-S. Yang · V. W. Hsu (✉)
Division of Rheumatology, Inflammation and Immunity, Brigham and Women's Hospital, and Department of Medicine, Harvard Medical School, Boston, MA, USA
e-mail: vhsu@bwh.harvard.edu

S.-Y. Park (✉)
Department of Life Sciences, Pohang University of Science and Technology (POSTECH), Pohang, Gyeongbuk, Republic of Korea
e-mail: seungpark@postech.ac.kr

A. Nakano, J. Saraste (eds.), *The Golgi Network, Volume II*, Subcellular Biochemistry 111, https://doi.org/10.1007/978-3-032-16833-7_2

Introduction

The Golgi complex plays a central role in the sorting of proteins and lipids for their proper distribution within the cell. This organelle is comprised of multiple stacks, consisting of cisternae. Following their synthesis in the endoplasmic reticulum (ER), cargo proteins are transported to the Golgi where they initially reach the cis cisterna followed by transit to the medial and then the trans cisternae. The Golgi cisternae also contain glycosylation enzymes. By being residents of different Golgi cisternae, these enzymes glycosylate cargo proteins progressively as they are transported through Golgi [1–3].

Because of its unusual morphology, being organized into cisternal stacks, the Golgi was originally proposed to transport cargoes through the movement of the cisternae, a process referred to as cisternal maturation/progression [4]. Subsequently, results from experiments that sought to reconstitute Golgi transport led to a different view, proposing instead that the Golgi cisternae are static, with transport across the stacks being mediated by vesicles [5, 6]. For the reconstitution studies, cells defective in a specific step of Golgi glycosylation were infected with the vesicular stomatitis virus (VSV) that expresses a glycosylated G protein (VSVG), which has been a model cargo for studies on anterograde transport through the secretory system. Golgi membrane was then isolated from these infected cells and also from uninfected cells that have intact Golgi glycosylation. When the two populations of Golgi membranes were incubated together in the presence of cytosol and adenosine triphosphate (ATP), the defect in VSVG with impaired glycosylation was corrected, suggesting that the protein had been transported from Golgi membranes with defective glycosylation to Golgi membranes that carry out normal glycosylation [7–10]. Further insight came from ways of inhibiting this reconstituted transport, which led to the observation of coated vesicles [11–14]. This finding suggested that Golgi transport involves not only vesicles shuttling among the stacks, but also that a coat complex is involved in generating these vesicles. The coat complex was subsequently identified as a multimeric complex [15], having α, β, β', γ, δ, ε, and ζ subunits [16–21]. The entire complex was named COP (Coat Protein) [17], and subsequently COPI, as another coat complex was identified that forms coated vesicles at the ER, which was named COPII [22]. ADP Ribosylation Factor 1 (ARF1), a small GTPase, was then found to recruit the cytosolic precursor of COPI, known as coatomer [16], to the Golgi membrane to initiate vesicle formation [23–25].

However, further studies led to renewed interest in the cisternal maturation model. Studies that examined the effects of Brefeldin A (BFA), a pharmacological agent that redistributes the Golgi enzymes to the ER through retrograde transport [26, 27], suggested that the Golgi cisternae are dynamic rather than static. Moreover, instead of binding to cargoes transported in the anterograde direction, COPI was found to bind retrograde cargoes [28, 29]. Accordingly, anterograde Golgi transport was proposed to be mediated by the Golgi cisternae acting as carriers, while COPI vesicles were proposed to transport cargoes in the retrograde direction. Retrograde cargoes include Golgi glycosylation enzymes, which in this way can maintain their distributions as residents of the cis, medial and trans Golgi cisternae [1, 2, 30]. The finding that cargoes which are too large to fit into small vesicles remained within the Golgi cisternae as they transit from the cis side to the trans side of the Golgi provided key support for this modified cisternal maturation model [31, 32]. In addition, Golgi glycosylation enzymes were detected in COPI vesicles [33–35]. Furthermore, studies on the yeast Golgi, in which the cis, medial, and trans cisternae are scattered and thus readily resolved by live cell imaging in light microscopy, revealed the dynamic movement of proteins that are residents of different Golgi cisternae [36, 37].

More recently, another mode of Golgi transport has been discovered, which involves the transport of cargoes through tubules that connect the Golgi cisternae. A temperature-sensitive mutation causes VSVG to accumulate in the ER at the non-permissive temperature and then become released as a synchronized population upon shift of cells to the permissive temperature. When the pulsed VSVG traversed through the Golgi, an increase in tubules that connect the Golgi cisternae was observed [38]. Moreover, VSVG was observed to be transported through these tubules [38]. Subsequently, a study performing detailed kinetic analysis concluded that VSVG is transported through the Golgi faster than the speed predicted by cisternal maturation [39]. To explain this finding, the study proposed that each cisterna of the Golgi exports cargo directly to the plasma membrane (PM), rather than cargoes transiting through all the Golgi cisternae before being delivered to the cell surface [39].

In recent years however, experimental support has accumulated for the tubule-based explanation for the fast mode of Golgi transport. Major support comes from an unexpected angle. Advancement in the understanding of how COPI vesicles are formed has led to the discovery that COPI also generates tubules. This revelation has, in turn, provided mechanistic insights into how the fast mode of transport through Golgi tubules occurs. In this chapter, we describe mechanistic studies on COPI transport over the years that have led to this unexpected turn of events.

COPI Vesicle Formation

Being a small GTPase, ARF1 cycles between its active (GTP-bound) state, which results in ARF1 being membrane-bound, and its inactive (GDP-bound) state, which results in cytosolic localization of ARF1. The cycling of ARF1 between these two states requires a guanine nucleotide exchange factor (GEF) that catalyzes ARF activation, and a GTPase activating protein (GAP) that catalyzes ARF deactivation [40, 41]. Because of this dynamic cycling, early studies used a non-hydrolyzable form of GTP (GTPγS) to lock ARF1 in its GTP-bound state, in order to stabilize ARF1 and COPI on the Golgi membrane [23–25]. Examination by electron microscopy (EM) revealed that this situation transformed the Golgi membrane into coated buds and vesicles, which led to the conclusion that ARF1 and coatomer are the only cytosolic factors needed to reconstitute COPI vesicle formation [42]. Moreover, because GTPγS acts by inhibiting the ability of the GAP to catalyze ARF deactivation, it was further concluded that the GAP acts after vesicle formation to uncoat COPI vesicles, which is needed for vesicles to fuse with their target compartment in completing a round of transport [43].

However, these views were upended by the subsequent identification of the GAP that acts in COPI transport, known as ARFGAP1, which enabled a more direct assessment of how the GAP acts. Whereas the addition of ARFGAP1 to the COPI vesicle reconstitution system, as expected, led to ARF1 being released from the Golgi membrane, ARFGAP1 was found unexpectedly to promote COPI vesicle formation by being a stoichiometric component of the coat [44]. In this regard, it is notable that the reconstitution of COPII vesicle formation from ER membrane, which transport cargoes from the ER to the Golgi, had revealed a similar role for the GAP that acts in COPII transport. Sar1 is the small GTPase that recruits the COPII complex to the ER membrane, and notably, Sec23 was found to act not only as a GAP that catalyzes Sar1 deactivation [45], but also as a component of the COPII coat [22]. As such, the finding that ARFGAP1 acts as a component of the COPI coat suggested a fundamentally conserved role for the GAP in vesicle biogenesis.

The subsequent refinement of the COPI vesicle reconstitution system led to the identification of another protein critical for COPI vesicle formation. To determine whether the Golgi membranes contain peripheral membrane protein(s) critical for COPI vesicle formation, they were washed with high stringency to release peripheral proteins. When the washed Golgi membranes were then used for the reconstitution system, the addition of ARF1, ARFGAP1, and coatomer was no longer sufficient to form COPI vesicles. This finding led to the subsequent identification of BFA-ADP-Ribosylated Substrate (BARS) as another factor critical for COPI vesicle formation [46]. BARS acts at the fission stage [46], when the neck of COPI buds undergo constriction, resulting in the release of vesicles from the Golgi membrane. This role of BARS was found to be regulated by two factors, with palmitoyl-coenzyme A (p-CoA) promoting, and nicotinamide adenine dinucleotide (reduced form, NADH) inhibiting, the role of BARS in fission [46]. The finding that p-CoA promotes BARS in vesicle fission also provided an explanation for a previous finding that p-CoA is needed to reconstitute COPI vesicles from Golgi membrane [47]; however, a mechanistic explanation had remained elusive.

BARS acting in COPI vesicle fission also suggested a puzzle. Prior to its discovery as a fission factor, BARS had been known as a transcription regulator, and was named C-terminal binding protein 1 (CtBP1) [48]. The CtBP family of transcription regulators are encoded by two genes, CtBP1 and CtBP2, and notably, the deletion of both genes still allowed embryonic fibroblasts to survive [49], which seemingly contradicted the essential role of COPI-mediated transport in cell viability. To elucidate a reconciling explanation, embryonic fibroblasts that had both CtBP genes deleted were found not to use BARS for COPI vesicle fission, but instead utilize endophilin B [50]. Notably, this role of endophilin B did not merely reflect developmental plasticity in response to the deletion of the CtBP genes, as additional studies revealed that some mammalian cells use BARS while others use endophilin B for COPI vesicle fission [50]. Such mechanistic flexibility in COPI vesicle formation had been unexpected, as all the key factors that had been discovered at the time had shown conserved function from yeast to man.

Further studies on how BARS acts in COPI vesicle fission led to another key finding. To examine whether BARS bends membrane, thus explaining its role in vesicle fission, liposomes were generated using defined lipids. When these liposomes were incubated with BARS, tubulation was observed, suggesting that BARS acts in vesicle fission by having a direct ability to bend membrane [51]. Notably, BARS could only deform liposomes in the presence of phosphatidic acid (PA) generated by phospholipase D2 (PLD2) [51]. The significance of this finding comes from the consideration that lipids had been known to promote vesicle formation by recruiting proteins to bend membrane, but whether lipids could participate more directly in membrane bending had been less clear. Thus, the finding that PA cooperates with BARS for COPI vesicle fission provided key evidence that a lipid can contribute directly to membrane bending in a physiologic event.

Subsequently, how PA directly promotes membrane bending for COPI vesicle fission was elucidated. The relative space occupied by the headgroup versus that occupied by the acyl chains of lipids had been proposed to explain how lipids could bend membrane. This explanation, based on lipid geometry, posits that when greater space is occupied by the acyl chains, a cone geometry is promoted. In contrast, when greater space is occupied by the headgroup, an inverted-cone geometry is promoted. In the context of vesicle formation, inverted-cone geometry promotes positive membrane curvature, which drives the budding stage, while cone geometry promotes negative membrane curvature, which drives the fission stage [52, 53] (Fig. 2.1). Supporting this lipid-based mechanism acting in vesicle biogenesis, the late stage of COPI vesicle fission was found to require the conversion of PA to diacylglycerol (DAG), which leads to a more cone-shaped lipid [54]. Notably, the further elucidation of how these key lipids act revealed that another type of lipid geometry is needed; namely, that both PA and DAG must have shortened acyl chains [54]. Explaining how these shorter lipids act, molecular simulation studies elucidated that they possess physical properties that are conducive to generating extreme negative membrane curvature that is needed to drive vesicle fission to completion. These include decreasing membrane thickness and rigidity, as well as increasing membrane fluidity [54].

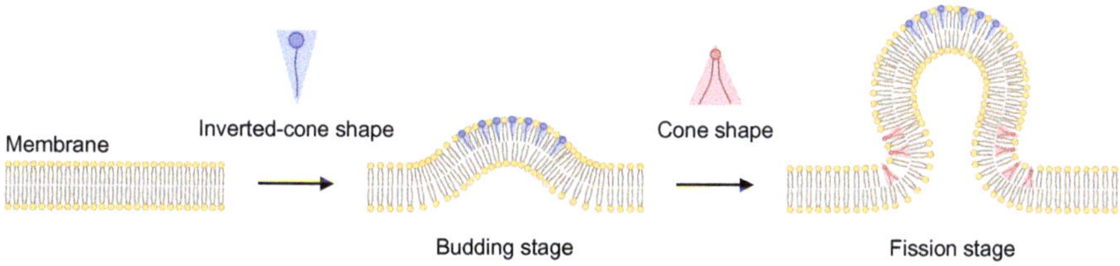

Fig. 2.1 How lipid geometry promotes vesicle formation. Inverted-cone shaped lipids promote the budding stage by inducing positive membrane curvature, while cone-shaped lipids promote the fission stage by inducing negative membrane curvature

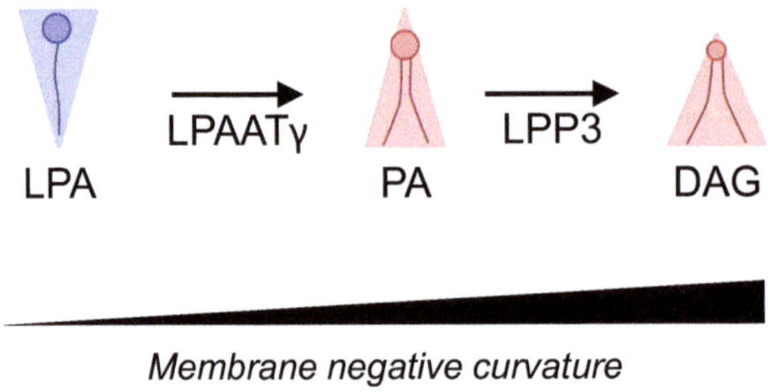

Fig. 2.2 Lipid geometry promoting the sub-stages of COPI vesicle fission. The conversion of LPA, which has one acyl chain, to PA, which has two acyl chains, results in a more cone-shaped lipid to promote early fission. Subsequently, the conversion of PA, which has a larger headgroup, to DAG, which has a smaller headgroup, results in an even more cone-shaped lipid to promote late fission

Besides PLD activity that converts phosphatidylcholine (PC) to PA [55], PA can also be generated from lysophosphatidic acid (LPA) through LPA acyltransferase (LPAAT) activity [56]. When a pharmacological inhibitor of LPAAT activity was added the COPI reconstitution system, vesicle formation was blocked. EM examination of the Golgi membranes revealed that COPI buds underwent tubulation [57]. This observation provided an explanation for why PLD and LPAAT activities have non-redundant roles in COPI vesicle formation, despite both activities resulting in the generation of PA. Because inhibition of PLD activity results in COPI buds with constricted necks [51], while inhibition of LPAAT activity result in COPI buds extending their necks to become tubules [57], these findings revealed that the LPAAT and PLD activities are needed for early and late fission, respectively. This conclusion is also consistent with how lipid geometry is predicted to promote vesicle formation. Besides altering the headgroup, changes in lipid geometry can also be achieved by altering the acyl chains. When compared to PA, LPA adopts a more inverted-cone shape by having one acyl chain rather than two. Thus, the transition from LPA to PA and then to DAG, which leads to progressively more cone-shaped lipids, is predicted by lipid geometry to increase the negative membrane curvature of the bud neck in driving vesicle fission (Fig. 2.2).

COPI Tubule Formation

The identification of key lipids needed to generate COPI vesicles also led to the subsequent revelation that COPI is involved in generating tubules that connect the Golgi cisternae for the faster mode of anterograde Golgi transport. After the initial discovery that tubules connecting the Golgi cisternae can transport cargo [38], a subsequent study identified the requirement of a member of the phospholipase A2 (PLA2) family of lipid enzymes, cytosolic PLA2 alpha (cPLA2α), in the formation

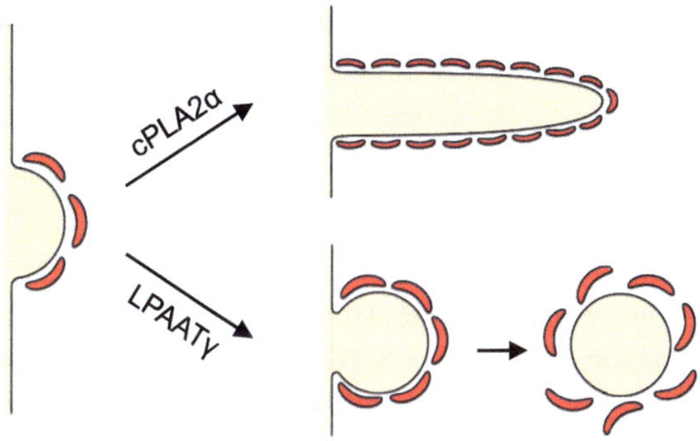

Fig. 2.3 Two opposing lipid enzyme activities determine whether COPI buds become vesicles or tubules. The conversion of LPA to PA by LPAAT activity directs COPI buds to become vesicles, while the opposite reaction that converts PA to LPA by PLA2 activity directs COPI buds to become tubules

of these tubules [58]. Furthermore, targeting against cPLA2α slowed VSVG transport through the Golgi [58]. Thus, because PLA2 activity opposes LPAAT activity by converting PA to LPA, an intriguing possibility was that COPI could be involved in generating cPLA2α-dependent Golgi tubules. Indeed, when cPLA2α was added to the COPI vesicle reconstitution system, COPI buds were found to become tubules [57]. Complementing this finding, the acute targeting against COPI in cells, by microinjecting an antibody against coatomer, was found to reduce the level of tubules that connect the Golgi cisternae and also slow VSVG transport through the Golgi [57]. These findings not only confirmed that COPI acts in generating cPLA2-α-dependent Golgi tubules for anterograde Golgi transport, but also revealed a mechanistic relationship between how COPI vesicles and tubules are formed. When LPAAT activity predominates, COPI buds are directed to become vesicles. In contrast, when PLA2 activity predominates, COPI buds are directed to become tubules (Fig. 2.3).

The revelation that COPI generates tubules for anterograde Golgi transport and vesicles for retrograde Golgi transport also raised a key mechanistic question: how can the same coat complex sort cargoes into the two opposing directions of Golgi transport? A hint was suggested by a previous study reporting that the ability of the small GTPase Cdc42 to induce cellular transformation requires its interaction with coatomer [59]. In this study, locking Cdc42 to either its GTP-bound or GDP-bound state inhibited VSVG transport, but a point mutation that enables Cdc42 to cycle faster between these two nucleotide-bound states was found to enhance VSVG transport and promote cellular transformation [59]. However, as COPI at the time had only been known to generate vesicles which operate in the retrograde direction, it had remained unclear how Cdc42 promotes VSVG transport, which occurs in the anterograde direction.

An explanation was subsequently elucidated by the revelation that Cdc42 has a role in COPI tubular transport. When the fast-cycling mutant of Cdc42 was added to the COPI reconstitution system, retrograde cargo was found to be prevented from being sorted into COPI tubules. Mechanistically, this finding was explained by Cdc42 having a dilysine motif that competes with the dilysine motif on retrograde cargoes for binding to coatomer [60]. The fast-cycling Cdc42 mutant was also found to promote COPI tubule formation, which was explained by Cdc42 having an intrinsic ability to induce membrane tubulation [60]. Thus, Cdc42 provided the specificity by which cargoes are sorted into COPI vesicles versus tubules, as well as modulating the ability of COPI to act in forming vesicles versus tubules.

These elucidated roles of Cdc42 also led to new insights into how Golgi transport occurs. The GTPase cycle of Cdc42 had been known to be regulated by intracellular signaling instigated by growth factors. This consideration led to the further elucidation that the ability of Cdc42 to modulate bidirectional Golgi transport is in turn regulated by growth-factor signaling [60]. Thus, Golgi transport does not act autonomously, but rather, is modulated by environmental cues received from outside of the cell.

Conclusion and Perspective

Our understanding of transport at the Golgi has evolved over the years. Initially, the cisternal maturation model was suggested to explain anterograde Golgi transport, which entails the Golgi cisternae acting as carriers. Subsequently, the vesicular transport model was proposed, which views Golgi cisternae as static compartments with COPI vesicles mediating anterograde transport across the cisternal stacks. The cisternal maturation model then regained support in explaining Golgi transport, which was modified to posit that anterograde Golgi transport is achieved by the Golgi cisternae acting as carriers, while COPI vesicles act in retrograde transport. In this modified model, COPI vesicles play a key role in maintaining the dynamic identity of the cis, medial, and trans cisternae of the Golgi, by transporting their residents in the retrograde direction in the face of the cisternae moving in the anterograde direction. More recently, the cisternal maturation model has been further modified to include a faster mode of anterograde transport mediated by tubules connecting the Golgi cisternae. Contributing to this evolving understanding of Golgi transport, advancements in the understanding of how COPI generates vesicles have led to the revelation that COPI also generates Golgi tubules that connect the Golgi cisternae. Thus, we anticipate that future studies on these two types of COPI carriers will result in an even more detailed understanding of how Golgi transport occurs.

Acknowledgements This work is funded by grants to VWH from the National Institutes of Health (R37GM058615) and to SYP from National Research Foundation of Korea (RS-2023-00208127, RS-2024-00344154) and Suh Kyungbae Foundation SUHF-24010035, Samsung Science and Technology Foundation SRFC-MA2402-13.

Compliance with Ethical Standards The authors declare no conflicts of interest.

References

1. Glick BS, Nakano A (2009) Membrane traffic within the Golgi apparatus. Annu Rev Cell Dev Biol 25:113–132
2. Nakano A, Luini A (2010) Passage through the Golgi. Curr Opin Cell Biol 22:471–478
3. Klumperman J (2011) Architecture of the mammalian Golgi. Cold Spring Harb Perspect Biol 3: a005181
4. Farquhar MG, Palade GE (1981) The Golgi apparatus (complex)-(1954–1981)—from artifact to center stage. J Cell Biol 91:77s–103s
5. Dunphy WG, Rothman JE (1985) Compartmental organization of the Golgi stack [Review]. Cell 42:13–21
6. Pfeffer SR, Rothman JE (1987) Biosynthetic protein transport and sorting by the endoplasmic reticulum and Golgi. Annu Rev Biochem 56:829–852
7. Fries E, Rothman JE (1980) Transport of vesicular stomatitis virus glycoprotein in a cell-free extract. Proc Natl Acad Sci USA 77:3870–3874
8. Rothman JE, Fries E (1981) Transport of newly synthesized vesicular stomatitis viral glycoprotein to purified Golgi membranes. J Cell Biol 89:162–168
9. Fries E, Rothman JE (1981) Transient activity of Golgi-like membranes as donors of vesicular stomatitis viral glycoprotein in vitro. J Cell Biol 90:697–704
10. Balch WE, Dunphy WG, Braell WA, Rothman JE (1984) Reconstitution of the transport of protein between successive compartments of the Golgi measured by the coupled incorporation of N-acetylglucosamine. Cell 39:405–416
11. Balch WE, Glick BS, Rothman JE (1984) Sequential intermediates in the pathway of intercompartmental transport in a cell-free system. Cell 39:525–536
12. Orci L, Glick BS, Rothman JE (1986) A new type of coated vesicular carrier that appears not to contain clathrin: its possible role in protein transport within the Golgi stack. Cell 46.171–184
13. Melancon P et al (1987) Involvement of GTP-binding "G" proteins in transport through the Golgi stack. Cell 51:1053–1062
14. Orci L, Malhotra V, Amherdt M, Serafini T, Rothman JE (1989) Dissection of a single round of vesicular transport: sequential intermediates for intercisternal movement in the Golgi stack. Cell 56:357–368
15. Malhotra V, Serafini T, Orci L, Shepherd JC, Rothman JE (1989) Purification of a novel class of coated vesicles mediating biosynthetic protein transport through the Golgi stack. Cell 58:329–336
16. Waters MG, Serafini T, Rothman JE (1991) "Coatomer": a cytosolic protein complex containing subunits of non-clathrin-coated Golgi transport vesicles. Nature 349:248–251
17. Serafini T et al (1991) A coat subunit of Golgi-derived non-clathrin-coated vesicles with homology to the clathrin-coated vesicle coat protein beta-adaptin. Nature 349:215–220
18. Stenbeck G et al (1992) Gamma-COP, a coat subunit of non-clathrin-coated vesicles with homology to Sec21p. FEBS Lett 314:195–198
19. Stenbeck G et al (1993) Beta prime COP, a novel subunit of coatomer. EMBO J 12:2841–2845
20. Kuge O et al (1993) Zeta-COP, a subunit of coatomer, is required for COP-coated vesicle assembly. J Cell Biol 123:1727–1734

21. Hara-Kuge S et al (1994) En bloc incorporation of coatomer subunits during the assembly of COP-coated vesicles [published erratum appears in J Cell Biol 1994 Jul;126(2):589]. J Cell Biol 124:883–892
22. Barlowe C et al (1994) COPII: a membrane coat formed by sec proteins that drive vesicle budding from the endoplasmic reticulum. Cell 77:895–907
23. Serafini T et al (1991) ADP-ribosylation factor is a subunit of the coat of Golgi-derived COP-coated vesicles: a novel role for a GTP-binding protein. Cell 67:239–253
24. Donaldson JG, Kahn RA, Lippincott-Schwartz J, Klausner RD (1991) Binding of ARF and beta-COP to Golgi membranes: possible regulation by a trimeric G protein. Science 254:1197–1199
25. Donaldson JG, Cassel D, Kahn RA, Klausner RD (1992) ADP-ribosylation factor, a small GTP-binding protein, is required for binding of the coatomer protein beta-COP to Golgi membranes. Proc Natl Acad Sci USA 89:6408–6412
26. Lippincott-Schwartz J, Yuan LC, Bonifacino JS, Klausner RD (1989) Rapid redistribution of Golgi proteins into the ER in cells treated with brefeldin A: evidence for membrane cycling from Golgi to ER. Cell 56:801–813
27. Donaldson JG, Lippincott-Schwartz J, Bloom GS, Kreis TE, Klausner RD (1990) Dissociation of a 110-kD peripheral membrane protein from the Golgi apparatus is an early event in brefeldin A action. J Cell Biol 111:2295–2306
28. Cosson P, Letourneur F (1994) Coatomer interaction with di-lysine endoplasmic reticulum retention motifs. Science 263:1629–1631
29. Letourneur F et al (1994) Coatomer is essential for retrieval of dilysine-tagged proteins to the endoplasmic reticulum. Cell 79:1199–1207
30. Glick BS, Luini A (2011) Models for Golgi traffic: a critical assessment. Cold Spring Harb Perspect Biol 3:a005215
31. Bonfanti L et al (1998) Procollagen traverses the Golgi stack without leaving the lumen of cisternae: evidence for cisternal maturation. Cell 95:993–1003
32. Mironov AA et al (2001) Small cargo proteins and large aggregates can traverse the Golgi by a common mechanism without leaving the lumen of cisternae. J Cell Biol 155:1225–1238
33. Love HD, Lin CC, Short CS, Ostermann J (1998) Isolation of functional Golgi-derived vesicles with a possible role in retrograde transport. J Cell Biol 140:541–551
34. Martinez-Menarguez JA et al (2001) Peri-Golgi vesicles contain retrograde but not anterograde proteins consistent with the cisternal progression model of intra-Golgi transport. J Cell Biol 155: 1213–1224
35. Gilchrist A et al (2006) Quantitative proteomics analysis of the secretory pathway. Cell 127: 1265–1281
36. Matsuura-Tokita K, Takeuchi M, Ichihara A, Mikuriya K, Nakano A (2006) Live imaging of yeast Golgi cisternal maturation. Nature 441:1007–1010
37. Losev E et al (2006) Golgi maturation visualized in living yeast. Nature 441:1002–1006
38. Trucco A et al (2004) Secretory traffic triggers the formation of tubular continuities across Golgi sub-compartments. Nat Cell Biol 6:1071–1081
39. Patterson GH et al (2008) Transport through the Golgi apparatus by rapid partitioning within a two-phase membrane system. Cell 133:1055–1067
40. D'Souza-Schorey C, Chavrier P (2006) ARF proteins: roles in membrane traffic and beyond. Nat Rev Mol Cell Biol 7:347–358
41. Donaldson JG, Jackson CL (2011) ARF family G proteins and their regulators: roles in membrane transport, development and disease. Nat Rev Mol Cell Biol 12:362–375
42. Orci L, Palmer DJ, Amherdt M, Rothman JE (1993) Coated vesicle assembly in the Golgi requires only coatomer and ARF proteins from the cytosol. Nature 364:732–734
43. Tanigawa G et al (1993) Hydrolysis of bound GTP by ARF protein triggers uncoating of Golgi-derived COP-coated vesicles. J Cell Biol 123:1365–1371
44. Yang JS et al (2002) ARFGAP1 promotes the formation of COPI vesicles, suggesting function as a component of the coat. J Cell Biol 159:69–78

45. Yoshihisa T, Barlowe C, Schekman R (1993) Requirement for a GTPase-activating protein in vesicle budding from the endoplasmic reticulum. Science 259:1466–1468
46. Yang JS et al (2005) A role for BARS at the fission step of COPI vesicle formation from Golgi membrane. EMBO J 24:4133–4143
47. Pfanner N et al (1989) Fatty acyl-coenzyme A is required for budding of transport vesicles from Golgi cisternae. Cell 59:95–102
48. Chinnadurai G (2002) CtBP, an unconventional transcriptional corepressor in development and oncogenesis. Mol Cell 9:213–224
49. Hildebrand JD, Soriano P (2002) Overlapping and unique roles for C-terminal binding protein 1 (CtBP1) and CtBP2 during mouse development. Mol Cell Biol 22:5296–5307
50. Yang JS et al (2006) Key components of the fission machinery are interchangeable. Nat Cell Biol 8:1376–1382
51. Yang JS et al (2008) A role for phosphatidic acid in COPI vesicle fission yields insights into Golgi maintenance. Nat Cell Biol 10:1146–1153
52. McMahon HT, Gallop JL (2005) Membrane curvature and mechanisms of dynamic cell membrane remodelling. Nature 438:590–596
53. Graham TR, Kozlov MM (2010) Interplay of proteins and lipids in generating membrane curvature. Curr Opin Cell Biol 22:430–436
54. Park SY et al (2019) The late stage of COPI vesicle fission requires short forms of phosphatidic acid and diacylglcerol. Nat Commun 10(1):3409
55. Frohman MA, Morris AJ (1999) Phospholipase D structure and regulation. Chem Phys Lipids 98:127–140
56. Bradley RM, Duncan RE (2018) The lysophosphatidic acid acyltransferases (acylglycerophosphate acyltransferases) family: one reaction, five enzymes, many roles. Curr Opin Lipidol 29:110–115
57. Yang JS et al (2011) COPI acts in both vesicular and tubular transport. Nat Cell Biol 13:996–1003
58. San Pietro E et al (2009) Group IV phospholipase A(2)alpha controls the formation of inter-cisternal continuities involved in intra-golgi transport. PLoS Biol 7:e1000194
59. Wu WJ, Erickson JW, Lin R, Cerione RA (2000) The gamma-subunit of the coatomer complex binds Cdc42 to mediate transformation. Nature 405:800–804
60. Park SY, Yang JS, Schmider AB, Soberman RJ, Hsu VW (2015) Coordinated regulation of bidirectional COPI transport at the Golgi by CDC42. Nature 521:529–532

Chapter 3
The Functions of RAB GTPases at the Golgi Complex

Bruno Goud

Abstract The RAB family in humans comprises approximately 70 members with about one third localizing to the Golgi apparatus and associated compartments. This significant localization reflects the number and diversity of transport pathways operating within this central cellular hub. We focus in this review on the molecular mechanisms by which RAB GTPases regulate membrane trafficking at the Golgi level.

Keywords RAB GTPases · Golgi complex · Intracellular transport · Golgins · Tethering complexes

In eukaryotic cells, RAB GTPases function as critical regulators of vesicular trafficking. Acting as molecular switches, they cycle between an active GTP-bound state and an inactive GDP-bound state. In their active conformation, they interact with a diverse range of effector proteins such as molecular motors, tethering complexes, lipid modifying enzymes (kinases and phosphatases), and signaling proteins [172, 55, 59].

The RAB family in humans comprises approximately 70 members with about one third localizing to the Golgi apparatus and associated compartments, such as the ERGIC (ER-Golgi intermediate compartment, also referred to as IC for intermediate compartment) and TGN (*trans*-Golgi network). This significant localization reflects the number and diversity of transport pathways operating within this central cellular hub.

The organization of the Golgi complex is tightly linked to vesicular transport processes, and disruptions in RAB function can profoundly impact its architecture, biogenesis and homeostasis. These aspects have been thoroughly reviewed elsewhere [75, 76, 173]. Here we focus on the molecular mechanisms by which RAB GTPases regulate membrane trafficking at the Golgi level.

B. Goud (✉)
Institut Curie, PSL Research University, Paris, France
e-mail: bruno.goud@curie.fr

A. Nakano, J. Saraste (eds.), *The Golgi Network, Volume II*, Subcellular Biochemistry 111, https://doi.org/10.1007/978-3-032-16833-7_3

Pre-Golgi Trafficking

The Golgi apparatus continuously receives a flow of molecules and membranes from the endoplasmic reticulum (ER) via anterograde transport. This forward flow is counterbalanced by a retrograde flow which facilitates the recycling of proteins (such as cargo receptors and components on the transport machinery) back to the ER. Among the best-characterized RAB proteins involved in pre-Golgi trafficking are RAB1 and RAB2. RAB1 and RAB2 are ubiquitously expressed as two isoforms (RAB1A and RAB1B, RAB2A and RAB2B) that share a high degree of sequence similarity (92 and 86%, respectively). At steady state, RAB1 and RAB2 predominantly associate with the ERGIC/IC, a network of pleiomorphic membranes situated between ER exit sites (ERES) and the Golgi apparatus. Additionally, they are also present on the first Golgi cisterna ("*cis*-Golgi") [14, 127].

RAB1 regulates both anterograde and retrograde membrane traffic between the ER and the *cis*-Golgi [112, 154]. Its role in the anterograde pathway is relatively well established, specifically in the targeting of ER-derived transport vesicles to *cis*-Golgi membranes. This process involves the TRAPIII complex, a RAB1 guanine-exchange factor (GEF) (for recent reviews on TRAPP complexes, see [5, 36]), and two RAB1 effectors, the tethering factor p115 (also known as USO1) [2, 3] and GM130 (also known as GOLGA2) [160]. GM130, a member of the golgin family, extended coiled-coil proteins that decorates the Golgi membranes, plays a dual role in maintaining Golgi architecture and capturing transport vesicles near the Golgi [44, 100]. This is achieved through tethering motifs located in the membrane-distal amino terminal region of the golgins [164]. RAB1 appears to also have a dual role in targeting ER-derived transport vesicles to Golgi membranes. First, it recruits p115 to COPII vesicles, which then interacts with the amino-terminal region of GM130 to facilitate vesicle capture [2, 101]. Second, RAB1 can directly interact with GM130 independently of p115 [97]. Notably, the RAB1-binding site of GM130 is located outside its amino-terminal region, suggesting that RAB1 acts downstream of vesicle capture. One possible function of RAB1 is to induce a conformational change in GM130, bending its structure to bring vesicles closer to the SNARE machinery required for their fusion with Golgi membranes. A similar mechanism has been demonstrated for RAB5 and the endosomal tether EEA1 [99]. As EEA1, GM130 has been shown to adopt a flexible rod-like structure [60].

The role of RAB1 in the Golgi-to-ER retrograde pathway is less well understood than its function in the anterograde pathway. RAB1 is known to localize to retrograde tubules originating from ERGIC/IC [86, 107, 122] and regulates retrograde transport of Golgi glycosyltransferases [35]. This anterograde function of RAB1 may be mediated by its recruitment of GBF1, the exchange factor of ARF1, a small GTPase essential for the formation of COPI vesicles [4].

Like RAB1, RAB2 is involved in bi-directional membrane traffic between ER and *cis*-Golgi [153, 154]. One of the RAB2's effectors is GMAP-210 (Golgi Microtubule-Associated Protein of 210 kDa, also known as TRP11), a golgin localized at the *cis*-Golgi [118]. In addition to its role in linking the Golgi apparatus to microtubules [119], GMAP-210 has been implicated in membrane trafficking [109, 120] although some studies have reported conflicting results [167]. Notably, GMAP-210 can capture ER- *or cis*-Golgi-derived vesicles both *in vivo* and *in vitro* [83, 164]. Unlike GM130, the capture mechanism relies on an ALPS (Amphipathic Lipid Packing Sensor) motif located at its N-terminus. This motif is capable of sensing the highly curved membranes characteristic of transport vesicles [27, 28].

In addition to targeting ER-derived transport vesicles to *cis*-Golgi membranes, RAB1 and RAB2 perform several other essential functions. First, they operate between the ER and the ERGIC/IC where p115, GM130 and GMAP-210 are localized [3, 87, 120], suggesting their involvement in capturing transport vesicles budding from the ER. Second, RAB1 and RAB2 help to organize membrane domains within the ERGIC/IC. Saraste and colleagues have proposed that RAB1 defines a stable sub-compartment of IC membranes known as the pericentrosomal intermediate compartment (pcIC) which accumulates around the centrosome under certain conditions [86, 124]. This compartment containing RAB1 and RAB2 effectors, along with COPI coats may serve as a functional link between pre Golgi compartments and the endosomal system, particularly RAB11-positive recycling endosomes [125]. It also contributes to the secretion of proteins via an unconventional pathway that bypasses the Golgi apparatus [123, 126]. Recently, RAB1 and RAB2 were identified as key regulators of a specialized sub-compartment within the IC/ERGIC that is essential for unconventional protein secretion (UPS). This function is mediated through their interactions with the *trans*-membrane Emp24 domain-containing protein (TMED10) and the microtubule-based motor KIF5B, respectively [152].

RAB1 and RAB2 also have roles beyond the interface between the IC/ERGIC and cis-Golgi. RAB1 is known to associate with early endosomes [98] and has recently been identified on a novel class of clathrin-associated carriers involved in rapid recycling [166]. Acute depletion of RAB1 results in perinuclear clustering of early endosomes and delays in transferrin recycling [50]. Despite these findings, the molecular mechanisms by which RAB1 regulates transport events in the early endocytic pathway remain unclear. RAB2 has been shown to interact *in vitro* with two golgins located at the *trans* side of the Golgi apparatus (GCC185 and Golgin-245) [52, 141]. Additionally, evidence suggests that RAB2 may play a role in Golgi exit, particularly in specific cell types such as neurons. For example, RAB2 regulates the biogenesis of pre-synaptic precursor vesicles in *Drosophila* motoneurons [43] and is involved in the biogenesis and transport of dense core vesicle in *C. elegans* [1, 29, 81].

Finally, RAB1 and RAB2 play critical roles in maintaining Golgi structure, likely through their interactions with various golgins and components of the COG complex. For RAB1, these include GM130, GMAP-210, the *cis*-golgin Golgin-84 and the *trans*-golgin TMF (TATA element modulatory factor, also known as ARA160), while RAB2 interacts with Golgin-245 [22, 138]. Both RAB1 and RAB2 are also critical regulators of autophagy, functioning in autophagosome biogenesis and maturation [24, 34, 47, 48, 57, 80, 96, 170]. Notably, RAB1A, but not RAB1B, has been shown to be specifically required for stress-induced autophagy [47]. The dual roles of RAB1 and RAB2 in secretion and autophagy depend on interactions with distinct sets of effectors, such as the HOPS complex and ATG9 [34, 66, 80]. However, the coordination between these processes remains poorly understood. Interestingly, in yeast, the RAB1 ortholog Ypt1 is activated by the same complex, TRAPIII, to regulate both secretion and autophagy, suggesting a potential mechanism for integration between these pathways [65].

In addition to RAB1 and RAB2, four other RAB proteins (RAB6, RAB18, RAB41 and RAB43) are implicated in ER to Golgi and Golgi to ER trafficking. Among these, RAB6, which predominantly associates with *trans*-Golgi and TGN membranes at steady state, also plays a role in intra-Golgi and post-Golgi trafficking (discussed below). RAB6 is central to a COPI-independent retrograde pathway [40, 161] which has been shown to mediate the transport of internalized Shiga toxin to the ER [64]. Although the endogenous cargoes of this pathway remain poorly characterized, evidence suggests they include Golgi-resident glycosylation enzymes that recycle through the ER [110, 131]. Additionally, this pathway may recycle anterograde cargoes lacking export signals [31]. A recent review by Dornan and Simpson provides a detailed discussion of the possible mechanisms underlying the RAB6-dependent retrograde pathway [26]. The main components of this pathway include: (i) non-muscle myosin II, which facilitates the fission of RAB6-positive transport carriers from Golgi membranes [94, 95], (ii) ZW10/RINT1, part of the NRZ (NAG-RINT1-ZW10) complex, proposed by the Storrie group to mediate the fusion of retrograde carriers with ER membranes [84, 151], (iii) Bicaudal D2, a dynein adaptor involved in the motility of these carriers [88, 90, 137, 139, 174], and (iv) SCFD1 (Sec1 family domain containing protein 1), recently identified by the Simpson group as an crucial regulator of the RAB6-dependent retrograde pathway [53]. This intricate network highlights the complex role of RAB6 in coordinating retrograde transport and maintaining Golgi-ER homeostasis.

RAB18 has been found in several cellular compartments but is predominantly associated with ER membranes (for a review, see [19]). It plays a key role in maintaining the normal structure of the ER [38]. Evidence suggests that RAB18 functions in a COPI-independent Golgi to ER retrograde pathway, likely overlapping with the pathway regulated by RAB6 [20]. The two RAB proteins may have complementary roles in tethering retrograde transport carriers to ER membranes. Supporting this idea, RAB18 has been shown to recruit the NRZ complex on lipid droplets (LDs), facilitating their tethering with ER membranes [106, 165]. This suggests that similar protein machinery might be shared between retrograde transport carriers and LDs. Furthermore, a functional link between RAB6 and RAB18 was also observed *in vivo*, as RAB6 depletion in luminal cells of the mammary gland affects the expression of RAB18 [13].

RAB41, a member of the RAB6 family, shares 82% sequence identity with RAB6B [108, 148]. While its precise localization and role in transport remain unclear, siRNA-mediated depletion of RAB41 in HeLa cells has been shown to partially inhibit ER to Golgi transport of VSV-G [77].

GFP-tagged RAB43 has been localized to multiple compartments, including the *cis*-Golgi [20] and the TGN [33]. Similar to RAB1, it can be activated by TRAPPIII

[49] and has been implicated in ER to Golgi trafficking [20], particularly in the trafficking of GPCRs [72]. RAB43 interacts with the COG complex [162] and regulates the sorting of specific membrane protein cargoes through the medial Golgi [18]. Additionally, it plays a role in the retrograde transport of Shiga toxin between endosomes and the Golgi apparatus [33].

Intra-Golgi Trafficking

Several models have been proposed to explain the mechanism of cargo transport across Golgi cisternae, and the maintenance of the Golgi apparatus's distinct compartments [8, 93, 103, 111]. While none of these models fully account for all aspects of Golgi functioning, substantial experimental evidence argues in favor of the cisternal maturation model. According to this model, COPI-coated retrograde transport vesicles mediate the backflow of Golgi-resident proteins, such as glycosylation enzymes and sugar transporters, to earlier cisternae, resulting in their maturation [17].

Golgi-resident vesicles are thought to be initially captured by golgins. In a study by Wong and Munro [164], three golgins—TMF, golgin-84 and GMAP-210—were shown to capture vesicles containing Golgi-resident proteins. Once captured, these vesicles could be tethered to Golgi membranes by the COG (Conserved Oligomeric Golgi) complex. Alternatively, golgins and COG complex may function in parallel, enhancing the efficiency and specificity of vesicle tethering [144, 162].

To date, two RAB proteins, RAB6 and RAB33B, have been directly implicated in intra-Golgi retrograde trafficking [175, 176, 177, 25, 156, 161]. Research from the Lupashin and Storrie groups has demonstrated that RAB6 initiates a retrograde pathway at the *trans*-Golgi involving the COG complex and COPI vesicles [142, 151]. RAB33B, localized on medial Golgi membranes, likely functions downstream of RAB6, potentially as part of a RAB cascade [115, 146].

The molecular mechanisms underlying the RAB6-dependent retrograde pathway remain poorly characterized. A likely key regulator is TMF, a protein that interacts with RAB6, its yeast counterpart Ypt6 in *S. cerevisiae*, and the COG complex [32, 92, 168]. Although the precise RAB6-binding site on TMF's C-terminus, has not been mapped, RAB6 depletion reduces Golgi-associated TMF, indicating that RAB6 facilitates TMF recruitment to Golgi membranes [168]. Notably, RAB6 depletion in HeLa cells increases the number of coated vesicles, including COPI vesicles, in the Golgi region [75]. A similar phenotype was observed in mouse embryonic fibroblast (MEFs) derived from *Rab6a* knockout mice [6]. Furthermore, the loss of RAB6 function epistatically suppresses Golgi dispersal caused by COG inactivation [151]. These findings suggest that RAB6 acts upstream of tethering processes mediated by TMF and the COG complex. Supporting this hypothesis, recent studies employing rapid depletion of COG proteins demonstrated that RAB6 remains associated with Golgi membranes even after COG depletion. Meanwhile, TMF remains localized to vesicles that depend on the COG complex [150]. Together, these observations highlight the upstream regulatory role of RAB6 in Golgi tethering and retrograde trafficking pathways.

In addition to TMF, RAB6 interacts with two other golgins, giantin and GORAB [30, 121]. The interaction between RAB6 and giantin is critical for Golgi recovery after alcohol-induced Golgi fragmentation, suggesting that these proteins play a role in the fusion of Golgi membranes [178]. Giantin also interacts with RAB1 [121]; however, whether RAB6- and RAB1-dependent transport pathways are functionally connected via giantin remains unclear. GORAB (also named SCYLBP1), whose mutations cause gerodermia osteodysplastica characterized by skin and bone disorders [54], is recruited on *trans*-Golgi membranes by RAB6 and ARF5 [30]. It functions as a COPI scaffolding factor required for efficient recycling of glycosyltransferases [163]. Whether RAB6's role is limited to recruiting GORAB to Golgi membranes, or if GORAB participates in the RAB6-dependent intra-Golgi retrograde transport pathway remains unknown. Finally, in addition to its role in retrograde transport, evidence exists that RAB6 is required for rapid anterograde transport between *medial*- and *trans*-Golgi [23].

Other RAB proteins have been identified as associated with Golgi membranes and classified as "bona fide" Golgi RABs in [173], including RAB19, RAB34, RAB36 and RAB39B. However, since no evidence supports a role of RAB19 in intra-Golgi traffic, we will consider this protein in the next section on post-Golgi trafficking.

Endogenous RAB34 was initially observed to associate with the Golgi apparatus in NRK cells and was found to regulate lysosome positioning through its interaction with RAB-interacting lysosomal protein (RILP), a RAB7 effector [157]. Recent studies have elucidated mechanisms underlying this process, identifying DENND6A, a RAB34 GEF, and Arl8b as key players [69]. While RAB34 has been implicated in intra-Golgi transport of VSV-G and constitutive secretion in response to hyperglycemia [41, 42], these roles have not been extensively explored. To date, the most well-characterized function of RAB34 is its involvement in ciliogenesis (Oguchi et al. 2020) [37, 105, 149].

RAB36 is closely related to RAB34 exhibiting 54% amino acid identity. Like RAB34, it also interacts with RILP and was shown to regulate the spatial distribution of lysosomes and the retrograde transport of melanosomes in melanocytes [15, 89]. While no evidence supports a role for RAB36 in intra-Golgi transport, its regulation by RUTBC2, a RAB9A effector, and GAP for both RAB36 and RAB34, suggests a possible involvement in late endosome-to-Golgi transport [104].

RAB39B, on the other hand, is predominantly expressed in brain tissues and has drawn significant attention due to its association with neurological disorders. Mutations in *RAB39B* are linked to X-linked intellectual disability with autism, epilepsy, and macrocephaly, as well as to early-onset Parkinson's disease [39, 71]. RAB39B has been implicated in ER-to-Golgi transport of GluA2/GluA3 α-amino-3-hydroxy-5-methyl-4-isoxazolepropionic acid receptor (AMPAR) [91].

Post-Golgi Trafficking

The TGN consists of a complex network of membranes located at the *trans* side of the Golgi apparatus. It functions as a trafficking platform to sort and package newly synthesized proteins into specific transport carriers destined for the plasma membrane (exocytosis) and other intracellular compartments. Additionally, the TGN plays a pivotal role in endocytic recycling. Unlike the well-defined Golgi stacks, the precise boundaries of the TGN—particularly with recycling endosomes—are difficult to delineate and can vary depending on the cellular context. Numerous RAB proteins, including RAB3B/D, RAB6, RAB7B, RAB8A/B, RAB9A/B, RAB10, RAB11A/B, RAB13, RAB14, RAB20, RAB29, RAB30, and RAB31/RAB22B have been identified or predicted to associate with the TGN. Based on recent insights into TGN organization and function across mammals, yeast, *Drosophila,* and plants, Aki Nakano proposed to divide this compartment into Golgi-associated and Golgi-independent TGN, the later including recycling endosomes (REs) [102]. Following this framework and the literature, we focus on RAB6, RAB29, and RAB30, as they localize to Golgi-associated TGN and are relevant to the scope of this review. RAB8A/B, although initially implicated in vesicular transport between the TGN and the basolateral plasma membrane in polarized MDCK cells [58], and later in the localization of apical proteins in intestinal cells [128, 129] is excluded here. Both endogenous and fluorescent protein (FP)-tagged RAB8 are predominantly associated with REs and cytoplasmic vesicles with minimal detection on the Golgi-associated TGN [46, 51]. The same applies to RAB10, which is closely related to RAB8 (66% identity) [108, 130], and shares many overlapping functions [16]. Similarly, RAB7B and RAB9, which are involved in late endosome-to-TGN transport, are not included due to their predominant localization to late endosomes and lysosomes under steady state conditions [79, 113].

Several molecular machineries have been identified for the sorting and packaging of cargoes at the TGN (for a review see [145]). One such mechanism involves RAB6 and its effectors, Myosin II and KIF20A [94]. Although RAB6 does not directly participate in the sorting process, it plays a critical role in the formation of its own transport carriers. These RAB6 positive carriers, which bud from TGN membranes contain a variety of secretory cargoes, positioning RAB6 as a potential master regulator of post-Golgi "constitutive" trafficking [45, 179]. Additionally, RAB6 associates with CARTS (CARriers of the TGN to the cell Surface), which represent a distinct class of secretory carriers. Unlike RAB6-positive carriers of the constitutive pathway, CARTS lack specific constitutive cargoes [180]. In both pathways, RAB6 cooperates with RAB8 to regulate the docking and fusion of secretory vesicles with the plasma membrane [46]. The recruitment of RAB8 to RAB6 positive transport carriers is mediated by ARHGEF10, a RHOA GEF [134]. The movement of RAB6 positive vesicles along microtubules is mediated by several motors, including KIF1C, KIF5B, KIF13B and the dynein-dynactin complex ensuring efficient transport to their destinations [9, 70, 132].

Several transport pathways connecting the endosomal compartments with the TGN membranes have been described (for a recent review see [11]). One such pathway, regulated by RAB6, particularly the RAB6A' isoform, likely functions at the level of tethering/docking of endosome-derived vesicles with the TGN [21, 85]. This pathway involves the retromer complex [181], early/recycling endosomal v-SNAREs, (VAMP3/cellubrevin and VAMP4), the TGN-localized t-SNAREs syntaxin 6, syntaxin 16 and Vti1a, and the GARP complex [85]. One of the RAB6 effectors, RAB6IP1 (also known as DENND5) interacts with sorting nexin 1 (SNX1), one the component of the retromer complex [159]. Additionally, RAB6 interacts with VPS52, a subunit of the GARP complex [68,74]. Knockout of VPS52 in MDCK cells phenocopies, at least in part, the effects of RAB6 knockout [55], although the precise role of the GARP complex in the RAB6-regulated retrograde pathway remains unproven. Several cargoes have been identified in this pathway, which likely corresponds to the one regulated by Ypt6 in yeast [140, 155]. These cargoes include Shiga toxin, beta1-integrin, and the transmembrane protein LAT (Linker of Activated T cells) in CD4+ lymphocytes [12, 64, 133]. Additionally, RAB6 has been shown to interact *in vitro* with several *trans*-golgins, including GCC88, Golgin-97, GCC185 and Golgin-245 [7, 141]. However, the functional significance of RAB6 interactions with GCC88, Golgin-97 and Golgin-245 is unclear. While it is debated whether RAB6 participates in recruiting of GCC185 to Golgi membranes, this interaction has not been definitively proven [10, 56].

RAB19 is a tissue-specific RAB [82] with a *Drosophila* counterpart (dRAB19). GFP-tagged dRAB19 localizes to the Golgi apparatus and interacts with the *trans*-golgin dGCC88 [141]. GCC88 has been implicated in endosome to TGN transport [73] and can capture endosome-derived vesicles at the TGN [135, 136]. This suggests that RAB19 could regulate a retrograde transport pathway between endosomes and Golgi; however, this hypothesis remains to be fully established. Additionally, recent studies have highlighted a role for RAB19 in ciliogenesis [61].

RAB29 has garnered significant attention due to its role in regulating the localization and activity of the leucine-rich repeat protein kinase (LRRK2), which is often mutated in patients suffering from Parkinson's disease [78, 114, 147]. While the trafficking role of RAB29 remains poorly understood, it is known to localize to the TGN and participate in retrograde trafficking between endosomes and TGN [158].

Exogenous GFP-tagged RAB30 associates with all Golgi cisternae in HeLa cells but siRNA-mediated depletion of RAB30 does not significantly affect anterograde or retrograde transport along the secretory pathway [67]. RAB30 also localizes to the TGN and REs in multiple cell types, with a documented role in endosome-to-Golgi transport [171]. RAB30 interacts *in vitro* with several TGN-associated golgins, including golgin-97, GCC88, GCC185 and golgin-245 [52, 141], although the functional significance of these interactions in RAB30-mediated retrograde trafficking remains unclear. Recently RAB30 expression was shown to increase in rat liver during fasting, suggesting a potential role for RAB30 in hepatic lipoprotein secretion [143].

Conclusions and Perspectives

Numerous studies over the past few decades have established that RAB proteins play essential roles in the regulation of intracellular transport at the Golgi complex. Figure 3.1 summarizes the localization of Golgi-associated RABs and the key transport pathways they regulate. Despite these insights, our understanding of their precise functions remains fragmented. Many findings have been derived from experiments involving the overexpression of dominant active or dominant negative forms of RAB proteins, or long-term depletion strategies. However, these approaches can introduce artifacts, particularly due to potential cross-talk and functional redundancy between RAB proteins. To clarify the exact functions of Golgi-associated RAB proteins, more refined methods, such as acute depletion approaches, are necessary. Furthermore, most data come from cultured cell models, highlighting the need for animal models to explore the physiological roles of these proteins *in vivo*.

In parallel, new paradigms in intracellular transport have emerged, particularly concerning the role of protein condensates formed by liquid-liquid phase separation (LLPS) in short range vesicular transport, such as between Golgi stacks. This concept was recently illustrated in the case of synaptic vesicle transport [116]. LLPS could serve as a mechanism for organizing RAB-mediated trafficking, concentrating RAB proteins and their effectors at key sites. This would be particularly crucial since RAB proteins are present in low abundance on membranes, with their effector-binding affinity generally in the low micromolar range. Notably,

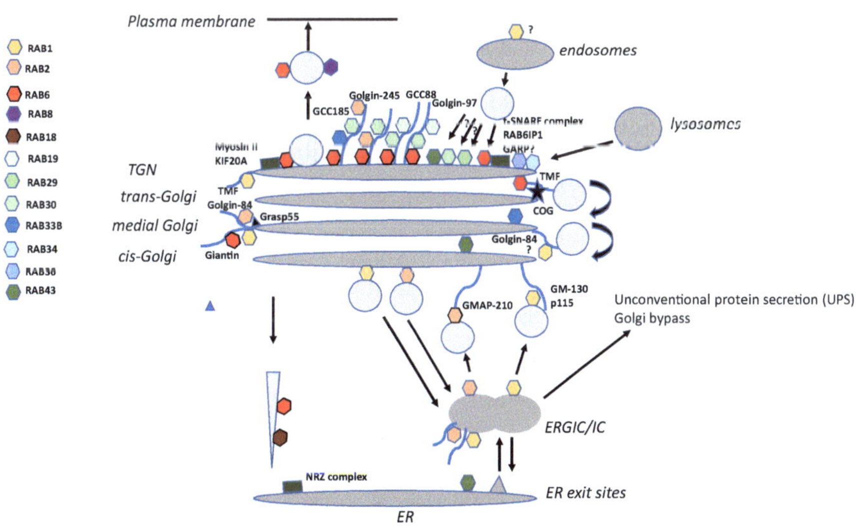

Fig. 3.1 Localization of Golgi-associated RABs and the key transport pathways they regulate. The question marks highlight insufficient data in the literature regarding RAB protein localization or regulated transport pathways (see text)

several RAB effectors exhibit LLPS properties, such as the RAB6 effector ELKS, whose condensates have been shown to capture RAB6-positive vesicles for efficient release at exocytic sites [63]. RAB effector condensates could also play a role in the formation of membrane domains [62]. Intriguingly, several golgin family proteins, including GM130 and GMAP-210, are also prone to phase separation [117, 169]. Further studies should investigate the role of phase separation mechanisms in the function of Golgi-associated RAB proteins.

Acknowledgements I thank Shaeri Mukherjee (USCF, USA) and Jaakko Saraste (University of Bergen, Norway) for critical reading of the manuscript.

References

1. Ailion M, Hannemann M, Dalton S, Pappas A, Watanabe S, Hegermann J, Liu Q, Han HF, Gu M, Goulding MQ, Sasidharan N, Schuske K, Hullett P, Eimer S, Jorgensen EM (2014) Two Rab2 interactors regulate dense-core vesicle maturation. Neuron 82(1):167–180. https://doi.org/10.1016/j.neuron.2014.02.017
2. Allan BB, Moyer BD, Balch WE (2000) Rab1 recruitment of p115 into a cis-SNARE complex: programming budding COPII vesicles for fusion. Science 289(5478):444–448. https://doi.org/10.1126/science.289.5478.444
3. Alvarez C, Fujita H, Hubbard A, Sztul E (1999) ER to Golgi transport: requirement for p115 at a pre-Golgi VTC stage. J Cell Biol 147(6):1205–1222. https://doi.org/10.1083/jcb.147.6.1205
4. Alvarez C, Garcia-Mata R, Brandon E, Sztul E (2003) COPI recruitment is modulated by a Rab1b-dependent mechanism. Mol Biol Cell 14(5):2116–2127. https://doi.org/10.1091/mbc.e02-09-0625
5. Bagde SR, Fromme JC (2023) The TRAPP complexes: discriminating GTPases in context. FEBS Lett 597(6):721–733. https://doi.org/10.1002/1873-3468.14557
6. Bardin S, Miserey-Lenkei S, Hurbain I, Garcia-Castillo D, Raposo G, Goud B (2015) Phenotypic characterisation of RAB6A knockout mouse embryonic fibroblasts. Biol Cell 107(12):427–439. https://doi.org/10.1111/boc.201400083
7. Barr FA (1999) A novel Rab6-interacting domain defines a family of Golgi-targeted coiled-coil proteins. Curr Biol 9(7):381–384. https://doi.org/10.1016/s0960-9822(99)80167-5
8. Boncompain G, Perez F (2013) The many routes of Golgi-dependent trafficking. Histochem Cell Biol 140(3):251–260. https://doi.org/10.1007/s00418-013-1124-7
9. Brault JB, Bardin S, Lampic M, Carpentieri JA, Coquand L, Penisson M, Lachuer H, Victoria GS, Baloul S, El Marjou F, Boncompain G, Miserey-Lenkei S, Belvindrah R, Fraisier V, Francis F, Perez F, Goud B, Baffet AD (2022) RAB6 and dynein drive post-Golgi apical transport to prevent neuronal progenitor delamination. EMBO Rep 23(10):e54605. https://doi.org/10.15252/embr.202254605
10. Burguete AS, Fenn TD, Brunger AT, Pfeffer SR (2008) Rab and Arl GTPase family members cooperate in the localization of the golgin GCC185. Cell 132(2):286–298. https://doi.org/10.1016/j.cell.2007.11.048
11. Buser DP, Spang A (2023) Protein sorting from endosomes to the TGN. Front Cell Dev Biol 11:1140605. https://doi.org/10.3389/fcell.2023.1140605
12. Carpier JM, Zucchetti AE, Bataille L, Dogniaux S, Shafaq-Zadah M, Bardin S, Lucchino M, Maurin M, Joannas LD, Magalhaes JG, Johannes L, Galli T, Goud B, Hivroz C (2018) Rab6-dependent retrograde traffic of LAT controls immune synapse formation and T cell activation. J Exp Med 215(4):1245–1265. https://doi.org/10.1084/jem.20162042

13. Cayre S, Faraldo MM, Bardin S, Miserey-Lenkei S, Deugnier MA, Goud B (2020) RAB6 GTPase regulates mammary secretory function by controlling the activation of STAT5. Development 147(19):dev190744. https://doi.org/10.1242/dev.190744
14. Chavrier P, Parton RG, Hauri HP, Simons K, Zerial M (1990) Localization of low molecular weight GTP binding proteins to exocytic and endocytic compartments. Cell 62(2):317–329. https://doi.org/10.1016/0092-8674(90)90369-p
15. Chen L, Hu J, Yun Y, Wang T (2010) Rab36 regulates the spatial distribution of late endosomes and lysosomes through a similar mechanism to Rab34. Mol Membr Biol 27(1): 23–30. https://doi.org/10.3109/09687680903417470
16. Chua CEL, Tang BL (2018) Rab 10-a traffic controller in multiple cellular pathways and locations. J Cell Physiol 233(9):6483–6494. https://doi.org/10.1002/jcp.26503
17. Cottam NP, Ungar D (2012) Retrograde vesicle transport in the Golgi. Protoplasma 249(4): 943–955. https://doi.org/10.1007/s00709-011-0361-7
18. Cox JV, Kansal R, Whitt MA (2016) Rab43 regulates the sorting of a subset of membrane protein cargo through the medial Golgi. Mol Biol Cell 27:1834–1844. https://doi.org/10.1091/mbc.e15-03-0123
19. Dejgaard SY, Presley JF (2019) Rab18: new insights into the function of an essential protein. Cell Mol Life Sci 76(10):1935–1945. https://doi.org/10.1007/s00018-019-03050-3
20. Dejgaard SY, Murshid A, Erman A, Kizilay O, Verbich D, Lodge R, Dejgaard K, Ly-Hartig TB, Pepperkok R, Simpson JC, Presley JF (2008) Rab18 and Rab43 have key roles in ER-Golgi trafficking. J Cell Sci 121(Pt 16):2768–2781. https://doi.org/10.1242/jcs.021808
21. Del Nery E, Miserey-Lenkei S, Falguières T, Nizak C, Johannes L, Perez F, Goud B (2006) Rab6A and Rab6A′ GTPases play non-overlapping roles in membrane trafficking. Traffic 7(4):394–407. https://doi.org/10.1111/j.1600-0854.2006.00395.x
22. Diao A, Rahman D, Pappin DJ, Lucocq J, Lowe M (2003) The coiled-coil membrane protein golgin-84 is a novel rab effector required for Golgi ribbon formation. J Cell Biol 160(2): 201–212. https://doi.org/10.1083/jcb.200207045
23. Dickson LJ, Liu S, Storrie B (2020) Rab6 is required for rapid, cisternal-specific, intra-Golgi cargo transport. Sci Rep 10(1):16604. https://doi.org/10.1038/s41598-020-73276-w
24. Ding X, Jiang X, Tian R, Zhao P, Li L, Wang X, Chen S, Zhu Y, Mei M, Bao S, Liu W, Tang Z, Sun Q (2019) RAB2 regulates the formation of autophagosome and autolysosome in mammalian cells. Autophagy 15(10):1774–1786. https://doi.org/10.1080/15548627.2019.1596478
25. Dong C, Wu G (2007) Regulation of anterograde transport of adrenergic and angiotensin II receptors by Rab2 and Rab6 GTPases. Cell Signal 19(11):2388–2399. https://doi.org/10.1016/j.cellsig.2007.07.017
26. Dornan LG, Simpson JC (2023) Rab6-mediated retrograde trafficking from the Golgi: the trouble with tubules. Small GTPases 14(1):26–44. https://doi.org/10.1080/21541248.2023.2238330
27. Drin G, Casella JF, Gautier R, Boehmer T, Schwartz TU, Antonny B (2007) A general amphipathic alpha-helical motif for sensing membrane curvature. Nat Struct Mol Biol 14(2): 138–146. https://doi.org/10.1038/nsmb1194
28. Drin G, Morello V, Casella JF, Gounon P, Antonny B (2008) Asymmetric tethering of flat and curved lipid membranes by a golgin. Science 320(5876):670–673. https://doi.org/10.1126/science.1155821
29. Edwards SL, Charlie NK, Richmond JE, Hegermann J, Eimer S, Miller KG (2009) Impaired dense core vesicle maturation in *Caenorhabditis elegans* mutants lacking Rab2. J Cell Biol 186(6):881–895. https://doi.org/10.1083/jcb.200902095
30. Egerer J, Emmerich D, Fischer-Zirnsak B, Chan WL, Meierhofer D, Tuysuz B, Marschner K, Sauer S, Barr FA, Mundlos S, Kornak U (2015) GORAB missense mutations disrupt RAB6 and ARF5 binding and Golgi targeting. J Invest Dermatol 135(10):2368–2376. https://doi.org/10.1038/jid.2015.192

31. Fossati M, Colombo SF, Borgese N (2014) A positive signal prevents secretory membrane cargo from recycling between the Golgi and the ER. EMBO J 33(18):2080–2097. https://doi.org/10.15252/embj.201488367
32. Fridmann-Sirkis Y, Siniossoglou S, Pelham HR (2004) TMF is a golgin that binds Rab6 and influences Golgi morphology. BMC Cell Biol 5:18. https://doi.org/10.1186/1471-2121-5-18
33. Fuchs E, Haas AK, Spooner RA, Yoshimura S, Lord JM, Barr FA (2007) Specific Rab GTPase-activating proteins define the Shiga toxin and epidermal growth factor uptake pathways. J Cell Biol 177(6):1133–1143. https://doi.org/10.1083/jcb.200612068
34. Fujita N, Huang W, Lin TH, Groulx JF, Jean S, Nguyen J, Kuchitsu Y, Koyama-Honda I, Mizushima N, Fukuda M, Kiger AA (2017) Genetic screen in Drosophila muscle identifies autophagy-mediated T-tubule remodeling and a Rab2 role in autophagy. Elife 6:e23367. https://doi.org/10.7554/eLife.23367
35. Galea G, Bexiga MG, Panarella A, O'Neill ED, Simpson JC (2015) A high-content screening microscopy approach to dissect the role of Rab proteins in Golgi-to-ER retrograde trafficking. J Cell Sci 128(13):2339–2349. https://doi.org/10.1242/jcs.167973
36. Galindo A, Munro S (2023) The TRAPP complexes: oligomeric exchange factors that activate the small GTPases Rab1 and Rab11. FEBS Lett 597(6):734–749. https://doi.org/10.1002/1873-3468.14553
37. Ganga AK, Kennedy MC, Oguchi ME, Gray S, Oliver KE, Knight TA, De La Cruz EM, Homma Y, Fukuda M, Breslow DK (2021) Rab34 GTPase mediates ciliary membrane formation in the intracellular ciliogenesis pathway. Curr Biol 31(13):2895–2905.e7. https://doi.org/10.1016/j.cub.2021.04.075
38. Gerondopoulos A, Bastos RN, Yoshimura S, Anderson R, Carpanini S, Aligianis I, Handley MT, Barr FA (2014) Rab18 and a Rab18 GEF complex are required for normal ER structure. J Cell Biol 205(5):707–720. https://doi.org/10.1083/jcb.201403026
39. Giannandrea M, Bianchi V, Mignogna ML, Sirri A, Carrabino S, D'Elia E, Vecellio M, Russo S, Cogliati F, Larizza L, Ropers HH, Tzschach A, Kalscheuer V, Oehl-Jaschkowitz B, Skinner C, Schwartz CE, Gecz J, Van Esch H, Raynaud M, Chelly J, de Brouwer AP, Toniolo D, D'Adamo P (2010) Mutations in the small GTPase gene RAB39B are responsible for X-linked mental retardation associated with autism, epilepsy, and macrocephaly. Am J Hum Genet 86(2):185–195. https://doi.org/10.1016/j.ajhg.2010.01.011
40. Girod A, Storrie B, Simpson JC, Johannes L, Goud B, Roberts LM, Lord JM, Nilsson T, Pepperkok R (1999) Evidence for a COP-I-independent transport route from the Golgi complex to the endoplasmic reticulum. Nat Cell Biol 1(7):423–430. https://doi.org/10.1038/15658
41. Goldenberg NM, Silverman M (2009) Rab34 and its effector munc13-2 constitute a new pathway modulating protein secretion in the cellular response to hyperglycemia. Am J Physiol Cell Physiol 297(4):C1053–C1058. https://doi.org/10.1152/ajpcell.00286.2009
42. Goldenberg NM, Grinstein S, Silverman M (2007) Golgi-bound Rab34 is a novel member of the secretory pathway. Mol Biol Cell 18(12):4762–4771. https://doi.org/10.1091/mbc.e06-11-0991
43. Götz TWB, Puchkov D, Lysiuk V, Lützkendorf J, Nikonenko AG, Quentin C, Lehmann M, Sigrist SJ, Petzoldt AG (2021) Rab2 regulates presynaptic precursor vesicle biogenesis at the trans-Golgi. J Cell Biol 220(5):e202006040. https://doi.org/10.1083/jcb.202006040
44. Goud B, Gleeson PA (2010) TGN golgins, Rabs and cytoskeleton: regulating the Golgi trafficking highways. Trends Cell Biol 20(6):329–336. https://doi.org/10.1016/j.tcb.2010.02.006
45. Grigoriev I, Splinter D, Keijzer N, Wulf PS, Demmers J, Ohtsuka T, Modesti M, Maly IV, Grosveld F, Hoogenraad CC, Akhmanova A (2007) Rab6 regulates transport and targeting of exocytotic carriers. Dev Cell 13(2):305–314. https://doi.org/10.1016/j.devcel.2007.06.010
46. Grigoriev I, Yu KL, Martinez-Sanchez E, Serra-Marques A, Smal I, Meijering E, Demmers J, Peränen J, Pasterkamp RJ, van der Sluijs P, Hoogenraad CC, Akhmanova A (2011) Rab6,

Rab8, and MICAL3 cooperate in controlling docking and fusion of exocytotic carriers. Curr Biol 21(11):967–974. https://doi.org/10.1016/j.cub.2011.04.030
47. Gyurkovska V, Murtazina R, Zhao SF, Shikano S, Okamoto Y, Segev N (2023) Dual function of Rab1A in secretion and autophagy: hypervariable domain dependence. Life Sci Alliance 6(5):e202201810. https://doi.org/10.26508/lsa.202201810
48. Haga K, Fukuda M (2024) Comprehensive knockout analysis of the RAB family small GTPases reveals an overlapping role of RAB2 and RAB14 in autophagosome maturation. Autophagy 21:21–36. https://doi.org/10.1080/15548627.2024.2374699
49. Harris NJ, Jenkins ML, Dalwadi U, Fleming KD, Nam SE, Parson MAH, Yip CK, Burke JE (2021) Biochemical insight into novel Rab-GEF activity of the mammalian TRAPPIII complex. J Mol Biol 433(18):167145. https://doi.org/10.1016/j.jmb.2021.167145
50. Hatoyama Y, Homma Y, Hiragi S, Fukuda M (2021) Establishment and analysis of conditional Rab1- and Rab5-knockout cells using the auxin-inducible degron system. J Cell Sci 134(24):jcs259184. https://doi.org/10.1242/jcs.259184
51. Hattula K, Furuhjelm J, Tikkanen J, Tanhuanpää K, Laakkonen P, Peränen J (2006) Characterization of the Rab8-specific membrane traffic route linked to protrusion formation. J Cell Sci 119(Pt 23):4866–4877. https://doi.org/10.1242/jcs.03275
52. Hayes GL, Brown FC, Haas AK, Nottingham RM, Barr FA, Pfeffer SR (2009) Multiple Rab GTPase binding sites in GCC185 suggest a model for vesicle tethering at the trans-Golgi. Mol Biol Cell 20(1):209–217. https://doi.org/10.1091/mbc.e08-07-0740
53. Heffernan LF, Suckrau PM, Banerjee T, Mysior MM, Simpson JC (2022) An imaging-based RNA interference screen for modulators of the Rab6-mediated Golgi-to-ER pathway in mammalian cells. Front Cell Dev Biol 10:1050190. https://doi.org/10.3389/fcell.2022.1050190
54. Hennies HC, Kornak U, Zhang H, Egerer J, Zhang X, Seifert W, Kühnisch J, Budde B, Nätebus M, Brancati F, Wilcox WR, Müller D, Kaplan PB, Rajab A, Zampino G, Fodale V, Dallapiccola B, Newman W, Metcalfe K, Clayton-Smith J, Tassabehji M, Steinmann B, Barr FA, Nürnberg P, Wieacker P, Mundlos S (2008) Gerodermia osteodysplastica is caused by mutations in SCYL1BP1, a Rab-6 interacting golgin. Nat Genet 40(12):1410–1412. https://doi.org/10.1038/ng.252
55. Homma Y, Hiragi S, Fukuda M (2021) Rab family of small GTPases: an updated view on their regulation and functions. FEBS J 288(1):36–55. https://doi.org/10.1111/febs.15453
56. Houghton FJ, Chew PL, Lodeho S, Goud B, Gleeson PA (2009) The localization of the Golgin GCC185 is independent of Rab6A/A′ and Arl1. Cell 138(4):787–794. https://doi.org/10.1016/j.cell.2009.05.048
57. Huang J, Birmingham CL, Shahnazari S et al (2011) Antibacterial autophagy occurs at PI(3)P-enriched domains of the endoplasmic reticulum and requires Rab1 GTPase. Autophagy 7(1):17–26. https://doi.org/10.4161/auto.7.1.13840
58. Huber LA, Pimplikar S, Parton RG, Virta H, Zerial M, Simons K (1993) Rab8, a small GTPase involved in vesicular traffic between the TGN and the basolateral plasma membrane. J Cell Biol 123(1):35–45. https://doi.org/10.1083/jcb.123.1.35
59. Hutagalung AH, Novick PJ (2011) Role of Rab GTPases in membrane traffic and cell physiology. Physiol Rev 91(1):119–149. https://doi.org/10.1152/physrev.00059.2009
60. Ishida R, Yamamoto A, Nakayama K, Sohda M, Misumi Y, Yasunaga T, Nakamura N (2015) GM130 is a parallel tetramer with a flexible rod-like structure and N-terminally open (Y-shaped) and closed (I-shaped) conformations. FEBS J 282(11):2232–2244. https://doi.org/10.1111/febs.13271
61. Jewett CE, Soh AWJ, Lin CH, Lu Q, Lencer E, Westlake CJ, Pearson CG, Prekeris R (2021) RAB19 directs cortical remodeling and membrane growth for primary ciliogenesis. Dev Cell 56(3):325–340.e8. https://doi.org/10.1016/j.devcel.2020.12.003
62. Jimenez AJ, Divoux S, Goud B, Perez F (2023) Phase partitioning rules Rab domain formation, growth and identity. bioRxiv 2023.04.17.537227. https://doi.org/10.1101/2023.04.17.537227

63. Jin G, Lin L, Li K, Li J, Yu C, Wei Z (2023) Structural basis of ELKS/Rab6B interaction and its role in vesicle capturing enhanced by liquid-liquid phase separation. J Biol Chem 299(6): 104808. https://doi.org/10.1016/j.jbc.2023.104808
64. Johannes L, Goud B (1998) Surfing on a retrograde wave: how does Shiga toxin reach the endoplasmic reticulum? Trends Cell Biol 8(4):158–162. https://doi.org/10.1016/s0962-8924(97)01209-9
65. Joiner AM, Phillips BP, Yugandhar K, Sanford EJ, Smolka MB, Yu H, Miller EA, Fromme JC (2021) Structural basis of TRAPPIII-mediated Rab1 activation. EMBO J 40(12):e107607. https://doi.org/10.15252/embj.2020107607
66. Kakuta S, Yamaguchi J, Suzuki C, Sasaki M, Kazuno S, Uchiyama Y (2017) Small GTPase Rab1B is associated with ATG9A vesicles and regulates autophagosome formation. FASEB J 31(9):3757–3773. https://doi.org/10.1096/fj.201601052R
67. Kelly EE, Giordano F, Horgan CP, Jollivet F, Raposo G, McCaffrey MW (2012) Rab30 is required for the morphological integrity of the Golgi apparatus. Biol Cell 104(2):84–101. https://doi.org/10.1111/boc.201100080
68. Khakurel A, Lupashin VV (2023) Role of GARP vesicle tethering complex in Golgi physiology. Int J Mol Sci 24(7):6069. https://doi.org/10.3390/ijms24076069
69. Kumar R, Khan M, Francis V, Aguila A, Kulasekaran G, Banks E, McPherson PS (2024) DENND6A links Arl8b to a Rab34/RILP/dynein complex, regulating lysosomal positioning and autophagy. Nat Commun 15(1):919. https://doi.org/10.1038/s41467-024-44957-1
70. Lee PL, Ohlson MB, Pfeffer SR (2015) Rab6 regulation of the kinesin family KIF1C motor domain contributes to Golgi tethering. Elife 4:e06029. https://doi.org/10.7554/eLife.06029
71. Lesage S, Bras J, Cormier-Dequaire F, Condroyer C, Nicolas A, Darwent L, Guerreiro R, Majounie E, Federoff M, Heutink P, Wood NW, Gasser T, Hardy J, Tison F, Singleton A, Brice A, French Parkinson's Disease Genetics Study Group (PDG) and the International Parkinson's Disease Genomics Consortium (IPDGC) (2015) Loss-of-function mutations in RAB39B are associated with typical early-onset Parkinson disease. Neurol Genet 1(1):e9. https://doi.org/10.1212/NXG.0000000000000009
72. Li C, Wei Z, Fan Y, Huang W, Su Y, Li H, Dong Z, Fukuda M, Khater M, Wu G (2017) The GTPase Rab43 controls the anterograde ER-Golgi trafficking and sorting of GPCRs. Cell Rep 21(4):1089–1101. https://doi.org/10.1016/j.celrep.2017.10.011
73. Lieu ZZ, Derby MC, Teasdale RD, Hart C, Gunn P, Gleeson PA (2007) The golgin GCC88 is required for efficient retrograde transport of cargo from the early endosomes to the trans-Golgi network. Mol Biol Cell 18(12):4979–4991. https://doi.org/10.1091/mbc.e07-06-0622. Epub 2007 Oct 3
74. Liewen H, Meinhold-Heerlein I, Oliveira V, Schwarzenbacher R, Luo G, Wadle A, Jung M, Pfreundschuh M, Stenner-Liewen F (2005) Characterization of the human GARP (Golgi associated retrograde protein) complex. Exp Cell Res 306(1):24–34. https://doi.org/10.1016/j.yexcr.2005.01.022
75. Liu S, Storrie B (2012) Are Rab proteins the link between Golgi organization and membrane trafficking? Cell Mol Life Sci 69:4093–4106
76. Liu S, Storrie B (2015) How Rab proteins determine Golgi structure. Int Rev Cell Mol Biol 315:1–22. https://doi.org/10.1016/bs.ircmb.2014.12.002
77. Liu S, Hunt L, Storrie B (2013) Rab41 is a novel regulator of Golgi apparatus organization that is needed for ER-to-Golgi trafficking and cell growth. PLoS One 8(8):e71886. https://doi.org/10.1371/journal.pone.0071886
78. Liu Z, Bryant N, Kumaran R, Beilina A, Abeliovich A, Cookson MR, West AB (2018) LRRK2 phosphorylates membrane-bound Rabs and is activated by GTP-bound Rab7L1 to promote recruitment to the trans-Golgi network. Hum Mol Genet 27(2):385–395. https://doi.org/10.1093/hmg/ddx410
79. Lombardi D, Soldati T, Riederer MA, Goda Y, Zerial M, Pfeffer SR (1993) Rab9 functions in transport between late endosomes and the trans Golgi network. EMBO J 12(2):677–682. https://doi.org/10.1002/j.1460-2075.1993.tb05701.x

80. Lőrincz P, Tóth S, Benkő P, Lakatos Z, Boda A, Glatz G, Zobel M, Bisi S, Hegedűs K, Takáts S, Scita G, Juhász G (2017) Rab2 promotes autophagic and endocytic lysosomal degradation. J Cell Biol 216(7):1937–1947. https://doi.org/10.1083/jcb.201611027
81. Lund VK, Lycas MD, Schack A, Andersen RC, Gether U, Kjaerulff O (2021) Rab2 drives axonal transport of dense core vesicles and lysosomal organelles. Cell Rep 35(2):108973. https://doi.org/10.1016/j.celrep.2021.108973
82. Lütcke A, Olkkonen VM, Dupree P, Lütcke H, Simons K, Zerial M (1995) Isolation of a murine cDNA clone encoding Rab19, a novel tissue-specific small GTPase. Gene 155(2): 257–260. https://doi.org/10.1016/0378-1119(94)00931-h
83. Magdeleine M, Gautier R, Gounon P, Barelli H, Vanni S, Antonny B (2016) A filter at the entrance of the Golgi that selects vesicles according to size and bulk lipid composition. Elife 5: e16988. https://doi.org/10.7554/eLife.16988
84. Majeed W, Liu S, Storrie B (2014) Distinct sets of Rab6 effectors contribute to ZW10- and COG-dependent Golgi homeostasis. Traffic 15(6):630–647. https://doi.org/10.1111/tra.12167
85. Mallard F, Tang BL, Galli T, Tenza D, Saint-Pol A, Yue X, Antony C, Hong W, Goud B, Johannes L (2002) Early/recycling endosomes-to-TGN transport involves two SNARE complexes and a Rab6 isoform. J Cell Biol 156(4):653–664. https://doi.org/10.1083/jcb.200110081
86. Maric M, Dale HA, Sannerud R, Saraste J (2009) The function of the intermediate compartment in pre-Golgi trafficking involves its stable connection with the centrosome. Mol Biol Cell 20(20):4458–4470. https://doi.org/10.1091/mbc.e08-12-1229
87. Marra P, Maffucci T, Daniele T, Tullio GD, Ikehara Y, Chan EK, Luini A, Beznoussenko G, Mironov A, De Matteis MA (2001) The GM130 and GRASP65 Golgi proteins cycle through and define a subdomain of the intermediate compartment. Nat Cell Biol 3(12):1101–1113. https://doi.org/10.1038/ncb1201-1101
88. Matanis T, Akhmanova A, Wulf P, Del Nery E, Weide T, Stepanova T, Galjart N, Grosveld F, Goud B, De Zeeuw CI, Barnekow A, Hoogenraad CC (2002) Bicaudal-D regulates COPI-independent Golgi-ER transport by recruiting the dynein-dynactin motor complex. Nat Cell Biol 4(12):986–992. https://doi.org/10.1038/ncb891
89. Matsui T, Ohbayashi N, Fukuda M (2012) The Rab interacting lysosomal protein (RILP) homology domain functions as a novel effector domain for small GTPase Rab36: Rab36 regulates retrograde melanosome transport in melanocytes. J Biol Chem 287(34): 28619–28631. https://doi.org/10.1074/jbc.M112.370544
90. Matsuto M, Kano F, Murata M (2015) Reconstitution of the targeting of Rab6A to the Golgi apparatus in semi-intact HeLa cells: a role of BICD2 in stabilizing Rab6A on Golgi membranes and a concerted role of Rab6A/BICD2 interactions in Golgi-to-ER retrograde transport. Biochim Biophys Acta 1853(10 Pt A):2592–2609. https://doi.org/10.1016/j.bbamcr.2015.05.005
91. Mignogna ML, Giannandrea M, Gurgone A, Fanelli F, Raimondi F, Mapelli L, Bassani S, Fang H, Van Anken E, Alessio M, Passafaro M, Gatti S, Esteban JA, Huganir R, D'Adamo P (2015) The intellectual disability protein RAB39B selectively regulates GluA2 trafficking to determine synaptic AMPAR composition. Nat Commun 6:6504. https://doi.org/10.1038/ncomms7504
92. Miller VJ, Sharma P, Kudlyk TA, Frost L, Rofe AP, Watson IJ, Duden R, Lowe M, Lupashin VV, Ungar D (2013) Molecular insights into vesicle tethering at the Golgi by the conserved oligomeric Golgi (COG) complex and the golgin TATA element modulatory factor (TMF). J Biol Chem 288(6):4229–4240. https://doi.org/10.1074/jbc.M112.426767
93. Mironov AA, Beznoussenko GV, Nicoziani P, Martella O, Trucco A, Kweon HS, Di Giandomenico D, Polishchuk RS, Fusella A, Lupetti P, Berger EG, Geerts WJ, Koster AJ, Burger KN, Luini A (2001) Small cargo proteins and large aggregates can traverse the Golgi by a common mechanism without leaving the lumen of cisternae. J Cell Biol 155(7): 1225–1238. https://doi.org/10.1083/jcb.200108073

94. Miserey-Lenkei S, Chalancon G, Bardin S, Formstecher E, Goud B, Echard A (2010) Rab and actomyosin-dependent fission of transport vesicles at the Golgi complex. Nat Cell Biol 12(7): 645–654. https://doi.org/10.1038/ncb2067
95. Miserey-Lenkei S, Bousquet H, Pylypenko O, Bardin S, Dimitrov A, Bressanelli G, Bonifay R, Fraisier V, Guillou C, Bougeret C, Houdusse A, Echard A, Goud B (2017) Coupling fission and exit of RAB6 vesicles at Golgi hotspots through kinesin-myosin interactions. Nat Commun 8(1):1254. https://doi.org/10.1038/s41467-017-01266-0
96. Mochizuki Y, Ohashi R, Kawamura T, Iwanari H, Kodama T, Naito M, Hamakubo T (2013) Phosphatidylinositol 3-phosphatase myotubularin-related protein 6 (MTMR6) is regulated by small GTPase Rab1B in the early secretory and autophagic pathways. J Biol Chem 288(2): 1009–1021. https://doi.org/10.1074/jbc.M112.395087
97. Moyer BD, Allan BB, Balch WE (2001) Rab1 interaction with a GM130 effector complex regulates COPII vesicle cis–Golgi tethering. Traffic 2(4):268–276. https://doi.org/10.1034/j.1600-0854.2001.1o007.x
98. Mukhopadhyay A, Quiroz JA, Wolkoff AW (2014) Rab1a regulates sorting of early endocytic vesicles. Am J Physiol Gastrointest Liver Physiol 306(5):G412–G424. https://doi.org/10.1152/ajpgi.00118.2013
99. Murray DH, Jahnel M, Lauer J, Avellaneda MJ, Brouilly N, Cezanne A, Morales-Navarrete H, Perini ED, Ferguson C, Lupas AN, Kalaidzidis Y, Parton RG, Grill SW, Zerial M (2016) An endosomal tether undergoes an entropic collapse to bring vesicles together. Nature 537(7618): 107–111. https://doi.org/10.1038/nature19326
100. Muschalik N, Munro S (2018) Golgins. Curr Biol 28(8):R374–R376. https://doi.org/10.1016/j.cub.2018.01.006
101. Nakamura N, Lowe M, Levine TP, Rabouille C, Warren G (1997) The vesicle docking protein p115 binds GM130, a cis-Golgi matrix protein, in a mitotically regulated manner. Cell 89(3): 445–455. https://doi.org/10.1016/s0092-8674(00)80225-1
102. Nakano A (2022) The Golgi apparatus and its next-door neighbors. Front Cell Dev Biol 10: 884360. https://doi.org/10.3389/fcell.2022.884360
103. Nakano A, Luini A (2010) Passage through the Golgi. Curr Opin Cell Biol 22(4):471–478. https://doi.org/10.1016/j.ceb.2010.05.003
104. Nottingham RM, Pusapati GV, Ganley IG, Barr FA, Lambright DG, Pfeffer SR (2012) RUTBC2 protein, a Rab9A effector and GTPase-activating protein for Rab36. J Biol Chem 287(27):22740–22748. https://doi.org/10.1074/jbc.M112.362558
105. Oguchi ME, Okuyama K, Homma Y, Fukuda M (2020) A comprehensive analysis of Rab GTPases reveals a role for Rab34 in serum starvation-induced primary ciliogenesis. J Biol Chem 295(36):12674–12685. https://doi.org/10.1074/jbc.RA119.012233
106. Ozeki S, Cheng J, Tauchi-Sato K, Hatano N, Taniguchi H, Fujimoto T (2005) Rab18 localizes to lipid droplets and induces their close apposition to the endoplasmic reticulum-derived membrane. J Cell Sci 118(Pt 12):2601–2611. https://doi.org/10.1242/jcs.02401
107. Palokangas H, Ying M, Väänänen K, Saraste J (1998) Retrograde transport from the pre-Golgi intermediate compartment and the Golgi complex is affected by the vacuolar H+-ATPase inhibitor bafilomycin A1. Mol Biol Cell 9(12):3561–3578. https://doi.org/10.1091/mbc.9.12.3561
108. Pereira-Leal JB, Seabra MC (2000) The mammalian Rab family of small GTPases: definition of family and subfamily sequence motifs suggests a mechanism for functional specificity in the Ras superfamily. J Mol Biol 301(4):1077–1087. https://doi.org/10.1006/jmbi.2000.4010
109. Pernet-Gallay K, Antony C, Johannes L, Bornens M, Goud B, Rios RM (2002) The overexpression of GMAP-210 blocks anterograde and retrograde transport between the ER and the Golgi apparatus. Traffic 3(11):822–832. https://doi.org/10.1034/j.1600-0854.2002.31107.x
110. Petrosyan A, Casey CA, Cheng PW (2016) The role of Rab6a and phosphorylation of non-muscle myosin IIA tailpiece in alcohol-induced Golgi disorganization. Sci Rep 6: 31962. https://doi.org/10.1038/srep31962

111. Pfeffer SR (2010) How the Golgi works: a cisternal progenitor model. Proc Natl Acad Sci USA 107(46):19614–19618. https://doi.org/10.1073/pnas.1011016107
112. Plutner H, Cox AD, Pind S, Khosravi-Far R, Bourne JR, Schwaninger R, Der CJ, Balch WE (1991) Rab1b regulates vesicular transport between the endoplasmic reticulum and successive Golgi compartments. J Cell Biol 115(1):31–43. https://doi.org/10.1083/jcb.115.1.31
113. Progida C, Cogli L, Piro F, De Luca A, Bakke O, Bucci C (2010) Rab7b controls trafficking from endosomes to the TGN. J Cell Sci 123(Pt 9):1480–1491. https://doi.org/10.1242/jcs.051474
114. Purlyte E, Dhekne HS, Sarhan AR, Gomez R, Lis P, Wightman M, Martinez TN, Tonelli F, Pfeffer SR, Alessi DR (2018) Rab29 activation of the Parkinson's disease-associated LRRK2 kinase. EMBO J 37(1):1–18. https://doi.org/10.15252/embj.201798099
115. Pusapati GV, Luchetti G, Pfeffer SR (2012) Ric1-Rgp1 complex is a guanine nucleotide exchange factor for the late Golgi Rab6A GTPase and an effector of the medial Golgi Rab33B GTPase. J Biol Chem 287(50):42129–42137. https://doi.org/10.1074/jbc.M112.414565
116. Qiu H, Wu X, Ma X, Li S, Cai Q, Ganzella M, Ge L, Zhang H, Zhang M (2024) Short-distance vesicle transport via phase separation. Cell 187(9):2175–2193.e21. https://doi.org/10.1016/j.cell.2024.03.003
117. Rebane AA, Ziltener P, LaMonica LC, Bauer AH, Zheng H, López-Montero I, Pincet F, Rothman JE, Ernst AM (2020) Liquid-liquid phase separation of the Golgi matrix protein GM130. FEBS Lett 594(7):1132–1144. https://doi.org/10.1002/1873-3468.13715
118. Rios RM, Tassin AM, Celati C, Antony C, Boissier MC, Homberg JC, Bornens M (1994) A peripheral protein associated with the cis-Golgi network redistributes in the intermediate compartment upon brefeldin A treatment. J Cell Biol 125(5):997–1013. https://doi.org/10.1083/jcb.125.5.997
119. Ríos RM, Sanchís A, Tassin AM, Fedriani C, Bornens M (2004) GMAP-210 recruits gamma-tubulin complexes to cis-Golgi membranes and is required for Golgi ribbon formation. Cell 118(3):323–335. https://doi.org/10.1016/j.cell.2004.07.012
120. Roboti P, Sato K, Lowe M (2015) The golgin GMAP-210 is required for efficient membrane trafficking in the early secretory pathway. J Cell Sci 128(8):1595–1606. https://doi.org/10.1242/jcs.166710
121. Rosing M, Ossendorf E, Rak A, Barnekow A (2007) Giantin interacts with both the small GTPase Rab6 and Rab1. Exp Cell Res 313(11):2318–2325. https://doi.org/10.1016/j.yexcr.2007.03.031
122. Sannerud R, Saraste J, Goud B (2003) Retrograde traffic in the biosynthetic-secretory route: pathways and machinery. Curr Opin Cell Biol 15(4):438–445. https://doi.org/10.1016/s0955-0674(03)00077-2
123. Sannerud R, Marie M, Nizak C, Dale HA, Pernet-Gallay K, Perez F, Goud B, Saraste J (2006) Rab1 defines a novel pathway connecting the pre-Golgi intermediate compartment with the cell periphery. Mol Biol Cell 17(4):1514–1526. https://doi.org/10.1091/mbc.e05-08-0792
124. Saraste J (2016) Spatial and functional aspects of ER-Golgi Rabs and tethers. Front Cell Dev Biol 4:28. https://doi.org/10.3389/fcell.2016.00028
125. Saraste J, Goud B (2007) Functional symmetry of endomembranes. Mol Biol Cell 18(4):1430–1436. https://doi.org/10.1091/mbc.e06-10-0933
126. Saraste J, Prydz K (2019) A new look at the functional organization of the Golgi ribbon. Front Cell Dev Biol 7:171. https://doi.org/10.3389/fcell.2019.00171
127. Saraste J, Lahtinen U, Goud B (1995) Localization of the small GTP-binding protein rab1p to early compartments of the secretory pathway. J Cell Sci 108(Pt 4):1541–1552. https://doi.org/10.1242/jcs.108.4.1541
128. Sato T, Mushiake S, Kato Y, Sato K, Sato M, Takeda N, Ozono K, Miki K, Kubo Y, Tsuji A, Harada R, Harada A (2007) The Rab8 GTPase regulates apical protein localization in intestinal cells. Nature 448(7151):366–369. https://doi.org/10.1038/nature05929
129. Sato T, Iwano T, Kunii M, Matsuda S, Mizuguchi R, Jung Y, Hagiwara H, Yoshihara Y, Yuzaki M, Harada R, Harada A (2014) Rab8a and Rab8b are essential for several apical

transport pathways but insufficient for ciliogenesis. J Cell Sci 127(Pt 2):422–431. https://doi.org/10.1242/jcs.136903

130. Schuck S, Gerl MJ, Ang A, Manninen A, Keller P, Mellman I, Simons K (2007) Rab10 is involved in basolateral transport in polarized Madin-Darby canine kidney cells. Traffic 8(1): 47–60. https://doi.org/10.1111/j.1600-0854.2006.00506.x
131. Sengupta P, Satpute-Krishnan P, Seo AY, Burnette DT, Patterson GH, Lippincott-Schwartz J (2015) ER trapping reveals Golgi enzymes continually revisit the ER through a recycling pathway that controls Golgi organization. Proc Natl Acad Sci USA 112(49):E6752–E6761. https://doi.org/10.1073/pnas.1520957112
132. Serra-Marques A, Martin M, Katrukha EA, Grigoriev I, Peeters CA, Liu Q, Hooikaas PJ, Yao Y, Solianova V, Smal I, Pedersen LB, Meijering E, Kapitein LC, Akhmanova A (2020) Concerted action of kinesins KIF5B and KIF13B promotes efficient secretory vesicle transport to microtubule plus ends. Elife 9:e61302. https://doi.org/10.7554/eLife.61302
133. Shafaq-Zadah M, Gomes-Santos CS, Bardin S, Maiuri P, Maurin M, Iranzo J, Gautreau A, Lamaze C, Caswell P, Goud B, Johannes L (2016) Persistent cell migration and adhesion rely on retrograde transport of β(1) integrin. Nat Cell Biol 18(1):54–64. https://doi.org/10.1038/ncb3287
134. Shibata S, Kawanai T, Hara T, Yamamoto A, Chaya T, Tokuhara Y, Tsuji C, Sakai M, Tachibana T, Inagaki S (2016) ARHGEF10 directs the localization of Rab8 to Rab6-positive executive vesicles. J Cell Sci 129(19):3620–3634. https://doi.org/10.1242/jcs.186817
135. Shin JJH, Gillingham AK, Begum F, Chadwick J, Munro S (2017) TBC1D23 is a bridging factor for endosomal vesicle capture by golgins at the trans-Golgi. Nat Cell Biol 19(12): 1424–1432. https://doi.org/10.1038/ncb3627
136. Shin JJH, Crook OM, Borgeaud AC, Cattin-Ortolá J, Peak-Chew SY, Breckels LM, Gillingham AK, Chadwick J, Lilley KS, Munro S (2020) Spatial proteomics defines the content of trafficking vesicles captured by golgin tethers. Nat Commun 11(1):5987. https://doi.org/10.1038/s41467-020-19840-4
137. Shomron O, Hirschberg K, Burakov A, Kamentseva R, Kornilova E, Nadezhdina E, Brodsky I (2021) Positioning of endoplasmic reticulum exit sites around the Golgi depends on BicaudalD2 and Rab6 activity. Traffic 22(3):64–77. https://doi.org/10.1111/tra.12774
138. Short B, Preisinger C, Körner R, Kopajtich R, Byron O, Barr FA (2001) A GRASP55-rab2 effector complex linking Golgi structure to membrane traffic. J Cell Biol 155(6):877–883. https://doi.org/10.1083/jcb.200108079
139. Short B, Preisinger C, Schaletzky J, Kopajtich R, Barr FA (2002) The Rab6 GTPase regulates recruitment of the dynactin complex to Golgi membranes. Curr Biol 12(20):1792–1795. https://doi.org/10.1016/s0960-9822(02)01221-6
140. Siniossoglou S, Pelham HR (2001) An effector of Ypt6p binds the SNARE Tlg1p and mediates selective fusion of vesicles with late Golgi membranes. EMBO J 20(21): 5991–5998. https://doi.org/10.1093/emboj/20.21.5991
141. Sinka R, Gillingham AK, Kondylis V, Munro S (2008) Golgi coiled-coil proteins contain multiple binding sites for Rab family G proteins. J Cell Biol 183(4):607–615. https://doi.org/10.1083/jcb.200808018
142. Smith RD, Willett R, Kudlyk T, Pokrovskaya I, Paton AW, Paton JC, Lupashin VV (2009) The COG complex, Rab6 and COPI define a novel Golgi retrograde trafficking pathway that is exploited by SubAB toxin. Traffic 10(10):1502–1517. https://doi.org/10.1111/j.1600-0854.2009.00965.x
143. Smith DM, Liu BY, Wolfgang MJ (2024) Rab30 facilitates lipid homeostasis during fasting. Nat Commun 15(1):4469. https://doi.org/10.1038/s41467-024-48959-x
144. Sohda M, Misumi Y, Yamamoto A, Nakamura N, Ogata S, Sakisaka S, Hirose S, Ikehara Y, Oda K (2010) Interaction of Golgin-84 with the COG complex mediates the intra-Golgi retrograde transport. Traffic 11(12):1552–1566. https://doi.org/10.1111/j.1600-0854.2010.01123.x

145. Stalder D, Gershlick DC (2020) Direct trafficking pathways from the Golgi apparatus to the plasma membrane. Semin Cell Dev Biol 107:112–125. https://doi.org/10.1016/j.semcdb.2020.04.001
146. Starr T, Sun Y, Wilkins N, Storrie B (2010) Rab33b and Rab6 are functionally overlapping regulators of Golgi homeostasis and trafficking. Traffic 11(5):626–636. https://doi.org/10.1111/j.1600-0854.2010.01051.x
147. Steger M, Diez F, Dhekne HS, Lis P, Nirujogi RS, Karayel O, Tonelli F, Martinez TN, Lorentzen E, Pfeffer SR, Alessi DR, Mann M (2017) Systematic proteomic analysis of LRRK2-mediated Rab GTPase phosphorylation establishes a connection to ciliogenesis. Elife 6:e31012. https://doi.org/10.7554/eLife.31012
148. Stein M, Pilli M, Bernauer S, Habermann BH, Zerial M, Wade RC (2012) The interaction properties of the human Rab GTPase family—comparative analysis reveals determinants of molecular binding selectivity. PLoS One 7(4):e34870. https://doi.org/10.1371/journal.pone.0034870
149. Stuck MW, Chong WM, Liao JC, Pazour GJ (2021) Rab34 is necessary for early stages of intracellular ciliogenesis. Curr Biol 31(13):2887–2894.e4. https://doi.org/10.1016/j.cub.2021.04.018
150. Sumya FT, Pokrovskaya ID, D'Souza Z, Lupashin VV (2023) Acute COG complex inactivation unveiled its immediate impact on Golgi and illuminated the nature of intra-Golgi recycling vesicles. Traffic 24(2):52–75. https://doi.org/10.1111/tra.12876
151. Sun Y, Shestakova A, Hunt L, Sehgal S, Lupashin V, Storrie B (2007) Rab6 regulates both ZW10/RINT-1 and conserved oligomeric Golgi complex-dependent Golgi trafficking and homeostasis. Mol Biol Cell 18(10):4129–4142. https://doi.org/10.1091/mbc.e07-01-0080
152. Sun Y, Tao X, Han Y, Lin X, Tian R, Wang H, Chang P, Sun Q, Ge L, Zhang M (2024) A dual role of ERGIC-localized Rabs in TMED10-mediated unconventional protein secretion. Nat Cell Biol 26(7):1077–1092. https://doi.org/10.1038/s41556-024-01445-4
153. Tisdale EJ, Balch WE (1996) Rab2 is essential for the maturation of pre-Golgi intermediates. J Biol Chem 271(46):29372–29379. https://doi.org/10.1074/jbc.271.46.29372
154. Tisdale EJ, Bourne JR, Khosravi-Far R, Der CJ, Balch WE (1992) GTP-binding mutants of rab1 and rab2 are potent inhibitors of vesicular transport from the endoplasmic reticulum to the Golgi complex. J Cell Biol 119(4):749–761. https://doi.org/10.1083/jcb.119.4.749
155. Tsukada M, Gallwitz D (1996) Isolation and characterization of SYS genes from yeast, multicopy suppressors of the functional loss of the transport GTPase Ypt6p. J Cell Sci 109 (Pt 10):2471–2481. https://doi.org/10.1242/jcs.109.10.2471
156. Valsdottir R, Hashimoto H, Ashman K, Koda T, Storrie B, Nilsson T (2001) Identification of rabaptin-5, rabex-5, and GM130 as putative effectors of rab33b, a regulator of retrograde traffic between the Golgi apparatus and ER. FEBS Lett 508(2):201–209. https://doi.org/10.1016/s0014-5793(01)02993-3
157. Wang T, Hong W (2002) Interorganellar regulation of lysosome positioning by the Golgi apparatus through Rab34 interaction with Rab-interacting lysosomal protein. Mol Biol Cell 13(12):4317–4332. https://doi.org/10.1091/mbc.e02-05-0280
158. Wang S, Ma Z, Xu X, Wang Z, Sun L, Zhou Y, Lin X, Hong W, Wang T (2014) A role of Rab29 in the integrity of the trans-Golgi network and retrograde trafficking of mannose-6-phosphate receptor. PLoS One 9(5):e96242. https://doi.org/10.1371/journal.pone.0096242
159. Wassmer T, Attar N, Harterink M, van Weering JR, Traer CJ, Oakley J, Goud B, Stephens DJ, Verkade P, Korswagen HC, Cullen PJ (2009) The retromer coat complex coordinates endosomal sorting and dynein-mediated transport, with carrier recognition by the trans-Golgi network. Dev Cell 17(1):110–122. https://doi.org/10.1016/j.devcel.2009.04.016
160. Weide T, Bayer M, Köster M, Siebrasse JP, Peters R, Barnekow A (2001) The Golgi matrix protein GM130: a specific interacting partner of the small GTPase rab1b. EMBO Rep 2(4):336–341. https://doi.org/10.1093/embo-reports/kve065

161. White J, Johannes L, Mallard F, Girod A, Grill S, Reinsch S, Keller P, Tzschaschel B, Echard A, Goud B, Stelzer EH (1999) Rab6 coordinates a novel Golgi to ER retrograde transport pathway in live cells. J Cell Biol 147(4):743–760. https://doi.org/10.1083/jcb.147.4.743
162. Willett R, Pokrovskaya I, Kudlyk T, Lupashin V (2014) Multipronged interaction of the COG complex with intracellular membranes. Cell Logist 4(1):e27888. https://doi.org/10.4161/cl.27888
163. Witkos TM, Chan WL, Joensuu M, Rhiel M, Pallister E, Thomas-Oates J, Mould AP, Mironov AA, Biot C, Guerardel Y, Morelle W, Ungar D, Wieland FT, Jokitalo E, Tassabehji M, Kornak U, Lowe M (2019) GORAB scaffolds COPI at the trans-Golgi for efficient enzyme recycling and correct protein glycosylation. Nat Commun 10:127. https://doi.org/10.1038/s41467-018-08044-6
164. Wong M, Munro S (2014) The specificity of vesicle traffic to the Golgi is encoded in the golgin coiled-coil proteins. Science 346(6209):1256898. https://doi.org/10.1126/science.1256898
165. Xu D, Li Y, Wu L, Li Y, Zhao D, Yu J, Huang T, Ferguson C, Parton RG, Yang H, Li P (2018) Rab18 promotes lipid droplet (LD) growth by tethering the ER to LDs through SNARE and NRZ interactions. J Cell Biol 217(3):975–995. https://doi.org/10.1083/jcb.201704184
166. Xu J, Liang Y, Li N, Dang S, Jiang A, Liu Y, Guo Y, Yang X, Yuan Y, Zhang X, Yang Y, Du Y, Shi A, Liu X, Li D, He K (2024) Clathrin-associated carriers enable recycling through a kiss-and-run mechanism. Nat Cell Biol 26(10):1652–1668. https://doi.org/10.1038/s41556-024-01499-4
167. Yadav S, Puri S, Linstedt AD (2009) A primary role for Golgi positioning in directed secretion, cell polarity, and wound healing. Mol Biol Cell 20(6):1728–1736. https://doi.org/10.1091/mbc.e08-10-1077
168. Yamane J, Kubo A, Nakayama K, Yuba-Kubo A, Katsuno T, Tsukita S, Tsukita S (2007) Functional involvement of TMF/ARA160 in Rab6-dependent retrograde membrane traffic. Exp Cell Res 313(16):3472–3485. https://doi.org/10.1016/j.yexcr.2007.07.010
169. Ziltener P, Rebane AA, Graham M, Ernst AM, Rothman JE (2020) The golgin family exhibits a propensity to form condensates in living cells. FEBS Lett 594(19):3086–3094. https://doi.org/10.1002/1873-3468.13884
170. Zoppino FCM, Militello RD, Slavin I et al (2010) Autophagosome formation depends on the small GTPase Rab1 and functional ER exit sites. Traffic 11(9):1246–1261. https://doi.org/10.1111/j.1600-0854.2010.01086.x
171. Zulkefli KL, Mahmoud IS, Williamson NA, Gosavi PK, Houghton FJ, Gleeson PA (2021) A role for Rab30 in retrograde trafficking and maintenance of endosome-TGN organization. Exp Cell Res 399(2):112442. https://doi.org/10.1016/j.yexcr.2020.112442
172. Stenmark H (2009) Rab GTPases as coordinators of vesicle traffic. Nat Rev Mol Cell Biol 10 (8):513–525. https://doi.org/10.1038/nrm2728
173. Goud B, Liu S, Storrie B (2018) Rab proteins as major determinants of the Golgi complex structure. SmallGTPases 9(1–2):66–75. https://doi.org/10.1080/21541248.2017.1384087
174. Young J, Stauber T, del Nery E, Vernos I, Pepperkok R, Nilsson T (2005) Regulation of microtubule-dependentrecycling at the trans-Golgi network by Rab6A and Rab6A'. Mol Biol Cell 16(1):162–177. https://doi.org/10.1091/mbc.e04-03-0260
175. Martinez O, Schmidt A, Salaméro J, Hoflack B, Roa M, Goud B (1994) The small GTP-binding protein rab6functions in intra-Golgi transport. J Cell Biol 127(6 Pt 1):1575–1588. https://doi.org/10.1083/jcb.127.6.1575
176. Martinez O, Antony C, Pehau-Arnaudet G, Berger EG, Salamero J, Goud B (1997) GTP-bound forms of rab6induce the redistribution of Golgi proteins into the endoplasmic reticulum. Proc Natl Acad Sci U S A 94(5):1828–1833. https://doi.org/10.1073/pnas.94.5.1828
177. Jiang S, Storrie B (2005) Cisternal rab proteins regulate Golgi apparatus redistribution in response to hypotonicstress. Mol Biol Cell 16(5):2586–2596. https://doi.org/10.1091/mbc.e04-10-0861

178. Casey CA, Thomes P, Manca S, Petrosyan A (2018) Giantin is required for post-alcohol recovery of Golgi in liver cells. Biomolecules 8(4):150. https://doi.org/10.3390/biom8040150
179. Fourriere L, Kasri A, Gareil N, Bardin S, Bousquet H, Pereira D, Perez F, Goud B, Boncompain G, Miserey-LenkeiS (2019) RAB6 and microtubules restrict protein secretion to focal adhesions. J Cell Biol 218(7):2215–2231. https://doi.org/10.1083/jcb.201805002
180. Wakana Y, van Galen J, Meissner F, Scarpa M, Polishchuk RS, Mann M, Malhotra V (2012) A new class of carriersthat transport selective cargo from the trans Golgi network to the cell surface. EMBO J 31(20):3976–90. https://doi.org/10.1038/emboj.2012.235
181. Seaman MN (2004) Cargo-selective endosomal sorting for retrieval to the Golgi requires retromer. J Cell Biol165(1):111–122. https://doi.org/10.1083/jcb.200312034

Chapter 4
The Functions of the Golgin Family of Coiled-Coil Proteins

Maraym Arab and Martin Lowe

Abstract The golgins are a family of ubiquitously expressed and evolutionary conserved Golgi-localised coiled-coil proteins. The golgins are particularly important for vesicle trafficking and structural organisation of the Golgi apparatus. The golgins mediate tethering, or long-range capture, of transport vesicles to Golgi membranes, a key step in vesicle traffic that ensures the specificity of vesicle transport at the Golgi. Following tethering, the golgins associate with other factors to ensure that vesicles undergo the transition to membrane fusion mediated by SNARE proteins. The golgins also contribute to the structural organisation of the Golgi apparatus, particularly in Golgi ribbon formation. This is achieved in several ways, which includes linking the Golgi membranes to the microtubule cytoskeleton, controlling the activity of cytoskeletal regulators and physically linking Golgi elements into the ribbon. Golgin-dependent interactions with the microtubule cytoskeleton also contribute to cell polarity, migration and division. In addition to regulating Golgi function and organisation, the golgins scaffold various factors to control the activity of these factors in a range of cellular processes which includes signaling, autophagy, spindle assembly and cell fate determination. In this book chapter we discuss current understanding of the golgins and their functional roles in cells.

Keywords Golgin · Membrane traffic · Vesicle tethering · Rab GTPase · Golgi structure · Golgi ribbon

---

M. Arab · M. Lowe (✉)
School of Biological Sciences, Faculty of Biology, Medicine and Health, University of Manchester, Manchester, UK
e-mail: martin.p.lowe@manchester.ac.uk

A. Nakano, J. Saraste (eds.), *The Golgi Network, Volume II*, Subcellular Biochemistry 111, https://doi.org/10.1007/978-3-032-16833-7_4

Introduction

The golgins are a conserved family of 11 Golgi-localised proteins that decorate the cytoplasmic face of the Golgi membrane [27, 66, 70, 117] (shown in Fig. 4.1). Many of the golgins were first identified as autoantigens [70]. They are homodimeric proteins comprised predominantly of coiled-coils and that are anchored to the Golgi membrane by their extreme carboxy-terminus. Because of these features, golgins are thought to project into the cytoplasm as extended rod-like proteins, although numerous short regions between the coiled-coil segments may function as flexible hinges to allow conformational changes, raising the possibility of more compact forms of the proteins [16, 36]. The golgins participate in various processes at the Golgi, which includes vesicle trafficking, structural organisation, and they can act as anchors for various proteins to regulate other processes in the cell. These varied functions will be discussed in this book chapter.

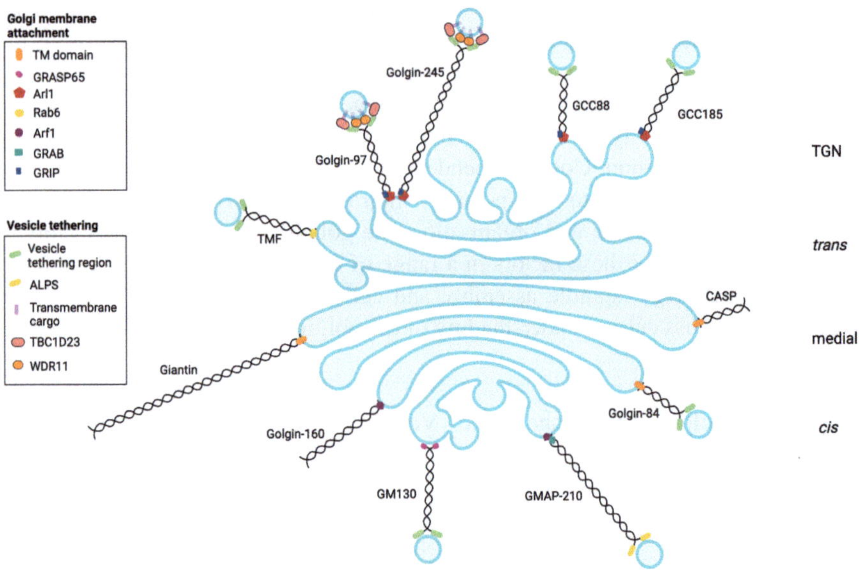

Fig. 4.1 Spatial organisation of golgins at the Golgi apparatus. The golgins are localised to different Golgi regions as indicated. Membrane attachment of the golgins is mediated by their extreme carboxy-terminus, either through a transmembrane domain (Golgin-84, CASP, giantin), binding to a small GTPase (GMAP-210, Golgin-160, TMF, GCC88, GCC185, Golgin-97, Golgin-245), or in the case of GM130 binding to GRASP65. Many of the golgins tether transport vesicles at their extreme amino-terminus as indicated. The localisation of the golgins correlates with their tethering of different vesicle types. However, in most cases it is not known how the golgins specifically recognise and tether transport vesicles. The GMAP-210 ALPS motif has been shown to bind directly to vesicle lipids to mediate vesicle tethering, whereas Golgin-97 and Golgin-245 bind to TBC1D23 and the WDR11 complex which can bind directly to the cytosolic tails of transmembrane vesicle cargo proteins

The different golgins are localised to distinct regions of the Golgi, consistent with them having distinct functions. In most cases, golgin targeting to different Golgi regions is mediated either by integral membrane tail anchors or through interactions with small GTPases [117]. The transmembrane golgins are giantin, CASP and golgin-84, which are located to the rims of the *cis*- and *medial*-Golgi [18, 28, 42, 48, 90]. Various small GTPases are involved in golgin recruitment. The *cis*-Golgi golgins GMAP210 and golgin-160 bind to Arf1 [20, 29, 121], which is mediated by a conserved GRAB domain in the case of GMAP-210 [29]. TMF, which resides at the rims of the *trans*-Golgi, binds to Rab6 [25, 122], whereas the *trans*-Golgi network (TGN) golgins GCC88, GCC185, golgin-97 and golgin-245 bind Arl1 via carboxy-terminal conserved GRIP domains [52, 72, 94], which in the case of GCC185 may be augmented by Rab6 binding [12]. Thus, the localisation and activation of the Arf1, Arl1 and Rab6 small GTPases is important for the correct targeting of the golgins to the Golgi. How the transmembrane golgins are targeted to distinct regions likely depends on the features of their transmembrane domains and possibly interactions with other proteins. In the case of GM130, membrane targeting is mediated by an interaction between the carboxy-terminal region of GM130 and the PDZ domain of the cis-Golgi localised protein GRASP65 [2, 4].

The golgins are conserved in evolution with five golgins apparently present in the last common eukaryotic ancestor [3]. Not all golgins are present across the entire eukaryotic domain however, with varying degrees of conservation [3, 117]. For example, giantin, which is the largest golgin, is restricted to vertebrates, whereas homologues of GM130 (Bug1), GMAP-210 (Rud3), CASP (Coy1), TMF1 (Sgm1) and the TGN golgins (Imh1) are present in the budding yeast *S. cerevisiae*. The emergence of additional golgins during evolution is believed to coincide with the additional trafficking complexity and Golgi organisation seen in 'higher' eukaryotes. In metazoan species, the golgins are widely expressed across different tissues, consistent with them playing a fundamental role in the correct functioning and organisation of the Golgi apparatus. This view is supported by functional studies in different model organisms and the association of golgin loss with human disease [50]. However, perhaps surprisingly, loss of a single golgin often manifests as a mild phenotype, or even lack of any apparent phenotype at all. This indicates redundancy between at least some of the golgins, which is supported by a recent study in *Drosophila* in which three golgins involved in intra-Golgi traffic (GMAP-210, golgin-84, TMF) were systematically knocked out in combination, resulting in progressively more severe phenotypes as more of the golgins were lost [73]. Interestingly, this study also revealed an inherent plasticity in Golgi trafficking which could contribute to the lack of phenotypes seen upon the single golgin knockouts. Thus far, only mutation of GMAP-210 has been shown to result in human disease, manifesting as a lethal skeletal dysplasia upon complete loss of the protein [99], or a milder skeletal disorder upon partial loss of expression [113]. A single study suggests loss of GM130 may also result in rare multisystemic disorder [95], but this needs to be verified in additional patients. Redundancy in golgin function or plasticity of Golgi trafficking may account for the apparent tolerance of humans to mutations in the other golgins.

Functions of the Golgins

Vesicle Transport

Tethering of Transport Vesicles

The predicted elongated conformation of the golgins projecting into the cytoplasm would appear ideal for the attachment or tethering of transport vesicles to the Golgi membrane, allowing for vesicle capture over relatively long distances (up to several hundred nanometres based upon predicted lengths of the golgins) (Fig. 4.1). The concept of golgins acting as 'tentacles' to tether vesicles [65] is certainly attractive and early studies were consistent with such a role, at least for those golgins analysed [9, 57, 67, 80, 103]. A role in vesicle tethering was further established in a landmark study published in 2014 which confirmed that golgins, when artificially relocated to mitochondria, are indeed sufficient to tether transport vesicles in cells [119]. Importantly, this study also showed that different golgins tether different populations of transport vesicles, indicating that golgin-mediated tethering can dictate the specificity of vesicle transport at the Golgi. The tethering of different vesicles correlated with the localisation of golgins at the Golgi; for example, the *cis*-Golgi golgins GM130 and GMAP-210 tethered ER-Golgi transport vesicles, whereas the TGN-localised GRIP-domain golgins tethered endosomal-derived vesicles mediating retrograde endosome to Golgi transport [119]. In contrast, golgins localised to the rims of the Golgi stack selectively tethered intra-Golgi transport vesicles. Surprisingly, not all the golgins could capture vesicles using the mitochondrial tethering assay, which may suggest those golgins (giantin, golgin-160, CASP, GCC185) do not tether vesicles, or it is possible the vesicles they tether were not analysed in the tethering assay used [119]. GCC185, for example, was shown previously to bind transport vesicles in vitro [16], consistent with a tethering function, and an earlier study suggested CASP contributes to vesicle tethering through binding to golgin-84 [57].

Consistent with the prediction of golgins acting over long distances to capture vesicles, it has been shown that they mediate tethering through their membrane-distal amino-termini [120]. In the case of GMAP-210, an amino-terminal ALPS motif, which is an extended amphipathic helix, mediates vesicle tethering by binding directly to the vesicle lipids [19, 20]. Binding is favoured by high curvature and unsaturated lipids, which induces lipid packing defects that promote amphipathic helix binding, allowing GMAP-210 to selectively bind small spherical transport vesicles at the cis-Golgi [54]. How the other golgins recognise and tether transport vesicles remains less well understood. The amino terminal regions of the golgins are conserved in evolution [120], as expected if they are functionally important, which may suggest binding to vesicle-associated protein partners to mediate tethering. For golgin-97 and golgin-245, which share highly conserved amino termini, this has been demonstrated. Both proteins bind to TBC1D23, a RabGAP-like protein that lacks catalytic activity, to mediate tethering of endosome-derived vesicles [96]. TBC1D23 can bind directly to the cytoplasmic tails of vesicle cargo proteins through a conserved acidic TLY motif to mediate tethering to the golgins [13]. TBC1D23 can also bind to the WASH complex found on endosome-derived vesicles and a tetrameric WDR11 complex found on both vesicles and the TGN [69, 96]. Interestingly, WDR11 can bind directly to vesicle cargo proteins via a super acidic cluster in the cargo protein cytoplasmic tails [17, 69]. Thus, golgin-mediated tethering of endosome-derived vesicles occurs via intermediary linker proteins, TBC1D23 and the WDR11 complex that each bind directly to the vesicle cargoes, providing an intuitive mechanism for capture in which the vesicle cargo dictates the site of vesicle delivery. These observations also indicate that signals in the tails of vesicle cargo proteins function not only in sorting of cargo into vesicles, but also contribute directly to the tethering that occurs at the target membrane, at least in the case of golgin-mediated tethering at the TGN. For the other golgins, including those in the Golgi stack and other TGN-localised golgins, the mechanisms underlying tethering remain obscure. GM130 at the *cis*-Golgi binds the vesicle docking protein p115, also called Uso1p, via its extreme amino-terminal region, and earlier studies suggested a tethering role for this interaction [67]. However, in the mitochondrial tethering assay, p115 binding appeared dispensable for tethering of ER to Golgi vesicles [119]. GM130 may therefore be able to capture vesicles using more than one mechanism. This also appears to be the case for GMAP-210, where mutation of the amino-terminal ALPS motif was shown to differentially affect tethering of different vesicle types [120]. Clearly, further work is required to understand how golgins tether transport vesicles, a key step in dictating the specificity of vesicle transport in the secretory pathway.

The Transition from Vesicle Tethering to Fusion

Following the initial long-range golgin-mediated tethering of a transport vesicle, the vesicle must be brought closer to the membrane to undergo membrane fusion and deliver its contents. SNARE proteins mediate the close apposition of vesicle and target membrane, in this case the Golgi membrane, which is referred to as docking, and they also catalyse the subsequent membrane fusion reaction [38]. How the transition from tethering to docking and then fusion occurs remains poorly understood. Interactions of golgins with additional proteins are likely to contribute to this process. The golgins can bind to Rab GTPases, often to more than one Rab, via binding sites along their length [35, 89, 90, 98]. This is not required for tethering per se but is likely to contribute to the transition to docking and fusion. This may occur either by inducing conformational changes that result in the collapse or bending of the golgin closer to the membrane, or by vesicle-associated Rabs 'hopping' between Rab-binding sites along the length of the same golgin or onto adjacent golgins able to bind the same Rab, moving the vesicle closer to the membrane where SNARE engagement can occur [66].

Some golgins can also bind to multi-subunit tethering complexes (MTCs), either to the COG complex in the Golgi stack [1, 64, 101], or the GARP complex at the TGN [15]. COG and GARP are structurally related and contain multiple subunits that interact with different components of the trafficking machinery, including coat proteins, SNARE proteins, members of the SM family of SNARE regulators and Rab GTPases, in addition to the golgins themselves [7, 40]. Both MTCs have been shown to adopt an extended conformation of 30–50 nm, which is less than the predicted lengths of the golgins when fully extended [105]. Based upon these features and additional functional data, it has been proposed that the MTCs orchestrate the transition from golgin-mediated tethering to SNARE-mediated membrane fusion, where they likely act as complex molecular chaperones to integrate the activity of the vesicle tethering and fusion machineries [118].

Structural Organisation of the Golgi

Golgi Ribbon Formation

In many species the Golgi is comprised of numerous individual Golgi stacks that are laterally connected by tubules to form a single copy ribbon [30]. This organisation was long thought to be restricted to vertebrates, but a recent study has shown the presence of Golgi ribbons in numerous non-vertebrate species [6]. Precisely how the ribbon is built is not fully understood, but the Golgi-localised proteins GRASP55 and GRASP65 function in this process [24, 32, 75, 126], likely by forming *trans*-oligomers through their PDZ domains that laterally connect the cisternae [2, 108] (Fig. 4.2). Golgin binding to the GRASPs, specifically golgin-45 (although strictly this protein does not conform to our definition of a golgin as outlined above and can be classed as golgin-like) binding to GRASP55 and GM130 binding to GRASP65, appear to be important in this process [2, 6, 75]. Mechanistically, how the golgins might play a role here remains unclear, but they could act by stabilising GRASP binding to the membrane in a conformation suitable for *trans*-oligomer formation.

The largest of the golgins, giantin, is localised to the Golgi rims where it could potentially tether transport vesicles, or mediate tethering between Golgi membranes to form lateral connections within the ribbon. Evidence exists for a tethering function [103], although recent studies are more consistent with a role in linking cisternae into a ribbon [91, 92]. Depletion of giantin in cells results in reduced connectivity between Golgi stacks, without the appearance of untethered transport vesicles in the vicinity of the Golgi, which might be expected if vesicle tethering was reduced [91, 92]. A lack of effect on vesicle tethering is also consistent with the mitochondrial tethering assay described above where giantin was unable to capture intra-Golgi vesicles, at least for those vesicles analysed in the assay [119]. Further support for giantin functioning in Golgi ribbon formation comes from the observation that giantin knockout in cultured cells leads to the breakdown of the ribbon into discrete 'units' [34]. This was accompanied by defects in glycosaminoglycan (GAG) synthesis, presumably due to the inability of GAG-synthesising enzymes to move between the now separated Golgi functional units. Mechanistically, how giantin might link Golgi membranes into a ribbon remains unclear. By analogy to the vesicle

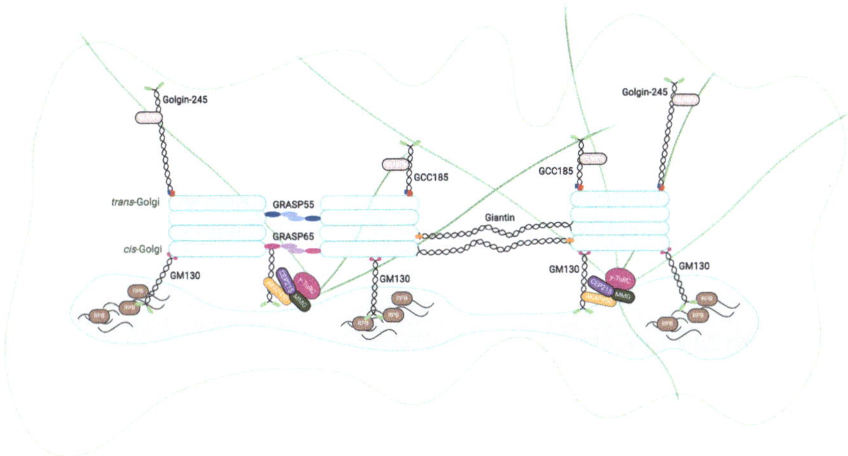

Fig. 4.2 Golgins and Golgi ribbon formation. Golgi cisternae are laterally connected to form the Golgi ribbon by the formation of GRASP55 and GRASP65 *trans*-oligomers. GM130 binds to GRASP65 and the golgin-like protein golgin-45 (not shown) binds to GRASP55 to help facilitate ribbon formation. Giantin could also potentially contribute to the formation of lateral connections between cisternae within the ribbon. Microtubules facilitate the movement and positioning of Golgi stacks and help maintain the Golgi ribbon. GM130 plays an important role in microtubule nucleation at the *cis*-Golgi by anchoring AKAP450 which binds via myomegalin (MMG) and CEP215 to γ-TURC (gamma-tubulin ring complex). At the TGN GCC185 and Golgin-245 associate with CLASPs stabilising the Golgi-nucleated microtubules. GM130 can also associate with poly(A) RNA containing bound RNA-binding proteins (RBP) to form a liquid phase condensate to stabilise the Golgi ribbon (darker green shading). There is also the possibility of a phase condensate more widely encompassing the Golgi ribbon, comprising other proteins and possibly other golgins (lighter green shading) Note that lateral tubules connecting cisternae within the ribbon are not shown. Similarly, actin, septin and intermediate filaments, which also contribute to ribbon integrity, are not shown

tethering golgins, it is tempting to suggest that giantin would bind to adjacent Golgi membranes via its membrane-distal amino-terminus, to promote lateral linking of Golgi stacks (Fig. 4.2). How any giantin-mediated tethering may be followed by Golgi membrane fusion to create a connected ribbon, also remains mysterious.

GM130 was shown several years ago to contribute to Golgi ribbon formation, which may be through interactions with GRASP65, as described above [2, 75]. However, a recent study has suggested another mechanism by which GM130 may fulfil its function in ribbon formation [127]. This novel mechanism requires GM130 to associate with poly(A) RNA to form a liquid phase condensate that structurally stabilises the Golgi ribbon [127] (Fig. 4.2). Interestingly, GM130 and other golgins were previously shown to undergo liquid phase separation, at least under specific experimental conditions [79, 128], suggesting this phenomenon may be more widespread at the Golgi membrane surface, although so far, GM130 is the only golgin shown to associate with RNA. Binding is mediated by a basic region in GM130 that is lacking from the other golgins [127], and this property may therefore be unique to GM130. Much remains to be discovered about the role of liquid phase separation at the Golgi, and how extensive and important it is for Golgi organisation and function, but a phase separated layer surrounding the Golgi could account for the long described 'zone of ribosome exclusion' around the Golgi that has been observed in numerous electron microscopy studies [84]. Interesting questions around the role of poly(A) RNA at the Golgi also arise, such as whether this localisation may contribute to localised protein translation, or conversely, the sequestration of certain RNAs away from the protein translation machinery.

Anchoring to the Cytoskeleton

Golgi Association with Microtubules

Golgi ribbon formation and its positioning near the centre of the cell are dependent upon association with the microtubule cytoskeleton and the microtubule motor protein dynein [81]. Dynein is a minus end-directed motor that drives the centripetal movement of Golgi elements towards the centrosome, the major microtubule organising centre in most cell types that is typically located adjacent to the nucleus. However, the Golgi itself plays an important role in regulating microtubule assembly and dynamics, and it too can function as a microtubule organising centre [14]. Microtubule nucleation at the Golgi is dependent upon GM130, which anchors the scaffolding protein AKAP450 to the *cis*-Golgi membrane, which in turn binds via myomegalin (MMG) and CEP215 to the γ-TURC (gamma-tubulin ring complex) microtubule nucleation complex [59, 83, 85]. The Golgi-nucleated microtubules are stabilised by association with CLASP1 and 2, which are microtubule plus end binding proteins. The CLASPs are bound to the TGN through association with the golgin GCC185 [21]. Thus, both a *cis*- and *trans*-Golgi golgin are required for the nucleation and assembly of Golgi-derived microtubules. The microtubules nucleated at the Golgi are important for maintaining the Golgi ribbon (Fig. 4.2), as well as for the polarised transport of cargo to the leading edge of migrating cells or to distinct regions of polarised cells [21, 33, 63, 86]. Interestingly, another TGN golgin, golgin-245, was also recently shown to contribute to CLASP localisation at the Golgi [26], indicating a role for several golgins in the regulation of Golgi microtubule dynamics.

As well as playing a role in nucleating microtubules at the Golgi in interphase, GM130 also contributes to microtubule nucleation during mitosis, with a role in mitotic spindle formation [115]. During mitosis, the Golgi ribbon is fragmented in preparation for its partitioning into the daughter cells during cell division [53]. Golgi fragmentation is triggered by the phosphorylation of several Golgi proteins [114], amongst which includes GM130 [51]. CDK1-mediation GM130 phosphorylation releases binding of p115, consistent with a reduced vesicle tethering function in mitosis [51]. The CDK1-mediated phosphorylation of GM130 also promotes binding to importin-α [115]. This releases importin-α from its binding partner TPX2, a spindle assembly factor, thereby promoting spindle assembly in the vicinity of the

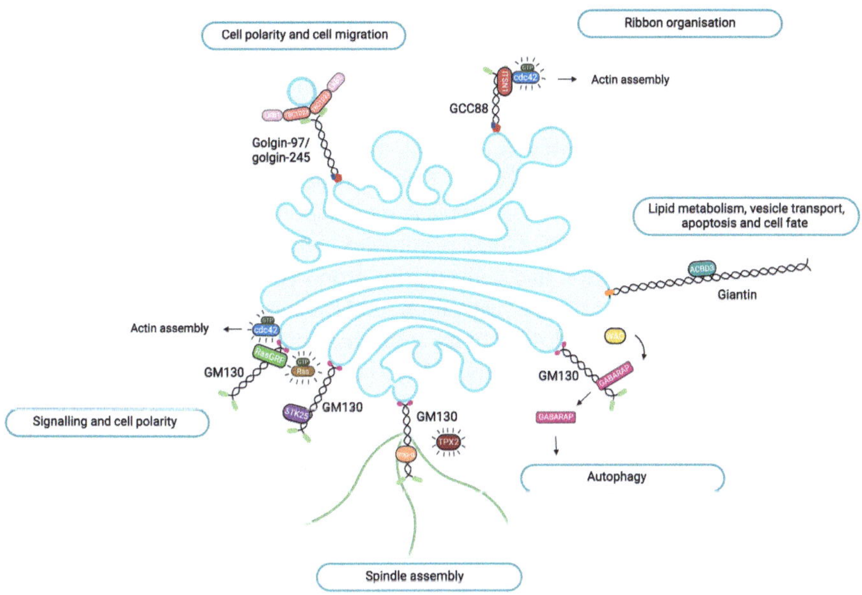

Fig. 4.3 Golgins as protein scaffolds. Golgins can act as scaffolds for the recruitment of different proteins to the Golgi membrane. GM130 in particular facilitates various processes such as signalling and Cdc42 activation (RasGRF), spindle assembly (importin-α), autophagy (WAC and GABARAP), and cell polarity and migration (STK25). ACBD3 interacts with giantin and golgin-45 (not shown) and its recruitment is important for various functions including lipid metabolism, apoptosis, asymmetric cell division and vesicle transport. GCC88 binds to the guanine nucleotide exchange factor ITSN1 which promotes Cdc42 activation at the TGN, which is important for Golgi ribbon maintenance. The TGN golgins Golgin-97 and Golgin-245 both bind to TBC1D23, which in addition to its role in vesicle tethering also binds the protein kinase LKB1, which is involved in the regulation of cell polarity and migration

Golgi membranes [115] (Fig. 4.3). Allied to this, GM130 can also capture the nascent microtubules, coupling the Golgi to the newly forming mitotic spindle. Hence, GM130 is a key microtubule regulator during both interphase and mitosis.

The mechanisms by which the dynein motor is recruited to the Golgi membrane remain poorly understood, but the dynactin and BICD2 complexes appear to be involved, which are able to associate with Golgi-localised Rab6 [60, 97]. A role for the golgin-160 protein in dynein recruitment to the Golgi has also been shown, which is important for ribbon formation and Golgi positioning in interphase cells [121]. Interestingly, golgin-160 dissociates from the Golgi during mitosis, which coincides with the breakup of the Golgi ribbon and its dispersion in the cytoplasm [121]. The mechanisms underlying golgin-160 binding to dynein in interphase and how this interaction is disrupted in mitosis remain to be determined, as does the relationship of golgin-160 to the other dynein-binding proteins at the Golgi.

Golgi Association with Other Cytoskeletal Elements

In addition to microtubules, the Golgi associates with other cytoskeletal elements, including actin, septin and intermediate filaments, which all contribute to the maintenance of Golgi organisation [22, 102, 111]. An important regulator of actin at the Golgi is the Rho family GTPase Cdc42, which exists in an active pool on Golgi membranes distinct from its typical localisation to the plasma membrane [23]. The role of Cdc42 at the Golgi is complex, with roles in Golgi trafficking, motility, ribbon organisation and cell polarity proposed [49]. Golgins function in controlling Golgi-localised Cdc42 activation. GM130 has been reported to bind the RhoGEF Tuba to promote Cdc42 activity at the *cis*-Golgi and in turn Golgi position and cell polarity [41], although others failed to detect an interaction between these proteins [5]. GM130 binds the negative regulator of Cdc42, RasGRF, whose sequestration by GM130 may also promote Cdc42 activation at the Golgi to influence cell polarity [5] (Fig. 4.3). The TGN golgin GCC88 binds to another protein with RhoGEF activity, intersectin-1 (ITSN1) [55] (Fig. 4.3). This interaction is thought to be important in maintaining the ribbon organisation of the Golgi through effects upon actin dynamics.

Septins are GTP-binding proteins that assemble in filaments, and a septin-1 scaffold has been found at the Golgi apparatus where it contributes to the structural integrity of the Golgi ribbon [102]. The Golgi-localised septin 1 scaffold functions in nucleating microtubules at the Golgi membrane, accounting for its role in Golgi organisation. Septin 1 localisation to the Golgi is dependent upon interaction with GM130, and both proteins can be found in a complex with γ-TURC and the centrosomal protein CEP170 [102]. How this complex relates to the previously described GM130-AKAP40 complex, which also associated with γ-TURC, remains to be determined, and it is possible they are part of the same larger assembly. In any case, it appears cooperation between the septin and microtubule cytoskeletons is necessary for Golgi-dependent microtubule nucleation, with GM130 playing a key role in these events.

Physical Coupling of ER Exit Sites to the Golgi Apparatus

In those species where the Golgi exists as a single or multiple discrete Golgi stacks, there is close apposition of the stack(s) to ER exit sites [68], where cargo proteins are sorted into carriers for delivery to the *cis*-Golgi [10, 56]. Even in those species that contain a Golgi ribbon, while some ERES are distributed throughout the cytoplasm, the majority lie in close physical proximity to the perinuclear Golgi ribbon [68]. The ER-Golgi-intermediate compartment lies between the ERES and *cis*-Golgi in vertebrates [87], and comparable structures have now been observed in yeast, plants and *Drosophila*, indicating this intermediate trafficking station between the ERES and Golgi is more universal than first thought [37, 107, 124]. The nature of the carriers operating at ERES and ERGIC remains a matter of debate, with recent studies showing tubular intermediates operating at these locations [93, 116, 123, 124]. In addition, what maintains the apparently tight coupling of the ERES, ERGIC and Golgi elements remains to be determined. The TANGO1 scaffolding protein can couple ERES to ERGIC [78], and phase separated condensates containing the protein TFG have also been proposed to organise the ERES/Golgi interface [76, 112]. However, it remains possible that golgins also contribute to the physical coupling of ERES, ERGIC and *cis*-Golgi. The golgins GMAP-210 and GM130 are present in the ERGIC in addition to the Golgi [58, 82], and may participate in tethering this compartment to the ER or *cis*-Golgi. Similarly, in plants the golgins AtCASP and Atgolgin-84A is present in a pre-Golgi compartment, suggesting a role in tethering at the ERES-ERGIC-Golgi interface [71, 110]. Indeed, loss of AtCASP decouples the Golgi from ER, consistent with a role in physically connecting these compartments [71].

Golgins and Exocytosis

Two of the TGN-localised golgins, golgin-97 and golgin-245, are required for the exocytic delivery of certain cargo proteins to the plasma membrane. For example, exocytosis of the basolateral cargo E-cadherin in epithelial cells is sensitive to loss of golgin-97 [129]. This role in exocytic traffic is dependent upon the mono-ADP-ribosylation of golgin-97 in a region within the coiled-coil domain, which is mediated by PARP12 following phosphorylation by protein kinase D [31], which was previously shown to regulate carrier formation at the TGN [47]. Golgin-97 is also important for the plasma membrane delivery of the potassium channel Kir2.1 [106]. In contrast, golgin-245 is required for the regulated secretion of TNF from macrophages [46]. These studies point to a role for the TGN golgins in exocytosis, but the mechanisms remain obscure. It remains possible that a role in TGN exit and subsequent transport to the plasma membrane is indirect, whereby defects in endosome to TGN transport due to defective vesicle tethering at the TGN impacts upon exocytotic transport of these cargo proteins. Alternatively, they may participate more directly in cargo sorting or export from the TGN by mechanisms unknown.

The Golgins as Protein Anchors

As described above, the golgins can associate with different proteins in the context of vesicle tethering and cytoskeletal organisation. However, it has also emerged that golgins can bind to proteins associated with other cellular activities, which includes signaling, autophagy regulation and cell fate determination, and anchor them to the Golgi membrane (Fig. 4.3). As mentioned above GM130 binds to RasGFF at the *cis*-Golgi, whose sequestration allows Tuba to activate Cdc42 at the Golgi

[5]. RasGRF also lies upstream of Ras and it has been proposed that GM130 binding may act to limit MAP kinase pathway activation and cellular differentiation [5]. GM130 can also bind to the STK25 (YSK1) protein kinase, which results in its activation at the Golgi [74]. STK25 is important for Golgi positioning, cell polarity, including dendrite specification in neurons, and cell migration [61, 62, 74, 77]. STK25 may act through phosphorylation of Golgi-localised 14-3-3ζ in this context [74]. STK25 lies downstream of the master kinase LKB1, which regulates various processes including cell polarity and migration, but also energy sensing via AMPK [8]. Interestingly, LKB1 recruitment and subsequent activation of LKB1 at the Golgi was recently shown to require binding to TBC1D23, the protein required for golgin-97 and golgin-245 mediated vesicle tethering at the TGN [109]. How the two functions of TBC1D23 at the TGN, vesicle tethering and LKB1 signaling, relate to one another requires further investigation.

In addition to the signaling role described above, GM130 also functions as a regulator of autophagy. GM130 can recruit the key autophagy factor GABARAP to the Golgi, whose sequestration there negatively regulates autophagy [39]. This interaction is competed by the WAC protein, releasing GABARAP from the Golgi and allowing it to promote autophagy in the pericentriolar region of the cell. While the role of GM130 is autophagy induction is important, the detailed molecular mechanisms await further study.

Another golgin-binding protein of interest is ACBD3, also known as GCP60, a scaffolding protein with numerous interaction partners [125]. ACBD3 appears to function in various cellular processes including lipid metabolism, apoptosis, neurite outgrowth, Golgi organisation and it has also reported to act as a cell fate determinant during cell division [125]. ACBD3 binds to giantin and golgin-45, amongst other proteins, which are required for ACBD3 recruitment to the Golgi membrane [100, 104]. Golgi-localised ACBD3 promotes recruitment of PI4-kinase (PI4KIIIβ) to generate PI4P at the Golgi membrane [88], which is important for lipid and vesicle transport there. Binding to giantin and golgin-45 appears to occur downstream from an initial interaction of ACBD3 with the SNARE Sec22B and its partner protein SCFD1, which may help coordinate ACBD3 recruitment with the arrival of incoming vesicles undergoing fusion events at the Golgi [104]. Further experiments are required to test this model, and it will also be interesting to determine how golgin interaction is coordinated functionally with the other interactions that ACBD3 undertakes.

Conclusion and Future Perspectives

The evolutionary conservation of the golgins indicates their importance, and in this book chapter we have described current knowledge regarding the various functions of these proteins. Their best characterised role is in vesicle tethering, where they capture transport vesicles in a specific manner at the Golgi membrane, a key process in the trafficking that occurs at the Golgi apparatus. However, many questions remain. These include the identity of the determinants on vesicles to which the golgins bind, which in the majority of cases is unknown. GMAP210 binds to lipids [19, 20], GM130 binds to p115 [67], whereas golgin-97 and golgin-245 are in a complex with the linker proteins TBC1D23 and the WDR11 complex that in turn bind to vesicle cargo proteins [13, 17, 69, 96]. Whether the other golgins also bind to vesicle lipids like GMAP210, or vesicle-associated proteins like GM130, golgin-97 and golgin-245, or even both, awaits further study. It is also unclear whether all the golgins function as vesicle tethers, or whether some function as tethers to other membranes, as may be the case for giantin in linking the Golgi ribbon. It will be interesting to analyse vesicle tethering in different cell types and model organisms, as has recently been performed in *Drosophila* for the golgins GMAP-210, golgin-84 and TMF1 [73], to obtain a better understanding of the requirements for the golgins in different cell types and tissues, and to better understand their specificity in the tethering process. Another question regards the contents of the vesicles tethered by the different golgins, and the extent to which these transport vesicles differ in their composition. Reconstitution of tethering combined with proteomics and lipidomics will provide important new insights in this regard [96]. A major unknown regards the events following golgin-mediated tethering, where the vesicle must manoeuvre closer to the membrane for SNARE engagement and membrane fusion. Various models have been proposed, discussed in this article, but experiments to test these models are currently lacking. It will be highly informative to reconstitute these events in vitro and perform a high-resolution structural analysis.

The golgins are important for the organisation of the Golgi ribbon and contribute to this process in different ways. Proposed mechanisms include stabilising *trans*-oligomers of the GRASP proteins (golgin-45 and GM130) [2, 6, 75], lateral tethering of Golgi membranes (giantin) [91, 92], liquid phase condensate formation (GM130) [127], and interacting with the microtubule cytoskeleton (GM130, GCC185 and golgin-245) [21, 26, 83] or the motor protein dynein (golgin-160) [121]. Interactions with actin machinery also likely contribute to this process, mediated through Cdc42 regulation by GM130 and GCC88 [5, 55]. It will be interesting to determine the extent to which these different mechanisms contribute to ribbon assembly and maintenance in different cell types, and how they are coordinated. As well as ribbon formation, the golgins contribute to other aspects of Golgi structural organisation; for example, they may contribute to cisternal stacking [45], and evidence points to a role in physical coupling of the Golgi to ERES and ERGIC [71]. In terms of stacking, the GRASP proteins were originally thought to mediate this process [130], although this view is somewhat controversial [11]. Nevertheless, it has been shown that golgins can functionally substitute for the GRASPs in stacking when over-expressed in cells, suggesting a possible role in this process [45]. The elongated nature of the golgins and their inherent ability to tether membranes would fit with possible roles in ribbon formation, cisternal stacking and ERES-Golgi coupling, as just described. Interestingly, when liver cells are undergoing steatosis, the Golgi is sequestered to the surface of lipid droplets, which accumulate in large numbers under these conditions, and this correlates with reduced secretory traffic [44]. Although the mechanisms by which the Golgi attaches to lipid droplets remain to be determined, it has been proposed the golgins fulfil this function, based upon their ability to tether membranes to the Golgi surface [43]. Interestingly, GMAP-210 binds to membranes with lipid packing defects [54], which are also seen on the surface of lipid droplets, potentially explaining how such tethering could work. Further experiments will reveal how the Golgi attaches to lipid droplets and whether the golgins indeed contribute to this process.

The golgins' major functions appear to be in Golgi trafficking and structural organisation, which is not surprising considering they are abundant Golgi-localised proteins. However, studies over the years have identified roles on other, non-Golgi processes. These include mitotic spindle formation, signaling, autophagy and asymmetric cell division. In this context, the golgins appear to act as scaffolds for the recruitment of different types of protein factors to the Golgi membrane. GM130 appears versatile in this regard, able to bind numerous factors to influence spindle formation (importin-α) [115], cell polarity and migration (STK25) [74], mitogenic signaling (RasGRF) [5] and autophagy (WAC and GABARAP) [39]. Why GM130 has evolved as a scaffold for so many factors, which also includes AKAP450 and Tuba or RasGRF for microtubule and Cdc42 regulation respectively [5, 41, 83], is an interesting question. It will also be interesting to explore the extent to which other golgins, which are elongated proteins with potentially numerous protein interaction sites along their length, bind to factors that regulate other cellular processes.

Acknowledgements Work in the Lowe lab is funded by the BBSRC (BB/T001984/1; BB/X006859/1), MRC (MR/Y014251/1), Leverhulme Trust (RPG-2024-031). Figures were made using BioRender (biorender.com).

Compliance with Ethical Standards The author declare no conflicts of interest.

References

1. Anderson NS, Mukherjee I, Bentivoglio CM, Barlowe C (2017) The Golgin protein Coy1 functions in intra-Golgi retrograde transport and interacts with the COG complex and Golgi SNAREs. Mol Biol Cell 28(20):2686–2700. https://doi.org/10.1091/mbc.E17-03-0137
2. Bachert C, Linstedt AD (2010) Dual anchoring of the GRASP membrane tether promotes trans pairing. J Biol Chem 285(21):16294–16301. https://doi.org/10.1074/jbc.M110.116129
3. Barlow LD, Nyvltova E, Aguilar M, Tachezy J, Dacks JB (2018) A sophisticated, differentiated Golgi in the ancestor of eukaryotes. BMC Biol 16(1):27. https://doi.org/10.1186/s12915-018-0492-9
4. Barr FA, Nakamura N, Warren G (1998) Mapping the interaction between GRASP65 and GM130, components of a protein complex involved in the stacking of Golgi cisternae. EMBO J 17(12):3258–3268. https://doi.org/10.1093/emboj/17.12.3258
5. Baschieri F, Confalonieri S, Bertalot G, Di Fiore PP, Dietmaier W, Leist M, Crespo P, Macara IG, Farhan H (2014) Spatial control of Cdc42 signalling by a GM130-RasGRF complex regulates polarity and tumorigenesis. Nat Commun 5:4839. https://doi.org/10.1038/ncomms5839
6. Benvenuto G, Leone S, Astoricchio E, Bormke S, Jasek S, D'Aniello E, Kittelmann M, McDonald K, Hartenstein V, Baena V, Escriva H, Bertrand S, Schierwater B, Burkhardt P,

Ruiz-Trillo I, Jekely G, Ullrich-Luter J, Luter C, D'Aniello S, Arnone MI, Ferraro F (2024) Evolution of the ribbon-like organization of the Golgi apparatus in animal cells. Cell Rep 43(3):113791. https://doi.org/10.1016/j.celrep.2024.113791
7. Blackburn JB, D'Souza Z, Lupashin VV (2019) Maintaining order: COG complex controls Golgi trafficking, processing, and sorting. FEBS Lett 593(17):2466–2487. https://doi.org/10.1002/1873-3468.13570
8. Bourouh M, Marignani PA (2022) The tumor suppressor kinase LKB1: metabolic nexus. Front Cell Dev Biol 10:881297. https://doi.org/10.3389/fcell.2022.881297
9. Brown FC, Schindelhaim CH, Pfeffer SR (2011) GCC185 plays independent roles in Golgi structure maintenance and AP-1-mediated vesicle tethering. J Cell Biol 194(5):779–787. https://doi.org/10.1083/jcb.201104019
10. Budnik A, Stephens DJ (2009) ER exit sites—localization and control of COPII vesicle formation. FEBS Lett 583(23):3796–3803. https://doi.org/10.1016/j.febslet.2009.10.038
11. Burd CG (2021) GRASPing for consensus about the Golgi apparatus. J Cell Biol 220(5): e202103117. https://doi.org/10.1083/jcb.202103117
12. Burguete AS, Fenn TD, Brunger AT, Pfeffer SR (2008) Rab and Arl GTPase family members cooperate in the localization of the golgin GCC185. Cell 132(2):286–298. https://doi.org/10.1016/j.cell.2007.11.048
13. Cattin-Ortola J, Kaufman JGG, Gillingham AK, Wagstaff JL, Peak-Chew SY, Stevens TJ, Boulanger J, Owen DJ, Munro S (2024) Cargo selective vesicle tethering: The structural basis for binding of specific cargo proteins by the Golgi tether component TBC1D23. Sci Adv 10(13):eadl0608. https://doi.org/10.1126/sciadv.adl0608
14. Chabin-Brion K, Marceiller J, Perez F, Settegrana C, Drechou A, Durand G, Pous C (2001) The Golgi complex is a microtubule-organizing organelle. Mol Biol Cell 12(7):2047–2060. https://doi.org/10.1091/mbc.12.7.2047
15. Chen YT, Wang IH, Wang YH, Chiu WY, Hu JH, Chen WH, Lee FS (2019) Action of Arl1 GTPase and golgin Imh1 in Ypt6-independent retrograde transport from endosomes to the trans-Golgi network. Mol Biol Cell 30(8):1008–1019. https://doi.org/10.1091/mbc.E18-09-0579
16. Cheung PY, Limouse C, Mabuchi H, Pfeffer SR (2015) Protein flexibility is required for vesicle tethering at the Golgi. Elife 4:e12790. https://doi.org/10.7554/eLife.12790
17. Deng H, Jia G, Li P, Tang Y, Zhao L, Yang Q, Zhao J, Wang J, Tu Y, Yong X, Zhang S, Mo X, Billadeau DD, Su Z, Jia D (2024) The WDR11 complex is a receptor for acidic-cluster-containing cargo proteins. Cell 187:4272–4288. https://doi.org/10.1016/j.cell.2024.06.024
18. Diao A, Rahman D, Pappin DJ, Lucocq J, Lowe M (2003) The coiled-coil membrane protein golgin-84 is a novel rab effector required for Golgi ribbon formation. J Cell Biol 160(2): 201–212. https://doi.org/10.1083/jcb.200207045
19. Drin G, Casella JF, Gautier R, Boehmer T, Schwartz TU, Antonny B (2007) A general amphipathic alpha-helical motif for sensing membrane curvature. Nat Struct Mol Biol 14(2): 138–146. https://doi.org/10.1038/nsmb1194
20. Drin G, Morello V, Casella JF, Gounon P, Antonny B (2008) Asymmetric tethering of flat and curved lipid membranes by a golgin. Science 320(5876):670–673. https://doi.org/10.1126/science.1155821
21. Efimov A, Kharitonov A, Efimova N, Loncarek J, Miller PM, Andreyeva N, Gleeson P, Galjart N, Maia AR, McLeod IX, Yates JR 3rd, Maiato H, Khodjakov A, Akhmanova A, Kaverina I (2007) Asymmetric CLASP-dependent nucleation of noncentrosomal microtubules at the trans-Golgi network. Dev Cell 12(6):917–930. https://doi.org/10.1016/j.devcel.2007.04.002
22. Egea G, Serra-Peinado C, Salcedo-Sicilia L, Gutierrez-Martinez E (2013) Actin acting at the Golgi. Histochem Cell Biol 140(3):347–360. https://doi.org/10.1007/s00418-013-1115-8
23. Erickson JW, Zhang C, Kahn RA, Evans T, Cerione RA (1996) Mammalian Cdc42 is a brefeldin A-sensitive component of the Golgi apparatus. J Biol Chem 271(43):26850–26854. https://doi.org/10.1074/jbc.271.43.26850

24. Feinstein TN, Linstedt AD (2008) GRASP55 regulates Golgi ribbon formation. Mol Biol Cell 19(7):2696–2707. https://doi.org/10.1091/mbc.e07-11-1200
25. Fridmann-Sirkis Y, Siniossoglou S, Pelham HR (2004) TMF is a golgin that binds Rab6 and influences Golgi morphology. BMC Cell Biol 5:18. https://doi.org/10.1186/1471-2121-5-18
26. Gareil N, Gervais A, Macaisne N, Chevreux G, Canman JC, Andreani J, Dumont J (2023) An unconventional TOG domain is required for CLASP localization. Curr Biol 33(16): 3522–3528. https://doi.org/10.1016/j.cub.2023.07.009
27. Gillingham AK, Munro S (2016) Finding the Golgi: golgin coiled-coil proteins show the way. Trends Cell Biol 26(6):399–408. https://doi.org/10.1016/j.tcb.2016.02.005
28. Gillingham AK, Pfeifer AC, Munro S (2002) CASP, the alternatively spliced product of the gene encoding the CCAAT-displacement protein transcription factor, is a Golgi membrane protein related to giantin. Mol Biol Cell 13(11):3761–3774. https://doi.org/10.1091/mbc.e02-06-0349
29. Gillingham AK, Tong AH, Boone C, Munro S (2004) The GTPase Arf1p and the ER to Golgi cargo receptor Erv14p cooperate to recruit the golgin Rud3p to the cis-Golgi. J Cell Biol 167(2):281–292. https://doi.org/10.1083/jcb.200407088
30. Gosavi P, Gleeson PA (2017) The function of the Golgi ribbon structure—an enduring mystery unfolds! Bioessays 39(11):1700063. https://doi.org/10.1002/bies.201700063
31. Grimaldi G, Filograna A, Schembri L, Lo Monte M, Di Martino R, Pirozzi M, Spano D, Beccari AR, Parashuraman S, Luini A, Valente C, Corda D (2022) PKD-dependent PARP12-catalyzed mono-ADP-ribosylation of Golgin-97 is required for E-cadherin transport from Golgi to plasma membrane. Proc Natl Acad Sci USA 119(1):e2026494119. https://doi.org/10.1073/pnas.2026494119
32. Grond R, Veenendaal T, Duran JM, Raote I, van Es JH, Corstjens S, Delfgou L, El Haddouti B, Malhotra V, Rabouille C (2020) The function of GORASPs in Golgi apparatus organization in vivo. J Cell Biol 219(9):e202004191. https://doi.org/10.1083/jcb.202004191
33. Hao H, Niu J, Xue B, Su QP, Liu M, Yang J, Qin J, Zhao S, Wu C, Sun Y (2020) Golgi-associated microtubules are fast cargo tracks and required for persistent cell migration. EMBO Rep 21(3):e48385. https://doi.org/10.15252/embr.201948385
34. Harada A, Kunii M, Kurokawa K, Sumi T, Kanda S, Zhang Y, Nadanaka S, Hirosawa KM, Tokunaga K, Tojima T, Taniguchi M, Moriwaki K, Yoshimura SI, Yamamoto-Hino M, Goto S, Katagiri T, Kume S, Hayashi-Nishino M, Nakano M, Miyoshi E, Suzuki KGN, Kitagawa H, Nakano A (2024) Dynamic movement of the Golgi unit and its glycosylation enzyme zones. Nat Commun 15(1):4514. https://doi.org/10.1038/s41467-024-48901-1
35. Hayes GL, Brown FC, Haas AK, Nottingham RM, Barr FA, Pfeffer SR (2009) Multiple Rab GTPase binding sites in GCC185 suggest a model for vesicle tethering at the trans-Golgi. Mol Biol Cell 20(1):209–217. https://doi.org/10.1091/mbc.e08-07-0740
36. Ishida R, Yamamoto A, Nakayama K, Sohda M, Misumi Y, Yasunaga T, Nakamura N (2015) GM130 is a parallel tetramer with a flexible rod-like structure and N-terminally open (Y-shaped) and closed (I-shaped) conformations. FEBS J 282(11):2232–2244. https://doi.org/10.1111/febs.13271
37. Ito Y, Uemura T, Nakano A (2018) The Golgi entry core compartment functions as a COPII-independent scaffold for ER-to-Golgi transport in plant cells. J Cell Sci 131(2):jcs203893. https://doi.org/10.1242/jcs.203893
38. Jahn R, Scheller RH (2006) SNAREs—engines for membrane fusion. Nat Rev Mol Cell Biol 7(9):631–643. https://doi.org/10.1038/nrm2002
39. Joachim J, Jefferies HB, Razi M, Frith D, Snijders AP, Chakravarty P, Judith D, Tooze SA (2015) Activation of ULK kinase and autophagy by GABARAP trafficking from the centrosome is regulated by WAC and GM130. Mol Cell 60(6):899–913. https://doi.org/10.1016/j.molcel.2015.11.018
40. Khakurel A, Lupashin VV (2023) Role of GARP vesicle tethering complex in Golgi physiology. Int J Mol Sci 24(7):6069. https://doi.org/10.3390/ijms24076069

41. Kodani A, Kristensen I, Huang L, Sutterlin C (2009) GM130-dependent control of Cdc42 activity at the Golgi regulates centrosome organization. Mol Biol Cell 20(4):1192–1200. https://doi.org/10.1091/mbc.e08-08-0834
42. Koreishi M, Gniadek TJ, Yu S, Masuda J, Honjo Y, Satoh A (2013) The golgin tether giantin regulates the secretory pathway by controlling stack organization within Golgi apparatus. PLoS One 8(3):e59821. https://doi.org/10.1371/journal.pone.0059821
43. Krahmer N, Mann M (2019) Catching lipid droplet contacts by proteomics. Contact 2: 2515256419859186. https://doi.org/10.1177/2515256419859186
44. Krahmer N, Najafi B, Schueder F, Quagliarini F, Steger M, Seitz S, Kasper R, Salinas F, Cox J, Uhlenhaut NH, Walther TC, Jungmann R, Zeigerer A, Borner GHH, Mann M (2018) Organellar proteomics and phospho-proteomics reveal subcellular reorganization in diet-induced hepatic steatosis. Dev Cell 47(2):205–221. https://doi.org/10.1016/j.devcel.2018.09.017
45. Lee I, Tiwari N, Dunlop MH, Graham M, Liu X, Rothman JE (2014) Membrane adhesion dictates Golgi stacking and cisternal morphology. Proc Natl Acad Sci USA 111(5): 1849–1854. https://doi.org/10.1073/pnas.1323895111
46. Lieu ZZ, Lock JG, Hammond LA, La Gruta NL, Stow JL, Gleeson PA (2008) A trans-Golgi network golgin is required for the regulated secretion of TNF in activated macrophages in vivo. Proc Natl Acad Sci USA 105(9):3351–3356. https://doi.org/10.1073/pnas.0800137105
47. Liljedahl M, Maeda Y, Colanzi A, Ayala I, Van Lint J, Malhotra V (2001) Protein kinase D regulates the fission of cell surface destined transport carriers from the trans-Golgi network. Cell 104(3):409–420. https://doi.org/10.1016/s0092-8674(01)00228-8
48. Linstedt AD, Hauri HP (1993) Giantin, a novel conserved Golgi membrane protein containing a cytoplasmic domain of at least 350 kDa. Mol Biol Cell 4(7):679–693. https://doi.org/10.1091/mbc.4.7.679
49. Long M, Kranjc T, Mysior MM, Simpson JC (2020) RNA interference screening identifies novel roles for RhoBTB1 and RhoBTB3 in membrane trafficking events in mammalian cells. Cells 9(5):1089. https://doi.org/10.3390/cells9051089
50. Lowe M (2019) The physiological functions of the golgin vesicle tethering proteins. Front Cell Dev Biol 7:94. https://doi.org/10.3389/fcell.2019.00094
51. Lowe M, Rabouille C, Nakamura N, Watson R, Jackman M, Jamsa E, Rahman D, Pappin DJ, Warren G (1998) Cdc2 kinase directly phosphorylates the cis-Golgi matrix protein GM130 and is required for Golgi fragmentation in mitosis. Cell 94(6):783–793. https://doi.org/10.1016/s0092-8674(00)81737-7
52. Lu L, Hong W (2003) Interaction of Arl1-GTP with GRIP domains recruits autoantigens Golgin-97 and Golgin-245/p230 onto the Golgi. Mol Biol Cell 14(9):3767–3781. https://doi.org/10.1091/mbc.e03-01-0864
53. Lucocq JM, Warren G (1987) Fragmentation and partitioning of the Golgi apparatus during mitosis in HeLa cells. EMBO J 6(11):3239–3246. https://doi.org/10.1002/j.1460-2075.1987.tb02641.x
54. Magdeleine M, Gautier R, Gounon P, Barelli H, Vanni S, Antonny B (2016) A filter at the entrance of the Golgi that selects vesicles according to size and bulk lipid composition. Elife 5: e16988. https://doi.org/10.7554/eLife.16988
55. Makhoul C, Gosavi P, Duffield R, Delbridge B, Williamson NA, Gleeson PA (2019) Intersectin-1 interacts with the golgin GCC88 to couple the actin network and Golgi architecture. Mol Biol Cell 30(3):370–386. https://doi.org/10.1091/mbc.E18-05-0313
56. Malhotra V (2024) Tailored assemblies of COPII proteins in secretion. J Cell Biol 223(8): e202404013. https://doi.org/10.1083/jcb.202404013
57. Malsam J, Satoh A, Pelletier L, Warren G (2005) Golgin tethers define subpopulations of COPI vesicles. Science 307(5712):1095–1098. https://doi.org/10.1126/science.1108061
58. Marra P, Maffucci T, Daniele T, Tullio GD, Ikehara Y, Chan EK, Luini A, Beznoussenko G, Mironov A, De Matteis MA (2001) The GM130 and GRASP65 Golgi proteins cycle through

and define a subdomain of the intermediate compartment. Nat Cell Biol 3(12):1101–1113. https://doi.org/10.1038/ncb1201-1101
59. Mascanzoni F, Iannitti R, Colanzi A (2022) Functional coordination among the Golgi complex, the centrosome and the microtubule cytoskeleton during the cell cycle. Cells 11(3):354. https://doi.org/10.3390/cells11030354
60. Matanis T, Akhmanova A, Wulf P, Del Nery E, Weide T, Stepanova T, Galjart N, Grosveld F, Goud B, De Zeeuw CI, Barnekow A, Hoogenraad CC (2002) Bicaudal-D regulates COPI-independent Golgi-ER transport by recruiting the dynein-dynactin motor complex. Nat Cell Biol 4(12):986–992. https://doi.org/10.1038/ncb891
61. Matsuki T, Matthews RT, Cooper JA, van der Brug MP, Cookson MR, Hardy JA, Olson EC, Howell BW (2010) Reelin and stk25 have opposing roles in neuronal polarization and dendritic Golgi deployment. Cell 143(5):826–836. https://doi.org/10.1016/j.cell.2010.10.029
62. Matsuki T, Chen J, Howell BW (2013) Acute inactivation of the serine-threonine kinase Stk25 disrupts neuronal migration. Neural Dev 8:21. https://doi.org/10.1186/1749-8104-8-21
63. Miller PM, Folkmann AW, Maia AR, Efimova N, Efimov A, Kaverina I (2009) Golgi-derived CLASP-dependent microtubules control Golgi organization and polarized trafficking in motile cells. Nat Cell Biol 11(9):1069–1080. https://doi.org/10.1038/ncb1920
64. Miller VJ, Sharma P, Kudlyk TA, Frost L, Rofe AP, Watson IJ, Duden R, Lowe M, Lupashin VV, Ungar D (2013) Molecular insights into vesicle tethering at the Golgi by the conserved oligomeric Golgi (COG) complex and the golgin TATA element modulatory factor (TMF). J Biol Chem 288(6):4229–4240. https://doi.org/10.1074/jbc.M112.426767
65. Munro S (2011) The golgin coiled-coil proteins of the Golgi apparatus. Cold Spring Harb Perspect Biol 3(6):a005256. https://doi.org/10.1101/cshperspect.a005256
66. Muschalik N, Munro S (2018) Golgins. Curr Biol 28(8):R374–R376. https://doi.org/10.1016/j.cub.2018.01.006
67. Nakamura N, Lowe M, Levine TP, Rabouille C, Warren G (1997) The vesicle docking protein p115 binds GM130, a cis-Golgi matrix protein, in a mitotically regulated manner. Cell 89(3): 445–455. https://doi.org/10.1016/s0092-8674(00)80225-1
68. Nakano A (2022) The Golgi apparatus and its next-door neighbors. Front Cell Dev Biol 10: 884360. https://doi.org/10.3389/fcell.2022.884360
69. Navarro Negredo P, Edgar JR, Manna PT, Antrobus R, Robinson MS (2018) The WDR11 complex facilitates the tethering of AP-1-derived vesicles. Nat Commun 9(1):596. https://doi.org/10.1038/s41467-018-02919-4
70. Nozawa K, Fritzler MJ, Chan EK (2005) Unique and shared features of Golgi complex autoantigens. Autoimmun Rev 4(1):35–41. https://doi.org/10.1016/j.autrev.2004.06.002
71. Osterrieder A, Sparkes IA, Botchway SW, Ward A, Ketelaar T, de Ruijter N, Hawes C (2017) Stacks off tracks: a role for the golgin AtCASP in plant endoplasmic reticulum-Golgi apparatus tethering. J Exp Bot 68(13):3339–3350. https://doi.org/10.1093/jxb/erx167
72. Panic B, Whyte JR, Munro S (2003) The ARF-like GTPases Arl1p and Arl3p act in a pathway that interacts with vesicle-tethering factors at the Golgi apparatus. Curr Biol 13(5):405–410. https://doi.org/10.1016/s0960-9822(03)00091-5
73. Park SY, Muschalik N, Chadwick J, Munro S (2022) In vivo characterization of Drosophila golgins reveals redundancy and plasticity of vesicle capture at the Golgi apparatus. Curr Biol 32(21):4549–4564. https://doi.org/10.1016/j.cub.2022.08.054
74. Preisinger C, Short B, De Corte V, Bruyneel E, Haas A, Kopajtich R, Gettemans J, Barr FA (2004) YSK1 is activated by the Golgi matrix protein GM130 and plays a role in cell migration through its substrate 14-3-3zeta. J Cell Biol 164(7):1009–1020. https://doi.org/10.1083/jcb.200310061
75. Puthenveedu MA, Bachert C, Puri S, Lanni F, Linstedt AD (2006) GM130 and GRASP65-dependent lateral cisternal fusion allows uniform Golgi-enzyme distribution. Nat Cell Biol 8(3):238–248. https://doi.org/10.1038/ncb1366

76. Qiu H, Wu X, Ma X, Li S, Cai Q, Ganzella M, Ge L, Zhang H, Zhang M (2024) Short-distance vesicle transport via phase separation. Cell 187(9):2175–2193. https://doi.org/10.1016/j.cell.2024.03.003
77. Rao S, Kirschen GW, Szczurkowska J, Di Antonio A, Wang J, Ge S, Shelly M (2018) Repositioning of somatic Golgi apparatus is essential for the dendritic establishment of adult-born hippocampal neurons. J Neurosci 38(3):631–647. https://doi.org/10.1523/JNEUROSCI.1217-17.2017
78. Raote I, Ortega-Bellido M, Santos AJ, Foresti O, Zhang C, Garcia-Parajo MF, Campelo F, Malhotra V (2018) TANGO1 builds a machine for collagen export by recruiting and spatially organizing COPII, tethers and membranes. Elife 7:e32723. https://doi.org/10.7554/eLife.32723
79. Rebane AA, Ziltener P, LaMonica LC, Bauer AH, Zheng H, Lopez-Montero I, Pincet F, Rothman JE, Ernst AM (2020) Liquid-liquid phase separation of the Golgi matrix protein GM130. FEBS Lett 594(7):1132–1144. https://doi.org/10.1002/1873-3468.13715
80. Reddy JV, Burguete AS, Sridevi K, Ganley IG, Nottingham RM, Pfeffer SR (2006) A functional role for the GCC185 golgin in mannose 6-phosphate receptor recycling. Mol Biol Cell 17(10):4353–4363. https://doi.org/10.1091/mbc.e06-02-0153
81. Rios RM, Bornens M (2003) The Golgi apparatus at the cell centre. Curr Opin Cell Biol 15(1):60–66. https://doi.org/10.1016/s0955-0674(02)00013-3
82. Rios RM, Tassin AM, Celati C, Antony C, Boissier MC, Homberg JC, Bornens M (1994) A peripheral protein associated with the cis-Golgi network redistributes in the intermediate compartment upon brefeldin A treatment. J Cell Biol 125(5):997–1013. https://doi.org/10.1083/jcb.125.5.997
83. Rivero S, Cardenas J, Bornens M, Rios RM (2009) Microtubule nucleation at the cis-side of the Golgi apparatus requires AKAP450 and GM130. EMBO J 28(8):1016–1028. https://doi.org/10.1038/emboj.2009.47
84. Rothman JE (2019) Jim's View: Is the Golgi stack a phase-separated liquid crystal? FEBS Lett 593(19):2701–2705. https://doi.org/10.1002/1873-3468.13609
85. Roubin R, Acquaviva C, Chevrier V, Sedjai F, Zyss D, Birnbaum D, Rosnet O (2013) Myomegalin is necessary for the formation of centrosomal and Golgi-derived microtubules. Biol Open 2(2):238–250. https://doi.org/10.1242/bio.20123392
86. Sanders AA, Kaverina I (2015) Nucleation and dynamics of Golgi-derived microtubules. Front Neurosci 9:431. https://doi.org/10.3389/fnins.2015.00431
87. Saraste J, Marie M (2018) Intermediate compartment (IC): from pre-Golgi vacuoles to a semi-autonomous membrane system. Histochem Cell Biol 150(5):407–430. https://doi.org/10.1007/s00418-018-1717-2
88. Sasaki J, Ishikawa K, Arita M, Taniguchi K (2012) ACBD3-mediated recruitment of PI4KB to picornavirus RNA replication sites. EMBO J 31(3):754–766. https://doi.org/10.1038/emboj.2011.429
89. Sato K, Roboti P, Mironov AA, Lowe M (2015) Coupling of vesicle tethering and Rab binding is required for in vivo functionality of the golgin GMAP-210. Mol Biol Cell 26(3):537–553. https://doi.org/10.1091/mbc.E14-10-1450
90. Satoh A, Wang Y, Malsam J, Beard MB, Warren G (2003) Golgin-84 is a rab1 binding partner involved in Golgi structure. Traffic 4(3):153–161. https://doi.org/10.1034/j.1600-0854.2003.00103.x
91. Satoh A, Hayashi-Nishino M, Shakuno T, Masuda J, Koreishi M, Murakami R, Nakamura Y, Nakamura T, Abe-Kanoh N, Honjo Y, Malsam J, Yu S, Nishino K (2019) The golgin protein giantin regulates interconnections between Golgi stacks. Front Cell Dev Biol 7:160. https://doi.org/10.3389/fcell.2019.00160
92. Satoh A, Hayashi-Nishino M, Nishino K (2023) An electron tomographic analysis of giantin-deficient Golgi proposes a new function of the golgin protein family. Methods Mol Biol 2557:235–246. https://doi.org/10.1007/978-1-0716-2639-9_15

93. Saxena S, Foresti O, Liu A, Androulaki S, Pena Rodriguez M, Raote I, Aridor M, Cui B, Malhotra V (2024) Endoplasmic reticulum exit sites are segregated for secretion based on cargo size. Dev Cell 59:2593–2608. https://doi.org/10.1016/j.devcel.2024.06.009
94. Setty SR, Shin ME, Yoshino A, Marks MS, Burd CG (2003) Golgi recruitment of GRIP domain proteins by Arf-like GTPase 1 is regulated by Arf-like GTPase 3. Curr Biol 13(5): 401–404. https://doi.org/10.1016/s0960-9822(03)00089-7
95. Shamseldin HE, Bennett AH, Alfadhel M, Gupta V, Alkuraya FS (2016) GOLGA2, encoding a master regulator of golgi apparatus, is mutated in a patient with a neuromuscular disorder. Hum Genet 135(2):245–251. https://doi.org/10.1007/s00439-015-1632-8
96. Shin JJH, Gillingham AK, Begum F, Chadwick J, Munro S (2017) TBC1D23 is a bridging factor for endosomal vesicle capture by golgins at the trans-Golgi. Nat Cell Biol 19(12): 1424–1432. https://doi.org/10.1038/ncb3627
97. Short B, Preisinger C, Schaletzky J, Kopajtich R, Barr FA (2002) The Rab6 GTPase regulates recruitment of the dynactin complex to Golgi membranes. Curr Biol 12(20):1792–1795. https://doi.org/10.1016/s0960-9822(02)01221-6
98. Sinka R, Gillingham AK, Kondylis V, Munro S (2008) Golgi coiled-coil proteins contain multiple binding sites for Rab family G proteins. J Cell Biol 183(4):607–615. https://doi.org/10.1083/jcb.200808018
99. Smits P, Bolton AD, Funari V, Hong M, Boyden ED, Lu L, Manning DK, Dwyer ND, Moran JL, Prysak M, Merriman B, Nelson SF, Bonafe L, Superti-Furga A, Ikegawa S, Krakow D, Cohn DH, Kirchhausen T, Warman ML, Beier DR (2010) Lethal skeletal dysplasia in mice and humans lacking the golgin GMAP-210. N Engl J Med 362(3):206–216. https://doi.org/10.1056/NEJMoa0900158
100. Sohda M, Misumi Y, Yamamoto A, Yano A, Nakamura N, Ikehara Y (2001) Identification and characterization of a novel Golgi protein, GCP60, that interacts with the integral membrane protein giantin. J Biol Chem 276(48):45298–45306. https://doi.org/10.1074/jbc.M108961200
101. Sohda M, Misumi Y, Yamamoto A, Nakamura N, Ogata S, Sakisaka S, Hirose S, Ikehara Y, Oda K (2010) Interaction of Golgin-84 with the COG complex mediates the intra-Golgi retrograde transport. Traffic 11(12):1552–1566. https://doi.org/10.1111/j.1600-0854.2010.01123.x
102. Song K, Gras C, Capin G, Gimber N, Lehmann M, Mohd S, Puchkov D, Rodiger M, Wilhelmi I, Daumke O, Schmoranzer J, Schurmann A, Krauss M (2019) A SEPT1-based scaffold is required for Golgi integrity and function. J Cell Sci 132(3):jcs225557. https://doi.org/10.1242/jcs.225557
103. Sonnichsen B, Lowe M, Levine T, Jamsa E, Dirac-Svejstrup B, Warren G (1998) A role for giantin in docking COPI vesicles to Golgi membranes. J Cell Biol 140(5):1013–1021. https://doi.org/10.1083/jcb.140.5.1013
104. Stalder D, Yakunin I, Pereira C, Eden J, Gershlick DC (2024) Recruitment of PI4KIIIbeta to the Golgi by ACBD3 is dependent on an upstream pathway of a SNARE complex and golgins. Mol Biol Cell 35(2):ar20. https://doi.org/10.1091/mbc.E23-09-0376
105. Stanton AE, Hughson FM (2023) The machinery of vesicle fusion. Curr Opin Cell Biol 83: 102191. https://doi.org/10.1016/j.ceb.2023.102191
106. Taneja TK, Ma D, Kim BY, Welling PA (2018) Golgin-97 targets ectopically expressed inward rectifying potassium channel, Kir2.1, to the trans-Golgi network in COS-7 cells. Front Physiol 9:1070. https://doi.org/10.3389/fphys.2018.01070
107. Tojima T, Suda Y, Jin N, Kurokawa K, Nakano A (2024) Spatiotemporal dissection of the Golgi apparatus and the ER-Golgi intermediate compartment in budding yeast. Elife 13: e92900. https://doi.org/10.7554/eLife.92900
108. Truschel ST, Sengupta D, Foote A, Heroux A, Macbeth MR, Linstedt AD (2011) Structure of the membrane-tethering GRASP domain reveals a unique PDZ ligand interaction that mediates Golgi biogenesis. J Biol Chem 286(23):20125–20129. https://doi.org/10.1074/jbc.C111.245324

109. Tu Y, Yang Q, Tang M, Gao L, Wang Y, Wang J, Liu Z, Li X, Mao L, Jia RZ, Wang Y, Tang TS, Xu P, Liu Y, Dai L, Jia D (2024) TBC1D23 mediates Golgi-specific LKB1 signaling. Nat Commun 15(1):1785. https://doi.org/10.1038/s41467-024-46166-2
110. Vieira V, Pain C, Wojcik S, Spatola Rossi T, Denecke J, Osterrieder A, Hawes C, Kriechbaumer V (2020) Living on the edge: the role of Atgolgin-84A at the plant ER-Golgi interface. J Microsc 280(2):158–173. https://doi.org/10.1111/jmi.12946
111. Vitali T, Sanchez-Alvarez R, Witkos TM, Bantounas I, Cutiongco MFA, Dudek M, Yan G, Mironov AA, Swift J, Lowe M (2023) Vimentin intermediate filaments provide structural stability to the mammalian Golgi complex. J Cell Sci 136(20):jcs260577. https://doi.org/10.1242/jcs.260577
112. Wegeng WR, Bogus SM, Ruiz M, Chavez SR, Noori KSM, Niesman IR, Ernst AM (2024) A hollow TFG condensate spatially compartmentalizes the early secretory pathway. bioRxiv. https://doi.org/10.1101/2024.03.26.586876
113. Wehrle A, Witkos TM, Unger S, Schneider J, Follit JA, Hermann J, Welting T, Fano V, Hietala M, Vatanavicharn N, Schoner K, Spranger J, Schmidts M, Zabel B, Pazour GJ, Bloch-Zupan A, Nishimura G, Superti-Furga A, Lowe M, Lausch E (2019) Hypomorphic mutations of TRIP11 cause odontochondrodysplasia. JCI Insight 4(3):e124701. https://doi.org/10.1172/jci.insight.124701
114. Wei JH, Seemann J (2009) Mitotic division of the mammalian Golgi apparatus. Semin Cell Dev Biol 20(7):810–816. https://doi.org/10.1016/j.semcdb.2009.03.010
115. Wei JH, Zhang ZC, Wynn RM, Seemann J (2015) GM130 regulates Golgi-derived spindle assembly by activating TPX2 and capturing microtubules. Cell 162(2):287–299. https://doi.org/10.1016/j.cell.2015.06.014
116. Weigel AV, Chang CL, Shtengel G, Xu CS, Hoffman DP, Freeman M, Iyer N, Aaron J, Khuon S, Bogovic J, Qiu W, Hess HF, Lippincott-Schwartz J (2021) ER-to-Golgi protein delivery through an interwoven, tubular network extending from ER. Cell 184(9):2412–2429. https://doi.org/10.1016/j.cell.2021.03.035
117. Witkos TM, Lowe M (2015) The Golgin family of coiled-coil tethering proteins. Front Cell Dev Biol 3:86. https://doi.org/10.3389/fcell.2015.00086
118. Witkos TM, Lowe M (2017) Recognition and tethering of transport vesicles at the Golgi apparatus. Curr Opin Cell Biol 47:16–23. https://doi.org/10.1016/j.ceb.2017.02.003
119. Wong M, Munro S (2014) Membrane trafficking. The specificity of vesicle traffic to the Golgi is encoded in the golgin coiled-coil proteins. Science 346(6209):1256898. https://doi.org/10.1126/science.1256898
120. Wong M, Gillingham AK, Munro S (2017) The golgin coiled-coil proteins capture different types of transport carriers via distinct N-terminal motifs. BMC Biol 15(1):3. https://doi.org/10.1186/s12915-016-0345-3
121. Yadav S, Puthenveedu MA, Linstedt AD (2012) Golgin160 recruits the dynein motor to position the Golgi apparatus. Dev Cell 23(1):153–165. https://doi.org/10.1016/j.devcel.2012.05.023
122. Yamane J, Kubo A, Nakayama K, Yuba-Kubo A, Katsuno T, Tsukita S, Tsukita S (2007) Functional involvement of TMF/ARA160 in Rab6-dependent retrograde membrane traffic. Exp Cell Res 313(16):3472–3485. https://doi.org/10.1016/j.yexcr.2007.07.010
123. Yan R, Chen K, Wang B, Xu K (2022) SURF4-induced tubular ERGIC selectively expedites ER-to-Golgi transport. Dev Cell 57(4):512–525. https://doi.org/10.1016/j.devcel.2021.12.018
124. Yang K, Liu M, Feng Z, Rojas M, Zhou L, Ke H, Pastor-Pareja JC (2021) ER exit sites in Drosophila display abundant ER-Golgi vesicles and pearled tubes but no megacarriers. Cell Rep 36(11):109707. https://doi.org/10.1016/j.celrep.2021.109707
125. Yue X, Qian Y, Gim B, Lee I (2019) Acyl-CoA-binding domain-containing 3 (ACBD3; PAP7; GCP60): a multi-functional membrane domain organizer. Int J Mol Sci 20(8):2028. https://doi.org/10.3390/ijms20082028

126. Zhang Y, Seemann J (2021) Rapid degradation of GRASP55 and GRASP65 reveals their immediate impact on the Golgi structure. J Cell Biol 220(1):e202007052. https://doi.org/10.1083/jcb.202007052
127. Zhang Y, Seemann J (2024) RNA scaffolds the Golgi ribbon by forming condensates with GM130. Nat Cell Biol 26(7):1139–1153. https://doi.org/10.1038/s41556-024-01447-2
128. Ziltener P, Rebane AA, Graham M, Ernst AM, Rothman JE (2020) The golgin family exhibits a propensity to form condensates in living cells. FEBS Lett 594(19):3086–3094. https://doi.org/10.1002/1873-3468.13884
129. Lock JG, Hammond LA, Houghton F, Gleeson PA, Stow JL (2005) E-cadherin transport from the trans-Golgi network in tubulovesicular carriers is selectively regulated by golgin-97. Traffic 6(12) Pages 1142–1156. https://doi.org/10.1111/j.1600-0854.2005.00349.x
130. Barr FA, Puype M, Vandekerckhove J, Warren G (1997)GRASP65, a protein involved in the stacking of Golgi cisternae Cell 91(2):253–262. https://doi.org/10.1016/s0092-8674(00)80407-9

Chapter 5
ARF GTPases Function at the Golgi

Petia Adarska, Luis Wong-Dilworth, and Francesca Bottanelli

Abstract ADP-ribosylation factors (ARFs) are small (~21 kDa) GTPases that play a central role in intracellular transport across eukaryotic cells. Structurally related to Ras GTPases, ARFs were initially discovered as allosteric activators of cholera toxin but later emerged as key regulators of membrane trafficking. ARFs function through a GDP-GTP cycle, where guanine nucleotide exchange factors (GEFs) promote the exchange of GDP for GTP, activating ARFs and facilitating their membrane association. Conversely, GTPase-activating proteins (GAPs) stimulate GTP hydrolysis, leading to ARF inactivation and membrane dissociation. This cycle is crucial for ARF-mediated recruitment of cargo adaptors, coat proteins which regulate cargo selection and trafficking pathways through a close interplay with lipid-based signaling.

Keywords ARF GTPases · GDP-GTP exchange factors · GTPase activating proteins · Golgi trafficking

Introduction: ARF GTPases—Master Regulators of Golgi Trafficking

ADP-ribosylation factor (ARF) are small (~21 kDa) GTPase and major regulators of intracellular transport, ubiquitously expressed in all eukaryotic cells. Recent studies suggest that ARFs originated from ArfRs, which are present in *Asgardarchaeota*, the closest known prokaryotic relatives of eukaryotes [99]. Interestingly, ArfRs possess many of the structural features of eukaryotic ARFs [99]. ARFs are structurally related to Ras GTPases and were initially identified as allosteric activators of cholera toxin, which inspired their name. Cholerae toxin acts by catalyzing the ADP-ribosylation of the alpha subunit of a trimeric G protein and the first ARF

P. Adarska · L. Wong-Dilworth · F. Bottanelli (✉)
Institute of Chemistry and Biochemistry, Freie Universität Berlin, Berlin, Germany
e-mail: bottanelli@zedat.fu-berlin.de

A. Nakano, J. Saraste (eds.), *The Golgi Network, Volume II*, Subcellular Biochemistry 111, https://doi.org/10.1007/978-3-032-16833-7_5

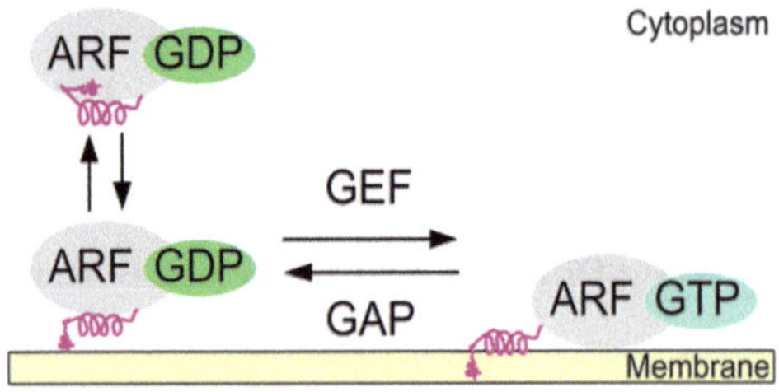

Fig. 5.1 The GDP-GTP cycle of ARFs to regulate membrane association. ARF-GDP reversibly binds to the membrane via its myristoylated N-terminal helix. Guanine nucleotide-exchange factors (GEFs) trigger the release of GDP, enabling ARFs to bind GTP which is present in the cytosol at higher quantifies. Upon GTP binding, ARFs undergo a conformational change, exposing the myristoylated N-terminal helix for stable membrane insertion. GTPase-activating proteins (GAPs) facilitate GTP hydrolysis that leads to the dissociation of ARFs from the membrane

was initially identified as a co-factor required for this reaction [37, 85]. ARF was later shown to have a high-affinity binding site for guanine nucleotides [38]. The first hint at a major cellular role for ARFs came from work in the yeast *Saccharomyces cerevisiae* (*S. cerevisiae*) by the Kahn lab [93], which showed that an *Arf1* null mutant displayed extensive trafficking defects and that Arf1 strongly localizes to the Golgi apparatus. Since the early 90s, ARFs have emerged as abundant and essential trafficking regulators at the Golgi.

ARFs cycle between an "active" GTP-bound state and an "inactive" GDP-bound state, where the nucleotide-sensitive conformational change is enabled by guanine nucleotide exchange factors (GEFs) and GTPase activating proteins (GAPs), respectively (Fig. 5.1).

The GTP-GDP cycle of ARFs is important for ARFs association with membranes and for the downstream recruitment of coat and effector proteins (i.e. adaptors responsible for cargo selection and lipid modifying enzymes). ARFs have a major role in the recruitment of coat proteins (COPI and clathrin) for the selection of cargoes to be loaded into different trafficking intermediates and transport pathways. ARFs are responsible for COPI recruitment at the Golgi, facilitating the retrieval of Golgi-resident enzymes and Golgi-to-endoplasmic reticulum (ER) transport. Additionally, ARFs regulate the recruitment of monomeric adaptors and adaptor protein complexes (APs) for the formation of different classes of post-Golgi carriers.

Many different ARF paralogues have been identified, performing a wide variety of cellular functions. 6 ARFs were identified mammals (5 in humans) and are classified into three groups based on their sequence homology: type I (ARF1 and ARF3; ARF2 not present in humans), type II (ARF4 and ARF5) and type III including only ARF6, which is the most divergent human ARF [34, 71] (Fig. 5.2). Only 3 ARFs exist in the yeast *S. cerevisiae*. Arf1 and Arf2 are homologous to type I ARFs, while yeast Arf3 is not essential, and it is closest to the mammalian ARF6. Type I and II ARFs all localize and function at the Golgi and will be the focus of this review.

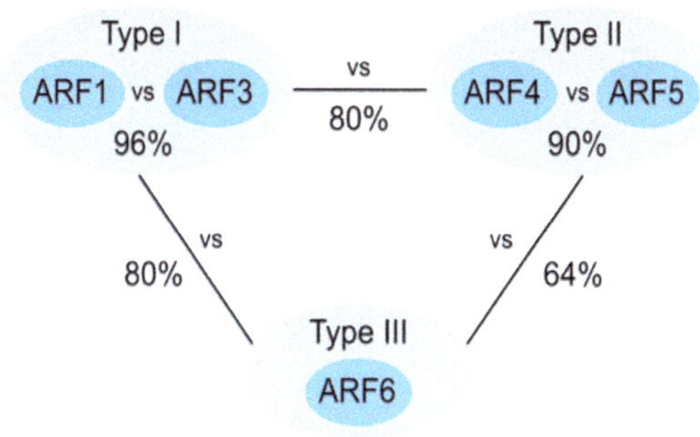

Fig. 5.2 Human ARF paralogues classification. Sequence homology between type I (ARF1 and ARF3), type II (ARF4 and ARF5) and type III (ARF6) are indicated. (Figure adapted from Nikolatou et al. [67])

Structural Organization of ARFs

ARFs paralogs are very similar in sequence and structure [39] and all comprise an N-terminal myristoylated amphipathic helix which is responsible for their membrane association, followed by canonical G domains with nucleotide-sensitive effector-binding switch regions (switch 1, interswitch and switch 2 regions) and C-terminal regions where nucleotide binding occurs (Fig. 5.3).

Membrane association of ARFs is provided by their N-terminal myristoylated amphipathic helix, which is shielded by the switch regions in the GDP-bound form [5, 46]. GDP-GTP exchange results in a conformational change and exposure of the helix which is then buried into the membrane [47, 106]. GTP binding by ARFs and the subsequent association of their N-terminal amphipathic helix with membranes are crucial for generating the membrane curvature necessary for vesicular and tubular carrier formation [7–9, 43, 49]. The membrane deformation function relies on the integrity of this helix, as mutations in the N-terminal amphipathic helix of ARF1 impair its ability to bend membranes [43]. ARF1 dimerization may be essential for inducing membrane curvature and facilitating scission at the neck of forming transport carriers [7, 8]. Additionally, the ARF effectors Arfaptin1 and 2 which are recruited to the *trans*-Golgi network (TGN) via binding to PI(4)P, harbor BAR (Bin/Amphiphysin/Rvs) domains that may support curvature generation and scission of transport carriers [17, 40].

Communication between the N-terminal helix and the nucleotide binding site at the other end of the protein is mediated by the interswitch region, which protrudes into the hydrophobic pocket occupied by the helix when ARF1 is bound to GDP [34]. The effector-binding Switch I and Switch II regions are nearly identical in all ARF paralogues hinting at identical three-dimensional organization [34, 71]. However, in cells, ARF paralogs interact with a wide range of different effectors, raising the question of where the specificity of effector binding originates. Small sequence differences outside of the switch regions may result in conformational differences in the effector-binding regions thus allowing ARFs to bind distinct effectors [57]. Importantly, it has been recently shown that ARF1 G domain populates different conformational membrane-associated or distal states, and this conformational dynamic may expose different interfaces allowing interaction with different

Fig. 5.3 Human ARF GTPases domain organization. Amphipathic helix (AH), switch 1 (SW1), interswitch (iSW) and switch 2 (SW2). (Figure adapted from Nikolatou et al. [67])

effectors [105]. Despite their high degree of structural similarity, the mechanisms by which ARFs interact with distinct effector sets remain poorly understood. This problem is especially notable for ARF1, given its well-characterized roles in early and post-Golgi intracellular trafficking [109].

Regulation of ARFs at the Golgi by GEFs and GAPs

Like all other regulatory GTPases, ARFs require the use of a guanine nucleotide exchange factor (GEFs) for their activation which facilitates GDP release for subsequent GTP binding. Importantly, GEFs are not able to activate ARF in absence of a membrane [34, 70, 80]. Thus, prior to GEF induced GDP release, ARF-GDP needs to be primed at a membrane to unlock ARFs myristolated N-terminal helix (for more details see [64]). The human genome encodes for ~15 ARF GEFs, divided in 6 sub-classes based on sequence and domain homology [95, 110]. All ARF GEFs possess a conserved Sec7 catalytic domain originally identified in the *S. cerevisea* protein Sec7p [1, 10, 73]. Three major Golgi-localized ARF GEFs have been identified in mammals; the Golgi Brefeldin A Resistant Guanine Nucleotide Exchange Factor 1 GBF1 [15] and the two Brefeldin A-inhibited guanine nucleotide exchange factors 1 and 2 (BIG1 and BIG2) [98]. Notably, the name GBF1 derives from its ability to confer brefeldin A (BFA) resistance to mammalian cells upon overexpression; however, GBF1 itself is a direct target of BFA, which inhibits its nucleotide exchange activity [14, 68]. The *S. cerevisea* homologue of BIG1 and BIG2 is Sec7 while two GEFs (Gea1 and Gea2) are homologues to GBF1. GBF1 localizes to the ER-Golgi intermediate compartment (ERGIC) and the cis-Golgi where it regulates the recruitment of COPI [41, 68, 107, 108]. BIG1 and BIG2 localize to late Golgi cisternae/the *trans*-Golgi network (TGN) and endosomal compartments where they regulate post-Golgi and endosomal traffic [88–90]. BIG1 and BIG2 support the recruitment of the adaptor protein complexes AP-1 and AP-3 via activation of ARF1 and ARF3 [18]. Interestingly, early Golgi GBF1 has been shown to regulate the recruitment of the downstream-acting GEFs BIG1 and BIG2 via activation of ARF4 and ARF5 [48], suggesting a regulatory cascade. Similarly to their mammalian counterparts, the yeast GEFs Gea1 has a role in COPI recruitment on *cis*-Golgi cisternae whereas Sec7 mediates post-Golgi trafficking. Notably, although first though to function redundantly to Gea1, recent studies have

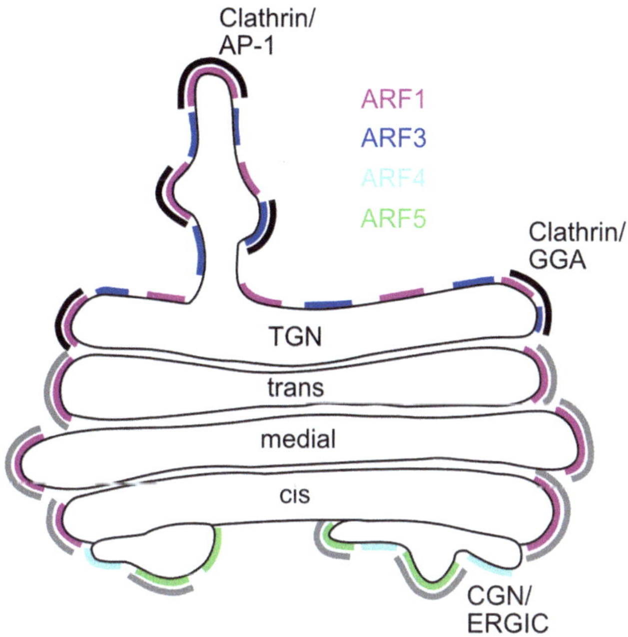

Fig. 5.4 Localization of ARF GTPases at the Golgi. TGN, trans-Golgi network; CGN, cis-Golgi network; ERGIC, ER-Golgi intermediate compartment

shown that Gea2 localizes to and functions at the trans Golgi [111, 112]. Recruitment of GEFs to a specific Golgi membrane may be regulated by their interaction with other small GTPases, such as Rab1 for GBF1 [59], Rab4 and the ARF-like protein ARL1 for BIG1 and BIG2 [12, 18, 30, 55] and through phosphoinositide binding [56]. Furthermore, the yeast BIG1/2 homolog Sec7 is recruited to the TGN through a positive feedback loop, positioning ARF1 as both a regulator and a product of its GEF [81].

As ARFs possess very low intrinsic GTP hydrolysis activity, termination of signaling is driven by GTPase activating proteins (GAPs) [77]. ARF GAPs are large family of regulators including over 31 members ubiquitously localized to all intracellular membranes in humans [95]. It is very interesting that the number of ARF GAPs is much higher than the number of GTPases they regulate (Fig. 5.4), suggesting roles beyond the simple regulation of the GDP-GTP cycle on ARFs. ARF GAPs harbor a conserved catalytic zinc-finger GAP domain which promotes GTP hydrolysis. ARF GAPs cellular functions extend beyond ARF inactivation and they can serve as effectors to propagate ARF signaling [91]. Many ARF GAPs modulate the recruitment of coats and sorting adaptors at the Golgi. Three different ARF GAPs (ARFGAP 1 and ARFGAP2/3) localize to the *cis*-Golgi in mammals, with their respective yeast homologues being Gcs1 and Glo3. All GAPs have been implicated in the orchestration of COPI recruitment on early-Golgi membranes [91, 103] but have been suggested to have diverging and yet elusive functions [6, 91]. ARFGAP2/

3 are recruited to the Golgi via direct interaction with the COPI coat [103]. Interestingly, ARFGAP2/3 activity is strongly enhanced upon binding. ARFGAP1 recruitment to the Golgi is curvature dependent [25] with preferential association to curved versus flat membranes, thus favoring GTP hydrolysis on highly curved transport carriers [26]. Structural modeling of the COPI coat architecture positioned ARF molecules in two distinct molecular environments within the coat [23], suggesting different regulation via different GAPs (ARFGAP2/3 vs ARFGAP1) [6].

Although ARF1 plays a crucial role in TGN export and the recruitment of adaptors and the clathrin coat, its regulation by GAPs at the TGN remains less studied compared to its function in COPI recruitment and early Golgi trafficking. The human GAP SMAP2 co-localizes together with the adaptor AP-1 to the TGN and is involved in AP-1-dependent retrieval from endosomes and late Golgi [29, 63]. The other SMAP paralogue SMAP1, which has an established ARF6-dependent role [96], was recently shown to be recruited to the TGN via AP-1 and to have a regulatory role in clathrin assembly [102]. Similarly, yeast Age2 (human SMAP2 homologue) localize to the late Golgi via an amphipathic helix to locally regulate levels of active ARF [53] and thus post-Golgi trafficking. AGAP1, was also shown to modulate TGN recruitment adaptors as overexpression disrupted AP-1 localization, while its paralogue AGAP2 regulated endocytic recycling on endosomes [66]. Interestingly, although ARFGAP3 binds coatomer, it has also been suggested to regulate the association of the monomeric adaptors GGAs to post-Golgi compartments [87].

Localization and Function of ARFs at the Golgi

ARF1 and type II ARFs ARF4 and ARF5 all localize to the *cis*-Golgi and the TGN [14, 21, 36, 41, 52, 54, 83], whereas Type I ARF3 localizes exclusively to the TGN [52] (Fig. 5.4). ARF1 and type II ARFs have also been observed on peripheral ERGICs [13]. Super-resolution microscopy of endogenously tagged ARFs could reveal that type II ARFs are particularly enriched in a population of ERGIC53-positive compartments closely associated to the cis-Golgi, most probably representing the cis-Golgi network (CGN, Fig. 5.4). Intriguingly, these Golgi-associated ERGICs/CGN compartments were devoid of ARF1, suggesting a functional differentiation with the *cis*-Golgi cisternae [104]. At the other end of the Golgi stack, Type I ARF1 and ARF3 were shown to localize to tubular-vesicular compartments emerging from the Golgi and transporting secretory cargoes [9, 94, 104]. Beyond the Golgi, Type I ARFs were also reported to localize to endosomes containing endocytosed transferrin (Tfn) [42] and on Rab4-positive endosomal compartments [18]. Interestingly, ERGIC-localized ARF5 was also found localizing to the plasma membrane (PM), where it is involved in cell migration [19, 60, 61].

Volpicelli-Daley and colleagues conducted the first study aimed at systematically dissecting the function of individual ARF members [100]. They proposed that ARFs function in pairs, as the siRNA-mediated knockdown of a single ARF did not produce detectable defects in coat recruitment, transport, or organelle morphology. However, knocking down ARF pairs resulted in specific phenotypes [100]. A possible function of pairs of ARFs in intracellular transport was re-assessed by Pennauer and colleagues, where they employed CRISPR-Cas9-mediated genomic deletion to generate knockout (KO) HeLa cells of type I and II ARFs, either in combination or individually [72]. They could show that cells lacking any ARF alone or in combination are viable and that ARF4 alone can sustain all essential cellular functions in the absence of all other ARF members [72]. Importantly, they reported that single KO cells displayed distinct cellular phenotypes. ARF1 KO cells displayed an enlarged Golgi and decreased COPI recruitment, while in ARF4 KO cells ER-resident proteins were secreted, suggesting a specific defect in COPI-dependent ER protein retrieval.

Studies aimed at dissecting the function of Golgi-associated ARFs were primarily focused on the role of ARF1 in the formation of COPI vesicles through *in vitro* reconstitution and hybrid approaches assessing COPI budding from purified Golgi membranes and semi-permeabilized cells [3, 69, 78]. ARF1, ARF4 and ARF5 all supported the formation of COPI vesicles from Golgi membranes in semi-permeabilized cells [75]. Further, the content of COPI vesicles generated with type I and II ARFs via this assay is similar [4]. These studies raised the possibility that ARFs may act redundantly on COPI vesicles formation, however, they lacked cellular spatial information. Further work showed that different ARF paralogues may drive COPI formation from different membranes along the secretory pathway to facilitate COPI-dependent retrieval from ER–Golgi intermediate compartment (ERGICs) (ARF4 and ARF5) and later early Golgi cisternae (ARF1) [104], hence the need of all ARFs to support COPI carriers formation. It remains so far unclear how ARFs associate to different intracellular membrane despite their high structural similarity.

Additional important functions of Golgi-localized ARFs have been identified in post-Golgi trafficking steps, including biosynthetic traffic from the Golgi to the PM and for Golgi-endosome communication [2]. ARFs play a major role in the recruitment of cargo adaptors to the TGN. Two classes of adaptors are responsible for cargo selection at the TGN; the monomeric Golgi-associated, γ-adaptin homologous, ARF-interacting proteins (GGAs) and adaptor protein complexes (APs). ARF1 supports the recruitment of AP-1 (and subsequently clathrin) to the TGN via conformationally unlocking the adaptor and allowing cargo binding [79, 92]. ARF-dependent AP-1 recruitment at the TGN and endosomes has long been considered crucial for bidirectional trafficking between these compartments [11, 28, 92, 101]. Recently, it has been shown that ARF1 defines tubular-vesicular compartments emerging from the TGN and carrying biosynthetic cargo and harboring AP-1/clathrin nanodomains [9, 82, 94]. It has been suggested that ARF1-dependent recruitment onto these compartments may have the sole function in retrieval of cargoes back to the Golgi [82] while anterograde transport of cargoes may be dependent on a GTPase-driven cascade and maturation of the compartments into recycling endosomes [94]. This is similar to what has been observed in yeast where ARF1 has been shown to modulate the Rab cascade driving intra-Golgi maturation [97]. ARF1 was shown to modulate an intra-Golgi Rab conversion via the recruitment of Gyp1 (a Rab1 GAP) and the TRAPPII complex (a GEF for Ypt32 [Rab11]) [97]. Further strengthening the idea that both Rab and ARF GTPases may act in concert in maturation cascades, Rab1 has been demonstrated to recruit the ARF GEF GBF1 to early secretory compartments [58].

ARF1 has been implicated in the recruitment of GGAs for trafficking of mannose 6-phosphate receptors (MPRs)-dependent soluble cargoes from the TGN to lysosomes in mammals and to the vacuole in yeast [20, 76]. The GAT domains of GGAs have been shown to bind to the switch regions of GTP bound ARF1 [16, 86]. ARF1 also regulates the association of AP-3 and AP-4 to the TGN for trafficking to the lysosome and to the phagophore respectively, has also shown to be regulated by ARF1. As the GEFs BIG1 and BIG2 seem to activate ARF1 to support recruitment of all adaptors, specific AP-dependent sorting steps may be instead regulated by the recruitment of a specific GAP. Supporting this model, the GAP AGAP1 was shown to associate to AP-3 [65] while AGAP2 to AP-1 [66]. While ARF1 function has been more widely investigated, ARF1 and ARF3 share the highest sequence similarity, making it challenging to distinguish their specific interactors and functions within the type I ARF family. ARF3, unlike the more broadly localized ARF1, localizes exclusively to the TGN. The two TGN-localized GEFs BIG1 and BIG2 were shown to act indiscriminatively on both ARF1 and ARF3, suggesting redundant roles in AP-1-dependent pathways [48]. However, an ARF3 knockdown did not affect AP-1 association to the TGN [52]. Interestingly, ARF3 was shown to regulate the PM levels of N-cadherin, either via modulating PM or turnover [84] suggesting ARF3-specific functions in post-Golgi trafficking. ARFs have also been implicated in specialized secretory functions across distinct cell types. For example, ARF4 plays a role in transporting rhodopsin to the primary cilium [21], with the associated GEFs and GAPs identified [22]. Additionally, ARF4 and ARF5 contribute to the transport of dense-core vesicles in rat PC12 cells [83].

Mechanisms of Golgi Membrane Targeting

Because of the high sequence and structural similarities among ARFs, the factors governing the specific association of ARF paralogues with distinct intracellular membranes and their differential interactions with effector proteins remain poorly understood. Recruitment of ARF GTPases is thought to be driven by the respective GEFs which induce the stable association of ARFs with the membrane via the mechanisms explained above. However, for example only one GEF is known to associate with ERGIC and cis-Golgi membranes while many different ARF paralogues have been showed to localize to segregate nanodomains on the same early secretory compartments, suggesting that other factors may be involved. ARF1 has been shown to be recruited to early secretory membranes by the SNARE membrin and SNAREs in general may play a role in the recruitment of ARFs [113]. ARF1 has also been shown to bind to the cytosolic tail of p24 in its GDP bound form. However, it is unclear what recruits ARF4 to different ERGIC membranes [35, 36, 50]. ARF activation is thought to be crucial for the recruitment of effectors and membrane association. However, GDP "inactive" ARFs have been shown to bind effectors and associate to membranes, raising the possibility that GDP-GTP exchange may act additionally as a functional switch. Studies using the fungal metabolite BFA, which binds to the catalytic Sec7 domain (Sec7d) of activating GEFs, preventing GDP displacement on ARFs, have suggested that ARF4 and ARF5 may be able to associate to membranes in their GDP-bound form, as they maintained their membrane association upon BFA treatment [13]. Subsequent investigations identified specific residues located in the N-terminal helix of type II ARFs as the structural determinants conferring BFA resistance and membrane association in their GDP-bound form, as the introduction of these type II-specific determinants into ARF1, which is BFA sensitive, conferred ARF4 membrane association properties [27]. Interestingly, in the case of the well studied PM-localized ARF6, it has been shown that it can also interact with effectors in the "inactive" GDP-bound state, suggesting that depending on the nucleotide bound, it would lead to activation of different signaling pathways [24]. GTP hydrolysis of ARFs does not only serve to dissociate ARFs from membranes to make them available for a next round of reactions, but is essential for their function, as a GTP-locked ARF1 mutant prevents cargo loading into COPI vesicles [51]. These observations suggest that membrane recruitment of ARFs may be mediated by the interaction with specific membrane receptors prior to GEF activation. In the case of ARF1, the ER-Golgi SNARE membrin was shown to recruit GDP-bound ARF1 to early Golgi cisternae via interactions with a specific 16 amino acid sequence located in the central domain of ARF1 [36]. ARF1-GDP is also targeted to Golgi membranes via its C-terminal interaction with p23, a member of the p24 family of transmembrane proteins [35, 50]. Similarly, ARF3 is targeted exclusively to the TGN via two unique determinants (A174 and K180) located at its C-terminus that are conserved among ARF3 homologs across species [52]. Interestingly, the N-terminal end of ARFs is responsible for the interaction with the catalytic Sec7 domain of their activating GEFs, suggesting the involvement of a yet unknown receptor for ARF3 responsible for its recruitment to TGN membranes.

Beyond Adaptors and Coat Recruitment: ARFs Role as Hubs for Effector Recruitment

ARFs interact with a wide variety of effectors to fulfil their many cellular sorting functions at the Golgi. Crosstalk between ARFs function and lipid-based signaling has emerged as a key regulatory node for the export of cargoes from the TGN. ARF1 mediates recruitment of PI(4)Kβ, thus stimulating the synthesis of PI(4)P [32]. PI(4)P regulates AP-1 recruitment to the Golgi, promoting the formation of sorting domains at the TGN to facilitate cargo enrichment and export [101]. Four-phosphate-adaptor protein 1 and 2 (FAPP1 and FAPP2) are also known to bind PI(4)P and ARF1 via their plekstrin homology (PH) domain to support secretory carrier formation at the TGN [33]. Interestingly, the PH domain of FAPP binds GTP-loaded active ARF1 and is thought to stabilize ARF1 on the membrane by interfering with GAP interaction [33]. The yeast ARF GEF Gea2p was also shown to interact with Drs2p, an ATPase required for phospholipid flippase activity [62]. The human lipid flippase ATP8A1 was also shown to bind to the GEFs BIG1 and BIG2 [74]. In addition to a direct role in carrier formation at the TGN, ARF1 was also shown to be critical for the Golgi localization of proteins responsible for lipid transfer across membranes at contact sites. ARF1 was shown to support the Golgi recruitment of oxysterol-binding proteins (OSBP) and ceramide transfer (CERT) via interaction with their PH domains [44]. Thus, a tight interplay between lipid-based signaling and ARFs is crucial for the regulation of carrier formation at the Golgi. Further, ARF1 has also been shown to support tethering of incoming vesicles on the *cis*-side of the Golgi by supporting membrane association of tethering molecules bearing long coiled-coiled domains responsible for vesicles capture. Both the yeast Golgin Rud3p [31] and the mammalian tether GMAP210 [26] were shown to be recruited to the Golgi via interaction with ARF proteins.

Conclusion and Future Perspective

We are just beginning to scratch the surface of our understanding of the diverse cellular functions of ARF GTPases. To gain deeper insight into the role of ARFs and their regulators in living cells, it is crucial to explore their regulatory networks and understand the functions of their numerous effectors and interactors. Recent work employing proximity-dependent proteomics has provided a discovery platform to discover ARF interactions with effectors and regulators [45]. However, many questions remain about the mechanisms governing ARF recruitment in living cells, especially given that their reported cellular functions far exceed the number of known GEFs. Conversely, ARF GAPs constitute a large protein family with over 30 members, potentially playing a regulatory role and contributing to the specificity of ARF functions within the cell. Thus, further investigation of ARF GAPs function in their cellular context may provide a better understanding of the many functions of ARFs. Due to its highly complex three-dimensional organization, the mammalian Golgi has been challenging to study using standard diffraction-limited light microscopy. However, advancements in microscopy hardware and image analysis are expected to enable the investigation of protein dynamics within the Golgi, allowing researchers to explore ARF functions at the level of individual Golgi cisternae.

References

1. Achstetter T, Franzusoff A, Field C, Schekman R (1988) SEC7 encodes an unusual, high molecular weight protein required for membrane traffic from the yeast Golgi apparatus. J Biol Chem 263:11711–11717
2. Adarska P, Wong-Dilworth L, Bottanelli F (2021) ARF GTPases and their ubiquitous role in intracellular trafficking beyond the Golgi. Front Cell Dev Biol 9:679046
3. Adolf F, Herrmann A, Hellwig A, Beck R, Brügger B, Wieland FT (2013) Scission of COPI and COPII vesicles is independent of GTP hydrolysis. Traffic 14:922–932
4. Adolf F, Rhiel M, Hessling B, Gao Q, Hellwig A, Béthune J, Wieland FT (2019) Proteomic profiling of mammalian COPII and COPI vesicles. Cell Rep 26:250–265
5. Antonny B, Beraud-Dufour S, Chardin P, Chabre M (1997) N-terminal hydrophobic residues of the G-protein ADP-ribosylation factor-1 insert into membrane phospholipids upon GDP to GTP exchange. Biochemistry 36:4675–4684
6. Arakel EC, Huranova M, Estrada AF, Rau EM, Spang A, Schwappach B (2019) Dissection of GTPase-activating proteins reveals functional asymmetry in the COPI coat of budding yeast. J Cell Sci 132:jcs232124
7. Beck R, Sun Z, Adolf F, Rutz C, Bassler J, Wild K, Sinning I, Hurt E, Brügger B, Béthune J, Wieland F (2008) Membrane curvature induced by Arf1-GTP is essential for vesicle formation. Proc Natl Acad Sci USA 105:11731–11736
8. Beck R, Prinz S, Diestelkötter-Bachert P, Röhling S, Adolf F, Hoehner K, Welsch S, Ronchi P, Brügger B, Briggs JA (2011) Coatomer and dimeric ADP ribosylation factor 1 promote distinct steps in membrane scission. J Cell Biol 194:765–777
9. Bottanelli F, Kilian N, Ernst AM, Rivera-Molina F, Schroeder LK, Kromann EB, Lessard MD, Erdmann RS, Schepartz A, Baddeley D, Bewersdorf J, Toomre D, Rothman JE (2017) A novel physiological role for ARF1 in the formation of bidirectional tubules from the Golgi. Mol Biol Cell 28:1676–1687
10. Chardin P, Paris S, Antonny B, Robineau S, Beraud-Dufour S, Jackson CL, Chabre M (1996) A human exchange factor for ARF contains Sec7- and pleckstrin-homology domains. Nature 384:481–484
11. Chi S, Cao H, Chen J, McNiven MA (2008) Eps15 mediates vesicle trafficking from the trans-Golgi network via an interaction with the clathrin adaptor AP-1. Mol Biol Cell 19:3564–3575
12. Christis C, Munro S (2012) The small G protein Arl1 directs the trans-Golgi-specific targeting of the Arf1 exchange factors BIG1 and BIG2. J Cell Biol 196:327–335
13. Chun J, Shapovalova Z, Dejgaard SY, Presley JF, Melançon P (2008) Characterization of class I and II ADP-Ribosylation factors (Arfs) in live cells: GDP-bound class II Arfs associate with the ER-Golgi intermediate compartment independently of GBF1. Mol Biol Cell 19:3488–3500
14. Claude A, Zhao B-P, Kuziemsky CE, Dahan S, Berger SJ, Yan J-P, Armold AD, Sullivan EM, Melançon P (1999a) Gbf1: a novel Golgi-associated Bfa-resistant guanine nucleotide exchange factor that displays specificity for Adp-ribosylation factor 5. J Cell Biol 146:71–84
15. Claude A, Zhao BP, Kuziemsky CE, Dahan S, Berger SJ, Yan JP, Armold AD, Sullivan EM, Melancon P (1999b) GBF1: a novel Golgi-associated BFA-resistant guanine nucleotide exchange factor that displays specificity for ADP-ribosylation factor 5. J Cell Biol 146:71–84
16. Collins BM, Watson PJ, Owen DJ (2003) The structure of the GGA1-GAT domain reveals the molecular basis for ARF binding and membrane association of GGAs. Dev Cell 4:321–332
17. Cruz-Garcia D, Ortega-Bellido M, Scarpa M, Villeneuve J, Jovic M, Porzner M, Balla T, Seufferlein T, Malhotra V (2013) Recruitment of arfaptins to the trans-Golgi network by PI(4)P and their involvement in cargo export. EMBO J 32:1717–1729
18. D'Souza RS, Semus R, Billings EA, Meyer CB, Conger K, Casanova JE (2014) Rab4 orchestrates a small GTPase cascade for recruitment of adaptor proteins to early endosomes. Curr Biol 24:1187–1198

19. D'Souza RS, Lim JY, Turgut A, Servage K, Zhang J, Orth K, Sosale NG, Lazzara MJ, Allegood J, Casanova JE (2020) Calcium-stimulated disassembly of focal adhesions mediated by an ORP3/IQSec1 complex. Elife 9:e54113
20. Dell'Angelica EC, Puertollano R, Mullins C, Aguilar RC, Vargas JD, Hartnell LM, Bonifacino JS (2000) GGAs: a family of ADP ribosylation factor-binding proteins related to adaptors and associated with the Golgi complex. J Cell Biol 149:81–94
21. Deretic D, Williams AH, Ransom N, Morel V, Hargrave PA, Arendt A (2005) Rhodopsin C terminus, the site of mutations causing retinal disease, regulates trafficking by binding to ADP-ribosylation factor 4 (ARF4). Proc Natl Acad Sci USA 102:3301–3306
22. Deretic D, Lorentzen E, Fresquez T (2021) The ins and outs of the Arf4-based ciliary membrane-targeting complex. Small GTPases 12:1–12
23. Dodonova SO, Aderhold P, Kopp J, Ganeva I, Rohling S, Hagen WJH, Sinning I, Wieland F, Briggs JAG (2017) 9A structure of the COPI coat reveals that the Arf1 GTPase occupies two contrasting molecular environments. Elife 6:e26691
24. Donaldson JG, Jackson CL (2011) ARF family G proteins and their regulators: roles in membrane transport, development and disease. Nat Rev Mol Cell Biol 12:362–375
25. Drin G, Casella JF, Gautier R, Boehmer T, Schwartz TU, Antonny B (2007) A general amphipathic alpha-helical motif for sensing membrane curvature. Nat Struct Mol Biol 14: 138–146
26. Drin G, Morello V, Casella JF, Gounon P, Antonny B (2008) Asymmetric tethering of flat and curved lipid membranes by a golgin. Science 320:670–673
27. Duijsings D, Lanke KH, Van Dooren SH, Van Dommelen MM, Wetzels R, De Mattia F, Wessels E, Van Kuppeveld FJ (2009) Differential membrane association properties and regulation of class I and class II Arfs. Traffic 10:316–323
28. Fölsch H, Ohno H, Bonifacino JS, Mellman I (1999) A novel clathrin adaptor complex mediates basolateral targeting in polarized epithelial cells. Cell 99:189–198
29. Funaki T, Kon S, Ronn RE, Henmi Y, Kobayashi Y, Watanabe T, Nakayama K, Tanabe K, Satake M (2011) Localization of SMAP2 to the TGN and its function in the regulation of TGN protein transport. Cell Struct Funct 36:83–95
30. Galindo A, Soler N, McLaughlin SH, Yu MM, Williams RL, Munro S (2016) Structural insights into Arl1-mediated targeting of the Arf-GEF BIG1 to the trans-Golgi. Cell Rep 16: 839–850
31. Gillingham AK, Tong AH, Boone C, Munro S (2004) The GTPase Arf1p and the ER to Golgi cargo receptor Erv14p cooperate to recruit the golgin Rud3p to the cis-Golgi. J Cell Biol 167: 281–292
32. Godi A, Pertile P, Meyers R, Marra P, Di Tullio G, Iurisci C, Luini A, Corda D, De Matteis MA (1999) ARF mediates recruitment of PtdIns-4-OH kinase-beta and stimulates synthesis of PtdIns(4,5)P2 on the Golgi complex. Nat Cell Biol 1:280–287
33. Godi A, Di Campli A, Konstantakopoulos A, Di Tullio G, Alessi DR, Kular GS, Daniele T, Marra P, Lucocq JM, De Matteis MA (2004) FAPPs control Golgi-to-cell-surface membrane traffic by binding to ARF and PtdIns(4)P. Nat Cell Biol 6:393–404
34. Goldberg J (1998) Structural basis for activation of ARF GTPase: mechanisms of guanine nucleotide exchange and GTP-myristoyl switching. Cell 95:237–248
35. Gommel DU, Memon AR, Heiss A, Lottspeich F, Pfannstiel J, Lechner J, Reinhard C, Helms JB, Nickel W, Wieland FT (2001) Recruitment to Golgi membranes of ADP-ribosylation factor 1 is mediated by the cytoplasmic domain of p23. EMBO J 20:6751–6760
36. Honda A, Al-Awar OS, Hay JC, Donaldson JG (2005) Targeting of Arf-1 to the early Golgi by membrin, an ER-Golgi SNARE. J Cell Biol 168:1039–1051
37. Kahn RA, Gilman AG (1984) Purification of a protein cofactor required for ADP-ribosylation of the stimulatory regulatory component of adenylate cyclase by cholera toxin. J Biol Chem 259:6228–6234
38. Kahn RA, Gilman AG (1986) The protein cofactor necessary for ADP-ribosylation of Gs by cholera toxin is itself a GTP binding protein. J Biol Chem 261:7906–7911

39. Kahn RA, Kern FG, Clark J, Gelmann EP, Rulka C (1991) Human ADP-ribosylation factors. A functionally conserved family of GTP-binding proteins. J Biol Chem 266:2606–2614
40. Kanoh H, Williger BT, Exton JH (1997) Arfaptin 1, a putative cytosolic target protein of ADP-ribosylation factor, is recruited to Golgi membranes. J Biol Chem 272:5421–5429
41. Kawamoto K, Yoshida Y, Tamaki H, Torii S, Shinotsuka C, Yamashina S, Nakayama K (2002) GBF1, a guanine nucleotide exchange factor for ADP-ribosylation factors, is localized to the cis-golgi and involved in membrane association of the COPI coat. Traffic 3:483–495
42. Kondo Y, Hanai A, Nakai W, Katoh Y, Nakayama K, Shin H-W (2012) ARF1 and ARF3 are required for the integrity of recycling endosomes and the recycling pathway. Cell Struct Funct 37:141–154
43. Krauss M, Jia J-Y, Roux A, Beck R, Wieland FT, De Camilli P, Haucke V (2008) Arf1-GTP-induced tubule formation suggests a function of Arf family proteins in curvature Acquisition at Sites of vesicle budding*. J Biol Chem 283:27717–27723
44. Levine TP, Munro S (2002) Targeting of Golgi-specific pleckstrin homology domains involves both PtdIns 4-kinase-dependent and -independent components. Curr Biol 12:695–704
45. Li FL, Wu Z, Gao YQ, Bowling FZ, Franklin JM, Hu C, Suhandynata RT, Frohman MA, Airola MV, Zhou H, Guan KL (2022) Defining the proximal interaction networks of Arf GTPases reveals a mechanism for the regulation of PLD1 and PI4KB. EMBO J 41:e110698
46. Liu Y, Kahn RA, Prestegard JH (2009) Structure and membrane interaction of myristoylated ARF1. Structure 17:79–87
47. Liu Y, Kahn RA, Prestegard JH (2010) Dynamic structure of membrane anchored Arf•GTP. Nat Struct Mol Biol 17:876–881
48. Lowery J, Szul T, Styers M, Holloway Z, Oorschot V, Klumperman J, Sztul E (2013) The Sec7 guanine nucleotide exchange factor GBF1 regulates membrane recruitment of BIG1 and BIG2 guanine nucleotide exchange factors to the trans-Golgi network (TGN). J Biol Chem 288: 11532–11545
49. Lundmark R, Doherty GJ, Vallis Y, Peter BJ, Mcmahon HT (2008) Arf family GTP loading is activated by, and generates, positive membrane curvature. Biochem J 414:189–194
50. Majoul I, Sohn K, Wieland FT, Pepperkok R, Pizza M, Hillemann J, Söling H-D (1998) KDEL receptor (Erd2p)-mediated retrograde transport of the cholera toxin a subunit from the Golgi involves COPI, p23, and the COOH terminus of Erd2p. J Cell Biol 143:601–612
51. Malsam J, Gommel D, Wieland FT, Nickel W (1999) A role for ADP ribosylation factor in the control of cargo uptake during COPI-coated vesicle biogenesis. FEBS Lett 462:267–272
52. Manolea F, Chun J, Chen DW, Clarke I, Summerfeldt N, Dacks JB, Melançon P (2010) Arf3 is activated uniquely at the trans-Golgi network by Brefeldin A inhibited guanine nucleotide exchange factors. Mol Biol Cell 21:1836–1849
53. Manzer KM, Fromme JC (2023) The Arf-GAP Age2 localizes to the late-Golgi via a conserved amphipathic helix. Mol Biol Cell 34:ar119
54. Mazelova J, Astuto-Gribble L, Inoue H, Tam BM, Schonteich E, Prekeris R, Moritz OL, Randazzo PA, Deretic D (2009) Ciliary targeting motif VxPx directs assembly of a trafficking module through Arf4. EMBO J 28:183–192
55. McDonold CM, Fromme JC (2014) Four GTPases differentially regulate the Sec7 Arf-GEF to direct traffic at the trans-Golgi network. Dev Cell 30:759–767
56. Meissner JM, Bhatt JM, Lee E, Styers ML, Ivanova AA, Kahn RA, Sztul E (2018) The ARF guanine nucleotide exchange factor GBF1 is targeted to Golgi membranes through a PIP-binding domain. J Cell Sci 131:jcs210245
57. Menetrey J, Macia E, Pasqualato S, Franco M, Cherfils J (2000) Structure of Arf6-GDP suggests a basis for guanine nucleotide exchange factors specificity. Nat Struct Biol 7:466–469
58. Monetta P, Slavin F, Romero N, Alvarez C (2007a) Rab1b interacts with GBF1 and modulates both ARR dynamics and COPI association. Mol Biol Cell 18:2400–2410
59. Monetta P, Slavin I, Romero N, Alvarez C (2007b) Rab1b interacts with GBF1 and modulates both ARF1 dynamics and COPI association. Mol Biol Cell 18:2400–2410

60. Moravec R, Conger KK, D'Souza R, Allison AB, Casanova JE (2012) BRAG2/GEP100/IQSec1 interacts with Clathrin and regulates α5β1 integrin endocytosis through activation of ADP Ribosylation factor 5 (Arf5)*. J Biol Chem 287:31138–31147
61. Nacke M, Sandilands E, Nikolatou K, Román-Fernández Á, Mason S, Patel R, Lilla S, Yelland T, Galbraith LCA, Freckmann EC, McGarry L, Morton JP, Shanks E, Leung HY, Markert E, Ismail S, Zanivan S, Blyth K, Bryant DM (2021) An ARF GTPase module promoting invasion and metastasis through regulating phosphoinositide metabolism. Nat Commun 12:1623
62. Natarajan P, Liu K, Patil DV, Sciorra VA, Jackson CL, Graham TR (2009) Regulation of a Golgi flippase by phosphoinositides and an ArfGEF. Nat Cell Biol 11:1421–1426
63. Natsume W, Tanabe K, Kon S, Yoshida N, Watanabe T, Torii T, Satake M (2006) SMAP2, a novel ARF GTPase-activating protein, interacts with clathrin and clathrin assembly protein and functions on the AP-1-positive early endosome/trans-Golgi network. Mol Biol Cell 17:2592–2603
64. Nawrotek A, Zeghouf M, Cherfils J (2016) Allosteric regulation of Arf GTPases and their GEFs at the membrane interface. Small GTPases 7:283–296
65. Nie Z, Boehm M, Boja ES, Vass WC, Bonifacino JS, Fales HM, Randazzo PA (2003) Specific regulation of the adaptor protein complex AP-3 by the Arf GAP AGAP1. Dev Cell 5:513–521
66. Nie Z, Fei J, Premont RT, Randazzo PA (2005) The Arf GAPs AGAP1 and AGAP2 distinguish between the adaptor protein complexes AP-1 and AP-3. J Cell Sci 118:3555–3566
67. Nikolatou K, Bryant DM, Sandilands E (2023) The ARF GTPase regulatory network in collective invasion and metastasis. Biochem Soc Trans 51:1559–1569
68. Niu TK, Pfeifer AC, Lippincott-Schwartz J, Jackson CL (2005) Dynamics of GBF1, a brefeldin A-sensitive Arf1 exchange factor at the Golgi. Mol Biol Cell 16:1213–1222
69. Orci L, Glick BS, Rothman JE (1986) A new type of coated vesicular carrier that appears not to contain clathrin: its possible role in protein transport within the Golgi stack. Cell 46:171–184
70. Paris S, Beraud-Dufour S, Robineau S, Bigay J, Antonny B, Chabre M, Chardin P (1997) Role of protein-phospholipid interactions in the activation of ARF1 by the guanine nucleotide exchange factor Arno. J Biol Chem 272:22221–22226
71. Pasqualato S, Menetrey J, Franco M, Cherfils J (2001) The structural GDP/GTP cycle of human Arf6. EMBO Rep 2:234–238
72. Pennauer M, Buczak K, Prescianotto-Baschong C, Spiess M (2021) Shared and specific functions of Arfs 1–5 at the Golgi revealed by systematic knockouts. J Cell Biol 221:e202106100
73. Peyroche A, Paris S, Jackson CL (1996) Nucleotide exchange on ARF mediated by yeast Gea1 protein. Nature 384:479–481
74. Pocognoni CA, Nawara T, Bhatt JM, Lee E, Jian X, Randazzo P, Sztul E (2024) The lipid flippase ATP8A1 regulates the recruitment of ARF effectors to the trans-Golgi network. Arch Biochem Biophys 758:110049
75. Popoff V, Langer JD, Reckmann I, Hellwig A, Kahn RA, Brügger B, Wieland FT (2011) Several ADP-ribosylation factor (Arf) isoforms support COPI vesicle formation. J Biol Chem 286:35634–35642
76. Puertollano R, Randazzo PA, Presley JF, Hartnell LM, Bonifacino JS (2001) The GGAs promote ARF-dependent recruitment of clathrin to the TGN. Cell 105:93–102
77. Randazzo PA, Kahn RA (1994) Gtp hydrolysis by Adp-Ribosylation factor is dependent on both an Adp-Ribosylation factor Gtpase-activating protein and acid phospholipids. (Vol 269, Pg 10758, 1994). J Biol Chem 269:16519–16519
78. Reinhard C, Schweikert M, Wieland FT, Nickel W (2003) Functional reconstitution of COPI coat assembly and disassembly using chemically defined components. Proc Natl Acad Sci USA 100:8253–8257
79. Ren X, Farias GG, Canagarajah BJ, Bonifacino JS, Hurley JH (2013) Structural basis for recruitment and activation of the AP-1 clathrin adaptor complex by Arf1. Cell 152:755–767

80. Renault L, Guibert B, Cherfils J (2003) Structural snapshots of the mechanism and inhibition of a guanine nucleotide exchange factor. Nature 426:525–530
81. Richardson BC, McDonold CM, Fromme JC (2012) The Sec7 Arf-GEF is recruited to the trans-Golgi network by positive feedback. Dev Cell 22:799–810
82. Robinson MS, Antrobus R, Sanger A, Davies AK, Gershlick DC (2024) The role of the AP-1 adaptor complex in outgoing and incoming membrane traffic. J Cell Biol 223:e202310071
83. Sadakata T, Shinoda Y, Sekine Y, Saruta C, Itakura M, Takahashi M, Furuichi T (2010) Interaction of calcium-dependent activator protein for secretion 1 (CAPS1) with the class II ADP-ribosylation factor small GTPases is required for dense-core vesicle trafficking in the trans-Golgi network. J Biol Chem 285:38710–38719
84. Sandilands E, Freckmann EC, Cumming EM, Roman-Fernandez A, McGarry L, Anand J, Galbraith L, Mason S, Patel R, Nixon C, Cartwright J, Leung HY, Blyth K, Bryant DM (2023) The small GTPase ARF3 controls invasion modality and metastasis by regulating N-cadherin levels. J Cell Biol 222:e202206115
85. Schleifer LS, Kahn RA, Hanski E, Northup JK, Sternweis PC, Gilman AG (1982) Requirements for cholera toxin-dependent ADP-ribosylation of the purified regulatory component of adenylate cyclase. J Biol Chem 257:20–23
86. Shiba T, Kawasaki M, Takatsu H, Nogi T, Matsugaki N, Igarashi N, Suzuki M, Kato R, Nakayama K, Wakatsuki S (2003) Molecular mechanism of membrane recruitment of GGA by ARF in lysosomal protein transport. Nat Struct Biol 10:386–393
87. Shiba Y, Kametaka S, Waguri S, Presley JF, Randazzo PA (2013) ArfGAP3 regulates the transport of cation independent mannose 6-phosphate receptor in the post-Golgi compartment. Curr Biol 23:1945–1951
88. Shin HW, Morinaga N, Noda M, Nakayama K (2004) BIG2, a guanine nucleotide exchange factor for ADP-ribosylation factors: its localization to recycling endosomes and implication in the endosome integrity. Mol Biol Cell 15:5283–5294
89. Shinotsuka C, Waguri S, Wakasugi M, Uchiyama Y, Nakayama K (2002) Dominant-negative mutant of BIG2, an ARF-guanine nucleotide exchange factor, specifically affects membrane trafficking from the trans-Golgi network through inhibiting membrane association of AP-1 and GGA coat proteins. Biochem Biophys Res Commun 294:254–260
90. Shinotsuka C, Yoshida Y, Kawamoto K, Takatsu H, Nakayama K (2002) Overexpression of an ADP-ribosylation factor-guanine nucleotide exchange factor, BIG2, uncouples brefeldin A-induced adaptor protein-1 coat dissociation and membrane tubulation. J Biol Chem 277: 9468–9473
91. Spang A, Shiba Y, Randazzo PA (2010) Arf GAPs: gatekeepers of vesicle generation. FEBS Lett 584:2646–2651
92. Stamnes MA, Rothman JE (1993) The binding of AP-1 clathrin adaptor particles to Golgi membranes requires ADP-ribosylation factor, a small GTP-binding protein. Cell 73:999–1005
93. Stearns T, Willingham MC, Botstein D, Kahn RA (1990) ADP-ribosylation factor is functionally and physically associated with the Golgi complex. Proc Natl Acad Sci USA 87:1238–1242
94. Stockhammer A, Adarska P, Natalia V, Heuhsen A, Klemt A, Bregu G, Harel S, Rodilla-Ramirez C, Spalt C, Ozsoy E, Leupold P, Grindel A, Fox E, Mejedo JO, Zehtabian A, Ewers H, Puchkov D, Haucke V, Bottanelli F (2024) ARF1 compartments direct cargo flow via maturation into recycling endosomes. Nat Cell Biol 26:1845–1859
95. Sztul E, Chen P-W, Casanova JE, Cherfils J, Dacks JB, Lambright DG, Lee F-JS, Randazzo PA, Santy LC, Schürmann A, Wilhelmi I, Yohe ME, Kahn RA (2019) ARF GTPases and their GEFs and GAPs: concepts and challenges. Mol Biol Cell 30:1249–1271
96. Tanabe K, Torii T, Natsume W, Braesch-Andersen S, Watanabe T, Satake M (2005) A novel GTPase-activating protein for ARF6 directly interacts with clathrin and regulates clathrin-dependent endocytosis. Mol Biol Cell 16:1617–1628
97. Thomas LL, Highland CM, Fromme JC (2021) Arf1 orchestrates Rab GTPase conversion at the trans-Golgi network. Mol Biol Cell 32:1104–1120

98. Togawa A, Morinaga N, Ogasawara M, Moss J, Vaughan M (1999) Purification and cloning of a brefeldin A-inhibited guanine nucleotide-exchange protein for ADP-ribosylation factors. J Biol Chem 274:12308–12315
99. Vargova R, Chevreau R, Alves M, Courbin C, Terry K, Legrand P, Elias M, Menetrey J, Dacks JB, Jackson CL (2025) The Asgard archaeal origins of Arf family GTPases involved in eukaryotic organelle dynamics. Nat Microbiol 10:495–508
100. Volpicelli-Daley LA, Li Y, Zhang C-J, Kahn RA (2005) Isoform-selective effects of the depletion of ADP-Ribosylation factors 1–5 on membrane traffic. Mol Biol Cell 16:4495–4508
101. Wang YJ, Wang J, Sun HQ, Martinez M, Sun YX, Macia E, Kirchhausen T, Albanesi JP, Roth MG, Yin HL (2003) Phosphatidylinositol 4 phosphate regulates targeting of clathrin adaptor AP-1 complexes to the Golgi. Cell 114:299–310
102. Wang S, Yao L, Zhang W, Cheng Z, Hu C, Liu H, Yan Y, Shi A (2021) AP-1 recruits SMAP-1/SMAPs to the trans-Golgi network to promote sorting in polarized epithelia. Front Cell Dev Biol 9:774401
103. Weimer C, Beck R, Eckert P, Reckmann I, Moelleken J, Brugger B, Wieland F (2008) Differential roles of ArfGAP1, ArfGAP2, and ArfGAP3 in COPI trafficking. J Cell Biol 183:725–735
104. Wong-Dilworth L, Rodilla-Ramirez C, Fox E, Restel SD, Stockhammer A, Adarska P, Bottanelli F (2023) STED imaging of endogenously tagged ARF GTPases reveals their distinct nanoscale localizations. J Cell Biol 222:e202205107
105. Zhang Y, Soubias O, Pant S, Heinrich F, Vogel A, Li J, Li Y, Clifton LA, Daum S, Bacia K, Huster D, Randazzo PA, Losche M, Tajkhorshid E, Byrd RA (2023a) Myr-Arf1 conformational flexibility at the membrane surface sheds light on the interactions with ArfGAP ASAP1. Nat Commun 14:7570
106. Zhang Y, Soubias O, Pant S, Heinrich F, Vogel A, Li J, Li Y, Clifton LA, Daum S, Bacia K, Huster D, Randazzo PA, Lösche M, Tajkhorshid E, Byrd RA (2023b) Myr-Arf1 conformational flexibility at the membrane surface sheds light on the interactions with ArfGAP ASAP1. Nat Commun 14:7570
107. Zhao X, Lasell TK, Melancon P (2002) Localization of large ADP-ribosylation factor-guanine nucleotide exchange factors to different Golgi compartments: evidence for distinct functions in protein traffic. Mol Biol Cell 13:119–133
108. Zhao X, Claude A, Chun J, Shields DJ, Presley JF, Melancon P (2006) GBF1, a cis-Golgi and VTCs-localized ARF-GEF, is implicated in ER-to-Golgi protein traffic. J Cell Sci 119:3743–3753
109. Yahara N, Ueda T, Sato K, Nakanom A (2001) Multiple roles of Arf1 GTPase in the yeast exocytic and endocytic pathways. Mol Biol Cell 12:221–238
110. Pipaliya SV, Schlacht A, Klinger CM, Kahn RA, Dacks J (2019) Ancient complement and lineage-specific evolution of the Sec7 ARF GEF proteins in eukaryotes. Mol Biol Cell 30:1846–1863
111. Tojima T, Suda Y, Jin N, Kurokawa K, Nakano A (2024) Spatiotemporal dissection of the Golgi apparatus and the ER-Golgi intermediate compartment in budding yeast. Elife 13
112. Yu CJ, Lee FJ (2017) Multiple activities of Arl1 GTPase in the trans-Golgi network. J Cell Sci 130:1691–1699
113. Rein et al (2002) ARF-GAP-mediated interaction between the ER-Golgi v-SNAREs and the COPI coat (PMID: 11970962. https://pubmed.ncbi.nlm.nih.gov/11970962/

Chapter 6
GARP Complex in Golgi Physiology

Amrita Khakurel, Walter S. Aragon-Ramirez, and Vladimir V. Lupashin

Abstract The Golgi Associated Retrograde Protein (GARP) complex, a member of the Complexes Associated with Tethering Containing Helical Rods (CATCHR) family, is proposed to tether vesicles arriving from endosomes to the *trans*-Golgi network (TGN). Discovered nearly 25 years ago, this protein complex is important for sorting vacuolar hydrolases and recycling membrane proteins from the endosomal/prevacuolar compartment to the TGN; however, its exact function, molecular partners, and the nature of GARP-dependent trafficking intermediates remain understudied. GARP-dependent transport route is utilized by various plasma membrane recycling proteins, lysosomal hydrolase receptors, and pathogens, including toxins. Mutations in GARP subunits have been associated with multiple neurological disorders, although the precise mechanisms by which these mutations lead to these conditions remain unclear. This chapter reviews the current understanding of GARP's structure, function, interacting partners, mutations, and associated pathologies in both humans and model organisms.

Keywords Endosome-TGN retrograde trafficking · GARP · Vesicle tethering · Trafficking machinery · Glycosylation

The GARP Complex

History and Discovery

Genetic and biochemical studies in yeast *S. cerevisiae* led to the discovery of the Golgi Associated Retrograde Protein (GARP) complex. Using a transposon-based mutagenesis approach, Stevens laboratory identified new genes involved in protein trafficking between the Golgi and pre-vacuolar (endosomal) compartments by

A. Khakurel · W. S. Aragon-Ramirez · V. V. Lupashin (✉)
Department of Physiology and Cell Biology, University of Arkansas for Medical Sciences, Little Rock, AR, USA
e-mail: vvlupashin@uams.edu

A. Nakano, J. Saraste (eds.), *The Golgi Network, Volume II*, Subcellular Biochemistry 111, https://doi.org/10.1007/978-3-032-16833-7_6

creating vacuolar protein sorting (vps) mutants [1]. These mutants, including *vps52*, *vps53*, and *vps54*, showed missorting of the vacuolar enzyme carboxypeptidase Y (CPY), slow growth at 30 °C, fragmented vacuolar morphology, and were sensitive to high concentrations of Zn^{2+}, Mn^{2+}, Cd^{2+}, and extreme pH [2]. *Vps54* mutants accumulated numerous small acidic vesicles, indicating the role of Vps54p in vesicle tethering and/or fusion. The complex was initially named Vps fifty-three (VFT) and later renamed GARP after the identification of a fourth subunit, Vps51p [3, 4]. Vps51p is essential for membrane trafficking and vacuole morphology, similar to the other subunits. Interestingly, in *vps51* mutants, the Vps52p/53p/54p complex can still assemble, though at reduced levels compared to wild-type cells [5]. The GARP complex is evolutionarily conserved with homologs in worms, flies, mammals, and plants [6–8].

Composition and Structure of the GARP Complex

GARP complex belongs to the CATCHR family of multisubunit tethering complexes (MTC) and is peripherally associated with the cytosolic face of TGN. It is an obligatory 1:1:1:1 heteromeric protein complex composed of four different subunits-VPS51, VPS52, VPS53, and VPS54 [7]. In most eukaryotes, the length of the GARP complex subunits ranges from 650 to 1700 amino acid residues whereas yeast Vps51p subunit is an exception as it is only 164 amino acids long [9].

In higher eukaryotes, in addition to GARP, the endolysosomal traffic is also regulated by the Endosomal Associated Retrograde Protein (EARP) complex. EARP is composed of VPS50/Syndentin, VPS51, VPS52 and VPS53 subunits. VPS51, VPS52, and VPS53 are common subunits with GARP complex and VPS50 is the unique subunit of EARP. EARP is involved in the recycling of a subset of plasma membrane receptors, such as transferrin receptors, from late endosomes back to the cell surface [10]. It is also essential for dense-core vesicle biogenesis [11].

In GARP complex, the amino-acid peptides in the N-terminus of each subunit are predicted to interact, forming a coiled-coil bundle facilitating GARP complex assembly. Hence, the complex is anticipated to have a core of amino terminals and four projecting arms corresponding to the subunit's carboxy-terminal regions (Fig. 6.1) [1, 5, 12].

Negative-stain electron microscopy of purified complex identified the GARP's structure in yeast as a Y-shaped complex with projecting legs labeled as legs A-C. Leg A is the shortest, while leg C is the longest. A hook was observed at the end of leg B, and a hinge in the middle of leg C. Double-GFP tagging revealed that legs A, B and C correspond to Vps52p, Vps53p and Vps54p, respectively. Vps51p was comparatively shorter and located at the connection point between three legs of a Y-shaped structure [13].

The predicted AlphaFold 3D structure of the human GARP complex is strikingly similar to the appearance of purified yeast GARP, with the exception of the extended VPS51 subunit (Fig. 6.1). In both yeast and human GARP complex, extended

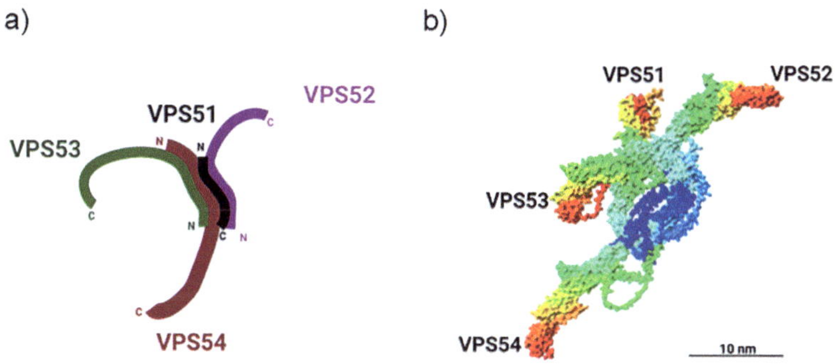

Fig. 6.1 Structure of the GARP complex. (**a**) Schematic representation of the proposed 3D structure of yeast GARP complex based on negative-stain electron microscopy [13]. (**b**) AlphaFold structure prediction of human GARP complex. The N-terminal regions of VPS51 (1–194), VPS52 (1–178), VPS53 (1–169), and VPS54 (1–310) were used to assemble a 3D core structure of the GARP complex using the ColabFold [98] plugin of the UCSF ChimeraX program [99]. AlphaFold predicted 3D structures of individual GARP subunits were fitted to the core structure using the "matchmaker" command of the UCSF ChimeraX program, and the resulting structures were colored using the Rainbow palette. Scale bar is 10 nm

C-terminal legs are widely (up to 35 nm) open, assembling docking platforms to tether 50–60 nm-wide trafficking intermediates. Importantly, GARP functions as a peripheral membrane protein, while the structural estimation was based on the purified soluble complex. This suggests that the functional conformation of GARP may differ from the 3D predictions. Related to this question, the functional homologs of GARP– Conserved Oligomeric Golgi (COG) and exocyst complexes are predicted to adhere to both "open" and "close" conformations [14, 15], indicating potential structural changes in membrane-bound and soluble forms of vesicle tethering complexes.

The complete 3D crystal structure of human GARP subunits is not yet available. However, in yeast, analysis of the crystal structure of Vps53p C- terminus (564–799 residues) at a resolution of 2.9 A revealed two alpha-helical bundles arranged in tandem. The C-terminus of Vps53p consists of a highly conserved surface patch that may take part in vesicle recognition [16]. Interestingly, the C-terminal region of VPS53 is predicted to contain the so-called MUN domain [17]. This domain was initially found in UNC13/Munc13–1 protein and predicted in many subunits of the CATCHR vesicle tether family (SEC6, EXO70, COG4, and RINT1). Deletion of Munc13-1 in mice causes severe defects in neurotransmitter release, indicating its fundamental role in synaptic transmission [18]. Like GARP, Munc13 controls priming of transport vesicles [19] and MUN domain is the minimal module responsible for Munc13-1/UNC13 activity [20].

The crystal structure of mouse VPS54 C-terminal 142 amino acid residues (836–977) at 1.7 A resolution, in conjunction with comparative sequence analysis, also revealed an α-helical bundle. In this structure, Leu967 is buried within the alpha-helical bundle through predominantly hydrophobic interactions critical for domain stability and folding in vitro. Mutation of Leu967 residue to glutamine in wobbler mice dramatically reduces the half-life and total protein level of VPS54, even though the integration of the affected subunit into the GARP complex is not prevented [21].

Comparison of the GARP complex with other tethering complexes of the CATCHR family uncovered a similarity in the subunit organization between GARP, COG, exocyst, and Dsl1 protein complexes. Comparisons of GARP complex with COG complex demonstrated that Vps51p structurally corresponds to Cog2p, Vps52p to Cog3p, Vps53p to Cog4p/Sec6/Dsl1, and Vps54p to Cog1p [13, 16, 22].

Localization of the GARP Complex to the Golgi Membrane

In both yeast and mammalian cells, the GARP complex is localized to *trans*-Golgi/TGN [1, 23, 24] (Fig. 6.2). The recruitment of the GARP in yeast was facilitated by the Rab GTPase Ypt6p, which is activated by the guanidine exchange factor (GEF) Ric1p-Rgp1p [25]. This activation of Ypt6p enables the tethering of vesicles containing SNARE protein Tlg1p to the TGN through GARP [25]. In human cells, the Ypt6p homolog RAB6 also interacts with VPS52, but this interaction is not critical for GARP complex association with the TGN, as RAB6 knockout (KO) does not disrupt GARP localization [26]. Instead, in mammalian cells, GARP complex recruitment to the TGN relies mainly on small GTPases ARL5 and ARFRP1. KO of ARL5 and ARFRP1 prevents GARP complex association with the Golgi, while re-expression of ARL5 restores GARP localization, highlighting the importance of these GTPases in GARP recruitment. The activated ARL5-GTP associates with the TGN membrane through the myristoylated N-terminal domain [26]. The specific GARP subunit that interacts directly with ARL5 is currently unknown. Similar interactions between ARL5 and the GARP complex have been observed in *Drosophila melanogaster*. In both flies and humans, ARL5 deletion leads to partial displacement of GARP from Golgi membranes and results in enlarged endosomal compartments [27].

While GARP and EARP complexes share VPS51, VPS52, and VPS53, the localization of these two complexes differs. VPS54, which is the unique component of the GARP complex, co-localizes with TGN46, indicating its association with the TGN (Fig. 6.2a). However, VPS50, a unique component of the EARP complex, is localized to RAB4-containing endosomes [10] and not in the Golgi (Fig. 6.2b). This differential localization underscores the distinct roles and spatial organization of the GARP and EARP complexes in intracellular trafficking.

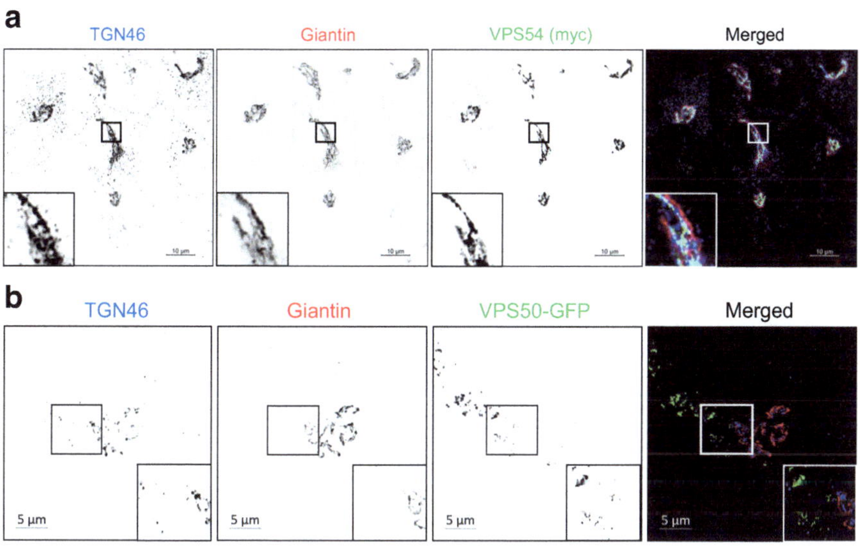

Fig. 6.2 GARP, but not EARP is localized in the Golgi in human cells. (**a**) hTERT- RPE1 VPS50-KO cells were transiently transfected with mVPS50-GFP and stained with antibodies against TGN46 and Giantin. Airyscan superresoution images (MIPs) were collected with 63x oil 1.4 objective on Zeiss LSM880. Enlarged (2x) Golgi region is shown in the bottom right corner insert. (**b**) hTERT-RPE1 VPS54-KO cells stably expressing VPS54–13 myc were stained with antibodies to the TGN protein TGN46, *cis/medial* Golgi protein Giantin and myc (VPS54). Airyscan superresolution images (Maximum Intensity Projections, MIPs) were collected with 63x oil 1.4 objective on Zeiss LSM880. Enlarged (5x) Golgi region is shown in the bottom left corner insert

GARP-Interacting Proteins

SNAREs

In *S. cerevisiae*, Vps51p interacts with Qc SNARE Tlg1p, through the amino-terminal regulatory Habc domain of Tlg1p. The binding site for Tlg1p has been mapped to the residues 18–30 at the N-terminus of Vps51p [3, 5]. The binding specificity of the Vps51p-Tlg1p was validated by deletion analysis and point mutations (Phe27 to Leu or Tyr28 to Ala). The N-terminal Vps51p peptide is partially alpha-helical and binds to a groove in the three-helix bundle formed by Tlg1p. Despite this interaction, point mutations that disrupt binding did not result in trafficking defects, suggesting that the Tlg1p-Vps51p interaction is not essential for GARP complex function [28].

Human VPS51 similarly interacts with the Habc domain of STX6, which is the human homolog of Tlg1p, highlighting the evolutionary conservation of this interaction. Crystal structure analysis of the complex between the STX6 Habc domain and the N-terminus of VPS51 reveals that VPS51 binds to a conserved hydrophobic groove in the STX6 Habc domain. This binding involves Tyr40 and Tyr41, suggesting that the VPS51 di-tyrosine motif is crucial for its interaction with the Habc domains of STX6 and STX10 [24, 28].

In vitro and *in vivo* studies in human cells demonstrated that the GARP complex interacts with the STX16/STX6/VTI1A/VAMP4 TGN SNARE complex. Microscopy analysis demonstrated colocalization of VPS54 with STX16, STX6, and VAMP4. Importantly, siRNA-mediated KD of VPS54 showed a reduction in STX16 SNARE complex formation and resulted in the redistribution of the TGN SNAREs. Interaction between the STX16 SNARE complex and GARP depends on the SNARE motif in SNARE proteins and N-terminal regions of VPS53 and VPS54 [29].

Interestingly, the COG complex, a related intra-Golgi vesicular MTC, interacts with the Qa SNARE STX5 [30–32]. Although the precise details of the GARP-SNARE interaction are not fully understood, the assembled GARP complex can bind to the STX16 SNARE complex, further elucidating the role of GARP in SNARE-mediated vesicle trafficking [29].

Small GTPases

In yeast, Vps52p interacts with the GTP-bound form of Rab GTPase Ypt6p. Similarly, mammalian RAB6 pulled down VPS52 indicating that GARP-RAB6 interaction is evolutionarily conserved [24]. Ypt6p is crucial for retrograde transport from endosomes to the TGN, and its mutation can lead to protein mislocalization and growth defects. Immunofluorescence imaging has shown that the depletion of Ypt6p causes a dispersion of the GARP signal from TGN. However, subcellular fractionation assays do not show a reduction in GARP association with the membrane fraction, suggesting that while Ypt6p is necessary for proper intracellular localization of GARP, it is not required for GARP's membrane association [5, 25]. Interestingly, defects in Ypt6p, such as impaired endosome-TGN transport and growth delays, can be partially compensated by overexpression of Arl1p or the coiled-coil Golgi protein Imh1p (homologue of human GOLGA1), indicating that some GARP functions can be restored independently of Ypt6 [33].

In *S. cerevisiae*, Vps53p interacts with the GTP-bound form of Arl1p [34]. However, the loss of Arl1p does not affect GARP localization to the TGN [3, 25]. In contrast, in mammals [26] and flies [27], GARP interacts with ARL5, another GTPase of the ARL subfamily, and this interaction is crucial for GARP's localization to the TGN.

In mammalian cells, VPS52 also interacts with ARF6, a small GTP-binding protein involved in endocytic trafficking. This interaction was demonstrated using a GTP-locked, constitutively active ARF6 Gln67Leu mutant. Mutation analysis indicated that the regions between amino acids 99–149 and 674–723, or the overall tertiary structure of VPS52, are necessary for binding ARF6. The VPS52ΔN149 mutant, which lacks the ARF6 interaction, was able to partially rescue some phenotypes associated with VPS52-KD. However, it could not rescue the increase in axonal length observed in hippocampal neurons [35]. This highlights the importance of the ARF6-VPS52 interaction in regulating neuronal morphogenesis. Since ARF6 is localized away from the Golgi, the VPS52-ARF6 binding is likely to occur within the context of an EARP-ARF6 interaction.

Coiled-Coil Tethers

In yeast, VPS53 binds to Golgi coiled-coil tether (CCT) Imh1. During ER stress, Vps53p –Imh1p interaction is enhanced to facilitate the recycling of SNAREs Snc1p and Tlg1p back to the late Golgi. The unfolded protein response activates the phosphorylation of Arl1p GEF Syt1p, which in turn recruits more Imh1p to the late Golgi. Imh1p is then phosphorylated at Ser25/Thr27 by MAPK Slt2p/Erk2p. This phosphorylation activates the GARP complex, enhancing the recycling of Tlg1p-positive vesicles to the late Golgi, and facilitating the transport of Snc1p to the Golgi and plasma membrane. It was proposed that, under ER stress, two distinct tethers, a coiled-coil golgin and the GARP, work together to promote the recycling of the Golgi SNAREs Snc1p and Tlg1p to late Golgi [36].

In mammalian cells, the TGN CCTs Golgin-97/GOLGA1, Golgin-245/GOLGA4, and GCC88/GCC1 are attached to the membrane via ARL1, while GARP membrane binding is facilitated by ARL5 [26, 27] (Fig. 6.3). Importantly, membrane association of both GTPases depends on the same machinery: the transmembrane protein SYS1 and the GTPase ARFRP1, indicating coordinated regulation of GARP and CCTs. Supporting this coordination, proximity biotinylation using endogenously expressed VPS54-TurboID revealed efficient labeling of GOLGA4, GOLGA1, and GCC1 (A.K., V.L., unpublished data).

Other Partners

VPS54-GFP stably expressed in human neuroglioma cells pulls down TSSC1, a WD40/β-propeller domain-containing protein, and confocal microscopy confirmed their colocalization. Disruption of TSSC1 affected both GARP and EARP, leading to delays in transport of the Shiga toxin B subunit to the TGN and the recycling of transferrin receptor to the plasma membrane. Additionally, localization of VPS54-GFP was partially impaired in TSSC1 knock-down (KD) cells, suggesting that TSSC1 aids in the recruitment of GARP to the TGN [37].

Yeast two-hybrid studies demonstrated that yeast dynamin Vps1p interacts with Vps51p, specifically through a 33 amino acid domain (99–132aa) near the C-terminal end of the latter. Highly conserved residues Glu127 and Tyr129 of this domain are involved in Vps1p-VPS51p interaction. Confocal microscopy supported the colocalization of Vps1p and Vps51p, and loss of Vps1p decreased the colocalization of Vps51p and Tlg1p at the *trans*-Golgi, suggesting that Vps1p assists the GARP tethering machinery in efficient tethering and fusion of trafficking intermediates at the TGN [38, 39].

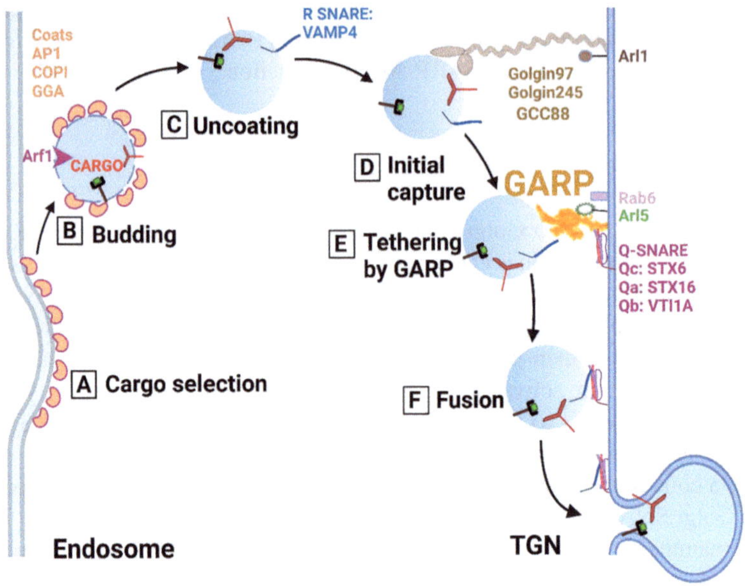

Fig. 6.3 Cartoon depicting the predicted role of the GARP complex and other trafficking components in the trafficking of endosome-derived vesicles to the TGN. (**a**) Cargo selection: Recycling soluble and transmembrane cargo molecules, such as mannose-6-phosphate receptors and TGN46, are selected and packaged in the trafficking intermediates (vesicle) by ARF1 GTPase and a vesicular coat (AP1, GGA or COPI). (**b**) Budding: After the vesicle is formed, it buds off the endosomal compartment. (**c**) Uncoating: ARF1 hydrolyzes GTP, and the coat gradually falls off. (**d**) Initial capture: Once the vesicle reaches proximity to the TGN, coiled-coil tethers (golgin-97, golgin-245, and GCC88) perform the initial capture. (**e**) Tethering by GARP: GARP complex binds coiled-coil tethers and SNAREs to control and coordinate docking of the vesicle with the TGN membrane. (**f**) Fusion: Following the tethering by the GARP complex, the trans-SNARE complex is formed, which results in vesicle fusion with the TGN and delivery of the recycling molecules

RNF41 (Ring Finger Protein 41), an E3 ubiquitin ligase involved in the intracellular sorting of various substrates was found to interact with VPS52 via high-throughput screening. Endogenous VPS52 was also shown to interact with RNF41 in HEK293T cells. Moreover, RNF41 could ubiquitinate and relocate VPS52 away from VPS53 towards so-called RNF41 bodies [40].

In mammalian cells, VPS52 also interacts with LRRK2, a serine/threonine-protein kinase, involved in vesicle trafficking, autophagy, and neuronal plasticity [41–43] Mutations in LRRK2, such as Gly2019Ser, are linked to Parkinson's disease. Studies in *C. elegans* show that KD of vps-54 exacerbates neuronal toxicity of LRRK2 mutations linking GARP to neuronal defects. Proteomics and imaging studies in HEK293T cells suggest that LRRK2 regulates endosome-TGN transport by scaffolding the GARP-SNARE interaction (STX16/STX6/VTI1A/VAMP4), suggesting a connection between GARP and LRRK2 in Parkinson's disease. Additionally, VPS54-KD led to a decreased viability of dopamine neurons, further supporting this link [44].

Functions of GARP Complex

Although the GARP is not essential for budding yeast survival [1], deletion of GARP subunits in this unicellular organism result in increased protein secretion, abnormal autophagy, and defects in vacuolar morphology [3, 4]. GARP deletion also causes infertility in worms [45], a shorter life span in flies [46], and embryonic lethality in mice [47]. In this section, we have summarized the major roles of the GARP complex and its individual subunits in cell physiology.

GARP as a Molecular Tether for the Endosome-Derived Vesicles

One of the most studied functions of the GARP complex is its role in retrograde vesicular transport from endosomes to the TGN, which enables the recycling of transmembrane proteins like vacuolar protein sorting receptor Vps10p in *S. cerevisiae*. Vps10p binds to the pro-carboxypeptidase (Pro-CPY) in the TGN. Pro-CPY is a vacuolar hydrolase precursor that is delivered to the vacuole, while the receptor Vps10p needs to be recycled back to TGN for further rounds of sorting. A mutation in any GARP subunit prevents Vps10p recycling to the TGN. Instead, Vps10p is missorted to the vacuole and degraded. Because of Vps10p degradation, GARP mutants missort precursors of vacuolar hydrolases, causing their secretion from the cell. Similarly, the recycling of the TGN protein Kex2p and the plasma membrane v-SNARE Snc1p is altered in GARP mutants, resulting in their vacuolar degradation [3].

In mammalian cells, cation-independent mannose 6-phosphate receptor (CI-MPR/IGF2R) performs a function similar to Vps10p. Knockdown of GARP subunits in HeLa cells prevented the retrieval of CI-MPR to the TGN, resulting in missorting and secretion of lysosomal hydrolases. Blockade in the sorting of lysosomal hydrolases gave rise to the development of enlarged lysosomes as a result of the accumulation of undegraded materials in the lumen of the organelle [12, 23].

Similarly, the retrieval of TGN46 is blocked in GARP mutants [23]. Moreover, the recycling Golgi resident proteins (GPP130, TMEM165, and TGN46) is significantly depleted in VPS54-KO and VPS53-KO cells [48]. In addition to endogenous proteins, the transport of the Shiga toxin is also inhibited in GARP-KD cells [21, 23].

Based on GARP mutant phenotypes and on the similarity between GARP and other CATCHR complexes, it was proposed that GARP serves as a TGN-located molecular tether for the recycling of endosome-derived vesicles (Fig. 6.3). In support of this model, yeast *vps54* mutant cells accumulate multiple vesicular structures [2], giving a phenotype which is similar to Golgi vesicle accumulation upon the acute depletion of COG complex [49]. In addition, siRNA depletion of GARP subunits in HeLa cells caused the redistribution of recycled CI-MPR to a vesicle-like "haze" [23]. It will be essential to isolate and characterize GARP-dependent membrane carriers and investigate the interplay between GARP and other players of TGN vesicle tethering/docking/fusion machinery.

GARP as a Regulator of SNARE Complexes

Like other MTC vesicular tethers [31, 50–53], GARP complex physically interacts with a specific subset of SNARE molecules. Mammalian GARP complex interacts with endosome-TGN STX16 SNARE complex (STX16/STX6/ VTI1A/VAMP4) (Fig. 6.3), and GARP-KD causes decreased formation and/or stability of that complex [29]. In addition, GARP has been shown to regulate the localization and stability of the intra-Golgi SNAREs: the Qb-SNARE GOSR1 and the Qc-SNARE BET1L. These v-SNAREs work in the STX5/GOSR1/BET1L/YKT6 SNARE complex to facilitate the fusion of intra-Golgi recycling vesicles. Immunofluorescence microscopy demonstrated the depletion of Golgi GOSR1 and BET1L signals in GARP-KOs. Consistently, the total protein abundance of v-SNAREs in the GARP-KOs was significantly decreased [54]. This indicates that the GARP is involved in the regulation of at least two different SNARE complexes, promoting the fusion of cargo vesicles not only in TGN, but also in the early-Golgi compartments.

Role of GARP Complex in the Maintenance of Golgi Glycosylation Machinery

Glycosylation of proteins and lipids contributes to a variety of biological processes, and abnormal glycosylation in humans is a hallmark of many diseases. Mutations in proteins that are directly or indirectly involved in the glycosylation process can cause congenital disorders of glycosylation [55, 56]. The glycosylation machinery is localized in the ER and Golgi [57]; therefore, it was initially assumed that post-Golgi trafficking does not significantly contribute to the glycosylation process. However, a recent publication described a 6-year-old patient with a neurodevelopmental disorder having a mutation in the VPS51 subunit of GARP/EARP complex and an abnormal pattern of glycosylation [58].

Similarly, a study from our lab revealed that KOs of GARP complex subunits VPS53 and VPS54 in different types of tissue culture cells (hTERT–RPE1, HEK293T and HeLa) [59] resulted in severe defects in both N- and O-linked protein glycosylation. Golgi glycosylation enzymes (MGAT1, B4GALT1, and ST6GAL1) were also significantly depleted. Retention using selective hooks (RUSH) assay showed that B4GALT1 was not retained at the Golgi complex in GARP-KO cells but was missorted to the endolysosomal compartment. Another *trans*-Golgi enzyme, ST6GAL1, was also not retained in the Golgi in GARP-impaired cells. The defect in Golgi enzyme retention/recycling depends entirely on GARP and not on the EARP complex. The malfunction, mislocalization, and reduced stabilization of Golgi enzymes associated with a glycosylation defect was rescued upon expression of the missing GARP subunit. This data indicates that GARP plays a crucial role in normal Golgi glycosylation by mediating the maintenance of the Golgi glycosylation machinery [48].

Role of the GARP Complex in Normal Golgi Physiology

Although the majority of GARP-sensitive proteins are predicted to reside in *trans*-Golgi compartments [48], the proteomic analysis of Golgi-enriched membranes in hTERT-RPE1 cells lacking GARP subunits revealed significant depletion of *cis- and medial*-Golgi proteins such as GLG1/MG-160, GALNT1, and MAN1A2 [54]. Subsequent experiments revealed that in human GARP-KO cells, COPI vesicular coat complex, which is primarily involved in intra-Golgi and Golgi-ER retrograde trafficking [60], is partially mislocalized to the ERGIC compartment. In GARP-KO cells, the COPI accessory proteins GOLPH3, ARFGAP1, and GBF1 were displaced from the membrane, while BIG1 was relocated to the endolysosomal compartment. In addition, transmission electron microscopy of GARP mutants revealed dysmorphic Golgi features. In VPS54-KO cells, the Golgi structure was enlarged, suggesting an increased distance between the cisternae. In VPS53-KO cells, the Golgi structure was round, swollen, and disrupted [54]. This suggests that GARP dysfunction impacts not just *trans*-Golgi compartments but also the entire Golgi complex (Fig. 6.4).

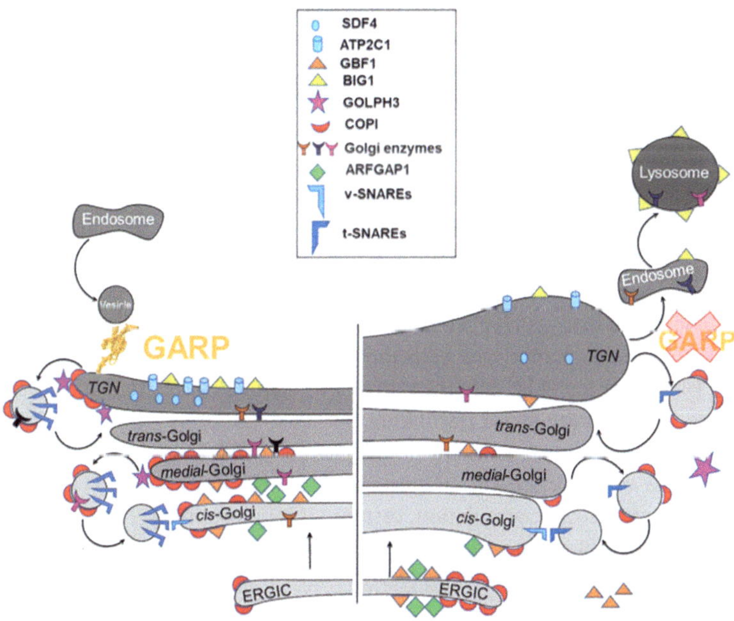

Fig. 6.4 GARP dysfunction results in the depletion and mislocalization of multiple Golgi proteins, altering Golgi homeostasis. (Left) A schematic of Golgi proteins in wild type cells. (Right) GARP deficient cells are depleted for multiple resident Golgi proteins. COPI coats and ARFGAP1 are mislocalized to the ERGIC, while BIG1 is relocated to endolysosomal compartments

GARP and Lipid Homeostasis

Fröhlich et al. identified a crucial role for the yeast GARP complex in lipid homeostasis. Deletion of any component of the GARP complex resulted in the build-up of sphingolipid intermediates and long-chain bases, whereas pharmacological inhibition of sphingolipid biosynthesis could repair all observed GARP-KO mutant abnormalities. This indicates that abnormal sphingolipid accumulation is the problem-causing factor in GARP-deficient cells [61]. Maintenance of the phospholipid's asymmetric distribution between the outer to the inner leaflet of the plasma membrane (PM) is carried out by phospholipid transporter flippases [62]. The mislocalization of PM localized flippases results in changes in lipid homeostasis [63]. Mutation in the GARP in yeast showed hypersensitivity to the phosphatidylethanolamine (PE) binding lantibiotic peptide Ro09–0198 (Ro) [63]. GARP mutants mislocalize PM flippases Dnf1p and Dnf2p to the vacuole, suggesting that remodeling of the lipid composition is one of the first defects occurring in GARP-depleted cells [64]. A study in flies showed that the GARP complex is involved in sterol transport, and KO of Vps54 resulted in reduced lifespan and impaired arborization of the *Drosophila melanogaster* [65].

The GARP complex is also involved as a regulator in the morphogenesis of *Candida albicans*. A peculiar feature of this opportunistic fungal pathogen is that it has the ability to undergo transition between yeast and filamentous hyphal form in response to cues. In GARP mutants, in response to cues such as Hsp90 inhibition, the transition from yeast to filamentous hyphal form is impaired, indicating that GARP is an important mediator of filamentation. This filamentation defect was found to be due to the disruption of lipid homeostasis in the GARP mutant [66, 67].

In human cells, dysfunction of the GARP complex resulted in the missorting of lysosomal resident protein NPC2, which is responsible for the egress of LDL-derived cholesterol from lysosomes. As a result, GARP-depleted cells accumulate cholesterol. This phenotype is similar to Niemann-Pick type C disease, a neurodegenerative disease characterized by a massive accumulation of free cholesterol and other lipids in almost all cells throughout the body, particularly in the brain and liver [68]. O'Brien et al. demonstrate that the GARP and EARP complexes both play a role in regulating sterol trafficking during dendrite development. In GARP-KO neurons, sterol accumulates specifically at the TGN. Oxysterol binding protein (OSBP) is one of the regulators of sterol and related lipids at the TGN. Targeting OSBP can rescue sterol accumulation, restore Golgi morphology, and promote adult dendrite growth, suggesting that GARP-dependent sterol transport is crucial for dendrite development in *Drosophila* [65].

Role of GARP Complex in Post-Golgi Secretory Pathway

Besides the endosomal-TGN trafficking function of the GARP complex and the promotion of SNARE function in membrane fusion, VPS52 has been implicated in the regulation of the secretory pathway. The secretion of model soluble secretory protein ss-sfGFP (Superfolder GFP fused to signal sequence) was significantly decreased in VPS52-KO MDCK cells with subsequent accumulation of ss-sfGFP in lysosomes. This phenotype was rescued by the expression of VPS52. The functional importance of the C-terminal region of VPS52 was assessed for its role in restoring secretory phenotypes and sorting lysosomal proteins. It was found that the region spanning amino acids 411 to 556 is necessary for ss-sfGFP sorting and secretion, while a longer region, from amino acids 411 to 723, is required for efficient lysosomal protein sorting [69]. Additional requirement for GARP in the secretory process was observed in HEK293T cells, where KO of GARP subunits caused severely defective anterograde transport of both GPI-anchored and transmembrane proteins from the TGN to the PM [70]. Overexpression of VAMP4, TMEM87A or its close homolog TMEM87B in VPS54-KO cells partially restored anterograde transport of GPI-anchored proteins in GARP deficient cells. The authors proposed that GARP-dependent endosome-to-TGN retrograde transport is required for the recycling of molecules critical for efficient post-Golgi anterograde transport of cell-surface proteins.

Hijacking of GARP by Intracellular Pathogens

A genome-wide insertional mutagenesis screen in HAP1 cells, aimed at identifying host components essential for monkeypox virus infection, identified the GARP complex as crucial for the generation of extracellular viruses. In VPS52-KO and VPS54-KO cells, the virus yield was drastically reduced, whereas the structure of the mature virion was unaffected. Interestingly, EM studies revealed a decrease in the membrane wrapping in GARP-KO cells, suggesting that GARP may play a role in the efficient membrane wrapping of the mature virions. Additionally, GARP-KO infected cells lacked actin tail production compared to controls, indicating a defect in one of the mechanisms viruses use to escape host cells. The role of Golgi trafficking machinery in the production of monkey pox and vaccinia virus is further supported by another study in COG-KO cells where COG-KO cells produced lower virus yield compared to WT [71, 72].

GARP Complex Mutants and Pathogenesis

GARP Complex Mutants

Null mutation: The Vps54 null mutation is embryonically lethal in mice. Vps54 null mutant embryos do not survive by embryonic day E11.5. These embryos show significant spinal cord underdevelopment and nearly complete absence of dorsal root ganglia. In addition, the atrial and ventricular myocardium exhibited significant hypoplasia, demonstrating the significance of VPS54 in early embryonic development [47]. Moreover, in *Drosophila melanogaster*, null mutants of the Vps54 homolog scattered (scat) have a defect in spermatogenesis [45].

Point mutation: The wobbler mouse is widely used animal model to study motor neuron disease. This mouse is characterized by its unsteady gait and exhibits muscular weakness, atrophy, and progressive motor neuron degeneration, mimicking human amyotrophic lateral sclerosis [73, 74]. Wobbler phenotype is caused by the point mutation in the exon 23 of VPS54 gene that results in destabilization of the protein [75, 76]. This point mutation contains an A-T transversion in the second position of codon 967, leading to the substitution of hydrophobic leucine for hydrophilic glutamine. As a result of this hydrophobic-to-hydrophilic amino acid substitution, the protein loses its stability, reducing its concentration in all tissues, including the spinal cord [76]. VPS53 levels also decrease, leading to a reduction in the entire GARP complex [47]. Proteomic analysis of wobbler brains showed increased glial fibrillary acidic protein and coronin, indicating astrogliosis [77]. Additionally, the wobbler mouse suffers from spermatogenesis deficiencies similar to human globozoospermia, where sperm lack an acrosome. Mass spectrometry revealed reduced FABP3 and increased HSP70 and HSP90 in wobbler testes, highlighting VPS54's role in fertility [78]. Expression of wild-type VPS54 in transgenic mice resulted in viable progeny and restored normal sperm morphology [47].

Pathogenesis of GARP Mutations

In Humans

VPS51 is essential for normal brain development. Whole exome sequencing of two siblings—one 9 years old and the other 30 months old—revealed a variant in the VPS51 gene, consisting of an in-frame deletion of the CTT codon in exon 5 (c.1419_1421del; p.Phe474del), detected in the homozygous state in both sisters. Both siblings exhibited delayed psychomotor development, absence of speech, severe intellectual disability, and postnatal microcephaly. Additionally, the 9-year-old had cerebellar atrophy, while the 30-month-old had a hypoplastic corpus callosum [79]. Similarly, another study reported a 6-year-old patient with severe global developmental delay and dysmorphic features, who carried compound heterozygous mutations in the VPS51 gene: c.1468C > T (p.Asp745Thrfs*93) and c.2232delC (p.Arg490Cys). The mutation in one allele causes a frameshift that produces a long but highly unstable protein that is degraded by the proteasome. In contrast, the other mutant allele produces a protein with a single amino acid substitution that is stable but assembles less efficiently with the other GARP/EARP subunits [58].

Mutation in VPS53 was shown to cause a progressive cerebello-cerebral atrophy 2 (PCCA2). The common features of PCCA2 are profound mental retardation, progressive microcephaly, spasticity, and early onset epilepsy. Whole exome sequencing identified c.2084A > G p. (Gln695Arg) missense mutation in exon 19 that replaces a neutral amino acid with a positively charged one within the second helix of the surface of VPS53's conserved C-terminal. Similarly, the c.1556 + 5G > A splice site mutation in the fifth nucleotide of intron 14 generates an unstable transcript that is predicted to encode a truncated protein, leaving the affected individuals with no functional VPS53 C-terminus and partially dysfunctional GARP [80]. Additionally, whole exome sequencing of two female siblings affected with complicated hereditary spastic paraparesis identified a mutation in the VPS53 subunit of the GARP complex (c.2084A > G: c.2084A > G, p. Gln695Arg) [81].

In Plants

In *Arabidopsis thaliana*, the homolog of VPS51, known as Unhinged/UNH-1, plays a crucial role in determining leaf shape and vein pattern by maintaining the polarity of PIN1 at the PM. The polarity of PIN1 is regulated through a process involving endocytosis, transcytosis, and recycling to the PM. PIN1 mutants exhibit leaves that are pointed and narrower, with fewer secondary and higher-order veins, and lack distal vein junctions compared to wild-type leaves. The UNH-1 leaf phenotype is mitigated by the pin1 mutation, indicating that the observed changes are due to increased PIN1 expression in the marginal epidermis. The study revealed that GARP is important for reducing PIN1 expression in margin cells, possibly by directing PIN1 to the lytic vacuole. Similar to its role in yeast, UNH-1 is crucial for vacuolar trafficking. Mutations in UNH-1 result in reduced vacuolar targeting and increased secretion of a vacuolar-targeted fluorophore [82]. Thus, VPS51 is crucial for the normal development of leaf and vein morphology in plants.

VPS52 has been shown to play a crucial role in protecting plants, such as *Arabidopsis thaliana* and potato, from aphid (*M. persicae*) infestation. In these species, the aphid effector Mp1 associates with VPS52 and reduces its levels by relocalizing the protein to prevacuolar compartments, which may lead to the degradation of the GARP subunit. Conversely, high levels of VPS52 in plants negatively impact aphid virulence, making VPS52 an important target for virulence suppression [83].

In Flies

Retrograde transport from the endosome to TGN is essential for the proper maturation of secretory granules and the exocytosis process during development in *Drosophila melanogaster*. Mutations in the GARP subunits Vps53 and Scattered (Vps54), impair the retrograde transport and this impairment leads to premature degradation of immature glue granules via crinophagy, bypassing the normal developmental process [84].

Downregulation of GARP in Cancer

VPS52 has been implicated in cancer biology, particularly in gastric cancer. It is found at lower levels in gastric cancer cell lines compared to normal gastric epithelial cells. In vitro studies show that reduced VPS52 leads to decreased cell viability and increased apoptosis. Conversely, VPS52 overexpression in vivo results in reduced tumor weight and volume. Genetic analysis of gastric cancer tissues revealed loss of heterozygosity and stop-gain mutations in VPS52, suggesting its role as a tumor suppressor [85].

Similarly, VPS53 is critical in suppressing colorectal cancer (CRC) malignancy. VPS53 levels are significantly lower in CRC tissues than in normal tissues. VPS53 depletion in CRC tissues showed a positive correlation with autophagy protein Beclin 1, indicating that autophagy induction could facilitate CRC cell death. Inhibiting autophagy mitigates VPS53's effects on cancer cell proliferation, invasion, and migration. Overexpression of VPS53 in CRC cells reduces cell growth by approximately 50%, likely due to increased apoptosis [86, 87]. VPS53 is also downregulated in medullary thyroid cancer [88] and prostate cancer [89, 90].

Conclusion and Future Perspectives in GARP Studies

GARP was first identified as a protein complex necessary for sorting vacuolar hydrolases and recycling membrane proteins from the endosomal/prevacuolar compartment to TGN. Despite studies carried out for over 25 years, many aspects of this evolutionarily conserved complex remain unclear. Mutant phenotypes in cells lacking GARP subunits exhibited not only defects in trafficking between endosomes and the TGN, but also alterations with glycosylation, intra-Golgi trafficking, and secretion.

The immediate defects of GARP dysfunction are not fully understood. Most research has used GARP subunit knockout or knockdown approaches, which take a long time to eliminate GARP complex activity. During this time, cells might adapt, potentially masking the immediate effect of GARP dysfunction. Moreover, prolonged depletion of GARP can lead to secondary defects that obscure the investigation. It is crucial to differentiate between primary defects caused directly by GARP dysfunction and secondary defects that develop over time due to prolonged dysfunction of TGN trafficking machinery.

GARP is a member of the CATCHR family of vesicle tethering complexes [16, 22]. However, we still lack understanding of how GARP and other CATCHR tethers interact with transport vesicles. Does GARP capture vesicles independently or together with coiled-coil tethers? Multiple proteins including M6PR [91], IGF2R [23, 92], ATP7A [93], GPP130 [94], TGN46 [95, 96], and TMEM87A [70] constantly cycle between endosomes and Golgi, most likely in GARP-dependent transport carriers. However, the specific set of protein and lipid cargoes that depend on the GARP complex for their transport is yet to be identified.

To address these questions, it is crucial to isolate and characterize GARP-dependent membrane carriers and investigate how GARP interacts with other components of the TGN vesicle tethering, docking, and fusion machinery. Although GARP is believed to facilitate the assembly of TGN SNAREs, the details of its interaction with SNAREs are not well understood. Does GARP interact exclusively with the STX16 SNARE complex? Does it bind and regulate the STX16-interacting

SM protein VPS45? Understanding all the proteins that interact with GARP is critical to determine which interactions are vital for its functions. Additionally, GARP is known to be involved not just in retrograde trafficking but also in regulating post-Golgi anterograde trafficking [70] and sphingolipid homeostasis [61, 64]. How exactly does GARP accomplish all these functions?

To understand the function of the GARP complex, it is equally essential to elucidate the dynamic nature of the GARP structure. Little is known regarding either the overall or the subunit 3D architecture of the mammalian GARP complex. The same applies for the free and membrane-bound forms of the complex? It is likely that GARP is switching from the "close" to the "open" conformation to tether inbound transport intermediates.

The GARP and EARP complexes share the VPS51/52/53 trimer, but their specific functions and localizations are different: GARP is thought to function at the TGN, while EARP functions on endosomes. The exact role of the shared VPS51/52/53 trimer in these processes is unclear. It is also not clear if the shared VPS51/52/53 trimer is dynamically switching between GARP and EARP conformations or, alternatively, if GARP and EARP complexes are stable and never exchange their subunits. Additionally, identifying other factors that distinguish GARP from EARP, such as SNAREs, coiled-coil tethers, and small GTPases, is essential for understanding GARP's unique functions. This information will help clarify how GARP functions differently from EARP and contribute to a more complete picture of its role in cellular processes.

The GARP complex is conserved across various species, including humans, mice, and plants, and its dysfunction can lead to significant pathologies. GARP deletions are not essential at the cellular level, but in mice, VPS54 null causes embryonic lethality. This raises important questions: How does GARP dysfunction lead to such severe outcomes during embryonal development? What are the specific roles of GARP in different types of cells? Furthermore, why does GARP deficiency have a particularly pronounced impact on neuronal function? Understanding these aspects is crucial for unraveling the complex role of GARP in development and disease.

GARP subunits are often downregulated in various cancers, and their overexpression can inhibit cancer cell proliferation, migration, invasion, and enhance apoptosis and autophagy. This raises several questions: How does the GARP complex influence autophagy? What mechanisms might be involved? What is the connection between GARP depletion and cancer development? Is this link related to abnormal glycosylation, or altered signaling pathways? Is this defect specific to certain tissues? Given the involvement of GARP in numerous cancers [86–90, 97], further investigation into the signaling pathways affected by GARP dysfunction is crucial.

Although initial studies focused on yeast and mammalian cell models, it is essential to explore GARP's role in various human tissues and across different species. Therefore, it is necessary to connect the dots and determine the GARP's role in Golgi physiology that eventually contributes to neurodevelopmental and other human disorders.

GARP plays a crucial role in cellular protein transport and is involved in various cellular processes, affecting the pathophysiology of several disorders. Despite extensive research over the past 25 years, much remains unknown about the GARP complex. What have we learned so far about GARP? Do we have a clear understanding of its specific functions? Can our current knowledge of GARP be applied to develop therapies for patients with GARP mutations? Comprehensive insights into GARP's functions are still lacking and to fully understand its role in cellular physiology, more detailed studies are essential.

References

1. Conibear E, Stevens TH (2000) Vps52p, Vps53p, and Vps54p form a novel multisubunit complex required for protein sorting at the yeast late Golgi. Mol Biol Cell 11(1):305–323
2. Conboy MJ, Cyert MS (2000) Luv1p/Rki1p/Tcs3p/Vps54p, a yeast protein that localizes to the late Golgi and early endosome, is required for normal vacuolar morphology. Mol Biol Cell 11(7):2429–2443
3. Conibear E, Cleck JN, Stevens TH (2003) Vps51p mediates the association of the GARP (Vps52/53/54) complex with the late Golgi t-SNARE Tlg1p. Mol Biol Cell 14(4):1610–1623
4. Reggiori F, Wang C-W, Stromhaug PE, Shintani T, Klionsky DJ (2003) Vps51 is part of the yeast Vps fifty-three tethering complex essential for retrograde traffic from the early endosome and Cvt vesicle completion. J Biol Chem 278(7):5009–5020
5. Siniossoglou S, Pelham HR (2002) Vps51p links the VFT complex to the SNARE Tlg1p. J Biol Chem 277(50):48318–48324
6. Oka T, Krieger M (2005) Multi-component protein complexes and Golgi membrane trafficking. J Biochem 137(2):109–114
7. Bröcker C, Engelbrecht-Vandré S, Ungermann C (2010) Multisubunit tethering complexes and their role in membrane fusion. Curr Biol 20(21):R943–R952
8. Koumandou VL, Dacks JB, Coulson RM, Field MC (2007) Control systems for membrane fusion in the ancestral eukaryote; evolution of tethering complexes and SM proteins. BMC Evol Biol 7:1–17
9. Bonifacino JS, Hierro A (2011) Transport according to GARP: receiving retrograde cargo at the trans-Golgi network. Trends Cell Biol 21(3):159–167
10. Schindler C, Chen Y, Pu J, Guo X, Bonifacino JS (2015) EARP is a multisubunit tethering complex involved in endocytic recycling. Nat Cell Biol 17(5):639–650
11. Topalidou I, Cattin-Ortolá J, Pappas AL, Cooper K, Merrihew GE, MacCoss MJ, Ailion M (2016) The EARP complex and its interactor EIPR-1 are required for cargo sorting to dense-core vesicles. PLoS Genet 12(5):e1006074
12. Pérez-Victoria FJ, Schindler C, Magadán JG, Mardones GA, Delevoye C, Romao M, Raposo G, Bonifacino JS (2010) Ang2/fat-free is a conserved subunit of the Golgi-associated retrograde protein complex. Mol Biol Cell 21(19):3386–3395
13. Chou H-T, Dukovski D, Chambers MG, Reinisch KM, Walz T (2016) CATCHR, HOPS and CORVET tethering complexes share a similar architecture. Nat Struct Mol Biol 23(8):761–763
14. Ha JY, Chou H-T, Ungar D, Yip CK, Walz T, Hughson FM (2016) Molecular architecture of the complete COG tethering complex. Nat Struct Mol Biol 23(8):758–760
15. Lepore DM, Martínez-Núñez L, Munson M (2018) Exposing the elusive exocyst structure. Trends Biochem Sci 43(9):714–725
16. Vasan N, Hutagalung A, Novick P, Reinisch KM (2010) Structure of a C-terminal fragment of its Vps53 subunit suggests similarity of Golgi-associated retrograde protein (GARP) complex to a family of tethering complexes. Proc Natl Acad Sci 107(32):14176–14181

17. Pei J, Ma C, Rizo J, Grishin NV (2009) Remote homology between Munc13 MUN domain and vesicle tethering complexes. J Mol Biol 391(3):509–517
18. Augustin I, Rosenmund C, Südhof TC, Brose N (1999) Munc13-1 is essential for fusion competence of glutamatergic synaptic vesicles. Nature 400(6743):457–461
19. Leitz J, Wang C, Esquivies L, Pfuetzner RA, Peters JJ, Couoh-Cardel S, Wang AL, Brunger AT (2024) Beyond the MUN domain, Munc13 controls priming and depriming of synaptic vesicles. Cell Rep 43(5):114026
20. Basu J, Shen N, Dulubova I, Lu J, Guan R, Guryev O, Grishin NV, Rosenmund C, Rizo J (2005) A minimal domain responsible for Munc13 activity. Nat Struct Mol Biol 12(11): 1017–1018
21. Pérez-Victoria FJ, Abascal-Palacios G, Tascón I, Kajava A, Magadán JG, Pioro EP, Bonifacino JS, Hierro A (2010) Structural basis for the wobbler mouse neurodegenerative disorder caused by mutation in the Vps54 subunit of the GARP complex. Proc Natl Acad Sci 107(29): 12860–12865
22. Whyte JR, Munro S (2001) The Sec34/35 Golgi transport complex is related to the exocyst, defining a family of complexes involved in multiple steps of membrane traffic. Dev Cell 1(4): 527–537
23. Pérez-Victoria FJ, Mardones GA, Bonifacino JS (2008) Requirement of the human GARP complex for mannose 6-phosphate-receptor-dependent sorting of cathepsin D to lysosomes. Mol Biol Cell 19(6):2350–2362
24. Liewen H, Meinhold-Heerlein I, Oliveira V, Schwarzenbacher R, Luo G, Wadle A, Jung M, Pfreundschuh M, Stenner-Liewen F (2005) Characterization of the human GARP (Golgi associated retrograde protein) complex. Exp Cell Res 306(1):24–34
25. Siniossoglou S, Pelham HR (2001) An effector of Ypt6p binds the SNARE Tlg1p and mediates selective fusion of vesicles with late Golgi membranes. EMBO J 20:5991–5998
26. Ishida M, Bonifacino JS (2019) ARFRP1 functions upstream of ARL1 and ARL5 to coordinate recruitment of distinct tethering factors to the trans-Golgi network. J Cell Biol 218(11): 3681–3696
27. Rosa-Ferreira C, Christis C, Torres IL, Munro S (2015) The small G protein Arl5 contributes to endosome-to-Golgi traffic by aiding the recruitment of the GARP complex to the Golgi. Biol Open 4(4):474–481
28. Fridmann-Sirkis Y, Kent HM, Lewis MJ, Evans PR, Pelham HR (2006) Structural analysis of the interaction between the SNARE Tlg1 and Vps51. Traffic 7(2):182–190
29. Pérez-Victoria FJ, Bonifacino JS (2009) Dual roles of the mammalian GARP complex in tethering and SNARE complex assembly at the trans-golgi network. Mol Cell Biol 29:5251–5263
30. Shestakova A, Suvorova E, Pavliv O, Khaidakova G, Lupashin V (2007) Interaction of the conserved oligomeric Golgi complex with t-SNARE Syntaxin5a/Sed5 enhances intra-Golgi SNARE complex stability. J Cell Biol 179(6):1179–1192
31. Willett R, Kudlyk T, Pokrovskaya I, Schönherr R, Ungar D, Duden R, Lupashin V (2013) COG complexes form spatial landmarks for distinct SNARE complexes. Nat Commun 4(1):1553
32. Laufman O, Kedan A, Hong W, Lev S (2009) Direct interaction between the COG complex and the SM protein, Sly1, is required for Golgi SNARE pairing. EMBO J 28(14):2006–2017
33. Chen Y-T, Wang I-H, Wang Y-H, Chiu W-Y, Hu J-H, Chen W-H, Lee F-JS (2019) Action of Arl1 GTPase and golgin Imh1 in Ypt6-independent retrograde transport from endosomes to the trans-Golgi network. Mol Biol Cell 30(8):1008–1019
34. Panic B, Whyte JR, Munro S (2003) The ARF-like GTPases Arl1p and Arl3p act in a pathway that interacts with vesicle-tethering factors at the Golgi apparatus. Curr Biol 13(5):405–410
35. Ibuchi K, Fukaya M, Shinohara T, Hara Y, Shiroshima T, Sugawara T, Sakagami H (2020) The Vps52 subunit of the GARP and EARP complexes is a novel Arf6-interacting protein that negatively regulates neurite outgrowth of hippocampal neurons. Brain Res 1745:146905

36. Wang Y-H, Chiu W-Y, Chen Y-T, Cai P-J, Wu Y-C, Wu J-L, Chen B-H, Liu Y-W, Yu C-J, Lee F-JS (2022) Golgin Imh1 and GARP complex cooperate to restore the impaired SNARE recycling transport induced by ER stress. Cell Rep 38(12):110488
37. Gershlick DC, Schindler C, Chen Y, Bonifacino JS (2016) TSSC1 is novel component of the endosomal retrieval machinery. Mol Biol Cell 27(18):2867–2878
38. Saimani U, Kim K (2017) Traffic from the endosome towards trans-Golgi network. Eur J Cell Biol 96(2):198–205
39. Lukehart J, Highfill C, Kim K (2013) Vps1, a recycling factor for the traffic from early endosome to the late Golgi. Biochem Cell Biol 91(6):455–465
40. Masschaele D, De Ceuninck L, Wauman J, Defever D, Stenner F, Lievens S, Peelman F, Tavernier J (2017) RNF41 interacts with the VPS52 subunit of the GARP and EARP complexes. PLoS One 12(5):e0178132
41. Zach S, Felk S, Gillardon F (2010) Signal transduction protein array analysis links LRRK2 to Ste20 kinases and PKC zeta that modulate neuronal plasticity. PLoS One 5(10):e13191
42. Gomez-Suaga P, Luzon-Toro B, Churamani D, Zhang L, Bloor-Young D, Patel S, Woodman PG, Churchill GC, Hilfiker S (2012) Leucine-rich repeat kinase 2 regulates autophagy through a calcium-dependent pathway involving NAADP. Hum Mol Genet 21(3):511–525
43. Penney J, Tsurudome K, Liao EH, Kauwe G, Gray L, Yanagiya A, Calderon MR, Sonenberg N, Haghighi AP (2016) LRRK2 regulates retrograde synaptic compensation at the Drosophila neuromuscular junction. Nat Commun 7(1):12188
44. Beilina A, Bonet-Ponce L, Kumaran R, Kordich JJ, Ishida M, Mamais A, Kaganovich A, Saez-Atienzar S, Gershlick DC, Roosen DA (2020) The Parkinson's disease protein LRRK2 interacts with the GARP complex to promote retrograde transport to the trans-Golgi network. Cell Rep 31(5):107614
45. Fabrizio JJ, Hime G, Lemmon SK, Bazinet C (1998) Genetic dissection of sperm individualization in Drosophila melanogaster. Development 125(10):1833–1843
46. Wilkinson EC, Starke EL, Barbee SA (2021) Vps54 regulates lifespan and locomotor behavior in adult Drosophila melanogaster. Front Genet 12:762012
47. Schmitt-John T, Drepper C, Mußmann A, Hahn P, Kuhlmann M, Thiel C, Hafner M, Lengeling A, Heimann P, Jones JM (2005) Mutation of Vps54 causes motor neuron disease and defective spermiogenesis in the wobbler mouse. Nat Genet 37(11):1213–1215
48. Khakurel A, Kudlyk T, Bonifacino JS, Lupashin VV (2021) The Golgi-associated retrograde protein (GARP) complex plays an essential role in the maintenance of the Golgi glycosylation machinery. Mol Biol Cell 32(17):1594–1610
49. Sumya FT, Pokrovskaya ID, Lupashin VV (2022) Rapid COG depletion in mammalian cell by auxin-inducible degradation system. In: Golgi: methods and protocols. Springer, pp 365–390
50. Peer M, Yuan H, Zhang Y, Korbula K, Novick P, Dong G (2022) Double NPY motifs at the N-terminus of the yeast t-SNARE Sso2 synergistically bind Sec3 to promote membrane fusion. Elife 11:e82041
51. Travis SM, DAmico K, Yu I-M, McMahon C, Hamid S, Ramirez-Arellano G, Jeffrey PD, Hughson FM (2020) Structural basis for the binding of SNAREs to the multisubunit tethering complex Dsl1. J Biol Chem 295(30):10125–10135
52. Torng T, Wickner W (2021) Phosphatidylinositol and phosphatidylinositol-3-phosphate activate HOPS to catalyze SNARE assembly, allowing small headgroup lipids to support the terminal steps of membrane fusion. Mol Biol Cell 32(21):ar19
53. Hong W, Lev S (2014) Tethering the assembly of SNARE complexes. Trends Cell Biol 24(1):35–43
54. Khakurel A, Kudlyk T, Pokrovskaya I, D'Souza Z, Lupashin VV (2022) GARP dysfunction results in COPI displacement, depletion of Golgi v-SNAREs and calcium homeostasis proteins. Front Cell Dev Biol 10:1066504
55. Smith RD, Lupashin VV (2008) Role of the conserved oligomeric Golgi (COG) complex in protein glycosylation. Carbohydr Res 343(12):2024–2031

56. Brasil S, Pascoal C, Francisco R, Marques-da-Silva D, Andreotti G, Videira PA, Morava E, Jaeken J, dos Reis Ferreira V (2018) CDG therapies: from bench to bedside. Int J Mol Sci 19(5): 1304
57. Stanley P (2011) Golgi glycosylation. Cold Spring Harb Perspect Biol 3(4):a005199
58. Gershlick DC, Ishida M, Jones JR, Bellomo A, Bonifacino JS, Everman DB (2019) A neurodevelopmental disorder caused by mutations in the VPS51 subunit of the GARP and EARP complexes. Hum Mol Genet 28(9):1548–1560
59. Khakurel A, Kudlyk T, Lupashin VV (2022) Generation and analysis of hTERT-RPE1 VPS54 knock-out and rescued cell lines. In: Golgi: methods and protocols. Springer, pp 349–364
60. Dell'Angelica EC, Bonifacino JS (2019) Coatopathies: genetic disorders of protein coats. Annu Rev Cell Dev Biol 35(1):131–168
61. Fröhlich F, Petit C, Kory N, Christiano R, Hannibal-Bach H-K, Graham M, Liu X, Ejsing CS, Farese JRV, Walther TC (2015) The GARP complex is required for cellular sphingolipid homeostasis. elife 4:e08712
62. Sakuragi T, Nagata S (2023) Regulation of phospholipid distribution in the lipid bilayer by flippases and scramblases. Nat Rev Mol Cell Biol 24(8):576–596
63. Takagi K, Iwamoto K, Kobayashi S, Horiuchi H, Fukuda R, Ohta A (2012) Involvement of Golgi-associated retrograde protein complex in the recycling of the putative Dnf aminophospholipid flippases in yeast. Biochem Biophys Res Commun 417(1):490–494
64. Eising S, Thiele L, Fröhlich F (2019) A systematic approach to identify recycling endocytic cargo depending on the GARP complex. Elife 8:e42837
65. O'Brien CE, Younger SH, Jan LY, Jan YN (2023) The GARP complex prevents sterol accumulation at the trans-Golgi network during dendrite remodeling. J Cell Biol 222(1): e202112108
66. Hossain S, Robbins N, Cowen LE (2022) The GARP complex is required for filamentation in Candida albicans. Genetics 222(4):iyac152
67. Hossain S, Veri AO, Liu Z, Iyer KR, O'Meara TR, Robbins N, Cowen LE (2021) Mitochondrial perturbation reduces susceptibility to xenobiotics through altered efflux in Candida albicans. Genetics 219(2):iyab095
68. Wei J, Zhang Y-Y, Luo J, Wang J-Q, Zhou Y-X, Miao H-H, Shi X-J, Qu Y-X, Xu J, Li B-L (2017) The GARP complex is involved in intracellular cholesterol transport via targeting NPC2 to lysosomes. Cell Rep 19(13):2823–2835
69. Homma Y, Fukuda M (2021) Knockout analysis of Rab6 effector proteins revealed the role of VPS52 in the secretory pathway. Biochem Biophys Res Commun 561:151–157
70. Hirata T, Fujita M, Nakamura S, Gotoh K, Motooka D, Murakami Y, Maeda Y, Kinoshita T (2015) Post-Golgi anterograde transport requires GARP-dependent endosome-to-TGN retrograde transport. Mol Biol Cell 26(17):3071–3084
71. Realegeno S, Priyamvada L, Kumar A, Blackburn JB, Hartloge C, Puschnik AS, Sambhara S, Olson VA, Carette JE, Lupashin V (2020) Conserved Oligomeric Golgi (COG) complex proteins facilitate orthopoxvirus entry, fusion and spread. Viruses 12(7):707
72. Realegeno S, Puschnik AS, Kumar A, Goldsmith C, Burgado J, Sambhara S, Olson VA, Carroll D, Damon I, Hirata T (2017) Monkeypox virus host factor screen using haploid cells identifies essential role of GARP complex in extracellular virus formation. J Virol 91(11). https://doi.org/10.1128/jvi.00011-00017
73. Boillée S, Peschanski M, Junier M-P (2003) The wobbler mouse: a neurodegeneration jigsaw puzzle. Mol Neurobiol 28:65–106
74. Bird MT, Shuttleworth E, Koestner A, Reinglass J (1971) The wobbler mouse mutant: an animal model of hereditary motor system disease. Acta Neuropathol 19:39–50
75. Kaupmann K, Simon-Chazottes D, Guénet J-L, Jockusch H (1992) Wobbler, a mutation affecting motoneuron survival and gonadal functions in the mouse, maps to proximal chromosome 11. Genomics 13(1):39–43
76. Moser JM, Bigini P, Schmitt-John T (2013) The wobbler mouse, an ALS animal model. Mol Gen Genomics 288:207–229

77. Murphy S, Schmitt-John T, Dowling P, Henry M, Meleady P, Swandulla D, Ohlendieck K (2023) Proteomic profiling of the brain from the wobbler mouse model of amyotrophic lateral sclerosis reveals elevated levels of the astrogliosis marker glial fibrillary acidic protein. Eur J Transl Myol 33(3):11555
78. Jockusch H, Holland A, Staunton L, Schmitt-John T, Heimann P, Dowling P, Ohlendieck K (2014) Pathoproteomics of testicular tissue deficient in the GARP component VPS54: the wobbler mouse model of globozoospermia. Proteomics 14(7–8):839–852
79. Uwineza A, Caberg J-H, Hitayezu J, Wenric S, Mutesa L, Vial Y, Drunat S, Passemard S, Verloes A, El Ghouzzi V (2019) VPS51 biallelic variants cause microcephaly with brain malformations: a confirmatory report. Eur J Med Genet 62(8):103704
80. Feinstein M, Flusser H, Lerman-Sagie T, Ben-Zeev B, Lev D, Agamy O, Cohen I, Kadir R, Sivan S, Leshinsky-Silver E (2014) VPS53 mutations cause progressive cerebello-cerebral atrophy type 2 (PCCA2). J Med Genet 51(5):303–308
81. Hausman-Kedem M, Ben-Shachar S, Menascu S, Geva K, Sagie L, Fattal-Valevski A (2019) VPS53 gene is associated with a new phenotype of complicated hereditary spastic paraparesis. Neurogenetics 20:187–195
82. Pahari S, Cormark RD, Blackshaw MT, Liu C, Erickson JL, Schultz EA (2014) Arabidopsis UNHINGED encodes a VPS51 homolog and reveals a role for the GARP complex in leaf shape and vein patterning. Development 141(9):1894–1905
83. Rodriguez PA, Escudero-Martinez C, Bos JI (2017) An aphid effector targets trafficking protein VPS52 in a host-specific manner to promote virulence. Plant Physiol 173(3):1892–1903
84. Csizmadia T, Dósa A, Farkas E, Csikos BV, Kriska EA, Juhász G, Lőw P (2022) Developmental program-independent secretory granule degradation in larval salivary gland cells of Drosophila. Traffic 23(12):568–586
85. Zhang J, Lin Y, Hu X, Wu Z, Guo W (2017) VPS52 induces apoptosis via cathepsin D in gastric cancer. J Mol Med 95:1107–1116
86. Peng H, Zheng J, Su Q, Feng X, Peng M, Gong L, Wu H, Pan X (2020) VPS53 suppresses malignant properties in colorectal cancer by inducing the autophagy signaling pathway. OncoTargets Ther 13:10667–10675
87. Kim TE, Kim YW, Hwang SY, Shin SM, Shin JW, Lee YH, Shin SY, Han KT, Lee JM, Namkoong SE (2002) Candidate tumor suppressor, HCCS-1, is downregulated in human cancers and induces apoptosis in cervical cancer. Int J Cancer 97(6):780–786
88. Shi L, Zhao S-m, Luo Y, Zhang A-w, Wei L-h, Xie Z-y, Li Y-y, Ma W (2017) MiR-375: a prospective regulator in medullary thyroid cancer based on microarray data and bioinformatics analyses. Pathol Res Pract 213(11):1344–1354
89. Rebbeck TR (2017) Prostate cancer genetics: variation by race, ethnicity, and geography. Semin Radiat Oncol 27(1):3–10
90. Goh CL, Eeles RA (2014) Germline genetic variants associated with prostate cancer and potential relevance to clinical practice. Recent Results Cancer Res 202:9–26
91. Hirst J, Futter CE, Hopkins CR (1998) The kinetics of mannose 6-phosphate receptor trafficking in the endocytic pathway in HEp-2 cells: the receptor enters and rapidly leaves multivesicular endosomes without accumulating in a prelysosomal compartment. Mol Biol Cell 9(4):809–816
92. Lin SX, Mallet WG, Huang AY, Maxfield FR (2004) Endocytosed cation-independent mannose 6-phosphate receptor traffics via the endocytic recycling compartment en route to the trans-Golgi network and a subpopulation of late endosomes. Mol Biol Cell 15(2):721–733
93. Horn N, Wittung-Stafshede P (2021) ATP7A-regulated enzyme metalation and trafficking in the menkes disease puzzle. Biomedicine 9(4):391
94. Natarajan R, Linstedt AD (2004) A cycling cis-Golgi protein mediates endosome-to-Golgi traffic. Mol Biol Cell 15(11):4798–4806
95. Buser DP, Spang A (2023) Protein sorting from endosomes to the TGN. Front Cell Dev Biol 11: 1140605
96. Makaraci P, Kim K (2018) Trans-Golgi network-bound cargo traffic. Eur J Cell Biol 97(3): 137–149

97. Zhang Z, Liu F, Xu Y, Huang H, Zou Y, Yang B, Luo Y, Zhang Q, Xiong A, Wang L (2018) GZFLW induces apoptosis of ectopic endometrial stromal cells via promoting VPS53 protein stability. Evid Based Complement Alternat Med 2018(1):1293630
98. Mirdita M, Schütze K, Moriwaki Y, Heo L, Ovchinnikov S, Steinegger M (2022) ColabFold: making protein folding accessible to all. Nat Methods 19(6):679–682
99. Pettersen EF, Goddard TD, Huang CC, Meng EC, Couch GS, Croll TI, Morris JH, Ferrin TE (2021) UCSF ChimeraX: structure visualization for researchers, educators, and developers. Protein Sci 30(1):70–82

Chapter 7
COG Complex in Golgi Trafficking and Glycosylation

Farhana Taher Sumya and Vladimir V. Lupashin

Abstract The Conserved Oligomeric Golgi (COG) complex, an evolutionary conserved octameric vesicular tether, is essential for maintaining Golgi function by ensuring accurate delivery of resident proteins to their specific locations. Mutations in human COG subunits result in severe multi-systemic diseases known as COG-Congenital Disorders of Glycosylation (COG-CDG). This review explores the current knowledge of COG complex structure, its dynamic behavior, interactions with partner proteins, and proposed models of its cellular functions. Furthermore, we will discuss the pathological implications of mutations in COG complex subunits, as observed in model organisms and human patients.

Keywords COG · Golgi · Vesicle tethering · SNARE · Congenital disorders of glycosylation

COG Structure

The individual COG subunits have been identified through a combination of biochemical and genetic approaches in mammals and yeast [148, 157]. COG is composed of eight subunits named COG1-COG8, each encoded by different genes. The COG complex is evolutionarily conserved and has been found in all eukaryotic species [74, 126]. Moreover, four COG subunits (COG3, COG4, COG6, and COG8) show more significant evolutionary conservation, as compared to the other four subunits [12, 44, 108, 158]. The total molecular weight of the complex varies from 590 to 750 kDa, depending on the species [74, 88]. The COG complex is divided into two or more stable subcomplexes. The subunit connectivity of COG has been described as a bi-lobed structure initially observed in electron microscopy (EM) studies of purified cross-linked bovine brain COG complex (Fig. 7.1). There

F. T. Sumya · V. V. Lupashin (✉)
Department of Physiology and Cell Biology, University of Arkansas for Medical Sciences, Little Rock, AR, USA
e-mail: vvlupashin@uams.edu

A. Nakano, J. Saraste (eds.), *The Golgi Network, Volume II*, Subcellular Biochemistry 111, https://doi.org/10.1007/978-3-032-16833-7_7

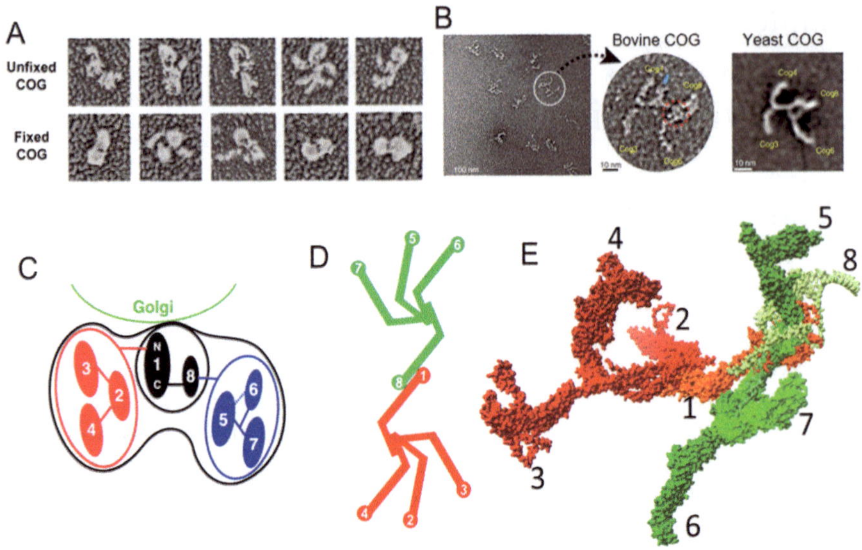

Fig. 7.1 COG structure. (**a**) Rotary-shadowed cryoelectron micrographs of purified bovine brain COG complex either untreated (top row) or after a fixation with glutardehyde (bottom row). Modified from [148]. (**b**) An image field of purified bovine COG, as well as a single particle, is shown at higher magnification. On the right, for comparison, is a representative class average for yeast COG; scale bar, 10 nm [59]. (**c**) 3-lobed model and subunit interaction map of mammalian complex [149]. (**d**) 2-lobed model of COG subunit interactions. Lobe A is shown in shades of red, and lobe B in shades of green [12]. (**e**) AlphaFold2 structure prediction of human COG complex. Note a similarity between AlfaFold and a single-particle EM image of an isolated COG complex in (**b**)

are several models for the intra-complex COG subunit connectivity. The "two-lobed" model postulates that COG subunits are grouped into two four-subunit subcomplexes named lobe A (COG1-4) and lobe B (COG5-8) [12, 158]. Lobes A and B are bridged through COG1-COG8 interactions [44, 158]. Another model postulates that the COG2/3/4 and COG5/6/7 subunits form two trimers, linked by a COG1–COG8 dimer [46, 149].

EM studies have revealed two primary conformations of the COG complex: one observed with purified, unfixed protein and the other following mild fixation with paraformaldehyde. The unfixed COG complex exhibits an extended and seemingly flexible structure, characterized by multiple elongated, curved arms with globular or hook-like ends, measuring approximately 50–75 nm. In contrast, the fixed COG complex appears more globular, about 37 nm in length, with rod-like connections between the two lobes (Fig. 7.1). It is currently unclear whether these two structural variants represent naturally occurring 'open' and 'closed' COG conformations. Observation of COG-related GARP [22] and exocyst [100] complexes in both open and closed conformations support the natural transition between the two forms. It is predicted that inside subcomplexes, COG subunits are connected by their N-terminal domains. In support of this model, the COG5–COG7 crystal structure demonstrates evolutionary conserved inter-subunit interaction through the formation of a rod-like structure [60].

There is a distant sequence homology between the subunits of the COG, exocyst, Dsl1, and GARP (Golgi-associated retrograde protein) complexes. Based on structural similarity, these complexes have been grouped into the CATCHR (Complexes Associated with Tethering Containing Helical Rods) subfamily of tethering complexes [88]. Despite structural and functional homology, the number of subunits in CATCHR complexes varies significantly. For example, the Dsl1/NRZ complex contains only three subunits [135]. The GARP is a tetramer comprising four different subunits [70]. Meanwhile, the COG and exocyst complexes each consist of an octamer, with two tetramers (COG1-4 and COG5-8, and CorEx1 and CorEx2) [78, 157]. The N-terminal domains of CATCHR subunits share similar primary sequences, while the central and C-terminal domains predominantly exhibit structural similarity by forming alpha-helical bundles [121, 152, 157]. Crystal structure information about individual COG subunits is limited, with details available for yeast COG2 [17], N-terminal domains of yeast COG5 and COG7 [60], and the C-terminal domain of human COG4 [121]. The X-ray structures indicate that the central and C-terminal parts of COG subunits adopt a CATCHR fold, that consists of a sequence of α-helical bundles, similar to the Dsl1 (Dsl1, Tip20), exocyst (Sec6, Sec15, Exo70, and Exo84), and GARP (Vps53, Vps54) complexes [17, 32, 103, 113, 119, 121, 130, 147, 152, 164]. In addition, The COG and several exocyst and GARP subunits (EXOC3, EXOC6, EXOC7, and VPS53) share a structural similarity with the MUN domain of a synaptic vesicle docking component UNC13 [92, 112]. The emerging structural similarities among CATCHR tethers provide strong evidence of a common evolutionary origin and may reflect shared mechanisms of action [121].

Single-particle EM of recombinant lobe A with GFP fiducial markers reveals that the N-termini of *S. cerevisiae* COG1, COG2, COG3, and COG4 wind around a proximal segment of one of its three legs. The EM data supports a model in which the N-terminal α-helices of COG1 and COG2 (aligned in one direction) combine with those of COG3 and COG4 (aligned in the opposite direction) to form a coiled-coil bundle. These studies also confirm that COG8 binds to the C-terminus of COG1, and in the absence of COG8, the C-terminus of COG1 becomes disordered [88].

Subsequent negative-stain EM and class averaging of lobe B purified from *Kluyveromyces lactis* reveal that the COG5-8 complex has an overall rod-like shape, measuring 27 ± 2 nm in length, with varying degrees of curvature. Additional GFP labeling shows that the two ends of the rod-like COG5-8 complex correspond to the C-termini of the two largest subunits, COG6 and COG8. The overall structure of the yeast COG complex is notable for its lack of compactness, featuring multiple (4–5) legs, each with reproducible curvature. The distance between the distal tips of COG3 and COG8 ranges from 7 to 38 nm, indicating significant flexibility in the complex's structure.

A similar open conformation has been observed in the COG complex purified from bovine brain. In mammals, the COG complex has additional legs, likely representing COG2, COG5, and COG7, which are larger in the mammalian complex. The authors conclude that the extended structure and apparent flexibility of the COG complex are well-suited to its role in vesicle capture and SNARE assembly [59].

COG Interaction with Membranes and Its Intracellular Localization

COG subunits do not have any predicted transmembrane or lipid-binding domains. Upon cell disruption, a significant fraction of COG is found to be soluble [71, 148]. This has been initially interpreted as evidence for the transient association of COG with membranes. At the same time, FRAP (Fluorescence Recovery after Photobleaching) experiments reveal that the on/off cycling of COG subunits from the Golgi membrane is very slow, and the rate of recovery is like the recovery of Golgi SNAREs [160]. Additionally, artificially membrane-glued COG4 and COG7 are found to be fully functional [25], and similar functionality has been observed for the membrane-attached exocyst subunits Sec3p and Exo70p [93]. This data indicates that COG is attached most of the time to Golgi membranes and/or transport intermediates via multipronged specific and highly regulated protein-protein interactions [160]. Several COG subunits are shown to interact with Golgi-localized Rabs, CCTs, and transmembrane SNAREs [85, 128, 132, 133, 160], but it is still unclear which single interaction is critical for COG membrane attachment.

COG complex subunits are primarily located at or near the tips and edges of the Golgi cisternae and on vesiculo-tubular structures found on both the *cis* and *trans* faces of the cisternal Golgi stacks. In some instances, they are found on vesicles containing COPI [153]. Immuno-EM and live cell super-resolution microscopy studies revealed that lobe A predominantly associates with the Golgi membrane and can be found on every cisterna [153], while lobe B preferentially interacts with vesicular membranes [158]. Interestingly, depletion of any lobe A subunits in human cells compromises lobe B association with the Golgi region [158, 178], indicating that Golgi localization of the lobe B subcomplex is determined by the functional lobe A subcomplex.

Interaction Partners of the COG Complex

COG acts as the primary CATCHR vesicle tethering complex at the Golgi. The COG complex interacts with various trafficking facilitators throughout the Golgi, including SNAREs, SNARE-interacting proteins, Rabs, coiled-coil tethers (CCT), and vesicular coats (Fig. 7.2). This network of interactions suggests that COG coordinates sequential processes triggered by activated Rab proteins, starting with vesicle attachment and ending with the assembly of SNARE complexes. Based on

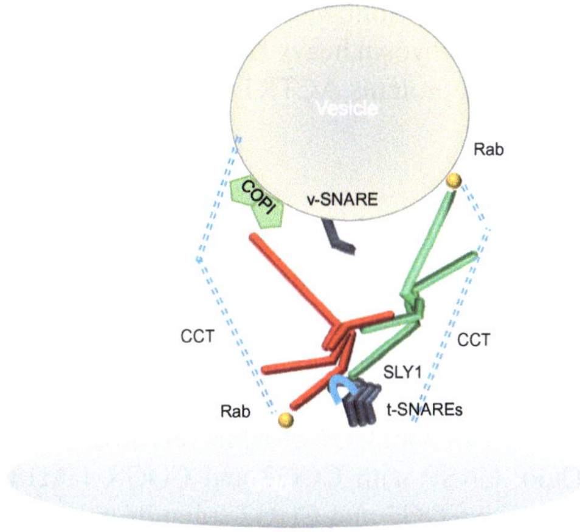

Fig. 7.2 Hypothetical model of the COG complex interacting with vesicle fusion and fission machinery. The schematic depicts vesicle tethering by the COG complex and its interactions with the Golgi trafficking machinery. Lobe A (red) primarily interacts with the COPI vesicular coat, CCT, Rabs, t-SNAREs, and the SNARE-associated protein SLY1/SCFD1, while lobe B (green) binds to Rabs, CCT, and SNAREs

protein-protein interaction network, different COG subunits could be designated as specific "interaction hubs": COG2 and COG7 act as "tether hubs," COG2, COG3, and COG4 act as "coat and motor hubs," COG4, COG5, and COG6 act as "Rab hubs," and COG4, COG6, and COG8 act as "SNARE hubs" [35, 71, 107, 117, 144, 150, 178].

COG Interaction with Vesicular Coats and Motor Proteins

Both yeast and human COG complexes co-immunoprecipitate (co-IP) with the COPI vesicular coat [144, 160]. COG4, being the only lobe A subunit that interacts with multiple SNARE and Rab proteins, also acts as a "coat hub" due to its potential interactions with the Gea2p (a yeast protein that facilitates guanine nucleotide exchange for the small GTPase ARF1p, regulating the formation of COPI-coated vesicles) and with clathrin heavy chain [19, 81]. Additional interaction with the clathrin coat has been observed for COG3, which was co-isolated with the clathrin adaptor protein AP3B1 [81]. A comprehensive yeast two-hybrid (Y2H) analysis of human COG-COPI interactions reveals that human βCOP/COPB1 can directly interact with three COG subunits: COG2, COG5, and COG8, with COG2 being the most vital βCOP partner [102, 161].

In addition to vesicular coats, COG is likely to interact with motor proteins. COG2 may function as a minor hub for kinesin motors [36, 91, 161]. Supporting the link between COG and kinesins, the depletion of two minus-end kinesins, KIFC3 and KIF25, suppresses Golgi fragmentation induced by COG3 depletion

[94]. Affinity capture of COG3 followed by mass spectrometry analysis reveals potential interactions with the myosin heavy (MYH14) and light (MYL12) chains, as well as with the actin-related proteins ACTR1A and ARP66B [55, 81].

COG Interactions with Coiled-Coil Tethers

COG-CCT interactions are believed to play a role in vesicle capture, facilitating SNARE assembly, and promoting membrane fusion [102, 132, 133, 160]. Original yeast studies reported a genetic interaction between *cis*-Golgi CCT USO1 and COG [150, 151]. Later, Y2H screen with CCT proteins revealed that TMF1 could interact with COG2 and COG6, CASP with COG2 and COG8, GM130 with COG2 and COG5, and Golgin-84 with COG7 and COG2 subunits respectively [102].

In yeasts, COG and USO1 mutants display synthetic lethality [151]. In mammalian cells, p115/USO1 binds to purified COG complex [160] and specifically to COG2. Mapping of COG2-p115 interaction reveals that HR2 domain of p115 (aa 200–247) and a 56-residue region of COG2 (residues 613–669) are sufficient for binding. Furthermore, the disruption of the COG2-p115 interaction leads to Golgi structural defects, strongly suggesting that this interaction plays a crucial role in Golgi biogenesis and maintenance [133].

GOLGA5/golgin-84 is a transmembrane CCT located on Golgi cisternae and on vesicles that accumulate in response to COG complex depletion [132, 141]. The COG complex co-IP with golgin-84, with the interaction being mediated by COG7 binding to the middle portion (amino acids 340–486) of CCT [132].

TMF1 is a CCT with reported *trans*-Golgi functions but a localization throughout the Golgi stack [38]. Y2H analysis shows that both COG2 and COG6 interacted with the two C-terminal TMF1 coils. A complete COG complex has two separate binding sites on TMF1: one in the C-terminal region and another in the N-terminal region of CCT [102].

Importantly, COG-CCT interactions are not unique and are observed for other members of the CATCHR family. Yeast exocyst interacts with long coiled-coil tether Sec2p, which is recruited to the Ypt31-positive Golgi compartment to function as a GEF for the Sec4p Rab GTPase [54]. GARP interacts with CCT Imh1p [156] via its Vps53 subunit, indicating that CATCHR-CCT interactions are universal protein-protein interactions that are likely to regulate the early stages of vesicle tethering.

COG Interaction with Rab GTPases

The majority of vesicular tethers are peripheral membrane proteins whose attachment to membranes is tightly regulated by small GTP-binding proteins, particularly Rab GTPases [63]. In the yeast cell, COG complex binds to Ypt1p and Ypt6p, homologs of mammalian RAB1 and RAB6. respectively [144]. Y2H and co-IP tests with mammalian proteins have identified eight COG-Rab interactions: COG4 with RAB30; COG5 with RAB2A and RAB39; and COG6 with RAB1A, 1B, 6A, 6B, and 41 [49, 102]. All observed interactions are with Rabs in GTP-bound form, indicating that COG acts as Rab effector. The mitochondrial relocalization assays have confirmed the interactions between COG5 and RAB2A, as well as between COG6 and RAB1A, RAB2A, RAB4A, and RAB6A. The molecular mechanism of COG-Rab interaction remains unclear, but interestingly, COG4-RAB30 interaction occurs via the N-terminal domain of COG4, which is near or overlaps with the mapped interaction regions with SCFD1 and STX5 (see below). This suggests potential coordination between COG-Rab and COG-SNARE interactions.

CATCHR-Rab interaction is a common theme in membrane trafficking. The exocyst complex is thought to be recruited to membranes by small GTPases [57, 176]. Recent *in vitro* assay has revealed that the exocyst can tether vesicles with a functional Rab GTPase, Sec4p, to vesicles lacking a functional Rab [122]. Moreover, the interaction between EXOC6/SEC15 and RAB33B contributes to cell migration [11], and the EXOC6-RAB11 interaction is involved in regulating lysosome exocytosis [40]. Regarding GARP, it has been reported that yeast Vps52p interacts with the GTP-bound form of Ypt6p. Similarly, in mammalian cells, RAB6 is found to pull down VPS52, suggesting an interaction between RAB6 and the GARP complex [62]. Additionally, RAB4B is implicated in VPS52-dependent endosome to TGN retrograde trafficking [53]. Interestingly, no Rab interaction has been reported for the Dsl1p complex [20].

COG Interactions with SNAREs and SM Proteins

Interactions between COG and SNAREs have been thoroughly established through *in vivo* and *in vitro* studies [34, 83, 87, 132, 144, 159, 178]. The COG complex is predicted to tether COPI-formed intra-Golgi carriers, which carry GS28/GOSR1 and GS15/BET1L v-SNAREs [141, 158, 178]. Indeed, the COG complex not only maintains the steady-state abundance and Golgi localization of GS28 and GS15 [107, 141], but is also necessary for the formation of intra-Golgi STX5/GS28/GS15/YKT6 SNARE complex [87, 128]. COG deficiency also impairs the formation of the *trans*-Golgi STX16/VTI1A/STX6/VAMP4 SNARE complex [85], highlighting the crucial role of the COG complex in all aspects of Golgi trafficking. Moreover, artificially anchoring COG4 and COG8 to the mitochondrial membrane is sufficient to reroute vesicles carrying STX5 and STX16, respectively, to COG-decorated mitochondria [159].

COG directly interacts with multiple Golgi SNAREs, and while the precise molecular details of these interactions remain unclear, some insights are beginning to emerge. COG4 directly binds to Qa SNARE STX5 via its N-terminal domain (aa 84–153) [87, 128]. In vitro COG-STX5 binding is relatively strong [160] and likely to be heavily regulated in vivo. One of the potential regulators is the STX5 partner protein SLY1/SCFD1, which binds to the N-terminal 84 amino acid domain of COG4. Disruption of the COG4-SLY1 interaction impairs the pairing of SNAREs involved in intra-Golgi transport, significantly affecting the stability of v-SNAREs and Golgi sensitivity to BFA treatment [87].

Y2H screen reveals that three COG subunits could directly interact with five Golgi-localized SNAREs [86, 159, 161]. In addition to COG4-STX5 interaction, COG6 can interact with STX5, STX6, GS27/GOSR2, and SNAP29, and COG8 can interact with STX5, GS27, and STX16. These COG-SNARE interactions are confirmed by a transient over-expression model [86, 87, 159], and with endogenous Golgi SNAREs [34, 128, 144, 160].

Interaction with SNAREs has been found for all other members of the CATCHR family and is likely to represent the most fundamental interaction between tethers and other components of fusion machinery. Exocyst subunit Sec3p facilitates the interaction between Q-SNAREs Sso2p and Sec9p, forming the binary t-SNARE complex and promoting subsequent membrane fusion. Another exocyst subunit Sec6p binds to Sec9p, facilitating formation of the Sec9p/Sso1p binary and Sec9p/Sso1p/Snc2p ternary SNARE complexes [99, 171].

In the case of GARP, Vps51p interacts with Qc-SNARE Tlg1p through the amino-terminal regulatory Habc domain of Tlg1p. In human cells, the GARP complex interacts with the STX16 TGN SNARE complex (STX16/STX6/VTI1A/VAMP4). These SNARE–GARP interactions rely on the SNARE motif in Golgi SNARE proteins and the N-terminal regions of VPS53 and VPS54 [70, 114].

Dsl1 complex is localized to the ER membrane through interactions with ER Q-SNAREs Ufe1/STX18, Use1p/USE1, and Sec20p/BNIP1 [80, 119]. Additionally, Dsl1 can bind to the R-SNAREs Sec22p and Ykt6p [101]. Recent cryo-EM structure of yeast Dsl1 complex bound to Use1p and Sec20p showed that N-terminal domains of the SNAREs form an integral part of the structure, stabilizing a Dsl1 complex configuration with unexpected similarities to the exocyst. Compromising the assembly of the Dsl1-SNARE complex is lethal in yeast, suggesting that the Dsl1 and Q-SNAREs function in intimate collaboration to help orchestrate vesicle capture, SNARE assembly, and membrane fusion [30]. The authors conclude that the structure of the SNARE-anchored Dsl1 complex reveals a fundamental principle of CATCHR function, suggesting that tethers and SNAREs can operate together as a single, integrated machine.

Functional Defects in COG-Deficient Cells

The COG complex has been studied in various organisms, ranging from single-celled *Saccharomyces cerevisiae* to more complex model organisms such as *Arabidopsis thaliana, Drosophila melanogaster, Caenorhabditis elegans*, mammalian, and human cells.

Yeast

Yeast COG mutants have been identified in several genetic screens for temperature-sensitive (ts) mutants that exhibit impairment in trafficking and glycosylation. At the restrictive temperature, mutants *sec34-1* (*cog3*), *sec35-1* (*cog2*), *tfi1/sec36/cod3* (*cog1*), *tfi2/cod2* (*cog6*), and *tfi3/cod1* (*cog4*) demonstrate deficiencies in N- and O-linked protein glycosylation [117, 136, 144, 157, 166]. Additionally, *cog1* mutant also amasses multiple membrane structures [117] and secretes vacuolar and ER resident proteins [144], indicating trafficking and sorting problems. The mutations in yeast COG also result in alterations in secretion [144, 166], and mislocalization of SNAREs Snc1p [157], Sec22p [144], and Bos1p [144]. Interestingly, when certain regulators of membrane trafficking are overexpressed—including Rab Ypt1p, SLY1 mutant Sly1pE532K/Sly1-20 [144, 151], the SNAREs Ykt6p, Bet1p, Sec22p, and the CCT Uso1p [151]—cell growth problems caused by the mutant COG are alleviated. This indicates that some functions of the COG complex in membrane trafficking can be compensated by the overproduction of other components of the Golgi docking-fusion machinery, suggesting that COG plays a catalytic and/or proofreading role in the vesicle tethering and fusion process.

A combination of COG mutations with mutations in COPI subunits [117, 144], USO1 [151], an αSNAP SEC17 [117], SLY1 [117], or a t-SNARE SED5 [144] result in synthetic lethality, indicating their involvement in the same intracellular process. Additional interactions between COG and Golgi Rabs Ypt1p and Ypt6p [144], and CCT Coy1p [5] are found in co-precipitation studies.

Deletions of lobe A subunits COG2, 3, and 4 are lethal, while COG1 knockout has a severe growth defect. Although lobe B deletions lack visible growth and structural defects, all COG mutants exhibit severe N- and O-linked glycosylation defects [144, 157]. CPY glycosylation defects in COG ts mutants were initially interpreted as a COG requirement in ER-Golgi trafficking [150, 151], but the lack of genetic interactions between COG and COPII components and data from higher eukaryotes [36] does not support COG involvement in the ER-Golgi trafficking step. Importantly, supporting COG's role in vesicle tethering, EM studies reveal a massive accumulation of 60-nm vesicles in COG ts cells [166].

Taking together, yeast cells exhibit a requirement for a functional COG complex to enable proper glycosylation, intra-Golgi trafficking, and protein sorting. Genetic studies suggest that the yeast COG complex is involved in intra-Golgi vesicular trafficking through its interactions with Golgi Rabs, CCTs, SNAREs, and the COPI coat.

Plants

Several studies demonstrate that the COG complex plays a crucial role in maintaining cell wall integrity and plant development. So far, COG studies have been conducted with *Arabidopsis thaliana* mutants of COG3, COG6, COG7, and COG8, which were generated through T-DNA insertions [21, 31, 67, 145]. Mutations in COG3, COG6, and COG8 led to defects in the secretion of cell wall components during pollen tube growth, resulting in slow and stunted growth and causing sterility in male plants. The *embryo yellow* (eye) mutation in COG7 leads to the misplacement of Golgi proteins and changes in Golgi's structure [67]. COG mutants mislocalize Golgi-ER recycling proteins ERD2 and EMP12, pectin biosynthesis enzyme GAUT14, and COPI coat. Knocking down COG1 and COG3 in barley plants increases their susceptibility to fungal infection, indicating impaired cell wall function [109]. Another plant, *Glycine max* has 2 paralogs of each COG gene, with one paralog of each gene family having a defense function to the parasitic nematode *Heterodera glycines* [75].

A recent study has reported that the COG7 mutant exhibits accelerated dark-induced carbon deprivation and senescence compared to wild-type plants. Multiomics and biochemical analyses uncover a faster induction of protein ubiquitination and autophagy in the COG7 mutant, along with an unexpected increase in protein N-linked glycosylation compared to the wild-type plants [21].

The plant COG complex plays a significant role in the structural integrity and development of cell walls, cell growth, expansion, and other processes by directly interacting with other trafficking components such as COPI, SNAREs [123], and the microtubule cytoskeleton. Beyond its evolutionarily conserved functions, the COG complex in plants also plays distinct roles in regulating ubiquitination and pathogen resistance.

Drosophila melanogaster

The first characterized multicellular animal with COG deficiency is *Drosophila melanogaster*. Like in yeasts, the COG complex in flies is involved in Golgi trafficking and physiology. COG3, COG5, and COG7 mutations render the male insect sterile due to defective spermatogenesis [9, 41]. The COG7 mutants display significant alterations in Golgi structure with a reduced number of total Golgi stacks in spermatocytes. The glycolipid GM1 and sialylated proteins are noticeably diminished throughout the cell, indicating a glycosylation disorder [9]. Moreover, the mislocalization of Golgi-associated proteins such as phosphatidylinositol transfer protein, Giotto, the GTPase Rab11, and t-SNARE STX16 suggest that the COG complex plays a crucial role in the localization of proteins involved in post-Golgi trafficking and cytokinesis [9, 48]. The lifespan of COG7 mutants is notably reduced compared to WT animals, and they also exhibited a range of neuromotor defects, including altered posture and reduced, erratic movement. N-glycan mass spectrometry analysis has shown that COG7 mutants display hyposialylation of N-glycans, similar to CHO COG mutants and COG-CDG patients [48]. Additionally, a separate study has reported that COG1 deficiency protected animals from copper diet induced death [27]. A detailed investigation reveals that the COG complex and the copper transporter ATP7A genetically interact in adult dopaminergic neurons, influencing copper-dependent mortality. Additionally, COG2 and COG6 mutants exhibit defective pain responses, suggesting potential nervous system defects in COG-deficient animals.

These studies collectively demonstrate that the COG complex in *Drosophila* plays a role in maintaining Golgi structure and facilitating trafficking within a network that encompasses numerous other trafficking proteins. This network of COG-dependent proteins also plays a crucial role in cytokinesis, which is particularly evident in spermatocytes, where cytokinesis has been demonstrated to rely on Golgi-derived vesicles. The *Drosophila* COG complex is also integral to neuronal function and the formation of neuromuscular junctions.

Caenorhabditis elegans

Another useful multicellular animal model for Golgi studies is *Caenorhabditis elegans* [69]. The first mutants of the COG complex in worms were discovered during a study of gonad formation defects [82]. COG mutants display altered gonadal morphology and a protruding vulva phenotype but remain viable and capable of proliferation, albeit at a slower rate than their wild-type counterparts. The proper migration of distal tip cells (DTC) and normal gonadal structure depend on the secretion of metalloprotease MIG-17. COG subunits COGC-3 and COGC-1 are essential for glycosylation and gonadal localization of MIG-17. Mutations in COGC-3 and COGC-1 cause aberrant DTC migration similar to that observed in mig-17 mutants [82]. Additionally, disruption to the COG complex affects the glycosylation of the Golgi protein MIG-23 [140]. N-glycan mass spectrometry analysis of glycosylation alterations in COG1-deficient *C. elegans* found that mutant animals have terminal fucoses on their complex glycans, as they lack sialylation [140]. In a subsequent study, it has been observed that combining a COG3 mutation with a GS28 mutant is embryonic lethal, a condition not found in either single mutant. Unlike COG mutants, GARP mutants do not exhibit synthetic lethality with Golgi SNAREs, highlighting distinct differences in their functions [96, 97]. Additionally, the *C. elegans* model for Saul-Wilson Syndrome (the rare human disease caused by COG4 G516R mutation, see section "COG-Associated Congenital Disorders of Glycosylation (CDGs)") was created. Surprisingly, *cogc-4* mutant exhibits no noticeable differences compared to wild-type worms [172], indicating organism-specific responses to COG point mutations.

The studies demonstrate the significance of the COG complex in the accurate glycosylation and development of *C. elegans*. Unlike the infertility observed in *Drosophila* and *Arabidopsis*, *C. elegans* COG mutants are viable and fertile. However, these mutants exhibit slower growth and dysmorphic features, along with changes in gonadal structure.

Mammalian Tissue Culture Cells

The mammalian COG complex subunits were initially identified in Monty Krieger's lab in the genetic screen for malfunctional LDL receptor (LDLR) in CHO cells [18, 73, 115]. Cells lacking COG1 (cog1/ldlB mutant) or COG2 (cog2/ldlC mutant) show decreased LDLR stability, likely due to its deficient glycosylation. COG mutants show similar hypoglycosylation patterns, such as immature N-, O-, and lipid-linked glycosylation, and reduced sialic acid residues in all glycan structures. It has been proposed that COG1 and COG2 may form a complex that regulates Golgi glycosylation. Subsequent studies in CHO, Vero, HeLa, HEK293, and hTERT-RPE1 cells confirm the existence of conserved Golgi-localized octameric COG complex [8, 108, 116, 148, 154, 159].

Biochemical studies in CHO cells were instrumental for understanding the composition and assembly of COG complex [108, 149]. These studies revealed that the depletion of COG subunits results in the destabilization of several Golgi proteins collectively known as "GEARs". GEAR proteins include SNAREs (GS28 and GS15), CCT/golgins (CASP, Giantin/GOLGB1, and Golgin-84), the glycosylation enzyme MAN2A1, and Golgi phosphoprotein GPP130/GOLIM4. Notably, the depletion of these proteins has been associated with disrupted COPI trafficking, as COG malfunction led to the destabilization of COPI, and depletion of COPI causes similar instability in the GEARs [107]. Furthermore, the depletion of COG subunits also affects enzymes SMS1/SGMS1 and CERT1 involved in sphingomyelin synthesis, thereby establishing a link between COG function and lipid homeostasis [137, 138].

COG studies in immortalized mammalian cells have been expanded by utilizing knockdown (KD), knock-sideways, and knockout (KO) techniques.

In one of the early studies, transient KD of the lobe A subunit COG3 [178] in HeLa cells resulted in reduced levels of COG1, COG2, and COG4, indicating their dependency on COG3. COG3 KD causes transient accumulation of COPI-positive vesicle-like intermediates named COG complex dependent (CCD) vesicles. Isolated CCD vesicles contain three GEAR proteins: GS15, GS28, and GPP130. This indicates that the COG dependency of GEARs may be connected to their presence in recycling vesicles that can no longer be properly tethered in the absence of the COG complex. Importantly, CCD vesicles could be isolated from cell lysate using COG-decorated agarose beads. Moreover, their re-attachment to the Golgi has been blocked by COG3-specific antibodies, supporting a direct role for the COG complex in vesicle tethering [129, 178]. The role of COG in glycosylation was further investigated by knocking down additional COG subunits. Analysis using lectin binding and N-glycan mass-spectrometry (MS) show decreased sialylation in COG3, COG4, and COG7 KD cells, while a minor increase in sialylation has been noted in COG6 and COG8 KD cells. Additionally, COG5 KD cells exhibit altered binding to WGA, PHA, ConA, and PNA lectins, indicating changes in sialic acid content, N-linked galactose, mannose, and O-linked sialylation, respectively [108, 116, 128, 129, 178]. Importantly, in COG KD cells, MAN2A1, MGAT1,

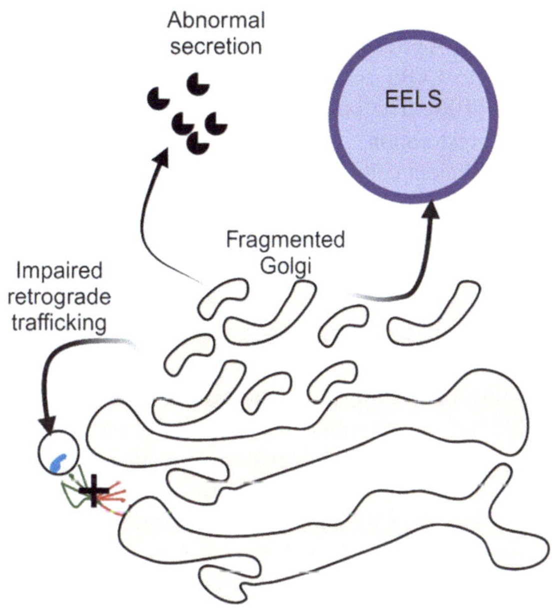

Fig. 7.3 Major cellular phenotypes associated with COG deficiency. The illustration depicts the impact of mammalian cell COG deletion, including improper retrograde trafficking, altered stability of glycosylation of enzymes, abnormal protein secretion, formation of enlarged endolysosomal structures (EELS), Golgi bloating and fragmentation

and GALNT2 are mislocalized to CCD vesicles, suggesting that glycosylation defects are primarily due to the mislocalization of Golgi glycosylation machinery. In addition, retrograde PM–Golgi–ER trafficking of Shiga and SubAB toxins are also greatly impaired in COG3 KD cells [131, 178].

To assess the role of each subunit in overall COG function without being confounded by variations in KD efficiencies, COG KOs have been generated in HEK293T cells using the CRISPR/Cas9 approach. All COG KO cells exhibit similar glycosylation, trafficking, and morphological defects, regardless of the lobe or subunit affected [8, 12, 14, 36].

Complete KO of individual COG subunits leads to morphological changes in the Golgi, including severe fragmentation and dilation of all Golgi cisternae. This results in additional glycosylation defects [8, 129], and affects the endolysosomal system, leading to the accumulation of Enlarged Endo-Lysosomal Structures (EELSs) [33]. It also causes delays in retrograde protein trafficking [131], protein missorting, and changes in the secretion profile [13] (Fig. 7.3). Although most of defects are common in all COG KO cells, minor differences between COG KO clones have been detected. Lobe A and COG6 KOs result in a more severely distorted Golgi structure, while COG 2, 3, 4, 5, and 7 KOs exhibit the most hypoglycosylated form of heavily N- and O-glycosylated LAMP2. In COG knockout cells, v-SNAREs GS15 and GS28 are depleted, suggesting that SNARE degradation contributes to trafficking and glycosylation defects. Surprisingly, the knockout of both Golgi SNAREs causes less damage to Golgi physiology compared to COG depletion, indicating that COG functions extend beyond assisting in SNARE complex formation [34].

Interestingly, COG-deficient cells demonstrate a delayed Golgi response to BFA (Brefeldin A) treatment [43, 116]. BFA inhibits the ARF1 exchange factor GBF1, leading to COPI coat dysfunction and an artificial fusion of Golgi and ER membranes [106]. The kinetics of both BFA-induced Golgi "collapse" and subsequent Golgi restoration following BFA wash-out are significantly delayed in COG-deficient cells [95, 124]. Resistance of COG-depleted cells to BFA treatment has been initially interpreted as COG's role in Golgi-ER retrograde transport. However, the real reason for BFA resistance was later linked to a partial displacement of GBF1 in COG-deficient cells [43].

Although CRISPR-based knockout and RNAi are powerful and widely used methods for studying COG function, they typically require a relatively long time (3–10 days) to manifest a mutant phenotype. Additionally, KO and KD experiments may result in incomplete silencing, off-target effects, and cellular adaptation, making it difficult to determine whether the observed phenotypes are direct or indirect consequences of COG depletion. It is also important to consider that Golgi membranes are highly dynamic, with transport through the Golgi in human cells typically taking less than 30 min [10, 15]. To overcome these challenges, a rapid degron-based strategy for COG4 degradation has been developed. COG4 depletion in hTERT-RPE1 cells effectively removes all other COG subunits from the Golgi, enabling the investigation of the primary defects associated with COG inactivation [141]. The acute COG inactivation causes a massive accumulation of CCD vesicles that carry the bulk of Golgi enzymes and resident proteins. Golgi stacks remained intact during the early stages of CCD vesicle accumulation but gradually fragmented as the number of CCD vesicles increased. This suggests that a defect in vesicle tethering is the primary issue associated with COG malfunction. The v-SNAREs (GS15, GS28) and v-tethers (giantin, golgin-84, and TMF1) are relocalized into CCD vesicles, while t-SNAREs (STX5, YKT6), t-tethers (GM130/GOLGA2, p115), and most of Rab proteins remain Golgi-associated. Superresolution microscopy and biochemical analysis reveal that different Golgi residents are segregated into different populations of CCD vesicles. Acute COG depletion significantly affects three Golgi-based vesicular coats—COPI, AP1, and GGA, suggesting that COG uniquely orchestrates the tethering of multiple types of intra-Golgi CCD vesicles, produced from different Golgi compartments by different coat machinery (Fig. 7.4).

In summary, in human cells, the deletion of any COG subunit renders the entire complex non-functional and leads to vesicular accumulation and subsequent degradation of several Golgi resident proteins including all tested Golgi glycosylation enzymes (MGAT1, MAN2A, B4GALT1, ST6GAL1) [116, 162], v-SNAREs (GS15/BET1L and GS28/GOSR1), and transmembrane golgins (giantin/GOLGB1 and golgin-84/GOLGA5) [107]. It is worth noting that not every Golgi protein is COG-sensitive. For instance, in COG-depleted cells, the levels and localization of the SNARE STX5, CCT GM130/GOLGA2, and many Golgi Rabs remain normal [8, 107, 141], suggesting the existence of a COG-independent Golgi recycling/retention mechanism.

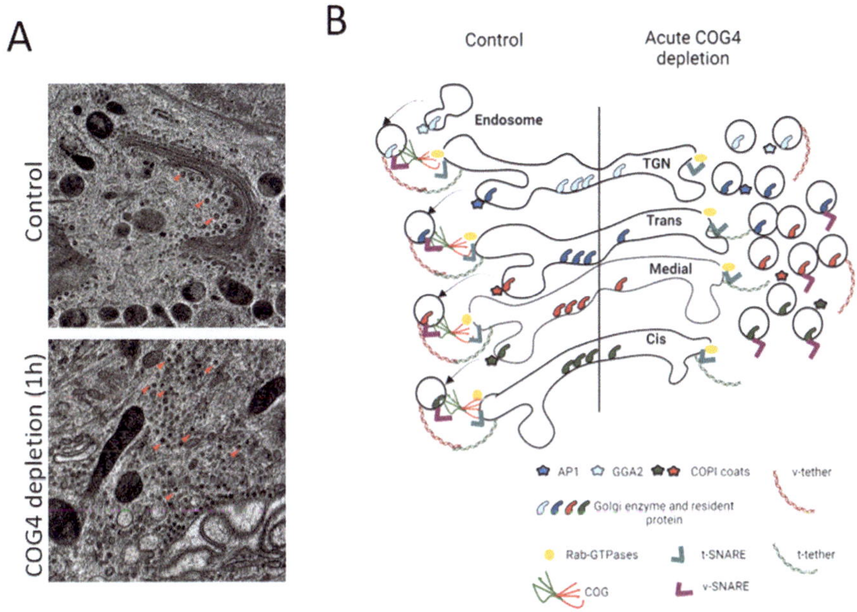

Fig. 7.4 The effect of acute COG depletion on Golgi homeostasis and vesicular trafficking. (**a**) Transmission electron micrographs depicting Golgi area in control and COG4 depleted hTERT-RPE1 cells. The scale bar is 1 μm. The red arrows indicate vesicles. (**b**) The model depicts the accumulation of multiple CCD vesicles in cells upon acute COG4 depletion

Models for COG Complex Function in Trafficking

The significance of COG-mediated trafficking lies in the cargo transported by the retrograde intra-Golgi vesicles, which primarily consists of glycosylation enzymes and other components of the Golgi glycosylation machinery. This vesicle transport route is crucial for maintaining the uneven distribution of glycosylation machinery across Golgi cisternae, a key factor in proper glycan biosynthesis. Interestingly, different COG mutations can lead to distinct glycosylation abnormalities [36]. Several models have been proposed to explain the function of the COG complex during vesicle tethering [12, 36, 88, 102, 145, 149, 153, 160]. In each model, the COG complex plays a central role in orchestrating membrane trafficking, but the specific stage at which it is involved varies. The models are shown in Fig. 7.5.

"Docking Station Assembly" Model

The initial proposed model was the "docking station assembly" model (Fig. 7.5a), which refers to the entire assembled COG complex. According to this model, distinct docking stations are assembled at different Golgi cisternae to recognize trafficking intermediates with different destination addresses. Docking complexes consisting of a specific combination of Rabs, CCTs, and t-SNAREs transiently assemble at the edges of Golgi cisternae. Upon COG priming, assembled docking complexes are

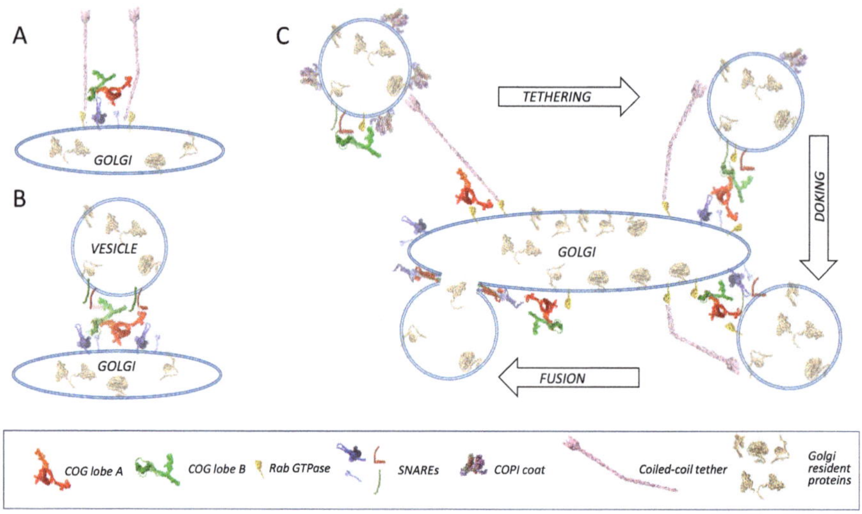

Fig. 7.5 Models for COG complex function. (**a**) In the "docking stations" model, the COG complex orchestrates the assembly of a pre-tethering protein complex through interactions with Rabs, coiled-coil tethers, and SNAREs. (**b**) In SNARE "stabilization model" COG subunits interact with one or more SNARE complexes, potentially contributing to its stabilization, proofreading and protection from NSF-mediated disassembly. (**c**) In the "dynamic tethering" model, COG subcomplexes initially localize to vesicles (Lobe B, green) and Golgi cisternae (Lobe A, red) through multiple weak interactions with Rabs, vesicular coat proteins, CCTs, and SNAREs. During the initial tethering, CCTs capture vesicles, allowing the COG complex to assemble via COG1-COG8 interactions. This assembly stabilizes and proofreads the tethering complex, leading to stable docking and the initial formation of the trans-SNARE complex. Additionally, the interaction between the COG complex and SLY1 further refines and finalizes SNARE complex formation, ultimately resulting in COG displacement and vesicle fusion

then ready to accept incoming retrograde transport vesicles. This model is supported by interactions of COG complex with Rabs, CCTs, and SNAREs and by *in vitro* reconstitution experiments [29, 154]. It is also in agreement with models proposed for the HOPS tethering complex [58, 61].

At the same time, the "docking model" fails to explain the existence of membrane-attached COG subcomplexes [66, 158] and the dispensability of lobe B subunits in yeast cells. Another problem with this model is the lack of information about the signal(s) that initiate the formation of the vesicle docking stations in the right place at the right time. Moreover, this model differs from the current models for the exocyst function in vesicle fusion [2, 4, 54, 122].

"SNARE Stabilization" Model

This model (Fig. 7.5b) has been developed based on observations from COG depletion studies that reveal a depletion of several Golgi SNAREs and a decrease in the stability of STX5 and STX16 SNARE complexes [84–87, 128]. These results led to the interpretation that the COG complex's interaction with v- and t-SNAREs may enhance SNARE stability and facilitate SNARE complex assembly. In this model, the COG complex does not directly tether incoming vesicles, but its primary function is to stabilize, catalyze, and possibly proofread the vesicle fusion machinery.

The COG complex or its sub-complexes may enhance the stability of SNAREs by binding to pre-assembled t-t and/or v-t SNARE complexes in multiple ways. The entire COG complex binds to the STX5/GS28/GS15 t-t SNARE complex through simultaneous interactions between COG4-STX5 [128], and COG7-GS28 [86]. Interactions between COG4 and the STX5 partner protein SLY1 [87] can stabilize the assembled SNARE complex and protect it from NSF or VCP-mediated disassembly. The model considers the MUN-like domain in COG4 [112], but does not explain the extended structural features of the COG complex and its multiple interactions with all classes of Golgi trafficking regulators. Apart from stabilization, the COG complex could also directly facilitate the formation of v-t SNARE complexes during vesicle docking. SNARE stabilization function has been proposed for other multisubunit tethering complexes, including HOPS [26] and GARP [114]. Surprisingly, the double knockout of GS28 and GS15 Golgi SNAREs causes less damage to Golgi physiology compared to COG depletion, indicating that COG functions extend beyond assisting in SNARE complex formation [34].

"Dynamic Tethering" Model

The "dynamic tethering" model seeks to reconcile the previous two models and explain the localization and dynamics of COG subcomplexes (Fig. 7.5c). According to this model, the lobe B subcomplex travels on recycling vesicles, while the lobe A subcomplex is predominantly associated with t-SNARE complexes on acceptor Golgi membranes. The binding of COG lobes to membranes is regulated by multiple COG-Rab interactions. Early in the tethering reaction, COG subcomplexes interact with coat remnants on vesicles and CCTs, which loosely attach trafficking intermediates at the appropriate Golgi cisternae. Multiple weak protein-protein interactions between the C-terminal extensions of COG subunits on vesicles and their corresponding partners on the acceptor membrane could act as a proofreading mechanism, ensuring correct vesicle delivery. Tight vesicle docking and initiation of SNARE priming for productive fusion occurs through the assembly of the COG complex via COG1-COG8 interaction. At this stage, COG may establish additional interactions with SLY1 and initiate or proofread the formation of the trans-SNARE complex. Subsequent to the formation of the trans-SNARE complex, COG is displaced from the vesicle-cisternae interface, enabling vesicle fusion to take place. In support of this model, COG binds to COPI coat components, Rabs, and CCTs [102, 117, 144, 178], signifying its direct involvement in recognizing COPI-coated trafficking intermediates within the Golgi. Additionally, if the efficient assembly or disassembly of the COG complex is inhibited, vesicle fusion is disrupted [158]. The modular nature of the COG complex may permit flexibility and the exchange of subunits with other CATCHR complexes. Several studies have reported such interactions [6, 28], which could explain the milder phenotypes observed in lobe B mutations in yeast and other model organisms. The dynamic tethering model for the COG complex is consistent with several models proposed for the exocyst [2, 105] and Dsl1 [30, 119, 146] complexes.

COG-Associated Congenital Disorders of Glycosylation (CDGs)

Over the last twenty years, more than a hundred individuals with COG mutations have been identified (Fig. 7.6). The accuracy of Golgi glycosylation heavily relies on the correct positioning of the glycosylation machinery. Disrupted intra-Golgi trafficking caused by COG malfunction results in glycosylation abnormalities [45]. Glycosylation disorders are part of a set of autosomal recessive disorders known as Congenital Disorders of Glycosylation (CDGs) which currently have no widely known cure. They are classified as type-I (CDG-I) or type-II (CDG-II) based on the affected protein. CDG-I involves defects in the synthesis and/or transfer of a lipid-linked oligosaccharide, while CDG-II involves abnormalities in the glycan modification and processing machinery, such as faulty glycosidases and glycosyltransferases. CDGs are multi-systemic disorders that significantly impact the quality of life. Recognizable symptoms include global developmental defects, dysmorphic features, microcephaly, and failure to thrive, often accompanied by varying degrees of liver and neurological involvement [36, 68, 111]. A CDG diagnosis relies on analyzing the isoelectric focusing (IEF) pattern of N-glycosylated transferrin. CDG-I is characterized by an abundance of asialylated and/or disialylated transferrin due to defects in glycan transfer and assembly.

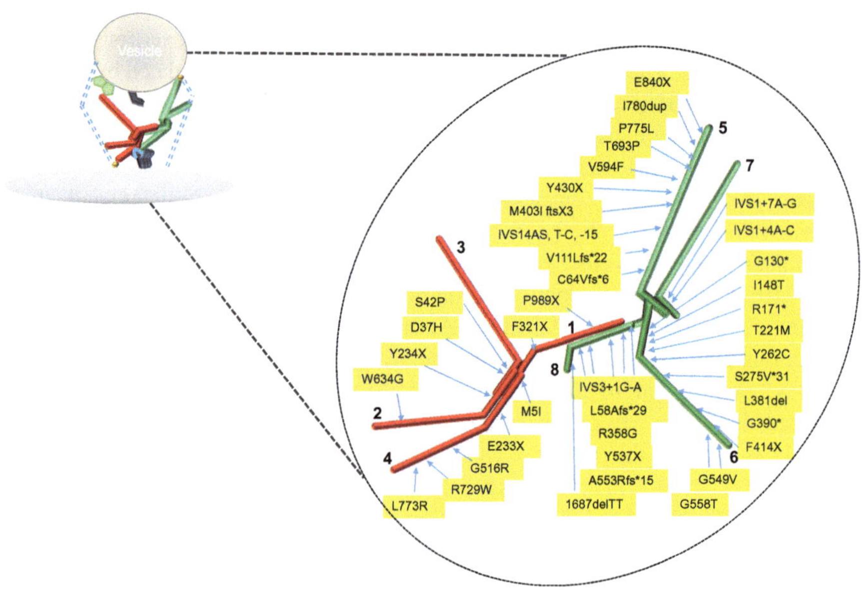

Fig. 7.6 The cartoon presents mutations found in COG-CDG patients. The diagram on the left shows COG's complex function during vesicle tethering. On the right side are the known human COG mutations in each subunit. Lobe A subunits COG1, COG2, COG3, and COG4 are in red, while lobe B subunits COG5, COG6, COG7, and COG8 are in green

Meanwhile, CDG-II is associated with defective glycan processing, leading to increased mono-, di-, tri-, and/or asialotransferrin. However, not all CDG-II patterns can be detected through transferrin's IEF, necessitating the use of glycan mass spectrometry and/or ApoCIII IEF as complementary testing methods [56, 89, 163]. The majority of CDG-II cases are caused by COG mutations. In COG-CDGs, the IEF pattern of serum transferrin indicates hyposialylated species, while IEF of apoC-III shows increased non-glycosylated apoC-III and decreased monosialo glycoforms [45]. Additional galactosylation and sialylation defects could be detected by glycan mass spectrometry. Following diagnosis, genotyping is typically performed to identify the mutations causing CDG. The following section summarizes the clinical presentations and molecular analyses of COG-CDG patients (Table 7.1) and aligns these findings with the current understanding of the cellular functions of the COG complex.

COG1-CDG

In 2006, the first study reported a case of CDG-causing mutation (a single nucleotide insertion, 2659-2660insC) in COG1 with feeding problems and failure to thrive. Upon clinical examination, the patient exhibited hypotonia, left ventricular hypertrophy, progressive microcephaly, mild hepatosplenomegaly, and slight cerebral and cerebellar atrophy [47]. The mutation resulted in a premature stop codon and the loss of 80 C-terminal amino acid residues of COG1 followed by the destabilization of COG1, all other lobe A subunits, and COG8 [47]. Additionally, the protein levels of glycosylation enzymes MAN2A and B4GALT1 were reduced. Two additional COG1 patients were documented with splice site intronic mutation c.1070 + 5G > A, which led to skipped exon 6 caused a premature stop codon in exon 7 [174]. As a consequence, a 659 amino acids long C-terminal fragment was lost. The patients present cerebrocosto-mandibular-like syndrome, growth retardation, moderate mental retardation, feeding difficulties, and cerebellar atrophy. The protein level of COG1 is reduced to 20% while COG8 is slightly reduced, and the expression of MAN2A remained unaffected [174].

COG2-CDG

In 2015, COG2 CDG case was documented [76] in a 6-year-old boy. The patient exhibited two mutations - a de novo c.701dup (p.Y234*) change in the paternal allele and an inherited c.1900 T > G (p.W634G) mutation in the maternal allele. At the protein level, COG2 is not detected, and COG3 and COG4 are also affected. The displayed symptoms were growth retardation and microcephaly, consistent with COG1-CDG patients. Additional symptoms include spastic quadriplegia, liver dysfunction, hypocupremia, hypocerulo-plasminemia, cerebral atrophy, and tonic seizures.

COG3-CDG

Two homozygous COG3 missense variant alleles c.124T > C, p. (S42P) and c.109G > C, p. (D37H) were identified in four affected individuals from two unrelated consanguineous families [37]. These variants caused changes in amino acid residues in a highly conserved alpha-helix structure near the N-terminus of

Table 7.1 Molecular and clinical presentations of COG complex mutations

Gene	Mutation	Effect on mutant subunit	Effect on COG complex	Clinical presentation
COG1	Homozygous 2659-2660insC (p. P888fsX900)	Truncated by 80aa at C-terminus due to premature stop codon Around 15% of mutant COG1 expressed and weakly localized to the Golgi	Depletion of all lobe A subunits and COG8	Feeding problems since birth, failure to thrive, hypotonia, psychomotor and growth retardation, short stature, microcephaly, liver and spleen enlargement, small hands and feet, straightened bitemporal space, antimongoloid eyelids, cardiac abnormalities including ventricular hypertrophy.
	Homozygous c.1070 + 5G > A	Mutant COG1 truncated by 321aa due to loss of exon 6 and new stop codon in exon 7. Truncated COG1 not expressed. Low protein levels of WT COG1 present	Decreased COG8 protein levels	Most severe COG1-CDG. Cerebrocostomandibular syndrome including mental retardation, cerebellar atrophy, renal and cardiac abnormalities, arthrogryposis, hearing loss and skeletal abnormalities.
COG2	Heterozygous mutations: c.701dup (p.Y234*) c.1900T > G (p.W634G)	c.701dup mRNA is not expressed. c.1900T > G mRNA expressed is low levels. COG2 protein is non-detected.	Decreased COG3, COG4 levels	Growth retardation, microcephaly, spastic quadriplegia, liver dysfunction, hypocupremia, hypoceruloplasminemia, cerebral atrophy and tonic seizures
COG3	Two Homozygous mutatons: c.124T > C, p.(S42P) and c.109G > C (p.D37H)	Alteration of amino acid residues in one highly conserved alpha-helix structure	The expression of COG3, COG4 is significantly reduced	Developmental delay, several motor and cognitive disorders, intellectual disability, severe microcephaly, dysmorphic facial features, visual impairment, etc.
COG4	c.2185C > T (p.R729W) 16q22 deletion	80% depletion of COG4 protein	Decreased COG1, COG2, COG3 COG5 levels	Mild psychomotor retardation, dysmorphia, epilepsia
	Heterozygous c.697G > T (p.E233X) [*de novo*]	Degradation of c.697G > T mRNA 70% depletion of COG4	No significant effect on other COG subunits	Developmental delay, hypotonia, failure to thrive, seizures, coagulopathy, liver cirrhosis, and recurrent infections

	or c.2318T > G (p.L773R)			Dysmorphia, Neurological defects, uncoordinated movements, bilateral cerebral atrophy of the frontotemporal regions, and recurrent respiratory infections
	Heterozygous *de novo* c.1546G > A or c.1546G > C (p.G516R)	–	No effect on other subunits	Skeletal dysplasia caused by Saul-Wilson Syndrome, developmental delay, and dysmorphic features
	Homozygous missense mutation c.15G > A; p.M5I			Developmental delay, spasticity, global dystonia, growth retardation
COG5	Homozygous c.1669-15T > C	Intronic substitution caused the skipping of exons 15 & 16. Truncated COG5 (by 58aa) not expressed. Low levels of WT COG5 expressed	–	Mild neurological phenotype including hypotonia, psychomotor retardation, delayed speech development, and truncal ataxia
	Heterozygous: c.556_560delAGTAAinsCT (p.S186_K187delinsL) c.1856T > C (p.I619T) [c.95T > G (p.M32R)]	Deletion is predicted to be in the frame. Affected M is in a highly conserved region. 75% depletion of COG5	About 90% depletion of COG5	Same as above with cirrhotic liver and hepatosplenomegaly
	Homozygous c.2518G > T (p.E840X)	75% depletion of COG5		Microcephaly, delayed speech and motor development, moderate to severe mental retardation, hypotonia, mild dysmorphia, microcephaly, short stature and walking difficulty
	Heterozygous c.189del.G (p.C64Vfs*6) c.2338_2340dupATT (p.I780dup)			Same as above with blindness and deafness
	Homozygous c.1780G > T (p.V594F)	Skipping of exon 16 75% depletion of COG5		Same as above with blindness and deafness

(continued)

Table 7.1 (continued)

Gene	Mutation	Effect on mutant subunit	Effect on COG complex	Clinical presentation
	Heterozygous c.1209delG (p.M403IfsX3)	Depletion of WT COG5. Low levels of mutant COG5 also present	–	Friedreich's-ataxia-like phenotype includes cerebellar atrophy, mild to moderate intellectual disability, and scoliosis.
	Heterozygous c.2324 C > T (p.P775L) c.330delT (p.V111Lfs*22)	P775 and V111 are highly conserved in mammals. Truncated p.V111Lfs*22 (by 690aa) non-functional	–	Most severe COG5-CDG phenotypes include severe mental retardation, delayed speech and motor development, microcephaly, cerebral and cerebellar atrophy, hypotonia, recurrent seizures, liver involvement, and small feet.
	Heterozygous c.1290C > A (p.Y430X) c.2077A > C (p.T693P)	T693 is highly conserved 80% depletion of WT COG5. Truncated COG5 present at very low levels	–	Neonatal jaundice, recurrent upper respiratory tract infections, hypohidrosis, hyperkeratosis, ulnar deviation, and delayed psychomotor development.
	Heterozygous c.2324C > T (p.P775L) c.1508dup (p.G505Wfs*3)	–	–	Neurodevelopmental disorders, non-syndromic intellectual disability
COG6	Homozygous c. 1646G > T (p.G549V)	80% depletion of COG6	Decreased COG5 and COG7 levels. Assembly of octameric COG is unaffected.	Fatal (2 siblings died within 2 months of birth.) Intractable focal seizures, vomiting, intracranial bleedings causing loss of consciousness, cholestasis, vitamin K deficiency, brain oedema, necrotizing enterocolitis with disseminated intravascular coagulation, Chronic inflammatory bowel disease, gastrointestinal malabsorption, liver cirrhosis, mild psychomotor retardation, and microcephaly, recurrent infections due to immunodeficiency, failure to thrive.

	c.1167-24A > G (p. G390FfsX6)	Intronic mutation created a new splice site. Mutant COG6 not expressed 30% of WT COG6 protein present	–	Shaheen syndrome (mild intellectual disability, hypohidrosis, abnormal teeth, and acquired microcephaly, palmoplantar hyperkeratosis)
		–	–	Shaheen syndrome, strabismus, splenomegaly, hypotonia, HLH
	Homozygous c.511C > T (p.R171*)		–	Fatal. Died a few days after birth. Respiratory failure extreme hypotonia
	Homozygous c.1746 + 2T gt; G	Splice site mutation	–	Fatal. (All patients died within one year) Developmental disability, chronic diarrhea with failure to thrive, splenomegaly, hypotonia, recurrent infections, bilateral sensorineural hearing loss, enlarged extra-axial CSF spaces, and asymmetric lateral ventricles.
	Homozygous c.1238_1239insA (p. F414Lfs*4)	–	–	Fatal. Died at the age of 1 year. Psychomotor disability, failure to thrive, chronic diarrhea and hepatosplenomegaly, recurrent infections from immune dysfunction, liver failure
	Heterozygous c.1646G > T (p.G549V) c.785A > G (p.Y262C)	–	–	Microcephaly, splenomegaly, and delay in motor development, gastroenteritis, high fever, and convulsions, recurrent infections from immune dysfunction
	Heterozygous c.511C > T (p.R171*) c.1746 + 2T > G.	–	–	Two siblings, one died at 15 months Diarrhea, fever at birth due to viral infection, hepatosplenomegaly, cirrhotic liver, psychomotor disability, hypotonia, scoliosis, microcephaly, dysmorphia, and growth retardation recurrent infections and hyperthermia
			–	

(continued)

Table 7.1 (continued)

Gene	Mutation	Effect on mutant subunit	Effect on COG complex	Clinical presentation
	Two missense mutation c.662C > T (p.T221M) and c.443T > C (p.I148T)			Progressive hepatosplenomegaly, jaundice, ascites, thrombocytopenia, microcephaly, hypotonia, hypohidrosis, hyperkeratosis, and recurrent hyperthermia, infection, and hypoglycemia.
	Two heterozygous mutations c.823delA (p.S275Vfs*31) and c.1141_1143delCTC (p.L381del)	Creates a frameshift, which will lead to a premature stop codon Creates an in-frame deletion of one amino acid.	–	Dysmorphism, severe neurological disability, corpus callosum dysgenesis, and a congenital recto-vaginal fistula
	Two heterozygous mutations c.1672C > T (p.G558T) and c.153 + 392A > G	Creates a premature termination codon, which may potentially lead to the production of a truncated form of COG6.	–	Abnormal transferase, liver cirrhosis, hemogram, coagulopathy, growth retardation, intellectual disability, frequent infections, and enamel hypoplasia
COG7	Homozygous IVS1 + 4 A → C aka c.169 + 4A > C	Intronic mutation in splice site 19 bp deletion due to altered splicing and premature stop codon Degradation of COG7 mRNA Less than 20% of protein expressed	Decreased COG5 (95%), COG6 (70%), and COG8 (71%) Lobe A is associated with the Golgi, not lobe B	Fatal. Cardiac failure and other complications within the first year of birth. Perinatal asphyxia, dysmorphia, skeletal deformities, hypotonia, hepatosplenomegaly, neonatal jaundice, recurrent epilepsies, infections
	Homozygous c.170-7A > G (p.56–57insAT)	Activation of upstream cryptic splice site inserted 6 bps between exon 1 and 2 About 70% of protein expressed	COG6 and COG8 levels decreased by 20% COG5 decreased by 60%	Milder clinical phenotype including growth retardation, failure to thrive, feeding problems, hypotonia, recurrent hyperthermia, and cerebral atrophy
COG8	Homozygous c.1611C > G (p.Y537X)	Loss of 76-C terminus amino acids due to a premature stop codon. 25% of truncated COG8 is expressed	Decreased COG1, COG2, COG3, COG6, COG7	Acute encephalopathy, loss of psychomotor abilities, mild dysmorphia, hypotonia, alternating esotropia, pseudo-ptosis,

			Golgi localization of COG subunits unaffected. Octameric COG assembly affected.	unregulated coagulation, mental retardation, cerebellar ataxia, PFAPA syndrome during infancy, oculomotor apraxia with dysinergia oculocephalica and pseudoptosis
	Heterozygous IVS3 + 1G > A TT 1687–1688	G > A in intron 3 alters splicing, leads to protein truncation by 306aa at the C-terminal. TT deletion in exon 5 creates a premature stop codon and truncates the protein by 76aa COG8 protein is undetected	Decreased COG1, COG5, COG6 and COG7 by 60–70% COG2, COG3 and COG4 associated with the Golgi	Severe growth retardation, Hypotonia, decreased reflexes, chronic axonal neuropathy, ventriculomegaly ex vacuo with atrophy, seizures, esotropia/amblyopia, keratosis pilaris, mild spasticity and contractures, and no bowel/bladder control.
	Heterozygous c.171dupG (p.L58Afs*29) and c.1656dupC (p.A553Rfs*15)	–	–	Psychomotor retardation, hypotonia, failure to thrive, microcephaly, skeletal deformities, and mild psychomotor retardation.
	Homozygous mutation c.1583-1G > A	G > A at the intron-exon boundary of intron 4	–	Fatal. Died 2 days after birth with severe neurological defects. Antenatal conditions included increased nuchal translucency, arthrogryposis multiplex congenita, Dandy-Walker malformation, and fetal growth restrictions.
	A missense mutation c.1073G > A (p.R358G)			Visual impairment, peripheral neuropathy, hallucinations, depression, delusion, dysmorphic hand, and difficulty in walking.

COG3. Analysis of serum transferrin revealed abnormal sialylation. Additionally, the protein levels of COG3 and COG4 in the patient were markedly reduced, while COG1 and COG2 remain unchanged compared to the controls and the parental sample. The affected patients were presenting developmental delays, several motor and cognitive disorders, intellectual disability, severe microcephaly, dysmorphic facial features, etc. In the case of a second family, a male-affected individual exhibited developmental delays, severe intellectual disability, absence of speech, microcephaly, drooling, muscle wasting, visual impairment, and gait issues.

COG4-CDG and Saul-Wilson Syndrome

Several patients have been reported to have CDG-causing mutations in COG4. Patient 1 had two mutations, de novo c.697G > T (p.E233X), c.2318 T > G (p. L773R), and Patient 2 carried a point mutation c.2185C > T (pR729W) in the COG4 gene. The protein level of COG4 was decreased significantly in both patients. COG4 subunit deficiency affects the stability of all remaining lobe A subunits and COG5 in patient 2 while all other subunits are unaltered in patient 1. In addition, glycan mass spectrometry revealed protein sialylation and galactosylation defects [120, 121]. Patient 1 exhibited severe symptoms, including developmental delay, hypotonia, failure to thrive, seizures, coagulopathy, liver cirrhosis, and recurrent infections. Patient 2 also displayed clinical signs such as dysmorphia, microcephaly, and moderate psychomotor delay, along with axial hypotonia, mild ataxia, uncoordinated movements, bilateral cerebral atrophy of the frontotemporal regions, and recurrent respiratory infections. Recently, a homozygous missense variant in COG4 (c.15G > A; p.M5I) has been reported [118]. The affected siblings presented symptoms including developmental delay, spasticity and global dystonia, growth retardation. The symptoms were evocative of inborn errors of metabolism.

Another identified COG4 mutation (c.1546G > A, or c.1546G > C, p.G516R) was associated with a rare skeletal dysplasia known as Saul-Wilson syndrome (SWS) [42]. Although not considered a CDG-causing mutation, individuals with SWS present with speech and developmental delays, dysmorphic features, skeletal dysplasia, and hearing loss. SWS fibroblasts accumulate glypicans, a group of heparan sulfate proteoglycans essential for growth and bone development through various signaling pathways.

Several model systems have been developed to study SWS. In a zebrafish model [168], glypican accumulation was observed, mirroring findings in SWS patients, and altered non-canonical Wnt signaling was proposed as a potential mechanism underlying SWS pathology. Another SWS model, using HEK293T and the chondrosarcoma cell line SW1353, revealed modest, cell-type-dependent changes in N-glycans [167]. Additionally, SW1353 cells carrying the G516R variant formed smaller spheroids and exhibited increased apoptosis, indicating impaired in vitro chondrogenesis.

The third model system utilized hTERT-RPE1 and HEK293T cells to compare SWS and COG4-CDGII (R729W) mutations [142]. Using a combination of CRISPR/Cas9 and lentiviral transduction technologies, both myc-tagged wild-type and mutant (G516R and R729W) COG4 proteins were expressed under the endogenous COG4 promoter. The analysis of constructed isogenic cell lines revealed similar stability and localization of COG complex subunits, wild-type cell growth, and normal Golgi morphology in both cell lines. Importantly, COG4-G516R cells demonstrated increased HPA lectin binding to the PM glycoconjugates, while COG4-R729W cells revealed high GNL lectin binding, indicating specific defects in O- and N-glycosylation. Both mutant cell lines express an elevated level of heparin sulfate proteoglycans. Moreover, a quantitative mass-spectrometry analysis of proteins secreted by COG-deficient cell lines revealed abnormal secretion of SIL1 and ERGIC-53/LMAN1 proteins by COG4-G516R cells.

COG5-CDG

Twelve cases of COG5-CDG have been reported over a period of 20 years. The first patient P1 (intronic homozygous mutation, c.1669-15 T > C) was documented in 2009 [110]. This mutation led to altered splicing resulting in the expression of truncated COG5 mRNA. P1 displayed mild neurological symptoms such as hypotonia, psychomotor retardation, delayed speech development, and truncal ataxia. The second patient (P2) was described in 2011 [50]. Mutational analysis of P2 revealed two heterozygous mutations: c.556_560delAGTAAinsCT (p.S186_K187delinsL) in the maternal allele and c.1856 T > C (p.I619T), c.95 T > G (p.M32R) in the paternal allele [50]. P2 exhibited a mild neurological phenotype, including mental retardation, hypotonia, delayed speech development, microcephaly, and liver cirrhosis with splenomegaly. Five more COG5-CDG patients have been reported in 2012 [124]. These included three female siblings (P3.1, P3.2, and P3.3) with a homozygous nonsense mutation c.2518G > T (p.E840X). P4 exhibited a heterozygous state for two mutations: c.189delG (p.C64Vfs*6) and c.2338_2340dupATT; (p.I780dup). P5 was identified as being homozygous for a missense mutation c.1780G > T (p.V594F). P3.1, P3.2, and P3.3 presented symptoms that include microcephaly, delayed speech, and motor development, moderate to severe mental retardation, hypotonia, mild dysmorphism, microcephaly, short stature, and walking difficulty. P4 and P5 had similar clinical presentations to the three siblings, with additional blindness and deafness. All these patients displayed about an 80% depletion of COG7 protein levels due to the COG5 mutation.

Another study reported on a family in which three siblings P6.1, P6.2, P6.3 (c.1209delG; p.M403IfsX3) revealed a heterozygous mutation (c.1209delG; p. M403IfsX3), resulting in severely depleted levels of the COG5 protein [72]. The patients were exhibiting symptoms resembling Friedreich's ataxia, including cerebellar atrophy, mild to moderate intellectual disability, and scoliosis. In 2019, a COG5-CDG patient (P7) was documented who exhibited severe neurological symptoms with smaller feet and liver involvement [170]. Another patient (P8) was found to be heterozygous for a missense mutation c.2324C > T (p.P775L) and a novel frameshift mutation c.330delT (p.V111Lfs*22) [155]. The patient exhibited clinical features including postural instability, walking difficulty, psychomotor delay, hypohidrosis, hyperkeratosis of the skin, macular hypoplasia, coagulation defects, liver lesions, and ulnar deviation of the right-hand fingers, characteristic of other COG5-CDG patients [155]. The authors suggested that COG5 compound heterozygous mutations may cause glycosylation defects similar to other COG5-CDGs. P8 also had two novel compound heterozygous mutations: a nonsense mutation c.1290C > A (p.Y430X) and a missense mutation c.2077A > C (p.T693P). Low levels of COG5 protein were seen, and the BFA-induced Golgi apparatus collapse to the endoplasmic reticulum was delayed.

COG6-CDG and Shaheen Syndrome

The first COG6-CDG patient carrying a homozygous mutation (c.1646G > T, pG549V) in COG6 was described in 2010 [95]. This mutation led to an 80% reduction in COG6 protein levels and a reduction of COG5 and COG7 protein levels, as well. The patient experienced a severe neurological disease, cholestasis, and vitamin K deficiency. The second patient [65] suffered from chronic inflammatory bowel disease, liver cirrhosis, mild psychomotor retardation, microcephaly, and recurrent infections. Furthermore, mass-spectrometry analysis of serum transferrin revealed hyposialylation. Another study [127] described a 12-year-old male with COG6-CDG who carried an intronic mutation c.1167-24A > G, resulting in a premature stop codon (p.G390FfsX6), causing a notable decrease in mutant COG6 mRNA. The patient had a history of intellectual disability, hypohidrosis, abnormal teeth, acquired microcephaly, palmoplantar hyperkeratosis, and hepatitis, collectively termed Shaheen syndrome. Another girl with COG6-CDG, also exhibiting Shaheen syndrome, was described in 2018 [3]. The patient's clinical presentation included a range of symptoms, such as global developmental disability, microcephaly, hypohidrosis, enamel hypoplasia, strabismus, splenomegaly, and hypotonia. Furthermore, a patient diagnosed with Hemophagocytic Lymphohistiocytosis (HLH) also carried the same mutation (c.1167-24A > G). In a study by Rymen and coworkers [125] four novel COG6 mutations were identified. One patient carried a homozygous c.511C > T (p.R171*) mutation, while another had a mutation (c.1746 + 2 T > G) resulting in altered splicing. The third patient had c.1238_1239insA (p.F414Lfs*4) mutation causing liver failure. Additionally, male siblings born to healthy non-consanguineous Moroccan parents both carried the c.785A > G (p.Y262C) mutation. One of the COG6-CDG (P5) cases was reported with growth retardation, developmental disability, microcephaly, liver and gastrointestinal disease, and recurrent infections [90]. Additionally, a retrospective study [98] reviewed 21 prenatal CDG cases with nonimmune hydrops fetalis (NIHF), with one case being COG6-CDG out of 17 live births. The authors proposed that developmental defects due to COG malfunction may contribute to NIHF. Two novel missense mutations of c.662C > T (p.T221M) in exon 7 and c.443 T > C (p.I148T) in exon 5 of COG6 were identified in 2021 [77]. The patient reported multi-system involvement, including progressive hepatosplenomegaly, jaundice, ascites, thrombocytopenia, microcephaly, hypotonia, hypohidrosis, hyperkeratosis, and recurrent hyperthermia, infection, and hypoglycemia. In 2022, a new variant of COG6-CDG was discovered [23]. The patient exhibited two heterozygous mutations: S275Vfs*31 (c.823delA) in exon 9, and L381del (c.1141_1143delCTC) in exon 12 of the COG6 gene. These novel variants were associated with dysmorphism, severe neurological disability, corpus callosum dysgenesis, and a congenital recto-vaginal fistula. Analysis of serum N-glycans using MALDI-MS revealed significant alterations, including severe hyposialylation and hypogalactosylation, as well as the presence of abnormal structures lacking antennary N-acetylglucosamine. Additionally, in 2024, a new COG6CDG patient was reported with a paternal variant c.1672C > T (G558T) and a maternal variant c.153 + 392A > G having a congenital disorder of glycosylation type IIL (CDG2L). The mutations created a premature termination codon may potentially lead to the production of a truncated form of COG6. The patient exhibited the symptoms including, liver cirrhosis, hemogram, coagulopathy, growth retardation, intellectual disability, frequent infections, and enamel hypoplasia [177].

The range of disorders associated with COG6-related pathologies varies in severity, but most patients exhibit common core features seen in all COG-CDGs. Within the first 2 years of life, 50% of patients had passed away, primarily due to recurrent infections, hyperthermia, and liver disease. The distinct characteristics of the COG6 clinical phenotype include hyperkeratosis, heart defects, and liver involvement [125].

COG7-CDG

The first identified mutation causing COG-CDG was found in the COG7 gene, which is recognized as the most severe form of the disorder. In 2004, two siblings (P1, male; P2, female) from a consanguineous Mediterranean family presented with typical CDG symptoms, including perinatal asphyxia, dysmorphia, skeletal deformities, hypotonia, and hepatosplenomegaly. They later developed jaundice, recurrent seizures, and infections and ultimately succumbed to cardiac failure [134, 165]. In the first two patients, sequencing of COG7 cDNA revealed a deletion of 19 bases in the first exon. Additionally, gDNA sequencing showed a homozygous intronic point mutation IVS1 + 4 A → C (c.169 + 4 A > C) leading to a significant reduction in mRNA and protein levels (95% in P1 and 85% in P2). Additionally, IEF patterns of serum transferrin and apolipoprotein C-III confirmed N- and O-glycosylation defects. Moreover, the BFA-induced collapse of the Golgi to the ER was delayed. Subsequent analysis of the fibroblasts from the patients revealed the mislocalization and depletion of COG-sensitive proteins, GOSR1 and BET1L. ERGIC-53 was also mislocalized to the Golgi, indicating impaired retrograde recycling from the ER to the Golgi [139]. In 2007, four children from three related Moroccan families were reported with a similar fatal COG7 mutation [173]. The affected children, two females, and two siblings, all died within their first year, exhibiting clinical features like stunted growth, severe microcephaly, hypotonia, adducted thumbs, gastrointestinal pseudo-obstruction, failure to thrive, cardiac abnormalities, wrinkled skin, and episodes of extreme hyperthermia [104]. Additionally, a different intronic mutation, c.170-7A > G in COG7, was identified in two COG7-CDG siblings with consanguineous Moroccan parents. N- and O-linked glycosylation effects were confirmed through IEF of serum transferrin and apoC-III. The protein levels of COG6, COG7, and COG8 were depleted by 20–30%, and COG5 was depleted by about 60%. The patients were suffering from delayed psychomotor development, hypotonia, areflexia of the lower limbs, and failure to thrive.

COG8-CDG

COG8-CDG patient carrying a nonsense mutation (c.1611C > G) in the COG8 gene, resulting in a premature stop codon at the C-terminus, was reported in 2007. The patient exhibited symptoms including acute encephalopathy, loss of psychomotor abilities, mental retardation, mild dysmorphia, hypotonia, unregulated coagulation, cerebellar ataxia, and PFAPA syndrome (Periodic Fever, Aphthous Stomatitis, Pharyngitis, Adenitis), oculomotor apraxia with dysinergia, ocular disorders [46]. This patient exhibited glycosylation defects as the loss of COG8 function affects the abundance of two Golgi glycosylation enzymes - MAN2A and B4GALT1. In the same year, another COG8-CDG patient was reported carrying two heterozygous mutations: IVS3 + 1G > A in intron 3, and a two-nucleotide deletion, TT1687–1688, in exon 5 [79]. Hyposialylation, due to reduced ST3GAL1 activity, was observed by IEF of serum transferrin, lectin binding, and mass spectrometry. In the patient's fibroblasts, the COG8 protein was significantly depleted and COG1, COG5, COG6, and COG7 were also depleted by 60–70% [84]. Additionally, the levels of GOSR1, BETL1, and VAMP4 were reduced, impairing the formation of the STX5 and STX16 SNARE complexes, which results in defective intra-Golgi and endosome to TGN retrograde trafficking. This patient presented hypotonia, decreased reflexes, chronic axonal neuropathy, ventriculomegaly ex vacuo with atrophy, seizures, and esotropia/amblyopia, a lack of bowel and bladder control. Another report [169] described another patient with two novel frameshift mutations, c.171dupG (L58Afs*29) and c.1656dupC (A553Rfs*15). The patient displayed similar clinical features, such as hypotonia, failure to thrive, elevated serum liver enzymes, microcephaly, skeletal deformities, and mild psychomotor retardation. There is a report [7] on a patient who was more severely affected compared to the three previous COG8-CDG patients. The patient was found to be homozygous for the c.1583-1G > A mutation located at the boundary of intron 4 and exon 5. The patient exhibited facial dysmorphia and neurological defects, including dilated ventricles and missing corpus callosum, vermian agenesis, and enlarged cisterna magna [7]. Recently, a new mutation c.1073G > A (R358G) in COG8 has been reported. The mutation was associated with a rare hereditary disease Cerebrotendinous xanthomatosis (CTX). The symptoms include visual impairment, peripheral neuropathy, hallucinations, depression, delusion, dysmorphic hands, and difficulty in walking [51]. Although no immunodeficiency has been previously reported in other COG-CDGs, recurrent infections have been documented in COG4 and COG8 CDGs [65].

Summary of COG-CDGs

The clinical presentations and cellular phenotypes of COG-CDGs vary, but patients typically exhibit a range of neurological, morphological, skeletal, and hepatic abnormalities [50, 125, 175].

The heterogeneity in symptoms may be attributed to the diverse effects of mutations on the stability and specific function of the mutant COG subunits and the COG complex. Reduced sialylation and galactosylation, indicating defects in Golgi glycosylation, are commonly observed in COG-CDG patients and KO cell lines.

Notably, the complete depletion of individual COG subunits in human cell lines resulted in more consistent and pronounced cellular phenotypes. To gain comprehensive insights into clinical phenotypes and develop effective treatment protocols, COG-CDG patients should undergo thorough testing for all COG-related abnormalities, particularly in cells obtained from biopsies of the most affected organ systems. Another approach is introducing COG-CDG patient mutations into an established hIPS cell line, differentiating these cells into the types most impacted by COG-CDGs, and analyzing the resulting phenotypic changes. Additionally, patient fibroblasts can be reprogrammed into iPSCs, which can then be differentiated into neuronal and bone cells, the most affected tissues in COG-CDGs.

Future Perspective and Conclusions

After over 35 years since the discovery of the first COG subunit, the COG complex has been established as a key player in Golgi trafficking. It has been extensively studied in various organisms, including single-celled yeast, fungi, worms, flies, plants, mammalian cell lines, and COG-CDG patients. The importance of COG for vesicle tethering, its interactions with the Golgi trafficking machinery, and the consequences of its deficiency for glycosylation and other Golgi functions are well documented.

At the same time, the molecular process of COG-mediated vesicle tethering has yet to be fully understood, and the specific sequence of events leading to the formation, docking, and fusion of retrograde membrane carriers at the Golgi apparatus remains speculative. Although we know that COG interacts with various components of the vesicle fusion machinery, the molecular details of COG-partner interactions and the precise temporal sequence of these interactions still need to be determined. It is anticipated that some of COG's interactions with the Golgi trafficking machinery occur in a sequential rather than simultaneous manner, as individual subunits interact with multiple trafficking partners. It is hoped that the *in vitro* reconstitution of the COG complex function will shed light on these mechanistic questions.

The assembly of the COG complex and its recruitment to the membranes and to the Golgi region is another topic of interest. It is unclear whether COG interacting Rabs, SNAREs, CCTs or Golgi-specific lipids are chiefly responsible for its recruitment to the Golgi rims. COG likely utilizes multiple membrane attachment strategies, as the deletion of individual Rabs or SNAREs does not result in the dislocation of COG to the cytosol [34, 94] and our unpublished observations).

Previous studies conducted by our lab and others have shown that malfunction or complete depletion of COG leads to the accumulation of non-tethered CCD vesicles [166, 178]. It has been suggested that these CCD vesicles are formed in a mostly COPI-dependent budding reaction and transport recycling molecules such as enzymes, transporters, cargo receptors, SNAREs, and Rabs [36, 128, 178]. However, the specific composition of CCD vesicles has not been thoroughly analyzed. It is unknown how heterogeneous CCD vehicles are. Are they all originating from *trans*-Golgi or from other Golgi subcompartments? What is the exact machinery for their production, movements and capturing? *In vitro* formation of several subtypes of COPI vesicles from Golgi was carefully studied and the content of these vesicles was analyzed [1, 39, 52], but it remains unclear how well these *in-vitro*-formed COPI vesicles represent CCD intermediates and how the COG complex identifies these vesicles. The development of an efficient COG-depletion strategy that allows for the rapid accumulation of CCD vesicles [141, 143] should aid in their detailed characterization. CCD vesicles accumulate several transmembrane proteins, including giantin, golgin-84, GS15, GS28, and TGN46/TGOLN2. Antibodies to these proteins are commercially available and can be used as tools for vesicle immunoisolation. Since these COG-dependent proteins are localized in different Golgi cisternae at steady state, it is reasonable to expect that they recycle through different types of membrane intermediates.

Furthermore, it is important to consider that some CCD intermediates may be formed by the AP1 or GGA1/2 vesicle coat machinery. Recent data from budding yeast suggests that a subset of Golgi and anterograde cargo uses an AP1-dependent mode of transportation for recycling at the TGN level [16]. Notably, COG deficiency affects not only the intra-Golgi STX5 SNARE complex but also the endosome-to-TGN STX16 SNARE complex [24].

The CATCHR complexes consist of three subunits (Dsl1), four subunits (GARP and EARP), or eight subunits (COG and exocyst). Despite having low sequence homology, these complexes share structural homology and possess common α-helical compact folds. This suggests that they originated from a common ancestor [64]. What led the exocyst and COG to acquire double the number of subunits? Do they serve additional functions compared to Dsl1 and GARP?

Is the function of the COG complex limited to intra-Golgi retrograde trafficking? Several studies propose that the COG complex subunits may have additional, potentially secondary, roles beyond the Golgi apparatus. Specifically, at the TGN, COG1 exhibits a specific interaction with RINT1, a component of the Dsl1/NRZ complex [6]. Additionally, COG6 interacts with the GARP partner SNARE STX6 [83], and we have observed multiple COG-GARP and COG-exocyst protein-protein interactions in cell lysates (own unpublished observations). Does this imply that CATCHR complexes can share their subunits to carry out new functions? Is there communication between the CATCHRs and other vesicle tethering factors?

At present, one of the primary challenges associated with COG-CDGs is underdiagnosis or misdiagnosis. To date, over a hundred individuals with more than 30 different COG mutations have been identified, each presenting diverse clinical features and cellular phenotypes. Various mutations in each subunit of COG have been documented and primarily studied in patients' fibroblasts. However, fibroblasts obtained from different patients may vary significantly in their genetic makeup, as they may only partially represent the organ systems and cell types most affected by COG mutations. Therefore, the development of a cellular COG disease model by using iPSC and CRISPR strategies could prove beneficial for comprehensively studying the disease and elucidating the detailed interplay between the COG complex and trafficking machinery.

References

1. Adolf F, Rhiel M, Hessling B, Gao Q, Hellwig A, Béthune J, Wieland FT (2019) Proteomic profiling of mammalian COPII and COPI vesicles. Cell Rep 26(1):250–265.e5. https://doi.org/10.1016/j.celrep.2018.12.041
2. Ahmed SM, Nishida-Fukuda H, Li Y, McDonald WH, Gradinaru CC, Macara IG (2018) Exocyst dynamics during vesicle tethering and fusion. Nat Commun 9(1):5140. https://doi.org/10.1038/s41467-018-07467-5
3. Althonaian N, Alsultan A, Morava E, Alfadhel M (2018) Secondary hemophagocytic syndrome associated with COG6 gene defect: report and review. JIMD Rep 42:105–111. https://doi.org/10.1007/8904_2018_88
4. An SJ, Rivera-Molina F, Anneken A, Xi Z, McNellis B, Polejaev VI, Toomre D (2021) An active tethering mechanism controls the fate of vesicles. Nat Commun 12(1):5434. https://doi.org/10.1038/s41467-021-25465-y
5. Anderson NS, Barlowe C (2019) Conserved juxtamembrane domains in the yeast golgin Coy1 drive assembly of a megadalton-sized complex and mediate binding to tethering and SNARE proteins. J Biol Chem 294(25):9690–9705. https://doi.org/10.1074/jbc.RA119.008107
6. Arasaki K, Takagi D, Furuno A, Sohda M, Misumi Y, Wakana Y, Inoue H, Tagaya M (2013) A new role for RINT-1 in SNARE complex assembly at the trans-Golgi network in coordination with the COG complex. Mol Biol Cell 24(18):2907–2917. https://doi.org/10.1091/mbc.E13-01-0014
7. Arora V, Puri RD, Bhai P, Sharma N, Bijarnia-Mahay S, Dimri N, Baijal A, Saxena R, Verma I (2019) The first case of antenatal presentation in COG8-congenital disorder of glycosylation with a novel splice site mutation and an extended phenotype. Am J Med Genet A 179(3):480–485. https://doi.org/10.1002/ajmg.a.61030
8. Bailey Blackburn J, Pokrovskaya I, Fisher P, Ungar D, Lupashin VV (2016) COG complex complexities: detailed characterization of a complete set of HEK293T cells lacking individual COG subunits. Front Cell Dev Biol 4:23. https://doi.org/10.3389/fcell.2016.00023
9. Belloni G, Sechi S, Riparbelli MG, Fuller MT, Callaini G, Giansanti MG (2012) Mutations in Cog7 affect Golgi structure, meiotic cytokinesis and sperm development during Drosophila spermatogenesis. J Cell Sci 125(Pt 22):5441–5452. https://doi.org/10.1242/jcs.108878
10. Beznoussenko GV, Parashuraman S, Rizzo R, Polishchuk R, Martella O, Di Giandomenico D, Fusella A, Spaar A, Sallese M, Capestrano MG, Pavelka M, Vos MR, Rikers YGM, Helms V, Mironov AA, Luini A (2014) Transport of soluble proteins through the Golgi occurs by diffusion via continuities across cisternae. elife 3. https://doi.org/10.7554/eLife.02009
11. Bjørnestad SA, Guadagno NA, Kjos I, Progida C (2022) Rab33b-exocyst interaction mediates localized secretion for focal adhesion turnover and cell migration. iScience 25(5):104250. https://doi.org/10.1016/j.isci.2022.104250
12. Blackburn JB, D'Souza Z, Lupashin VV (2019) Maintaining order: COG complex controls Golgi trafficking, processing, and sorting. FEBS Lett 593(17):2466–2487. https://doi.org/10.1002/1873-3468.13570
13. Blackburn JB, Kudlyk T, Pokrovskaya I, Lupashin VV (2018) More than just sugars: conserved oligomeric Golgi complex deficiency causes glycosylation-independent cellular defects. Traffic Cph Den 19(6):463–480. https://doi.org/10.1111/tra.12564
14. Blackburn JB, Lupashin VV (2016) Creating knockouts of conserved oligomeric Golgi complex subunits using CRISPR-mediated gene editing paired with a selection strategy based on glycosylation defects associated with impaired COG complex function. Methods Mol Biol Clifton NJ 1496:145–161. https://doi.org/10.1007/978-1-4939-6463-5_12
15. Boncompain G, Divoux S, Gareil N, de Forges H, Lescure A, Latreche L, Mercanti V, Jollivet F, Raposo G, Perez F (2012) Synchronization of secretory protein traffic in populations of cells. Nat Methods 9(5):493–498. https://doi.org/10.1038/nmeth.1928

16. Casler JC, Papanikou E, Barrero JJ, Glick BS (2019) Maturation-driven transport and AP-1-dependent recycling of a secretory cargo in the Golgi. J Cell Biol 218(5):1582–1601. https://doi.org/10.1083/jcb.201807195
17. Cavanaugh LF, Chen X, Richardson BC, Ungar D, Pelczer I, Rizo J, Hughson FM (2007) Structural analysis of conserved oligomeric Golgi complex subunit 2. J Biol Chem 282(32): 23418–23426. https://doi.org/10.1074/jbc.M703716200
18. Chatterton JE, Hirsch D, Schwartz JJ, Bickel PE, Rosenberg RD, Lodish HF, Krieger M (1999) Expression cloning of LDLB, a gene essential for normal Golgi function and assembly of the ldlCp complex. Proc Natl Acad Sci USA 96(3):915–920. https://doi.org/10.1073/pnas.96.3.915
19. Chen S, Cai H, Park S-K, Menon S, Jackson CL, Ferro-Novick S (2011) Trs65p, a subunit of the Ypt1p GEF TRAPPII, interacts with the Arf1p exchange factor Gea2p to facilitate COPI-mediated vesicle traffic. Mol Biol Cell 22(19):3634–3644. https://doi.org/10.1091/mbc.E11-03-0197
20. Chia PZC, Gleeson PA (2014) Membrane tethering. F1000prime Rep 6:74. https://doi.org/10.12703/P6-74
21. Choi H-S, Bjornson M, Liang J, Wang J, Ke H, Hur M, De Souza A, Kumar KS, Mortimer JC, Dehesh K (2023) COG-imposed Golgi functional integrity determines the onset of dark-induced senescence. Nat Plants 9(11):1890–1901. https://doi.org/10.1038/s41477-023-01545-3
22. Chou H-T, Dukovski D, Chambers MG, Reinisch KM, Walz T (2016) CATCHR, HOPS and CORVET tethering complexes share a similar architecture. Nat Struct Mol Biol 23(8): 761–763. https://doi.org/10.1038/nsmb.3264
23. Cirnigliaro L, Bianchi P, Sturiale L, Garozzo D, Mangili G, Keldermans L, Rizzo R, Matthijs G, Fiumara A, Jaeken J, Barone R (2022) COG6-CDG: novel variants and novel malformation. Birth Defects Res 114(5–6):165–174. https://doi.org/10.1002/bdr2.1981
24. Climer LK, Dobretsov M, Lupashin V (2015) Defects in the COG complex and COG-related trafficking regulators affect neuronal Golgi function. Front Neurosci 9:405. https://doi.org/10.3389/fnins.2015.00405
25. Climer LK, Pokrovskaya ID, Blackburn JB, Lupashin VV (2018) Membrane detachment is not essential for COG complex function. Mol Biol Cell 29(8):964–974. https://doi.org/10.1091/mbc.E17-11-0694
26. Collins KM, Thorngren NL, Fratti RA, Wickner WT (2005) Sec17p and HOPS, in distinct SNARE complexes, mediate SNARE complex disruption or assembly for fusion. EMBO J 24(10):1775–1786. https://doi.org/10.1038/sj.emboj.7600658
27. Comstra HS, McArthy J, Rudin-Rush S, Hartwig C, Gokhale A, Zlatic SA, Blackburn JB, Werner E, Petris M, D'Souza P, Panuwet P, Barr DB, Lupashin V, Vrailas-Mortimer A, Faundez V (2017) The interactome of the copper transporter ATP7A belongs to a network of neurodevelopmental and neurodegeneration factors. elife 6:e24722. https://doi.org/10.7554/eLife.24722
28. Costanzo M, VanderSluis B, Koch EN, Baryshnikova A, Pons C, Tan G, Wang W, Usaj M, Hanchard J, Lee SD, Pelechano V, Styles EB, Billmann M, van Leeuwen J, van Dyk N, Lin Z-Y, Kuzmin E, Nelson J, Piotrowski JS, Srikumar T, Bahr S, Chen Y, Deshpande R, Kurat CF, Li SC, Li Z, Usaj MM, Okada H, Pascoe N, San Luis B-J, Sharifpoor S, Shuteriqi E, Simpkins SW, Snider J, Suresh HG, Tan Y, Zhu H, Malod-Dognin N, Janjic V, Przulj N, Troyanskaya OG, Stagljar I, Xia T, Ohya Y, Gingras A-C, Raught B, Boutros M, Steinmetz LM, Moore CL, Rosebrock AP, Caudy AA, Myers CL, Andrews B, Boone C (2016) A global genetic interaction network maps a wiring diagram of cellular function. Science 353(6306): aaf1420. https://doi.org/10.1126/science.aaf1420
29. Cottam NP, Wilson KM, Ng BG, Körner C, Freeze HH, Ungar D (2014) Dissecting functions of the conserved oligomeric Golgi tethering complex using a cell-free assay. Traffic Cph Den 15(1):12–21. https://doi.org/10.1111/tra.12128

30. DAmico KA, Stanton AE, Shirkey JD, Travis SM, Jeffrey PD, Hughson FM (2024) Structure of a membrane tethering complex incorporating multiple SNAREs. Nat Struct Mol Biol 31(2): 246–254. https://doi.org/10.1038/s41594-023-01164-8
31. Dell'Aglio E (2021) A COG in the machine: how good Golgi makes good pollen. Plant Physiol 186(1):201–203. https://doi.org/10.1093/plphys/kiab078
32. Dong G, Hutagalung AH, Fu C, Novick P, Reinisch KM (2005) The structures of exocyst subunit Exo70p and the Exo84p C-terminal domains reveal a common motif. Nat Struct Mol Biol 12(12):1094–1100. https://doi.org/10.1038/nsmb1017
33. D'Souza Z, Blackburn JB, Kudlyk T, Pokrovskaya ID, Lupashin VV (2019) Defects in COG-mediated Golgi trafficking Alter Endo-lysosomal system in human cells. Front Cell Dev Biol 7:118. https://doi.org/10.3389/fcell.2019.00118
34. D'Souza Z, Pokrovskaya I, Lupashin VV (2023) Syntaxin-5's flexibility in SNARE pairing supports Golgi functions. Traffic Cph Den 24(8):355–379. https://doi.org/10.1111/tra.12903
35. D'Souza Z, Sumya FT, Khakurel A, Lupashin V (2021) Getting sugar coating right! The role of the Golgi trafficking machinery in glycosylation. Cells 10(12):3275. https://doi.org/10.3390/cells10123275
36. D'Souza Z, Taher FS, Lupashin VV (2020) Golgi inCOGnito: from vesicle tethering to human disease. Biochim Biophys Acta Gen Subj 1864(11):129694. https://doi.org/10.1016/j.bbagen.2020.129694
37. Duan R, Marafi D, Xia Z-J, Ng BG, Maroofian R, Sumya FT, Saad AK, Du H, Fatih JM, Hunter JV, Elbendary HM, Baig SM, Abdullah U, Ali Z, Efthymiou S, Murphy D, Mitani T, Withers MA, Jhangiani SN, Coban-Akdemir Z, Calame DG, Pehlivan D, Gibbs RA, Posey JE, Houlden H, Lupashin VV, Zaki MS, Freeze HH, Lupski JR (2023) Biallelic missense variants in COG3 cause a congenital disorder of glycosylation with impairment of retrograde vesicular trafficking. J Inherit Metab Dis. https://doi.org/10.1002/jimd.12679
38. Elkis Y, Bel S, Rahimi R, Lerer-Goldstein T, Levin-Zaidman S, Babushkin T, Shpungin S, Nir U (2015) TMF/ARA160 governs the dynamic spatial orientation of the Golgi apparatus during sperm development. PLoS One 10(12):e0145277. https://doi.org/10.1371/journal.pone.0145277
39. Elsner M, Hashimoto H, Simpson JC, Cassel D, Nilsson T, Weiss M (2003) Spatiotemporal dynamics of the COPI vesicle machinery. EMBO Rep 4(10):1000–1004. https://doi.org/10.1038/sj.embor.embor942
40. Escrevente C, Bento-Lopes L, Ramalho JS, Barral DC (2021) Rab11 is required for lysosome exocytosis through the interaction with Rab3a, Sec15 and GRAB. J Cell Sci 134(11): jcs246694. https://doi.org/10.1242/jcs.246694
41. Farkas RM, Giansanti MG, Gatti M, Fuller MT (2003) The Drosophila Cog5 homologue is required for cytokinesis, cell elongation, and assembly of specialized Golgi architecture during spermatogenesis. Mol Biol Cell 14(1):190–200. https://doi.org/10.1091/mbc.e02-06-0343
42. Ferreira CR, Xia Z-J, Clément A, Parry DA, Davids M, Taylan F, Sharma P, Turgeon CT, Blanco-Sánchez B, Ng BG, Logan CV, Wolfe LA, Solomon BD, Cho MT, Douglas G, Carvalho DR, Bratke H, Haug MG, Phillips JB, Wegner J, Tiemeyer M, Aoki K, Nordgren A, Hammarsjö A, Duker AL, Rohena L, Hove HB, Ek J, Adams D, Tifft CJ, Onyekweli T, Weixel T, Macnamara E, Radtke K, Powis Z, Earl D, Gabriel M, Russi AHS, Brick L, Kozenko M, Tham E, Raymond KM, Phillips JA, Tiller GE, Wilson WG, Hamid R, Malicdan MCV, Nishimura G, Grigelioniene G, Jackson A, Westerfield M, Bober MB, Gahl WA, Freeze HH (2018) A recurrent De Novo heterozygous COG4 substitution leads to Saul-Wilson syndrome, disrupted vesicular trafficking, and altered proteoglycan glycosylation. Am J Hum Genet 103(4):553–567. https://doi.org/10.1016/j.ajhg.2018.09.003
43. Flanagan-Steet H, Johnson S, Smith RD, Bangiyeva J, Lupashin V, Steet R (2011) Mislocalization of large ARF-GEFs as a potential mechanism for BFA resistance in COG-deficient cells. Exp Cell Res 317(16):2342–2352. https://doi.org/10.1016/j.yexcr.2011.06.005

44. Fotso P, Koryakina Y, Pavliv O, Tsiomenko AB, Lupashin VV (2005) Cog1p plays a central role in the organization of the yeast conserved oligomeric Golgi complex. J Biol Chem 280(30):27613–27623. https://doi.org/10.1074/jbc.M504597200
45. Foulquier F (2009) COG defects, birth and rise! Biochim Biophys Acta 1792(9):896–902. https://doi.org/10.1016/j.bbadis.2008.10.020
46. Foulquier F, Ungar D, Reynders E, Zeevaert R, Mills P, García-Silva MT, Briones P, Winchester B, Morelle W, Krieger M, Annaert W, Matthijs G (2007) A new inborn error of glycosylation due to a Cog8 deficiency reveals a critical role for the Cog1-Cog8 interaction in COG complex formation. Hum Mol Genet 16(7):717–730. https://doi.org/10.1093/hmg/ddl476
47. Foulquier F, Vasile E, Schollen E, Callewaert N, Raemaekers T, Quelhas D, Jaeken J, Mills P, Winchester B, Krieger M, Annaert W, Matthijs G (2006) Conserved oligomeric Golgi complex subunit 1 deficiency reveals a previously uncharacterized congenital disorder of glycosylation type II. Proc Natl Acad Sci 103(10):3764–3769. https://doi.org/10.1073/pnas.0507685103
48. Frappaolo A, Sechi S, Kumagai T, Robinson S, Fraschini R, Ghahnavieh AK, Belloni G, Piergentili R, Tiemeyer KH, Tiemeyer M, Giansanti MG (2017) COG7 deficiency in *Drosophila* generates multifaceted developmental, behavioral, and protein glycosylation phenotypes. J Cell Sci jcs.209049. https://doi.org/10.1242/jcs.209049
49. Fukuda M, Kanno E, Ishibashi K, Itoh T (2008) Large scale screening for novel Rab effectors reveals unexpected broad Rab binding specificity. Mol Cell Proteomics 7(6):1031–1042. https://doi.org/10.1074/mcp.M700569-MCP200
50. Fung CW, Matthijs G, Sturiale L, Garozzo D, Wong KY, Wong R, Wong V, Jaeken J (2012) COG5-CDG with a mild neurohepatic presentation. JIMD Rep 3:67–70. https://doi.org/10.1007/8904_2011_61
51. Ghoshouni H, Sarmadian R, Irilouzadian R, Biglari HN, Gilani A (2023) A rare case of Cerebrotendinous Xanthomatosis associated with a mutation on COG8 gene. J Investig Med High Impact Case Rep 11:23247096231168109. https://doi.org/10.1177/23247096231168109
52. Gilchrist A, Au CE, Hiding J, Bell AW, Fernandez-Rodriguez J, Lesimple S, Nagaya H, Roy L, Gosline SJC, Hallett M, Paiement J, Kearney RE, Nilsson T, Bergeron JJM (2006) Quantitative proteomics analysis of the secretory pathway. Cell 127(6):1265–1281. https://doi.org/10.1016/j.cell.2006.10.036
53. Gilleron J, Chafik A, Lacas-Gervais S, Tanti J-F, Cormont M (2024) Golgi-associated retrograde protein (GARP) complex-dependent endosomes to trans Golgi network retrograde trafficking is controlled by Rab4b. Cell Mol Biol Lett 29(1):54. https://doi.org/10.1186/s11658-024-00574-w
54. Gingras RM, Sulpizio AM, Park J, Bretscher A (2022) High-resolution secretory timeline from vesicle formation at the Golgi to fusion at the plasma membrane in S. cerevisiae. eLife 11: e78750. https://doi.org/10.7554/eLife.78750
55. Giot L, Bader JS, Brouwer C, Chaudhuri A, Kuang B, Li Y, Hao YL, Ooi CE, Godwin B, Vitols E, Vijayadamodar G, Pochart P, Machineni H, Welsh M, Kong Y, Zerhusen B, Malcolm R, Varrone Z, Collis A, Minto M, Burgess S, McDaniel L, Stimpson E, Spriggs F, Williams J, Neurath K, Ioime N, Agee M, Voss E, Furtak K, Renzulli R, Aanensen N, Carrolla S, Bickelhaupt E, Lazovatsky Y, DaSilva A, Zhong J, Stanyon CA, Finley RL, White KP, Braverman M, Jarvie T, Gold S, Leach M, Knight J, Shimkets RA, McKenna MP, Chant J, Rothberg JM (2003) A protein interaction map of Drosophila melanogaster. Science 302(5651):1727–1736. https://doi.org/10.1126/science.1090289
56. Guillard M, Morava E, van Delft FL, Hague R, Körner C, Adamowicz M, Wevers RA, Lefeber DJ (2011) Plasma N-glycan profiling by mass spectrometry for congenital disorders of glycosylation type II. Clin Chem 57(4):593–602. https://doi.org/10.1373/clinchem.2010.153635

57. Guo W, Roth D, Walch-Solimena C, Novick P (1999) The exocyst is an effector for Sec4p, targeting secretory vesicles to sites of exocytosis. EMBO J 18(4):1071–1080. https://doi.org/10.1093/emboj/18.4.1071
58. Guo Z, Johnston W, Kovtun O, Mureev S, Bröcker C, Ungermann C, Alexandrov K (2013) Subunit organisation of in vitro reconstituted HOPS and CORVET multisubunit membrane tethering complexes. PLoS One 8(12):e81534. https://doi.org/10.1371/journal.pone.0081534
59. Ha JY, Chou H-T, Ungar D, Yip CK, Walz T, Hughson FM (2016) Molecular architecture of the complete COG tethering complex. Nat Struct Mol Biol 23(8):758–760. https://doi.org/10.1038/nsmb.3263
60. Ha JY, Pokrovskaya ID, Climer LK, Shimamura GR, Kudlyk T, Jeffrey PD, Lupashin VV, Hughson FM (2014) Cog5-Cog7 crystal structure reveals interactions essential for the function of a multisubunit tethering complex. Proc Natl Acad Sci USA 111(44):15762–15767. https://doi.org/10.1073/pnas.1414829111
61. Ho R, Stroupe C (2015) The HOPS/class C Vps complex tethers membranes by binding to one Rab GTPase in each apposed membrane. Mol Biol Cell 26(14):2655–2663. https://doi.org/10.1091/mbc.E14-04-0922
62. Homma Y, Fukuda M (2021) Knockout analysis of Rab6 effector proteins revealed the role of VPS52 in the secretory pathway. Biochem Biophys Res Commun 561:151–157. https://doi.org/10.1016/j.bbrc.2021.05.009
63. Homma Y, Hiragi S, Fukuda M (2021) Rab family of small GTPases: an updated view on their regulation and functions. FEBS J 288(1):36–55. https://doi.org/10.1111/febs.15453
64. Hughson FM, Reinisch KM (2010) Structure and mechanism in membrane trafficking. Curr Opin Cell Biol 22(4):454–460. https://doi.org/10.1016/j.ceb.2010.03.011
65. Huybrechts S, De Laet C, Bontems P, Rooze S, Souayah H, Sznajer Y, Sturiale L, Garozzo D, Matthijs G, Ferster A, Jaeken J, Goyens P (2012) Deficiency of subunit 6 of the Conserved Oligomeric Golgi Complex (COG6-CDG): second patient, different phenotype. JIMD Rep 4: 103–108. https://doi.org/10.1007/8904_2011_79
66. Ishii M, Lupashin VV, Nakano A (2018) Detailed analysis of the interaction of yeast COG complex. Cell Struct Funct 43(2):119–127. https://doi.org/10.1247/csf.18014
67. Ishikawa T, Machida C, Yoshioka Y, Ueda T, Nakano A, Machida Y (2008) EMBRYO YELLOW gene, encoding a subunit of the conserved oligomeric Golgi complex, is required for appropriate cell expansion and meristem organization in Arabidopsis thaliana. Genes Cells Devoted Mol Cell Mech 13(6):521–535. https://doi.org/10.1111/j.1365-2443.2008.01186.x
68. Jaeken J (2013) Congenital disorders of glycosylation. Handb Clin Neurol 113:1737–1743. https://doi.org/10.1016/B978-0-444-59565-2.00044-7
69. Kaletta T, Hengartner MO (2006) Finding function in novel targets: C. elegans as a model organism. Nat Rev Drug Discov 5(5):387–398. https://doi.org/10.1038/nrd2031
70. Khakurel A, Lupashin VV (2023) Role of GARP vesicle tethering complex in Golgi physiology. Int J Mol Sci 24(7):6069. https://doi.org/10.3390/ijms24076069
71. Kim DW, Sacher M, Scarpa A, Quinn AM, Ferro-Novick S (1999) High-copy suppressor analysis reveals a physical interaction between Sec34p and Sec35p, a protein implicated in vesicle docking. Mol Biol Cell 10(10):3317–3329. https://doi.org/10.1091/mbc.10.10.3317
72. Kim YO, Yun M, Jeong JH, Choi SM, Kim SK, Yoon W, Park C, Hong Y, Woo YJ (2017) A mild form of COG5 defect showing early-childhood-onset Friedreich's-Ataxia-Like phenotypes with isolated cerebellar atrophy. J Korean Med Sci 32(11):1885–1890. https://doi.org/10.3346/jkms.2017.32.11.1885
73. Kingsley DM, Kozarsky KF, Segal M, Krieger M (1986) Three types of low density lipoprotein receptor-deficient mutant have pleiotropic defects in the synthesis of N-linked, O-linked, and lipid-linked carbohydrate chains. J Cell Biol 102(5):1576–1585. https://doi.org/10.1083/jcb.102.5.1576
74. Klinger CM, Klute MJ, Dacks JB (2013) Comparative genomic analysis of multi-subunit tethering complexes demonstrates an ancient pan-eukaryotic complement and sculpting in Apicomplexa. PLoS One 8(9):e76278. https://doi.org/10.1371/journal.pone.0076278

75. Klink VP, Darwish O, Alkharouf NW, Lawaju BR, Khatri R, Lawrence KS (2021) Conserved oligomeric Golgi (COG) complex genes functioning in defense are expressed in root cells undergoing a defense response to a pathogenic infection and exhibit regulation my MAPKs. PLoS One 16(8):e0256472. https://doi.org/10.1371/journal.pone.0256472
76. Kodera H, Ando N, Yuasa I, Wada Y, Tsurusaki Y, Nakashima M, Miyake N, Saitoh S, Matsumoto N, Saitsu H (2015) Mutations in COG2 encoding a subunit of the conserved oligomeric golgi complex cause a congenital disorder of glycosylation. Clin Genet 87(5): 455–460. https://doi.org/10.1111/cge.12417
77. Kong C, Zheng T, Wang S, Li Q, Hua S (2021) Congenital disorder of glycosylation caused by compound heterozygous mutation of the COG6 gene: a case report and literature review. Chin J Perinat Med 12:113–119
78. Koumandou VL, Dacks JB, Coulson RMR, Field MC (2007) Control systems for membrane fusion in the ancestral eukaryote; evolution of tethering complexes and SM proteins. BMC Evol Biol 7:29. https://doi.org/10.1186/1471-2148-7-29
79. Kranz C, Ng BG, Sun L, Sharma V, Eklund EA, Miura Y, Ungar D, Lupashin V, Winkel RD, Cipollo JF, Costello CE, Loh E, Hong W, Freeze HH (2007) COG8 deficiency causes new congenital disorder of glycosylation type IIh. Hum Mol Genet 16(7):731–741. https://doi.org/10.1093/hmg/ddm028
80. Kraynack BA, Chan A, Rosenthal E, Essid M, Umansky B, Waters MG, Schmitt HD (2005) Dsl1p, Tip20p, and the novel Dsl3(Sec39) protein are required for the stability of the Q/t-SNARE complex at the endoplasmic reticulum in yeast. Mol Biol Cell 16(9):3963–3977. https://doi.org/10.1091/mbc.e05-01-0056
81. Kristensen AR, Gsponer J, Foster LJ (2012) A high-throughput approach for measuring temporal changes in the interactome. Nat Methods 9(9):907–909. https://doi.org/10.1038/nmeth.2131
82. Kubota Y, Sano M, Goda S, Suzuki N, Nishiwaki K (2006) The conserved oligomeric Golgi complex acts in organ morphogenesis via glycosylation of an ADAM protease in C. elegans. Dev Camb Engl 133(2):263–273. https://doi.org/10.1242/dev.02195
83. Kudlyk T, Willett R, Pokrovskaya ID, Lupashin V (2013) COG6 interacts with a subset of the Golgi SNAREs and is important for the Golgi complex integrity. Traffic Cph Den 14(2): 194–204. https://doi.org/10.1111/tra.12020
84. Laufman O, Freeze HH, Hong W, Lev S (2013a) Deficiency of the Cog8 subunit in normal and CDG-derived cells impairs the assembly of the COG and Golgi SNARE complexes. Traffic Cph Den 14(10):1065–1077. https://doi.org/10.1111/tra.12093
85. Laufman O, Hong W, Lev S (2011) The COG complex interacts directly with Syntaxin 6 and positively regulates endosome-to-TGN retrograde transport. J Cell Biol 194(3):459–472. https://doi.org/10.1083/jcb.201102045
86. Laufman O, Hong W, Lev S (2013b) The COG complex interacts with multiple Golgi SNAREs and enhances fusogenic assembly of SNARE complexes. J Cell Sci 126(Pt 6): 1506–1516. https://doi.org/10.1242/jcs.122101
87. Laufman O, Kedan A, Hong W, Lev S (2009) Direct interaction between the COG complex and the SM protein, Sly1, is required for Golgi SNARE pairing. EMBO J 28(14):2006–2017. https://doi.org/10.1038/emboj.2009.168
88. Lees JA, Yip CK, Walz T, Hughson FM (2010) Molecular organization of the COG vesicle tethering complex. Nat Struct Mol Biol 17(11):1292–1297. https://doi.org/10.1038/nsmb.1917
89. Lefeber DJ, Morava E, Jaeken J (2011) How to find and diagnose a CDG due to defective N-glycosylation. J Inherit Metab Dis 34(4):849–852. https://doi.org/10.1007/s10545-011-9370-0
90. Li G, Xu Y, Hu X, Li N, Yao R, Yu T, Wang X, Guo W, Wang J (2019) Compound heterozygous variants of the COG6 gene in a Chinese patient with deficiency of subunit 6 of the conserved oligomeric Golgi complex (COG6-CDG). Eur J Med Genet 62(1):44–46. https://doi.org/10.1016/j.ejmg.2018.04.017

91. Li S, Armstrong CM, Bertin N, Ge H, Milstein S, Boxem M, Vidalain P-O, Han J-DJ, Chesneau A, Hao T, Goldberg DS, Li N, Martinez M, Rual J-F, Lamesch P, Xu L, Tewari M, Wong SL, Zhang LV, Berriz GF, Jacotot L, Vaglio P, Reboul J, Hirozane-Kishikawa T, Li Q, Gabel HW, Elewa A, Baumgartner B, Rose DJ, Yu H, Bosak S, Sequerra R, Fraser A, Mango SE, Saxton WM, Strome S, Van Den Heuvel S, Piano F, Vandenhaute J, Sardet C, Gerstein M, Doucette-Stamm L, Gunsalus KC, Harper JW, Cusick ME, Roth FP, Hill DE, Vidal M (2004) A map of the interactome network of the metazoan C. elegans. Science 303(5657):540–543. https://doi.org/10.1126/science.1091403
92. Li W, Ma C, Guan R, Xu Y, Tomchick DR, Rizo J (2011) The crystal structure of a Munc13 C-terminal module exhibits a remarkable similarity to vesicle tethering factors. Struct Lond Engl 1993 19(10):1443–1455. https://doi.org/10.1016/j.str.2011.07.012
93. Liu D, Li X, Shen D, Novick P (2018) Two subunits of the exocyst, Sec3p and Exo70p, can function exclusively on the plasma membrane. Mol Biol Cell 29(6):736–750. https://doi.org/10.1091/mbc.E17-08-0518
94. Liu S, Majeed W, Grigaitis P, Betts MJ, Climer LK, Starkuviene V, Storrie B (2019) Epistatic analysis of the contribution of Rabs and Kifs to CATCHR family dependent Golgi organization. Front Cell Dev Biol 7:126. https://doi.org/10.3389/fcell.2019.00126
95. Lübbehusen J, Thiel C, Rind N, Ungar D, Prinsen BHCMT, de Koning TJ, van Hasselt PM, Körner C (2010) Fatal outcome due to deficiency of subunit 6 of the conserved oligomeric Golgi complex leading to a new type of congenital disorders of glycosylation. Hum Mol Genet 19(18):3623–3633. https://doi.org/10.1093/hmg/ddq278
96. Luo L, Hannemann M, Koenig S, Hegermann J, Ailion M, Cho M-K, Sasidharan N, Zweckstetter M, Rensing SA, Eimer S (2011) The Caenorhabditis elegans GARP complex contains the conserved Vps51 subunit and is required to maintain lysosomal morphology. Mol Biol Cell 22(14):2564–2578. https://doi.org/10.1091/mbc.E10-06-0493
97. Maekawa M, Inoue T, Kobuna H, Nishimura T, Gengyo-Ando K, Mitani S, Arai H (2009) Functional analysis of GS28, an intra-Golgi SNARE, in Caenorhabditis elegans. Genes Cells Devoted Mol Cell Mech 14(8):1003–1013. https://doi.org/10.1111/j.1365-2443.2009.01325.x
98. Makhamreh MM, Cottingham N, Ferreira CR, Berger S, Al-Kouatly HB (2020) Nonimmune hydrops fetalis and congenital disorders of glycosylation: a systematic literature review. J Inherit Metab Dis 43(2):223–233. https://doi.org/10.1002/jimd.12162
99. Mei K, Guo W (2018) The exocyst complex. Curr Biol CB 28(17):R922–R925. https://doi.org/10.1016/j.cub.2018.06.042
100. Mei K, Li Y, Wang S, Shao G, Wang J, Ding Y, Luo G, Yue P, Liu J-J, Wang X, Dong M-Q, Wang H-W, Guo W (2018) Cryo-EM structure of the exocyst complex. Nat Struct Mol Biol 25(2):139–146. https://doi.org/10.1038/s41594-017-0016-2
101. Meiringer CTA, Rethmeier R, Auffarth K, Wilson J, Perz A, Barlowe C, Schmitt HD, Ungermann C (2011) The Dsl1 protein tethering complex is a resident endoplasmic reticulum complex, which interacts with five soluble NSF (N-ethylmaleimide-sensitive factor) attachment protein receptors (SNAREs): implications for fusion and fusion regulation. J Biol Chem 286(28):25039–25046. https://doi.org/10.1074/jbc.M110.215327
102. Miller VJ, Sharma P, Kudlyk TA, Frost L, Rofe AP, Watson IJ, Duden R, Lowe M, Lupashin VV, Ungar D (2013) Molecular insights into vesicle tethering at the Golgi by the Conserved Oligomeric Golgi (COG) complex and the Golgin TATA Element Modulatory Factor (TMF). J Biol Chem 288(6):4229–4240. https://doi.org/10.1074/jbc.M112.426767
103. Miller VJ, Ungar D (2012) Re'COG'nition at the Golgi. Traffic Cph Den 13(7):891–897. https://doi.org/10.1111/j.1600-0854.2012.01338.x
104. Morava E, Zeevaert R, Korsch E, Huijben K, Wopereis S, Matthijs G, Keymolen K, Lefeber DJ, De Meirleir L, Wevers RA (2007) A common mutation in the COG7 gene with a consistent phenotype including microcephaly, adducted thumbs, growth retardation, VSD and episodes of hyperthermia. Eur J Hum Genet EJHG 15(6):638–645. https://doi.org/10.1038/sj.ejhg.5201813

105. Moskalenko S, Tong C, Rosse C, Mirey G, Formstecher E, Daviet L, Camonis J, White MA (2003) Ral GTPases regulate exocyst assembly through dual subunit interactions. J Biol Chem 278(51):51743–51748. https://doi.org/10.1074/jbc.M308702200
106. Niu T-K, Pfeifer AC, Lippincott-Schwartz J, Jackson CL (2005) Dynamics of GBF1, a Brefeldin A-sensitive Arf1 exchange factor at the Golgi. Mol Biol Cell 16(3):1213–1222. https://doi.org/10.1091/mbc.e04-07-0599
107. Oka T, Ungar D, Hughson FM, Krieger M (2004) The COG and COPI complexes interact to control the abundance of GEARs, a subset of Golgi integral membrane proteins. Mol Biol Cell 15(5):2423–2435. https://doi.org/10.1091/mbc.e03-09-0699
108. Oka T, Vasile E, Penman M, Novina CD, Dykxhoorn DM, Ungar D, Hughson FM, Krieger M (2005) Genetic analysis of the subunit organization and function of the conserved oligomeric golgi (COG) complex: studies of COG5- and COG7-deficient mammalian cells. J Biol Chem 280(38):32736–32745. https://doi.org/10.1074/jbc.M505558200
109. Ostertag M, Stammler J, Douchkov D, Eichmann R, Hückelhoven R (2013) The conserved oligomeric Golgi complex is involved in penetration resistance of barley to the barley powdery mildew fungus. Mol Plant Pathol 14(3):230–240. https://doi.org/10.1111/j.1364-3703.2012.00846.x
110. Paesold-Burda P, Maag C, Troxler H, Foulquier F, Kleinert P, Schnabel S, Baumgartner M, Hennet T (2009) Deficiency in COG5 causes a moderate form of congenital disorders of glycosylation. Hum Mol Genet 18(22):4350–4356. https://doi.org/10.1093/hmg/ddp389
111. Péanne R, de Lonlay P, Foulquier F, Kornak U, Lefeber DJ, Morava E, Pérez B, Seta N, Thiel C, Van Schaftingen E, Matthijs G, Jaeken J (2018) Congenital disorders of glycosylation (CDG): Quo vadis? Eur J Med Genet 61(11):643–663. https://doi.org/10.1016/j.ejmg.2017.10.012
112. Pei J, Ma C, Rizo J, Grishin NV (2009) Remote homology between Munc13 MUN domain and vesicle tethering complexes. J Mol Biol 391(3):509–517. https://doi.org/10.1016/j.jmb.2009.06.054
113. Pérez-Victoria FJ, Abascal-Palacios G, Tascón I, Kajava A, Magadán JG, Pioro EP, Bonifacino JS, Hierro A (2010) Structural basis for the wobbler mouse neurodegenerative disorder caused by mutation in the Vps54 subunit of the GARP complex. Proc Natl Acad Sci USA 107(29):12860–12865. https://doi.org/10.1073/pnas.1004756107
114. Pérez-Victoria FJ, Bonifacino JS (2009) Dual roles of the mammalian GARP complex in tethering and SNARE complex assembly at the trans-golgi network. Mol Cell Biol 29(19):5251–5263. https://doi.org/10.1128/MCB.00495-09
115. Podos SD, Reddy P, Ashkenas J, Krieger M (1994) LDLC encodes a brefeldin A-sensitive, peripheral Golgi protein required for normal Golgi function. J Cell Biol 127(3):679–691. https://doi.org/10.1083/jcb.127.3.679
116. Pokrovskaya ID, Willett R, Smith RD, Morelle W, Kudlyk T, Lupashin VV (2011) Conserved oligomeric Golgi complex specifically regulates the maintenance of Golgi glycosylation machinery. Glycobiology 21(12):1554–1569. https://doi.org/10.1093/glycob/cwr028
117. Ram RJ, Li B, Kaiser CA (2002) Identification of Sec36p, Sec37p, and Sec38p: components of yeast complex that contains Sec34p and Sec35p. Mol Biol Cell 13(5):1484–1500. https://doi.org/10.1091/mbc.01-10-0495
118. Raynor A, Vincent-Delorme C, Alaix A-S, Cholet S, Dupré T, Vuillaumier-Barrot S, Fenaille F, Besmond C, Bruneel A (2021) Normal transferrin patterns in congenital disorders of glycosylation with Golgi homeostasis disruption: apolipoprotein C-III at the rescue! Clin Chim Acta Int J Clin Chem 519:285–290. https://doi.org/10.1016/j.cca.2021.05.016
119. Ren Y, Yip CK, Tripathi A, Huie D, Jeffrey PD, Walz T, Hughson FM (2009) A structure-based mechanism for vesicle capture by the multisubunit tethering complex Dsl1. Cell 139(6):1119–1129. https://doi.org/10.1016/j.cell.2009.11.002
120. Reynders E, Foulquier F, Leão Teles E, Quelhas D, Morelle W, Rabouille C, Annaert W, Matthijs G (2009) Golgi function and dysfunction in the first COG4-deficient CDG type II patient. Hum Mol Genet 18(17):3244–3256. https://doi.org/10.1093/hmg/ddp262

121. Richardson BC, Smith RD, Ungar D, Nakamura A, Jeffrey PD, Lupashin VV, Hughson FM (2009) Structural basis for a human glycosylation disorder caused by mutation of the COG4 gene. Proc Natl Acad Sci USA 106(32):13329–13334. https://doi.org/10.1073/pnas.0901966106
122. Rossi G, Puller GC, Brennwald P (2024) Asymmetric tethering by exocyst in vitro requires a Rab GTPase, an R-SNARE and a Sac1-sensitive phosphoinositide lipid. Mol Biol Cell 35(3): br8. https://doi.org/10.1091/mbc.E23-08-0311
123. Rui Q, Tan X, Liu F, Li Y, Liu X, Li B, Wang J, Yang H, Qiao L, Li T, Fang S, Gao R, Wang W, Bednarek SY, Bao Y (2021) Syntaxin of plants31 (SYP31) and SYP32 is essential for Golgi morphology maintenance and pollen development. Plant Physiol 186(1):330–343. https://doi.org/10.1093/plphys/kiab049
124. Rymen D, Keldermans L, Race V, Régal L, Deconinck N, Dionisi-Vici C, Fung C, Sturiale L, Rosnoblet C, Foulquier F, Matthijs G, Jaeken J (2012) COG5-CDG: expanding the clinical spectrum. Orphanet J Rare Dis 7(1):94. https://doi.org/10.1186/1750-1172-7-94
125. Rymen D, Winter J, Van Hasselt PM, Jaeken J, Kasapkara C, Gokçay G, Haijes H, Goyens P, Tokatli A, Thiel C, Bartsch O, Hecht J, Krawitz P, Prinsen HCMT, Mildenberger E, Matthijs G, Kornak U (2015) Key features and clinical variability of COG6-CDG. Mol Genet Metab 116(3):163–170. https://doi.org/10.1016/j.ymgme.2015.07.003
126. Santana-Molina C, Gutierrez F, Devos DP (2021) Homology and modular evolution of CATCHR at the origin of the eukaryotic endomembrane system. Genome Biol Evol 13(7): evab125. https://doi.org/10.1093/gbe/evab125
127. Shaheen R, Ansari S, Alshammari MJ, Alkhalidi H, Alrukban H, Eyaid W, Alkuraya FS (2013) A novel syndrome of hypohidrosis and intellectual disability is linked to COG6 deficiency. J Med Genet 50(7):431–436. https://doi.org/10.1136/jmedgenet-2013-101527
128. Shestakova A, Suvorova E, Pavliv O, Khaidakova G, Lupashin V (2007) Interaction of the conserved oligomeric Golgi complex with t-SNARE Syntaxin5a/Sed5 enhances intra-Golgi SNARE complex stability. J Cell Biol 179(6):1179–1192. https://doi.org/10.1083/jcb.200705145
129. Shestakova A, Zolov S, Lupashin V (2006) COG complex-mediated recycling of Golgi glycosyltransferases is essential for normal protein glycosylation. Traffic Cph Den 7(2): 191–204. https://doi.org/10.1111/j.1600-0854.2005.00376.x
130. Sivaram MVS, Furgason MLM, Brewer DN, Munson M (2006) The structure of the exocyst subunit Sec6p defines a conserved architecture with diverse roles. Nat Struct Mol Biol 13(6): 555–556. https://doi.org/10.1038/nsmb1096
131. Smith RD, Willett R, Kudlyk T, Pokrovskaya I, Paton AW, Paton JC, Lupashin VV (2009) The COG complex, Rab6 and COPI define a novel Golgi retrograde trafficking pathway that is exploited by SubAB toxin. Traffic Cph Den 10(10):1502–1517. https://doi.org/10.1111/j.1600-0854.2009.00965.x
132. Sohda M, Misumi Y, Yamamoto A, Nakamura N, Ogata S, Sakisaka S, Hirose S, Ikehara Y, Oda K (2010) Interaction of Golgin-84 with the COG complex mediates the intra-Golgi retrograde transport. Traffic Cph Den 11(12):1552–1566. https://doi.org/10.1111/j.1600-0854.2010.01123.x
133. Sohda M, Misumi Y, Yoshimura S-I, Nakamura N, Fusano T, Ogata S, Sakisaka S, Ikehara Y (2007) The interaction of two tethering factors, p115 and COG complex, is required for Golgi integrity. Traffic Cph Den 8(3):270–284. https://doi.org/10.1111/j.1600-0854.2006.00530.x
134. Spaapen LJM, Bakker JA, van der Meer SB, Sijstermans HJ, Steet RA, Wevers RA, Jaeken J (2005) Clinical and biochemical presentation of siblings with COG-7 deficiency, a lethal multiple O- and N-glycosylation disorder. J Inherit Metab Dis 28(5):707–714. https://doi.org/10.1007/s10545-005-0015-z
135. Spang A (2012) The DSL1 complex: the smallest but not the least CATCHR. Traffic Cph Den 13(7):908–913. https://doi.org/10.1111/j.1600-0854.2012.01362.x

136. Spelbrink RG, Nothwehr SF (1999) The yeast GRD20 gene is required for protein sorting in the trans-Golgi network/endosomal system and for polarization of the actin cytoskeleton. Mol Biol Cell 10(12):4263–4281. https://doi.org/10.1091/mbc.10.12.4263
137. Spessott W, Uliana A, Maccioni HJF (2010a) Defective GM3 synthesis in Cog2 null mutant CHO cells associates to mislocalization of lactosylceramide sialyltransferase in the Golgi complex. Neurochem Res 35(12):2161–2167. https://doi.org/10.1007/s11064-010-0319-8
138. Spessott W, Uliana A, Maccioni HJF (2010b) Cog2 null mutant CHO cells show defective sphingomyelin synthesis. J Biol Chem 285(53):41472–41482. https://doi.org/10.1074/jbc.M110.150011
139. Steet R, Kornfeld S (2006) COG-7-deficient human fibroblasts exhibit altered recycling of Golgi proteins. Mol Biol Cell 17(5):2312–2321. https://doi.org/10.1091/mbc.e05-08-0822
140. Struwe WB, Reinhold VN (2012) The conserved oligomeric Golgi complex is required for fucosylation of N-glycans in Caenorhabditis elegans. Glycobiology 22(6):863–875. https://doi.org/10.1093/glycob/cws053
141. Sumya FT, Pokrovskaya ID, D'Souza Z, Lupashin VV (2023a) Acute COG complex inactivation unveiled its immediate impact on Golgi and illuminated the nature of intra-Golgi recycling vesicles. Traffic Cph Den 24(2):52–75. https://doi.org/10.1111/tra.12876
142. Sumya FT, Pokrovskaya ID, Lupashin V (2021) Development and initial characterization of cellular models for COG complex-related CDG-II diseases. Front Genet 12:733048. https://doi.org/10.3389/fgene.2021.733048
143. Sumya FT, Pokrovskaya ID, Lupashin VV (2023b) Rapid COG depletion in mammalian cell by Auxin-inducible degradation system. Methods Mol Biol Clifton NJ 2557:365–390. https://doi.org/10.1007/978-1-0716-2639-9_23
144. Suvorova ES, Duden R, Lupashin VV (2002) The Sec34/Sec35p complex, a Ypt1p effector required for retrograde intra-Golgi trafficking, interacts with Golgi SNAREs and COPI vesicle coat proteins. J Cell Biol 157(4):631–643. https://doi.org/10.1083/jcb.200111081
145. Tan X, Cao K, Liu F, Li Y, Li P, Gao C, Ding Y, Lan Z, Shi Z, Rui Q, Feng Y, Liu Y, Zhao Y, Wu C, Zhang Q, Li Y, Jiang L, Bao Y (2016) Arabidopsis COG complex subunits COG3 and COG8 modulate Golgi morphology, vesicle trafficking homeostasis and are essential for pollen tube growth. PLoS Genet 12(7):e1006140. https://doi.org/10.1371/journal.pgen.1006140
146. Travis SM, DAmico K, Yu I-M, MCmahon C, Hamid S, Ramirez-Arellano G, Jeffrey PD, Hughson FM (2020) Structural basis for the binding of SNAREs to the multisubunit tethering complex Dsl1. J Biol Chem 295(30):10125–10135. https://doi.org/10.1074/jbc.RA120.013654
147. Tripathi A, Ren Y, Jeffrey PD, Hughson FM (2009) Structural characterization of Tip20p and Dsl1p, subunits of the Dsl1p vesicle tethering complex. Nat Struct Mol Biol 16(2):114–123. https://doi.org/10.1038/nsmb.1548
148. Ungar D, Oka T, Brittle EE, Vasile E, Lupashin VV, Chatterton JE, Heuser JE, Krieger M, Waters MG (2002) Characterization of a mammalian Golgi-localized protein complex, COG, that is required for normal Golgi morphology and function. J Cell Biol 157(3):405–415. https://doi.org/10.1083/jcb.200202016
149. Ungar D, Oka T, Vasile E, Krieger M, Hughson FM (2005) Subunit architecture of the conserved oligomeric Golgi complex. J Biol Chem 280(38):32729–32735. https://doi.org/10.1074/jbc.M504590200
150. VanRheenen SM, Cao X, Lupashin VV, Barlowe C, Gerard Waters M (1998) Sec35p, a novel peripheral membrane protein, is required for ER to Golgi vesicle docking. J Cell Biol 141(5):1107–1119. https://doi.org/10.1083/jcb.141.5.1107
151. VanRheenen SM, Cao X, Sapperstein SK, Chiang EC, Lupashin VV, Barlowe C, Waters MG (1999) Sec34p, a protein required for vesicle tethering to the yeast Golgi apparatus, is in a complex with Sec35p. J Cell Biol 147(4):729–742
152. Vasan N, Hutagalung A, Novick P, Reinisch KM (2010) Structure of a C-terminal fragment of its Vps53 subunit suggests similarity of Golgi-associated retrograde protein (GARP) complex

to a family of tethering complexes. Proc Natl Acad Sci USA 107(32):14176–14181. https://doi.org/10.1073/pnas.1009419107
153. Vasile E, Oka T, Ericsson M, Nakamura N, Krieger M (2006) IntraGolgi distribution of the Conserved Oligomeric Golgi (COG) complex. Exp Cell Res 312(16):3132–3141. https://doi.org/10.1016/j.yexcr.2006.06.005
154. Walter DM, Paul KS, Waters MG (1998) Purification and characterization of a novel 13 S hetero-oligomeric protein complex that stimulates in vitro Golgi transport. J Biol Chem 273(45):29565–29576. https://doi.org/10.1074/jbc.273.45.29565
155. Wang X, Han L, Wang X-Y, Wang J-H, Li X-M, Jin C-H, Wang L (2020) Identification of two novel mutations in COG5 causing congenital disorder of glycosylation. Front Genet 11:168. https://doi.org/10.3389/fgene.2020.00168
156. Wang Y-H, Chiu W-Y, Chen Y-T, Cai P-J, Wu Y-C, Wu J-L, Chen B-H, Liu Y-W, Yu C-J, Lee F-JS (2022) Golgin Imh1 and GARP complex cooperate to restore the impaired SNARE recycling transport induced by ER stress. Cell Rep 38(12):110488. https://doi.org/10.1016/j.celrep.2022.110488
157. Whyte JR, Munro S (2001) The Sec34/35 Golgi transport complex is related to the exocyst, defining a family of complexes involved in multiple steps of membrane traffic. Dev Cell 1(4): 527–537. https://doi.org/10.1016/s1534-5807(01)00063-6
158. Willett R, Blackburn JB, Climer L, Pokrovskaya I, Kudlyk T, Wang W, Lupashin V (2016) COG lobe B sub-complex engages v-SNARE GS15 and functions via regulated interaction with lobe A sub-complex. Sci Rep 6:29139. https://doi.org/10.1038/srep29139
159. Willett R, Kudlyk T, Pokrovskaya I, Schönherr R, Ungar D, Duden R, Lupashin V (2013a) COG complexes form spatial landmarks for distinct SNARE complexes. Nat Commun 4:1553. https://doi.org/10.1038/ncomms2535
160. Willett R, Pokrovskaya I, Kudlyk T, Lupashin V (2014) Multipronged interaction of the COG complex with intracellular membranes. Cell Logist 4(1):e27888. https://doi.org/10.4161/cl.27888
161. Willett R, Ungar D, Lupashin V (2013b) The Golgi puppet master: COG complex at center stage of membrane trafficking interactions. Histochem Cell Biol 140(3):271–283. https://doi.org/10.1007/s00418-013-1117-6
162. Willett RA, Pokrovskaya ID, Lupashin VV (2013c) Fluorescent microscopy as a tool to elucidate dysfunction and mislocalization of Golgi glycosyltransferases in COG complex depleted mammalian cells. Methods Mol Biol Clifton NJ 1022:61–72. https://doi.org/10.1007/978-1-62703-465-4_6
163. Wopereis S, Morava É, Grünewald S, Adamowicz M, Huijben KMLC, Lefeber DJ, Wevers RA (2005) Patients with unsolved congenital disorders of glycosylation type II can be subdivided in six distinct biochemical groups. Glycobiology 15(12):1312–1319. https://doi.org/10.1093/glycob/cwj017
164. Wu S, Mehta SQ, Pichaud F, Bellen HJ, Quiocho FA (2005) Sec15 interacts with Rab11 via a novel domain and affects Rab11 localization in vivo. Nat Struct Mol Biol 12(10):879–885. https://doi.org/10.1038/nsmb987
165. Wu X, Steet RA, Bohorov O, Bakker J, Newell J, Krieger M, Spaapen L, Kornfeld S, Freeze HH (2004) Mutation of the COG complex subunit gene COG7 causes a lethal congenital disorder. Nat Med 10(5):518–523. https://doi.org/10.1038/nm1041
166. Wuestehube LJ, Duden R, Eun A, Hamamoto S, Korn P, Ram R, Schekman R (1996) New mutants of Saccharomyces cerevisiae affected in the transport of proteins from the endoplasmic reticulum to the Golgi complex. Genetics 142(2):393–406. https://doi.org/10.1093/genetics/142.2.393
167. Xia Z-J, Mahajan S, Paul Daniel EJ, Ng BG, Saraswat M, Campos AR, Murad R, He M, Freeze HH (2022) COG4 mutation in Saul-Wilson syndrome selectively affects secretion of proteins involved in chondrogenesis in chondrocyte-like cells. Front Cell Dev Biol 10:979096. https://doi.org/10.3389/fcell.2022.979096

168. Xia Z-J, Zeng X-XI, Tambe M, Ng BG, Dong PDS, Freeze HH (2021) A dominant heterozygous mutation in COG4 causes Saul–Wilson syndrome, a Primordial Dwarfism, and disrupts zebrafish development via Wnt signaling. Front Cell Dev Biol 9:2450. https://doi.org/10.3389/fcell.2021.720688
169. Yang A, Cho SY, Jang J-H, Kim J, Kim SZ, Lee BH, Yoo H-W, Jin D-K (2017) Further delineation of COG8-CDG: a case with novel compound heterozygous mutations diagnosed by targeted exome sequencing. Clin Chim Acta Int J Clin Chem 471:191–195. https://doi.org/10.1016/j.cca.2017.06.010
170. Yin S, Gong L, Qiu H, Zhao Y, Zhang Y, Liu C, Jiang H, Mao Y, Kong L-Y, Liang B, Lv Y (2019) Novel compound heterozygous COG5 mutations in a Chinese male patient with severe clinical symptoms and type IIi congenital disorder of glycosylation: a case report. Exp Ther Med 18(4):2695–2700. https://doi.org/10.3892/etm.2019.7834
171. Yu I-M, Hughson FM (2010) Tethering factors as organizers of intracellular vesicular traffic. Annu Rev Cell Dev Biol 26(1):137–156. https://doi.org/10.1146/annurev.cellbio.042308.113327
172. Zafra I, Nebenfuehr B, Golden A (2021) Saul-Wilson syndrome Missense Allele does not show obvious Golgi defects in a C. elegans model. MicroPubl Biol 2021(03). https://doi.org/10.17912/micropub.biology.000373
173. Zeevaert R, Foulquier F, Cheillan D, Cloix I, Guffon N, Sturiale L, Garozzo D, Matthijs G, Jaeken J (2009a) A new mutation in COG7 extends the spectrum of COG subunit deficiencies. Eur J Med Genet 52(5):303–305. https://doi.org/10.1016/j.ejmg.2009.06.006
174. Zeevaert R, Foulquier F, Dimitrov B, Reynders E, Van Damme-Lombaerts R, Simeonov E, Annaert W, Matthijs G, Jaeken J (2009b) Cerebrocostomandibular-like syndrome and a mutation in the conserved oligomeric Golgi complex, subunit 1. Hum Mol Genet 18(3):517–524. https://doi.org/10.1093/hmg/ddn379
175. Zeevaert R, Foulquier F, Jaeken J, Matthijs G (2008) Deficiencies in subunits of the Conserved Oligomeric Golgi (COG) complex define a novel group of Congenital Disorders of Glycosylation. Mol Genet Metab 93(1):15–21. https://doi.org/10.1016/j.ymgme.2007.08.118
176. Zhang X-M, Ellis S, Sriratana A, Mitchell CA, Rowe T (2004) Sec15 is an effector for the Rab11 GTPase in mammalian cells. J Biol Chem 279(41):43027–43034. https://doi.org/10.1074/jbc.M402264200
177. Zhang X-Y, Zhang J, Lu Y (2024) COG6-CDG: two novel variants and milder phenotype in a Chinese patient. Hum Mutat 2024:1–7. https://doi.org/10.1155/2024/9857442
178. Zolov SN, Lupashin VV (2005) Cog3p depletion blocks vesicle-mediated Golgi retrograde trafficking in HeLa cells. J Cell Biol 168(5):747–759. https://doi.org/10.1083/jcb.200412003

Chapter 8
Lipid Metabolism and Trafficking in the Golgi Apparatus

Jan Parolek, Yeongho Kim, and Christopher G. Burd

Abstract In organelle biogenesis pathways, the Golgi apparatus plays a central role in synthesizing and distributing complex sphingolipids, glycerolipids, and cholesterol, which form the membranes of organelles. The concept of 'lipid flow' describes the circulation of lipids throughout the endomembrane system, mediated by vesicle-based carriers and lipid transfer proteins that directly transfer lipids at sites of organelle-organelle contact. The synthesis and trafficking of lipids are closely interconnected, as they undergo successive enzymatic transformations that begin in the endoplasmic reticulum and are completed in the Golgi apparatus before being distributed to other organelles. A phosphatidylinositol 4-phosphate signaling network that converges on the *trans* cisterna and *trans* Golgi network compartments regulates lipid flow to, though, and out of the Golgi apparatus, acting as a rheostat that tunes the rates and specificities of lipid trafficking by controlling phosphatidylinositol 4-phosphate levels in Golgi membranes. The integration of glycolipid and glycoprotein processing reactions within the Golgi with lipid sorting and trafficking pathways suggests that the Golgi apparatus serves as a lipid-based sorting station at the heart of organelle biogenesis pathways.

Keywords Lipid synthesis · Lipid transport · Golgi apparatus · Cholesterol · Sphingolipids · PI4P signaling

J. Parolek · Y. Kim · C. G. Burd (✉)
Department of Cell Biology, Yale School of Medicine, New Haven, CT, USA
e-mail: jan.parolek@yale.edu; yeongho.kim@yale.edu; christopher.burd@yale.edu

A. Nakano, J. Saraste (eds.), *The Golgi Network, Volume II*, Subcellular Biochemistry 111, https://doi.org/10.1007/978-3-032-16833-7_8

Introduction

The membrane of each organelle of the endomembrane network is composed of a unique collection of lipids and integral membrane proteins. Whereas the mechanisms of protein sorting and trafficking are well defined, knowledge of how lipid metabolism, sorting and trafficking contribute to organelle biogenesis is relatively rudimentary. The Golgi apparatus plays a foundational role in linking each of these aspects of lipid cell biology. Organelle membrane homeostasis pathways, which maintain the lipid compositions of organelle membranes, ensure organelle resilience by maintaining the barrier function of the membrane. This is achieved by the presence of integral membrane proteins and that catalyze the formation of distinct membrane lipids at specific sites within the cell. Indeed, the structures and functions of integral membrane proteins, including ion channels, signaling receptors, inter-organelle trafficking factors and other classes of proteins, have co-evolved with organelle membrane properties to optimize and regulate membrane properties that contribute to the unique function(s) of each integral membrane protein [5, 14, 42, 43]. Lipid synthesis and sorting pathways within the Golgi apparatus are vital for the biogenesis of every membrane-delimited organelle. Organelle membrane homeostasis pathways are also influenced by lipid degradation pathways, exemplified best by the lysosome storage disorders such as Niemann-Pick and Gaucher diseases, in which sphingolipids that are synthesized in the Golgi apparatus fail to be degraded properly in the lysosome, resulting in lipid accumulation and consequent loss of lysosome function.

The Membrane Landscape of Animal Cells

The distinctions between the lipid compositions of different organelle membranes center on three questions: Which lipids comprise the organelle membrane? Where are the lipids produced? How are the lipids sorted to the appropriate organelle (s) from their site(s) of synthesis? Organelle membranes are composed of three major classes of lipids: glycerolipids, sphingolipids and sterols (principally cholesterol in human cells). A comparison of the membrane compositions of the different cellular organelles reveals an evolutionarily conserved dichotomy between the organelles of the secretory pathway and of the endo–lysosome network (Fig. 8.1). The organelle membranes of the endo–lysosome network contain relatively high amounts of sphingolipids and cholesterol, which associate with each other and pack at high relative density within the membrane. These membranes are also asymmetric, with sphingolipids restricted to the exofacial leaflet (facing the outside of the cell) and aminophospholipids (phosphatidylserine (PS), and phosphatidylethanolamine (PE)) restricted to the cytoplasmic leaflet. In contrast, the membranes of early secretory pathway organelles contain small proportions of complex sphingolipids and cholesterol, and these membrane bilayers are symmetric (Fig. 8.2). Yet, these

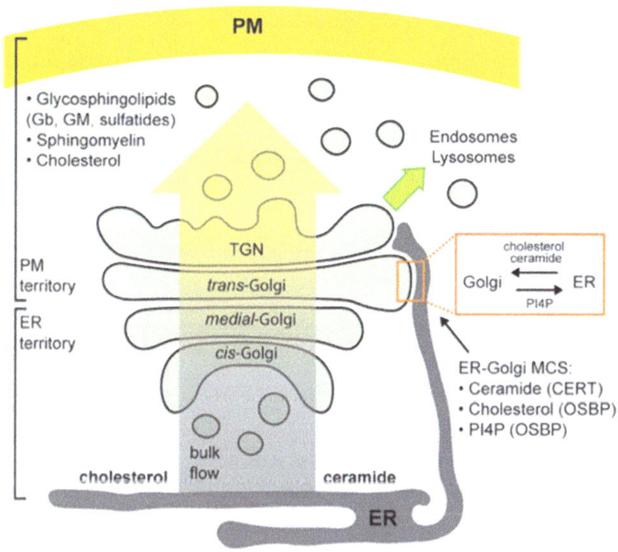

Fig. 8.1 Lipid synthesis through the Golgi apparatus. Membrane lipids are predominantly synthesized in the endoplasmic reticulum (ER), including ceramide and cholesterol. Lipids are trafficked to the Golgi apparatus trafficked either by bulk flow (ER-derived vesicles) or by lipid-transfer proteins (LTP) that function at the ER-TGN membrane-contact sites. In the case of ceramide and cholesterol, the transfer is made by ceramide transfer protein (CERT) and oxysterol binding protein (OSBP), respectively. Lipids synthesized within the Golgi apparatus, namely sphingomyelin (SM), globosides (Gb) and gangliosides (GM), are transported to the plasma membrane (PM) by bulk flow in vesicle carriers. Carrier vesicles with cargo for other organelles (lysosomal or endosomal proteins) are sorted through various mechanisms after leaving the Golgi apparatus

lipids are synthesized via sequential enzymatic transformations that are initiated in the ER and completed in the Golgi apparatus.

The location of individual membrane lipids within the cell is highly dynamic, such that the lipids of organelle membranes are rapidly exchanged with lipids of nearly all other organelle membranes, undermining the concept of a resident organelle for particular lipid species. 'Lipid flow' is a concept that describes the trafficking of lipids between organelles within the entire endomembrane network. Bulk membrane flows out of the ER due to mass action; there is a net addition of lipids to the ER network via synthesis. The Golgi apparatus receives the bulk of ER membrane via vesicle- and non-vesicular transport mechanisms (Fig. 8.1). Although membrane lipids are the most abundant molecules of transport vesicles, the membrane lipid composition of nearly all types of transport vesicles is unknown. To date, there is little evidence for sorting of specific lipids upon export from the ER in human cells. In human cells, non-vesicular transport of ceramide is mediated by ceramide transfer protein (CERT) [27] and non-vesicular cholesterol transport is primarily mediated by Oxysterol Binding Protein (OSBP) [48].

Based upon the general membrane features described in the preceding paragraph, organelles are assigned to one of two membrane 'territories', which are defined as

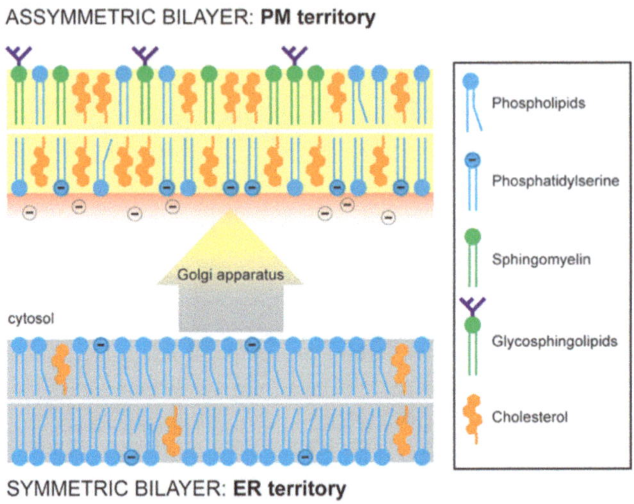

Fig. 8.2 Structure of ER versus PM lipid territory. Based on their composition, organelle membranes of animal cells can be roughly divided into two lipid territories. The first one is the plasma membrane with relatively high content of cholesterol and sphingolipids. This increases the tightness of the plasma membrane, resulting in a good barrier against extracellular elements. Phosphatidylserine (PS) is enriched in the inner PM leaflet resulting in negative net charge, while glycosphingolipids are present in the outer PM leaflet as a part of the glycocalyx. The second membrane territory is the endoplasmic reticulum membrane with relatively high fluidity and low content of cholesterol. Golgi apparatus facilitates the transition zone between these two territories by lipid synthesis and trafficking

the sets of organelles whose membranes are similar in composition and topology [30]. The membranes of organelles within the same territory have similar lipid compositions and hence the physical properties of the organelle membranes within a territory are similar and distinct from the organelle membranes of the other territory [4] (Fig. 8.2). One membrane territory, which we refer to as the "plasma membrane territory" [36], is comprised of the plasma membrane and the organelles of the endo-lysosome network that exchange membrane with plasma membrane via endocytosis and plasma membrane recycling pathways. The other membrane territory is comprised of organelles of the early secretory pathway—the endoplasmic reticulum (ER), ER-Golgi intermediate compartment (ERGIC) and *cis* Golgi cisternae. We refer to this as the 'endoplasmic reticulum territory' and it is where most lipid synthesis occurs. The lipids in ER territory membranes are loosely packed compared to PM territory membranes, which is thought to provide a favorable molecular environment to support the insertion of newly synthesized integral membrane proteins. The Golgi receives biosynthetic lipid intermediates, such as diacylglycerol and ceramide, from the ER and further processes them into glycerolipids and complex sphingolipids, respectively. Given the segregation of lipid metabolic steps between the ER and Golgi apparatus, a fundamental question is, how is lipid metabolism linked to lipid and membrane transport into and out of the Golgi apparatus?

The interface of the ER and PM membrane territories is the *trans* Golgi network (TGN), as it is within the membranes of the *trans* cisterna and the TGN (herein denoted, '*trans*/TGN'), where complex sphingolipids are produced and where membrane cholesterol content is abruptly increased. The change in organelle membrane lipid composition between lipid territories that occurs in the *trans*/TGN emerges from two lipid trafficking pathways. One traffics ceramide and cholesterol from the ER to the *trans*/TGN to remodel its membrane lipid content to roughly match that of PM territory organelles. The second pathway is the 'recycling' of membrane between organelles within each territory, which maintains characteristic membrane composition by preventing the mixing of membrane from organelles of the ER and PM territories [36]. Membrane bilayer asymmetry, a characteristic of PM territory organelle membranes, arises from constraints imposed during sphingolipid synthesis; the active sites of sphingolipid biosynthetic enzymes in the Golgi apparatus lie within the exofacial membrane leaflet, and their products (principally, sphingomyelins (SMs) and glycolipids) are consequently present at the exofacial leaflet of Golgi membranes (Figs. 8.1 and 8.2). The zwitterionic nature of the headgroup of SM and net charge of glycolipid head groups minimizes spontaneous flipping between membrane leaflets and hence, SM and glycolipids are fixed in the exofacial leaflet. For phosphatidylserine, however, flippase activities in the membranes of PM territory organelles maintain PS localization to the cytoplasmic leaflet by catalyzing transfer of PS from the luminal leaflet to the cytoplasmic leaflet of the membrane (i.e., flipping) (Fig. 8.2). Notably, the *trans*/TGN is the first organelle within the anterograde secretory pathway where PS was observed to accumulate in the cytoplasmic membrane leaflets [21]. Thus, asymmetric membrane, a hallmark of PM territory organelle membranes, is transformed from symmetric to asymmetric membrane in the Golgi apparatus.

Lipids of the Golgi Apparatus

The major lipid groups of Golgi membranes in metazoan cells are glycerolipids, sphingolipids and cholesterol. These lipid species are either made in the Golgi apparatus (glycerolipids, sphingolipids) or traffic in large proportion through the Golgi apparatus (*e.g.*, cholesterol) prior to sorting to other organelles. Recent innovations in organelle purification approaches have facilitated the most comprehensive, quantitative analyses of Golgi membrane lipids to date. Of the glycerolipids, phosphatidylcholine, phosphatidylinositol, and phosphatidylserine are the most abundant in immune-purified Golgi organelles and glycosphingolipids are enriched [23]. Further, the analysis of Golgi-enriched fraction confirmed that glycosphingolipids and their lipid- and sugar-precursors are enriched within the Golgi. Presumably, lipid metabolism in the Golgi apparatus membrane homeostasis pathways contributes to lipid trafficking and synthesis to maintain suitable proportions of each lipid in Golgi membranes.

Sphingolipids

Sphingolipid synthesis is initiated in the ER by the formation of the sphingoid base, which is the building block of complex sphingolipids. The sphingoid base is converted to a ceramide by one of six ER-resident ceramide synthases and the ceramides are then transported to the Golgi apparatus where headgroups are added by Golgi-resident enzymes [55]. The segregation of sphingolipid synthesis between the ER and Golgi apparatus indicates that the rates of complex sphingolipid synthesis are determined not only by the catalytic properties of the Golgi-resident enzymes, but also on the rates of ceramide trafficking from the ER to the *trans*/TGN.

There are two trafficking pathways that transport ceramide from the ER to the Golgi apparatus: direct lipid transfer by a cytosolic lipid transfer protein (LTP) called ceramide transfer protein (CERT, also called STARD11) [55] and by bulk vesicle-mediated membrane trafficking [26, 37, 61]. Interestingly, the ceramide binding pocket of CERT can only accommodate ceramides with $C \leq 20$ *N*-linked fatty acyl chains, termed "long chain (LC)" ceramide species [39, 41], such that synthesis of LC complex sphingolipids relies on CERT, but synthesis of "very long chain (VLC)" sphingolipids ($C > 20$) does not. Rather, bulk vesicular trafficking of VLC-ceramides from the ER to the Golgi requiring the coatomer II (COPII) vesicle coat protein complex that mediates bulk membrane export from the ER [37, 45]. In principle, long chain ceramides could also be trafficked to the Golgi by the COPII network, though this has yet to be definitively demonstrated. The different ER-to-Golgi routes of LC- versus VLC-sphingolipids suggest that LC- and VLC-complex sphingolipid synthesis might be distinctly regulated, which could ensure that homeostatic proportions of LC and VLC sphingolipids are synthesized in the Golgi. Consequently, sphingolipid homeostasis in the Golgi apparatus coordinates sphingolipid trafficking and metabolism to maintain appropriate amounts of ceramides and sphingosine.

Ceramides are converted to sphingomyelin or glycosphingolipids (GSL) in the Golgi apparatus. Sphingomyelin is a bulk phospholipid of the plasma membrane that is synthesized by two sphingomyelin synthases, SMS1 and SMS2, that catalyze the transfer of the headgroup of PC (phosphocholine) to ceramide, thereby producing an equivalent amount of sphingomyelin and diacylglycerol (DAG) products (Fig. 8.3). Sphingomyelin synthases are not known to possess specificity towards ceramide species with different length fatty acids (*i.e.*, LC- vs. VLC-ceramides). Ongoing SM synthesis seems to be required for efficient secretion as genetic depletion or ablation of SM production slows anterograde trafficking of VSVG and influenza glycoproteins to the PM [62, 63] and it causes mis-sorting of certain secreted proteins [19, 20]. Decreased SM production has also been shown to impair glucose-regulated insulin segregation [62].

The production of glycosphingolipids is more intricate. Ceramides can serve as substrates for ceramide galactosyltransferase and glucosylceramide synthase, producing galactosylceramide and glucosylceramide, respectively. In neurons, the headgroup of galactosylceramide is further sulfated in the Golgi by cerebroside

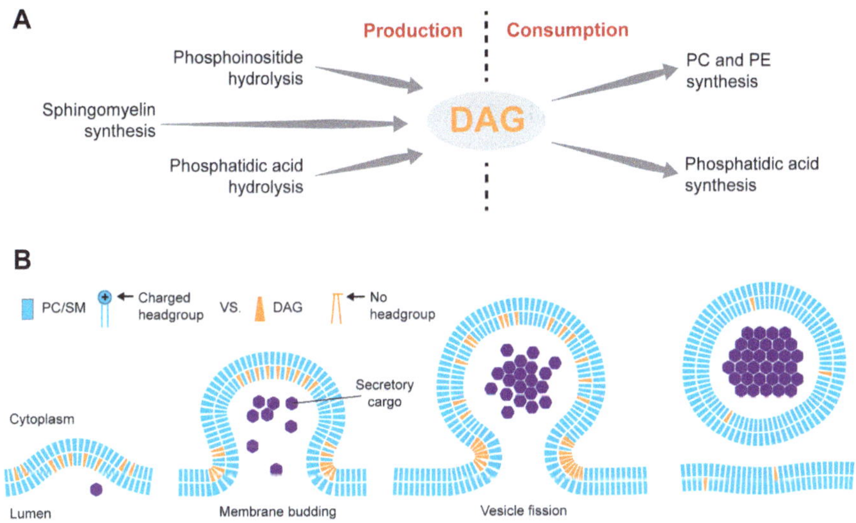

Fig. 8.3 Diacylglycerol metabolism promotes vesicle budding and signal transduction in the Golgi apparatus. (**a**) Diacylglycerol production in the Golgi apparatus by phosphoinositide hydrolysis (phospholipase C), sphingomyelin synthesis (sphingomyelin synthase) and during phosphatidic acid hydrolysis (phosphatide phosphatase). Diacylglycerol is consumed during phosphatidylcholine and phosphatidylethanolamine synthesis (choline- or ethanolamine-phosphotransferase) and during phosphatidic acid synthesis (diacylglycerol kinase). (**b**) Diacylglycerol molecule has a converted cone shape, and it can promote the formation of positive membrane curvature. Diacylglycerol has been proposed to promote formation of membrane buds and fission of vesicles from the *trans*-Golgi/TGN that are addressed to the PM

sulfotransferase, creating sulfatides, that are a key component of the myelin sheath. Glucosylceramides are further modified in the Golgi apparatus with addition of various carbohydrates, yielding more complex sphingolipids, lactosylceramides, globosides (Gb) and gangliosides (GM). Many of these complex sphingolipids are either cancer-associated or developmentally regulated antigens [24]. Globosides and gangliosides present in the PM play a key role in cell-cell communication, adhesion and defining different blood types, some of these glycolipids can be exploited by bacteria or viral pathogens to enter and infect the cells carrying them. For instance, Gb3 is one of the targets of Shiga toxin of *Escherichia coli*, whereas gangliosides are targeted by haemagglutinin of influenza virus.

The synthesis of glucosylceramide (GC) may be a peculiar case of sphingolipid synthesis that impacts the synthesis of all glycolipids because GC is the core of the glycolipids with larger, more complex head groups. The membrane topology of the initial step of GC synthesis - the transfer of glucosyl headgroup onto the ceramide - has not yet been firmly established. While GC is present solely in the exofacial leaflet, where it undergoes further processing of the glycan headgroup, early studies that were based on metabolic turnover of a fluorescent GC analog proposed that the active site of GC synthase (GCS) is localized at the cytosolic leaflet of the Golgi membrane [8, 25]. This begged the question of how GC gets translocated to the Golgi lumen. To date, there has not been identified any GC-specific lipid floppase (an integral membrane protein that can transfer lipids from the cytoplasmic leaflet to the luminal leaflet) within the Golgi. It was suggested that newly synthesized GC is transported back to the ER where it is flopped by ATP-independent lipid floppase activity [8, 25]. Importantly, more recent studies showed that GCS localizes to the *trans*-Golgi, where SMS1 also resides [7].

Glycerolipids

Pioneering genetic studies of secretion using *Saccharomyces cerevisiae* (budding yeast) cells pointed to critical roles for diacylglycerol (DAG), sterol, and phosphoinositide metabolism during secretion [2, 11, 22, 35, 53]. Diacylglycerol is a neutral lipid species that is produced in the Golgi apparatus during sphingomyelin synthesis in which the choline headgroup of PC is transferred to ceramide, making equivalent amounts of sphingomyelin and DAG. In addition to SM synthesis, there are two other sources of DAG in the Golgi membrane: phosphatidic acid metabolism and PI(4,5)P_2 hydrolysis.

The SM synthesis pathway represents an intersection between glycerolipid and sphingolipid pathways through production of sphingomyelin and DAG from ceramide and PC, respectively (Fig. 8.3). PC and PE are produced in the Golgi apparatus from diacylglycerol (DAG) and CDP-choline and CDP-ethanolamine by cholinephosphotransferase-1 (CHPT1) and ethanolaminephosphotransferase-1 (EPT1), respectively [31–34]. Phosphatidic acid is a short-lived signaling lipid species and is not known to be enriched at a particular organelle(s). It is synthesized by diacylglycerol kinases while it is degraded by phosphatidate phosphatases (Fig. 8.3). Propranolol, a pharmacological inhibitor of beta receptors, is known to deplete the DAG pool in the Golgi, presumably by blocking phosphatidate phosphohydrolases which regenerate DAG by dephosphorylation of PA. This suggests that despite its relatively low concentration in cells, phosphatidic acid turnover plays an important role in maintaining the diacylglycerol content in the Golgi apparatus [3].

Three functions are attributed to DAG in Golgi membranes: as an intermediate in glycerophospholipid synthesis, in transport vesicle budding and fission, and as a component of signal transduction pathways. Transport carriers were proposed to bud from the *trans*/TGN by higher DAG content which would facilitate membrane deformations required to produce a vesicle. Due to its inverted cone shape, DAG can promote the formation of positive membrane curvature (*i.e.*, membrane bending into the cytoplasm) [57] (Fig. 8.3). Treatment of cells with a small molecule that causes depletion of DAG, propranolol, elicits an accumulation of membrane buds from Golgi membranes, suggesting that DAG depletion arrests transport vesicle formation at the stage of vesicle growth (*i.e.*, budding or fission). Interestingly, no vesicle coat proteins that would be candidate sorting factors have been identified on secretory vesicles budding from Golgi, highlighting the hypothesis that lipids, in particular DAG, drive the process of membrane budding and vesicle formation.

The functions of DAG in signal transduction pathways that converge on the Golgi apparatus is in the recruitment of protein kinase C (e.g. PKC isoform epsilon) and protein kinase D (PKD isoform 1) via their C1ab structural domains that directly bind DAG [3, 12]. PKD has been also proposed to regulate the fission of transport vesicles from the TGN based on the observation that elongated tubules (>5 μm) containing secretory cargo emanate from the Golgi in cells expressing a kinase-dead dominant-negative mutant of PKD [3, 44, 47]. This was interpreted to indicate that PKD regulates fission of secretory carriers, however a relevant Golgi-localized PKD signaling effector(s) that mediates vesicle fission has yet to be identified, and this remains a major question for the field.

Cholesterol

Cholesterol is the major sterol of animal cells. It is obtained from dietary sources and is synthesized by ER-resident enzymes when needed and then distributed to other organelles via lipid transfer at organelle membrane contact sites and by vesicular trafficking. Cellular cholesterol homeostasis is controlled in large part by the sterol responsive element binding protein (SREBP) sensor pathway that elicits increased cholesterol synthesis when ER membrane cholesterol content falls below a 5 mol% threshold (CHO cells) [46, 56], and suppresses cholesterol synthesis when cholesterol content of ER membrane is above the threshold.

In the Golgi apparatus, cholesterol forms a *cis*-to-*trans* (low to high) gradient across the stack of cisternae [54], in which early/*cis* Golgi cisterna have relatively low amounts and late Golgi cisternae have relatively high amounts. High relative cholesterol content is characteristic of organelles of the PM territory. Oxysterol-binding protein (OSBP) was identified as the key LTP that transfers cholesterol from the ER to the *trans*/TGN and other organelles. OSBP is the prototype of a family of structurally related LTPs that transfers cholesterol to numerous organelles, but most prominently, the Golgi apparatus and plasma membrane. Two other OSBP family members, ORP9 (Oxysterol Binding Protein-related Protein 9) and ORP10 (Oxysterol Binding Protein-related Protein 10), form a heterodimer that is required for tethering of ER and *trans*/TGN membranes to form the ER-Golgi MCS [29, 67] (Fig. 8.4).

A large proportion of cholesterol in the PM [approx. 2/3 in cultured human fibroblasts [16]] is bound to proteins or lipids, sphingomyelin being the predominant bound lipid species. The unbound or "free" cholesterol pool is the portion of cholesterol available for transport by OSBP and other cholesterol LTPs. The amount of sphingomyelin in organelle membranes is therefore a critical determinant of membrane cholesterol content. In addition, inhibition of OSBP by the small molecule, OSW-1 causes large changes in lipid order by a parallel increase of cholesterol and also in the saturation level of phospholipids from the ER to the PM [4, 52]. With

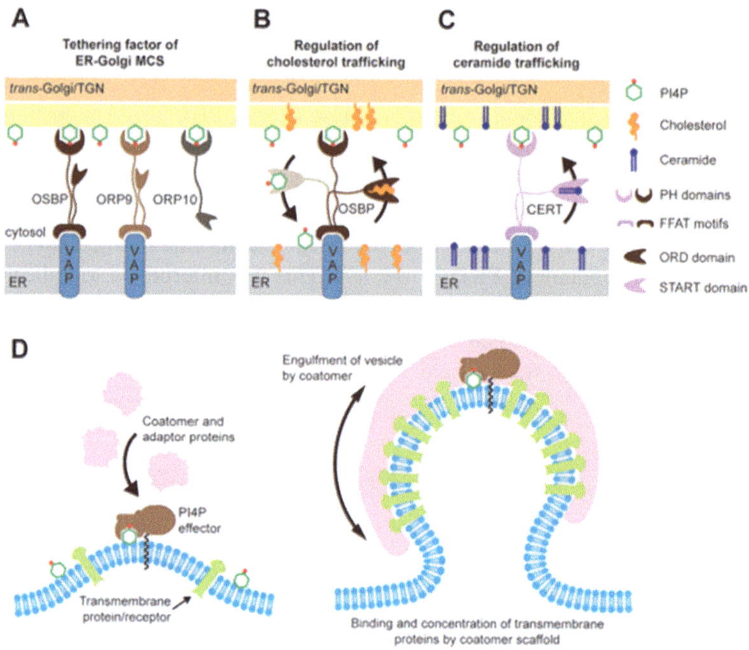

Fig. 8.4 PI4P signaling network regulates lipid trafficking into and out of the Golgi apparatus. (**a**) PI4P is responsible for creating ER-TGN membrane contact sites by recruitment of tethering proteins OSBP, ORP9 and ORP10. (**b**) OSBP acts as cholesterol-PI4P antiporter between the ER and PM, where a molecule of cholesterol is exchanged for a molecule of PI4P. In the ER, PI4P is hydrolyzed into PI that will be reused in the Golgi for making PI4P (**c**) CERT transfers ceramide from the ER-TGN at the ER-TGN MCS. It utilizes PI4P at the Golgi membrane as anchor by its PH domain. The concentration of ceramide in the Golgi is kept low by converting ceramide to sphingomyelin. (**d**) PI4P recruits its effectors, such as AP-1, GGAs and GOLPH3, to the budding membrane. This interaction is recruiting protein adaptors and protein coatomer I (COPI) for formation of the vesicle scaffold. When the vesicle fills with sorted cargo, it buds of and is fully covered with coatomer

its cholesterol transfer activity between the ER and TGN, OSBP is recognized as a key regulator of lipid packing of membranes of the secretory pathway before and after Golgi complex [49].

Lipid Trafficking Between the Golgi Apparatus and Other Organelles

Membrane lipids are trafficked between organelles via two mechanisms: by vesicles containing millions of lipids and by soluble LTPs that transport one lipid at a time at MCSs. In vesicle-mediated transport, a vesicle that fuses with a target organelle and, upon fusion of the vesicle and the membrane of the acceptor organelle, membrane topology is preserved. Non-vesicular lipid transport by soluble LTPs, on the other hand, is intrinsically asymmetric; lipids are transferred between the cytoplasmic leaflets of opposed organelle membranes at membrane contact sites, resulting in more lipid in the cytoplasmic leaflet compared to the luminal leaflet. Discrepancies in the numbers of lipids in each leaflet of a bilayer are resolved by lipid scramblases which facilitate free exchange of lipids between two membrane leaflets. Hence, the implications of vesicle-mediated versus LTP-mediated transport have fundamentally distinct consequences to membrane structure. Lipids are trafficked to the Golgi apparatus by both vesicle- and LTP-mediated trafficking, but vesicle-mediated trafficking is the major mode of lipid export from the Golgi apparatus.

Regarding vesicle-mediated trafficking pathways, the Golgi apparatus is the 'distribution hub' of the endomembrane organelle network for lipids and integral membrane proteins. Given the abrupt, tightly regulated remodeling of *trans*/TGN membrane composition, much attention has focused on investigating sphingolipid- and/or cholesterol-based sorting mechanisms that could segregate lipids and proteins within the *trans*/TGN membrane. Formation of lipid nanodomains in the *trans*/TGN membrane that are rich in cholesterol, glycosphingolipids and sphingomyelin, could function as platforms for sorting of lipids and proteins that have avidity for the nanodomain lipids surrounding them. According to this principle, 'hydrophobic matching' between the transmembrane domain of proteins and bilayer lipids can serve as a mechanism by which membranes organize localization and behavior of proteins. For instance, hydrophobic mismatch between the length of transmembrane domains of SNARE proteins (syntaxins 1 and 4) have been proposed to induce self-clustering dependent on the surrounding lipids [50]. Nanodomains containing sphingolipids and cholesterol were proposed to constitute a 'lipid raft' [58] that could capture integral membrane proteins with high avidity for sphingolipid-cholesterol membrane nanodomains would concentrate these proteins within the membrane. Although the existence of static lipid raft structures remains a matter of debate, it has been well described that local phase condensation of lipids and proteins play key roles in protein sorting.

PI4P Signaling Controls Lipid Flow into and Out of the Golgi

Phosphatidylinositol-phosphates ('PIPs') are phosphorylated forms of phosphatidylinositol, in which the 3, 4 and 5 positions of the inositol headgroup are reversibly phosphorylated to generate transient, local pools of mono, di- and tri-phosphorylated species. PIPs are present solely in the cytoplasmic leaflet of organelle membranes and, in general, each PIP localizes to a distinct or subset of organelles, leading to the concept of an organelle-specific 'PIP code' that directs the formation of organelle-specific protein complexes that participate in organelle biogenesis. The core components of PIP signaling networks are phosphoinositide kinases and phosphoinositide phosphatases. A phosphatidylinositol 4-phosphate (PI4P) signaling network regulates nearly all aspects of Golgi lipid dynamics

(Fig. 8.4). It is necessary for tethering ER and *trans*/TGN membranes to each other at the ER-Golgi MCS [66, 67] and it is used as a 'flag' by peripheral membrane proteins to identify Golgi membranes.

In contrast to the restricted localization of most PIPs to one or a few organelles, PI4P is present in the membranes of many organelles: plasma membrane, Golgi apparatus, endosome, lysosome and ER. In the Golgi apparatus, PI4P is produced by phosphatidylinositol 4-kinase III-β (PI4KIIII-β), but in the absence of active PI4KIIII-β, the Golgi acquires PI4P that is produced in the endosome membrane by phosphatidylinositol 4-kinase IIα (PI4KIIα), probably by vesicle-mediated endosome-to-Golgi retrograde membrane trafficking [49, 69]. PI4KIIII-β is a peripheral membrane protein whose enzymatic output (i.e., location and amount of product, PI4P) is subject to two points of regulation: targeting of the enzyme to *trans*/TGN membranes and the enzyme's intrinsic phospho-transfer activity. Targeting of PI4KIIII-β to the Golgi requires binding to acyl-coenzyme A binding domain containing protein 3 (ACBD3), another peripheral Golgi membrane protein of unknown cellular function [38, 60]. Localization of ACBD3 itself to the Golgi apparatus is mediated by binding to the Golgi integral proteins, giantin [59] and the SNARE protein, Sec22b [10, 60]. The most significant PIP phosphatase that terminates PI4P signaling in the Golgi apparatus is an ER resident integral membrane protein called Sac1.

The mechanisms of PI4KIIII-β regulation are not well understood. The enzyme is phosphorylated by PKD1 and PKD2, which was reported to increase the amount of cellular PI4P [7, 28], though the increase was due to protection of the unphosphorylated enzyme from degradation, rather than an increase in the specific activity of the enzyme [28]. Phosphorylated PI4KIIII-β is bound by 14-3-3 proteins, but binding does not affect the intrinsic enzymatic activity [9]. In experiments that used pure proteins, the activity of membrane-associated PI4KIIII-β was increased, suggesting that the membrane environment enhances PI4KIIII-β activity [38]. The potential for the lipid composition of the *trans*/TGN of PI4K activity has not yet been comprehensively explored.

Ongoing PI4P synthesis and trafficking from the Golgi apparatus to the ER is required to support the OSBP functional cycle, which provides the chemical energy that drives OSBP-mediated cholesterol transfer from the ER to the *trans*/TGN, against its concentration gradient (low in the ER, high in the *trans*/TGN) (Fig. 8.1). The foundation of the OSBP cholesterol transfer mechanism lies in its lipid transfer cycle. OSBP has the capacity to extract cholesterol from the ER membrane, transfer it to the *trans*/TGN membrane and exchange cholesterol for a molecule of PI4P which is then transferred to the ER membrane [1, 17, 48]. The cycle would continue to run until the amounts of PI4P in Golgi and ER membranes equilibrate. However, the PIP phosphatase, Sac1, converts PI4P to PI in ER membrane, driving ER-to-Golgi vectoral transport of cholesterol. Thus, the difference in the amounts of PI4P in *trans*/TGN membranes versus the ER membrane drives the directionality of OSBP-mediated cholesterol trafficking from the ER to the Golgi. In immortalized retinal pigmental epithelial cells (hTERT-RPE1), OSBP consumes about half of the total cellular pool of PI4P, and this consumption depends on the amount of transported cholesterol [49]. The reduction of PI4P levels by OSBP eventually reduces the activities of other PI4P effectors, including CERT [7]. Notably, six different LTPs have been reported to operate at the ER-Golgi MCS [65] and their localization to the Golgi membrane also relies on PI4P recognition, however, to date it has not been definitively shown that Golgi-associated LTPs other than OSBP use PI4P counter transport to drive lipid transport.

In contrast to the OSBP counter-transport cycle, CERT does not use an analogous ceramide/PI4P counter-transport cycle to drive ceramide into the Golgi apparatus. Rather, ceramide is converted to a complex sphingolipid in the Golgi apparatus, thereby maintaining an ER to Golgi gradient that drives ceramide transport to the Golgi apparatus. This mechanism was termed a “ceramide metabolic trap” and it ensures that LC-ceramides do not accumulate in *trans*/TGN membrane [15].

The activities of OSBP and CERT are regulated by their phosphorylation. PKD phosphorylates OSBP and inhibits its targeting to the ER-Golgi MCS [51]. PKD also phosphorylates CERT and this modification inhibits its ceramide transfer activity [40]. Collectively, OSBP-mediated changes in amounts of PI4P and PKD phosphorylation-mediated regulation of LTP activities constituted a rheostat-like mechanism that coordinates the level of cholesterol, ceramide and PS in the TGN via the PI4P signaling network.

PI4P Promotes Transport Vesicle Biogenesis

At the *trans*/TGN, PI4P is involved in the sorting of macromolecules into transport vesicles that ferry cargo from the Golgi to the endo-lysosome network (anterograde trafficking), and in intra-Golgi retrograde trafficking from late to early cisternae. Although these pathways direct the formation of transport vesicles that traffic in opposite directions, PI4P functions in both vesicle biogenesis pathways by facilitating capture of integral membrane protein cargo that will be incorporated into a nascent transport vesicle. The Golgi associated clathrin coated vesicle adapters AP1 [69], GGA1 and GGA2 [18, 68] recognize both PI4P and activated ARF1, a global GTPase regulator of Golgi functions, to target to the *trans*/TGN membrane via a coincidence detection mechanism. Regarding vesicle cargo capture, AP1 selects Golgi resident proteins that require retrograde trafficking from late Golgi-to-early Golgi cisternae to maintain residence in the Golgi apparatus. The GGA1 and GGA2 cargo adapters sort lysosome resident pro-enzymes from the Golgi apparatus to the late endosome and then ultimately to the lysosome.

Intra-Golgi transport vesicles coated with coatomer I (COPI) proteins mediate retrograde trafficking of Golgi enzymes which is required to maintain residence in the Golgi apparatus. An effector of PI4P signaling is a peripheral membrane protein GOLPH3 that binds PI4P, COPI and Golgi-resident proteins simultaneously, thereby mediating the retention of Golgi residents via retrograde trafficking [64, 70, 71]. A relatively small body of evidence provided precedent for the sorting of specific lipids within the cisternae of the Golgi apparatus. Sphingomyelin (C18) binds to a subunit of the COPI network, p24 [13], and regulates the function of the COPI coatomer, although lower concentrations of sphingomyelin have been observed in COPI vesicles [6].

Conclusions and Perspectives

Organelle biogenesis requires huge quantities of lipids that are produced in organelles of the early secretory pathway—the ER territory—and then distributed to the PM and endo-lysosome network—the PM territory. Trafficking of lipid metabolic intermediates from the ER to the Golgi channels these metabolic intermediates to enzymes that generate the head groups of sphingolipids. In the context of vesicle-mediated and LTP-mediated lipid transport, the concept of lipid flow explains how the PM and organelles of the endo-lysosome network become enriched in sphingolipids and cholesterol, relative to the membranes of the ER and Golgi apparatus. The Golgi apparatus is the principal sorting station within the endomembrane organelle network. A PI4P-based signaling network regulates the assembly of and flow of lipid across the ER-Golgi membrane contact site and it also participates in vesicle-mediated lipid and protein trafficking pathways from the Golgi apparatus. As such, co-ordinate regulation of lipid synthesis and lipid and protein trafficking pathways by the PI4P signaling network that can be thought of as a rheostat, where the abundance of PI4P regulates the flux of ceramides and cholesterol into and PI4P out of the *trans*/TGN. With the general outline of lipid metabolism and trafficking to, through and out of the Golgi apparatus in hand, the major gap in our knowledge of these pathways asks, what are the lipid sensing mechanisms that form the basis of homeostasis mechanisms that maintain the amounts of each lipid that is made in and/or traffics through the Golgi apparatus?

Acknowledgments We are grateful to colleagues for discussions and critical reading of the manuscript. The authors declare no conflicts of interest. This work is supported by funds from the National Institute of General Medical Sciences of the National Institutes of Health under award number GM144096 to CGB.

References

1. Antonny B, Bigay J, Mesmin B (2018) The oxysterol-binding protein cycle: burning off PI(4)P to transport cholesterol. Annu Rev Biochem 87:809–837
2. Bankaitis VA, Aitken JR, Cleves AE, Dowhan W (1990) An essential role for a phospholipid transfer protein in yeast Golgi function. Nature 347:561–562
3. Baron CL, Malhotra V (2002) Role of diacylglycerol in PKD recruitment to the TGN and protein transport to the plasma membrane. Science 295:325–328
4. Bigay J, Antonny B (2012) Curvature, lipid packing, and electrostatics of membrane organelles: defining cellular territories in determining specificity. Dev Cell 23:886–895

5. Botelho AV, Huber T, Sakmar TP, Brown MF (2006) Curvature and hydrophobic forces drive oligomerization and modulate activity of rhodopsin in membranes. Biophys J 91:4464–4477
6. Brugger B, Sandhoff R, Wegehingel S, Gorgas K, Malsam J, Helms JB, Lehmann WD, Nickel W, Wieland FT (2000) Evidence for segregation of sphingomyelin and cholesterol during formation of COPI-coated vesicles. J Cell Biol 151:507–518
7. Capasso S, Sticco L, Rizzo R, Pirozzi M, Russo D, Dathan NA, Campelo F, van Galen J, Holtta-Vuori M, Turacchio G, Hausser A, Malhotra V, Riezman I, Riezman H, Ikonen E, Luberto C, Parashuraman S, Luini A, D'Angelo G (2017) Sphingolipid metabolic flow controls phosphoinositide turnover at the trans-Golgi network. EMBO J 36:1736–1754
8. Chalat M, Menon I, Turan Z, Menon AK (2012) Reconstitution of glucosylceramide flip-flop across endoplasmic reticulum: implications for mechanism of glycosphingolipid biosynthesis. J Biol Chem 287:15523–15532
9. Chalupska D, Eisenreichova A, Rozycki B, Rezabkova L, Humpolickova J, Klima M, Boura E (2017) Structural analysis of phosphatidylinositol 4-kinase IIIbeta (PI4KB) - 14-3-3 protein complex reveals internal flexibility and explains 14-3-3 mediated protection from degradation in vitro. J Struct Biol 200:36–44
10. Chalupska D, Rozycki B, Klima M, Boura E (2019) Structural insights into Acyl-coenzyme A binding domain containing 3 (ACBD3) protein hijacking by picornaviruses. Protein Sci 28: 2073–2079
11. Cleves AE, McGee TP, Whitters EA, Champion KM, Aitken JR, Dowhan W, Goebl M, Bankaitis VA (1991) Mutations in the CDP-choline pathway for phospholipid biosynthesis bypass the requirement for an essential phospholipid transfer protein. Cell 64:789–800
12. Colon-Gonzalez F, Kazanietz MG (2006) C1 domains exposed: from diacylglycerol binding to protein-protein interactions. Biochim Biophys Acta 1761:827–837
13. Contreras FX, Ernst AM, Haberkant P, Bjorkholm P, Lindahl E, Gonen B, Tischer C, Elofsson A, von Heijne G, Thiele C, Pepperkok R, Wieland F, Brugger B (2012) Molecular recognition of a single sphingolipid species by a protein's transmembrane domain. Nature 481: 525–529
14. Cornelius F (2001) Modulation of Na,K-ATPase and Na-ATPase activity by phospholipids and cholesterol. I. Steady-state kinetics. Biochemistry 40:8842–8851
15. Dadsena S, Hassan DG, Holthuis JCM (2019) Unraveling the molecular principles by which ceramides commit cells to death. Cell Stress 3:280–283
16. Das A, Brown MS, Anderson DD, Goldstein JL, Radhakrishnan A (2014) Three pools of plasma membrane cholesterol and their relation to cholesterol homeostasis. elife 3:e02882
17. de Saint-Jean M, Delfosse V, Douguet D, Chicanne G, Payrastre B, Bourguet W, Antonny B, Drin G (2011) Osh4p exchanges sterols for phosphatidylinositol 4-phosphate between lipid bilayers. J Cell Biol 195:965–978
18. Demmel L, Gravert M, Ercan E, Habermann B, Muller-Reichert T, Kukhtina V, Haucke V, Baust T, Sohrmann M, Kalaidzidis Y, Klose C, Beck M, Peter M, Walch-Solimena C (2008) The clathrin adaptor Gga2p is a phosphatidylinositol 4-phosphate effector at the Golgi exit. Mol Biol Cell 19.1991–2002
19. Deng Y, Pakdel M, Blank B, Sundberg EL, Burd CG, von Blume J (2018) Activity of the SPCA1 calcium pump couples Sphingomyelin synthesis to sorting of secretory proteins in the trans-Golgi network. Dev Cell 47(464–478):e468
20. Deng Y, Rivera-Molina FE, Toomre DK, Burd CG (2016) Sphingomyelin is sorted at the trans Golgi network into a distinct class of secretory vesicle. Proc Natl Acad Sci USA 113:6677–6682
21. Fairn GD, Schieber NL, Ariotti N, Murphy S, Kuerschner L, Webb RI, Grinstein S, Parton RG (2011) High-resolution mapping reveals topologically distinct cellular pools of phosphatidylserine. J Cell Biol 194:257–275
22. Fang M, Kearns BG, Gedvilaite A, Kagiwada S, Kearns M, Fung MK, Bankaitis VA (1996) Kes1p shares homology with human oxysterol binding protein and participates in a novel regulatory pathway for yeast Golgi-derived transport vesicle biogenesis. EMBO J 15:6447–6459

23. Fasimoye R, Dong W, Nirujogi RS, Rawat ES, Iguchi M, Nyame K, Phung TK, Bagnoli E, Prescott AR, Alessi DR, Abu-Remaileh M (2023) Golgi-IP, a tool for multimodal analysis of Golgi molecular content. Proc Natl Acad Sci USA 120:e2219953120
24. Hakomori SI (2008) Structure and function of glycosphingolipids and sphingolipids: recollections and future trends. Biochim Biophys Acta 1780:325–346
25. Halter D, Neumann S, van Dijk SM, Wolthoorn J, de Maziere AM, Vieira OV, Mattjus P, Klumperman J, van Meer G, Sprong H (2007) Pre- and post-Golgi translocation of glucosylceramide in glycosphingolipid synthesis. J Cell Biol 179:101–115
26. Hammond AT, Glick BS (2000) Dynamics of transitional endoplasmic reticulum sites in vertebrate cells. Mol Biol Cell 11:3013–3030
27. Hanada K, Kumagai K, Yasuda S, Miura Y, Kawano M, Fukasawa M, Nishijima M (2003) Molecular machinery for non-vesicular trafficking of ceramide. Nature 426:803–809
28. Hausser A, Link G, Hoene M, Russo C, Selchow O, Pfizenmaier K (2006) Phospho-specific binding of 14-3-3 proteins to phosphatidylinositol 4-kinase III beta protects from dephosphorylation and stabilizes lipid kinase activity. J Cell Sci 119:3613–3621
29. He R, Liu F, Wang H, Huang S, Xu K, Zhang C, Liu Y, Yu H (2023) ORP9 and ORP10 form a heterocomplex to transfer phosphatidylinositol 4-phosphate at ER-TGN contact sites. Cell Mol Life Sci 80:77
30. Holthuis JC, Menon AK (2014) Lipid landscapes and pipelines in membrane homeostasis. Nature 510:48–57
31. Horibata Y, Ando H, Sugimoto H (2020) Locations and contributions of the phosphotransferases EPT1 and CEPT1 to the biosynthesis of ethanolamine phospholipids. J Lipid Res 61:1221–1231
32. Horibata Y, Elpeleg O, Eran A, Hirabayashi Y, Savitzki D, Tal G, Mandel H, Sugimoto H (2018) EPT1 (selenoprotein I) is critical for the neural development and maintenance of plasmalogen in humans. J Lipid Res 59:1015–1026
33. Horibata Y, Hirabayashi Y (2007) Identification and characterization of human ethanolaminephosphotransferase1. J Lipid Res 48:503–508
34. Horibata Y, Sugimoto H (2021) Differential contributions of choline phosphotransferases CPT1 and CEPT1 to the biosynthesis of choline phospholipids. J Lipid Res 62:100100
35. Kearns BG, McGee TP, Mayinger P, Gedvilaite A, Phillips SE, Kagiwada S, Bankaitis VA (1997) Essential role for diacylglycerol in protein transport from the yeast Golgi complex. Nature 387:101–105
36. Kim Y, Burd CG (2023) Lipid sorting and organelle identity. Cold Spring Harb Perspect Biol 15:a041397
37. Kim Y, Mavodza G, Senkal CE, Burd CG (2023) Cholesterol-dependent homeostatic regulation of very long chain sphingolipid synthesis. J Cell Biol 222:e202308055
38. Klima M, Toth DJ, Hexnerova R, Baumlova A, Chalupska D, Tykvart J, Rezabkova L, Sengupta N, Man P, Dubankova A, Humpolickova J, Nencka R, Veverka V, Balla T, Boura E (2016) Structural insights and in vitro reconstitution of membrane targeting and activation of human PI4KB by the ACBD3 protein. Sci Rep 6:23641
39. Kudo N, Kumagai K, Tomishige N, Yamaji T, Wakatsuki S, Nishijima M, Hanada K, Kato R (2008) Structural basis for specific lipid recognition by CERT responsible for nonvesicular trafficking of ceramide. Proc Natl Acad Sci USA 105:488–493
40. Kumagai K, Kawano M, Shinkai-Ouchi F, Nishijima M, Hanada K (2007) Interorganelle trafficking of ceramide is regulated by phosphorylation-dependent cooperativity between the PH and START domains of CERT. J Biol Chem 282:17758–17766
41. Kumagai K, Yasuda S, Okemoto K, Nishijima M, Kobayashi S, Hanada K (2005) CERT mediates intermembrane transfer of various molecular species of ceramides. J Biol Chem 280: 6488–6495
42. Lee AG (2003) Lipid-protein interactions in biological membranes: a structural perspective. Biochim Biophys Acta 1612:1–40

43. Levental I, Lyman E (2023) Regulation of membrane protein structure and function by their lipid nano-environment. Nat Rev Mol Cell Biol 24:107–122
44. Liljedahl M, Maeda Y, Colanzi A, Ayala I, Van Lint J, Malhotra V (2001) Protein kinase D regulates the fission of cell surface destined transport carriers from the trans-Golgi network. Cell 104:409–420
45. Loizides-Mangold U, David FP, Nesatyy VJ, Kinoshita T, Riezman H (2012) Glycosylphosphatidylinositol anchors regulate glycosphingolipid levels. J Lipid Res 53: 1522–1534
46. Luo J, Yang H, Song BL (2020) Mechanisms and regulation of cholesterol homeostasis. Nat Rev Mol Cell Biol 21:225–245
47. Maeda Y, Beznoussenko GV, Van Lint J, Mironov AA, Malhotra V (2001) Recruitment of protein kinase D to the trans-Golgi network via the first cysteine-rich domain. EMBO J 20: 5982–5990
48. Mesmin B, Bigay J, Moser von Filseck J, Lacas-Gervais S, Drin G, Antonny B (2013) A four-step cycle driven by PI(4)P hydrolysis directs sterol/PI(4)P exchange by the ER-Golgi tether OSBP. Cell 155:830–843
49. Mesmin B, Bigay J, Polidori J, Jamecna D, Lacas-Gervais S, Antonny B (2017) Sterol transfer, PI4P consumption, and control of membrane lipid order by endogenous OSBP. EMBO J 36: 3156–3174
50. Milovanovic D, Honigmann A, Koike S, Gottfert F, Pahler G, Junius M, Mullar S, Diederichsen U, Janshoff A, Grubmuller H, Risselada HJ, Eggeling C, Hell SW, van den Bogaart G, Jahn R (2015) Hydrophobic mismatch sorts SNARE proteins into distinct membrane domains. Nat Commun 6:5984
51. Nhek S, Ngo M, Yang X, Ng MM, Field SJ, Asara JM, Ridgway ND, Toker A (2010) Regulation of oxysterol-binding protein Golgi localization through protein kinase D-mediated phosphorylation. Mol Biol Cell 21:2327–2337
52. Niko Y, Didier P, Mely Y, Konishi G, Klymchenko AS (2016) Bright and photostable push-pull pyrene dye visualizes lipid order variation between plasma and intracellular membranes. Sci Rep 6:18870
53. Ohashi M, Jan de Vries K, Frank R, Snoek G, Bankaitis V, Wirtz K, Huttner WB (1995) A role for phosphatidylinositol transfer protein in secretory vesicle formation. Nature 377:544–547
54. Orci L, Montesano R, Meda P, Malaisse-Lagae F, Brown D, Perrelet A, Vassalli P (1981) Heterogeneous distribution of filipin--cholesterol complexes across the cisternae of the Golgi apparatus. Proc Natl Acad Sci USA 78:293–297
55. Pothukuchi P, Agliarulo I, Pirozzi M, Rizzo R, Russo D, Turacchio G, Nuchel J, Yang JS, Gehin C, Capolupo L, Hernandez-Corbacho MJ, Biswas A, Vanacore G, Dathan N, Nitta T, Henklein P, Thattai M, Inokuchi JI, Hsu VW, Plomann M, Obeid LM, Hannun YA, Luini A, D'Angelo G, Parashuraman S (2021) GRASP55 regulates intra-Golgi localization of glycosylation enzymes to control glycosphingolipid biosynthesis. EMBO J 40:e107766
56. Radhakrishnan A, Goldstein JL, McDonald JG, Brown MS (2008) Switch-like control of SREBP-2 transport triggered by small changes in ER cholesterol: a delicate balance. Cell Metab 8:512–521
57. Shemesh T, Luini A, Malhotra V, Burger KN, Kozlov MM (2003) Prefission constriction of Golgi tubular carriers driven by local lipid metabolism: a theoretical model. Biophys J 85:3813–3827
58. Simons K, Ikonen E (1997) Functional rafts in cell membranes. Nature 387:569–572
59. Sohda M, Misumi Y, Yamamoto A, Yano A, Nakamura N, Ikehara Y (2001) Identification and characterization of a novel Golgi protein, GCP60, that interacts with the integral membrane protein giantin. J Biol Chem 276:45298–45306
60. Stalder D, Yakunin I, Pereira C, Eden J, Gershlick DC (2024) Recruitment of PI4KIIIbeta to the Golgi by ACBD3 is dependent on an upstream pathway of a SNARE complex and golgins. Mol Biol Cell 35:ar20

61. Stephens DJ, Lin-Marq N, Pagano A, Pepperkok R, Paccaud JP (2000) COPI-coated ER-to-Golgi transport complexes segregate from COPII in close proximity to ER exit sites. J Cell Sci 113(Pt 12):2177–2185
62. Subathra M, Qureshi A, Luberto C (2011) Sphingomyelin synthases regulate protein trafficking and secretion. PLoS One 6:e23644
63. Tafesse FG, Sanyal S, Ashour J, Guimaraes CP, Hermansson M, Somerharju P, Ploegh HL (2013) Intact sphingomyelin biosynthetic pathway is essential for intracellular transport of influenza virus glycoproteins. Proc Natl Acad Sci USA 110:6406–6411
64. Tu L, Tai WC, Chen L, Banfield DK (2008) Signal-mediated dynamic retention of glycosyltransferases in the Golgi. Science 321:404–407
65. Venditti R, Masone MC, De Matteis MA (2020) ER-Golgi membrane contact sites. Biochem Soc Trans 48:187–197
66. Venditti R, Masone MC, Rega LR, Di Tullio G, Santoro M, Polishchuk E, Serrano IC, Olkkonen VM, Harada A, Medina DL, La Montagna R, De Matteis MA (2019a) The activity of Sac1 across ER-TGN contact sites requires the four-phosphate-adaptor-protein-1. J Cell Biol 218:783–797
67. Venditti R, Rega LR, Masone MC, Santoro M, Polishchuk E, Sarnataro D, Paladino S, D'Auria S, Varriale A, Olkkonen VM, Di Tullio G, Polishchuk R, De Matteis MA (2019b) Molecular determinants of ER-Golgi contacts identified through a new FRET-FLIM system. J Cell Biol 218:1055–1065
68. Wang J, Sun HQ, Macia E, Kirchhausen T, Watson H, Bonifacino JS, Yin HL (2007) PI4P promotes the recruitment of the GGA adaptor proteins to the trans-Golgi network and regulates their recognition of the ubiquitin sorting signal. Mol Biol Cell 18:2646–2655
69. Wang YJ, Wang J, Sun HQ, Martinez M, Sun YX, Macia E, Kirchhausen T, Albanesi JP, Roth MG, Yin HL (2003) Phosphatidylinositol 4 phosphate regulates targeting of clathrin adaptor AP-1 complexes to the Golgi. Cell 114:299–310
70. Welch LG, Peak-Chew SY, Begum F, Stevens TJ, Munro S (2021) GOLPH3 and GOLPH3L are broad-spectrum COPI adaptors for sorting into intra-Golgi transport vesicles. J Cell Biol 220:e202106115
71. Wood CS, Schmitz KR, Bessman NJ, Setty TG, Ferguson KM, Burd CG (2009) PtdIns4P recognition by Vps74/GOLPH3 links PtdIns 4-kinase signaling to retrograde Golgi trafficking. J Cell Biol 187:967–975

Chapter 9
Cargo Sorting Zones in Plant *Trans*-Golgi Network

Yutaro Shimizu and Tomohiro Uemura

Abstract Membrane trafficking is crucial for various aspects of plant growth, development, and responses to environmental stimuli. The *trans*-Golgi network (TGN) in plants serves as a central sorting station for membrane trafficking and operates at the intersection of the biosynthetic and endocytic pathways for both plasma membrane/extracellular molecules and vacuolar proteins. Although the molecular machinery that controls membrane trafficking is evolutionarily conserved, how plant TGNs perform their multiple sorting operations still remains poorly understood. Multitasking of plant TGNs manifests as a partition of the molecular machinery within a single organelle. This chapter presents the current state of knowledge on plant TGNs, explores emerging concepts in the field, and highlights key questions and challenges for future research.

Keywords Plant · *Arabidopsis thaliana* · Membrane trafficking · *Trans*-Golgi network · Cargo sorting · Organelle zone

Introduction

Membrane trafficking is a fundamental cellular process in eukaryotes that involves the movement of proteins, lipids, and other molecules between different membrane compartments in both anterograde and retrograde directions using membrane carriers, such as vesicles and tubules. In plants, membrane trafficking is crucial for various processes for growth and development, including seed development [1],

---

Y. Shimizu
RIKEN Center for Sustainable Resource Science, Saitama, Japan

Present Address: Laboratoire de Biogenèse Membranaire, University of Bordeaux, Villenave d'Ornon, France
e-mail: yutaro.shimizu@u-bordeaux.fr

T. Uemura (✉)
Natural Science Division, Faculty of Core Research, Ochanomizu University, Tokyo, Japan
e-mail: uemura.tomohiro@ocha.ac.jp

A. Nakano, J. Saraste (eds.), *The Golgi Network, Volume II*, Subcellular Biochemistry 111, https://doi.org/10.1007/978-3-032-16833-7_9

reproduction [2, 3], hormone distribution [4, 5], nutrient uptake [6, 7], and immune responses [8, 9]. This trafficking system consists of conserved processes, including two major steps: carrier formation accompanied by cargo packing at the donor membrane, which is mediated by coat proteins, and carrier fusion with the target membrane, mediated by soluble N-ethylmaleimide sensitive factor attachment protein receptors (SNAREs). Coat proteins, which form the coat around vesicles, are fundamental in cargo protein sorting and vesicle formation at the donor membrane [10–12]. SNAREs can be classified as Q-SNAREs and R-SNAREs. Accurate fusion of the carrier with the target membrane is mediated by SNARE complexes formed by specific combinations of two or three Q-SNAREs and one R-SNARE [13, 14]. Two classes of small GTPases, SAR/ARFs and RABs, are key regulators of membrane trafficking. SAR/ARFs recruit coat proteins to the donor membrane, whereas RABs mainly contribute to the targeting and tethering of the carrier to the target membrane [12, 14, 15].

The endomembrane system in plant cells comprises the endoplasmic reticulum (ER), Golgi apparatus, *trans*-Golgi network (TGN), multivesicular body/prevacuolar compartment (MVB/PVC), vacuoles, and the plasma membrane (PM) [16, 17] (Fig. 9.1a). Among these organelles, the TGN is a key hub in post-Golgi membrane trafficking of plant cells [17, 18]. It functions as a central cargo-sorting station that receives cargo from the Golgi apparatus and directs it to different destinations, including the PM and vacuoles. In plants, the TGN also functions as an early endosome that receives endocytosed material from the PM [17, 18]; the endocytosed cargo proteins are then recycled to the PM or redirected to vacuoles via the MVB/PVC for degradation [17, 18]. Thus, the plant TGN operates at the intersection of secretory and vacuolar trafficking routes for both biosynthetic and endocytic cargo proteins and has been recognized as an important sorting platform in post-Golgi membrane trafficking (Fig. 9.1a). However, how the TGN sorts a variety of cargo proteins to different destinations remains unclear. The focus on TGN in plant cell biology research has increased recently, and studies have revealed its complex organization, dynamic behavior, and critical roles in plant growth and development, and a wide range of other physiological processes. In this chapter, we focus on the cargo sorting mechanisms of plant TGN and highlight recent advances and emerging concepts in plant TGN research.

Morphology, Dynamics, and Key Features of the Plant TGN

The term TGN was coined by Griffiths and Simons [19] based on mammalian research and was then introduced to the plant field [20]. In contrast to the flattened cisternae of the Golgi apparatus, the plant TGN is characterized as a tubulovesicular structure that forms transport carriers, such as clathrin-coated vesicles (CCVs) and secretory vesicles [21, 22]. At the protein level, the plant TGN is characterized by Qa-SNARE SYP4s [23, 24], Qc-SNARE SYP61 [23, 24], vacuolar H^+-ATPase subunit VHAa1 [25], and clathrin [21, 22, 26]; all of these are absent from the

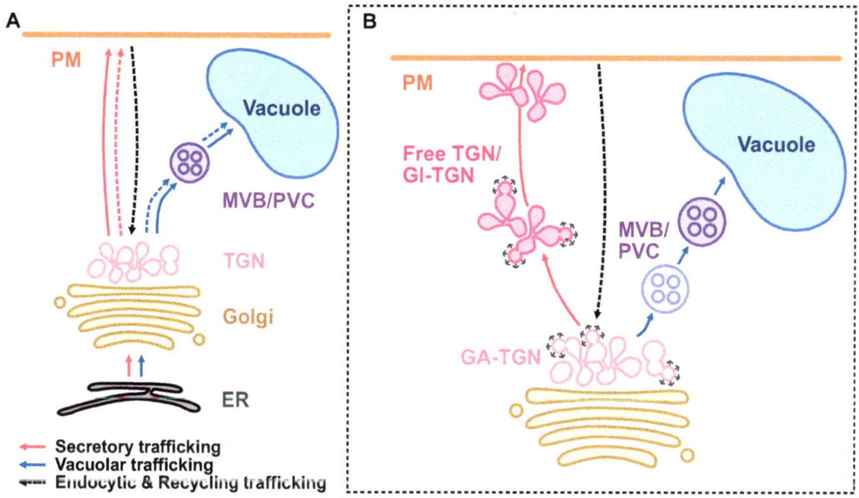

Fig. 9.1 Schematic of plant endomembrane system. (**a**) TGN stands at the intersection of different trafficking pathways. Secretory/PM or vacuolar cargo proteins are synthesized at the ER and transported to their destinations through the Golgi and TGN. The TGN also receives endocytic cargo and redirects it to the PM or vacuole. (**b**) The post-Golgi trafficking is mediated by the organelles that can mature. The Free-TGN/GI-TGN is derived from the GA-TGN, moves independently of the Golgi, and mediates secretory cargo transport to the PM. The free-TGN/GI-TGN just after being generated and the GA-TGN harbors clathrin-coated vesicle buds. The MVB/PVC is the intermediate organelle in the vacuolar trafficking pathway. The MVB/PVC could mature from the Rab5-positive MVB/PVC (pale violet) to the Rab7-positive MVB/PVC (dark violet) on the way to the vacuole. ER Endoplasmic reticulum, TGN Trans-Golgi network, GA-TGN Golgi-associated TGN, GI-TGN Golgi-released independent TGN, PM Plasma membrane, MVB/PVC Multivesicular body/prevacuolar compartment

main body of the Golgi apparatus, highlighting the functional differences between the Golgi and TGN. SYP4s and SYP61 are plant orthologs of yeast Tlg2/mammalian Syntaxin16 and yeast Tlg1/mammalian Syntaxin6/10, respectively [27]. These proteins are typically used as TGN markers in various organisms. In support of the model in which the TGN functions as a branch of the secretory and vacuolar trafficking pathways, at least two different types of R-SNARE vesicle associated membrane proteins (VAMPs) that mediate membrane fusion at the PM or vacuolar membrane can localize to the TGN. VAMP721/722 localizes to both the TGN and PM and functions in secretory trafficking [28–30]. In contrast, VAMP727 localizes to the TGN, MVB/PVC, and vacuolar membrane and mainly functions in the vacuolar trafficking pathway [31, 32].

The TGN is a membrane compartment that is generally associated with the *trans* face of the Golgi apparatus. However, both 3D electron tomography and live-cell imaging revealed that the plant TGN can be associated with the Golgi apparatus or move independently from it [21, 22, 33, 34], termed as Golgi-associated TGN (GA-TGN) or free TGN/Golgi-independent TGN (GI-TGN), respectively (Fig. 9.1b). Live imaging of TGN markers, such as SYP43 and SYP61, demonstrated that a portion of the GA-TGN detach from the Golgi apparatus to form free TGN/GI-TGN, whereas the other portion remains associated with the Golgi apparatus [32, 34]. Thus, the TGN can mature by converting GA-TGN to free TGN/GI-TGN (Fig. 9.1b). Toyooka et al. also reported the presence of free TGN/GI-TGN-like structures in tobacco BY-2 cells and termed them secretory vesicle clusters (SVCs) because the vesicle buds in the SVCs contained secretory markers, such as secretory GFP and pectin [35]. Currently, some SVCs are considered equivalent to the free TGN/GI-TGN. Free TGN/GI-TGN is abound with VAMP721, supporting its role in secretory trafficking to the PM [36]. Although some SVCs contain clathrin coats, like free TGN/GI-TGN, the SVCs close to and almost fusing with the PM do not contain clathrin coats [35]. Thus, the free TGN/GI-TGNs may further mature due to CCV budding on their way to the PM. The generation of free TGN/GI-TGN is apparently coupled with physiological functions because the proportion of free TGN/GI-TGN increases with the maturation of root epidermal cells [34]. Free TGN/GI-TGN also mediate the secretion of anti-pathogenic molecules into the apoplast in response to infection with powdery mildew fungi or bacterial pathogens [36, 37].

Jia et al. identified a *loss of tgn (lot)* mutant that leads to overstacking of the Golgi cisternae, loss of obvious TGN-like membranes, and dispersed cytosolic distributions of TGN-localized proteins, including Qb-SNARE VTI12, RabA4b, and RabA5d [38, 39]. LOT encodes a putative Golgi-localized guanine nucleotide exchange factor (GEF) for RabH/Rab6 [38]. These observations partially support the "maturation" process in which GA-TGN is derived and generated from the Golgi apparatus, as observed in budding yeast [40, 41]. However, observations of pharmacologically-induced regeneration of Golgi and TGN using brefeldin A (BFA), a fungal toxin that binds to a subset of ARF–GEF complexes, suggest that the TGN can be generated independently of the Golgi [42]. An additional yet unidentified mechanism might exist to generate the TGN in plants. The biogenesis or transport of free TGN/GI-TGN requires a TGN-associated microtubule-binding protein, designated TGNap1 [43], suggesting that secretory trafficking is partially dependent on the microtubule cytoskeleton. *tgnap1* mutants showed an increase in the number and size of vesicles in the TGN region, suggesting the inhibition of the peeling of GA-TGN from the Golgi [43]. TGNap1 also interacts with RabH/Rab6 and YPT/RAB GTPase interacting protein 4 (YIP4) [43]. YIP4 interacts and recruits various key regulators of membrane trafficking, such as SYP61, VHAa1, and ECHIDNA as well as TGNap1, to the TGN [43, 44]. Nevertheless, *yip4* and *echidna* mutants seem to exhibit a phenotype opposite to that of *tgnap1* mutants, namely, significant dissociation of GA-TGN from the Golgi [44, 45]. These findings indicate that TGN differentiation is intricately regulated by protein-interaction networks.

Electron microscopy has frequently shown that some of the MVBs, which also function as late endosomes or PVCs [46, 47], are located in close proximity to the TGN in plants [48, 49]. Although the mechanism is not fully understood, application of the V-ATPase inhibitor concanamycin A reduces the number of MVBs, as they aggregate with TGN [48]. Scheuring et al. discovered that the MVB structure was connected to TGN-like structures and proposed that MVB matured from the TGN subdomain [48]. MVB is generally characterized by the presence of intraluminal vesicles formed by invaginating the limiting membrane of the MVB through endosomal sorting complex required for transport (ESCRT) machinery [47]. Four separate ESCRT complexes, namely ESCRT-0, -I, -II, and -III, sequentially function on the limiting membrane of endosomes, recognizing ubiquitinated membrane proteins and sorting them into the intraluminal vesicle for protein degradation after the MVB fuses with the vacuole [47]. Scheuring et al. also demonstrated that the ESCRT-I subunit VACUOLAR PROTEIN SORTING 28 (VPS28), ESCRT-II subunit VPS22, and ESCRT-III subunit VPS28 were distributed in a continuous manner between the TGN and MVB as a gradient [48]. Thereafter, in plants, target of myb1-like (TOL) proteins were characterized as the functional counterparts of ESCRT-0 and colocalized with TGN and PM markers [50, 51]. This suggests that ubiquitinated cargo proteins are recognized by the ESCRT machinery during the initial stages of the endocytic vacuolar pathway. The sequential recruitment of Rab5 and Rab7 is a unique feature of endosomal maturation from the early to late endosomes in mammalian cells. However, in plant cells, conversion from Rab5 to Rab7 appears to occur on the MVB/PVC/late endosome [52–54], indicating that the MVB/PVC/late endosome matures further after formation (Fig. 9.1b).

Regulation of Endocytic and Recycling Trafficking by the TGN

The dual nature of the plant TGN as an early endosome was discovered by observing the endocytic marker FM4-64 [25, 55, 56]. In animal cells, the endocytosed molecules are first received by the early endosome, which is spatially separated from the TGN. However, in plants endocytosed FM4-64 reaches the TGN (within 5 min) earlier than other membrane compartments, indicating that the plant TGN also acts as an early endosome [25, 55, 56]. Specifically, FM4-64 initially labels the GA-TGN rather than the free TGN/GI-TGN, suggesting that GA-TGN serves as an early endosome [36]. Live imaging using total internal reflection fluorescence microscopy indicated that partially or fully coated CCVs on the PM could be picked up directly and/or through the actin cytoskeleton by the GA-TGN [57]. Consequently, the plant TGN is often designated as the TGN/early endosome, and, across the eukaryotic kingdom, only plants were believed to display a TGN that can directly handle retrograde transport. However, this view has recently changed, as recent studies on receptor endocytosis in yeast have shown that budding yeasts also have the same endosome system as plants [58, 59].

As an early endosome, the plant TGN regulates the fate of PM proteins by recycling (sometimes in a polarized manner) them to the PM or directing them to the vacuole. In plants, PM-localized mineral transporters, including the borate exporter REQUIRES HIGH BORON 1 (BOR1), phosphate importer PHOSPHATE TRANSPORTER 1 (PHT1), iron importer IRON-REGULATED TRANSPORTER

1 (IRT1), and divalent metal transporter NATURAL RESISTANCE ASSOCIATED MACROPHAGE PROTEIN 1 (NRAMP1), undergo vacuolar degradation in response to an increase in their substrate concentration, which prevents excess mineral toxicity [60–66]. Under control conditions (e.g., low-boron conditions for BOR1), these transporters predominantly localize to both the TGN and PM and are recycled between these compartments through endocytosis and TGN-mediated plasma membrane delivery [62–64, 67]. When the specific minerals are present in excess (e.g., under high boron conditions for BOR1), the endocytosed transporters are rapidly ubiquitinated and subsequently transported into the vacuole via the TGN and then the MVBs for protein degradation [33, 65–70].

Plant cells utilize the polar localization of PM proteins to drive the directional transport of nutrients or hormones. For example, boron transporters exhibit an asymmetric distribution along the lateral axis. The boric acid influx channel nodulin 26-like intrinsic protein 5;1 (NIP5;1) and the boric acid efflux transporter BOR1 are localized to the outer and inner lateral PM, respectively, in root epidermal cells [33, 71, 72]. This asymmetric distribution of boric acid transporters contributes to the directional flow of boron from the soil to plant xylem [6]. Clathrin-mediated endocytosis is essential for the strong polar localization of NIP5;1 and BOR1, suggesting that constitutive recycling between the TGN and PM plays an important role in maintaining the polar localization of boric acid transporters [71–74]. Concentration gradients of the phytohormone auxin, which are induced by its directional transport, play a crucial role in diverse aspects of plant growth and development, such as embryogenesis, organ formation, and tropism [75, 76]. The directional auxin transport predominantly relies on the apical–basal localization of auxin efflux carriers named PIN-FORMED (PIN) proteins [77, 78]. Arabidopsis PIN2 is the most studied polarly localized protein, which exhibits apical localization patterns on the PM of root epidermal cells. In root epidermal cells, PIN2 acquires the apical polarity immediately after cytokinesis by its removal from the basal membrane via endocytosis [79, 80]. Evidence from the model plant *Arabidopsis thaliana* indicates that PIN2 polarity is maintained through a combination of endocytosis, TGN-mediated recycling, and restriction of lateral diffusion [81–84]. Interestingly, the basal polar localization of PIN1 in stele cells, but not the apical localization of PIN2 in epidermal cells, is sensitive to treatment with BFA, an inhibitor of ARF-GEFs that regulate endocytic recycling [85]. This highlights the fact that different factors are involved in different polar secretions. It has been suggested that ARF-GEF GNOM regulates the targeted recycling of PIN1 to the basal PM but only partially contributes to the recycling of apically localized PIN2 [86]. In contrast, the compound Endosidin 16 targets RabA2a/Rab11 and interferes with the trafficking of apically localized PM proteins without affecting the basal polarity [87]. However, the mechanisms involved in the recycling and polar targeting of PM proteins from the TGN remain poorly understood.

The polar localization of auxin transporters or boric acid transporters have also been experimentally confirmed in other plant clades, such as bryophyte and monocot, suggesting that the polar localization of proteins on the PM is a common feature in plants [88–90]. However, in contrast to well-characterized examples of polarity formation at the PM in *Arabidopsis*, endocytosis inhibition does not perturb the polar localization of the boric acid transporters OsNIP2;1 and OsBOR1 or silicon transporters [89, 90]. In *Physcomitrella patens* and *Marchantia polymorpha*, endogenous PIN2s are polarly localized to the apical tip of filamentous cells, whereas exogenously expressed *Arabidopsis* PIN2 is distributed at the PM in a non-polar manner with significant accumulation in intracellular compartments [88]. In *Arabidopsis* root epidermal cells, exogenous *Physcomitrella* or *Marchantia* PIN2 is localized in the PM without apical polarity [88]. The disruption of microtubules by oryzalin treatment disturbed the polar localization of PIN2 in *Physcomitrella* and *Marchantia* filamentous cells but did not abolish the polar localization of PIN2 in *Arabidopsis* root hair cells [88]. These findings suggest that the mechanisms underlying the establishment of polarity have evolved in different cell types and/or species. How much the biosynthetic secretory trafficking machinery contributes to the establishment of PM polarity remains unknown because of the lack of reliable techniques to investigate biosynthetic secretion in plants.

Clathrin and Its Adaptor Proteins (APs) in the TGN

Clathrin and AP complexes are key coat proteins involved in post-Golgi trafficking. AP complexes are generally heterotetramers consisting of two large subunits (β and α/γ/δ/ε/ζ), a medium subunit (μ), and a small subunit (σ) [11, 91]. AP complexes sort their cargo proteins by recognizing and binding to cytoplasmic sorting motifs or signals, such as tyrosine and dileucine motifs [10]. *Arabidopsis* genome encodes all the components of five AP complexes, except for AP5σ [92]. Among the five plant AP complexes, AP-1 and AP-4 localize to the TGN [93–96]. In the history of the identification of AP complexes from bovine brains, AP-1 and AP-2 complexes have been shown to interact with clathrin [97, 98]. Plant AP-1 and AP-2 complexes also interact with clathrin [93, 99–101]. As in mammalian cells and yeasts, AP-2/clathrin-mediated endocytosis is a well-established process in plants [99–102]. However, the exact function of AP-1 and the interaction of AP-4 with clathrin at the TGN in plants remain controversial.

The virtual impossibility to maintain *ap1μ1 ap1μ2* [93] and *ap1γ1 ap1γ2* [103, 104] double homozygous mutants revealed that AP-1 complex is essential for the survival and reproduction of *Arabidopsis*. The partial functional loss of AP-1, such as *ap1μ2* mutants (*ap1μ2* is the most abundant isoform of AP-1 μ subunits), causes severe defects in seedling growth and fertility [93–95, 104–107]. A crucial aspect is that plant AP-1 appears to function in two independent trafficking pathways: secretory trafficking to the PM and vacuolar trafficking [108] (Fig. 9.2). For example, genetic analyses demonstrated that the recycling of PIN2, BRASSINOSTEROID INSENSITIVE1, and RARE COLD-INDUCIBLE 2A to the PM was impaired in *ap1μ2* mutant root cells [95, 109]. The secretory trafficking of cytokinesis-specific Qa-SNARE KNOLLE/SYP111 to the cell plate was also blocked in *ap1μ2* mutant root cells, reflecting the failure of normal cell division and

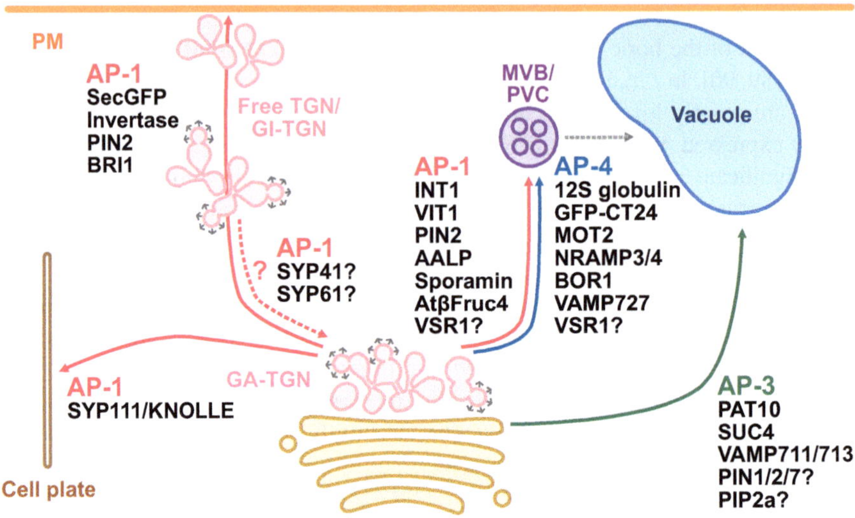

Fig. 9.2 Roles of Golgi- and TGN-localized AP complexes in plants. AP-1 is involved in the transport of proteins destined for the apoplast, PM, cell plate, and vacuole. In the *ap-1*-mutant cells, TGN-localized Q-SNAREs, SYP41 and SYP61, were mislocalized to the PM, suggesting that AP-1 retains or aids in the retrograde trafficking of TGN-resident proteins. Both AP-1 and AP-4 can interact with VSR1. AP-3 and AP-4 sort distinct vacuolar cargo proteins. AP-4 functions at the TGN, whereas AP-3 is suggested to function at the Golgi. Loss of function of AP-3 results in intracellular aggregation of PINs and PIP2a, suggesting possible involvement of AP-3 in the trafficking of these PM-localized proteins

severe growth defects in *ap1μ2* mutants [93, 94]. Furthermore, the delivery of degradative PIN2 to the vacuole was impaired in *ap1μ2* root cells, and RNA interference of *AP1γ1* in *AP1γ2*-mutant cells led to defects in vacuolar targeting of VACUOLAR ION TRANSPORTER 1 and INOSITOL TRANSPORTER 1 [103]. Although the precise mechanisms remain unclear, AP-1 seems to affect cargo exit from the ER because the soluble apoplastic protein secretory invertase and soluble vacuolar protein *Arabidopsis* aleurain-like protease (AALP) are retained in the ER in *ap1m2*-mutant protoplasts [93]. AP-1 is also required for the secretion of polysaccharides in *Arabidopsis* seed coat cells [110]; however, little is known about the molecular mechanism by which non-amino acid components are sorted by the plant AP complex. AP-1 is recruited to the TGN by ARF1 and its GEFs, designated as BFA-inhibited GEFs (BIGs), and impairing the function BIGs also resulted in defects in both the secretory and vacuolar trafficking pathways [111]. However, how AP-1 sorts a variety of different cargo proteins remains an open question. A possible explanation is that AP-1 utilizes different unknown interactors depending on the trafficking pathways. However, we speculate that AP-1 sorts cargo proteins that should be retrogradely transported from the TGN to earlier compartments. Notably, TGN-resident Q-SNAREs SYP41 and SYP61 were significantly mislocalized from the TGN to the PM in *ap1μ2*-mutant root cells, whereas VHAa1 localized similarly to the TGN in both *ap1μ2* mutant and wild-type cells [109]. This suggests that the function of the TGN was compromised in the *ap1μ2* mutant because of the loss of TGN-resident proteins, including Q-SNAREs. Both *ap1μ2* and *syp4* mutants exhibited an abnormally curved Golgi/TGN, indicating dysfunction of the Golgi/TGN [93, 112]. Similar to AP-1 component mutants, *syp4* mutants also exhibit pleiotropic phenotypes, including severe defects in growth, fertilization competence and both secretory and vacuolar trafficking pathways [112]. As clathrin and AP-1 remain in the TGN during maturation from GA-TGN to free TGN/GI-TGN and dissociate from free TGN/GI-TGN [22, 32, 35], AP-1/clathrin-coated vesicles may transport Golgi- and/or GA-TGN-resident proteins from free TGN/GI-TGN. AP-1/clathrin-mediated retrograde trafficking from the TGN has been recently reported in yeast [58, 113–115] and mammals [116]. In *Arabidopsis* protoplasts, Shao et al. recently demonstrated that clathrin mediates retrograde recycling of VACUOLAR SORTING RECEPTOR 7 (VSR7) from the TGN [117].

The functional loss of AP-4 results in dwarfing, reduced male fertility, and hyperbranching of leaf trichomes; however, the phenotypic changes were less severe than those in AP-1-related mutants [96, 118, 119]. In contrast to the complexity in understanding AP-1 functions, research related to AP-4 has revealed that it mostly functions in the vacuolar trafficking pathway (Fig. 9.2). Using a forward genetic screen, Fuji et al. identified AP-4 subunits to be responsible for the vacuolar sorting of the artificial cargo protein GFP-CT24 [96]. Analyses of loss-of-function mutants of AP-4 revealed the impaired vacuolar sorting of 12S globulin, molybdate transporter 2, and vacuolar metal transporters NRAMP3 and 4 [96, 118]. *ap4b* and *ap4m* mutants did not affect the polar localization of BOR1 but delayed its high-boron-induced vacuolar sorting [70], suggesting that AP-4 is involved in the vacuolar targeting of endocytosed cargo proteins. The requirement of clathrin for AP-4-coated vesicles in plants has been debated. Biochemical analyses have shown an interaction between clathrin and AP-4 [32, 96, 120], but super-resolution imaging demonstrated that AP-4 did not completely colocalize with clathrin at the TGN, unlike AP-1 which tended to colocalize with clathrin [32]. The segregation between AP-4 and clathrin is also supported by the fact that the MODIFIED TRANSPORT TO THE VACUOLE1 (MTV1), an accessory protein of AP-4, is located further from clathrin than EPSIN1, an accessory protein of AP-1 [121]. Further studies, using genetic approaches, are needed to reveal the relationship between AP-4 and clathrin.

Plant AP-3 also sorts a subset of vacuolar proteins independent of the MVB/PVC-mediated trafficking pathway, as observed in yeast (Fig. 9.2). Forward genetic screening based on GFP-tagged cargo protein imaging showed that the mutations in AP-3 β and σ genes resulted in aberrant intracellular accumulation of proteins, such as PIN1, PIN2, PIN7, and the aquaporin PLASMA MEMBRANE INTRINSIC PROTEIN 2a [122–124]. Vacuolar membrane targeting of VAMP711, VAMP713, SUCROSE TRANSPORTER 4, and PROTEIN S-ACYL TRANSFERASE10 is compromised in *ap-3* mutants [53, 125]. Notably, SUCROSE TRANSPORTER 4 and PROTEIN S-ACYL TRANSFERASE10 colocalized with the Golgi marker rather than the TGN marker in *ap-3* mutants, suggesting that AP-3 functions at the Golgi apparatus and not the TGN [122, 125]. Pertl-Obermeyer et al. analyzed putative cargo proteins of AP-3 and AP-4 using a comparative proteomic approach by employing the wild-type cells and *ap3β* and *ap4β* mutants [119]. Despite large differences between the macroscopic phenotypes of *ap3β* and *ap4β* mutants, highly overlapping results were obtained between AP-3 and AP-4 cargo candidates, suggesting possible redundancy [119]. The expression levels or affinities of AP complexes for the same cargo protein may differ between tissues and cells, leading to differences in the degree of distinct phenotypes. Indeed, a yeast two-hybrid assay demonstrated that both μ subunits of AP-3 and AP-4 interacted with tyrosine motifs of BOR1 [73]. However, neither polar localization of BOR1 nor its targeting to the vacuole was significantly affected in *ap3m* or *ap3b* mutants [70]. Furthermore, vacuolar targeting of BOR1 was comparable between *ap4m* and *ap3m ap4m* double mutants [70]. Thus, the interaction ability does not necessarily reflect *in vivo* functions.

In addition to AP complexes, monomeric accessory proteins containing an N-terminal homology (ENTH) domain are presumed to coordinately sort the cargo proteins [126, 127]. The ENTH protein EPSIN1 localizes to the TGN and binds to clathrin and AP-1 [121, 128]. Similar to AP-1, EPSIN1 is involved in the delivery of vacuolar proteins [128] and a subset of PM proteins [129]. Another TGN-localized ENTH protein is MTV1, which is phylogenetically distant from other ENTH protein members. However, it is well-conserved and mostly present as a single-copy gene in plants [130] MTV1 is primarily involved in vacuolar cargo sorting [121, 130]. Both *epsin1* and *mtv1* single mutants exhibited no obvious effects on plant growth; however, *epsin1 mtv1* double mutants exhibited a dwarf phenotype, suggesting functional redundancy [121]. Genetic analyses have also revealed that AP-4 is required to recruit MTV1, but not EPSIN1, to the TGN and functions in the same trafficking pathway as that of MTV1 [121]. Orthologs of Golgi-localizing, γ-adaptin ear homology domain, ADP-ribosylation factor-binding proteins (GGAs), which are well-studied monomeric accessory proteins functioning as the adaptor for CCVs at the TGN in yeast and mammalian cells, appear to be absent in plants [11].

The existence of crosstalk mechanisms between secretory trafficking from the TGN to the PM and the clathrin-mediated endocytosis has been suggested [109, 131]. In *ap-2* complex mutants, the exocytosis and recruitment of AP-1 and clathrin to the TGN were impaired [109]. In contrast, the recruitment of AP-2 and clathrin to the PM, as well as clathrin-mediated endocytosis, was compromised in *ap-1* mutants without changes in the expression or protein levels of clathrin and its accessory proteins [109].

Upon comprehensive interactome analyses of clathrin adaptor proteins in *Arabidopsis,* the protein designated as p34, a putative plant homolog of the human α- and γ-adaptin binding protein, was identified as the a key regulator of post-Golgi trafficking [132]. *Arabidopsis* p34 directly interacts with the subunits of AP-1, AP-2, and AP-4 but not with those of AP-3, AP-5, or clathrin [132]. Protein abundance levels, but not the transcript expression levels, of AP-1, AP-2, and AP-4 were reduced in *p34* mutants [132]. In contrast, for AP-3 and AP-5, the protein abundance, transcript levels, and subcellular localization were not altered in *p34* mutants [132]. The mutants of *p34* influence various post-Golgi trafficking pathways, including reduced endocytosis, abnormal polar localization of PIN2, mistargeting of SYP61, ectopic KNOLLE/SYP111 localization, and recruitment of MTV1, which is reminiscent of the known mutant phenotypes of *ap-1*, *ap-2*, or *ap-4*. These results suggest that p34 functions as a common regulator of these AP complexes [132].

Soluble Cargo Protein Receptors and Their Retrograde Trafficking

Soluble cargo proteins are co-translationally translocated to the lumen of the ER because of the presence of signal peptides. They were originally thought to be targeted to the PM by default in plant membrane trafficking, because when the signal peptide was fused to non-secretory proteins, the resulting chimeric proteins were secreted into the extracellular space [133]. In contrast, soluble cargo proteins with vacuolar-targeting signals are actively sorted into the vacuoles. Soluble cargo proteins do not directly interact with coat adaptors and thus utilize mediators for sorting. Transmembrane receptors bind to both cargo proteins and coat adaptors to aid in the sorting of soluble cargo proteins. VSRs are well-studied cargo receptors in plants. They are the functional equivalents of mannose-6-phosphate receptors (MPRs) in mammalian cells and VPS10 in yeast [134]. In mammalian cells, post-translational glycan modification of cargo proteins serves as a signal for lysosomal targeting [135]. MPRs recognize mannose-6-phosphate residues on the N-linked oligosaccharides present in lysosomal enzymes [135]. By contrast, plant VSRs sort its cargo proteins, such as major seed storage proteins 12S globulin and 2S albumin, by binding to amino acid sequences encoded within the vacuolar cargo protein [136–139]. Despite the difference in their vacuolar/lysosomal sorting signals, MPRs, VSRs, and VPS10 share a common mechanism of receptor-ligand interaction: the receptors bind and release their ligands at relatively neutral and acidic pH conditions, respectively [134]. Mammalian MPR and yeast VPS10 bind cargo proteins in the TGN and release them in the post-TGN compartment, whereas the receptors are recycled back to the TGN for another round of sorting [140, 141].

In 1994, Kirsch et al. first identified VSRs, originally named BP-80 (binding protein of approximately 80 kDa), in the CCVs of peas [142]. Subcellular localization analyses of VSR family proteins demonstrated that they localize to both the Golgi/TGN and MVB/PVC [117, 137, 143–145]. Subsequent studies have demonstrated the following interactions between VSRs and TGN-localized adaptors: VSR1 and AP-1 [93], VSR1 and AP-4 [96], VSR1 and EPSIN1 [128], VSR2 and AP-1 or AP-4 [146], and VSR4 and AP-1 [147]. Importantly, VSRs seem to localize primarily to the limiting membrane, rather than to the intraluminal vesicles, of the MVB/PVC, in contrast to other cargo proteins targeted to the vacuole [33, 148], indicating that these receptors could be recycled from the MVB/PVC for another round of anterograde trafficking. Many studies have consistently shown that the components of the core retromer complex (VPS26, VPS29, and VPS35), which is the coat complex involving retrograde trafficking, mainly localize to the MVB [149–

155]. Co-immunoprecipitation analysis showed that VPS35 interacts with VSR1 [149], and a knockdown mutant of VPS29 failed to recycle VSR1 from the MVB/PVC to the TGN [156]. Consequently, a model similar to that in mammalian cells and yeast was proposed, according to which VSRs sort cargo proteins into CCVs at the TGN, release cargo at the MVB/PVC, and cycle between the TGN and MVB [157, 158]. Notably, although the localization of plant SNX-BAR, another coat protein complex recently suggested to function independently of the core retromer complex in plants, is controversial, it also plays a role in retrograde trafficking [158–160].

In contrast, Pimpl group developed a nanobody-fused reporter to monitor the *in vivo* interaction between VSRs and ligands and proposed a new model for VSR-mediated vacuolar sorting. They demonstrated that VSRs bind to vacuolar cargo proteins in the ER and Golgi, but not in the TGN, suggesting that they are required for the transport of their cargo from the ER and/or Golgi to the TGN [117, 161, 162]. Therefore, soluble vacuolar cargo proteins are thought to be released in the TGN and transported from the TGN to the vacuole by default without no further role for VSRs [117, 161, 162]. This default pathway involves organelle maturation, by which the TGN is converted to the MVB [48]. Undoubtedly, both models support the role of plant TGN in the trafficking of VSRs and soluble vacuolar cargo proteins; however, further studies are needed to fully elucidate the mechanism of VSR-mediated trafficking of vacuolar cargo proteins.

Zones of Plant TGN

TGN-localized membrane traffic regulators, including Rab GTPases, SNAREs, and coat proteins, are spatially segregated in plants [23, 32, 43–45, 56, 96, 121, 163–165], suggesting that the multiple functions of the TGN rely on its partitioning into subdomains. Importantly, advanced imaging techniques have allowed researchers to clearly visualize the key subdomains for cargo sorting in plant TGN. The Nakano group at RIKEN developed super-resolution confocal live-imaging microscopy (SCLIM) and revealed that the plant TGN consists of at least two subdomains characterized by either 1) VAMP721, clathrin, and AP-1 or 2) VAMP727 and AP-4 [32]. Owing to the secretory trafficking function of VAMP721 and the vacuolar trafficking function of VAMP727, the subdomains are designated as the secretory and vacuolar trafficking zones. Using Zeiss Airyscan imaging and genetic and biochemical analyses, the Sauer group demonstrated that the distinct TGN subdomains were composed of AP-1 with EPSIN1 or AP-4 with MTV1 [121]. Taken together, these findings indicate that the plant TGN is further divided into two subdomains, AP-1/EPSIN1/VAMP721 and AP-4/MTV1/VAMP727.

Besides the composition of proteins, lipid composition also differs between TGN subdomains [166]. This was demonstrated by lipidomic analyses, following the immunopurification of TGN subdomains [164, 167]. The SYP61 and RabA2a subdomains were analyzed, demonstrating that the SYP61 subdomain is more enriched in the final form of sphingolipids containing α-hydroxylated very long-chain fatty acids, such as GIPC and GlcCer, as compared to the RabA2a subdomain. Consistent with this finding, INOSITOLPHOSPHORYLCERAMID SYNTHASE 2, a key lipid enzyme in production of the final form of sphingolipids from ceramides, colocalizes well with SYP61 [168] and the SYP61 subdomain-localized protein ECHIDNA [169]. In contrast, ceramides are thought to be enriched in the RabA2a subdomain of the TGN [166, 170]. Interestingly, a pharmacological or genetic reduction in the acyl chain length of the SYP61 subdomain-abundant sphingolipids GIPC and GlcCer perturbed the secretory sorting of PIN2 to the apical polar domain of the PM [164, 169]. However, this pharmacological treatment inhibited neither the basal localization of PIN1 nor the nonpolar localization of PM proteins, including AUX1, aquaporin PIP1;4, and Qb-SNARE NPSN12 [164]. In contrast, the secretory trafficking of PIN1 and AUX1, but not PIN2, to the basal PM membrane was impaired when the acyl chain length of the ceramides was shortened [170]. The TGN protein ECHIDNA is localized to the SYP61 subdomain and is involved in the secretory sorting of AUX1, but not of PIN1 or PIN2. These results show that even within one subdomain, further functional partitioning occurs. Thus, the TGN might differentiate secretory trafficking according to specialized nanodomains or vesicles with specific regulators, despite the seemingly single path to the PM. Otherwise, the *de novo* secretion or endocytic recycling of PM proteins may be differentially regulated by distinct molecules within the same subdomains.

Conclusion and Future Perspectives

Functional studies have highlighted the central role of the TGN in a wide range of physiological processes in plants, such as reproduction [93, 104–107], defense responses against pathogens [36, 37], and adaptive growth decisions under abiotic stress conditions [171]. Hence, understanding the mechanisms and functions of membrane trafficking in plants is not only of fundamental scientific interest but also has important implications for agriculture and biotechnology, as membrane trafficking is essential for traits, such as crop yield, quality, and stress resilience. Research on plant membrane trafficking has been influenced to some extent by the canonical models established in animals and yeast. However, findings using advanced technology do not fully account for these established simple models and suggest complicated regulations of cargo transport. These discoveries highlight both the conservation of core trafficking machinery across eukaryotes and the unique mechanisms that have evolved in plants. Thus, a comparison of plant studies with those from other kingdoms is beneficial for understanding the underlying mechanism of membrane trafficking in eukaryotes. However, we need to re-consider the experimental results independent of other model systems. Future studies using cutting-edge techniques, such as super-resolution imaging and correlative light and electron microscopy, can further our understanding of post-TGN membrane trafficking in plants. For example, conflicting models of the sorting of vacuolar-soluble cargo proteins have recently tried to be unified according to the differential distribution of membrane trafficking machinery on the way to the vacuole from the TGN [160]. Importantly, the dynamic behavior and subdomains of the TGN identified thus far may reflect the actual mechanisms by which the TGN efficiently sorts cargo proteins. Despite significant progress in understanding the structure, function, and suborganellar organization of plant TGN, many questions remain unanswered. How is the TGN generated and organized at the molecular level? How are the TGN subdomains formed and maintained, and how do these subdomains contribute to TGN's functions? What are the underlying mechanisms that regulate the biosynthetic and endocytic pathways at the TGN? How does the TGN dynamically interact with other organelles, such as the Golgi and MVB/PVC? Answering these questions requires a combination of advanced imaging and genetic and biochemical approaches. Our knowledge of plant membrane trafficking is primarily derived from the model plant *Arabidopsis*, which is quite different from crop plants. Hence, future studies investigating membrane trafficking in crops would be beneficial for animal health and food production.

Acknowledgements We are grateful to all members of the Live Cell Super-Resolution Imaging Research Team and Molecular Bioregulation Research Team at RIKEN and the Uemura lab at Ochanomizu University for the fruitful and stimulating discussions. We apologize to our colleagues whose work could not be cited or discussed in depth here because of the focus and scope of this chapter. Y.S. was supported by the Japan Society for the Promotion of Science Research Fellowship for Young Scientists (22KJ3129). T.U. was supported by Grants-in-Aid for Scientific Research from the Ministry of Education, Culture, Sports, Science and Technology (18H04857), the Japan Science and Technology Agency (JPMJCR20E5), and a Grant for Basic Science Research Project from the Sumitomo Foundation.

Compliance with Ethical Standards The authors have no conflicts of interest to disclose.

References

1. Zheng P, Zheng C, Otegui MS, Li F (2022) Endomembrane mediated-trafficking of seed storage proteins: from Arabidopsis to cereal crops. J Exp Bot 73:1312–1326. https://doi.org/10.1093/jxb/erab519
2. Li H-J, Meng J-G, Yang W-C (2018) Multilayered signaling pathways for pollen tube growth and guidance. Plant Reprod 31:31–41. https://doi.org/10.1007/s00497-018-0324-7
3. Hao G-J, Zhao X-Y, Zhang M-M et al (2022) Vesicle trafficking in Arabidopsis pollen tubes. FEBS Lett 596:2231–2242. https://doi.org/10.1002/1873-3468.14343
4. Richter S, Anders N, Wolters H et al (2010) Role of the GNOM gene in Arabidopsis apical-basal patterning--from mutant phenotype to cellular mechanism of protein action. Eur J Cell Biol 89:138–144. https://doi.org/10.1016/j.ejcb.2009.11.020
5. Paez Valencia J, Goodman K, Otegui MS (2016) Endocytosis and endosomal trafficking in plants. Annu Rev Plant Biol 67:309–335. https://doi.org/10.1146/annurev-arplant-043015-112242
6. Yoshinari A, Takano J (2017) Insights into the mechanisms underlying boron homeostasis in plants. Front Plant Sci 8:1951. https://doi.org/10.3389/fpls.2017.01951

7. Spielmann J, Fanara S, Cotelle V, Vert G (2023) Multilayered regulation of iron homeostasis in Arabidopsis. Front Plant Sci 14:1250588. https://doi.org/10.3389/fpls.2023.1250588
8. Inada N, Ueda T (2014) Membrane trafficking pathways and their roles in plant-microbe interactions. Plant Cell Physiol 55:672–686. https://doi.org/10.1093/pcp/pcu046
9. Gu Y, Zavaliev R, Dong X (2017) Membrane trafficking in plant immunity. Mol Plant 10: 1026–1034. https://doi.org/10.1016/j.molp.2017.07.001
10. Arora D, Van Damme D (2021) Motif-based endomembrane trafficking. Plant Physiol 186: 221–238. https://doi.org/10.1093/plphys/kiab077
11. Dacks JB, Robinson MS (2017) Outerwear through the ages: evolutionary cell biology of vesicle coats. Curr Opin Cell Biol 47:108–116. https://doi.org/10.1016/j.ceb.2017.04.001
12. Singh MK, Jürgens G (2018) Specificity of plant membrane trafficking - ARFs, regulators and coat proteins. Semin Cell Dev Biol 80:85–93. https://doi.org/10.1016/j.semcdb.2017.10.005
13. Lipka V, Kwon C, Panstruga R (2007) SNARE-ware: the role of SNARE-domain proteins in plant biology. Annu Rev Cell Dev Biol 23:147–174. https://doi.org/10.1146/annurev.cellbio.23.090506.123529
14. Saito C, Ueda T (2009) Functions of RAB and SNARE proteins in plant life. Int Rev Cell Mol Biol 274:183–233. https://doi.org/10.1016/S1937-6448(08)02004-2
15. Ito E, Uemura T (2022) RAB GTPases and SNAREs at the trans-Golgi network in plants. J Plant Res 135:389–403. https://doi.org/10.1007/s10265-022-01392-x
16. Aniento F, de Medina S, Hernández V, Dagdas Y et al (2022) Molecular mechanisms of endomembrane trafficking in plants. Plant Cell 34:146–173. https://doi.org/10.1093/plcell/koab235
17. Ito Y, Boutté Y (2020) Differentiation of trafficking pathways at Golgi entry core compartments and post-Golgi subdomains. Front Plant Sci 11:609516. https://doi.org/10.3389/fpls.2020.609516
18. Nakano A (2022) The Golgi Apparatus and its next-door neighbors. Front Cell Dev Biol 10: 884360. https://doi.org/10.3389/fcell.2022.884360
19. Griffiths G, Simons K (1986) The trans Golgi network: sorting at the exit site of the Golgi complex. Science 234:438–443. https://doi.org/10.1126/science.2945253
20. Staehelin LA, Giddings TH Jr, Kiss JZ, Sack FD (1990) Macromolecular differentiation of Golgi stacks in root tips of Arabidopsis and Nicotiana seedlings as visualized in high pressure frozen and freeze-substituted samples. Protoplasma 157:75–91. https://doi.org/10.1007/BF01322640
21. Staehelin LA, Kang B-H (2008) Nanoscale architecture of endoplasmic reticulum export sites and of Golgi membranes as determined by electron tomography. Plant Physiol 147:1454–1468. https://doi.org/10.1104/pp.108.120618
22. Kang B-H, Nielsen E, Preuss ML et al (2011) Electron tomography of RabA4b- and PI-4Kβ1-labeled trans Golgi network compartments in Arabidopsis. Traffic 12:313–329. https://doi.org/10.1111/j.1600-0854.2010.01146.x
23. Bassham DC, Sanderfoot AA, Kovaleva V et al (2000) AtVPS45 complex formation at the trans-Golgi network. Mol Biol Cell 11:2251–2265. https://doi.org/10.1091/mbc.11.7.2251
24. Uemura T, Ueda T, Ohniwa RL et al (2004) Systematic analysis of SNARE molecules in Arabidopsis: dissection of the post-Golgi network in plant cells. Cell Struct Funct 29:49–65. https://doi.org/10.1247/csf.29.49
25. Dettmer J, Hong-Hermesdorf A, Stierhof Y-D, Schumacher K (2006) Vacuolar H+-ATPase activity is required for endocytic and secretory trafficking in Arabidopsis. Plant Cell 18:715–730. https://doi.org/10.1105/tpc.105.037978
26. Ito E, Fujimoto M, Ebine K et al (2012) Dynamic behavior of clathrin in Arabidopsis thaliana unveiled by live imaging. Plant J 69:204–216. https://doi.org/10.1111/j.1365-313X.2011.04782.x
27. Sanderfoot A (2007) Increases in the number of SNARE genes parallels the rise of multicellularity among the green plants. Plant Physiol 144:6–17. https://doi.org/10.1104/pp.106.092973

28. Kwon C, Neu C, Pajonk S et al (2008) Co-option of a default secretory pathway for plant immune responses. Nature 451:835–840. https://doi.org/10.1038/nature06545
29. Zhang L, Zhang H, Liu P et al (2011) Arabidopsis R-SNARE proteins VAMP721 and VAMP722 are required for cell plate formation. PLoS One 6:e26129. https://doi.org/10.1371/journal.pone.0026129
30. El Kasmi F, Krause C, Hiller U et al (2013) SNARE complexes of different composition jointly mediate membrane fusion in Arabidopsis cytokinesis. Mol Biol Cell 24:1593–1601. https://doi.org/10.1091/mbc.e13-02-0074
31. Ebine K, Okatani Y, Uemura T et al (2008) A SNARE complex unique to seed plants is required for protein storage vacuole biogenesis and seed development of Arabidopsis thaliana. Plant Cell 20:3006–3021. https://doi.org/10.1105/tpc.107.057711
32. Shimizu Y, Takagi J, Ito E et al (2021) Cargo sorting zones in the trans-Golgi network visualized by super-resolution confocal live imaging microscopy in plants. Nat Commun 12: 1–14. https://doi.org/10.1038/s41467-021-22267-0
33. Viotti C, Bubeck J, Stierhof Y-D et al (2010) Endocytic and secretory traffic in Arabidopsis merge in the trans-Golgi network/early endosome, an independent and highly dynamic organelle. Plant Cell 22:1344–1357. https://doi.org/10.1105/tpc.109.072637
34. Uemura T, Suda Y, Ueda T, Nakano A (2014) Dynamic behavior of the trans-golgi network in root tissues of Arabidopsis revealed by super-resolution live imaging. Plant Cell Physiol 55: 694–703. https://doi.org/10.1093/pcp/pcu010
35. Toyooka K, Goto Y, Asatsuma S et al (2009) A mobile secretory vesicle cluster involved in mass transport from the Golgi to the plant cell exterior. Plant Cell 21:1212–1229. https://doi.org/10.1105/tpc.108.058933
36. Uemura T, Nakano RT, Takagi J et al (2019) A Golgi-released subpopulation of the Trans-Golgi network mediates protein secretion in Arabidopsis. Plant Physiol 179:519–532. https://doi.org/10.1104/pp.18.01228
37. Bhandari DD, Ko DK, Kim S-J et al (2023) Defense against phytopathogens relies on efficient antimicrobial protein secretion mediated by the microtubule-binding protein TGNap1. Nat Commun 14:6357. https://doi.org/10.1038/s41467-023-41807-4
38. Jia P-F, Xue Y, Li H-J, Yang W-C (2018) Golgi-localized LOT regulates trans-Golgi network biogenesis and pollen tube growth. Proc Natl Acad Sci USA 115:12307–12312. https://doi.org/10.1073/pnas.1809206115
39. Jia P-F, Xue Y, Li H-J, Yang W-C (2019) LOT regulates TGN biogenesis and Golgi structure in plants. Plant Signal Behav 14:e1573100. https://doi.org/10.1080/15592324.2019.1573100
40. Losev E, Reinke CA, Jellen J et al (2006) Golgi maturation visualized in living yeast. Nature 441:1002–1006. https://doi.org/10.1038/nature04717
41. Matsuura-Tokita K, Takeuchi M, Ichihara A et al (2006) Live imaging of yeast Golgi cisternal maturation. Nature 441:1007–1010. https://doi.org/10.1038/nature04737
42. Ito Y, Toyooka K, Fujimoto M et al (2017) The trans-Golgi network and the Golgi stacks behave independently during regeneration after brefeldin A treatment in tobacco BY-2 cells. Plant Cell Physiol 58:811–821. https://doi.org/10.1093/pcp/pcx028
43. Renna L, Stefano G, Slabaugh E et al (2018) TGNap1 is required for microtubule-dependent homeostasis of a subpopulation of the plant trans-Golgi network. Nat Commun 9:5313. https://doi.org/10.1038/s41467-018-07662-4
44. Gendre D, McFarlane HE, Johnson E et al (2013) Trans-Golgi network localized ECHIDNA/Ypt interacting protein complex is required for the secretion of cell wall polysaccharides in Arabidopsis. Plant Cell 25:2633–2646. https://doi.org/10.1105/tpc.113.112482
45. Gendre D, Oh J, Boutté Y et al (2011) Conserved Arabidopsis ECHIDNA protein mediates trans-Golgi-network trafficking and cell elongation. Proc Natl Acad Sci USA 108:8048–8053. https://doi.org/10.1073/pnas.1018371108
46. Tse YC, Mo B, Hillmer S et al (2004) Identification of multivesicular bodies as prevacuolar compartments in Nicotiana tabacum BY-2 cells. Plant Cell 16:672–693. https://doi.org/10.1105/tpc.019703

47. Cui Y, Shen J, Gao C et al (2016) Biogenesis of plant prevacuolar multivesicular bodies. Mol Plant 9:774–786. https://doi.org/10.1016/j.molp.2016.01.011
48. Scheuring D, Viotti C, Krüger F et al (2011) Multivesicular bodies mature from the trans-Golgi network/early endosome in Arabidopsis. Plant Cell 23:3463–3481. https://doi.org/10.1105/tpc.111.086918
49. Cui Y, Cao W, He Y et al (2019) A whole-cell electron tomography model of vacuole biogenesis in Arabidopsis root cells. Nat Plants 5:95–105. https://doi.org/10.1038/s41477-018-0328-1
50. Korbei B, Moulinier-Anzola J, De-Araujo L et al (2013) Arabidopsis TOL proteins act as gatekeepers for vacuolar sorting of PIN2 plasma membrane protein. Curr Biol 23:2500–2505. https://doi.org/10.1016/j.cub.2013.10.036
51. Moulinier-Anzola J, Schwihla M, De-Araújo L et al (2020) TOLs function as ubiquitin receptors in the early steps of the ESCRT pathway in higher plants. Mol Plant 13:717–731. https://doi.org/10.1016/j.molp.2020.02.012
52. Cui Y, Zhao Q, Gao C et al (2014) Activation of the Rab7 GTPase by the MON1-CCZ1 complex is essential for PVC-to-vacuole trafficking and plant growth in Arabidopsis. Plant Cell 26:2080–2097. https://doi.org/10.1105/tpc.114.123141
53. Ebine K, Inoue T, Ito J et al (2014) Vacuolar trafficking occursvacuolar trafficking occurs through Distinctly Regulated Pathways.distinctly regulated pathways. Curr Biol 24:1375–1382. https://doi.org/10.1016/j.cub.2014.05.004
54. Singh MK, Krüger F, Beckmann H et al (2014) Protein delivery to vacuole requires SAND protein-dependent Rab GTPase conversion for MVB-vacuole fusion. Curr Biol 24:1383–1389. https://doi.org/10.1016/j.cub.2014.05.005
55. Lam SK, Siu CL, Hillmer S et al (2007) Rice SCAMP1 defines clathrin-coated, trans-Golgi-located tubular-vesicular structures as an early endosome in tobacco BY-2 cells. Plant Cell 19:296–319. https://doi.org/10.1105/tpc.106.045708
56. Chow C-M, Neto H, Foucart C, Moore I (2008) Rab-A2 and Rab-A3 GTPases define a trans-golgi endosomal membrane domain in Arabidopsis that contributes substantially to the cell plate. Plant Cell 20:101–123. https://doi.org/10.1105/tpc.107.052001
57. Narasimhan M, Johnson A, Prizak R et al (2020) Evolutionarily unique mechanistic framework of clathrin-mediated endocytosis in plants. elife 9:e52067. https://doi.org/10.7554/eLife.52067
58. Day KJ, Casler JC, Glick BS (2018) Budding yeast has a minimal endomembrane system. Dev Cell 44:56–72. https://doi.org/10.1016/j.devcel.2017.12.014
59. Toshima JY, Tsukahara A, Nagano M et al (2023) The yeast endocytic early/sorting compartment exists as an independent sub-compartment within the trans-Golgi network. elife 12:e84850. https://doi.org/10.7554/eLife.84850
60. Huang T-K, Han C-L, Lin S-I et al (2013) Identification of downstream components of ubiquitin-conjugating enzyme PHOSPHATE2 by quantitative membrane proteomics in Arabidopsis roots. Plant Cell 25:4044–4060. https://doi.org/10.1105/tpc.113.115998
61. Kerkeb L, Mukherjee I, Chatterjee I et al (2008) Iron-induced turnover of the Arabidopsis IRON-REGULATED TRANSPORTER1 metal transporter requires lysine residues. Plant Physiol 146:1964–1973. https://doi.org/10.1104/pp.107.113282
62. Barberon M, Zelazny E, Robert S et al (2011) Monoubiquitin-dependent endocytosis of the iron-regulated transporter 1 (IRT1) transporter controls iron uptake in plants. Proc Natl Acad Sci USA 108:E450–E458. https://doi.org/10.1073/pnas.1100659108
63. Bayle V, Arrighi J-F, Creff A et al (2011) Arabidopsis thaliana high-affinity phosphate transporters exhibit multiple levels of posttranslational regulation. Plant Cell 23:1523–1535. https://doi.org/10.1105/tpc.110.081067
64. Agorio A, Giraudat J, Bianchi MW et al (2017) Phosphatidylinositol 3-phosphate-binding protein AtPH1 controls the localization of the metal transporter NRAMP1 in Arabidopsis. Proc Natl Acad Sci USA 114:E3354–E3363. https://doi.org/10.1073/pnas.1702975114

65. Takano J, Miwa K, Yuan L et al (2005) Endocytosis and degradation of BOR1, a boron transporter of Arabidopsis thaliana, regulated by boron availability. Proc Natl Acad Sci USA 102:12276–12281. https://doi.org/10.1073/pnas.0502060102
66. Kasai K, Takano J, Miwa K et al (2011) High boron-induced ubiquitination regulates vacuolar sorting of the BOR1 borate transporter in Arabidopsis thaliana. J Biol Chem 286:6175–6183. https://doi.org/10.1074/jbc.M110.184929
67. Yoshinari A, Korbei B, Takano J (2018) TOL proteins mediate vacuolar sorting of the borate transporter BOR1 in Arabidopsis thaliana. Soil Sci Plant Nutr 64:598–605. https://doi.org/10.1080/00380768.2018.1504322
68. Takano J, Tanaka M, Toyoda A et al (2010) Polar localization and degradation of Arabidopsis boron transporters through distinct trafficking pathways. Proc Natl Acad Sci USA 107:5220–5225. https://doi.org/10.1073/pnas.0910744107
69. Yoshinari A, Hosokawa T, Beier M et al (2020) Transport-coupled ubiquitination of the borate transporter BOR1 for its boron-dependent degradation. Plant Cell 33:420–438. https://doi.org/10.1093/plcell/koaa020
70. Yoshinari A, Shimizu Y, Hosokawa T et al (2024) Rapid vacuolar sorting of the borate transporter BOR1 requires the adaptor protein complex AP-4 in Arabidopsis. Plant Cell Physiol pcae096. https://doi.org/10.1093/pcp/pcae096
71. Yoshinari A, Fujimoto M, Ueda T et al (2016) DRP1-dependent endocytosis is essential for polar localization and boron-induced degradation of the borate transporter BOR1 in Arabidopsis thaliana. Plant Cell Physiol 57:1985–2000. https://doi.org/10.1093/pcp/pcw121
72. Wang S, Yoshinari A, Shimada T et al (2017) Polar localization of the NIP5;1 boric acid channel is maintained by endocytosis and facilitates boron transport in Arabidopsis roots. Plant Cell 29:824–842. https://doi.org/10.1105/tpc.16.00825
73. Yoshinari A, Hosokawa T, Amano T et al (2019) Polar localization of the borate exporter BOR1 requires AP2-dependent endocytosis. Plant Physiol 179:1569–1580. https://doi.org/10.1104/pp.18.01017
74. Yoshinari A, Toda Y, Takano J (2021) GNOM-dependent endocytosis maintains polar localisation of the borate exporter BOR1 in Arabidopsis. Biol Cell 113:264–269. https://doi.org/10.1111/boc.202000106
75. Yu Z, Zhang F, Friml J, Ding Z (2022) Auxin signaling: research advances over the past 30 years. J Integr Plant Biol 64:371–392. https://doi.org/10.1111/jipb.13225
76. Friml J (2022) Fourteen stations of auxin. Cold Spring Harb Perspect Biol 14:a039859. https://doi.org/10.1101/cshperspect.a039859
77. Muroyama A, Bergmann D (2019) Plant cell polarity: creating diversity from inside the box. Annu Rev Cell Dev Biol 35:309–336. https://doi.org/10.1146/annurev-cellbio-100818-125211
78. Hammes UZ, Murphy AS, Schwechheimer C (2022) Auxin transporters-A biochemical view. Cold Spring Harb Perspect Biol 14:a039875. https://doi.org/10.1101/cshperspect.a039875
79. Men S, Boutté Y, Ikeda Y et al (2008) Sterol-dependent endocytosis mediates post-cytokinetic acquisition of PIN2 auxin efflux carrier polarity. Nat Cell Biol 10:237–244. https://doi.org/10.1038/ncb1686
80. Glanc M, Fendrych M, Friml J (2018) Mechanistic framework for cell-intrinsic re-establishment of PIN2 polarity after cell division. Nat Plants 4:1082–1088. https://doi.org/10.1038/s41477-018-0318-3
81. Yang K, Wang L, Le J, Dong J (2020) Cell polarity: regulators and mechanisms in plants. J Integr Plant Biol 62:132–147. https://doi.org/10.1111/jipb.12904
82. Marhava P (2022) Recent developments in the understanding of PIN polarity. New Phytol 233:624–630. https://doi.org/10.1111/nph.17867
83. Sauer M, Grebe M (2021) Plant cell biology: PIN polarity maintained. Curr Biol 31:R449–R451. https://doi.org/10.1016/j.cub.2021.03.070

84. Kleine-Vehn J, Wabnik K, Martinière A et al (2011) Recycling, clustering, and endocytosis jointly maintain PIN auxin carrier polarity at the plasma membrane. Mol Syst Biol 7:540. https://doi.org/10.1038/msb.2011.72
85. Kleine-Vehn J, Dhonukshe P, Sauer M et al (2008) ARF GEF-dependent transcytosis and polar delivery of PIN auxin carriers in Arabidopsis. Curr Biol 18:526–531. https://doi.org/10.1016/j.cub.2008.03.021
86. Geldner N, Anders N, Wolters H et al (2003) The Arabidopsis GNOM ARF-GEF mediates endosomal recycling, auxin transport, and auxin-dependent plant growth. Cell 112:219–230. https://doi.org/10.1016/S0092-8674(03)00003-5
87. Li R, Rodriguez-Furlan C, Wang J et al (2017) Different endomembrane trafficking pathways establish apical and basal polarities. Plant Cell 29:90–108. https://doi.org/10.1105/tpc.16.00524
88. Tang H, Lu K-J, Zhang Y et al (2024) Divergence of trafficking and polarization mechanisms for PIN auxin transporters during land plant evolution. Plant Commun 5:100669. https://doi.org/10.1016/j.xplc.2023.100669
89. Huang S, Konishi N, Yamaji N et al (2022) Boron uptake in rice is regulated post-translationally via a clathrin-independent pathway. Plant Physiol 188:1649–1664. https://doi.org/10.1093/plphys/kiab575
90. Konishi N, Huang S, Yamaji N, Ma JF (2022) Cell-type-dependent but CME-independent polar localization of silicon transporters in rice. Plant Cell Physiol 63:699–712. https://doi.org/10.1093/pcp/pcac032
91. Hirst J, Barlow LD, Francisco GC et al (2011) The fifth adaptor protein complex. PLoS Biol 9: e1001170. https://doi.org/10.1371/journal.pbio.1001170
92. Larson RT, Dacks JB, Barlow LD (2019) Recent gene duplications dominate evolutionary dynamics of adaptor protein complex subunits in embryophytes. Traffic 20:961–973. https://doi.org/10.1111/tra.12698
93. Park M, Song K, Reichardt I et al (2013) Arabidopsis μ-adaptin subunit AP1M of adaptor protein complex 1 mediates late secretory and vacuolar traffic and is required for growth. Proc Natl Acad Sci USA 110:10318–10323. https://doi.org/10.1073/pnas.1300460110
94. Teh O-K, Shimono Y, Shirakawa M et al (2013) The AP-1 μ adaptin is required for KNOLLE localization at the cell plate to mediate cytokinesis in Arabidopsis. Plant Cell Physiol 54:838–847. https://doi.org/10.1093/pcp/pct048
95. Wang J-G, Li S, Zhao X-Y et al (2013) HAPLESS13, the Arabidopsis μ1 adaptin, is essential for protein sorting at the trans-Golgi network/early endosome. Plant Physiol 162:1897–1910. https://doi.org/10.1104/pp.113.221051
96. Fuji K, Shirakawa M, Shimono Y et al (2016) The adaptor complex AP-4 regulates vacuolar protein sorting at the trans-Golgi network by interacting with VACUOLAR SORTING RECEPTOR1. Plant Physiol 170:211–219. https://doi.org/10.1104/pp.15.00869
97. Pearse BM, Robinson MS (1984) Purification and properties of 100-kd proteins from coated vesicles and their reconstitution with clathrin. EMBO J 3:1951–1957. https://doi.org/10.1002/j.1460-2075.1984.tb02075.x
98. Keen JH (1987) Clathrin assembly proteins: affinity purification and a model for coat assembly. J Cell Biol 105:1989–1998. https://doi.org/10.1083/jcb.105.5.1989
99. Di Rubbo S, Irani NG, Kim SY et al (2013) The clathrin adaptor complex AP-2 mediates endocytosis of brassinosteroid insensitive1 in Arabidopsis. Plant Cell 25:2986–2997. https://doi.org/10.1105/tpc.113.114058
100. Kim SY, Xu Z-Y, Song K et al (2013) Adaptor protein complex 2-mediated endocytosis is crucial for male reproductive organ development in Arabidopsis. Plant Cell 25:2970–2985. https://doi.org/10.1105/tpc.113.114264
101. Yamaoka S, Shimono Y, Shirakawa M et al (2013) Identification and dynamics of Arabidopsis adaptor protein-2 complex and its involvement in floral organ development. Plant Cell 25: 2958–2969. https://doi.org/10.1105/tpc.113.114082

102. Fan L, Hao H, Xue Y et al (2013) Dynamic analysis of Arabidopsis AP2 σ subunit reveals a key role in clathrin-mediated endocytosis and plant development. Development 140:3826–3837. https://doi.org/10.1242/dev.095711
103. Wang X, Cai Y, Wang H et al (2014) Trans-Golgi network-located AP1 gamma adaptins mediate dileucine motif-directed vacuolar targeting in Arabidopsis. Plant Cell 26:4102–4118. https://doi.org/10.1105/tpc.114.129759
104. Wang J-G, Feng C, Liu H-H et al (2017) AP1G mediates vacuolar acidification during synergid-controlled pollen tube reception. Proc Natl Acad Sci USA 114:E4877–E4883. https://doi.org/10.1073/pnas.1617967114
105. Wang J-G, Feng C, Liu H-H et al (2016) HAPLESS13-mediated trafficking of STRUBBELIG is critical for ovule development in Arabidopsis. PLoS Genet 12:e1006269. https://doi.org/10.1371/journal.pgen.1006269
106. Zhou Y, Fang W, Pang Z et al (2022) AP1G2 affects mitotic cycles of female and male gametophytes in Arabidopsis. Front Plant Sci 13:924417. https://doi.org/10.3389/fpls.2022.924417
107. Yin G-M, Fang Y-R, Wang J-G et al (2024) Arabidopsis HAPLESS13/AP-1μ is critical for pollen sac formation and tapetal function. Plant Sci 341:111998. https://doi.org/10.1016/j.plantsci.2024.111998
108. Robinson DG, Pimpl P (2014) Clathrin and post-Golgi trafficking: a very complicated issue. Trends Plant Sci 19:134–139. https://doi.org/10.1016/j.tplants.2013.10.008
109. Yan X, Wang Y, Xu M et al (2021) Cross-talk between clathrin-dependent post-Golgi trafficking and clathrin-mediated endocytosis in Arabidopsis root cells. Plant Cell 33:3057–3075. https://doi.org/10.1093/plcell/koab180
110. Shimada T, Kunieda T, Sumi S et al (2018) The AP-1 complex is required for proper mucilage formation in Arabidopsis seeds. Plant Cell Physiol 59:2331–2338. https://doi.org/10.1093/pcp/pcy158
111. Richter S, Kientz M, Brumm S et al (2014) Delivery of endocytosed proteins to the cell-division plane requires change of pathway from recycling to secretion. elife 3:e02131. https://doi.org/10.7554/eLife.02131.018
112. Uemura T, Kim H, Saito C et al (2012) Qa-SNAREs localized to the trans-Golgi network regulate multiple transport pathways and extracellular disease resistance in plants. Proc Natl Acad Sci USA 109:1784–1789. https://doi.org/10.1073/pnas.1115146109
113. Casler JC, Papanikou E, Barrero JJ, Glick BS (2019) Maturation-driven transport and AP-1-dependent recycling of a secretory cargo in the Golgi. J Cell Biol 218:1582–1601. https://doi.org/10.1083/jcb.201807195
114. Casler JC, Glick BS (2020) A microscopy-based kinetic analysis of yeast vacuolar protein sorting. elife 9:e56844. https://doi.org/10.7554/eLife.56844
115. Casler JC, Johnson N, Krahn AH et al (2021) Clathrin adaptors mediate two sequential pathways of intra-Golgi recycling. J Cell Biol 221:e202103199. https://doi.org/10.1101/2021.03.17.435835
116. Robinson MS, Antrobus R, Sanger A et al (2024) The role of the AP-1 adaptor complex in outgoing and incoming membrane traffic. J Cell Biol 223:e202310071. https://doi.org/10.1083/jcb.202310071
117. Shao X, Xu H, Pimpl P (2023) Nanobody-based VSR7 tracing shows clathrin-dependent TGN to Golgi recycling. Nat Commun 14:6926. https://doi.org/10.1038/s41467-023-42331-1
118. Müdsam C, Wollschläger P, Sauer N, Schneider S (2018) Sorting of Arabidopsis NRAMP3 and NRAMP4 depends on adaptor protein complex AP4 and a dileucine-based motif. Traffic 19:503–521. https://doi.org/10.1111/tra.12567
119. Pertl-Obermeyer H, Wu XN, Schrodt J et al (2016) Identification of cargo for adaptor protein (AP) complexes 3 and 4 by sucrose gradient profiling. Mol Cell Proteomics 15:2877–2889. https://doi.org/10.1074/mcp.M116.060129

120. Dahhan DA, Reynolds GD, Cárdenas JJ et al (2022) Proteomic characterization of isolated Arabidopsis clathrin-coated vesicles reveals evolutionarily conserved and plant-specific components. Plant Cell 34:2150–2173. https://doi.org/10.1093/plcell/koac071
121. Heinze L, Freimuth N, Rößling A-K et al (2020) EPSIN1 and MTV1 define functionally overlapping but molecularly distinct trans-Golgi network subdomains in Arabidopsis. Proc Natl Acad Sci USA 117:25880–25889. https://doi.org/10.1073/pnas.2004822117
122. Wolfenstetter S, Wirsching P, Dotzauer D et al (2012) Routes to the tonoplast: the sorting of tonoplast transporters in Arabidopsis mesophyll protoplasts. Plant Cell 24:215–232. https://doi.org/10.1105/tpc.111.090415
123. Feraru E, Paciorek T, Feraru MI et al (2010) The AP-3 β adaptin mediates the biogenesis and function of lytic vacuoles in Arabidopsis. Plant Cell 22:2812–2824. https://doi.org/10.1105/tpc.110.075424
124. Zwiewka M, Feraru E, Möller B et al (2011) The AP-3 adaptor complex is required for vacuolar function in Arabidopsis. Cell Res 21:1711–1722. https://doi.org/10.1038/cr.2011.99
125. Feng Q-N, Song S-J, Yu S-X et al (2017) Adaptor protein-3-dependent vacuolar trafficking involves a subpopulation of COPII and HOPS tethering proteins. Plant Physiol 174:1609–1620. https://doi.org/10.1104/pp.17.00584
126. Zouhar J, Sauer M (2014) Helping hands for budding prospects: ENTH/ANTH/VHS accessory proteins in endocytosis, vacuolar transport, and secretion. Plant Cell 26:4232–4244. https://doi.org/10.1105/tpc.114.131680
127. Feng Y, Hiwatashi T, Minamino N et al (2022) Membrane trafficking functions of the ANTH/ENTH/VHS domain-containing proteins in plants. FEBS Lett 596:2256–2268. https://doi.org/10.1002/1873-3468.14368
128. Song J, Lee MH, Lee G-J et al (2006) Arabidopsis EPSIN1 plays an important role in vacuolar trafficking of soluble cargo proteins in plant cells via interactions with clathrin, AP-1, VTI11, and VSR1. Plant Cell 18:2258–2274. https://doi.org/10.1105/tpc.105.039123
129. Collins CA, LaMontagne ED, Anderson JC et al (2020) EPSIN1 modulates the plasma membrane abundance of FLAGELLIN SENSING2 for effective immune responses. Plant Physiol 182:1762–1775. https://doi.org/10.1104/pp.19.01172
130. Sauer M, Delgadillo MO, Zouhar J et al (2013) MTV1 and MTV4 encode plant-specific ENTH and ARF GAP proteins that mediate clathrin-dependent trafficking of vacuolar cargo from the trans-Golgi network. Plant Cell 25:2217–2235. https://doi.org/10.1105/tpc.113.111724
131. Larson ER, Van Zelm E, Roux C et al (2017) Clathrin heavy chain subunits coordinate endo- and exocytic traffic and affect stomatal movement. Plant Physiol 175:708–720. https://doi.org/10.1104/pp.17.00970
132. Wang P, Siao W, Zhao X et al (2023) Adaptor protein complex interaction map in Arabidopsis identifies P34 as a common stability regulator. Nat Plants 9:355–371. https://doi.org/10.1038/s41477-022-01328-2
133. Denecke J, Botterman J, Deblaere R (1990) Protein secretion in plant cells can occur via a default pathway. Plant Cell 2:51–59. https://doi.org/10.1105/tpc.2.1.51
134. de Marcos LC, Denecke J (2016) Lysosomal and vacuolar sorting: not so different after all! Biochem Soc Trans 44:891–897. https://doi.org/10.1042/BST20160050
135. Ghosh P, Dahms NM, Kornfeld S (2003) Mannose 6-phosphate receptors: new twists in the tale. Nat Rev Mol Cell Biol 4:202–212. https://doi.org/10.1038/nrm1050
136. Kirsch T, Saalbach G, Raikhel NV, Beevers L (1996) Interaction of a potential vacuolar targeting receptor with amino- and carboxyl-terminal targeting determinants. Plant Physiol 111:469–474. https://doi.org/10.1104/pp.111.2.469
137. Paris N, Rogers SW, Jiang L et al (1997) Molecular cloning and further characterization of a probable plant vacuolar sorting receptor. Plant Physiol 115:29–39. https://doi.org/10.1104/pp.115.1.29

138. Shimada T, Kuroyanagi M, Nishimura M, Hara-Nishimura I (1997) A pumpkin 72-kDa membrane protein of precursor-accumulating vesicles has characteristics of a vacuolar sorting receptor. Plant Cell Physiol 38:1414–1420. https://doi.org/10.1093/oxfordjournals.pcp.a029138
139. Tsao H-E, Lui SN, Lo AH-F et al (2022) Structural insights into how vacuolar sorting receptors recognize the sorting determinants of seed storage proteins. Proc Natl Acad Sci USA 119:e2111281119. https://doi.org/10.1073/pnas.2111281119
140. Braulke T, Bonifacino JS (2009) Sorting of lysosomal proteins. Biochim Biophys Acta 1793: 605–614. https://doi.org/10.1016/j.bbamcr.2008.10.016
141. Bowers K, Stevens TH (2005) Protein transport from the late Golgi to the vacuole in the yeast Saccharomyces cerevisiae. Biochim Biophys Acta 1744:438–454. https://doi.org/10.1016/j.bbamcr.2005.04.004
142. Kirsch T, Paris N, Butler JM et al (1994) Purification and initial characterization of a potential plant vacuolar targeting receptor. Proc Natl Acad Sci USA 91:3403–3407. https://doi.org/10.1073/pnas.91.8.3403
143. Hinz G, Hillmer S, Baumer M, Hohl II (1999) Vacuolar storage proteins and the putative vacuolar sorting receptor BP-80 exit the golgi apparatus of developing pea cotyledons in different transport vesicles. Plant Cell 11:1509–1524. https://doi.org/10.1105/tpc.11.8.1509
144. Miao Y, Yan PK, Kim H et al (2006) Localization of green fluorescent protein fusions with the seven Arabidopsis vacuolar sorting receptors to prevacuolar compartments in tobacco BY-2 cells. Plant Physiol 142:945–962. https://doi.org/10.1104/pp.106.083618
145. Yang L, Jiang L (2023) The seven rice vacuolar sorting receptors localize to prevacuolar compartments. J Plant Physiol 291:154137. https://doi.org/10.1016/j.jplph.2023.154137
146. Gershlick DC, de Lousa CM, Foresti O et al (2014) Golgi-dependent transport of vacuolar sorting receptors is regulated by COPII, AP1, and AP4 protein complexes in tobacco. Plant Cell 26:1308–1329. https://doi.org/10.1105/tpc.113.122226
147. Nishimura K, Matsunami E, Yoshida S et al (2016) The tyrosine-sorting motif of the vacuolar sorting receptor VSR4 from Arabidopsis thaliana, which is involved in the interaction between VSR4 and AP1M2, μ1-adaptin type 2 of clathrin adaptor complex 1 subunits, participates in the post-Golgi sorting of VSR4. Biosci Biotechnol Biochem 80:694–705. https://doi.org/10.1080/09168451.2015.1116925
148. Otegui MS, Herder R, Schulze J et al (2006) The proteolytic processing of seed storage proteins in Arabidopsis embryo cells starts in the multivesicular bodies. Plant Cell 18:2567–2581. https://doi.org/10.1105/tpc.106.040931
149. Oliviusson P, Heinzerling O, Hillmer S et al (2006) Plant retromer, localized to the prevacuolar compartment and microvesicles in Arabidopsis, may interact with vacuolar sorting receptors. Plant Cell 18:1239–1252. https://doi.org/10.1105/tpc.105.035907
150. Jaillais Y, Santambrogio M, Rozier F et al (2007) The retromer protein VPS29 links cell polarity and organ initiation in plants. Cell 130:1057–1070. https://doi.org/10.1016/j.cell.2007.08.040
151. Yamazaki M, Shimada T, Takahashi H et al (2008) Arabidopsis VPS35, a retromer component, is required for vacuolar protein sorting and involved in plant growth and leaf senescence. Plant Cell Physiol 49:142–156. https://doi.org/10.1093/pcp/pcn006
152. Zelazny E, Santambrogio M, Pourcher M et al (2013) Mechanisms governing the endosomal membrane recruitment of the core retromer in Arabidopsis. J Biol Chem 288:8815–8825. https://doi.org/10.1074/jbc.M112.440503
153. Munch D, Teh O-K, Malinovsky FG et al (2015) Retromer contributes to immunity-associated cell death in Arabidopsis. Plant Cell 27:463–479. https://doi.org/10.1105/tpc.114.132043
154. Rodriguez-Furlan C, Domozych D, Qian W et al (2019) Interaction between VPS35 and RABG3f is necessary as a checkpoint to control fusion of late compartments with the vacuole. Proc Natl Acad Sci USA 116:21291–21301. https://doi.org/10.1073/pnas.1905321116

155. Hu S, Li B, Wu F et al (2022) Plant ESCRT protein ALIX coordinates with retromer complex in regulating receptor-mediated sorting of soluble vacuolar proteins. Proc Natl Acad Sci USA 119:e2200492119. https://doi.org/10.1073/pnas.2200492119
156. Kang H, Kim SY, Song K et al (2012) Trafficking of vacuolar proteins: the crucial role of Arabidopsis vacuolar protein sorting 29 in recycling vacuolar sorting receptor. Plant Cell 24: 5058–5073. https://doi.org/10.1105/tpc.112.103481
157. Kang H, Hwang I (2014) Vacuolar sorting receptor-mediated trafficking of soluble vacuolar proteins in plant cells. Plants 3:392–408. https://doi.org/10.3390/plants3030392
158. Ivanov R, Robinson DG (2020) EMAC, Retromer, and VSRs: do they connect? Protoplasma 257:1725–1729. https://doi.org/10.1007/s00709-020-01543-8
159. Heucken N, Ivanov R (2018) The retromer, sorting nexins and the plant endomembrane protein trafficking. J Cell Sci 131:jcs203695. https://doi.org/10.1242/jcs.203695
160. Zouhar J, Cao W, Shen J, Rojo E (2023) Retrograde transport in plants: circular economy in the endomembrane system. Eur J Cell Biol 102:151309. https://doi.org/10.1016/j.ejcb.2023.151309
161. Künzl F, Früholz S, Fäßler F et al (2016) Receptor-mediated sorting of soluble vacuolar proteins ends at the trans-Golgi network/early endosome. Nat Plants 2:16017. https://doi.org/10.1038/nplants.2016.17
162. Früholz S, Fäßler F, Kolukisaoglu Ü, Pimpl P (2018) Nanobody-triggered lockdown of VSRs reveals ligand reloading in the Golgi. Nat Commun 9:643. https://doi.org/10.1038/s41467-018-02909-6
163. Boutté Y, Jonsson K, McFarlane HE et al (2013) ECHIDNA-mediated post-Golgi trafficking of auxin carriers for differential cell elongation. Proc Natl Acad Sci USA 110:16259–16264. https://doi.org/10.1073/pnas.1309057110
164. Wattelet-Boyer V, Brocard L, Jonsson K et al (2016) Enrichment of hydroxylated C24- and C26-acyl-chain sphingolipids mediates PIN2 apical sorting at trans-Golgi network subdomains. Nat Commun 7:12788. https://doi.org/10.1038/ncomms12788
165. Ravikumar R, Kalbfuß N, Gendre D et al (2018) Independent yet overlapping pathways ensure the robustness and responsiveness of trans-Golgi network functions in Arabidopsis. Development 145:dev169201. https://doi.org/10.1242/dev.169201
166. Fougère L, Mongrand S, Boutté Y (2024) The function of sphingolipids in membrane trafficking and cell signaling in plants, in comparison with yeast and animal cells. Biochim Biophys Acta Mol Cell Biol Lipids 1869:159463. https://doi.org/10.1016/j.bbalip.2024.159463
167. Ito Y, Grison M, Esnay N et al (2020) Immunopurification of intact endosomal compartments for lipid analyses in Arabidopsis. Methods Mol Biol 2177:119–141. https://doi.org/10.1007/978-1-0716-0767-1_11
168. Wang W, Yang X, Tangchaiburana S et al (2008) An inositolphosphorylceramide synthase is involved in regulation of plant programmed cell death associated with defense in Arabidopsis. Plant Cell 20:3163–3179. https://doi.org/10.1105/tpc.108.060053
169. Ito Y, Esnay N, Platre MP et al (2021) Sphingolipids mediate polar sorting of PIN2 through phosphoinositide consumption at the trans-Golgi network. Nat Commun 12:4267. https://doi.org/10.1038/s41467-021-24548-0
170. Markham JE, Molino D, Gissot L et al (2011) Sphingolipids containing very-long-chain fatty acids define a secretory pathway for specific polar plasma membrane protein targeting in Arabidopsis. Plant Cell 23:2362–2378. https://doi.org/10.1105/tpc.110.080473
171. Wiese C, Abele M, Al B et al (2024) Regulation of adaptive growth decisions via phosphorylation of the TRAPPII complex in Arabidopsis. J Cell Biol 223:e202311125. https://doi.org/10.1083/jcb.202311125

Part II
Golgi Functions in Glycosylation and Processing

Chapter 10
Siamese Twins: The Multimodular Mechanisms of Golgi Maturation and Glycan Synthesis Are Coupled at Their Core

Domenico Russo, Seetharaman Parashuraman, and Alberto Luini

Abstract The Golgi apparatus functions as the central processing hub for proteins and lipids, orchestrating glycosylation, sorting, and secretion. Once viewed as a passive site of bulk enzyme recycling, the Golgi is now recognized as a highly dynamic, multimodular system in which intra-Golgi trafficking and glycan synthesis operate as tightly interdependent "Siamese twins." The classical cisternal maturation model, initially based on uniform COPI-mediated recycling, has evolved into the multimodular cisternal maturation (MCM) framework, revealing the coexistence of multiple, specialized recycling modules. Distinct sets of glycosyltransferases form coherent enzymatic modules, each maintained by dedicated retrograde pathways—some COPI-dependent, others COPI-independent—that ensure compartmental fidelity and enable differential regulation of glycan synthesis. These pathways are coordinated by adaptors, retainers, and lipid identity cues that collectively sustain Golgi polarity and adaptability to cellular and metabolic states. Disruption of this modular recycling logic leads to enzyme mislocalization, defective glycosylation, and disease, ranging from congenital disorders of glycosylation to oncogenic transformation. The transition from bulk to multimodular recycling thus redefines the Golgi as an integrated regulatory platform linking membrane trafficking to metabolic and signaling networks.

Keywords Golgi apparatus · Cisternal maturation · Multimodular recycling · Glycosylation · Congenital disorders of glycosylation · Cancer

D. Russo (✉) · S. Parashuraman (✉) · A. Luini (✉)
Institute of Endotypes in Oncology, Metabolism and Immunology (IEOMI), National Research Council of Italy, Naples, Italy
e-mail: domenico.russo@cnr.it; seetharaman.parashuraman@cnr.it; alberto.luini@ieos.cnr.it

A. Nakano, J. Saraste (eds.), *The Golgi Network, Volume II*, Subcellular Biochemistry 111, https://doi.org/10.1007/978-3-032-16833-7_10

Introduction

Proteins and lipids, once synthesized in the endoplasmic reticulum (ER), are delivered to the Golgi complex, where they undergo extensive chemical processing, primarily through glycosylation, and are then sorted into specialized carriers that deliver them to their final destinations, including the basolateral or apical plasma membrane and other cellular compartments such as the endo-lysosomal system. The Golgi apparatus has often been described as the cellular glycosylation factory, as it is the organelle where most glycosylation reactions occur, although some steps also take place in other compartments. The processes of glycosylation and cargo sorting represent the main specializations of the Golgi complex and are involved in a vast number of physiological processes, as well as in many pathological conditions where their regulation is altered. The processes of intra-Golgi transport and glycosylation can be regarded as Siamese twins because glycan synthesis is regulated during cargo transport through the Golgi, and, crucially, by the core mechanisms that mediate Golgi transport.

Glycan synthesis, or glycosylation, is a fundamental biological process involved in a vast array of functions across all forms of life, including eukaryotic cells. It is a remarkably complex and demanding mechanism for a cell to execute. Glycans are abundant cellular components in both bacterial and eukaryotic systems, accounting for approximately 5–10% of the cell dry weight and exhibit extraordinary structural diversity. Moreover, rather than being a single homogeneous process, glycosylation comprises more than fourteen distinct biosynthetic pathways that give rise to chemically and functionally diverse classes of glycans [116]. It is a template-independent process, proceeding through the progressive assembly of sugar chains onto protein or lipid scaffolds within the Golgi lumen as cargo transits through the organelle, and occurs via a series of sequential reactions catalyzed by distinct, though sometimes partially overlapping, sets of glycoenzymes, primarily glycosyltransferases and glycosidases, amounting to over 200 enzymes in the human genome, organized into specific metabolic routes [116].

How can such a multitude of synthetic pathways be executed independently and yet with such a remarkable degree of precision in one organelle? In addition, beyond its complexity, another fundamental challenge for the glycosylation machinery lies in its regulation. Glycosylation can be regarded as one of the main metabolic processes of the cell, like protein or DNA synthesis, as well as the anabolic and catabolic pathways of general metabolism. Like these other biosynthetic networks, it must continuously adapt to the changing functional and metabolic demands of the cell. The regulation of other metabolic processes like metabolism and protein and DNA synthesis is provided directly by a rich and sophisticated network of hundreds of different signals that ensure rapid and precise responses to changing cellular needs. In this regard, by contrast, glycosylation has a unique problem, namely, its topology. Glycosylation takes place within the lumen of the Golgi or other transport organelles, and the vast majority of its products, whether glycoproteins or glycolipids, are ultimately displayed on the outer surface of the plasma membrane (topologically equivalent to organelle lumen) or secreted into the extracellular space. The Golgi must therefore employ mechanisms capable of translating the regulatory information arriving from the cell surface or intracellular organelles into signals that can reach the lumen of its compartments, where glycosylation reactions take place.

The major group of "signal transducers" that achieve this are the membrane trafficking regulators that include small GTPases (Rabs) of the Ras superfamily, adaptors, soluble N-ethylmaleimide-sensitive factor attachment protein receptors (SNARE) and tethering molecules. These molecules are cytosolic and therefore susceptible to regulation by the signaling machinery. When they bind to the Golgi membranes they can transduce the signal into regulation of membrane trafficking events at the Golgi. This regulatory principle appears to apply primarily to the glycosylation machinery itself. However, emerging evidence indicates that certain cargo proteins can also be regulated to modulate their rate of passage through the Golgi apparatus, presumably to control the extent and pattern of their glycosylation.

The Golgi complex represents an evolutionary response to this challenge of glycosylation regulation, since a key feature of the organelle is the generation of anatomically distinct and compositionally specialized membrane compartments, or processing stations, where specific steps can occur under optimal conditions. Structurally, the Golgi apparatus is organized into distinct cis, medial, and trans compartments, each with a characteristic morphology and a specific composition of glycosylation enzymes [62]. The fidelity of glycan biosynthesis depends on this ordered distribution, ensuring that each enzyme acts precisely on the intermediate produced by its predecessor [57]. Therefore, the mechanisms that establish and maintain this spatial organization are of fundamental importance.

Decades of morphological and biochemical studies have converged toward the cisternal maturation model, which envisions Golgi cisternae as dynamic entities that progressively mature from cis to trans stages. During this process, cargo is carried forward within the maturing cisternae, while resident enzymes are recycled backward through COPI-coated vesicles to preserve the polarized enzymatic architecture [16, 17]. This model explains how glycosylation enzymes remain within the Golgi apparatus even as cargo continues its anterograde journey.

However, a single, uniform retrograde route mediated by COPI cannot fully account for the precise compartmentalization of hundreds of distinct glycosylation enzymes into specific cis, medial, and trans domains. Recent studies have demonstrated that multiple retrograde pathways coexist within the Golgi, many still COPI-dependent but operating through different adaptor systems, each specialized for enzymatic modules belonging to a specific pathway. This realization has led to a refinement of the original paradigm to the multimodular recycling model, which proposes that enzyme localization and recycling are governed by a network of specialized adaptors and distinct trafficking modules [114].

Within this framework, two major classes of adaptors operate in parallel: those mediating enzyme recycling, such as COPI subunits themselves itself and COPI-associated factors like GOLPH3, GOLPH3L, FAM114A, LEPROT, Vps74, and Erd1, and those promoting enzyme retention within specific cisternae, notably GRASP55. Each adaptor recognizes subsets of enzymes through cytosolic motifs, lipid identity, or local cisternal context [35, 44, 102, 109, 115, 128, 129, 135, 136]. Together, these mechanisms sustain the enzymatic gradients that define Golgi polarity while allowing the organelle to dynamically adapt to changes in cargo load, metabolic demand, or developmental conditions.

In addition, there are COPI-independent pathways, including those mediated by clathrin adaptors (AP-1, Ent5, Dop1), that further expand the repertoire of recycling mechanisms, contributing to the robustness and versatility of Golgi function [15, 16, 148]. These findings highlight the intricate complexity of intra-Golgi dynamics and suggest that enzyme localization is controlled by a network of overlapping and specialized recycling routes rather than by a single constitutive pathway.

In this context, Golgi organization can be viewed as the outcome of a multimodular recycling network consisting of multiple COPI-dependent and COPI-independent retrograde transport pathways. These pathways operate in concert on functionally coherent enzymatic modules or subsets of different glycosylation pathways to modify the glycan output composition in response to cellular needs, and maintain compartmentalization and polarity. Disruption of this finely tuned organization, by genetic mutations or functional alterations, leads to defective glycosylation and contributes to a variety of human diseases.

Structural and Morphodynamic Organization of the Golgi Apparatus

The architecture of Golgi is defined by stacks of flattened, fenestrated cisternae arranged in a polarized fashion, with the cis-face oriented toward the ER and the trans-face directed toward the plasma membrane or other organelles [62, 79]. Each Golgi stack typically comprises 4–11 cisternae and surrounded by a ribosome-free proteinaceous matrix known as the Golgi matrix. This matrix contains structural proteins that stabilize the stack and may regulate its morphodynamics and spatial organization [62]. In mammalian cells, the stacks are laterally connected by tubular networks forming the Golgi ribbon, a continuous membrane system whose volume and complexity vary significantly depending on cell type, developmental stage, and metabolic activity. For example, in insulin-producing β cells, the noncompact zone contributes substantially to total Golgi volume, reflecting its importance in high-throughput secretory processes [62, 80].

The ER-Golgi intermediate compartment (ERGIC) is found on the cis-side of the Golgi apparatus. It appears as a structurally distinct layer of polymorphic elements situated between the ER and the cis-most Golgi cisterna. These elements may represent vesicular tubular clusters (VTCs) stalled in transit or in the process of forming new cisternae [67]. Their presence supports models in which Golgi cisternae are continuously renewed through membrane influx from the ER and ERGIC. This dynamic renewal process is essential for maintaining the functional integrity of the Golgi and for accommodating fluctuations in cargo load.

Morphologically, the Golgi exhibits a high degree of variation in structure. At the cis- face, VTCs mediate ER-to-Golgi transport and retrograde trafficking via COPI-coated vesicles [62]. These vesicles are concentrated at cisternal rims and fenestrated zones, with the highest density observed on cis cisternae. The central VTC flattens into cisternal membranes that fuse with or deposit parallel to the cis-most Golgi cisterna, facilitating the integration of newly synthesized cargo into the Golgi stack. At the trans-face, the trans-Golgi network (TGN), a network of tubules and vesicles, directs cargo to the plasma membrane, endosomes, lysosomes, or secretory granules. Its structure varies significantly among cell types, being extensive in cells with elaborate lysosomal systems and minimal or absent in cells forming large secretory granules [22]. Advances in high-voltage electron microscopy (HVEM) and tomographic reconstruction have provided unprecedented insights into the three-dimensional organization of the Golgi. In mammalian cells, the Golgi ribbon consists of compact zones of stacked cisternae interspersed with noncompact regions rich in vesicles and polymorphic membranous elements [67]. Tubular bridges connect cisternae at equivalent levels across stacks. Quantitative analyses reveal that their volumes and fenestration patterns of cisternae vary along the cis-trans axis, suggesting functional compartmentalization and specialization.

The TGN, with a network of cisterna-like elements and tubules is traditionally viewed as a singular sorting station. It is now understood to comprise multiple trans cisternae (C5–C7), each with distinct structural and functional properties [84]. C7 typically produces clathrin-coated vesicles for lysosomal delivery, while C6 generates lace-coated vesicles destined for the plasma membrane [66, 68]. These coat types are mutually exclusive and arise from distinct tubular elements, indicating that molecular sorting occurs prior to carrier formation. These findings support a refined model of Golgi export, where distinct trans cisternae specialize in packaging cargo for different destinations. In neuroendocrine PC12 cells, a specialized trans-Golgi subcompartment enriched in Golgin97 and VAMP4 transiently hosts synaptic and granule proteins during synchronized trafficking events. Live-cell imaging and temperature-block experiments revealed that VAChT, synaptophysin, and chromogranin A colocalize in this compartment before segregating into synaptic protein transport vesicles (SPTVs) and large dense-core vesicles (LDCVs) [98]. The recruitment of SNARE proteins such as VAMP2 and SNAP25 suggests that this compartment serves as a platform for assembling exocytic machinery. Similar sorting behaviour in rat cortical neurons supports the conservation of this mechanism across neuronal cell types. These findings highlight the existence of specialized sorting domains within the Golgi that coordinate the packaging and dispatch of functionally distinct cargoes.

The Golgi structural plasticity is evident across a wide range of cell types. In Sertoli cells, the organelle forms an extended ribbon-like network that spans from the juxtanuclear region toward the cell apex, while in spermatids, it adopts a hemispherical configuration with a cortical zone composed of stacked cisternae and a medullary zone rich in vesicular and tubular profiles [106]. In skeletal muscle cells, Golgi elements reorganize during myogenic differentiation, transitioning from mobile carriers in early stages to dispersed structures flanking ER exit sites in mature myofibers [74]. These architectural variations underscore the ability of Golgi apparatus to adapt its morphology to specific cellular contexts and functional requirements. Moreover, the Golgi undergoes dynamic reorganization in response to physiological stimuli. Under conditions of increased metabolic demand, such as glucose stimulation in pancreatic β-cells or increased transport following viral infection direct continuities can form between cisternae at different levels of the stack, bypassing intervening cisternae and potentially facilitating rapid transit of newly synthesized proteins [80, 127].

Collectively, these findings portray the Golgi apparatus as a morphodynamically regulated organelle whose structural complexity and spatial compartmentalization are tightly coupled to its diverse functional roles. Its architecture is not static but adapts dynamically to the physiological state, cargo load, and developmental stage of the cell. The variability in Golgi organization across cell types and conditions underscores its centrality in maintaining cellular homeostasis and efficient secretion. Crucially, the ability of cisternae to undergo structural remodelling, functional specialization, and coordinated fragmentation lays the foundation for understanding the mechanisms of cisternal maturation, a process that integrates membrane dynamics, enzymatic progression, and cargo sorting, and will be explored in detail in the following section.

Cisternal Maturation: Mechanism, Discovery, and Supporting Evidence

The concept of cisternal maturation has evolved through several distinct stages, reflecting the progressive refinement of our understanding of membrane traffic within the Golgi complex. The earliest view is derived from the pioneering ultrastructural work of Grassé and Morré [50, 85]. In their membrane flow or progression model, developed in the 1950s–1970s, Mollenhauer and Morré envisioned the Golgi as a flow-through system in which membranes continuously move from the ER via the Golgi to the plasma membrane [85, 88]. In this progression model, newly formed cis cisternae originated by the fusion of ER-derived carriers, while trans cisternae fragmented into secretory vesicles. The Golgi stack was thus seen as a conveyor belt, with cisternae themselves acting as the carriers of cargo.

Subsequent biochemical and morphological observations revealed that adjacent Golgi cisternae possess distinct and stable enzyme compositions, incompatible with a simple progression mechanism [30]. This led to the vesicular transport model, in which stable cisternae exchanged cargo via small anterograde vesicles. The discovery of coated vesicles and of machinery—COPI, COPII, ARF, and SNAREs—required for vesicle formation provided molecular support for this view [112]. In this case, the Golgi was conceptualized as a stack of static compartments connected by vesicular trafficking, with directionality ensured by coat and tethering complexes [112].

More recently, inconsistencies in the vesicular model, particularly, the transport in algae of large cargoes (scales) that were never observed in vesicles, and were also too large to fit into vesicles, revived the idea of cisternal progression [4]. This type of evidence had been available early on, but scales had been explained as products of anomalous transport formulae in remote and evolutionary ancient mechanisms [37]. This argument was later undermined by the increasing awareness that fundamental cellular mechanisms are conserved [33]. Finally, the observation that large protein aggregates in mammalian cells that could not fit in small vesicles were able to traverse the Golgi at normal rates, made it evident that a modified cisternal progression model was necessary [11]. The modern progression maturation model introduced the key notion that cisternae themselves progress forward but change composition over time through retrograde recycling of resident enzymes [82]. This reconciliation coupled anterograde membrane flow with enzyme recycling in lockstep, explaining how cisternae can maintain distinct enzymatic identities while cargo moves forward within the same lumen. The main experimental support for cisternal maturation as a universal transport mechanism therefore came from the demonstration that procollagen moves through intact cisternae without leaving their lumen [11] and from live-cell imaging in yeast showing sequential transformation of cis, medial, and trans cisternae, confirming maturation as a dynamic process [73, 81].

According to the cisterna maturation model, new cisternae are formed at the cis face of the Golgi from ER-derived COPII vesicles and progressively mature into medial and trans cisternae by acquiring and losing specific sets of resident enzymes. During this maturation process, cargo remains within the same cisterna, while Golgi-resident enzymes are recycled back in a retrograde fashion via COPI-coated vesicles. Eventually, the maturing cisterna becomes part of the TGN, from which cargo is sorted and exported to its final destination [36, 46, 47]. Live-cell imaging in yeast provided the first direct, dynamic evidence for cisternal maturation. In *Saccharomyces cerevisiae,* where Golgi cisternae are unstacked and spatially resolvable, Glick, Nakano and their colleagues [73, 81] used fluorescently tagged resident proteins to show that individual cisternae undergo a stereotyped sequence of maturation, acquiring and losing specific markers over time. These transitions occurred over 2–3 min and matched the known transit time of secretory cargo, supporting the idea that Golgi cisternae mature in cis to trans direction. Further support for cisternal maturation came from Rizzo et al. [108], who engineered polymerizable Golgi-resident enzymes to block their recycling. These constructs remained confined within cisternae and moved from cis to trans, providing direct evidence for enzyme recycling in synchrony with cisternal progression also in mammalian cells.

Mechanistically, cisternal maturation depends on the continuous retrograde recycling of resident Golgi enzymes. COPI-coated vesicles are the primary carriers mediating this recycling, retrieving enzymes from more mature cisternae and returning them to earlier ones [48]. Functional evidence for this process came from temperature-sensitive COPI mutants (ret1-1), which severely impaired cisternal maturation at restrictive temperatures [55]. Auxin-inducible degradation of COPI subunits (Ret1/α-COP and Sec21/γ-COP) further confirmed that depletion of these proteins blocks cis-to-medial and medial-to-trans transitions, with cisternae retaining early markers indefinitely [55].

The specificity and directionality of vesicle fusion events during retrograde transport are regulated by SNARE proteins, Rab GTPases, and tethering factors. SNAREs such as Syntaxin 5 and GOS28 are distributed throughout the Golgi and mediate intra-Golgi fusion events [131]. Rab GTPases act in cascades to enforce temporal order during maturation: early-acting Rabs recruit GEFs for downstream Rabs, while later-acting Rabs recruit GAPs to inactivate upstream Rabs [124]. For example, in yeast, Ypt32 recruits a GTPase activating protein (GAP) to inactivate Ypt1, driving the transition from early to late Golgi identity [96]. Arf1, a small GTPase, plays a pivotal role in recruiting COPI and clathrin adaptors, organizing membrane curvature, and coordinating lipid composition [56]. Its activity is also tightly regulated by GEFs and GAPs, ensuring spatial and temporal control of vesicle formation and fusion. Tethering factors such as Golgins also contribute to the specificity of retrograde transport. Wong and Munro [139] demonstrated that Golgins like TMF, Golgin-84, and GMAP-210 selectively capture distinct intra-Golgi carriers containing resident enzymes. Their rerouting experiments showed that these tethers contribute in an important manner to the spatial organization of retrograde trafficking, reinforcing the cisternal maturation model.

Later studies have refined the model by incorporating observations from live-cell imaging and biochemical synchronization. Kurokawa et al. [65] used super-resolution confocal live imaging microscopy (SCLIM) to show that cargo remains within a single cisterna as it matures from cis to trans identity. Casler et al. [16] developed a regulatable fluorescent cargo in yeast and observed a burst of intra-Golgi recycling during the early-to-late transition, mediated by the clathrin adaptor AP-1. Finally, Pirozzi et al. [101] revisited the model using synchronized cargo transport and high-resolution imaging. They showed that cargoes move vectorially from cis to trans compartments and accumulate in the TGN before exiting. Their data suggest that the TGN functions as a distinct, rate-limiting compartment, and that the final step of intra-Golgi transport may involve fusion of cargo-containing cisternae with the TGN, a mechanism distinct from cisternal maturation. This hybrid model preserves the core tenets of cisternal maturation while acknowledging the structural and functional distinctiveness of the TGN.

In conclusion, the cisternal maturation model of intra-Golgi trafficking is supported by a robust body of morphological, biochemical, and kinetic evidence (Fig. 10.1). Subsequently this facilitated the extension of the maturation principle to other systems, such as early-to-late endosome conversion [107], establishing the general concept of compartment maturation and Rab conversion as universal mechanisms of organelle identity. It provides a coherent framework for understanding how the Golgi apparatus maintains its polarized structure, ensures the fidelity of glycosylation, and coordinates the export of cargo. While alternative models may operate in parallel, particularly for small soluble proteins (see below), cisternal maturation remains the most comprehensive and experimentally validated explanation for intra-Golgi transport. This dynamic model of intra-Golgi transport not only explains how secretory cargo is conveyed through the stack but also provides a

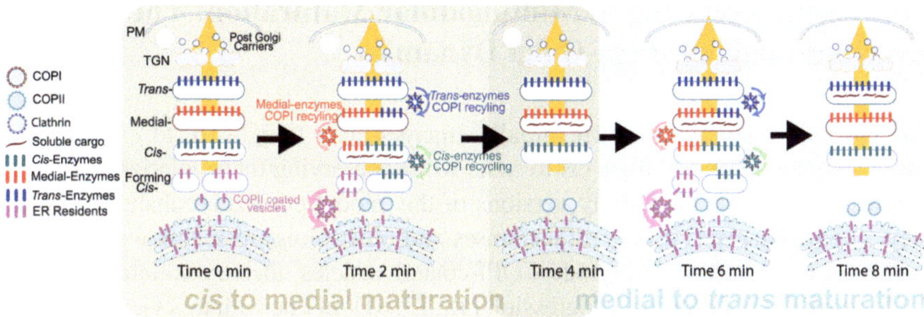

Fig. 10.1 The process of cisternal progression and interconversion. Once cargo-containing carriers reach the cis-Golgi, they fuse to generate new cis cisternae. Trafficking then proceeds through the process of cisternal progression and interconversion. Coloured bars indicate the distinct enzymatic composition of each compartment: ER and forming cis cisterna (pink), cis (green), medial (red), and trans (blue). Transmembrane cargoes are represented as short vertical bars and correspond to transported molecules not belonging to the trafficking or glycosylation machinery. The figure illustrates two sequential rounds of cisternal interconversion. During the cis-to-medial transition, newly formed cis cisternae arise from ER-derived membranes and progressively mature through retrograde vesicular transport of resident enzymes between adjacent cisternae. Each cisterna acquires enzymes from the distal and transfers others to the proximal cisterna, thus shifting its position within the stack (forming → cis → medial → trans). The enzymatic identity of each cisterna therefore correlates with its position. The trans cisterna behaves differently: while its enzymes are recycled back to the medial cisterna, it disassembles into transport carriers containing cargo proteins destined for their final compartments. Throughout this process, enzymes experience both an anterograde flow driven by cisternal progression and a retrograde flow mediated by vesicular recycling. The equilibrium between these opposing fluxes determines the steady-state localization of glycoenzymes—including those involved in glycosphingolipid metabolism—and thereby defines the outcome of competing glycosylation reactions. Enzymes that fail to recycle rapidly enough are carried forward into post-Golgi carriers and exit the stack, whereas cargo proteins and lipids remain within the maturing cisternae. Upon disassembly of the trans cisterna, cargoes are packaged into anterograde carriers targeting the plasma membrane or the endo-lysosomal system. Through successive rounds of interconversion, cargoes are sequentially exposed to the distinct enzymatic environments of each cisterna, each acting as a discrete reaction chamber that ensures ordered glycosylation and a defined metabolic output. The entire progression–interconversion cycle, from the formation of a new cis cisterna to its conversion into a trans element, occurs approximately every 3–5 min, so that complete cis-to-trans transit requires 15–20 min depending on cell type. This schematic representation omits certain molecular complexities, such as the retrograde transport of some yeast cargoes from the TGN to the Golgi (see [148] for details)

mechanistic basis for the spatial organization of glycosylation enzymes. The continuous retrograde recycling of resident enzymes ensures their polarized distribution across cis, medial, and trans cisternae, thereby establishing the enzymatic gradients necessary for sequential glycan processing. As a result, the mechanisms underlying cisternal maturation are directly responsible for shaping the glycosylation landscape of the cell. The implications of this model for the spatial organization for glycosylation enzymes and cellular function will be explored in the following sections.

From Bulk Recycling to Multimodular Maturation: The Evolving Logic of Intra-Golgi Dynamics

The cisternal maturation model initially emerged as a framework to explain how Golgi cisternae progress from cis to trans while preserving their polarized distribution of resident enzymes. Early versions of this model posited a constitutive, bulk recycling mechanism: glycosyltransferases and other Golgi residents were thought to recycle uniformly in retrograde COPI-coated vesicles, thereby counterbalancing the forward flow of cisternal progression [48, 73, 81]. In line with this view, Liu et al. [72] provided compelling evidence that cis-Golgi glycosyltransferases directly bind to the δ- and ζ-subunits of COPI via a conserved N-terminal motif (ϕ-(K/R)-X-L-X-(K/R)). Using proximity labelling, pulldown assays, and biophysical binding studies, they showed that this interaction is essential for Golgi retention and is mediated by the μ-homology domain of δ-COP, structurally analogous to cargo-binding domains in clathrin adaptors. However, this direct binding mechanism applies only to a subset of glycosyltransferases bearing the consensus motif, immediately revealing a specificity that contradicts a uniform recycling paradigm. While this view provided a partial solution for enzyme retention within the Golgi, it soon faced a conceptual obstacle. If all enzymes rely on the same COPI machinery for retrieval, whether through direct binding or other mechanisms, how is their precise localization into cis-, medial-, and trans-compartments maintained? A uniform recycling pathway could not adequately explain the ordered distribution of distinct enzymes across the stack, hinting at a more complex, multimodular organization as discussed recently [114].

Several published studies have revealed that cisternal maturation is far more multimodular than initially anticipated. Rather than depending solely on direct COPI binding or a single bulk pathway, different groups of Golgi residents engage specialized recycling routes, often with the help of dedicated adaptor proteins that bridge between residents lacking direct COPI interaction motifs and the COPI machinery itself. Yeast studies first highlighted this principle with the discovery of Vps74, a peripheral Golgi protein that binds a conserved pentameric motif (F/L-L/I/V-XX-R/K) in the cytosolic tails of glycosyltransferases and links them to COPI vesicles, thereby ensuring their retention in early Golgi compartments [128, 129, 141]. Vps74 forms a tetrameric complex that interacts with COPI, facilitating the selective incorporation of its client enzymes into retrograde vesicles. This adaptor-mediated mechanism allows residents that lack direct COPI-binding motifs to access the COPI recycling machinery in a cargo-selective manner. Loss of Vps74 disrupts enzyme localization and impairs glycosylation, highlighting its essential role in compartment-specific recycling [128, 129].

In mammalian cells, the Vps74 homolog GOLPH3 functions as an adaptor for enzymes involved in glycosphingolipid (GSL) biosynthesis, incorporating them into retrograde COPI vesicles [109]. Eckert et al. [35] showed that GOLPH3 selectively binds to specific cytosolic motifs (e.g., LLRRR) on enzymes such as C2GnT and SiaTI, mediating their incorporation into COPI vesicles through direct interaction with the coatomer complex. Importantly, not all glycosyltransferases interact with GOLPH3, as B4GalT1, for instance, lacks the binding motif and is excluded from GOLPH3-mediated recycling, highlighting the functional specificity of adaptors. Additionally, Welch et al. [135] identified GOLPH3 and its paralogue GOLPH3L as broad-spectrum COPI adaptors that bind to short, positively charged cytoplasmic tails of glycosylation enzymes. Their deletion leads to the mislocalization and degradation of a wide array of Golgi enzymes, resulting in global glycosylation defects.

More recently, Lesniak et al. [70] identified the client-binding site on Vps74, revealing that this interaction involves two membrane-facing loops and a PI4P-binding site, and is mutually exclusive with Arf1-GTP and Sac1 binding, suggesting a regulated mechanism of cargo selection. Wood et al. [140] further demonstrated that Vps74-mediated recycling is controlled by local phosphatidylinositol 4-phosphate [PI(4)P] signalling in the Golgi apparatus, with the Sac1 phosphoinositide phosphatase modulating this process. Thus, the localization of GOLPH3 clients is dictated by the lipid identity of the cisterna, particularly the presence of PI(4)P at the trans-Golgi, which recruits GOLPH3 and initiates client recycling [28, 105, 141]. This system exemplifies how cisternal identity markers and adaptors cooperate to enforce enzyme localization.

Other adaptors extend this paradigm. Sardana et al. [115] identified Erd1 as a transmembrane adaptor in yeast that links subsets of enzymes to COPI through Vps74. Erd1 interacts directly with both the cytosolic receptor Vps74 and its client enzymes, stabilizing their incorporation into COPI vesicles. Loss of Erd1 phenocopied Vps74 deletion, resulting in mislocalization of glycosyltransferases to the vacuole and severe glycosylation defects. Notably, Erd1 itself requires Vps74 for its recycling, suggesting a cooperative mechanism in which transmembrane and cytosolic adaptors act in concert to mediate selective retrieval. More recently, FAM114A proteins were identified as adaptors for enzyme recycling in mammalian cells [136]. These transmembrane adaptors bind to similar polybasic motifs and interact directly with COPI subunits, exhibiting cargo specificity and differential localization within the Golgi stack, suggesting that distinct COPI containing complexes operate in different cisternal contexts.

In parallel, mammalian cells also employ LEPROT and LEPROTL1 as transmembrane adaptors that recognize long transmembrane domains in a pH-dependent manner and interact directly with COPI subunits, further expanding the repertoire of cargo-specific recycling modules [44]. GORAB and Scyl1 scaffold COPI at the trans-Golgi to promote the retrieval of ST6GAL1 and other trans-Golgi enzymes [138]. Witkos et al. demonstrated that GORAB recruits COPI to discrete domains at the trans-Golgi via interaction with Scyl1, stabilizing COPI assembly and facilitating enzyme recycling. These domains define a functionally distinct pool of COPI vesicles, spatially segregated from those operating at the cis-Golgi and ERGIC, supporting the multimodular organization of COPI-mediated transport.

Together, these findings demonstrate that COPI recycling itself is heterogeneous, mediated by multiple overlapping but distinct pathways. Structural and biochemical analyses reinforce this view: COPI isoforms show differential distribution across cisternae [83], with γ1/ζ1-COP isoforms enriched in the early Golgi and γ2/ζ2-COP isoforms localized to the trans-Golgi. This spatial segregation suggests that COPI vesicles form distinct subpopulations tailored to specific transport routes and cargo sets [94]. Proteomics confirms compositional diversity [1], although different isoforms generate vesicles with similar cargo composition in vitro, indicating that their differential localization in vivo may reflect spatial regulation rather than intrinsic cargo selectivity. Cryo-electron microscopy reveals that COPI vesicles vary in size, membrane thickness, and luminal density depending on their cisternal origin, consistent with specialized functions [14].

Critically, the multimodular Golgi architecture extends beyond COPI-dependent mechanisms. COPI-independent recycling routes have been described, especially at the late Golgi, further demonstrating the complexity of intra-Golgi transport. In yeast, Zhao et al. [148] uncovered a COPI-independent retrograde pathway operating at the TGN, identifying Dop1 as a conserved protein essential for the retrieval of glycosyltransferases such as Och1 from the TGN back to the early Golgi. In Dop1-deficient cells, Och1 accumulates in the TGN and fails to return to the cis-Golgi, leading to defective N- and O-glycosylation. This pathway is independent of Vps74 and COPI, as shown by the absence of vacuolar degradation of Och1 in Dop1-deficient cells, in contrast to vps74Δ mutants where mislocalized enzymes are degraded. Additional studies identified sequential pathways mediated by AP-1 and Ent5 clathrin adaptors, which organize the polarized distribution of enzyme classes in the Golgi [15]. Using live-cell 4D confocal microscopy Casler et al. [15] demonstrated that clathrin adaptors mediate two sequential intra-Golgi recycling routes, each defining a distinct kinetic class of Golgi proteins. Some transmembrane proteins, such as Kex2 and Drs2, recycle via AP-1/Ent5-dependent pathways, while others like Vps10 follow GGA-mediated routes through endosomes, suggesting temporal modularity in intra-Golgi trafficking. Multi-pass proteins such as TM9SF2 and LAPTM4A also participate in the localization of subsets of enzymes [142, 144], though whether their action is COPI-dependent remains unresolved. Most recently, Krahn et al. [64] used live-cell imaging combined with vesicle capture assays to demonstrate the existence of at least three kinetically segregated intra-Golgi recycling routes in yeast: an early COPI(b) pathway recycling Vrg4 and Kre2, an intermediate COPI(b′) pathway recycling Tmn1 and Gnt1, and a late clathrin-mediated route dependent on AP-1 and Ent5 adaptors recycling Kex2, Ste13, and Stv1. Notably, some proteins such as Sys1 and Aur1 appear to partition between multiple pathways, suggesting plasticity in cargo selection.

Thus, cisternal maturation involves a complex interplay of COPI-dependent and independent machineries, reinforcing the multimodular nature of intra-Golgi dynamics. In parallel, retainer molecules help lock enzymes into position once they reach their cisternal destination. GRASP55, for example, prevents further recycling of specific glycosyltransferases, stabilizing their steady-state localization and ensuring ordered compartmentalization [102]. Pothukuchi et al. demonstrated that GRASP55 functions not as a recycling COPI adaptor but as a retainer, a molecule that binds specific glycosylation enzymes (e.g., glucosylceramide synthase or GCS) via PDZ-binding motifs in their cytosolic tails and actively prevents their entry into

COPI vesicles, anchoring them in the trans-Golgi. Loss of GRASP55 leads to a shift of these enzymes toward the cis-Golgi, increasing their access to upstream substrates and altering glycosphingolipid biosynthesis. This demonstrates that retainers act as counterforces to recycling adaptors, shaping the spatial distribution of enzymes by limiting their retrograde flux. These retention mechanisms operate alongside the constant balance of anterograde cisternal progression and retrograde retrieval, reinforcing the concept that enzyme distribution is the product of both recycling pathways and retainers within a multimodular framework.

Cisternal maturation also influences enzyme abundance. At steady state, a fraction of glycoenzymes inevitably escapes recycling and is targeted for degradation, either via lysosomal delivery or ectodomain shedding at the plasma membrane [122]. Recycling efficiency therefore dictates enzyme levels: impaired recycling increases degradation, reducing Golgi enzyme pools, while enhanced recycling stabilizes enzyme abundance, sometimes leading to overaccumulation [109]. Rizzo et al. showed that GOLPH3 not only regulates enzyme localization but also their stability and degradation. In the absence of GOLPH3, client enzymes fail to recycle and are misrouted to lysosomes for degradation, leading to altered GSL biosynthesis. Conversely, GOLPH3 overexpression enhances enzyme retention, increases downstream GSLs (e.g., Gb3), and promotes mitogenic signalling via the PI3K-Akt-mTOR axis. This mechanism is clinically relevant: GOLPH3 is amplified in several human cancers, and its expression correlates with increased levels of LCS (B4GALT5) and poor prognosis in non-small cell lung cancer [109]. This dual control over localization and abundance highlights how tightly cisternal maturation regulates the enzymatic landscape of the Golgi.

Additional molecular evidence supports the multimodular view. The number of Rabs and Golgins involved in Golgi trafficking far exceeds the minimum required for a three-step maturation process, suggesting that multiple distinct pathways coexist. Indeed, more than twenty Rabs are localized to the Golgi [3, 124], and different Golgin tethers mediate distinct vesicle targeting events, such as GM130 for ER-derived vesicles, or TMF, GMAP210, and Golgin-84 for COPI vesicles from medial and late cisternae respectively [17, 139]. This redundancy and specialization are consistent with a multimodular maturation process rather than a unitary pathway. Papanikou et al. [97] used an anchor-away strategy to acutely inactivate COPI and observed that early Golgi proteins such as Vrg4 failed to recycle, while late Golgi proteins like Sec7 and Kex2 continued to cycle dynamically, leading to the formation of hybrid cisternae containing both early and late markers. This indicated that COPI is selectively required for the recycling of early residents, while late residents utilize alternative, COPI-independent routes, further supporting the multimodular paradigm.

In addition to the study of retrograde transport, the study of anterograde transport of cargo has also provided evidence for a multimodular cisternal maturation mechanism. Classical cisternal maturation mechanism predicts that all cargoes traverse the Golgi at the same velocity. Contrary to this prediction, experimental observations have shown that the cargo velocity across the Golgi can vary [6, 25, 39, 126] suggesting that there are likely multiple ways to traverse the Golgi and/or cargo proteins can also be subjected to retrograde transport, which is not predicted by the classical cisternal maturation model.

Taken together, these insights paint a picture of cisternal maturation as a coordinated, multimodular network of recycling routes. Instead of a single constitutive pathway, intra-Golgi transport relies on multiple, partially independent mechanisms: direct COPI binding by specific enzyme subsets, adaptor-mediated COPI vesicles bridging residents with shared cytosolic information to the coat machinery, COPI-independent retrieval routes, lipid identity markers, retainer proteins, and a vast array of tethers and Rabs. This multimodularity explains how enzymes achieve stable but diverse intra-Golgi distributions while maintaining dynamic control of their abundance. The implications extend beyond cell biology. A shift from viewing cisternal maturation as a uniform housekeeping mechanism to recognizing its multimodular nature reframes how alterations in specific pathways can affect disease. Errors in enzyme recycling or retention can lead to aberrant glycosylation patterns, with consequences ranging from congenital disorders of glycosylation (CDG) to cancer and other pathologies [43, 52, 89, 133]. Day et al. [27] demonstrated that budding yeast has a minimal endomembrane system, yet even this simplified system exhibits the complexity of multiple recycling pathways. Sahu et al. [114] reviewed the role of the multimodular cisternal maturation machinery in glycan synthesis and oncogenesis, emphasizing how COPI adaptors and their cargo recognition contribute to spatial organization and disease processes. In this light, the Golgi emerges not only as a central hub of trafficking but also as a regulatory platform whose multimodular maturation machinery safeguards glycosylation homeostasis.

Polarized Distribution and Modular Organization of Glycosyltransferases in the Golgi Apparatus

Glycosylation is mediated by a large and diverse repertoire of glycosyltransferases, with more than 200 distinct enzymes encoded in the human genome [13, 92, 116]. These enzymes act sequentially to elongate glycan chains and produce the extraordinary structural diversity that characterizes glycoconjugates. A defining feature of glycosyltransferase biology is their polarized distribution along the cis-to-trans axis of the Golgi apparatus. This arrangement mirrors the sequential order of reactions during glycan maturation, with early-acting enzymes, such as mannosidases and N-acetylglucosaminyltransferases, enriched in cis and medial cisternae, and late-acting enzymes, such as galactosyltransferases, fucosyltransferases, and sialyltransferases, localized in trans cisternae and the TGN [30, 76, 104, 111]. Classical biochemical fractionation studies demonstrated that proteins such as the vesicular stomatitis virus glycoprotein undergo trimming in early Golgi compartments and terminal modification in later ones [29, 31, 32], thereby establishing the concept of functionally ordered compartmentation. Immunocytochemical and electron microscopy approaches confirmed that specific glycosyltransferases are codistributed with reference markers of defined cisternae [104, 111]. Together, these findings provided the first evidence for the strict spatial compartmentalization that is now recognized as a hallmark of Golgi architecture.

The functional relevance of this organization lies in its ability to ensure that each enzyme acts on the appropriate intermediate generated by its predecessor, thereby preserving the fidelity of glycan biosynthesis. Importantly, enzymes that catalyze consecutive reactions are frequently co-localized within the same cisternae, forming functional modules [104, 128]. These modules enhance biosynthetic efficiency by promoting substrate channelling and reducing the diffusion of intermediates. Evidence for such modularity comes from the observation that GlcNAc-transferase I and mannosidase II co-localize in medial/trans cisternae, while galactosyltransferase and sialyltransferase cluster in trans cisternae and the TGN [61, 104]. Similarly, enzymes of glycolipid biosynthesis, such as GalNAc-transferase and GalT2, were shown to physically interact and form multi-protein complexes in the Golgi [45, 77] . These complexes facilitate substrate channelling and enhance the efficiency of ganglioside synthesis, further supporting the concept of enzymatic modularity. On the contrary, enzymes that catalyse reactions belonging to different biosynthetic pathways are physically segregated within the same cisternae, forming distinct glycosylation units. This segregation ensures that non-sequential metabolic routes do not interfere with each other, thereby maintaining the specificity and control of the glycosylation process [54].

Another important need for such polarized organisation arises from the need to control competitive reactions that happen in the Golgi [103]. The glycosylation enzymes are known for their promiscuity and often compete with each other for the available substrates, i.e. the proteins and lipids that pass through the Golgi apparatus [57]. Moreover, differences in enzyme specificity and promiscuity affect glycan output: while certain enzymes, such as MGAT2 and MGAT5, display strict substrate specificity, others exhibit broader reactivity, generating microheterogeneity [7]. Regulation of the products produced by the competing glycosylation enzymes is achieved by controlling their access to the substrates. Recently it was shown that GRASP55 regulates the intra-Golgi localization of glucosylceramide synthase (GCS), a key enzyme in glycosphingolipid biosynthesis [102]. GCS competes with sphingomyelin synthase 1 (SMS1) for the same substrate *viz* ceramide. Under normal conditions both GCS and SMS1 colocalize in the trans-Golgi where they compete for ceramide. In the absence of GRASP55, GCS shifts to the cis-Golgi where it gets preferential access to ceramide thus making more glucosylceramide. Thus, the control of intra-Golgi localization of an enzyme controls substrate flux across competing pathways.

The importance of recycling pathways in regulating the glycan output has been further substantiated by computational and imaging approaches. Modelling studies have shown that defects in enzyme trafficking perturb glycosylation fidelity, leading to reduced branching and altered glycan profiles [38, 57]. Thus, compartmentalization mitigates the consequences of enzyme promiscuity by restricting enzymatic action to defined compartments, thereby shaping glycan profiles in a controlled manner [57]. The maintenance of polarized distribution and modular organization depends on dynamic recycling pathways. COPI-dependent retrograde transport plays a central role in retrieving enzymes from distal to proximal cisternae, ensuring that cisternal identity is preserved as cisternae mature from cis to trans [91, 95, 129, 141]. Cytoplasmic sorting motifs and interactions with adaptors such as Vps74p in yeast or GOLPH3 in mammals mediate the recognition and retrieval of glycosyltransferases [109, 128]. In addition to COPI-mediated retrieval, COPI-independent pathways contribute to enzyme homeostasis. The conserved oligomeric Golgi (COG) complex functions as a tethering machinery required for retrograde transport of Golgi enzymes, and mutations in COG subunits cause severe congenital disorders of glycosylation due to enzyme mislocalization and impaired glycan processing [40, 130].

Spatial enzyme organization is further modulated by intrinsic structural and biophysical properties. Transmembrane domain length and composition influence the partitioning of enzymes into lipid domains of different bilayer thickness, thereby contributing to their localization within the Golgi [34, 90, 91]. Super-resolution imaging techniques such as FLASH-PAINT have revealed that cis-to-trans polarization and enzyme clustering are preserved even under conditions of Golgi fragmentation, underscoring the robustness of modular organization [54, 117].

In summary, the Golgi apparatus is not a passive conduit for glycosylation reactions but a highly dynamic and spatially organized platform. The polarized distribution of over 200 glycosyltransferases along the cis-trans axis ensures the sequential and faithful maturation of glycans, while the modular clustering of enzymes catalysing consecutive reactions enhances biosynthetic efficiency. These modules are actively maintained through a combination of COPI-dependent and independent recycling pathways, structural determinants of enzyme localization, and adaptor-mediated retention. Together, these mechanisms establish a mosaic organization that supports the remarkable versatility and adaptability of glycosylation (Fig. 10.2). In the following section, we will examine in more detail the specific contributions of COPI-dependent and independent pathways to the maintenance of Golgi enzyme localization and their implications for physiology and disease.

Golgi-Dependent Regulation of Glycosylation Enzymes and Its Implications in Disease

The Golgi apparatus is a central hub of post-translational modification, where glycosylation, sorting, and trafficking of proteins and lipids are precisely coordinated to ensure cellular homeostasis. Among the various modifications that occur in this organelle, glycosylation is one of the most structurally complex and functionally consequential. It is also tightly linked to Golgi architecture and trafficking dynamics [103]. It modulates protein folding, stability, secretion, and receptor signalling [69, 132] and its fidelity depends on the correct localization and activity of

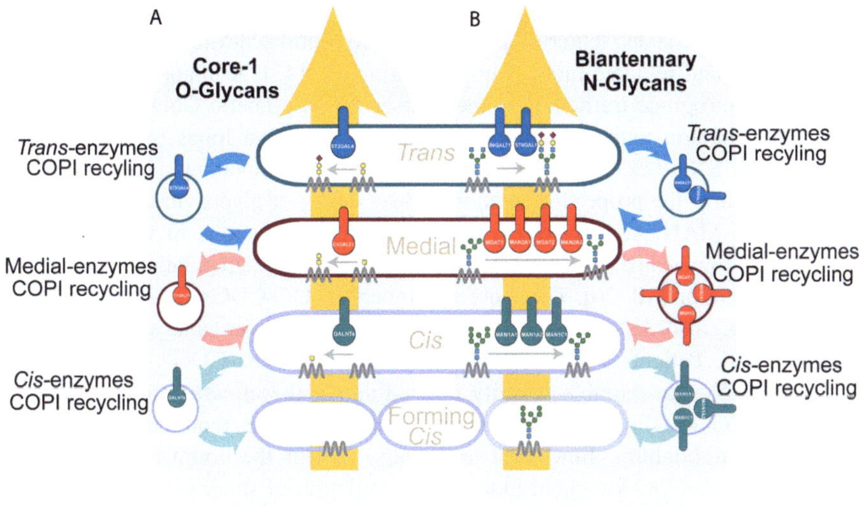

Fig. 10.2 Cisternal maturation orchestrates the spatial organization of distinct subsets of glycoenzymes within Golgi cisternae. The biosynthetic pathways for Core-1 O-glycans (**a**) and biantennary N-glycans (**b**) are illustrated. The figure depicts how cisternal maturation mechanisms, whether COPI-dependent or COPI-independent, regulate enzyme localization across the cis-, medial-, and trans-Golgi. By modulating the recycling of sequentially acting enzymes, they ultimately shape the final glycosylation pattern. Enzymes that act in sequence are known to physically interact and form functional enzymatic modules within the same cisterna [61]. Most enzymes are enriched within one or two cisternae, although the sharpness of their localization can vary [102, 125]. For simplicity, enzymes are shown within a single cisterna. Briefly, O-glycosylation (**a**) takes place exclusively in the Golgi apparatus. It initiates at the cis-Golgi with the transfer of GalNAc to Ser/Thr residues of cargo proteins. In the medial-Golgi, depending on the available glycosyltransferases, galactose or GlcNAc residues are added; in the case of Core-1 O-glycans, galactose addition occurs. Finally, at the trans-Golgi, sialic acid is incorporated by sialyltransferases. In contrast, N-glycan biosynthesis (**b**) begins in the ER. Proteins entering the Golgi carry a preassembled oligosaccharide of 14 sugars, four of which have already been trimmed. In the cis-Golgi, three additional mannose residues are removed by alpha-mannosidases (MAN1A1, MAN1A2, and MAN1C1). In the medial-Golgi, after the removal of two more mannoses by MAN2 enzymes, GlcNAc residues are added by MGAT1 and MGAT2. Finally, upon reaching the trans-Golgi, galactose and terminal sialic acids are attached, completing the maturation of biantennary N-glycans

glycosyltransferases, sugar nucleotide transporters, and regulatory and structural proteins [87, 116].

The organization of these enzymes within the Golgi cisternae is dynamically maintained through recycling and retention mechanisms that balance their retrograde retrieval and forward transport. When these pathways are perturbed by mutations, altered expression, or structural disorganization, they lead to aberrant glycosylation, a hallmark of both congenital disorders of glycosylation (CDGs) [52] and cancer [49, 59, 119, 146].

As per the cisternal maturation model, as the Golgi cisternae containing cargoes mature the resident enzymes are continuously recycled retrogradely to maintain their compartmental positioning [12, 114]. The fidelity of this process is maintained by coat protein complexes, tethering factors, and structural scaffolds that stabilize the Golgi ribbon and regulate intra-Golgi trafficking [5, 143]. A major orchestrator of intra-Golgi retrograde trafficking is the Conserved Oligomeric Golgi (COG) complex, an octameric tethering complex organized into two lobes (A and B) that interact with SNAREs, Rab GTPases, and coatomer components [10]. The COG complex ensures the proper localization and recycling of glycosyltransferases such as MGAT1, MAN2A, B4GALT1, and ST6GAL1. Mutations in COG subunits disrupt tethering, causing vesicle accumulation, Golgi fragmentation, and severe hypoglycosylation [10, 26]. These defects underlie COG-CDGs, which manifest as multisystemic disorders affecting the nervous system, liver, and skeletal tissues [21, 41, 149]. Patients with nearly seventy different COG-CDG mutations have been reported, with variable severity depending on the affected subunit: lobe A mutations (COG1–4) often produce more severe phenotypes than lobe B mutations (COG5–8), highlighting functional asymmetry within the complex [53, 78]. In cellular models, COG knockout induces accumulation of small Golgi-derived vesicles termed COG complex–dependent (CCD) vesicles, which are enriched in recycling enzymes and fail to tether properly with the cistenae [137, 150]. The resulting enzyme mislocalization alters both N- and O-glycosylation, causing incomplete sialylation and galactosylation [8, 9]. The COG complex also coordinates the trafficking of sugar nucleotide transporters and Golgi milieu regulators such as TMEM165, which maintains manganese homeostasis essential for galactosylation [86]. TMEM165 mutations or COG dysfunction reduce Mn^{2+} availability, leading to defective galactosylation that can be partially rescued by galactose supplementation, further emphasizing the tight coupling between trafficking, ion balance, and enzymatic activity [26].

Among recycling adaptors, GOLPH3 (Golgi phosphoprotein 3) has emerged as a key factor linking enzyme localization to oncogenic signalling. GOLPH3 binds PI(4)P at the TGN via its pleckstrin homology (PH) domain and interacts with the COPI coatomer to retrieve specific glycosyltransferases bearing Lxx(R/K) motifs [129, 141]. Its clients include enzymes of the glycosphingolipid (GSL) biosynthetic pathway, such as lactosylceramide synthase (B4GalT5), Gb3 synthase (A4GALT1), and GM3 synthase (ST3GAL5) [114]. GOLPH3 depletion causes their lysosomal degradation, whereas its overexpression enhances recycling and shifts enzyme distribution towards the cis-Golgi, conferring a positional advantage in competitive reactions [109]. GOLPH3 which is amplified in multiple cancers—8% of lung adenocarcinomas, 6% of ovarian, and 6% of bladder tumours—promotes Golgi dispersal [18, 24], increased secretion of pro-tumorigenic factors, and enhanced mTOR signalling [118, 120]. Mechanistically, its overexpression alters the glycosylation flux toward oncogenic GSL species that modulate receptor tyrosine kinase signalling and cell adhesion, thereby promoting metastasis [63, 109, 113]. The discovery that GOLPH3 couples structural remodelling of the Golgi with oncogenic glycosylation changes, positions it as a bridge between trafficking control and tumorigenesis [2, 146].

The complementing recycling adaptor, GRASP55, functions as a retainer that maintains specific enzymes within their resident cisternae. This Golgi stacking protein also mediates cisternal tethering and ribbon formation through regulated oligomerization [5, 147]. GRASP55 directly binds subsets of glycosyltransferases, preventing their incorporation into COPI vesicles and stabilizing them in maturing cisternae [102]. Reduction in GRASP55 levels mislocalizes these enzymes and alters flux across competing glycosylation reactions [102]. Loss of GRASP55 and GRASP65 function also increases vesicle budding and shortens cargo residence time, leading to incomplete glycan synthesis. Consistently, unstacked Golgi caused by GRASP depletion shows reduced glycan diversity and aberrant glycoprotein profiles [51, 143]. Altered GRASP expression and phosphorylation have also been linked to Golgi fragmentation in cancer, contributing to enhanced cell migration, secretion, and drug resistance [23].

Golgi structural proteins such as GM130, Golgins, and GRASPs are not only architectural elements but also active participants in enzyme targeting and disease. Their isoform-specific tethering functions help maintain compartmentalization and enzyme fidelity [58]. GM130 interacts with GRASP65 and p115 to maintain Golgi ribbon continuity and directly mediates localization of C1GalT1, a key enzyme in mucin-type O-glycosylation [99]. Loss of GM130 results in accumulation of Tn and sialyl-Tn antigens, frequently seen in carcinomas [19, 100]. Alterations in Golgi structure, fragmentation, dispersal, or ribbon loss, are among the most consistent features of tumor cells and correlate with increased secretion, anoikis resistance, and metastasis [2, 23].

The recruitment of trafficking proteins such as GOLPH3 and COPI depends on the phosphoinositide composition of Golgi membranes. The lipid PI(4)P, generated by PI4KIIIβ and PI4KIIα, defines Golgi identity and membrane curvature [134]. Amplification of PI4KIIIβ occurs in approximately 4% of cancers, often co-occurring with KRAS mutations in lung adenocarcinomas [123]. Elevated PI(4)P levels enhance GOLPH3 recruitment, thereby promoting malignant glycosylation patterns. Similarly, the phosphatidylinositol transfer protein PITPNC1, which delivers PI to the Golgi, is frequently amplified in breast, gastric, and colorectal cancers [151, 152]. Its overexpression enhances secretion of matrix metalloprotease 9 (MMP9) and platelet-derived growth factor (PDGF), fostering angiogenesis and metastasis [123]. Thus, lipid mediated signalling at the Golgi directly regulates enzyme recycling and tumour-associated glycosylation remodelling. Alterations in upstream signalling pathways like hyperactive Src–Arf1–GBF1 signalling can drive relocation of GALNTs from Golgi to ER [20] resulting in truncated O-glycans such as Tn and sialyl-Tn antigens, that activate EGFR and MMP14 [110], reinforcing oncogenic signalling [42, 93].

Interestingly, Golgi stress responses have recently been recognized as contributors to tumour survival under chemotherapeutic pressure. In colorectal cancer, DNA-damaging agents induce Golgi dispersal and activation of a PKCα/GSK3α/TFE3 signalling axis, which enhances trafficking and increases surface expression of the drug efflux transporter ABCG2 and modulates P-gp/MRP1 levels via ST6GAL1

[71, 75]. This Golgi stress response, requiring GOLPH3, illustrates how organelle remodelling and glycosylation control intersect to promote chemoresistance.

Collectively, these observations establish the Golgi as both a metabolic and signalling hub, where genetic amplification, mutation, or overexpression of trafficking and structural components reprogram glycosylation to favour malignant transformation [121]. Conversely, loss-of-function mutations in the same machinery produce congenital glycosylation disorders. Understanding these dual roles highlights the Golgi's regulatory plasticity and suggests that targeting glycoenzyme trafficking could provide novel therapeutic opportunities in both congenital and neoplastic diseases [145].

Conclusions

The Golgi apparatus functions as a dynamic and polarized organelle whose structural complexity and compartmental specialization are tightly linked to its multiple functional roles. The multimodular cisternal maturation (MCM) model provides a comprehensive and experimentally supported framework for understanding intra-Golgi transport and enzyme compartmentalization. Based on extensive morphological, biochemical, and kinetic evidence, this model proposes that Golgi cisternae progressively mature from cis to trans, while resident enzymes are recycled in a retrograde direction through a complex, multimodular network of trafficking pathways including COPI-dependent and independent routes [46, 47, 114]. The presence of these multiple mechanisms enhances the robustness and regulatory potential of the Golgi, allowing it to adapt to fluctuations in cargo load, metabolic demand, and developmental cues [15, 148]. The coexistence of multiple recycling modules, each with distinct kinetic and spatial properties, underscores the multimodular architecture of cisternal maturation.

This spatial organization and trafficking complexity are fundamental for maintaining the fidelity of protein and lipid glycosylation, a critical post-translational modification that influences protein folding, stability, receptor signalling, and intercellular communication [87, 103]. The polarized distribution of glycosyltransferases, maintained through recycling and retention mechanisms, ensures sequential glycan processing and supports the formation of functional enzyme modules that enhance biosynthetic efficiency. Indeed, perturbations in Golgi trafficking and enzyme localization contribute to a wide spectrum of pathological conditions.

Understanding the multimodular logic of cisternal maturation and enzyme recycling offers new perspectives on the molecular basis of glycosylation and its deregulation in disease. This knowledge not only deepens our understanding of Golgi biology but also informs therapeutic strategies aimed at restoring glycosylation fidelity and cellular function. Targeting specific adaptors, lipid regulators, or structural scaffolds may provide novel avenues for intervention in both congenital and neoplastic diseases, positioning the Golgi as a central regulatory platform in cell physiology and pathology.

Looking ahead, several promising directions emerge for future research. First, the molecular dissection of individual recycling modules, including their cargo specificity, regulatory inputs, and spatial dynamics, will be essential to fully understand how enzyme localization is fine-tuned in different cellular contexts. The understanding that membrane trafficking and glycosylation are interconnected as "Siamese twins" provides a rationale for devising strategies to dissect these modules. By using the regulation of enzyme localization as a probe, molecular machinery that controls membrane trafficking can be dissected. High-resolution live-cell imaging, proximity labelling, and cryo-electron tomography will likely play a pivotal role in mapping these pathways with unprecedented detail.

Second, the integration of Golgi trafficking with cellular signalling networks remains an underexplored frontier. Given the role of glycosylation in modulating receptor activity and immune responses, future studies should investigate how Golgi adaptors and lipid regulators interface with oncogenic and stress-responsive pathways, particularly in cancer and inflammation.

Third, the development of pharmacological tools that selectively modulate Golgi trafficking, such as small molecules targeting GOLPH3, PI4KIIIβ, or GRASP55, could offer new therapeutic strategies to correct glycosylation defects or inhibit tumour-promoting glycan remodelling. These approaches may be especially valuable in cancers characterized by Golgi fragmentation and altered glycan profiles.

Finally, expanding our understanding of Golgi dynamics in non-mammalian systems and specialized cell types, such as neurons, immune cells, and plants, will reveal conserved and divergent principles of glycosylation control. Comparative studies across species and tissues may uncover novel adaptors or retention mechanisms that could be harnessed for biotechnological or therapeutic applications.

In sum, the Golgi apparatus lies at the crossroads of membrane trafficking, enzymatic regulation, and disease biology. Continued investigation into its modular architecture, and into the unique mechanisms by which cytosolic signals regulate the localization and activity of both the glycosylation machinery and of the cargo proteins undergoing glycosylation, promises to provide transformative insights into the principles of cellular organization and the molecular basis of pathophysiology.

References

1. Adolf F, Rhiel M, Hessling B, Gao Q, Hellwig A, Béthune J, Wieland FT (2016) Proteomic profiling of mammalian COPII and COPI vesicles. Cell Rep 15(6):1338–1351
2. Bajaj R, Warner AN, Fradette JF, Gibbons DL (2022) Dance of the Golgi: understanding Golgi dynamics in cancer metastasis. Cells 11(9):1484
3. Barr FA (2009) Membrane traffic: Golgi stumbles over cilia. Curr Biol 19(6):R253–R255
4. Becker B, Bölinger B, Melkonian M (1995) Anterograde transport of algal scales through the Golgi complex is not mediated by vesicles. Trends Cell Biol 5(8):305–307

5. Bekier ME, Wang L, Li J, Huang H, Tang D, Zhang X, Wang Y (2017) Knockout of the Golgi stacking proteins GRASP55 and GRASP65 impairs Golgi structure and function. Mol Biol Cell 28(21):2833–2842
6. Beznoussenko GV, Parashuraman S, Rizzo R, Polishchuk R, Martella O, Di Giandomenico D, Fusella A, Spaar A, Sallese M, Capestrano MG, Pavelka M, Vos MR, Rikers YGM, Helms V, Mironov AA, Luini A (2014) Transport of soluble proteins through the Golgi occurs by diffusion via continuities across cisternae. elife 3:e02009
7. Biswas A, Thattai M (2020) Promiscuity and specificity of eukaryotic glycosyltransferases. Biochem Soc Trans 48(3):891–900
8. Blackburn JB, Kudlyk T, Pokrovskaya I, Lupashin VV (2018) More than just sugars: conserved oligomeric Golgi complex deficiency causes glycosylation-independent cellular defects. Traffic 19(6):463–480
9. Blackburn JB, Pokrovskaya I, Fisher P, Ungar D, Lupashin VV (2016) COG complex complexities: detailed characterization of a complete set of HEK293T cells lacking individual COG subunits. Front Cell Dev Biol 4:23
10. Blackburn JB, D'Souza Z, Lupashin VV (2019) Maintaining order: COG complex controls Golgi trafficking, processing, and sorting. FEBS Lett 593(17):2466–2487
11. Bonfanti L, Mironov AA Jr, Martínez-Menárguez JA, Martella O, Fusella A, Baldassarre M, Buccione R, Geuze HJ, Mironov AA, Luini A (1998) Procollagen traverses the Golgi stack without leaving the lumen of cisternae: evidence for cisternal maturation. Cell 95(7):993–1003
12. Bonifacino JS, Glick BS (2004) The mechanisms of vesicle budding and fusion. Cell 116(2): 153–166
13. Breton C, Šnajdrová L, Jeanneau C, Koča J, Imberty A (2006) Structures and mechanisms of glycosyltransferases. Glycobiology 16(2):29R–37R
14. Bykov YS, Schaffer M, Dodonova SO, Albert S, Plitzko JM, Baumeister W, Engel BD, Briggs JAG (2017) The structure of the COPI coat determined within the cell. elife 6:e32493
15. Casler JC, Johnson N, Krahn AH, Pantazopoulou A, Day KJ, Glick BS (2021) Clathrin adaptors mediate two sequential pathways of intra-Golgi recycling. J Cell Biol 221(1): e202103199
16. Casler JC, Papanikou E, Barrero JJ, Glick BS (2019) Maturation-driven transport and AP-1-dependent recycling of a secretory cargo in the Golgi. J Cell Biol 218(5):1582–1601
17. Cattin-Ortolá J, Kaufman JGG, Gillingham AK, Wagstaff JL, Peak-Chew S-Y, Stevens TJ, Boulanger J, Owen DJ, Munro S (2024) Cargo selective vesicle tethering: the structural basis for binding of specific cargo proteins by the Golgi tether component TBC1D23. Sci Adv 10(13):eadl0608
18. Cerami E, Gao J, Dogrusoz U, Gross BE, Sumer SO, Aksoy BA, Jacobsen A, Byrne CJ, Heuer ML, Larsson E, Antipin Y, Reva B, Goldberg AP, Sander C, Schultz N (2012) The cBio cancer genomics portal: an open platform for exploring multidimensional cancer genomics data. Cancer Discov 2(5):401–404
19. Cheng PW, Davidson S, Bhat G (2020) Markers of malignant prostate cancer cells: Golgi localization of alpha-mannosidase 1A at GM130-GRASP65 site and appearance of high mannose N-glycans on cell surface. Biochem Biophys Res Commun 527(2):406–410
20. Chia J, Wang S-C, Wee S, Gill DJ, Tay F, Kannan S, Verma CS, Gunaratne J, Bard FA (2021) Src activates retrograde membrane traffic through phosphorylation of GBF1. elife 10:e68678
21. Climer LK, Dobretsov M, Lupashin V (2015) Defects in the COG complex and COG-related trafficking regulators affect neuronal Golgi function. Front Neurosci 9:405
22. Clermont Y, Rambourg A, Hermo L (1995) Trans-Golgi network (TGN) of different cell types: three-dimensional structural characteristics and variability. Anat Rec 242(3):289–301
23. Colanzi A, Parashuraman S, Reis CA, Ungar D (2022) Editorial: does the Golgi complex enable oncogenesis? Front Cell Dev Biol 10:1000946
24. Curtis C, Shah SP, Chin S-F, Turashvili G, Rueda OM, Dunning MJ, Speed D, Lynch AG, Samarajiwa S, Yuan Y, Gräf S, Ha G, Haffari G, Bashashati A, Russell R, McKinney S, METABRIC Group, Langerød A, Green A, Provenzano E, Wishart G, Pinder S, Watson P,

Markowetz F et al (2012) The genomic and transcriptomic architecture of 2,000 breast tumours reveals novel subgroups. Nature 486(7403):346–352
25. D'Angelo G, Prencipe L, Iodice L, Beznoussenko G, Savarese M, Marra P, Di Tullio G, Martire G, De Matteis MA, Bonatti S (2009) GRASP65 and GRASP55 sequentially promote the transport of C-terminal valine-bearing cargos to and through the Golgi complex. J Biol Chem 284(50):34849–34860
26. D'Souza Z, Taher FS, Lupashin VV (2020) Golgi inCOGnito: from vesicle tethering to human disease. Biochim Biophys Acta 1864(11):129694
27. Day KJ, Casler JC, Glick BS (2018) Budding yeast has a minimal endomembrane system. Dev Cell 44(1):56–72.e4
28. Dippold HC, Ng MM, Farber-Katz SE, Lee S-K, Kerr ML, Peterman MC, Sim R, Wiharto PA, Galbraith KA, Madhavarapu S, Fuchs GJ, Meerloo T, Farquhar MG, Zhou H, Field SJ (2009) GOLPH3 bridges phosphatidylinositol-4-phosphate and actomyosin to stretch and shape the Golgi to promote budding. Cell 139(2):337–351
29. Dunphy WG, Fries E, Urbani LJ, Rothman JE (1981) Early and late functions associated with the Golgi apparatus reside in distinct compartments. Proc Natl Acad Sci USA 78(12): 8255–8259
30. Dunphy WG, Rothman JE (1983) Compartmentation of asparagine-linked oligosaccharide processing in the Golgi apparatus. J Cell Biol 97(1):270–275
31. Dunphy WG, Rothman JE (1985) Compartmental organization of the Golgi stack. Cell 42(1): 13–21
32. Dunphy WG, Brands R, Rothman JE (1985) Attachment of terminal N acetylglucosamine to asparagine-linked oligosaccharides occurs in central cisternae of the Golgi stack. J Cell Biol 101(6):1881–1891
33. Dunphy WG, Pfeffer SR, Clary DO, Wattenberg BW, Glick BS, Rothman JE (1986) Yeast and mammals utilize similar cytosolic components to drive protein transport through the Golgi complex. Proc Natl Acad Sci USA 83(6):1622–1626
34. Dünser K, Schoberer J (2025) Trafficking and localization of Golgi-resident N-glycan processing enzymes in plants. Front Plant Sci 16:1624949
35. Eckert ESP, Reckmann I, Hellwig A, Röhling S, El-Battari A, Wieland FT, Popoff V (2014) Golgi phosphoprotein 3 triggers signal-mediated incorporation of glycosyltransferases into coatomer-coated (COPI) vesicles. J Biol Chem 289(45):31319–31329
36. Emr S, Glick BS, Linstedt AD, Lippincott-Schwartz J, Luini A, Malhotra V, Marsh BJ, Nakano A, Pfeffer SR, Rabouille C, Rothman JE, Warren G, Wieland FT (2009) Journeys through the Golgi—taking stock in a new era. J Cell Biol 187(4):449–453
37. Farquhar MG, Palade GE (1998) The Golgi apparatus: 100 years of progress and controversy. Trends Cell Biol 8(1):2–10
38. Fisher P, Thomas-Oates J, Wood AJ, Ungar D (2019) The N-glycosylation processing potential of the mammalian Golgi apparatus. Front Cell Dev Biol 7:157
39. Fossati M, Colombo SF, Borgese N (2014) A positive signal prevents secretory membrane cargo from recycling between the Golgi and the ER. EMBO J 33(18):2080–2097
40. Foulquier F, Vasile E, Schollen E, Callewaert N, Raemaekers T, Quelhas D, Jaeken J, Mills P, Winchester B, Krieger M, Annaert W, Matthijs G (2006) Conserved oligomeric Golgi complex subunit 1 deficiency reveals a previously uncharacterized congenital disorder of glycosylation type II. Proc Natl Acad Sci USA 103(10):3764–3769
41. Foulquier F (2009) COG defects, birth and rise! Biochim Biophys Acta Mol basis Dis 1792(9): 896–902
42. Freitas D, Campos D, Gomes J, Pinto F, Macedo JA, Matos R et al (2019) O-glycans truncation modulates gastric cancer cell signaling and transcription leading to a more aggressive phenotype. EBioMedicine 40:349–362
43. Freeze HH (2007) Genetic defects in the human glycome. Nat Rev Genet 8:537–551

44. Gao J, Cong C, Xiang Y, Wu Y, Yan B, Jia Q, Yao Z-P, Guo Y, Hu J (2025) A class of COPI adaptors regulates processing of transmembrane receptors by reinforcing their Golgi retention. bioRxiv. Preprint
45. Giraudo CG, Daniotti JL, Maccioni HJF (2001) Physical and functional association of glycolipid N-acetyl-galactosaminyl and galactosyl transferases in the Golgi apparatus. Glycobiology 11(7):539–548
46. Glick BS, Nakano A (2009a) Membrane traffic. Two faces of the Golgi. Cell 137(1):16–22
47. Glick BS, Nakano A (2009b) Membrane traffic within the Golgi apparatus. Annu Rev Cell Dev Biol 25:113–132
48. Glick BS, Elston T, Oster G (1997) A cisternal maturation mechanism can explain the asymmetry of the Golgi stack. FEBS Lett 414:177–181
49. Granovsky M, Fata J, Pawling J, Muller WJ, Khokha R, Dennis JW (2000) Suppression of tumor growth and metastasis in Mgat5-deficient mice. Nat Med 6:306–312
50. Grassé PP, Carasso N (1957) Ultra-structure of the Golgi apparatus in Protozoa and Metazoa (somatic and germinal cells). J Biophys Biochem Cytol 3(1):7–24
51. Grond R, Veenendaal T, Duran JM, Raote I, Van Es JH, Corstjens S et al (2020) The function of GORASPs in Golgi apparatus organization in vivo. J Cell Biol 219(6):e202004191
52. Grunewald S, Matthijs G, Jaeken J (2002) Congenital disorders of glycosylation: a review. Pediatr Res 52(5):618–624
53. Haijes HA, Jaeken J, Foulquier F, van Hasselt PM (2018) Hypothesis: lobe A COG1–4-CDG causes a more severe phenotype than lobe B COG5–8-CDG. J Med Genet 55(3):137–142
54. Harada A, Kunii M, Kurokawa K, Sumi T, Kanda S, Zhang Y, Nadanaka S, Hirosawa KM, Tokunaga K, Tojima T, Taniguchi M, Moriwaki K, Yoshimura S, Yamamoto-Hino M, Goto S, Katagiri T, Kume S, Hayashi-Nishino M, Nakano M, Miyoshi E, Suzuki KGN, Kitagawa H, Nakano A (2024) Dynamic movement of the Golgi unit and its glycosylation enzyme zones. Nat Commun 15:5678
55. Ishii M, Suda Y, Kurokawa K, Nakano A (2016) COPI is essential for Golgi cisternal maturation and dynamics. J Cell Sci 129(17):3251–3261
56. Jackson CL, Bouvet S (2014) Arfs at a glance. J Cell Sci 127(Pt 19):4103–4109
57. Jaiman A, Thattai M (2020) Golgi compartments enable controlled biomolecular assembly using promiscuous enzymes. elife 9:e49573
58. Jarvela T, Linstedt AD (2014) Isoform-specific tethering links the Golgi ribbon to maintain compartmentalization. Mol Biol Cell 25(1):133–144
59. Ji Y, Wei S, Hou J, Zhang C, Xue P, Wang J et al (2017) Integrated proteomic and N-glycoproteomic analyses of doxorubicin sensitive and resistant ovarian cancer cells reveal glycoprotein alteration in protein abundance and glycosylation. Oncotarget 8:13413–13427
60. Jungmann J, Munro S (1998) Multi-protein complexes in the cis Golgi of Saccharomyces cerevisiae with alpha-1,6-mannosyltransferase activity. EMBO J 17(2):423–434
61. Khoder-Agha F, Harrus D, Brysbaert G, Lensink MF, Harduin-Lepers A, Glumoff T, Kellokumpu S (2019) Assembly of B4GALT1/ST6GAL1 heteromers in the Golgi membranes involves lateral interactions via highly charged surface domains. J Biol Chem 294(39): 14383–14393
62. Klumperman J (2011) Architecture of the mammalian Golgi. Cold Spring Harb Perspect Biol 3(7):a005181
63. Kovbasnjuk O, Mourtazina R, Baibakov B, Wang T, Elowsky C, Choti MA et al (2005) The glycosphingolipid globotriaosylceramide in the metastatic transformation of colon cancer. Proc Natl Acad Sci USA 102(52):19087–19092
64. Krahn AH, Pantazopoulou A, Austin J II, Johnson N, Lee-Smith CF, Glick BS (2025) COPI-dependent intra-Golgi recycling at an intermediate stage of cisternal maturation. bioRxiv. Preprint
65. Kurokawa K, Osakada H, Kojidani T, Waga M, Suda Y, Asakawa H, Haraguchi T, Nakano A (2019) Visualization of secretory cargo transport within the Golgi apparatus. J Cell Biol 218(5):1602–1618

66. Ladinsky MS, Kremer JR, Furcinitti PS, McIntosh JR, Howell KE (1994) HVEM tomography of the trans-Golgi network: structural insights and identification of a lace-like vesicle coat. J Cell Biol 127(1):29–38
67. Ladinsky MS, Mastronarde DN, McIntosh JR, Howell KE, Staehelin LA (1999) Golgi structure in three dimensions: functional insights from the normal rat kidney cell. J Cell Biol 144(6):1135–1149
68. Ladinsky MS, Wu CC, McIntosh S, McIntosh JR, Howell KE (2002) Structure of the Golgi and distribution of reporter molecules at 20°C reveals the complexity of the exit compartments. Mol Biol Cell 13(8):2810–2825
69. Lau KS, Partridge EA, Grigorian A, Silvescu CI, Reinhold VN, Demetriou M, Dennis JW (2007) Complex N-glycan number and degree of branching cooperate to regulate cell proliferation and differentiation. Cell 129(1):123–134
70. Lesniak AM, Ye Z, Banfield DK (2025) Identification of the client-binding site on the Golgi membrane protein adaptor Vps74/yGOLPH3. iScience 28(10):113494
71. Li Y, Mu L, Li Y, Mi Y, Hu Y, Li X, Tao D, Qin J (2024) Golgi dispersal in cancer stem cells promotes chemoresistance of colorectal cancer via the Golgi stress response. Cell Death Dis 15:417
72. Liu L, Doray B, Kornfeld S (2018) Recycling of Golgi glycosyltransferases requires direct binding to coatomer. Proc Natl Acad Sci USA 115(36):8984–8989. https://doi.org/10.1073/pnas.1810291115
73. Losev E, Reinke CA, Jellen J, Strongin DE, Bevis BJ, Glick BS (2006) Golgi maturation visualized in living yeast. Nature 441(7096):1002–1006
74. Lu Z, Joseph D, Bugnard E, Zaal KJ, Ralston E (2001) Golgi complex reorganization during muscle differentiation: visualization in living cells and mechanism. Mol Biol Cell 12(4):795–808
75. Ma H, Cheng L, Hao K, Li Y, Song X, Zhou H, Jia L (2014) Reversal effect of ST6GAL1 on multidrug resistance in human leukemia by regulating the PI3K/Akt pathway and the expression of P-gp and MRP1. PLoS One 9(1):e85113
76. Maccioni HJF (2007) Glycosylation of glycolipids in the Golgi complex. J Neurochem 103(4):1–15
77. Maccioni HJF, Quiroga R, Ferrari ML (2004) Cellular and molecular biology of glycosphingolipid glycosylation. Biochim Biophys Acta 1680(1):1–17
78. Marques-da-Silva D, Dos Reis Ferreira V, Monticelli M, Janeiro P, Videira PA, Witters P et al (2017) Liver involvement in congenital disorders of glycosylation (CDG). J Inherit Metab Dis 40(2):195–207
79. Marsh BJ, Mastronarde DN, Buttle KF, Howell KE, McIntosh JR (2001) Organellar relationships in the Golgi region of the pancreatic beta cell line, HIT-T15, visualized by high resolution electron tomography. Proc Natl Acad Sci USA 98(5):2399–2406
80. Marsh BJ, Volkmann N, McIntosh JR, Howell KE (2004) Direct continuities between cisternae at different levels of the Golgi complex in glucose-stimulated mouse islet beta cells. Proc Natl Acad Sci USA 101(15):5565–5570
81. Matsuura-Tokita K, Takeuchi M, Ichihara A, Mikuriya K, Nakano A (2006) Live imaging of yeast Golgi cisternal maturation. Nature 441(7096):1007–1010
82. Mironov AA, Weidman P, Luini A (1997) Variations on the intracellular transport theme: maturing cisternae and trafficking tubules. J Cell Biol 138(3):481–484
83. Moelleken J, Malsam J, Betts MJ, Movafeghi A, Reckmann I, Meissner I, Hellwig A, Russell RB, Söllner T, Brügger B, Wieland FT (2007) Differential localization of coatomer complex isoforms within the Golgi apparatus. Proc Natl Acad Sci USA 104(11):4425–4430
84. Mogelsvang S, Marsh BJ, Ladinsky MS, Howell KE (2004) Predicting function from structure: 3D structure studies of the mammalian Golgi complex. Traffic 5(5):338–345
85. Mollenhauer HH, Morré DJ (1966) Golgi apparatus and plant secretion. Annu Rev Plant Physiol 17:1–24

86. Morelle W, Potelle S, Witters P, Wong S, Climer L, Lupashin V, Matthijs G, Gadomski T, Jaeken J, Cassiman D, Morava E, Foulquier F (2017) Galactose supplementation in patients with TMEM165-CDG rescues the glycosylation defects. J Clin Endocrinol Metabol 102(4): 1375–1386
87. Moremen KW, Tiemeyer M, Nairn AV (2012) Vertebrate protein glycosylation: diversity, synthesis and function. Nat Rev Mol Cell Biol 13(7):448–462
88. Morré DJ, Franke WW, Deumling B, Nyquist SE, Ovtracht L (1974) Golgi apparatus function in membrane flow and differentiation: origin of plasma membrane from endoplasmic reticulum. In: Biomembranes. Springer, pp 95–104
89. Munkley J, Elliott DJ (2016) Hallmarks of glycosylation in cancer. Oncotarget 7:35478–35489
90. Munro S (1995a) A comparison of the transmembrane domains of Golgi and plasma membrane proteins. Biochem Soc Trans 23(3):527–530
91. Munro S (1995b) An investigation of the role of transmembrane domains in Golgi protein retention. EMBO J 14(19):4695–4704
92. Narimatsu Y, Joshi HJ, Nason R, Van Coillie J, Karlsson R, Sun L et al (2019) An atlas of human glycosylation pathways enables display of the human glycome by gene-engineered cells. Mol Cell 75(2):394–407
93. Nguyen AT, Chia J, Ros M, Hui KM, Saltel F, Bard F (2017) Organelle-specific O-glycosylation drives MMP14 activation, tumor growth, and metastasis. Cancer Cell 32(5):639–653.e6
94. Orci L, Stamnes M, Ravazzola M, Amherdt M, Perrelet A, Söllner TH, Rothman JE (1997) Bidirectional transport by distinct populations of COPI-coated vesicles. Cell 90(2):335–349
95. Pantazopoulou A, Glick BS (2019) A kinetic view of membrane traffic pathways can transcend the classical view of Golgi compartments. Front Cell Dev Biol 7:153
96. Papanikou E, Glick BS (2014) Golgi compartmentation and identity. Curr Opin Cell Biol 29: 74–81
97. Papanikou E, Day KJ, Austin J II, Glick BS (2015) COPI selectively drives maturation of the early Golgi. elife 4:e13232
98. Park JJ, Gondré-Lewis MC, Eiden LE, Loh YP (2011) A distinct trans-Golgi network subcompartment for sorting of synaptic and granule proteins in neurons and neuroendocrine cells. J Cell Sci 124(Pt 5):735–744
99. Petrosyan A, Ali MF, Cheng PW (2012) Glycosyltransferase-specific Golgi-targeting mechanisms. J Biol Chem 287(45):37621–37627
100. Petrosyan A, Holzapfel MS, Muirhead DE, Cheng PW (2014) Restoration of compact Golgi morphology in advanced prostate cancer enhances susceptibility to galectin-1-induced apoptosis by modifying mucin O-glycan synthesis. Mol Cancer Res 12(12):1704–1716
101. Pirozzi M, Agliarulo I, Di Martino R, Marzullo VM, Quarto M, Turacchio G, Beznoussenko GV, Russo D, Luini A, Mironov AA, Parashuraman S (2025) Cargoes move from cis- to trans-Golgi compartments and concentrate in the TGN before exiting. EMBO Rep 26:19
102. Pothukuchi P, Agliarulo I, Pirozzi M, Rizzo R, Russo D, Turacchio G, Nüchel J, Yang J-S, Gehin C, Capolupo L, Hernandez-Corbacho MJ, Biswas A, Vanacore G, Dathan N, Nitta T, Henklein P, Thattai M, Inokuchi J-I, Hsu VW, Plomann M, Obeid LM, Hannun YA, Luini A, D'Angelo G, Parashuraman S (2021) GRASP55 regulates intra-Golgi localization of glycosylation enzymes to control glycosphingolipid biosynthesis. EMBO J 40:e107766
103. Pothukuchi P, Agliarulo I, Russo D, Rizzo R, Russo F, Parashuraman S (2019) Translation of genome to glycome: role of the Golgi apparatus. FEBS Lett 593(17):2390–2411
104. Rabouille C, Hui N, Hunte F, Kieckbusch R, Berger EG, Warren G, Nilsson T (1995) Mapping the distribution of Golgi enzymes involved in the construction of complex oligosaccharides. J Cell Sci 108(Pt 4):1617–1627
105. Rahajeng J, Kuna RS, Makowski SL, Tran TTT, Buschman MD, Li S, Cheng N, Ng MM, Field SJ (2019) Efficient Golgi forward trafficking requires GOLPH3-driven, PI4P-dependent membrane curvature. Mol Biol Cell 30(7):792–805

106. Rambourg A, Clermont Y, Hermo L (1981) Three-dimensional structure of the Golgi apparatus. Methods Cell Biol 23:155–166
107. Rink J, Ghigo E, Kalaidzidis Y, Zerial M (2005) Rab conversion as a mechanism of progression from early to late endosomes. Cell 122(5):735–749
108. Rizzo R, Parashuraman S, Mirabelli P, Puri C, Lucocq J, Luini A (2013) The dynamics of engineered resident proteins in the mammalian Golgi complex relies on cisternal maturation. J Cell Biol 201(7):1027–1036
109. Rizzo R, Russo D, Kurokawa K, Sahu P, Lombardi B, Supino D et al (2021) Golgi maturation-dependent glycoenzyme recycling controls glycosphingolipid biosynthesis and cell growth via GOLPH3. EMBO J 40(8):e107238
110. Ros M, Nguyen AT, Chia J, Le Tran S, Le Guezennec X, McDowall R et al (2020) ER-resident oxidoreductases are glycosylated and trafficked to the cell surface to promote matrix degradation by tumour cells. Nat Cell Biol 22(11):1371–1381
111. Roth J, Berger EG (1982) Immunocytochemical localization of galactosyltransferase in HeLa cells: codistribution with thiamine pyrophosphatase in trans-Golgi cisternae. J Cell Biol 93(1): 223–229
112. Rothman JE (1996) The protein machinery of vesicle budding and fusion. Protein Sci 5(2): 185–194
113. Russo D, Parashuraman S, D'Angelo G (2016) Glycosphingolipid-protein interaction in signal transduction. Int J Mol Sci 17(11):1732
114. Sahu P, Balakrishnan A, Di Martino R, Luini A, Russo D (2022) Role of the mosaic cisternal maturation machinery in glycan synthesis and oncogenesis. Front Cell Dev Biol 10:842448
115. Sardana R, Highland CM, Straight BE, Chavez CF, Fromme JC, Emr SD (2021) Golgi membrane protein Erd1 is essential for recycling a subset of Golgi glycosyltransferases. elife 10:e70774
116. Schjoldager KT, Narimatsu Y, Joshi HJ, Clausen H (2020) Global view of human protein glycosylation pathways and functions. Nat Rev Mol Cell Biol 21:729–749
117. Schueder F, Rivera-Molina F, Su M, Marin Z, Kidd P, Rothman JE, Toomre D, Bewersdorf J (2024) Unraveling cellular complexity with transient adapters in highly multiplexed super-resolution imaging. Cell 187(7):1769–1784.e18
118. Sechi S, Frappaolo A, Karimpour-Ghahnavieh A, Piergentili R, Giansanti MG (2020) Oncogenic roles of GOLPH3 in the physiopathology of cancer. Int J Mol Sci 21(3):933
119. Spano D, Colanzi A (2022) Golgi complex: a signaling hub in cancer. Cells 11(13):1990
120. Scott KL, Kabbarah O, Liang M-C, Ivanova E, Anagnostou V, Wu J, Dhakal S, Wu M, Chen S, Feinberg T, Huang J, Saci A, Widlund HR, Fisher DE, Xiao Y, Rimm DL, Protopopov A, Wong K-K, Chin L (2009) GOLPH3 modulates mTOR signalling and rapamycin sensitivity in cancer. Nature 459(7250):1085–1090
121. Seales EC, Jurado G, Singhal A, Bellis SL (2005) Overexpression of alpha2,6-sialyltransferase in human carcinoma cells enhances tumor cell motility. J Biol Chem 280(32):28369–28377
122. Sun X, Mahajan D, Chen B, Song Z, Lu L (2021) A quantitative study of the Golgi retention of glycosyltransferases. J Cell Sci 134(20):jcs258564
123. Tan X, Banerjee P, Pham EA, Rutaganira FUN, Basu K, Bota-Rabassedas N, Guo H-F, Grzeskowiak CL, Liu X, Yu J, Shi L, Peng DH, Rodriguez BL, Zhang J, Zheng V, Duose DY, Solis LM, Mino B, Raso MG, Behrens C et al (2020) PI4KIIIβ is a therapeutic target in chromosome 1q-amplified lung adenocarcinoma. Sci Transl Med 12(527):eaax3772
124. Thomas LL, Fromme JC (2020) Extensive GTPase crosstalk regulates Golgi trafficking and maturation. Curr Opin Cell Biol 65:1–7
125. Tie HC, Ludwig A, Sandin S, Lu L (2018) The spatial separation of processing and transport functions to the interior and periphery of the Golgi stack. elife 7:e41301
126. Tie HC, Wang H, Mahajan D, Lam HYI, Sun X, Chen B, Mu Y, Lu L (2025) Quantitative intra-Golgi transport and organization data suggest the stable compartment nature of the Golgi. elife 14:e98582

127. Trucco A, Polishchuk RS, Martella O, Di Pentima A, Fusella A, Di Giandomenico D, San Pietro E, Beznoussenko GV, Polishchuk EV, Baldassarre M, Buccione R, Geerts WJC, Koster AJ, Burger KNJ, Mironov AA, Luini A (2004) Secretory traffic triggers the formation of tubular continuities across Golgi sub-compartments. Nat Cell Biol 6(11):1071–1081
128. Tu L, Banfield DK (2010) Localization of Golgi-resident glycosyltransferases. Cell Mol Life Sci 67(1):29–41
129. Tu L, Tai WCS, Chen L, Banfield DK (2008) Signal-mediated dynamic retention of glycosyltransferases in the Golgi. Science 321(5887):404–407
130. Ungar D, Oka T, Krieger M, Hughson FM (2006) Retrograde transport on the COG railway. Trends Cell Biol 16(2):113–120
131. Volchuk A, Ravazzola M, Perrelet A, Eng WS, Di Liberto M, Varlamov O, Fukasawa M, Engel T, Söllner TH, Rothman JE, Orci L (2004) Countercurrent distribution of two distinct SNARE complexes mediating transport within the Golgi stack. Mol Biol Cell 15(4): 1506–1518. https://doi.org/10.1091/mbc.e03-08-0625. Epub 2004 Jan 23
132. Varki A (2017) Biological roles of glycans. Glycobiology 27(1):3–49
133. Vajaria BN, Patel PS (2017) Glycosylation: a hallmark of cancer? Glycoconj J 34(2):147–156
134. Waugh MG (2019) Phosphatidylinositol 4-kinases, phosphatidylinositol 4-phosphates and the Golgi apparatus: building a phosphoinositide toolkit. Adv Biol Regul 72:51–61
135. Welch LG, Peak-Chew S-Y, Begum F, Stevens TJ, Munro S (2021) GOLPH3 and GOLPH3L are broad-spectrum COPI adaptors for sorting into intra-Golgi transport vesicles. J Cell Biol 220(10):e202106115
136. Welch DR, Wang Y, Glick BS (2024) FAM114A proteins are transmembrane COPI adaptors with cargo specificity. J Cell Biol 223(4):e202306064
137. Willett R, Blackburn JB, Climer L, Pokrovskaya I, Kudlyk T, Wang W, Lupashin V (2016) COG lobe B sub-complex engages v-SNARE GS15 and functions via regulated interaction with lobe A sub-complex. Sci Rep 6:29139
138. Witkos TM, Chan WL, Joensuu M, Rhiel M, Pallister E, Thomas-Oates J, Mould AP, Mironov AA, Biot C, Guerardel Y, Morelle W, Ungar D, Wieland FT, Jokitalo E, Tassabehji M, Kornak U, Lowe M (2019) GORAB scaffolds COPI at the trans-Golgi for efficient enzyme recycling and correct protein glycosylation. Nat Commun 10:127
139. Wong M, Munro S (2014) Membrane trafficking: the specificity of vesicle traffic to the Golgi is encoded in the golgin coiled-coil proteins. Science 346(6209):1256898
140. Wood CS, Hung C-S, Huoh Y-S, Mousley CJ, Stefan CJ, Bankaitis V, Ferguson KM, Burd CG (2012) Local control of phosphatidylinositol 4-phosphate signaling in the Golgi apparatus by Vps74 and Sac1 phosphoinositide phosphatase. Mol Biol Cell 23(13):2527–2536
141. Wood CS, Schmitz KR, Bessman NJ, Gangi Setty T, Ferguson KM, Burd CG (2009) PtdIns4P recognition by Vps74/GOLPH3 links PtdIns 4-kinase signaling to retrograde Golgi trafficking. J Cell Biol 187(7):967–975
142. Woo CH, Gao C, Yu P, Tu L, Meng Z, Banfield DK, Yao X, Jiang L (2015) Conserved function of the lysine-based KXD/E motif in Golgi retention for endomembrane proteins among different organisms. Mol Biol Cell 26(23):4280–4293
143. Xiang Y, Zhang X, Nix DB, Katoh T, Aoki K, Tiemeyer M, Wang Y (2013) Regulation of protein glycosylation and sorting by the Golgi matrix proteins GRASP55/65. Nat Commun 4: 1659
144. Yamaji T, Sekizuka T, Tachida Y, Sakuma C, Morimoto K, Kuroda M, Hanada K (2019) A CRISPR screen identifies LAPTM4A and TM9SF proteins as glycolipid-regulating factors. iScience 11:409–424
145. Ye Z, Liu J, Zhang H (2019) The plasticity of Golgi apparatus and its implications in diseases. Cell Biol Int 43(11):1315–1322
146. Zhang X (2021) Alterations of Golgi structural proteins and glycosylation defects in cancer. Front Cell Dev Biol 9:665289
147. Zhang Y, Seemann J (2021) Rapid degradation of GRASP55 and GRASP65 reveals their immediate impact on the Golgi structure. J Cell Biol 220(1):e202007052

148. Zhao S-B, Suda Y, Nakanishi H, Wang N, Yoko-O T, Gao X-D, Fujita M (2019) Yeast Dop1 is required for glycosyltransferase retrieval from the trans-Golgi network. Biochim Biophys Acta Gen Subj 1863(6):1147–1157
149. Zeevaert R, Foulquier F, Jaeken J, Matthijs G (2008) Deficiencies in subunits of the conserved oligomeric Golgi COG complex define a novel group of congenital disorders of glycosylation. Mol Genet Metab 93(1):15–21
150. Zolov SN, Lupashin VV (2005) Cog3p depletion blocks vesicle-mediated Golgi retrograde trafficking in HeLa cells. J Cell Biol 168(5):747–759
151. Halberg N, Sengelaub CA, Navrazhina K, Molina H, Uryu K, Tavazoie SF (2016) PITPNC1 recruits RAB1B to the Golgi network to drive malignant secretion. Cancer Cell 29 (3):339–353. https://doi.org/10.1016/j.ccell.2016.02.013
152. Ashlin TG, Blunsom NJ, Cockcroft S (2021) Courier service for phosphatidylinositol: PITPs deliver on demand. Biochim Biophys Acta Mol Cell Biol Lipids 1866(9):158985. https://doi.org/10.1016/j.bbalip.2021.158985

Chapter 11
Regulation of Glycan Modification by Glycosyltransferases in the Golgi Apparatus

Hirokazu Yagi, Seigo Tateo, and Koichi Kato

Abstract Protein glycosylation, a fundamental modification critical for diverse biological functions, is well orchestrated within the Golgi apparatus, a central organelle in the secretory pathway. This review explores the mechanisms by which the Golgi governs protein-specific glycan modification, emphasizing the spatial organization of glycosyltransferases within the Golgi and the regulated transport of glycoproteins through its compartments. Specificity arises from the strategic localization of glycosyltransferases within the Golgi, enabling them to recognize both glycan structures and protein domains on their substrates. Advanced imaging reveals a complex and dynamic organization within the Golgi, challenging traditional models and highlighting specialized zones. Regulated transport of glycoproteins through the Golgi is critical for controlling their access to modifying enzymes. Specific sequence motifs within glycoproteins can act as 'passport sequence', directing them to distinct Golgi regions and influencing glycosylation by altering their proximity to specific enzymes and accessory proteins. These mechanisms support the hypothesis that sequence-dependent signals within substrate proteins influence their trafficking, enabling precise glycosylation as they transit the Golgi. Understanding these processes provides new insights into how glycan diversity is generated and regulated within the Golgi, highlighting the interplay between enzyme localization, protein trafficking, and the dynamic organization of this essential organelle.

Keywords Glycosylation · Golgi apparatus · Glycosyltransferase · Protein-specific glycan modification · Substrate recognition · Glycoprotein

---

H. Yagi (✉) · S. Tateo · K. Kato
Graduate School of Pharmaceutical Sciences, Nagoya City University, Nagoya, Japan

Exploratory Research Center on Life and Living Systems (ExCELLS), National Institutes of Natural Sciences, Okazaki, Japan
e-mail: hyagi@phar.nagoya-cu.ac.jp

A. Nakano, J. Saraste (eds.), *The Golgi Network, Volume II*, Subcellular Biochemistry 111, https://doi.org/10.1007/978-3-032-16833-7_11

Introduction

Glycosylation, the modification of proteins with glycans, is a prevalent phenomenon in nature, with glycoproteins playing indispensable roles in biological processes such as tissue formation and immune defense [4]. Glycan biosynthesis is shaped through the sequential addition and removal of sugar residues mediated by glycosyltransferases and glycosidases. This process primarily occurs within the Golgi apparatus, where distinct sets of glycosyltransferases localized in specific compartments contribute to the formation of diverse glycan structures. Additionally, the supply of nucleotide sugars, which serve as donor substrates for glycosyltransferases, are known to regulate glycan expression patterns. Glycan modification in the Golgi apparatus is a highly regulated process and represents a critical factor contributing to glycan structural diversity [4, 37]. Unlike proteins, glycan sequences are not directly encoded by the genome, making their structures difficult to predict and characterized by significant heterogeneity. However, accumulating evidence suggests that apparently stochastic glycan modification processes in the Golgi are governed by certain mechanisms that control glycan structural diversity. For instance, the *N*-glycans modifying the Fc region of immunoglobulin G (IgG), a representative serum glycoprotein, exhibit terminal structural variations depending on physiological and pathological conditions, yet consistently maintain a biantennary complex-type structure [53]. Furthermore, matriglycans involved in cell-cell adhesion are expressed exclusively at specific sites on a particular protein, α-dystroglycan [16, 56]. Our previous studies have elucidated differences in glycan expression patterns before and after neural stem cell differentiation, suggesting that these differences actively contribute to cellular differentiation processes and the maintenance of stemness [46, 51, 52]. We identified that specific N-glycans containing Lewis X [Galβ1-4(Fucα1-3)GlcNAc] and HNK-1 [HSO_3-3GlcAβ1-3Galβ1-4GlcNAc-] are expressed in undifferentiated neural stem cells and are modified exclusively on specific proteins (Fig. 11.1) [46, 51, 52]. Thus, these examples imply that glycan modifications, not directly controlled by the genome, occur in a protein-specific manner, as if following a blueprint, mediated by molecular mechanisms within the Golgi apparatus. However, the systematic understanding of the mechanisms underlying such regulation remains incomplete.

In this article, we aim to discuss the mechanisms of protein-specific glycan modification, focusing on the Golgi apparatus, including findings from our previous research.

Glycosyltransferases

Glycan modification in the Golgi apparatus is closely regulated, largely due to the activity of glycosyltransferases, which ensure the sequential addition of sugar residues onto proteins and lipids. According to the UniProt and GlyCosmos Portal

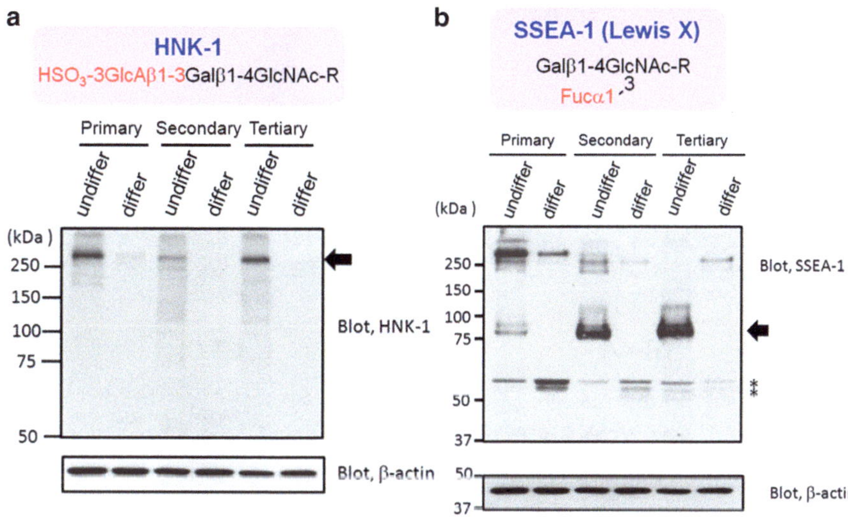

Fig. 11.1 Specific glycotopes expressed on the specific proteins in neural stem cells. The lysates prepared from primary, secondary, and tertiary neurospheres and cells differentiated from primary, secondary and tertiary neurospheres were analyzed by Western blotting with (**a**) anti-HNK-1 antibody and (**b**) anti-LewisX antibody or anti-β-actin antibody (Blot). (This figure was adapted from the literature with permission [51, 52])

databases [41, 54], there are 215 types of glycosyltransferases in humans, most of which localize to the endoplasmic reticulum (ER) and Golgi apparatus, participating in the synthesis of complex carbohydrates [47]. The catalytic domains of these enzymes are classified into three structural classes: GT-A, GT-B, and GT-C (Fig. 11.2). Among them, eight glycosyltransferases possess two catalytic domains (GT-AA, GT-AB). GT-C enzymes are multipass membrane glycosyltransferases with their active sites located within the transmembrane domains. Most glycosyltransferases, except for GT-C, are type II membrane proteins composed of a catalytic domain, a stem region, a transmembrane domain, and a cytoplasmic region. Some glycosyltransferases also contain additional domains adjacent to the catalytic domain (add-on domains), which are involved in substrate recognition and complex formation [47]. The specificity of glycosyltransferases in recognizing their substrates plays a crucial role in protein-specific glycan modification, a process that will be explored in the following sections.

Protein-Specific Glycan Modification Through Direct Interaction Between Substrate Glycoproteins and Glycosyltransferases

Certain glycosyltransferases determine specific glycan structures by recognizing both sugar moieties and protein substrates. To illustrate this principle with a familiar example, consider protein phosphorylation: specific kinases recognize the amino

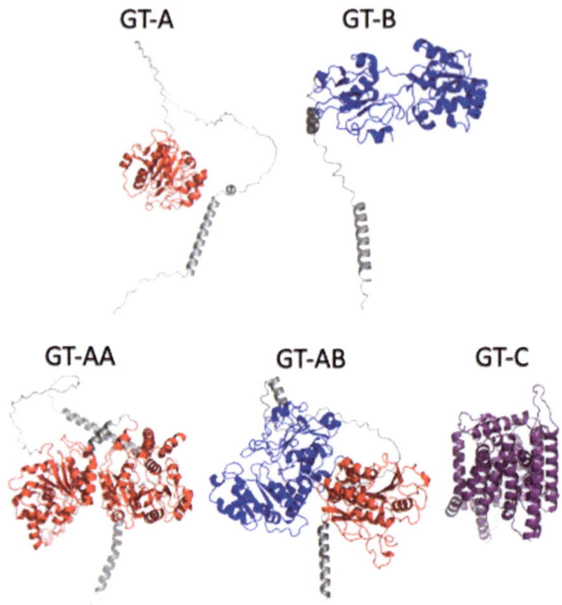

Fig. 11.2 Protein folds of catalytic domains of human glycosyltransferases. This figure illustrates five typical folds: GT-A, GT-B, GT-AA, GT-AB, and GT-C. Each structure reflects a unique arrangement of catalytic domains, with GT-A and GT-B featuring a single catalytic domain, GT-AA and GT-AB comprising two catalytic domains, and GT-C incorporating a membrane-integral domain. The 3D structural models presented here are sourced from the AlphaFold Protein Structure Database, representing prominent glycosyltransferases: (GT-A) B4GALT1, (GT-B) FUT9, (GT-AA) LARGE1, (GT-AB) EXT1, and (GT-C) ALG10. Color; N-terminal CTS region, gray; GT-A catalytic domain, red; GT-B catalytic domain, blue; GT-C membrane-integral catalytic domain, purple. (This figure was adapted from the literature with permission [47])

acid sequences within substrate proteins, leading to protein-specific phosphorylation [13]. This highlights how enzyme-substrate recognition can drive specificity. In the context of glycosylation, certain glycosyltransferases directly attach single sugar residues (such as glucose (Glc), fucose (Fuc), or *N*-acetylgalactosamine (GalNAc)) to proteins, exhibiting sequence specificity through recognition of protein regions [5, 11, 20]. It is conceivable that glycosyltransferases involved in glycan chain elongation—those that mediate the transfer of monosaccharides to existing sugar chains–may also recognize not only the glycan portions but also the protein regions of substrate glycoproteins, thereby exhibiting protein specificity. Indeed, enzymes involved in the biosynthesis of matriglycans, and HNK-1 and LewisX containing glycans are believed to recognize both the glycan and protein portions of substrate glycoproteins, facilitating specific glycan modifications [46, 51, 52].

Matriglycan

Deficiencies in matriglycans, which are involved in cell-cell adhesion, are known to cause muscular dystrophy. Over the past decade, glycosyltransferases involved in the synthesis of matriglycans have been identified, and in 2016, their overall structure was elucidated [14]. This structure comprises a phosphorylated trisaccharide (ManP-GlcNAc-GalNAc) linked via phosphodiester bonds to tandemly connected ribitol phosphate (RboP), followed by a repeating sequence of xylose (Xyl) and glucuronic acid (GlcA) at the non-reducing end, catalyzed by the bifunctional enzyme LARGE. Interestingly, this glycan is exclusively expressed on α-dystroglycan (α-DG), specifically at Thr317/Thr319 sites [9]. We identified that POMGNT2, an enzyme responsible for synthesizing the GlcNAc moiety of the phosphorylated trisaccharide structure, modifies mannose (Man) residues at the specific positions on α-DG with *N*-acetylglucosamine (GlcNAc) [45]. This indicates that POMGNT2 is a critical enzyme determining the sites of matriglycan formation on α-DG. Indeed, X-ray crystallography has shown that POMGNT2 recognizes peptides proximal to Thr317/Thr319 [12]. Furthermore, even LARGE, which is responsible for the final extension of the outer glycan structure, 'reaches back' to directly interact with the N-terminal domain of α-DG, highlighting the crucial role of protein-enzyme interactions in ensuring specificity [15].

Thus, matriglycans undergo finely tuned control through the recognition of substrate proteins by multiple enzymes, leading to glycan modifications at specific amino acid residues on the specific substrates. In similar way, the sialyltransferase (ST8SiaV), which is involved in polysialic acid synthesis, directly interacted with carrier proteins (NCAM and NRP-2), thereby acquiring specific protein glycosylation [1, 7, 23].

HNK-1 Glycans

HNK-1 glycans are characterized by a unique structure where sulfate is attached to the 3-position of glucuronic acid (GlcA). They are expressed on both *N*- and *O*-linked glycans [28, 35]. Glycosyltransferases involved in HNK-1 modification include GlcA transferases (GlcAT-P, GlcAT-S) and a sulfotransferase (HNK-1ST), which form complexes to efficiently construct the glycan structure [17]. HNK-1 glycans are specifically expressed on adhesion molecules with immunoglobulin domains such as NCAM, L1, and myelin-associated glycoprotein (MAG), as well as extracellular matrix proteins like Tenascin-R, where they regulate neural cell adhesion, migration, and neurite outgrowth [38].

Our study has shown that in neural stem cells, glycans containing HNK-1 are specifically expressed on Tenascin-C (TNC) [52]. Moreover, these glycans regulate the expression levels of the epidermal growth factor receptor (EGFR), thereby activating the mitogen-activated protein kinase (MAPK) pathway and contributing to the maintenance of stemness in neural stem cells [46, 52]. Interestingly, HNK-1 glycans on TNC are expressed only on specific spliced domains. As differentiation progresses, the expression of TNC isoforms containing these spliced domains decreases, leading to the loss of HNK-1 expression [52]. This study has indicated that HNK-1 glycan expression is regulated by the splicing process without affecting the expression levels of glycosyltransferases involved in the HNK-1biosynthesis. Although all six spliced domains of TNC share the same fibronectin type III (FNIII) domain structure, HNK-1 glycans are expressed only on specific domains. This suggests that glycosyltransferases such as GlcAT recognize distinct amino acid sequences characteristic of these domains to synthesize domain-specific HNK-1 glycans.

Lewis X Glycans

We have previously reported that neural stem cells abundantly express Lewis X glycans and that glycoproteins bearing Lewis X include LAMP-1 [51]. The glycosyltransferase responsible for the biosynthesis of Lewis X glycans is FUT9. Interestingly, when FUT9 was expressed in cultured mammalian cells, Lewis X modification was observed specifically on LAMP-1 [34]. This suggests the presence of a molecular code within the LAMP-1 molecule that determines its interaction with FUT9. To identify this molecular code, we investigated the regions of LAMP-1 that regulate Lewis X modification by FUT9. The luminal region of LAMP-1 consists of two homologous immunoglobulin-like domains, both containing multiple *N*-glycosylation sites. By evaluating Lewis X expression in deletion mutants and systematic domain swap mutants of LAMP-1, we found that a specific 29-residue segment within the N-terminal domain (referred to as the L29 sequence) is essential for Lewis X glycan modification [34]. Furthermore, when the L29 sequence was fused to model glycoproteins such as erythropoietin (EPO) and fetuin in cells expressing FUT9, Lewis X modification was observed on these proteins. Additionally, *in vitro* glycosylation reactions using recombinant model glycoproteins and FUT9 demonstrated Lewis X modification on proteins with the L29 sequence, consistent with cellular results (Fig. 11.3). These findings reveal that the 29-residue segment within the N-terminal domain of LAMP-1 functions as a molecular code that governs direct interaction with FUT9.

While direct interactions between glycosyltransferases and their substrates are of key importance in glycan modification, the spatial organization of these enzymes within the Golgi apparatus also significantly impacts glycosylation. This will be explored in the following section.

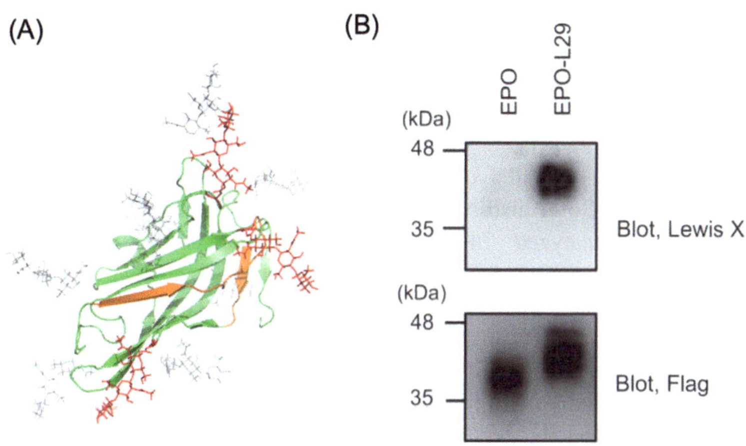

Fig. 11.3 Interaction of FUT9 with EPO dependent upon the L29 sequence. (**a**) Three-dimensional structural model of the N-domain of LAMP-1 with multiple glycan chains, highlighting the L29 sequence (IKTVESITDIRADIDKKYRCVSGTQVEMN) in orange. (**b**) FUT9-catalyzed Lewis X formation on desialylated EPO under cell-free conditions. After confirming complete desialylation, EPO with or without the L29 sequence (EPO and EPO-L29, respectively) was incubated with FUT9. The reaction mixtures were analyzed by immunoblotting using anti-Lewis X and anti-Flag antibodies. (This figure was adapted from the literature with permission [34])

Glycosylation Control Via Intra-Golgi Transport Pathways

As mentioned above, some glycosyltransferases are believed to achieve protein-specific glycosylation through direct interactions with the protein portion of the substrate. Within the secretory pathway, glycoproteins synthesized in the endoplasmic reticulum undergo a series of glycan modifications as they traverse the Golgi apparatus, encountering a diverse array of glycosidases and glycosyltransferases [4, 37]. The sequential nature of this transit suggests that the accessibility of substrate proteins to these modifying enzymes is tightly regulated, thereby enabling specific glycosylation patterns to be established. In this section, we will explore the specific localization of glycosyltransferases within the Golgi and how the regulated transport of substrates affects their modification.

Sub-Golgi Localization of Glycosyltransferases

Elucidating the mechanisms governing protein-specific glycosylation hinges on understanding glycosyltransferase localization. The spatial organization of these enzymes within the Golgi apparatus dictates glycan biosynthesis and, consequently, substrate specificity, thereby influencing glycoprotein function. The Golgi apparatus is composed of a series of cisternae, organized into *cis*, *medial*, and *trans* compartments that define the secretory pathway [8, 22, 25]. Glycosyltransferases are believed to be strategically compartmentalized within distinct sub-Golgi regions, enabling them to mediate specific glycan modifications; however, the exact distribution within the Golgi apparatus remains largely unclear. Emerging research indicates a heterogeneous distribution of glycosyltransferases even within individual Golgi cisterna [10, 42, 48].

Super-resolution confocal live imaging microscopy (SCLIM) has been utilized to investigate the sub-Golgi distribution of glycosyltransferases [19, 40]. Using this technique, we analyzed the three-dimensional colocalization of six glycosyltransferases (B3GALT6, ST3GAL3, ST3GAL4, ST6GAL1, FUT9, and B4GALT1) in pairs (Fig. 11.4a) [48]. B3GALT6, a β1,3-galactosyltransferase involved in glycosaminoglycan synthesis, localizes to the *medial*-Golgi. In contrast, the remaining five enzymes localize to the *trans*-Golgi, where they contribute to modification of glycoprotein glycans. Specifically, this region houses α2,3-sialyltransferases (ST3GAL3 and ST3GAL4) and α2,36-sialyltransferase (ST6GAL1) as well as α1,3-fucosyltransferase (FUT9) and β1,4-galactosyltransferase (B4GALT1). Glycosyltransferase distribution reveals a clear separation between *medial*-Golgi (B3GALT6) and *trans*-Golgi enzymes (Fig. 11.4b). Notably, even within the *trans*-Golgi, glycosyltransferases exhibit differential colocalization patterns, suggesting a complex organization that extends beyond the traditional *cis*, *medial*, and *trans* compartments (Fig. 11.4c, d). Interestingly, even isoforms of the same α2,3-sialyltransferase, such as ST3GAL3 and ST3GAL4, showed distinct localizations. The findings challenge the traditional view of distinct *cis*, *medial*, and *trans*-Golgi compartments, revealing a more complex subcompartmental organization of glycosyltransferases. Consistent with this, a SCLIM-based live-imaging demonstrated that the Golgi apparatus is composed of dynamic compartments with distinct zones of glycosyltransferase localization [10]. In this study, the localization of glycosyltransferases within nocodazole-induced Golgi ministacks was examined. These ministacks have also been utilized in a side-averaging approach to visualize the cisternal organization of Golgi-localized proteins [39]. This method offers a superior quantitative and systematic framework for analyzing the vertical localization of glycosyltransferases (Fig. 11.5).

These studies suggest that the spatial organization of glycosyltransferases within the Golgi is regulated by mechanisms far more precise than previously assumed.

Mechanistic Insights into Glycosyltransferase Localization

The sub-Golgi localization of glycosyltransferases is likely determined by their interactions with other proteins and membrane components. For example, GlcNAc-1-phosphotransferase, which is responsible for synthesizing mannose-6-phosphate glycans involved in lysosomal transport, is retained in the Golgi through interactions with LYSET (TMEM251) [30, 32]. In LYSET-deficient cells, this enzyme fails to localize to the Golgi, leading to defective mannose-6-phosphate

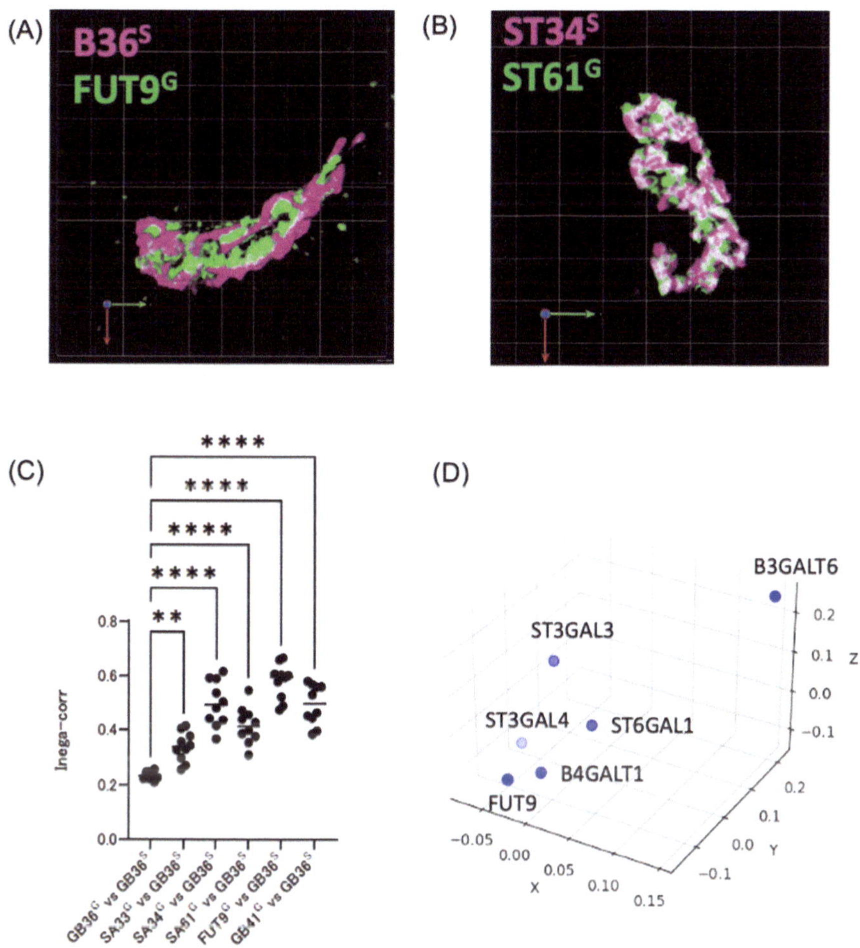

Fig. 11.4 Distribution of glycosyltransferases in the Golgi Apparatus. (**a**, **b**) Three-dimensional imaging of glycosyltransferases using super-resolution confocal live imaging microscopy (SCLIM). (**a**) 3D SCLIM images of the Golgi ribbon region in Expi293F cells expressing B3GALT6 (GB36) fused with mScarlet-I (^{S}) and FUT9 fused with mNeonGreen (^{G}). (**b**) 3D SCLIM images of Expi293F cells expressing ST3GAL4 fused with mScarlet-I (ST34^{S}) and ST6GAL1 fused with mNeonGreen (ST61^{G}). (**c**) Quantification of glycosyltransferase distribution differences within the Golgi apparatus using the normalized mean deviation product. Statistical significance (p-values) was indicated as follows: **$p < 0.01$, ****$p < 0.0001$. (**d**) Dimensional reduction mapping to visualize the spatial relationships of glycosyltransferases based on their colocalization patterns. The mapping was generated using multidimensional scaling (MDS) based on Pearson correlation coefficients representing the localization of each glycosyltransferase. The distances between data points reflect differences in enzyme localization as assessed by imaging analysis. Abbreviations: GB41: B4GALT1, GB36: B3GALT6, SA33: ST3GAL3, SA34: ST3GAL4, SA36: ST3GAL6, SA61: ST6GAL1, G: mNeonGreen, S: mScarlet-I. (This figure was adapted from the literature with permission [48])

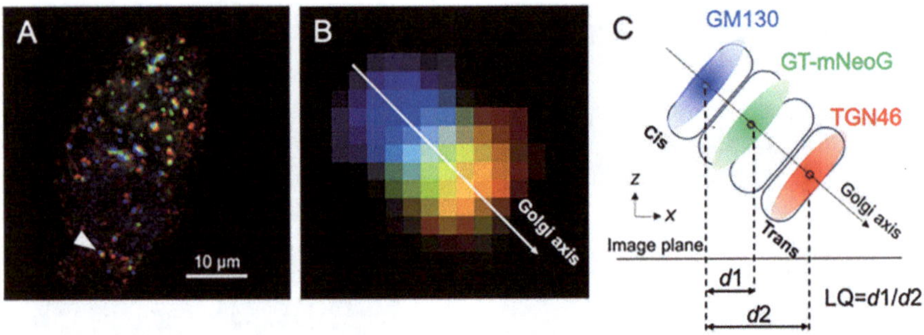

Fig. 11.5 Quantification of glycosyltransferase distribution in mini-Golgi stacks. (**a**, **b**) Immunofluorescence imaging of mini-Golgi stacks using confocal microscopy. HEK293 cells expressing β1,4-glucuronyltransferase 1 (B4GAT1) fused with mNeoGreen (mNeoG-GT) were immunostained for endogenous GM130 (blue) and TGN46 (red). (**a**) An overview of mini-Golgi stacks, with an arrowhead indicating the region magnified in panel (**b**). (**b**) Enlarged view of the indicated mini-Golgi stack, showing the centers of mass along the Golgi axis (white arrowhead). The color-coded representation highlights spatial distribution along the Golgi axis. (**c**) Schematic representation corresponding to the image in panel (**b**). The diagram illustrates the Golgi axis orientation, axial angle, and the measurement of distances ($d1$ and $d2$) used to calculate the localization quotient ($LQ = d1/d2$). The mini-Golgi stack, originally in three-dimensional space, is projected onto the image plane, providing a view along the xz-plane

glycosylation of lysosomal proteins. We have previously shown that glycosyltransferases involved in synthesizing matriglycan form complexes with each other [44]. Thus, glycosyltransferases may localize within the Golgi through interactions with transporters and other glycosyltransferases.

Electron microscopy reveals that the membrane thickness of the Golgi apparatus increases from the *cis*- to the *trans*-side [3]. Bioinformatics analyses indicate that glycosyltransferases localized on the *trans*-side tend to have longer transmembrane regions compared to those localized in the endoplasmic reticulum or *cis*-Golgi [36]. This suggests that transmembrane domain length, interacting with the lipid membrane, dictates glycosyltransferase localization based on membrane thickness, leading to specific distributions.

The cytoplasmic regions of certain glycosyltransferases can interact with COPI coat subunits and COPI adaptors (such as GOLPH3/Vps74), contributing to their retention within the Golgi [42]. For instance, the cytoplasmic region of GALNT6 contains the sequence "MRLLRR," which corresponds to the φ-(K/R)-X-L-X-(K/R) motif, a consensus sequence for binding δ-COP, suggesting direct interaction with δ-COP and localization to the *cis*-Golgi [21]. In contrast, ST3GAL4 and ST6GAL1 interact with GOLPH3 through sequences (L-X-X-K/R) in their cytoplasmic regions [31]. Recently, COPI adaptor molecules with functions similar to GOLPH3 have been identified, highlighting their different functions in the localization of glycosyltransferases [43]. We analyzed the colocalization of recombinant enzymes in which the cytoplasmic, transmembrane, and stem (CTS) regions of B3GALT6

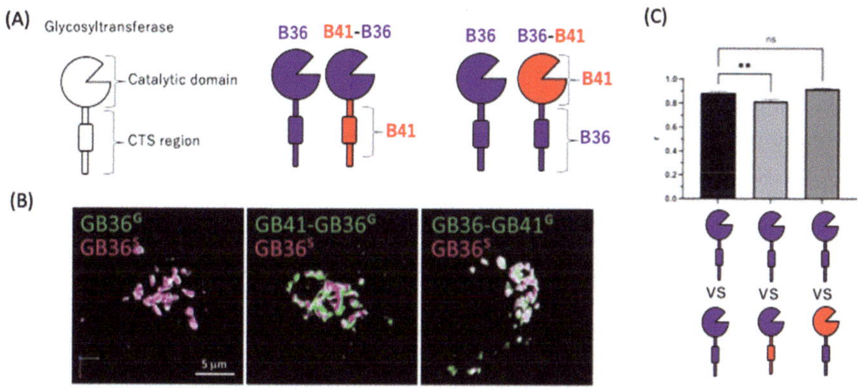

Fig. 11.6 3D images of chimeric glycosyltransferase mutants. (**a**) Schematic representation of chimeric glycosyltransferases. (**b**) Dual-color 3D images of the Golgi ribbon area of Expi293F cells expressing B3GALT6S versus B3GALT6G, B4GALT1-B3GALT6G, or B3GALT6-B4GALT1G. Bars, 5 μm. (**c**) Colocalization analysis quantified based on Pearson's correlation coefficient values. Statistical significance (p) was indicated as follows: $**p < 0.01$. Key: GB41, B4GALT1; GB36, B3GALT6; GB41-GB36, B4GALT1-B3GALT6; GB36-GB41, B3GALT6-B4GALT1. (This figure was adapted from the literature with permission [48])

and B4GALT1 were swapped. The recombinant enzymes sharing CTS regions showed higher colocalization with the original enzymes, suggesting that localization is influenced more by the CTS regions than by the catalytic domains (Fig. 11.6) [48]. However, enzymes with identical CTS regions did not exhibit completely identical localizations, implying a role for the luminal catalytic domains as well. These findings suggest that glycosyltransferase localization is determined by complex interactions involving multiple regions.

Proximity-dependent labeling is illuminating the intricate molecular networks associated with individual glycosyltransferases [50]. Integrating this with proteomic approaches provides a comprehensive understanding of sub-Golgi localization and glycosylation regulation.

Control of Glycosylation Modifications Through Glycoprotein Specific Transport

Glycan modification specificity hinges on both the specific localization of glycosyltransferases within the Golgi and the transport pathways of substrate glycoproteins. Research in *Drosophila* demonstrated that different classes of glycoprotein-related enzymes (i.e. Fringe-Connection, sulfateless, and Rhomboid) transverse distinct Golgi compartments, suggesting the formation of specialized Golgi zones tailored to specific glycosylation requirements [55]. This suggests that substrate proteins are efficiently glycosylated by the appropriate enzymes within these specialized Golgi zones, according to their transport pathway.

A notable example is contactin-1, which plays a critical role in axon-glia junctions within the paranodal region [33]. Contactin-1 is known to follow different secretory pathways depending on whether it forms a complex with its interacting partner, paranodin, within the cell [2]. Specifically, when complexed with paranodin, contactin-1 carries high-mannose glycans, whereas in the absence of this complex, it exhibits complex-type glycans. This indicates that contactin-1 undergoes distinct glycosylation modifications depending on its secretory route.

As another example of how transport pathway control can influence glycan modification, the interaction between cargo receptor and blood coagulation factor VIII (FVIII) can be considered. Efficient secretion of FVIII relies on the function of specific cargo receptor, notably ERGIC-53 and MCFD2, which facilitate the transport of FVIII from the endoplasmic reticulum to the Golgi apparatus [26, 27, 57]. Dysfunction of these cargo receptors, often due to genetic mutations, leads to impaired FVIII secretion and can result in bleeding disorders. Within the FVIII glycoprotein, we identified a 10-amino acid sequence that is specifically recognized by the MCFD2 cargo receptor. This sequence is crucial for the interaction between FVIII and MCFD2, initiating the transport processes [49]. This MCFD2-recognized sequence is not only essential for efficient FVIII secretion but also, surprisingly, enhances the secretion of other proteins. Introducing this sequence into erythropoietin (EPO) resulted in a significant increase in EPO secretion. This observation led to the designation of the sequence as the 'passport sequence', reflecting its ability to facilitate protein trafficking. Furthermore, it was found that incorporating the passport sequence into EPO not only enhanced its secretion but also significantly improved the maturation of its *N*-glycans [50]. Specifically, a marked increase in both galactosylation and sialylation of EPO's *N*-linked glycans was observed, suggesting that the passport sequence promotes access to or enhances the activity of glycosylation enzymes. Proximity-dependent labeling revealed that EPO harboring the passport sequence exhibits altered localization within the Golgi, residing in close proximity to the galactosyltransferase B4GALT1 and the protein NUCB1. Gene suppression experiments demonstrated that NUCB1 facilitates the biosynthesis of glycans containing galactose and sialic acid in a passport sequence-dependent manner, while NMR analysis demonstrated a direct interaction between NUCB1 and the passport sequence. Furthermore, EPO containing the passport sequence was found to colocalize with B4GALT1 and NUCB1 in specific Golgi zones, suggesting that this sequence enhances glycosylation efficiency through interactions with this galactosyltransferase (Fig. 11.7). Thus, the passport sequence functions as a molecular code that improves glycosylation efficiency by directing substrate proteins to specific Golgi regions.

Considering these findings, it is hypothesized that specific amino acid sequences embedded within substrate proteins can dictate their intracellular transport pathways. This mechanism potentially directs substrate glycoproteins to distinct regions of the Golgi apparatus where specific glycosyltransferases are localized, thereby enabling

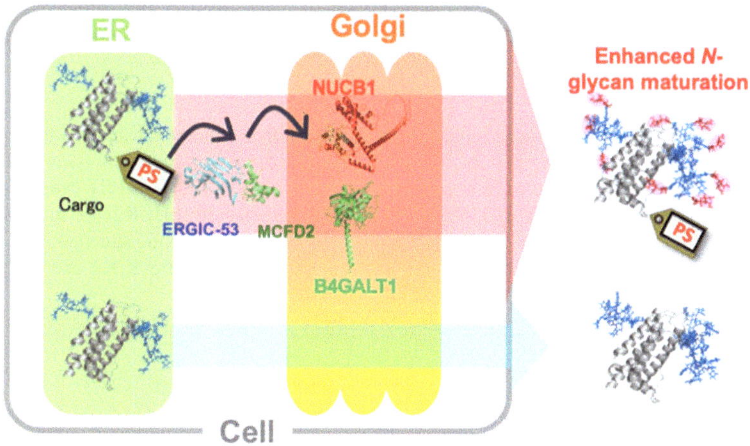

Fig. 11.7 The passport sequence functions as a molecular code that improves glycosylation efficiency by directing substrate proteins to specific Golgi regions. Glycoproteins carrying the passport sequence followed a specific route in the secretory pathway through the interaction with MCFD2, and that galactose modification was promoted through the interaction with the NUCB1/B4GALT1 complex on the route. (This figure was adapted from the literature with permission [50])

precise, protein-specific glycosylation modifications. Understanding these transport mechanisms provides new insights into how glycan diversity is regulated in the Golgi.

Conclusion

Our understanding of glycosylation in the Golgi apparatus is rapidly evolving. Traditionally, glycosyltransferases in the Golgi were thought to have relatively simple distributions and functions. However, recent studies reveal that these enzymes exhibit highly organized and dynamic localization patterns, interacting with specific substrates to perform precise glycosylation modifications.

Furthermore, discoveries such as identification of the 'passport sequence' have unveiled a novel layer of regulation, indicating that substrate-specific sequences and intracellular transport pathways are critical factors in glycosylation control. These findings challenge the traditional view of glycan biosynthesis as a passive process, highlighting that the spatially and functionally defined accessibility of glycosyltransferases in the Golgi determines protein-specific glycan modifications.

Moving forward, emerging technologies, including super-resolution imaging, proximity-dependent labeling, and omics-based approaches, hold immense potential to reveal the molecular mechanisms governing Golgi glycosylation [6, 18, 24, 29, 39, 50]. A deeper understanding of glycosyltransferase localization, substrate recognition, and trafficking pathways promises to revolutionize our understanding of glycan-related diseases and cellular processes, leading to innovative therapeutic strategies and advancements in glycobiology and cell biology.

Acknowledgments We express gratitude to our collaborators for their contributions to this study. This work was supported in part by Grants-in-Aid for Scientific Research (Grant Numbers JP21H02625 and JP25K02410 to H·Y. and JP24H00599 to K·K.), JST FOREST Program (Grant Number JPMJFR2255 to H·Y.), JST-CREST (Grant Number JP MJCR21E3 to K.K.), the ExCELLS Advanced Co-creation Platform (Spatiotemporal atlas of dynamic structure and function of organelles, 23EXC601 to H.Y.), ExCELLS "Golgi Atlas Project" (to K.K.) and the Human Glycome Atlas Project.

References

1. Bhide GP, Fernandes NRJ, Colley KJ (2016) Sequence requirements for neuropilin-2 recognition by ST8SiaIV and polysialylation of its O-glycans. J Biol Chem 291:9444–9457. https://doi.org/10.1074/jbc.M116.714329
2. Bonnon C, Bel C, Goutebroze L et al (2007) PGY repeats and N-glycans govern the trafficking of paranodin and its selective association with contactin and neurofascin-155. Mol Biol Cell 18: 229–241. https://doi.org/10.1091/mbc.E06-06-0570
3. Bykov YS, Schaffer M, Dodonova SO et al (2017) The structure of the COPI coat determined within the cell. elife 6. https://doi.org/10.7554/eLife.32493
4. Varki A, Cummings RD, Esko JD, Stanley P, Hart GW, Aebi M, Mohnen D, Kinoshita T, Packer NH, Prestegard JH, Schnaar RL, Seeberger PH (2022) Essentials of glycobiology, 4th edn. Cold Spring Harbor Laboratory Press, New York
5. Fahie KMM, Papanicolaou KN, Zachara NE (2022) Integration of O-GlcNAc into stress response pathways. Cells 11:3509. https://doi.org/10.3390/cells11213509
6. Fasimoye R, Dong W, Nirujogi RS et al (2023) Golgi-IP, a tool for multimodal analysis of Golgi molecular content. Proc Natl Acad Sci USA 120:e2219953120. https://doi.org/10.1073/pnas.2219953120
7. Foley DA, Swartzentruber KG, Colley KJ (2009) Identification of sequences in the polysialyltransferases ST8Sia II and ST8Sia IV that are required for the protein-specific polysialylation of the neural cell adhesion molecule, NCAM. J Biol Chem 284:15505–15516. https://doi.org/10.1074/jbc.M809696200
8. Glick BS (2000) Organization of the Golgi apparatus. Curr Opin Cell Biol 12:450–456. https://doi.org/10.1016/s0955-0674(00)00116-2
9. Hara Y, Kanagawa M, Kunz S et al (2011) Like-acetylglucosaminyltransferase (LARGE)-dependent modification of dystroglycan at Thr-317/319 is required for laminin binding and arenavirus infection. Proc Natl Acad Sci USA 108:17426–17431. https://doi.org/10.1073/pnas.1114836108
10. Harada A, Kunii M, Kurokawa K et al (2024) Dynamic movement of the Golgi unit and its glycosylation enzyme zones. Nat Commun 15:4514. https://doi.org/10.1038/s41467-024-48901-1
11. Holdener BC, Haltiwanger RS (2019) Protein O-fucosylation: structure and function. Curr Opin Struct Biol 56:78–86. https://doi.org/10.1016/j.sbi.2018.12.005
12. Imae R, Kuwabara N, Manya H et al (2021) The structure of POMGNT2 provides new insights into the mechanism to determine the functional O-mannosylation site on α-dystroglycan. Genes Cells 26:485–494. https://doi.org/10.1111/gtc.12853

13. Johnson JL, Yaron TM, Huntsman EM et al (2023) An atlas of substrate specificities for the human serine/threonine kinome. Nature 613:759–766. https://doi.org/10.1038/s41586-022-05575-3
14. Kanagawa M, Kobayashi K, Tajiri M et al (2016) Identification of a post-translational modification with Ribitol-phosphate and its defect in muscular dystrophy. Cell Rep 14:2209–2223. https://doi.org/10.1016/j.celrep.2016.02.017
15. Kanagawa M, Saito F, Kunz S et al (2004) Molecular recognition by LARGE is essential for expression of functional dystroglycan. Cell 117:953–964. https://doi.org/10.1016/j.cell.2004.06.003
16. Kanagawa M, Toda T (2018) Ribitol-phosphate-a newly identified posttranslational glycosylation unit in mammals: structure, modification enzymes and relationship to human diseases. J Biochem 163:359–369. https://doi.org/10.1093/jb/mvy020
17. Kizuka Y, Matsui T, Takematsu H et al (2006) Physical and functional association of glucuronyltransferases and sulfotransferase involved in HNK-1 biosynthesis. J Biol Chem 281:13644–13651. https://doi.org/10.1074/jbc.M601453200
18. Koga D, Kusumi S, Yagi H, Kato K (2024) Three-dimensional analysis of the intracellular architecture by scanning electron microscopy. Microscopy 73:215–225. https://doi.org/10.1093/jmicro/dfad050
19. Kurokawa K, Miyashiro D, Nakano A (2021) Development of super-resolution confocal live imaging microscopy, SCLIM, which elucidates the mechanism of cargo protein transport within the Golgi apparatus by 4D multicolor live cell imaging. Rev Laser Eng 49:271. https://doi.org/10.2184/lsj.49.5_271
20. Larsen ISB, Narimatsu Y, Clausen H et al (2019) Multiple distinct O-Mannosylation pathways in eukaryotes. Curr Opin Struct Biol 56:171–178. https://doi.org/10.1016/j.sbi.2019.03.003
21. Liu L, Doray B, Kornfeld S (2018) Recycling of Golgi glycosyltransferases requires direct binding to coatomer. Proc Natl Acad Sci 115:8984–8989. https://doi.org/10.1073/pnas.1810291115
22. Makhoul C, Gosavi P, Gleeson PA (2019) Golgi dynamics: the morphology of the mammalian Golgi apparatus in health and disease. Front Cell Dev Biol 7:112. https://doi.org/10.3389/fcell.2019.00112
23. Mendiratta SS, Sekulic N, Lavie A, Colley KJ (2005) Specific amino acids in the first fibronectin type III repeat of the neural cell adhesion molecule play a role in its recognition and polysialylation by the polysialyltransferase ST8Sia IV/PST. J Biol Chem 280:32340–32348. https://doi.org/10.1074/jbc.M506217200
24. Miyashiro D, Tojima T, Nakano A (2024) Extremely high spatiotemporal resolution microscopy for live cell imaging by single photon counting, noise elimination, and a novel restoration algorithm based on probability calculation. Front Cell Dev Biol 12:1324906. https://doi.org/10.3389/fcell.2024.1324906
25. Nakano A (2022) The Golgi apparatus and its next-door neighbors. Front Cell Dev Biol 10:884360. https://doi.org/10.3389/fcell.2022.884360
26. Nishio M, Kamiya Y, Mizushima T et al (2010) Structural basis for the cooperative interplay between the two causative gene products of combined factor V and factor VIII deficiency. Proc Natl Acad Sci USA 107:4034–4039. https://doi.org/10.1073/pnas.0908526107
27. Nyfeler B, Zhang B, Ginsburg D et al (2006) Cargo selectivity of the ERGIC-53/MCFD2 transport receptor complex. Traffic 7:1473–1481. https://doi.org/10.1111/j.1600-0854.2006.00483.x
28. Oka S (2019) HNK-1 carbohydrate, an attractive unique glyco-Epitope. Trends Glycosci Glycotechnol 31:SE12–SE14. https://doi.org/10.4052/tigg.1905.2se
29. Pasquarelli RR, Quan JJ, Cheng ES et al (2024) Characterization and functional analysis of Toxoplasma Golgi-associated proteins identified by proximity labeling. MBio 15:e0238024. https://doi.org/10.1128/mbio.02380-24

30. Pechincha C, Groessl S, Kalis R et al (2022) Lysosomal enzyme trafficking factor LYSET enables nutritional usage of extracellular proteins. Science 378:eabn5637. https://doi.org/10.1126/science.abn5637
31. Qi F, Isaji T, Duan C et al (2020) ST3GAL3, ST3GAL4, and ST3GAL6 differ in their regulation of biological functions via the specificities for the α2,3-sialylation of target proteins. FASEB J 34:881–897. https://doi.org/10.1096/fj.201901793R
32. Richards CM, Jabs S, Qiao W et al (2022) The human disease gene LYSET is essential for lysosomal enzyme transport and viral infection. Science 378:eabn5648. https://doi.org/10.1126/science.abn5648
33. Rios JC, Melendez-Vasquez CV, Einheber S et al (2000) Contactin-associated protein (Caspr) and contactin form a complex that is targeted to the paranodal junctions during myelination. J Neurosci 20:8354–8364. https://doi.org/10.1523/jneurosci.20-22-08354.2000
34. Saito T, Yagi H, Kuo C-W et al (2022) An embeddable molecular code for Lewis X modification through interaction with fucosyltransferase 9. Commun Biol 5:676. https://doi.org/10.1038/s42003-022-03616-1
35. Schachner M, Martini R (1995) Glycans and the modulation of neural-recognition molecule function. Trends Neurosci 18:183–191. https://doi.org/10.1016/0166-2236(95)93899-9
36. Sharpe HJ, Stevens TJ, Munro S (2010) A comprehensive comparison of transmembrane domains reveals organelle-specific properties. Cell 142:158–169. https://doi.org/10.1016/j.cell.2010.05.037
37. Stanley P (2011) Golgi glycosylation. Cold Spring Harb Perspect Biol 3:a005199–a005199. https://doi.org/10.1101/cshperspect.a005199
38. Sytnyk V, Leshchyns'ka I, Schachner M (2021) Neural glycomics: the sweet side of nervous system functions. Cell Mol Life Sci 78:93–116. https://doi.org/10.1007/s00018-020-03578-9
39. Tie HC, Mahajan D, Chen B et al (2016) A novel imaging method for quantitative Golgi localization reveals differential intra-Golgi trafficking of secretory cargoes. Mol Biol Cell 27:848–861. https://doi.org/10.1091/mbc.E15-09-0664
40. Tojima T, Miyashiro D, Kosugi Y, Nakano A (2023) Super-resolution live imaging of cargo traffic through the Golgi apparatus in mammalian cells. Methods Mol Biol 2557:127–140. https://doi.org/10.1007/978-1-0716-2639-9_10
41. UniProt Consortium (2023) UniProt: the universal protein knowledgebase in 2023. Nucleic Acids Res 51:D523–D531. https://doi.org/10.1093/nar/gkac1052
42. Welch LG, Munro S (2019) A tale of short tails, through thick and thin: investigating the sorting mechanisms of Golgi enzymes. FEBS Lett 593:2452–2465. https://doi.org/10.1002/1873-3468.13553
43. Welch LG, Peak-Chew S-Y, Begum F et al (2021) GOLPH3 and GOLPH3L are broad-spectrum COPI adaptors for sorting into intra-Golgi transport vesicles. J Cell Biol 220. https://doi.org/10.1083/jcb.202106115
44. Yagi H, Kuo C-W, Obayashi T et al (2016) Direct mapping of additional modifications on phosphorylated O-glycans of α-Dystroglycan by mass spectrometry analysis in conjunction with knocking out of causative genes for dystroglycanopathy. Mol Cell Proteomics 15:3424–3434. https://doi.org/10.1074/mcp.M116.062729
45. Yagi H, Nakagawa N, Saito T et al (2013) AGO61-dependent GlcNAc modification primes the formation of functional glycans on α-dystroglycan. Sci Rep 3:3288. https://doi.org/10.1038/srep03288
46. Yagi H, Saito T, Yanagisawa M et al (2012) Lewis X-carrying N-glycans regulate the proliferation of mouse embryonic neural stem cells via the Notch signaling pathway. J Biol Chem 287:24356–24364. https://doi.org/10.1074/jbc.M112.365643
47. Yagi H, Takagi K, Kato K (2024a) Exploring domain architectures of human glycosyltransferases: highlighting the functional diversity of non-catalytic add-on domains. Biochim Biophys Acta Gen Subj 1868:130687. https://doi.org/10.1016/j.bbagen.2024.130687

48. Yagi H, Tateo S, Saito T et al (2024b) Deciphering the sub-Golgi localization of glycosyltransferases via 3D super-resolution imaging. Cell Struct Funct advpub:24008. https://doi.org/10.1247/csf.24008
49. Yagi H, Yagi-Utsumi M, Honda R et al (2020) Improved secretion of glycoproteins using an N-glycan-restricted passport sequence tag recognized by cargo receptor. Nat Commun 11:1368. https://doi.org/10.1038/s41467-020-15192-1
50. Yagi H, Yamada R, Saito T et al (2024c) Molecular tag for promoting *N*-glycan maturation in the cargo receptor-mediated secretion pathway. iScience 27:in revision. https://doi.org/10.1016/j.isci.2024.111457
51. Yagi H, Yanagisawa M, Kato K, Yu RK (2010a) Lysosome-associated membrane protein 1 is a major SSEA-1-carrier protein in mouse neural stem cells. Glycobiology 20:976–981. https://doi.org/10.1093/glycob/cwq054
52. Yagi H, Yanagisawa M, Suzuki Y et al (2010b) HNK-1 epitope-carrying tenascin-C spliced variant regulates the proliferation of mouse embryonic neural stem cells. J Biol Chem 285:37293–37301. https://doi.org/10.1074/jbc.M110.157081
53. Yagi H, Yanaka S, Kato K (2021) Structural and functional roles of the N-glycans in therapeutic antibodies. In: Comprehensive Glycoscience. Elsevier, pp 534–542
54. Yamada I, Shiota M, Shinmachi D et al (2020) The GlyCosmos portal: a unified and comprehensive web resource for the glycosciences. Nat Methods 17:649–650. https://doi.org/10.1038/s41592-020-0879-8
55. Yano H, Yamamoto-Hino M, Abe M et al (2005) Distinct functional units of the Golgi complex in *Drosophila* cells. Proc Natl Acad Sci 102:13467–13472. https://doi.org/10.1073/pnas.0506681102
56. Yoshida-Moriguchi T, Campbell KP (2015) Matriglycan: a novel polysaccharide that links dystroglycan to the basement membrane. Glycobiology 25:702–713. https://doi.org/10.1093/glycob/cwv021
57. Zhang B, Cunningham MA, Nichols WC et al (2003) Bleeding due to disruption of a cargo-specific ER-to-Golgi transport complex. Nat Genet 34:220–225. https://doi.org/10.1038/ng1153

Part III
New Golgi Functions

Chapter 12
Insights into Microtubule Nucleation at the Golgi Apparatus

Rosa M. Rios and María P. Gavilán

Abstract The arrangement of microtubule (MT) networks is essential for various cellular processes including chromosome segregation during cell division, organelle positioning and transport, and the maintenance of cell polarity and morphology. The configuration of MT arrays heavily depends on the sites of MT assembly and anchoring of their minus ends. These sites are often organized into specialized structures known as microtubule-organizing centers (MTOCs). The nucleation activity of MTOCs relies on MT-nucleating factors, such as γ-tubulin-containing complexes, their associated activators and receptors, and proteins that anchor and stabilize MT minus ends.

Traditionally, the centrosome has been considered the primary MTOC in animal cells. However, recent research has revealed significant contributions from other subcellular structures in shaping MT networks. Notably, the Golgi Apparatus (GA) has emerged as a prominent alternative MTOC, operating in a diverse range of organisms from *Drosophila* to humans and across various cell types, from undifferentiated cells to highly specialized ones. This novel activity expands the functional repertoire of the GA, complementing its central role in secretion with essential contributions to MT cytoskeleton configuration.

This chapter offers a comprehensive overview of the current understanding of MT formation at the GA, including how this process is coordinated with the centrosome. It also explores the specific functions of Golgi-nucleated MTs and their participation in generating the complex MT networks of specialized cells, such as neurons, muscle cells, and pancreatic cells.

R. M. Rios (✉)
Centro Andaluz de Biología Molecular y Medicina Regenerativa (CABIMER), Universidad de Sevilla-CSIC-Universidad Pablo de Olavide, Seville, Spain
e-mail: rosa.rios@cabimer.es

M. P. Gavilán
Centro Andaluz de Biología Molecular y Medicina Regenerativa (CABIMER), Universidad de Sevilla-CSIC-Universidad Pablo de Olavide, Seville, Spain

Departamento de Citología e Histología Normal y Patológica, Facultad de Medicina, Universidad de Sevilla, Seville, Spain

A. Nakano, J. Saraste (eds.), *The Golgi Network, Volume II*, Subcellular Biochemistry 111, https://doi.org/10.1007/978-3-032-16833-7_12

Keywords Golgi Apparatus · Microtubule Nucleation · Microtubule Stabilization · AKAP450 · Golgi integrity · Cell polarity and migration · Neurons · Muscle cells · Pancreatic β-cells

Introduction

The GA is the central organelle of the secretory pathway, acting as a factory where proteins from the endoplasmic reticulum (ER) undergo modifications before being dispatched to their final destinations. It plays an essential role in protein processing, sorting, and transport within the cell. The GA is also key for carbohydrate and glycolipid synthesis and serves as an intermediate storage site for products derived from the ER. Structurally, the GA consists of a series of membrane-bound compartments, each with specific functions in the modification and sorting of cellular components [6, 40].

A distinctive feature of GA organization is the presence of flat, perforated cisternae. In simpler organisms, such as the budding yeast *Saccharomyces cerevisiae*, the GA consists of dispersed cisternae or isolated tubular networks that support efficient secretory activity. Other species of budding yeasts, unicellular green algae, and many protozoa organize the GA into a single stack of flattened cisternae aligned in parallel, commonly referred to as a Golgi stack, which represents the basic structural unit of the GA. In most invertebrate animals, including *Caenorhabditis elegans* and *Drosophila*, as well as in plant cells and many fungi, scattered Golgi stacks are positioned near the ER exit sites, forming a secretory unit. In contrast, vertebrate cells typically contain several stacks interconnected by narrow tubules, creating a continuous membrane system known as the Golgi ribbon [70, 96]. Notably, Golgi ribbons have also been recently described in certain taxa of invertebrates [13].

Golgi stacks feature two distinct faces: the cis-face (entry side) and the trans-face (exit side). Adjacent to these faces there are specialized compartments known as the cis-Golgi network (CGN) and the trans-Golgi network (TGN). Both networks are composed of interconnected tubular structures and cisternal chambers, primarily involved in sorting proteins and lipids [75]. Materials transported from the ER move to the CGN via tubular clusters. As molecules progress through the Golgi stack, they undergo a complex series of covalent modifications. Finally, modified proteins, lipids, and polysaccharides are transported to the TGN, which serves as a central sorting and distribution hub directing molecules towards lysosomes, the plasma membrane or extracellular spaces [97, 102].

In most non-differentiated vertebrate cells, the Golgi ribbon localizes near the nucleus and surrounds the centrosome (Fig. 12.1a, c; [38, 39, 119, 173]). Curiously, time-lapse microscopy studies have shown that the overall three-dimensional arrangement of the GA as a pericentrosomal juxtanuclear ribbon remains stable despite intense membrane trafficking [110]. This stability may be mediated by its close relationship with the cytoskeleton. Indeed, both Golgi ribbon integrity and positioning depend on MTs [146]. The key piece of evidence linking the GA to the

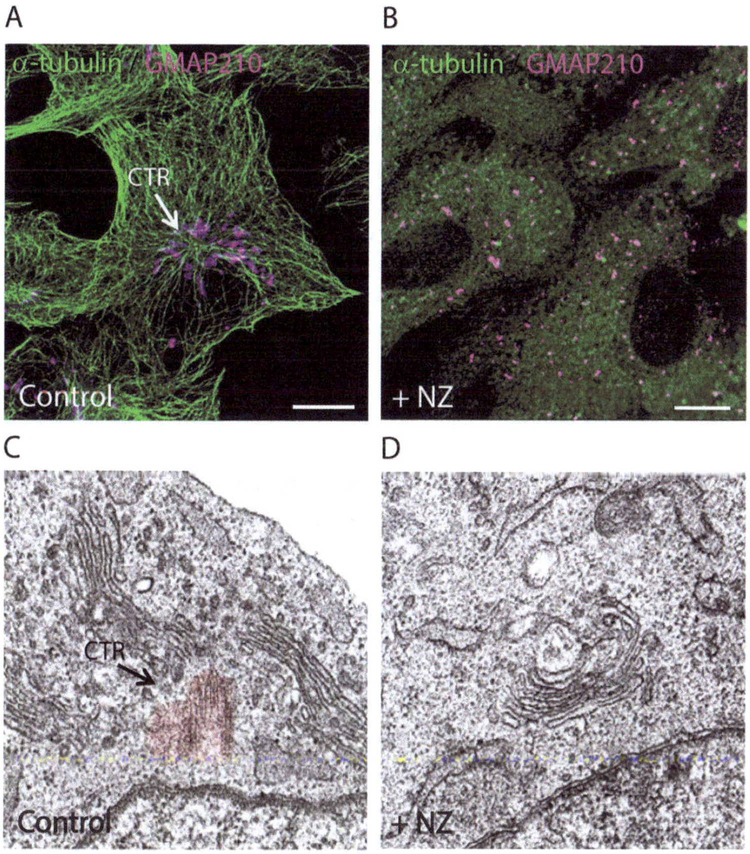

Fig. 12.1 (**a, c**) The Golgi ribbon in mammalian cells is typically located near the nucleus and surrounds the centrosome. (**b, d**) Its integrity and positioning depend on MTs. Panels (**a, b**) display immunofluorescence images of RPE-1 cells double-labeled for α-tubulin (green) and the cis-Golgi protein GMAP210 (magenta). In panel (**a**) (control cells), MTs are intact, while in panel (**b**), MTs have been depolymerized through nocodazole treatment. Panels (**c, d**) present electron microscopy images of control (**c**) and nocodazole-treated (**d**) cells. In panel (**c**), the centrosome, shaded in magenta, is surrounded by the characteristic ribbon-like arrangement of adjacent Golgi stacks. After nocodazole treatment (**d**), the Golgi ribbon breaks into numerous discrete elements, although they maintain their typical stacked morphology. Arrows indicate the centrosome (CTR). Scale bars represent 10 μm. ((**c**) and (**d**) Partial reproduction of Egea and Rios [38]. © *2008, Springer-Verlag/ Wien (with permissions).* (**a**) and (**b**) our unpublished images)

microtubular cytoskeleton is the observation that treatment with MT-depolymerizing agents leads to the dispersion of the perinuclear Golgi ribbon into scattered stacks, which typically co-localize with ER exit sites (Fig. 12.1b, d; [30, 38]). Following drug washout, ministacks are transported back toward the cell center along the newly formed MT network, eventually reassembling into a new Golgi ribbon [60].

MTs are long, stiff polymeric filaments that serve as tracks for intracellular transport and organelle positioning, contribute to cell shape determination, and regulate cell polarity and motility. During cell division, MTs drive chromosome segregation, while in post-mitotic cells, they form the axonemes of cilia and flagella [6]. MTs assemble from heterodimers of α- and β-tubulin, arranged in a head-to-tail fashion to form linear, structurally polarized protofilaments (Fig. 12.2a; [72]). Thirteen of these protofilaments associate laterally, giving rise to a hollow tube with an outer diameter of approximately 25 nm. The end of each protofilament exposing the α-tubulin subunits is known as the minus end (−), whereas the β-tubulin subunits are exposed at the plus end (+). In cells, the plus ends correspond to the primary sites of MT elongation (fast-growing ends), while the minus ends are typically stabilized and anchored at specific subcellular locations (slow-growing ends). MT ends undergo stochastic transitions between phases of growth and shrinkage, a process known as dynamic instability (Fig. 12.2b; [5]). This nonequilibrium process is driven by the GTPase activity of tubulin. Tubulin heterodimers are incorporated into the MT polymer with GTP bound to their β-tubulin subunits, which is subsequently hydrolyzed, resulting in a polymer lattice mainly composed of GDP-β-tubulin. However, due to a delay in GTP hydrolysis, a growing MT retains a structurally distinct stabilizing 'cap' of GTP-tubulin at its end. The transitions between growth and shrinkage (termed "catastrophe") and between shrinkage and growth (known as 'rescue') are regulated by the loss and regaining of this stabilizing cap. MT dynamic instability is controlled by a complex and interconnected network of MT-associated proteins that act on both MT ends and the lattice itself [5, 36, 160].

MTs control GA localization and organization in animal cells through various mechanisms. In non-differentiated dividing cells, the centrosome typically localizes near the nucleus and organizes a symmetric radial MT array by anchoring the minus ends while the growing plus ends point toward the plasma membrane [16]. In such cells, the minus end–directed motor dynein mediates the retrograde transport toward the cell center, while the plus end–directed motor kinesins facilitate outward, anterograde transport [137]. The GA is maintained in a pericentrosomal location through dynein-mediated centripetal transport along the radial MTs emanating from the centrosome [8, 24, 34, 58]. Indeed, both depletion or removal of dynein from membranes lead to the fragmentation and scattering of Golgi stacks throughout the cytoplasm [26, 57].

Additionally, the GA itself can nucleate a significant number of MTs [27, 37, 120]. These Golgi-derived MTs are thought to contribute to proper Golgi ribbon formation and to maintain the integrity and position of the GA within the cell [66, 95]. Furthermore, Golgi-derived MTs are believed to enhance cell polarity by increasing MT density in the region where the GA is located. Notably, due to the perinuclear location of the GA, this arrangement preserves the overall plus-end-out orientation of the MT network imposed by the centrosome [118]. However, Golgi-derived MTs can be distinguished from centrosomal MTs based on differences in their geometry, stability, and specific post-translational modifications.

Traditionally, research on MT organization has focused on non-differentiated dividing cells grown in two-dimensional (2D) cultures, where MTs typically form radial arrays, as mentioned above. However, in many cell types -especially in

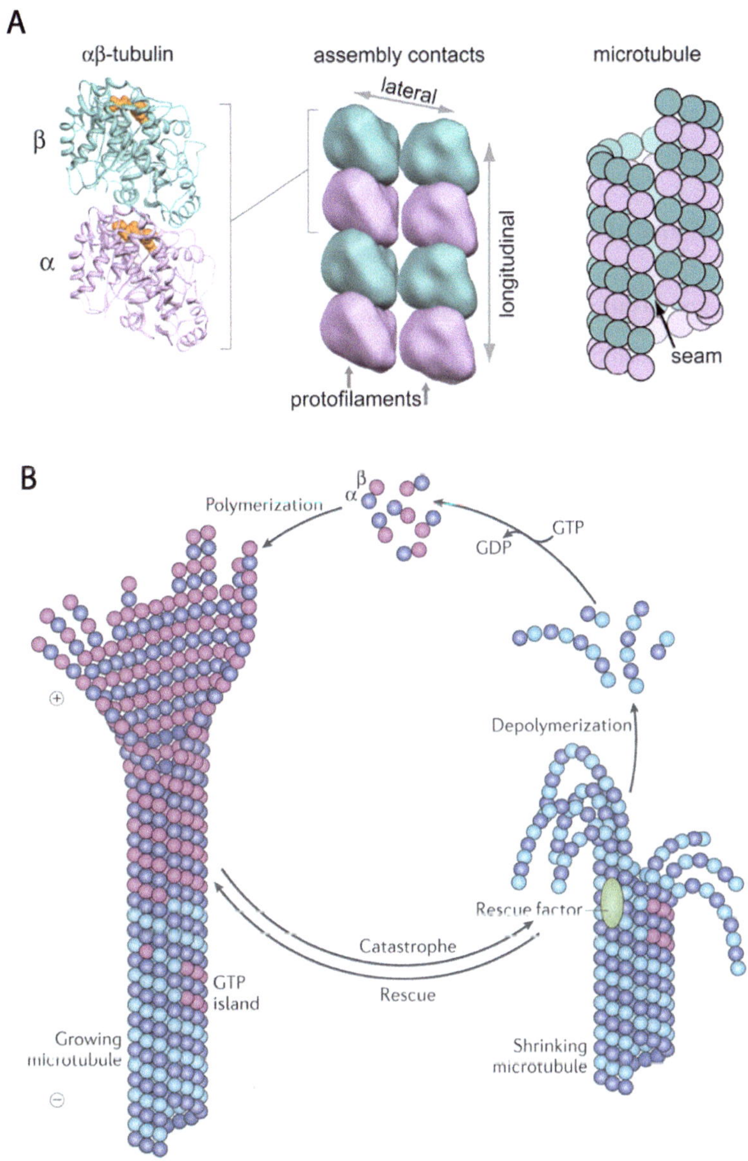

Fig. 12.2 (**a**) The α-tubulin–β-tubulin (αβ-tubulin) heterodimer serves as the essential repeating unit of MTs. When bound to GTP (shown in orange in the left image), heterodimers assemble through two types of interactions (illustrated by double-headed arrows): GTP-mediated longitudinal contacts between α-tubulin and β-tubulin that form protofilaments, and lateral contacts between protofilaments mediated by α-tubulin-α-tubulin and β-tubulin-β-tubulin interactions. This assembly results in a hollow structure that constitutes the MT. Additionally, a 'seam' is created due to lateral interactions between α-tubulin and β-tubulin in adjacent protofilaments, introducing slight asymmetry into the MT structure. (**b**) The cycle of MT polymerization and depolymerization is driven by the hydrolysis of GTP bound to β-tubulin. After the incorporation of GTP-bound tubulin dimers at the polymerizing ends, GTP hydrolysis occurs with a delay, leaving a stabilizing GTP cap at the growing ends of the MTs. Loss of this cap triggers MT catastrophe, leading to rapid

differentiated animal cells- MT arrays exhibit diverse geometries and densities [4, 126]. Likewise, the GA adopts different organizational forms depending on the cell type [96, 152]. These structural variations underscore how both MT networks and the GA adapt to meet the unique physiological demands of each cell type.

Mechanisms of Microtubule Nucleation

- **The gamma-tubulin ring complex**

MTs can be assembled *in vitro* from purified tubulin dimers by adding GTP and magnesium ions. However, despite the simplicity of this setup, the process faces significant kinetic barriers that results in a relatively slow MT formation rate. Additionally, spontaneous MT nucleation requires much higher tubulin concentrations than those needed for the elongation of pre-existing MTs. In contrast, MTs are formed with remarkable efficiency within cells, even though the tubulin concentration is below the threshold required for spontaneous nucleation. This efficiency is achieved by their ability to assemble a stable initial oligomer, which serves as a template for MT elongation. The formation of this initial oligomer is facilitated by a ring-shaped structure known as the γ-tubulin ring complex (γ-TuRC) composed of γ-tubulin and various accessory proteins [43, 123, 141, 148].

In budding yeasts, γ-tubulin self-assembles with Spc97 and Spc98 proteins to form the γ-tubulin small complex (γ-TuSC). Spc97 and Spc98 bind two copies of yeast γ-tubulin, forming a V-shaped tetramer where Spc97/98 interact laterally through their N-terminal regions while keeping the two γ-tubulin copies apart (Fig. 12.3a; [72, 73, 79]). At the spindle pole body, the yeast equivalent of the animal centrosome, seven γ-TuSC tetramers assemble into a left-handed spiral ring (referred to as yeast γ-TuSC ring), with exposed γ-tubulin molecules at the top, resembling the structure of a MT tube (Fig. 12.3b,c; [71, 73]). Recent cryogenic electron microscopy studies at near-atomic resolution have revealed that yeast γ-TuSC ring typically includes 6.5 γ-TuSC molecules per turn, resulting in 13 γ-tubulin molecules per ring, similar to a 13-protofilament MT (Fig. 12.3b, d; [35, 54, 188]).

In metazoans, γ-tubulin complex proteins (GCPs) 2 and 3, analogous to Spc97/98 in yeast, interact with two copies of γ-tubulin to form the γ-TuSC. Additional γ-tubulins and GCPs 4–6 proteins combine with γ-TuSCs to form the γ-TuRC, which has a molecular weight of approximately 2.2 MDa [51, 143, 187]. In recent

Fig. 12.2 (continued) depolymerization and subsequent shrinkage of the MTs. However, this shrinkage can be interrupted by rescue events, potentially facilitated by the presence of 'GTP islands' within the MT lattice or specific 'rescue factors' that promote MT stability and regrowth. ((**a**) Section extracted from [72]. © *2011, Springer Nature Limited (with permissions).* (**b**) Segment of figure taken from Akhmanova and Steinmetz [5]. © *2015, Springer Nature Limited* (*with permissions*))

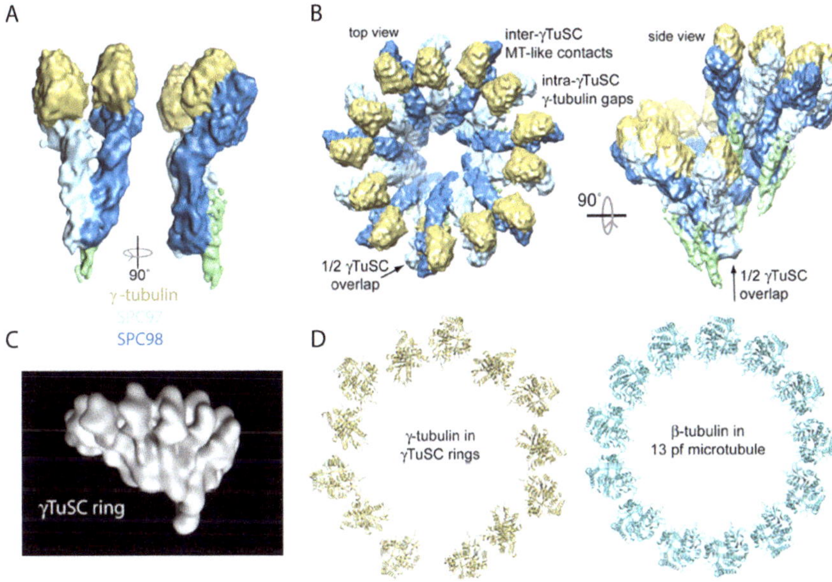

Fig. 12.3 In panel (**a**), the 8 Å cryo-electron microscopy structure of the γ-tubulin small complex (γTuSC) from *Saccharomyces cerevisiae* is depicted. This complex, which is bound to the spindle pole body component 110 (Spc110, shown in green), represents a single subunit of a larger γTuSC oligomer (refer to panel B). In this view, the amino termini of SPC97 and SPC98 are positioned at the bottom, while their carboxy-terminal domains are near the top, interacting with γ-tubulin. Panel (**b**) presents top-down and side views of the γTuSC ring. This ring is composed of six and a half γTuSCs per turn, resulting from a half-γTuSC overlap between the first and seventh subunits in the ring, as illustrated in the side view. This arrangement leads to 13 γ-tubulins per turn, which corresponds to the number of protofilaments found *in vivo* in MTs. In panel (**c**), a low-resolution negative-stain EM reconstruction of the γTuSC ring, as shown in panel B, is presented. Panel (**d**) compares the positions of γ-tubulin in γTuSC rings with the arrangement of α-tubulin–β-tubulin in MTs, revealing a mismatch in geometry with alternating contacts and gaps in the arrangement of γ-tubulin. (Adapted from [72]. © *2011, Springer Nature Limited* (*with permissions*))

years, high-resolution cryo-electron microscopy structures of purified native or recombinant γ-TuRCs have transformed our understanding of γ-TuRC assembly and regulation [32, 80, 166, 167, 172, 185]. The γ-TuRC consists of 14 γ-tubulin/GCP spokes, which are ordered (GCP2–3)4, GCP4-GCP5, GCP4-GCP6, GCP2–3, forming a cone-shaped structure (Fig. 12.4; [187]). The base of the cone is composed of GCPs while γ-tubulins are positioned near the nucleation interface. Each γ-tubulin molecule binds directly to an αβ-tubulin dimer to promote new MT assembly. Additionally, the γ-TuRC structure includes a stabilizing scaffold within the luminal space at the base of the cone, comprising GCP3 and GCP6 N-termini, along with two copies of the Mozart1 protein and one copy of globular actin (Fig. 12.4; [187]). The resolved structure of γ-TuRC supports a model of MT nucleation in which γ-tubulins recruit αβ-tubulin dimers and promote their lateral interactions during the initial stages of MT assembly [144].

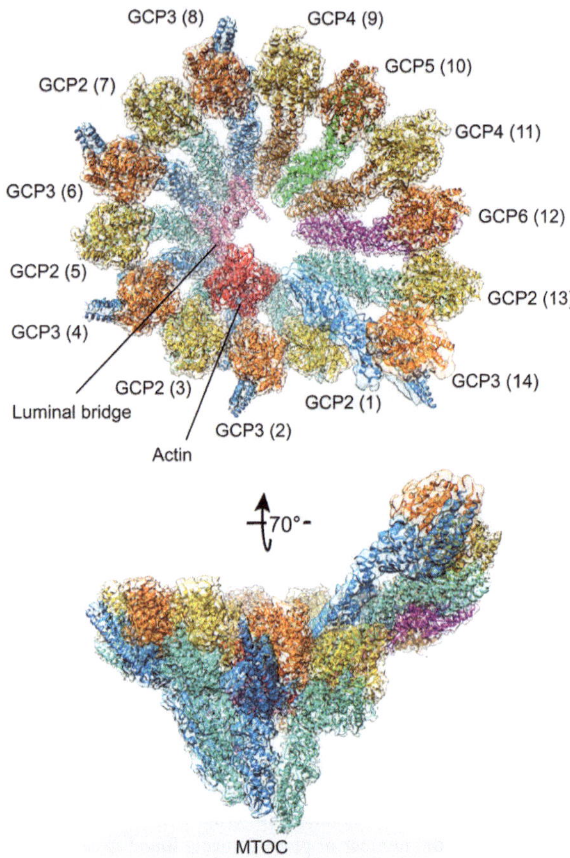

Fig. 12.4 The overall architecture of the left-handed γ-TuRC spiral, as revealed by cryo-electron microscopy single-particle analysis, is illustrated. This structure includes various components: γ-tubulins are depicted in yellow and orange, while GCP2 is shown in aquamarine, GCP3 in blue, GCP4 in brown, GCP5 in green, and GCP6 in purple. Actin is represented in red, and the luminal bridge is illustrated in pink. The spoke numbers are provided in brackets for reference. In the tilted view, the approximate position of the MTOC is indicated. (Partial reproduction of [187].)

Considering that MTs typically comprise 13 protofilaments, the resolved structures have revealed a significant mismatch between the γ-TuRC template and the geometry of the MT that it nucleates. This points to a mechanism where cytosolic γ-TuRC assembly is associated with a low-activity open conformation that requires activation to enhance its MT nucleation capacity. Supporting this, purified γ-TuRCs exhibits relatively low nucleation *in vitro* activity, which does not correspond to their potent *in vivo* activity. Additionally, γ-TuRC is widespread throughout the cytosol, yet ectopic MT nucleation is rare [11, 99]. The mechanisms underlying γ-TuRC activation remain unclear but may involve conformational changes that better align the γ-TuRC with the 13-protofilament MT structure [2, 21, 86, 157]. Promising candidates that can modify γ-TuRC structure or promote αβ-tubulin addition have been identified.

- **Activation of the gamma-tubulin ring complex**

The most well-known γ-TuRC-activating proteins are those containing the centrosomin motif 1 (CM1) at their N terminus [79]. This ~60–70 amino acid motif has been identified at the N terminus of orthologous spindle pole body and centrosomal proteins from yeasts to humans [20, 132, 161, 181]. *Drosophila* centrosomin and its homolog in mammals CDK5Rap2 contain the CM1 domain and have been extensively studied. The first half of CM1 consists of a ~30-residue sequence that specifically interacts with γ-TuRCs and enhances their MT-nucleating activity. This sequence is referred to as the γTuRC-mediated nucleation activator (γTuNA) domain. Numerous studies have demonstrated that γTuNA robustly binds to γ-TuRCs and activates them both *in vitro* and *in vivo* [29, 114]. In fact, its activation *in vivo* is so strong that the γTuNA domain is autoinhibited to prevent undesired nucleation activity. Thus, the CM1 motif contains a sequence that promotes MT nucleation from γ-TuRCs, while another sequence inhibits this activation [149, 175]. Data from *in vitro* assays support the idea that CDK5Rap2 induces a conformational change or enhances the conformational flexibility of the γ-TuRC structure, facilitating a better alignment with the 13-protofilament MT structure [114].

A significant breakthrough in the field was the discovery that XMAP215 (in *Xenopus*) and ch-TOG (in humans), a well-characterized MT polymerase, also functions as a γ-TuRC co-nucleation factor essential for MT nucleation [46, 54, 142]. XMAP215/ch-TOG was initially described as a plus-end binding protein that stimulates MT polymerization by promoting αβ-tubulin addition or preventing their loss [23, 165]. More recently, it was found that XMAP215/ch-TOG directly binds to γ-TuRC, and both components form a module that synergistically nucleates MTs in the cell [74, 142]. Single-molecule experiments combined with computational modeling suggest that ch-TOG indirectly induces the open-to-closed conformation change of purified γ-TuRCs by increasing the likelihood of protofilament formation, with lateral contacts between protofilaments promoting γ-TuRC closure [93].

Another highly conserved family of TOG-domain-containing proteins includes cytoplasmic linker-associated proteins, CLASP1 and CLASP2. CLASPs stabilize MTs by suppressing catastrophe and promoting rescue at plus ends, thereby facilitating MT outgrowth from stabilized seeds [1]. *In vitro* studies have now revealed a role for CLASPs as activators of γ-TuRCs, where CLASP2 shows an activation effect comparable in magnitude to that of CDK5Rap2 [113].

- **Microtubule Organizing Centers**

In animal cells, the arrangement, quantity, and orientation of MT networks are essential for determining cellular architecture, polarity, and proliferation. The assembly and remodeling of these arrays require careful spatial and temporal regulation of new MT formation. This regulation is facilitated by proteins that tether γ-TuRCs to specific subcellular structures. These structures, which accumulate γ-TuRC-tethering factors, are collectively known as Microtubule-Organizing Centers (MTOCs; [127, 171]). By regulating MTOC activity, cells achieve precise control over MT nucleation. MTOCs also influence MT orientation by anchoring the minus ends of MTs [158]. This anchoring capability is crucial, as it determines the polarity and organization of MT arrays, ultimately shaping the overall structure and behavior of the MT cytoskeleton.

For many years, the centrosome has been considered the primary MTOC in animal cells. The centrosome consists of two centrioles arranged orthogonally, surrounded by a dense protein matrix termed the pericentriolar material (PCM), where MT-nucleating capacity resides [16, 17, 76, 169]. Centrosomes are self-replicating structures that duplicate once per cell cycle, specifically during the S phase, a process driven by the PLK4 kinase. As the cell progresses to the G2/M transition, the duplicated centrosomes separate and undergo a substantial increase in size—from approximately 500 nm in diameter during interphase to several microns during mitosis. This expansion, known as centrosome maturation, is primarily due to the PLK1 kinase-mediated recruitment of additional PCM components. Centrosome maturation enhances the MT-nucleating capacity of the centrosome, ensuring proper spindle pole formation during mitosis (Fig. 12.5a; [15, 31, 155]).

PCM components include a diverse array of large coiled-coil proteins organized as a toroid around the centriole pair. During interphase, most of the PCM is associated with the older of the two centrioles, known as the mother centriole. Certain PCM components, such as Pericentrin and Cep152, extend radially from the centriole surface to form a proximal layer, which incorporates additional proteins to create a branched matrix that provides binding sites for γ-TuRCs and their adaptor proteins (Fig. 12.5b; [47]). PCM proteins Pericentrin and its paralog AKAP450 (formally known as AKAP9), as well as CDK5Rap2 and its paralog Myomegalin have the ability to bind to γ-TuRCs, potentially serving as anchors at the centrosome [79, 118]. Additionally, CDK5Rap2 and possibly Myomegalin can enhance MT nucleation via the CM1 activator domain. However, recent findings indicate that while these proteins are essential for centrosome maturation, they are largely dispensable for MT nucleation at the centrosome during interphase [52]. Another PCM scaffold protein, Cep192, plays a more significant role in the recruitment of γ-TuRCs to the centrosome. Intriguingly, even in the absence of all these proteins, the centrosome retains 50% of its γ-tubulin and corresponding MT-nucleating capacity [52]. This indicates that γ-TuRC recruitment and activation at the centrosome in human cells depend on a multitude of adaptors and targeting factors, some of which are yet to be identified. Evidence supporting this includes the discovery that the MT polymerase XMAP215/ch-TOG is crucial for centrosome-based MT nucleation [7].

Notably, some of these PCM proteins can interact with different cellular structures, leading to the formation of non-centrosomal MTOCs in various locations, such as the GA, endosomes, mitochondria, nuclear envelope, kinetochores, and the base of cilia [127, 171].

Microtubule Nucleation at the Golgi Apparatus

- **Mechanisms of MT nucleation at the GA**

The idea that the GA might exhibit MT nucleation ability first emerged in the early 2000s. In a pioneering study, Christian Pous' group reported that in hepatic cells,

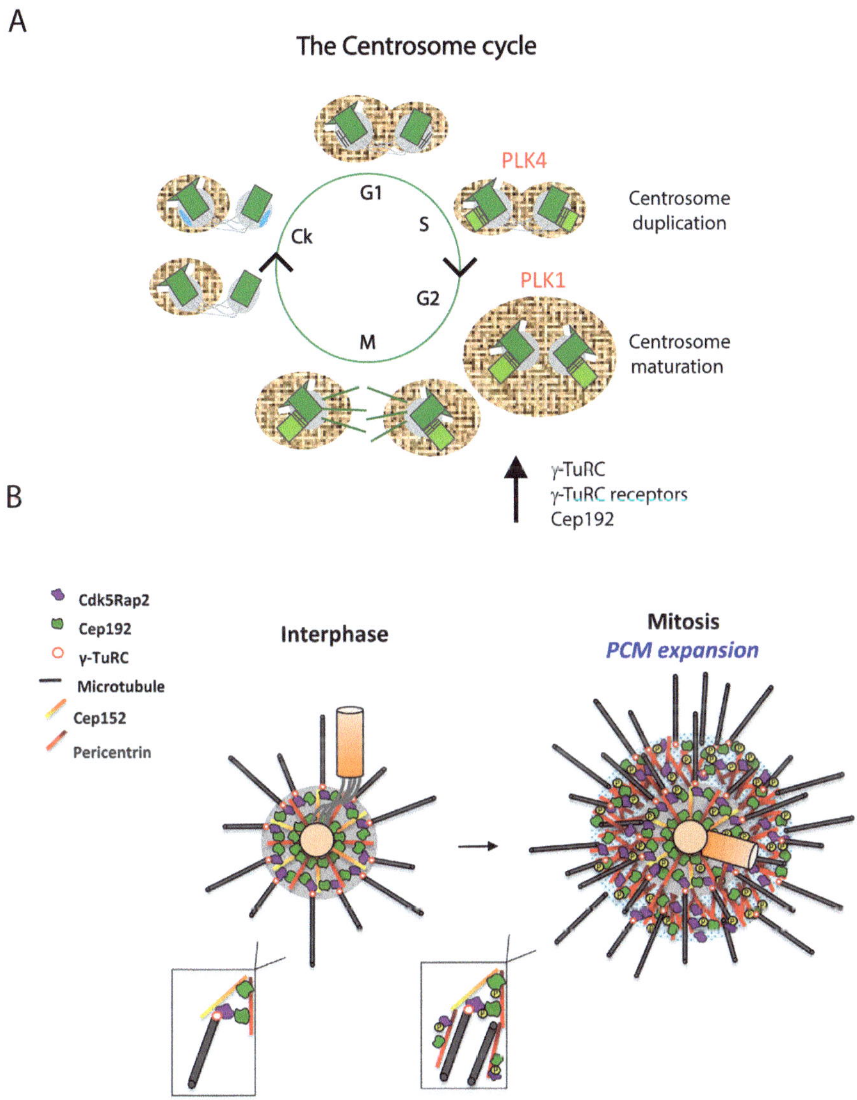

Fig. 12.5 (**a**) The centrosome–centriole duplication–segregation cycle is typically divided into distinct phases. During G1, the two centrioles are loosely connected. In the subsequent S phase, the PLK4 kinase promotes the duplication of both centrioles, resulting in two centrosomes, each comprising two closely associated centrioles. As the cycle progresses into G2, centrosome maturation occurs through the PLK1-mediated accumulation of PCM proteins. This maturation process enhances the recruitment of γ-TuRCs, increasing the MT nucleation activity required for effective spindle assembly. Centrosome separation occurs in the late G2 phase, leading to the formation of the spindle poles. Panel (**b**) provides a simplified overview of PCM organization. During interphase, most of the PCM is associated with one of the two centrioles, represented as orange cylinders. Certain PCM components, such as Pericentrin and Cep152, extend radially from the centriole surface to create a proximal layer (gray circle). This layer incorporates additional proteins, including CDK5Rap2 and Cep192, which together create a branched matrix that provides binding sites for γ-TuRCs and their adaptor proteins. At mitotic entry, centrosome maturation is marked by

Golgi membranes could assemble and stabilize MTs during recovery from drug-induced MT depolymerization [27]. Furthermore, purified Golgi membranes were found to contain γ-tubulin and to facilitate MT assembly. Building on this, Kaverina's group definitively showed that a subset of MTs extends directly from Golgi membranes by tracking polymerizing MTs in human retinal pigmented epithelial RPE-1 cells [37]. Additional significant findings of this work included the observation that siRNA-mediated depletion of γ-tubulin inhibits MT formation at both the GA and the centrosome and that laser ablation of the centrosome does not impact the number of MTs formed at the GA. It was estimated that up to 50% of MTs in RPE-1 interphase cells are generated by the GA.

Since then, significant advances have been made in understanding the organization of MTs at the GA, with notable contributions from the Rios and Akhmanova groups [52, 66, 120, 170]. Their research has uncovered key mechanisms of MT nucleation at the GA, including the identification of specific proteins and complexes involved in this process. MT nucleation can be readily observed after depolymerizing MTs using cold or nocodazole treatments, followed by a brief warming or washout of the drug (Fig. 12.6a, c; [120]). Cold treatment preserves the integrity of the Golgi ribbon, whereas nocodazole induces the fragmentation of the ribbon into scattered Golgi stacks as mentioned above. The critical factor controlling MT nucleation is AKAP450, which localizes at both the GA and the centrosome (Fig. 12.6b; [120]). AKAP450 is a large, multiply spliced protein that serves as a scaffold for several kinases, phosphatases, and other enzymes. Structurally, it is a long, likely dimeric, protein that contains several coiled coil domains as well as unstructured and disordered regions. AKAP450 is targeted to the cis-GA through the interaction of its N-terminal coiled-coil domain, termed the AK1B domain, with the cis-Golgi matrix protein GM130 [66]. This targeting restricts MT nucleation activity to the cis-face of the GA.

Depletion of AKAP450 in RPE-1 cells abolishes MT nucleation at the GA (Fig. 12.6c; [52, 120]). Similarly, GM130 knockdown or AK1B domain overexpression leads to the dissociation of AKAP450, thereby hindering MT nucleation at the GA [66, 120]. Interestingly, treatment with the drug brefeldin A, which perturbs Golgi organization, results in the relocation of GM130 and AKAP450 to the ER exit sites, which concomitantly acquire AKAP450-dependent MT nucleation capacity. Through its C-terminal region, Golgi-bound AKAP450 directly binds to the γ-TuRC and recruits the γ-TuRC activator CDK5Rap2, as well as the Golgi-specific isoform Myomegalin 8, both of which also associate with γ-TuRCs [66, 125, 162, 170]. This mechanism appears to be evolutionarily conserved, as the

Fig. 12.5 (continued) PCM expansion, a process facilitated by the phosphorylation of multiple proteins, such as Pericentrin, Cdk5Rap2, and Cep192. This expansion leads to the formation of an outer layer with gel-like properties (blue hatched circle), which, while less structured, incorporates scaffolds to enhance MT nucleation capacity, crucial for proper spindle assembly. ((**b**) Modified from [47].

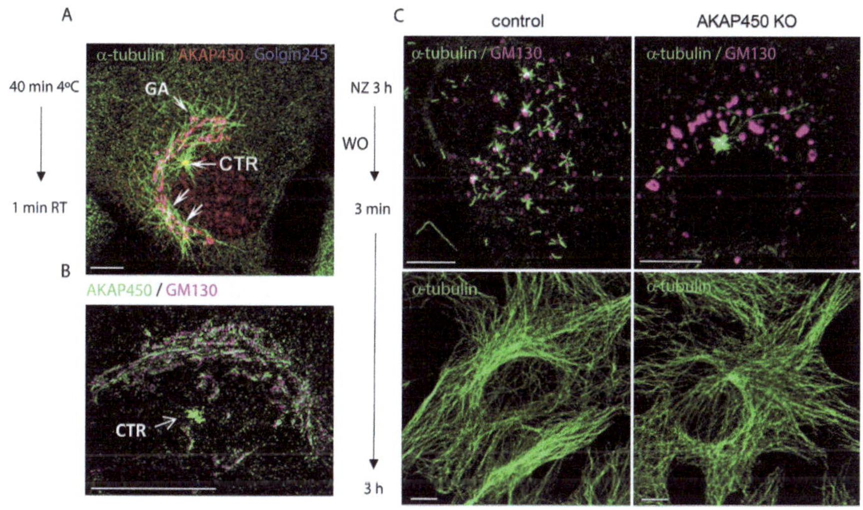

Fig. 12.6 MT nucleation at the GA. (**a**) MT repolymerization after cold treatment and rewarming is illustrated. Cells were fixed and stained for α-tubulin (green), AKAP450 (red) and the trans-Golgi marker Golgin 245 (blue). Note that MTs extend radially from the centrosome (depicted in yellow) and tangentially from Golgi membranes, with arrows indicating the centrosome (CTR) and the GA. (**b**) RPE-1 cells stained for AKAP450 (green) and the cis-Golgi marker GM130 (magenta) were visualized using an OMX super-resolution microscope. An arrow indicates the centrosome. (**c**) MT regrowth experiments were performed in RPE-1 cells following nocodazole treatment, comparing control cells (left panels) with AKAP450-depleted cells (right panels). Shortly after drug washout, MTs are observed growing from dispersed Golgi elements. Three hours post-washout, control cells exhibit a reformed asymmetric MT network, while in the absence of Golgi-nucleated MTs, the AKAP450-depleted cells develop a radial network centered around the centrosome. α-Tubulin is shown in green and GM130 in magenta. Scale bars represent 10 μm. ((**a**) Portion of figure taken from [120]. © *2009, European Molecular Biology Organization (with permissions)*. (**b**) and (**c**) our unpublished images)

Drosophila homologs of AKAP450 and CDK5Rap2 have been shown to control MT outgrowth from Golgi elements, known as Golgi outposts, in fly neurons [108, 174].

The minimal sequence of AKAP450 necessary for MT nucleation at Golgi membranes comprises the N-terminal quarter of the protein (amino acids 1-1029), which targets the protein to the GA, fused with the C-terminal segment (amino acids 2873-3907), responsible for recruiting γ-TuRCs and their activators [170]. Surprisingly, cells deficient in CDK5Rap2 exhibit only a 30% reduction in MT nucleation at the GA while Myomegalin 8 knockout cells show no decrease in nucleation [52, 170]. Even in the absence of both CDK5Rap2 and Myomegalin 8, 70% of Golgi MT nucleation activity persists, indicating that additional factors or activation mechanisms may be involved at the GA. In RPE-1 cells, similar to centrosomal MT nucleation, MT nucleation from the Golgi also involves the MT polymerase ch-TOG, though the exact mechanism requires further investigation [7].

- **Mechanisms of MT anchoring and stabilization at the GA**

AKAP450 also provides cis-Golgi membranes with MT-minus end anchoring or stabilizing activities through the recruitment of additional factors. Co-immunoprecipitation experiments and proteomics identified an interaction between the N-terminus of AKAP450 and the DCTN1/p150^{glued} component of dynactin ([66], our unpublished data). DCTN1/p150^{glued} is the largest subunit of dynactin, a massive protein complex necessary for most, if not all, types of dynein activity in eukaryotes [133]. DCTN1/p150^{glued} assembles as an elongated dimer that interacts with MTs through a well-characterized motif located in its N-terminal end, known as the CAP-Gly domain [133]. It is worth mentioning that overexpression of a dominant-negative mutant of DCTN1/p150^{glued}, which has been reported to fragment the Golgi ribbon into isolated stacks, likely by disrupting dynein's function [112], inhibits MT nucleation from the GA [120].

Another line of research has identified an important role for CAMSAP2 in MT generation at the GA, without directly influencing MT nucleation [170]. It was previously believed that MT minus ends do not grow in cells; however, studies on CAMSAP proteins have shown that they can recognize and protect free MT minus ends formed through minus-end polymerization. This recognition is mediated by a C-terminal domain common to CAMSAP proteins. When MTs are depolymerized, CAMSAPs are diffusely distributed, but as polymerization progresses, they bind to the minus ends of MTs that are released from nucleation sites [3].

CAMSAP proteins exhibit distinct functions depending on the specific family member: CAMSAP1 tracks growing MT minus ends, while CAMSAP2 and CAMSAP3 specifically bind to MT lattice segments formed by minus-end growth. In RPE-1 cells, the majority of CAMSAP2 stretches, extending 1 to 3 microns along MT minus ends, colocalize with the GA [170]. Accordingly, most Golgi-anchored MTs are lost in cells lacking CAMSAP2, although GA-associated MT nucleation remains unaffected. Interestingly, the stretches decorated with CAMSAP2 are recruited to the GA by AKAP450 through the intermediary protein Myomegalin 8, which in turn recruits EB1/EB3 proteins, the core components of MT plus end tracking protein complexes. The C-terminal domain of AKAP450 interacts with the N-terminal domain of Myomegalin 8, forming a complex that effectively binds to the C-terminal region of CAMSAP2. Loss of Myomegalin 8 results in the detachment of CAMSAP2 stretches from the GA (Fig. 12.7; [170]).

Effective MT formation at the GA also depends on CLASP proteins. CLASPs are MT plus-end binding proteins that additionally localize at the centrosome and the GA [77]. At the GA, CLASPs are specifically recruited by the TGN-associated protein GCC185 [37]. Initially, CLASPs were thought to anchor MTs at the TGN; however, more recent studies indicate that they actually stabilize CAMSAP2-decorated stretches at the GA [170]. Cells lacking CLASPs exhibit a significant reduction or complete loss of CAMSAP2 stretches, leading to the absence of Golgi-derived MTs.

Finally, the MT-crosslinking proteins MTCL1 and its paralog MTCL2 have been shown to promote MT elongation on both sides of the Golgi stacks

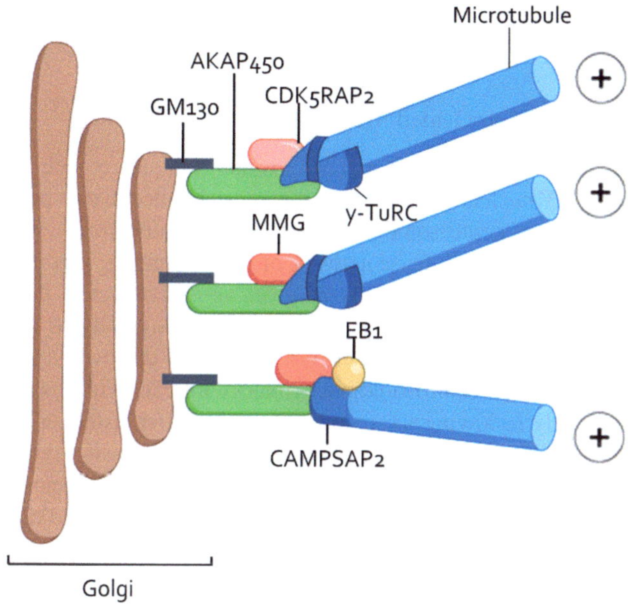

Fig. 12.7 The mechanism of MT generation at the cis-Golgi depends on the recruitment of AKAP450 to the GA by the cis-Golgi protein GM130. AKAP450 binds to γ-TuRC either directly or through CDK5Rap2 or its homologue Myomegalin, thereby promoting MT nucleation. Additionally, a complex involving AKAP450, Myomegalin, EB proteins and CAMSAP2 helps anchor and stabilize MT minus ends at the GA independently of γ-TuRC. This dual mechanism highlights how the GA supports MT organization both through γ-TuRC-mediated nucleation and γ-TuRC-independent stabilization. (Adapted with modifications from Akhmanova and Kapitein [4]. © *2022, Springer Nature Limited* (*with permissions*))

[92, 131]. MTCL1 is recruited to Golgi membranes through interactions with either CLASPs or AKAP450. When MTCL1 is knocked down, the perinuclear MT network that normally colocalizes with the GA is disrupted, impairing the tethering and extension of the Golgi ribbon. MTCL1 is a long coiled-coil protein with two MT-binding domains located at its N- and C-terminal regions. The C-terminal domain has a unique ability to stabilize MTs. Thus, MTCL1 stabilizes Golgi-associated MTs via its C-terminal MT-binding domain while crosslinking them using its N-terminal domain [131].

In summary, Golgi-dependent MT nucleation involves multiple factors, with AKAP450 fulfilling three key roles (Fig. 12.7; [4]): directly nucleating MTs, potentially through γ-TuRC binding; recruiting MT nucleation activators, such as CDK5Rap2, and stabilizers like MTCLs; and anchoring CAMSAP2-bound MT minus ends via interaction with Myomegalin 8. These findings highlight that the GA functions as an MT-nucleating site by recruiting AKAP450, which utilizes both γ-TuRC-dependent nucleation and γ-TuRC-independent stabilization to organize Golgi-derived MTs. Furthermore, despite the presence of large cytoplasmic pools of AKAP450, CDK5Rap2, Myomegalin 8, CAMSAP2, and γ-TuRCs, MT formation does not typically occur in the cytoplasm. This suggests that AKAP450-based complexes acquire MT-nucleating capacity when assembled at cis-Golgi membranes.

- **Properties of Golgi-nucleated Microtubules**

One notable feature of Golgi-associated MTs is their unique geometry. Long before the discovery of MT nucleation at the GA, it was established that a specific subset of MTs, distinct from centrosome-anchored MTs, colocalized with the GA [134, 146]. These MTs form a dense network of short filaments, seemingly anchored at both ends to the GA. Approximately 20 years ago, Marsh et al. employed high-pressure freezing, freeze-substitution, and electron tomography to investigate the spatial relationship between MTs and the GA [88]. They discovered that these MTs do not originate from the centrosome but primarily interact with the cis-face of the Golgi stacks, closely following the membranes of the first cisternae. Occasionally, these MTs make direct contact with the cisternae and traverse the Golgi stacks through openings in the membranes. In contrast, the medial and trans cisternae are not associated with MTs. Thus, while centrosomal MTs are arranged radially, Golgi-associated MTs primarily adopt a tangential orientation.

A potential mechanism for the formation of this intra-Golgi network is that AKAP450-nucleated MTs elongate from the cis-face of the GA stacks, with their plus ends capped by TGN-associated CLASPs. It has been suggested that CAMSAP2-decorated MT lattices, which can be relatively long (1–3 microns), may span across different Golgi compartments and become anchored or stabilized at both ends [170]. This arrangement would result in short MTs predominantly parallel to the GA ribbon, potentially facilitating the lateral linkage and merging of GA stacks into a continuous ribbon. Additionally, another subset of Golgi-derived MTs extends from the TGN toward the leading edge of motile cells [37, 95]. In this context, CLASPs could stabilize cis-Golgi-nucleated MTs that grow toward the TGN.

Another distinct characteristic of GA-nucleated MTs is their stability, clearly demonstrated by their resistance to MT-depolymerizing drugs [145]. This increased stability can arise from two mechanisms: the association with MT-stabilizing proteins and the capping of their plus ends. Both mechanisms are active at the GA, as evidenced by the presence of MTCLs and CLASPs, which help stabilize and anchor MTs (see above). Stability is also linked to specific post-translational modifications, such as tubulin de-tyrosination and acetylation [68]. Immunofluorescence microscopy analysis in RPE-1 cells shows that MTs enriched in acetylated tubulin often co-localize with the GA (Fig. 12.8a; [120]), both under normal conditions and during recovery from MT depolymerization (Fig. 12.8b; [120]). Notably, AKAP450-dependent Golgi-derived MTs become acetylated shortly after their formation, suggesting a rapid acquisition of stability-associated modifications [27, 120].

The effects of post-translational tubulin modifications on Golgi organization and function remain incompletely understood. The positioning and shaping of the GA are typically governed by a dynamic equilibrium between the opposing forces of the minus-end-directed dynein motor and plus-end-directed kinesin motors [8, 24, 44,

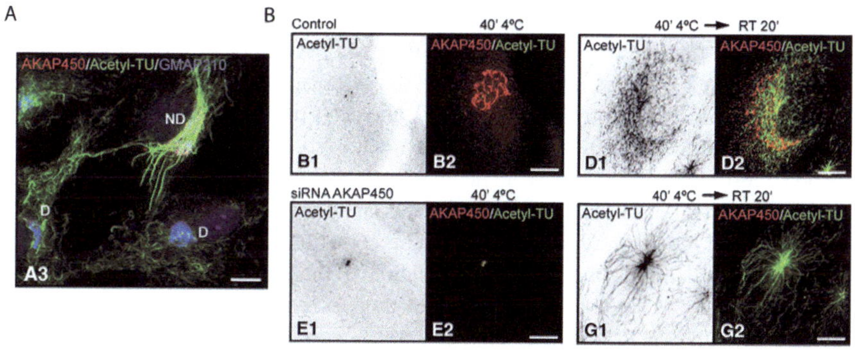

Fig. 12.8 In (**a**), a dense network of acetylated MTs is observed colocalizing with the GA in RPE-1 cells. These MTs are confirmed to be nucleated at the GA, as they are absent in cells depleted of AKAP450. For comparison, acetylated MTs are visible in AKAP450-non-depleted (ND) cells but not in AKAP450-depleted (D) cells. Acetylated tubulin is shown in green, AKAP450 in red, and GMAP210 in blue. In panel (**b**), MT repolymerization is analyzed after cold treatment followed by rewarming in both AKAP450-non-depleted (top gallery) and AKAP450-depleted (bottom gallery) cells. Single-label images for acetylated tubulin are displayed alongside triple-label images as seen in panel (**a**). Notably, a strong acetylation of Golgi-derived MTs is observed shortly after their polymerization. (Derived from [120]. © *2009, European Molecular Biology Organization (with permissions)*)

87]. Tubulin acetylation has been proposed to enhance kinesin-1 transport efficiency along MTs [9, 117]. This leads to a plausible mechanism whereby the acetylation of Golgi-derived MTs facilitates the extension of the GA around the nuclear envelope. In this scenario, kinesin-1-driven movement counterbalances the inward-pulling forces generated by dynein towards the MT minus ends, which are primarily anchored at the centrosome. Interestingly, Kelliher et al. [69] demonstrated that the motor activity of kinesin-1 influences the spatial distribution of Golgi outposts in neurons, thereby affecting overall neuronal function.

- **Regulation of MT nucleation activity of the GA**

Current evidence supports that MT nucleation at the GA is regulated on at least three different levels, with the first being spatial. MT nucleation does not occur uniformly across Golgi membranes; rather, it is specifically localized to the cis-side of Golgi stacks. This restriction is likely imposed by the distribution of the AKAP450 receptor GM130 [120]. Moreover, MT nucleation activity is not randomly distributed along the cis-Golgi cisternae. In MT-repolymerization experiments following nocodazole treatment, MTs are observed to regrow from specific, localized regions of isolated GA stacks (see Fig. 12.6c). At these regions, the MT minus ends colocalize with AKAP450 [52, 120]. Super-resolution microscopy and quantitative modeling have confirmed that MT nucleation at the GA is confined to distinct hotspots within the Golgi. These hotspots are spherical regions of approximately 0.4 μm in diameter, containing both Golgi membrane markers and multiple MT minus ends [129]. While these hotspots are predominantly found in

AKAP450-positive fragments, not all AKAP450-positive areas qualify as hotspots. This indicates that although AKAP450 is necessary for hotspot formation, it is not sufficient on its own. The formation of these hotspots also requires activation through the γTUNA motif of CDK5Rap2 and Myomegalin 8 [129]. Conversely, cytoplasmic aggregates of AKAP450 that do not colocalize with Golgi stacks in nocodazole-treated cells fail to nucleate MTs, confirming that only Golgi-bound AKAP450 can assemble the complexes necessary for MT nucleation [120].

The second level of regulation implies the coordination between MT nucleation activity of the GA and other MTOCs, particularly the centrosome. The presence of two MTOCs in the same cell suggests that their activities must be finely tuned to maintain cellular MT homeostasis. Studies on cell lines with γ-TuRC receptor proteins knocked out, along with the use of the PLK4 inhibitor centrinone has provided valuable insights into this coordination [28, 52, 170]. PLK4 is a serine-threonine kinase essential for centriole duplication. After few cell divisions in the presence of centrinone, cells progressively lose their centrioles leading to cells that lack centrosomes. Centrinone can also be used to generate cells with extra centrosomes, as its removal induces a wave of centriole duplication [168].

These studies have established the centrosome as the primary regulator of MT nucleation coordination. Inhibiting MT nucleation at the GA by knocking out AKAP450 does not affect centrosomal MT nucleation activity. Conversely, removing the centrosome significantly increases MT nucleation at the GA [52, 170]. Repolymerization experiments demonstrated a marked increase in both the number of Golgi elements capable of nucleating MTs and the quantity of MTs produced by each element after centrosome depletion (Fig. 12.9a; [52]). In contrast, cells with extra centrosomes following centrinone washout displayed reduced Golgi-associated MT nucleation activity in a centriole number-dependent manner (Fig. 12.9b; [52]). Notably, the total MT density in cells lacking centrosomes was double that of untreated cells, while MT density remained unchanged in cells with extra centrosomes (Fig. 12.9c). These findings indicate that the centrosome regulates MT numbers in cells by downregulating the MT nucleation activity of the GA [52].

The mechanism by which the centrosome regulates MT nucleation at the GA remains not fully understood. Depletion of centrosomes results in increased recruitment of AKAP450, CDK5Rap2, and γ-tubulin to Golgi membranes [52, 170]. Under these conditions, CAMSAP2 is no longer required for MT anchoring at the GA, suggesting that the Golgi itself enhances its capacity to anchor MT minus ends [170]. Notably, centrinone treatment also promotes the association of the PCM protein Pericentrin with Golgi membranes. Given the significant cytoplasmic pools of all these proteins, their availability does not appear to be the limiting factor; rather, the centrosomal pool may play a crucial role in regulating this process. Supporting this notion, depletion of the PCM proteins Cep192 or Pericentrin has been shown to increase MT nucleation at the GA, indicating a regulatory role for both proteins in this process [52, 106].

Finally, MT formation at the Golgi is tightly regulated throughout the cell cycle. Extensive MT nucleation during interphase is sharply reduced at prometaphase and completely inhibited during metaphase when the duplicated centrosomes become the

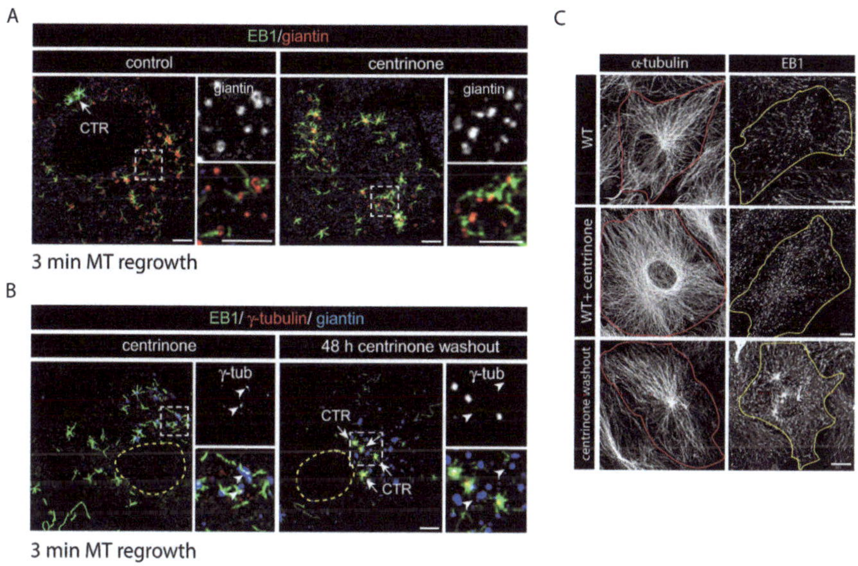

Fig. 12.9 Centrosome depletion effects on MT nucleation at the GA. (**a**) Control and centrinone-treated RPE1 cells were subjected to a 3-min MT regrowth assay following nocodazole treatment. Cells were then fixed and labeled for EB1 (green), a marker of growing MTs, and giantin (red), a marker of GA elements. Insets display magnified images of boxed regions. (**b**) RPE-1 cells were treated with centrinone for 7 days (left) or for 5 days followed by a 48-h recovery in the absence of the drug prior to fixation (right). Representative images of triple immunofluorescence staining for EB1 (green), γ-tubulin (red), and giantin (blue) are presented. Insets highlight MT nucleation from Golgi elements (left) or from multiple centrioles (right), with enlargements provided on the right. Arrowheads indicate the position of Golgi elements, while arrows indicate centrosomes. (**c**) Left panels: Confocal images show MT networks of control cells (top), cells lacking centrosomes (middle), and cells containing extra centrosomes (bottom), all stained for α-tubulin. Right panels: Growing MTs are visualized by labeling with EB1 under identical conditions. The red and yellow lines delineate the contours of the cells. (Recreated from [52]. © *2018, Gavilan et al (with permissions)*)

primary sites of MT nucleation for mitotic spindle formation. As the cell transitions into telophase, MT nucleation at the Golgi gradually reactivates, whereas the centrosomal MT nucleation activity strongly decreases [85].

These changes in MT nucleation activity are accompanied by notable remodeling of Golgi membranes. The continuous and interconnected structure of the Golgi ribbon poses challenges for its equal segregation during cell division. To address this, the Golgi ribbon is disassembled during mitosis. This disassembly process begins in late G2/early prophase and involves severing the lateral connections between cisternae, followed by the unstacking and vesiculation of Golgi elements. This results in tubular-vesicular membranes that are partitioned between daughter cells [33, 91, 163]. At telophase, the tubulovesicles fuse and reform intact and functional stacks. During cytokinesis, the reforming Golgi elements in daughter cells initially organize into two asymmetrical membrane clusters on either side of the nucleus. The smaller cluster, known as the "twin Golgi," is positioned near the intercellular bridge. This cluster subsequently migrates to the opposite side of the nucleus, merging with the larger pericentrosomal Golgi cluster, thus contributing to the reformation of the interphase Golgi ribbon [50].

There are ongoing debates regarding whether Golgi-nucleated MTs at G2/prophase contribute to spindle formation or if mitotic Golgi elements rely on the mitotic spindle for accurate partitioning between daughter cells. GM130, the Golgi receptor of AKAP450, has been reported to localize to the mitotic spindle and play a direct role in spindle formation by activating the MT nucleator TPX2 and stimulating astral MT growth [55, 164]. However, GM130 or AKAP450 knockout cells do not exhibit significant defects in spindle formation or cell cycle progression [52, 147]. This, along with the observed cell cycle-dependent modulation of MT nucleation at the Golgi, supports the notion that Golgi-derived MTs likely fulfill specific roles within the interphase/prophase MT network rather than directly contributing to the mitotic spindle.

While the molecular mechanisms driving Golgi disassembly—primarily involving phosphorylation and ubiquitination—are well understood [64], little is known about the processes that inactivate the machinery responsible for MT nucleation at mitotic entry and subsequently reactivate it. Interestingly, Cdk1-dependent phosphorylation of PTTG1/securin reduces its binding to Golgi membranes, delaying MT regrowth after nocodazole washout. Given that PTTG1 forms a complex with proteins involved in MT nucleation, such as GM130, AKAP450, and γ-tubulin, this mechanism may contribute to the downregulation of Golgi-derived MT nucleation at the onset of mitosis [98]. Other factors that could play a role include the mitotic phosphorylation of GM130 [81] and CLASPs [84]. Additionally, phosphoproteomics analyses have revealed that AKAP450 is phosphorylated during mitosis, suggesting further layers of regulation.

Functions of Golgi-Derived Microtubules

- **Golgi ribbon formation**

One of the primary functions of Golgi-derived MTs is to maintain the morphology and integrity of the Golgi ribbon, which is crucial as the GA undergoes substantial structural changes in response to various physiological demands [130, 186]. For instance, during cell migration, the Golgi ribbon temporarily unlinks and repositions itself according to migration cues [14, 94, 170]. Similarly, the Golgi ribbon fragments during the G2/M phase and reforms at mitotic exit, as previously discussed.

Before the discovery of Golgi-associated MT nucleation, it was believed that Golgi assembly was primarily driven by the radial MT array originating from the centrosome. However, it is now evident that Golgi-derived MTs also play a significant role in Golgi reassembly following cell division or experimentally induced fragmentation. Live-cell imaging of Golgi ribbon formation following nocodazole

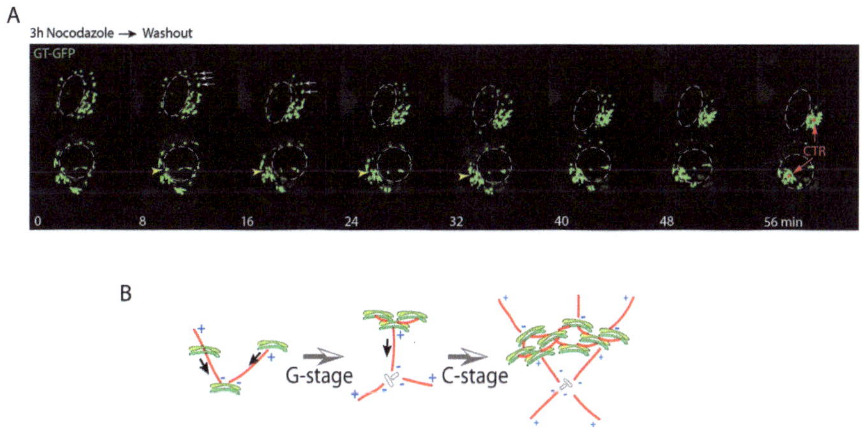

Fig. 12.10 Reassembly of the GA after nocodazole washout. (**a**) RPE-1 cells expressing the Golgi protein Galactosyl-Transferase fused with the fluorescent protein mCherry (GT-mcherry) were treated with nocodazole and cold for 4 h to depolymerized MTs and fragment the GA. After washout, cells were recorded using video microscopy, with the time elapsed since nocodazole removal indicated. (**b**) A cartoon illustrating the two-step GA reassembly process as proposed by Kaverina. Golgi elements partially fuse at the cell periphery before translocating toward the centrosome. Peripheral Golgi fusion relies on Golgi-derived MTs, while translocation to the cell center is driven by centrosomal MTs. ((**a**) Based on [66]. © *2011, Hurtado* et al. (**b**) Section extracted from [95]. © *2009, Springer Nature Limited (with permissions)*)

washout in RPE-1 cells has led to the proposal of a two-step mechanism (Fig. 12.10a, b; [66, 95]). Initially, Golgi ministacks cluster and fuse at the cell periphery. Quantitative analysis of these images, combined with data from fixed immunostained samples, shows that Golgi elements double in size during this phase. This growth is mediated by Golgi-nucleated MTs that frequently orient tangentially to the Golgi stacks. Subsequently, the Golgi elements relocate to the cell center along radial centrosomal MTs to complete Golgi ribbon formation around the centrosome (Fig. 12.10a, b; [66, 95]). Both processes are dependent on dynein.

While the two-stage model of Golgi ribbon assembly is now widely accepted, studies that interfere with MT nucleation have revealed unexpected flexibility in this process. For instance, AKAP450 knockout cells exhibit an intact and compacted pericentrosomal Golgi ribbon, despite the absence of Golgi MT nucleation [52, 170]. Conversely, a centrally located Golgi ribbon that extends around the nuclear envelope can still form when centrosomes are depleted using centrinone [52, 170]. Surprisingly, in cells lacking both AKAP450 and centrosomes—resulting in the absence of both Golgi- and centrosome-derived MT nucleation, yet still possessing a robust MT network from *de novo* cytoplasmic MTOCs—a Golgi ribbon persists, albeit in a more randomly positioned manner (Fig. 12.11; [52]). These observations collectively suggest that neither Golgi nor centrosomal MTs are strictly necessary for Golgi ribbon assembly, highlighting the inherent resilience of

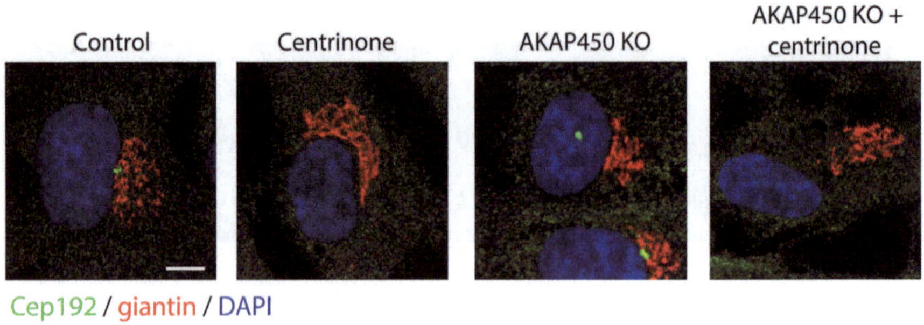

Fig. 12.11 Golgi integrity and positioning are not strictly dependent on either centrosomes or Golgi-nucleated MTs. Control, centrinone-treated, AKAP450 knockout, and AKAP450 knockout centrinone-treated cells were double-stained for Cep192, a centrosomal marker (green), and giantin, a Golgi marker (red). DNA was counterstained with DAPI and is shown in blue. Scale bars represent 10 μm. Our unpublished images

Golgi stacks in merging into a ribbon. The key factors for this process appear to be the presence of a functional MT network and dynein activity.

Interestingly, different methods of centrosome depletion, such as centrinone treatment and laser ablation, yield distinct outcomes for Golgi complex dynamics. Laser ablation of the centrosome does not disrupt existing Golgi complexes but prevents cells from reassembling a functional Golgi complex after nocodazole washout [159]. In contrast, centrosome depletion via centrinone still permits Golgi ribbon formation. This discrepancy may stem from the fact that laser ablation removes both centrioles and PCM proteins, while centrinone specifically targets centrioles without affecting PCM proteins, allowing these proteins to redistribute to the GA.

- **Golgi ribbon morphology**

The compact pericentrosomal morphology of the GA in cells lacking Golgi-nucleated MTs, combined with the extended Golgi ribbon configuration observed in centrosome-less cells, suggests that Golgi-derived MTs play a role in shaping Golgi ribbons. In the absence of tangential MTs nucleated at the GA, which probably facilitate lateral movements and interactions between stacks, centripetal movements driven by centrosome-anchored MTs become more prominent. Conversely, in cells lacking centrosomes, increased MT nucleation at the GA may enhance lateral movement, resulting in more extended Golgi ribbon configurations. This balance between the two MT subsets likely determines the steady-state morphology of the Golgi ribbon.

It is worth reminding that studies on Golgi ribbon localization, assembly, and dynamics have primarily focused on cells at early G1 or mitosis, reinforcing the conventional view that the GA surrounds the centrosome during interphase and disassembles at the onset of mitosis. However, recent research has shown that the localization, morphology, and association of the Golgi ribbon with the centrosome undergo significant changes throughout interphase [48]. During this phase of the cell cycle, the GA transitions between two distinct configurations: a highly compact morphology surrounding the centrosome (C mode) and a perinuclear extension away from the centrosome (E mode). C mode is characteristic of G1 cells that have just exited mitosis, while Golgi extension around the nucleus increases during the S phase, reaching maximum extension by late G2 or early prophase. This is followed by a rapid transition back to high compactness near the separated, duplicated centrosomes, typically just before nuclear envelope breakdown, which indicates entry into mitosis. These Golgi rearrangements are supported by Golgi-derived MTs and likely reflect modifications in the balance of membrane-bound dynein versus kinesin activities, as mentioned previously [48]. Understanding how this balance is regulated and how the GA maintains its perinuclear organization even in the absence of the centrosome warrants further investigation.

- **Cell polarity and migration**

Cell migration is a dynamic and intricately regulated process that is essential for various developmental activities, including organ and tissue formation, as well as wound repair. The first requirement for directional migration is the polarization of the cell into a leading edge and a trailing edge. Cell polarization is established through the tightly regulated asymmetric remodeling of the actin cytoskeleton, uneven formation and turnover of focal adhesions—anchorage sites for the cell—and directed membrane trafficking toward the protruding lamella. Polarization also involves GA and centrosome reorientation toward the direction of movement, along with the rearragenment of the MT network.

A key step in motility is the initiation of symmetry breaking, where cells establish their front and rear along the migration axis. MTs are crucial in this process, as they facilitate polarized trafficking, reinforce protrusive activity, and organize signaling to regulate the actin machinery responsible for membrane extensions. The MT network in migrating cells is characterized by a high concentration of MTs at the leading edge, imparting both structural and functional asymmetry (Fig. 12.12a; [37]). Since GA-nucleated MTs contribute to this asymmetry, their role in cell migration was quickly investigated. Depletion of AKAP450 [66, 120], CLASPs [95, 159], or CAMSAP2 [170] demonstrated that while Golgi-derived MTs are not required for random cell migration, they are essential for maintaining directionality. Notably, cells can polarize and migrate in the absence of centrosomes, but inhibiting Golgi MT formation impairs this process [90, 159]. Furthermore, post-Golgi trafficking orientation is disrupted in the absence of Golgi-derived MTs, though overall membrane trafficking remains unaffected. These findings highlight the importance of Golgi-derived MTs in maintaining polarized trafficking from the Golgi ribbon, which is essential for directed cell movement [66, 90, 159, 170].

Most data on cell polarization and directional motility come from studies of 2D cell migration, typically using wound-healing assays. However, examining cell invasion in three-dimensional (3D) matrices provides a more physiologically relevant context for understanding these processes. In 3D environments, cells are embedded in a matrix, leading to the formation of adhesions not only at the cell tips but also distributed across the entire cell surface. These adhesions cluster at the

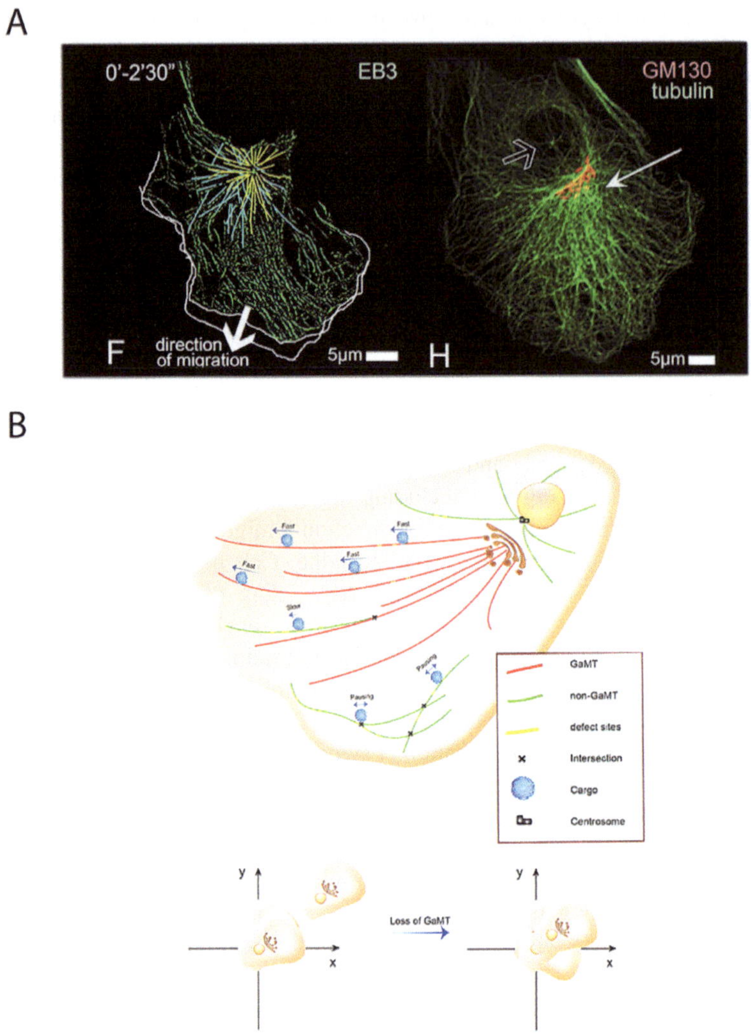

Fig. 12.12 (**a**) Directionality of Golgi-nucleated MTs in a motile cell. Time-lapse imaging of RPE-1 cells expressing mCherry-EB3, a marker of growing MTs, reveals the orientation of Golgi-nucleated MTs. Tracks of mCherry-EB3 were recorded over 2.5 min, overlaid (false-colored green), and displayed alongside centrosomal (yellow) and Golgi-associated (cyan) tracks at the cell center. The arrow indicates the direction of cell migration. The cell's protruding front, captured in the first and last frames of the 5-min video, is outlined with white lines (left panel). A prominent MT array (thin arrow) in a polarized cell is more closely associated with the GA (red) than with the centrosome (hollow arrow). Immunostaining shows α-tubulin in green and GM130 in red. (**b**) Cartoon illustrating Golgi MTs as fast tracks for post-Golgi cargo. Golgi-derived MTs, compared to non-Golgi MTs, are more stable, morphologically polarized toward the cell's leading edge, and exhibit significantly fewer intersections and defect repair sites, where cargo often pauses. These Golgi MTs enhance the persistence of cell migration without altering the directionality of motile cells. ((**a**) Portion of figure adapted from [37]. © *2007, Elsevier Inc. All rights reserved (with permissions)*. (**b**) Extracted from [56]. © *2020, Hao et al (with permissions)*)

contact points between the cell and the matrix, particularly in the anterior, neck-like regions of cell protrusions. Notably, the loss of Golgi-anchored MTs impairs the ability of cells to undergo invasive movement in 3D even more severely than in 2D migration [90, 170].

The prevailing view is that polarized membrane trafficking stabilizes protrusion sites and serves as a mechanism through which Golgi-associated MTs support and orient cell motility. Recent studies combining *in vivo* and super-resolution microscopy techniques have strongly supported this perspective (Fig. 12.12b; [56]). These analyses focused on a subset of MTs with their minus ends localized near the GA, extending toward the leading edge of migrating cells. The results indicated that Golgi-anchored MTs are more stable and morphologically polarized than non-Golgi MTs. Additionally, post-Golgi cargos moved significantly faster along Golgi-anchored MTs compared to non-Golgi MTs, suggesting that Golgi MTs serve as "fast tracks" for the anterograde trafficking of post-Golgi cargos (Fig. 12.12b; [56]).

Efficient cell migration also relies on the mechanical interplay between the cell and its environment, where forces generated at adhesion sites drive cytoskeletal dynamics. A specific type of migration, known as durotaxis, involves cells moving along stiffness gradients in the substrate and is critical in both development and disease. Recent studies have demonstrated that Golgi-derived MTs are essential for durotaxis, while centrosomal MTs are dispensable. Selective depletion of Golgi-associated MTs resulted in an increase in cell area, accompanied by a reduction in both the assembly/disassembly rate of focal adhesions and the frequency of MT-focal adhesion contacts. These findings indicate that Golgi-derived MTs are primary regulators of focal adhesion dynamics and may play a key role in establishing and maintaining cell polarization on stiffness gradient substrates [122].

Microtubule Nucleation in Differentiated Cells

As cells differentiate to fulfill organ- or tissue-specific functions, their MT networks undergo reorganization to accommodate new functional demands. During this reorganization, the centrosomal MTOC function diminishes, and MTs are primarily organized from non-centrosomal MTOCs [127]. The transition from centrosomal to non-centrosomal MTOCs occurs mainly in cells that, under physiological conditions, permanently exit the cell cycle during differentiation. Prominent examples are differentiated epithelial cells, cardiomyocytes, skeletal myotubes, and neurons [22, 45, 78, 135, 180]. Given the role of the centrosome in cell division, it is logical that cells that reduce their centrosomal MTOC also withdraw from a proliferative state. Conversely, exiting the cell cycle may be a prerequisite that allows cells to diminish the dominance of the centrosomal MTOC in order to establish non-centrosomal MTOCs. Neurons, striated muscle cells (i.e., skeletal and cardiac) and pancreatic beta-cells retain minimal MTOC activity at centrosomes, if any, while the predominant MTOC function is carried out by non-centrosomal sites. Among these, the GA emerges as the major non-centrosomal MTOC in differentiated cells.

- **Neurons**

Neurons exhibit a unique polarized morphology characterized by the distinct separation of axons and dendrites. The establishment and maintenance of this polarity critically depend on the precise organization and regulation of MTs [83, 109]. In axons, MTs are predominantly arranged in parallel bundles with uniform polarity, with their plus ends oriented away from the soma. In contrast, MT polarity in dendrites is more variable, influenced by factors such as neuron type, model organism, and developmental stage (Fig. 12.13a; [83, 121, 153]). Nevertheless, dendrites consistently differ from axons by the presence of minus-end-out MTs, a characteristic observed in both mammals and *Drosophila* [136, 177]. This intricate organization enables motor-dependent, compartment-specific sorting of cargo, facilitating efficient transport and distribution within the cell [138, 182]. Similarly, the GA in neurons exhibits a complex organization, featuring a convoluted structure within the soma and discrete Golgi outposts in dendrites (Fig. 12.13b; [183]). Although these outposts share common markers with the perinuclear Golgi body, they are smaller (typically 1–2 μm in diameter or length) and are exclusively localized to dendrites, where they contribute to local secretory networks [62]. Their prominence increases during periods of rapid dendritic growth and synaptogenesis [61, 152, 179].

As in other cell types undergoing differentiation, the centrosome gradually diminishes its role as the primary MTOC during neuronal development [83, 105, 135]. While proliferating cells, such as neuroprogenitors, utilize the centrosome to organize MTs, neurons exhibit a non-centrosomal organization of their MT network. In fact, studies involving laser or genetic ablation of the centrosome in *Drosophila* neurons have demonstrated that proper MT organization, axon specification, and neuronal development can occur independently of the centrosome [10, 105]. Notably, even after centrosome inactivation, the neuronal cytoplasm retains nucleation activity [105, 128, 178].

Since the discovery that acentrosomal MT nucleation is crucial for the complex morphology of neurons, significant efforts have been made to understand the role of Golgi-nucleated MTs in the development and maintenance of neuronal processes, including axons and dendrites. Time-lapse imaging in *Drosophila* class IV da neurons, known for their complex dendritic arbors, showed that many growing MTs originate from Golgi outposts at dendritic branch points [108]. This process requires both γ-tubulin and CP309/Plp (the *Drosophila* homolog of AKAP450). Ori-McKenney et al. also demonstrated that isolated Golgi membranes can nucleate MTs in a γ-tubulin-dependent manner, suggesting a key role for Golgi-associated MT nucleation in dendritic morphogenesis. Subsequent research revealed that the somatic Golgi in *Drosophila* da neurons consists of typical cis, medial, and trans compartments, while dendrites contain two distinct types of Golgi outposts: one with single compartments (either medial or trans) and another with all three compartments at dendritic branch points (Fig. 12.13b; [183]). Multicompartmental Golgi outposts were identified as primary initiation sites for MT growth in dendrites [183]. Supporting this, centrosomin (the *Drosophila* homolog of CDK5Rap2) was

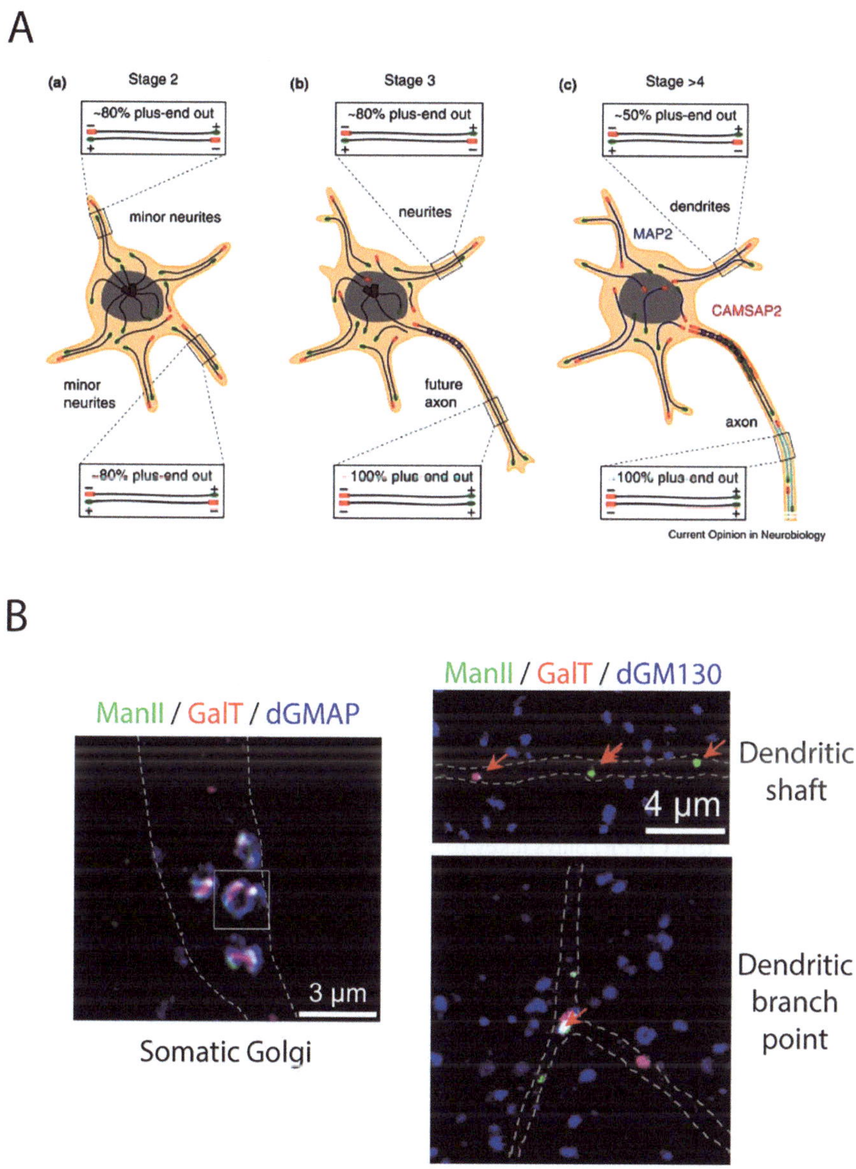

Fig. 12.13 (**a**) MT reorganization during neuronal polarization and axon formation can be observed in cultured hippocampal neurons. Shortly after plating, these neurons extend lamellipodia around the cell body, followed by the development of several minor neurites. Axon formation marks the first step in breaking the neuron's morphological symmetry. Finally, the remaining minor neurites differentiate into dendrites. Changes in MT organization throughout these developmental stages is represented by the percentage of plus-end-out MTs. The MT-associated protein MAP2 becomes enriched in the somatodendritic compartment, while the MT minus-end binding protein CAMSAP2 accumulates in the proximal region of the developing axon. (**b**) Distinct Golgi architectures in the soma and dendrites of *Drosophila* da neurons. The somatic Golgi consists of ring-shaped structures that contain cis- (dGMAP, blue), medial- (ManII, green), and trans- (GalT, magenta) Golgi markers. In contrast, Golgi compartments in the dendritic shafts are often

found to localize at the cis-side of Golgi outposts and promote minus-end-out MT nucleation [174].

However, these observations were challenged by another study that found γ-tubulin accumulation at dendritic branch points with limited co-localization with Golgi outposts. Analysis of *Drosophila* class IV and class I da neurons revealed that Golgi outposts predominantly exist in large dendritic arbors, whereas minus-end-out MTs also appear in smaller arbors. Importantly, the removal of Golgi outposts using kinesin-Golgi fusions did not affect γ-tubulin concentration at dendritic branch points [104]. More recent research suggests that Golgi outposts are not essential for the formation or maintenance of a dendrite-specific cytoskeleton. However, alterations in their organization or localization disrupt MT polarity in both dendrites and axons [176].

Interestingly, the cis-face of the Golgi body in the soma was identified as the primary site of γ-tubulin localization in both *Drosophila* class IV and I da neurons. The somatic Golgi actively nucleates MTs, which are directed into the axon and away from dendrites in a kinesin-2-dependent manner, thereby contributing to plus-end-out MT polarity in axons [100].

In summary, despite some contradictory results, multicompartmental Golgi outposts appear to be favored sites for γ-tubulin recruitment and MT nucleation at dendritic branch points, employing molecular mechanisms similar to those observed in non-neuronal cells. The somatic Golgi also functions as an active MTOC nucleating MTs that contribute to the plus-end-out MT pool in axons. Further research is needed to clarify some issues, identify the molecular actors involved, and extend these findings to vertebrate models. It is worth mentioning that Golgi-associated MT nucleation is not unique to neurons within nervous tissue. Golgi outposts have recently been identified in oligodendrocytes, where the tubulin polymerization-promoting protein (TPPP) plays a key role in MT nucleation. These Golgi-nucleated MTs have been shown to be essential for the myelination process [49].

- **Muscle cells**

Skeletal muscle contractile cells, known as myofibers, are elongated multinucleated syncytia formed by the fusion of proliferative myoblasts. In humans, myofibers can measure hundreds of microns in diameter and several millimeters in length. Although myofibers are non-mitotic, eliminating the need for MTs in cell division, MTs play an essential role in myofiber development and function. During myogenesis, the process by which skeletal muscle develops, mononucleated

Fig. 12.13 (continued) disconnected, as indicated by the separation of ManII (green) and GalT (magenta) labelings. The cis-Golgi protein dGM130 (blue) is not detectable in these regions. Red arrows highlight Golgi compartments containing either ManII or GalT. In dendritic branch points, Golgi outposts exhibit multiple compartments. Red arrows point to Golgi compartments containing ManII-GFP, GalT-RFP, and dGM130. ((**a**) Reproduced from van Beuningen and Hoogenraad [153]. (**b**) Modified from [183].

myoblasts stop proliferating and elongate to become fusion-competent. The fusion of myoblasts to form multinucleated myofibers involves the relocation of nuclei to the cell periphery, reorganization of the GA, loss of centrosomal MTOC activity, and a dramatic rearrangement of the MT cytoskeleton [12, 82, 115].

In myoblasts, the MT network is organized into a single radial array originating from the juxtanuclear centrosome, which is surrounded by the GA. As myoblasts fuse to form myotubes, which then elongate, MTs reorganize into stable bundles that encircle the nuclei and form grid-like networks aligned with the longitudinal axis of the myofiber. Concurrently, the GA dissociates from the centrosome, reconfiguring into a belt-like ring structure around each myotube nucleus and forming numerous small, static Golgi elements known as Golgi outposts, similar to those found in neurons (Fig. 12.14a; [107, 116, 152]). These changes in cellular architecture are accompanied by the redistribution of MT-nucleating capacity from the centrosome to non-centrosomal MTOCs.

Nearly four decades ago, pioneering studies by the Bornens group revealed that the nuclear envelope serves as the primary MTOC in striated muscle cells [139]. They observed that the GA in myotubes adopts a perinuclear distribution, closely aligned with this perinuclear MTOC, with the cis-side consistently positioned at a fixed distance from the nuclear envelope. This finding suggested a structural association between the GA and the nuclear periphery [140]. Further research indicated that PCM proteins, including AKAP450, PCM1, Pericentrin, CDK5Rap2, and γ-tubulin, localize to the nuclear envelope during myotube differentiation [25, 41, 42, 53].

Recent studies have begun to clarify the molecular mechanisms underlying MTOC assembly at the nuclear envelope. In skeletal muscle cells, the outer nuclear membrane protein Nesprin-1α recruits AKAP450, PCM1, and PCNT to the nuclear envelope, with only AKAP450 involved in MT nucleation [53]. Furthermore, AKAP450-mediated MT nucleation from the nuclear envelope is essential for nuclear spreading in myotubes. In rat cardiomyocytes, the muscle-specific protein AKAP6 localizes to the nuclear envelope and acts as an adaptor between Nesprin-1α and either PCNT or AKAP450, forming a protein platform that connects the Golgi to the nucleus. Unlike myotubes, cardiomyocytes exhibit MTOC activity at both the Golgi and the nuclear envelope, utilizing either AKAP450 or the PCNT-AKAP450 complex, respectively [156].

Actually, the close proximity of the nuclear envelope and surrounding Golgi belt make it challenging to distinguish whether MTs originate from the nuclear envelope or the adjacent Golgi. By examining gaps in the perinuclear GA, where the nuclear envelope was exposed, Ide et al. [67] clearly demonstrated that the GA serves as the primary site of MT nucleation during and after skeletal muscle cell differentiation (Fig. 12.14b–d). They identified CDK5Rap2-rich regions on the GA, near the nuclear envelope, as the key MT nucleating sites in skeletal myotubes (Fig. 12.14e). Furthermore, disrupting the perinuclear GA with brefeldin A significantly impaired MT regrowth around the nucleus, indicating that the nuclear envelope has minimal MT nucleating activity on its own.

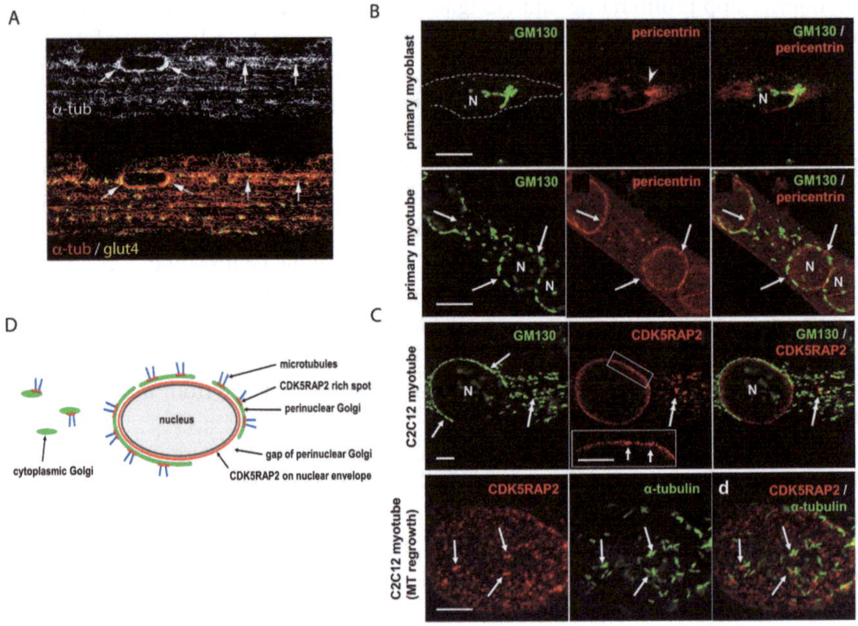

Fig. 12.14 (**a**) MT organization in a rat type I muscle fiber was examined through double staining for α-tubulin and GLUT4, a marker for Golgi elements. The top panel shows single labeling of α-tubulin, revealing MTs as dense bundles surrounding the nuclei and extending in long lines between them. Golgi elements are localized in both regions. (**b**) Dynamics of the GA and the centrosome during skeletal muscle cell differentiation are illustrated. In myoblasts, the GA is typically positioned near the centrosome (arrowhead), adjacent to the nucleus (top panels). During differentiation, both the GA and the pericentriolar material (PCM) translocate to the nuclear envelope, marking the onset of muscular differentiation (bottom panels). Gaps in the perinuclear GA can be observed (arrows). The GA was stained for GM130, and the centrosome for pericentrin. (**c**) The PCM protein CDK5Rap2 co-localizes with the GA, labeled for GM130, in C2C12 myotubes (top panels). Arrows indicate the perinuclear GA, while the double arrow highlights CDK5Rap2-positive cytoplasmic Golgi outposts. The inset provides an enlarged view of CDK5Rap2-rich spots (top panels). MT repolymerization experiments show MTs growing from CDK5Rap2-rich spots, as indicated by arrows (bottom panels). Scale bars represent 5 μm. (**d**) A cartoon depicts MT nucleation in myotubes. MTs are nucleated at CDK5Rap2-rich spots (red) on the perinuclear GA and at CDK5Rap2-positive cytoplasmic Golgi outposts (green). ((**a**) Taken from [116]. © *1999, Society for Neuroscience.* (**b–d**) Derived from [67]. © *2021, Ide* et al. *under exclusive licence to Springer-Verlag GmbH Germany, part of Springer Nature*)

In addition to the perinuclear GA, MTs are also nucleated from cytoplasmic Golgi outposts, forming bundles that grow orthogonally to pre-existing MTs and eventually interconnect, creating a lattice-like organization [107]. This lattice overlays the orthogonal network formed by dystrophin, the protein that is absent in Duchenne muscular dystrophy [111]. Furthermore, Golgi outposts themselves create a grid and are anchored along both the transverse and longitudinal components of dystrophin. Interestingly, PCNT, but not AKAP450, was detected at Golgi outposts via immunofluorescence microscopy. This finding led to the hypothesis that Golgi outpost-dependent nucleation involves PCNT in conjunction with γ-tubulin [107]. However, additional evidence, including depletion studies, is necessary to validate this hypothesis.

- **Pancreatic β-cells**

Pancreatic β-cells play a crucial role in maintaining glucose homeostasis by regulating insulin release, which is vital for human health. Excessive insulin secretion can lead to hypoglycemia, resulting in damage to the brain and other tissues [103], while insufficient secretion contributes to the development of diabetes [65]. After synthesis, insulin is stored in secretory granules and transported to the plasma membrane, where it is held in readiness for release upon demand [18]. Despite the presence of thousands of insulin granules in the cytoplasm, only 1–2% undergo exocytosis in response to glucose stimulation [124], indicating the presence of mechanisms that restrain their release.

Over the years, various studies have demonstrated that MTs serve as tracks for insulin granule transport and precisely regulate their subcellular positioning [19, 59, 150, 184]. Initially, MTs were believed to solely facilitate the transport of insulin granules from the Golgi to the plasma membrane [154]. However, research primarily conducted by Kaverina's group has shown that in functional β-cells, MTs play a more significant role in restricting rather than promoting insulin secretion. In β-cells, the MT cytoskeleton is organized into an internal meshwork of cross-linked and bent MTs, along with a population of MTs situated beneath and parallel to the plasma membrane [18, 101]. Insulin granules cluster near the subset of MTs oriented laterally to the plasma membrane [101]. Experimental depolymerization of MTs causes insulin granules to accumulate at the cell periphery, suggesting that MTs limit the number of readily releasable insulin granules at secretion sites [19, 63].

Live-cell imaging and nocodazole-recovery experiments have demonstrated that most MTs in β-cells are nucleated by the GA, while the centrosome exhibits very low MT nucleation activity [184]. In β-cells, the GA is notably large, spanning a significant portion of the cytoplasm and providing multiple nucleation sites for the generation of a dense, multi-directional MT network [89, 101]. Interestingly, the dynamics of this complex MT network vary substantially depending on glucose concentration (Fig. 12.15; [18]). Glucose stimulation induces MT remodeling by dramatically enhancing nucleation from the Golgi while favoring the depolymerization of pre-existing MTs [151, 184]. These changes release the grip of MTs on insulin granules, facilitating their secretion [18, 184]. Under low-glucose conditions, insulin granules move randomly along the non-directional tracks of the MT meshwork, resulting in a decreased number of granules available for secretion (Fig. 12.15; [18]). Thus, by limiting the trafficking of insulin vesicles, Golgi-derived MTs play a critical role in fine-tuning glucose-stimulated insulin secretion.

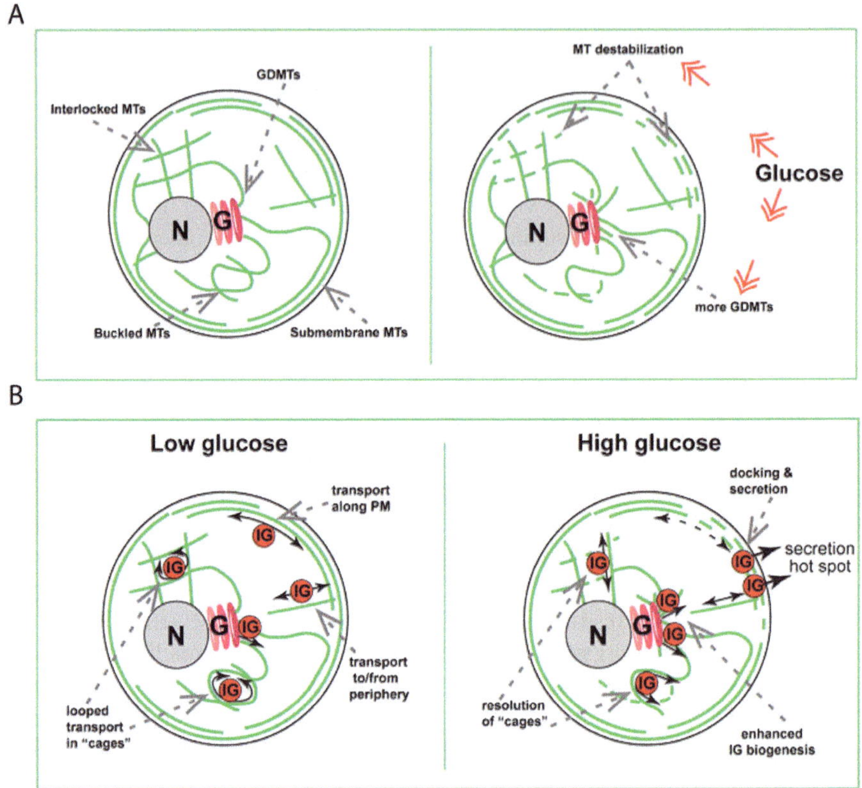

Fig. 12.15 (**a**) Illustration of MT sub-populations in a pancreatic beta cell. Under steady-state basal glucose conditions, new MTs are nucleated by the GA. Additionally, a submembrane MT array and an inner MT meshwork, formed by interlocking and buckled MTs, are present (left panel). When glucose levels rise, MT remodeling occurs through two processes: MT destabilization and increased MT nucleation at the GA (right panel). (**b**) MT-driven insulin granule transport in a pancreatic beta cell. During basal glucose conditions, few nascent MTs are generated at the GA, while many insulin granules are confined within MT cages in the inner cytoplasm, where they exhibit looped movements. At the periphery, submembrane and randomly oriented MTs enable bi-directional transport of insulin granules (left panel). Under high glucose conditions, partial destabilization of the inner MT meshwork allows for the release of insulin granules from their cages, leading to more directional transport. Simultaneously, local destabilization of submembrane MTs disrupts the retrieval of granules from active secretion sites, facilitating their docking and release (right panel). GA (G), Nucleus (N), Insulin Granules (IG). (Figures from [18]. © *2022 Bracey, Gu and Kaverina*)

Conclusions and Perspectives

Since the discovery of the MTOCs, research has predominantly focused on the centrosome, which organizes a radial array of MTs ideally suited for assembling the mitotic spindle. Indeed, nearly all animal cells, with very few exceptions, rely on centrosomes for mitosis. However, non-radial MT arrangements are better adapted for physiological processes that depend on cell polarity and support the specialized functions of differentiated cells. To generate such MT arrays, cells can reassign MTOC activity to non-centrosomal sites, with the GA being the most common alternative.

Over the past two decades, significant progress has been made in understanding the molecular mechanisms underlying the MTOC activity of the GA. These studies have revealed notable similarities with the centrosome. Current evidence suggests that the GA acquires MTOC activity by recruiting a 'nucleating module' composed of well-known and highly conserved components, including AKAP450 (and possibly PCNT), CDK5Rap2, Myomegalin 8, and γ-TuRCs. Intriguingly, despite sharing molecular components, the MTOC activities of the GA and the centrosome are inversely regulated during cell cycle progression and differentiation. How this regulation occurs remains a key question warranting further investigation. For instance, the mechanisms behind the shutdown of GA MTOC activity upon mitotic entry—while centrosomal activity increases—are still unclear. It is also unknown whether this process contributes to Golgi-associated mitotic checkpoints or the disassembly of the organelle during mitosis. Moreover, the mechanisms of centrosome downregulation or loss during cell differentiation are poorly understood. Future research will undoubtedly elucidate how cells utilize post-translational modifications, express specific regulators, and engage in competitive interactions between MTOCs to control different types of MT arrays.

Golgi-derived MTs are essential for cell polarization, adhesion, and migration in cycling cells, and they play a particularly significant role in the function of differentiated cells, such as neurons and muscle cells. Studies on other differentiated cell types and tissues, as well as on biological processes like apoptosis and autophagy, are expected to broaden our understanding of Golgi-nucleated MT functions. Additionally, recent advances in manipulating and visualizing MT dynamics in three-dimensional cultures are anticipated to enhance our understanding of these processes in more physiological environments. Continued progress in high-resolution three-dimensional microscopy, along with more efficient methods for gene knockout and endogenous protein tagging in specific tissues and organs, will propel the field forward and illuminate new research directions.

Research on GA MTOC activity holds significant medical relevance. Beyond its central role in protein sorting and trafficking, the mammalian Golgi's distinctive structure, arranged as an MT-dependent ribbon, is crucial for non-classical roles such as intracellular signaling and stress sensing, which are vital for maintaining cellular homeostasis. Golgi ribbon fragmentation and MT alterations are common hallmarks of neurodegenerative diseases and cancer. Fragmentation has been observed in model systems of conditions such as Creutzfeldt–Jakob disease, Alzheimer's disease, spinocerebellar ataxia type 2, Parkinson's disease, amyotrophic lateral sclerosis, Huntington's disease, corticobasal degeneration, and spinal muscular atrophy, typically occurring before other cellular disturbances. Such fragmentation may contribute to disease pathomechanisms by disrupting protein processing and trafficking to the synapse. Conversely, defective axonal transport observed in several neurodegenerative diseases may lead to Golgi disassembly due to protein accumulation in the soma. Given its proposed role in the formation of axons and dendrites, impaired MTOC activity of the GA could significantly impact the integrity and function of these structures.

In the context of cancer, unrestrained proliferation, along with aberrant survival and migration, increases the need for Golgi-dependent sensing and adaptation to intrinsic and extrinsic stressors. Investigating how Golgi fragmentation enhances cancer cell proliferation and metastasis will be a compelling area of future research, as will the role of Golgi-derived MTs in the dissemination of cancer cells. Furthermore, efforts may focus on developing small molecules or molecular tools aimed at rescuing the Golgi structure in various diseases, which could potentially delay disease progression.

Acknowledgements We thank C. Marcozzi for critically reading the manuscript. Work in Rios' lab is supported by FEDER/Ministerio de Ciencia e Innovación–Agencia Estatal de Investigación (PGC2018-095057-B-I00; PID2022-141680NB-I00) and Consejo Superior de Investigaciones Científicas (CSIC PIE Ref 202080E095).

References

1. Aher A, Kok M, Sharma A, Rai A, Olieric N, Rodriguez-Garcia R, Katrukha EA, Weinert T, Olieric V, Kapitein LC, Steinmetz MO, Dogterom M, Akhmanova A (2018) CLASP suppresses microtubule catastrophes through a single TOG domain. Dev Cell 46:40–58.e8. https://doi.org/10.1016/j.devcel.2018.05.032
2. Aher A, Urnavicius L, Xue A, Neselu K, Kapoor TM (2024) Structure of the γ-tubulin ring complex-capped microtubule. Nat Struct Mol Biol 31:1124–1133. https://doi.org/10.1038/s41594-024-01264-z
3. Akhmanova A, Hoogenraad CC (2015) Microtubule minus-end-targeting proteins. Curr Biol 25:R162–R171. https://doi.org/10.1016/j.cub.2014.12.027
4. Akhmanova A, Kapitein LC (2022) Mechanisms of microtubule organization in differentiated animal cells. Nat Rev Mol Cell Biol 23:541–558. https://doi.org/10.1038/s41580-022-00473-y
5. Akhmanova A, Steinmetz MO (2015) Control of microtubule organization and dynamics: two ends in the limelight. Nat Rev Mol Cell Biol 16:711–726. https://doi.org/10.1038/nrm4084
6. Alberts B, Heald R, Johnson A, Morgan D, Raff M, Roberts K, Walter P (2022) Molecular biology of the cell, 7th edn. W. W. Norton & Company. ISBN 13: 9780393884821
7. Ali A, Vineethakumari C, Lacasa C, Lüders J (2023) Microtubule nucleation and γTuRC centrosome localization in interphase cells require ch-TOG. Nat Commun 14:289. https://doi.org/10.1038/s41467-023-35955-w
8. Allan VJ, Thompson HM, McNiven MA (2002) Motoring around the Golgi. Nat Cell Biol 4: E236–E242. https://doi.org/10.1038/ncb1002-e236
9. Balabanian L, Berger CL, Hendricks AG (2017) Acetylated microtubules are preferentially bundled leading to enhanced Kinesin-1 motility. Biophys J 113:1551–1560. https://doi.org/10.1016/j.bpj.2017.08.009
10. Basto R, Lau J, Vinogradova T, Gardiol A, Woods CG, Khodjakov A, Raff JW (2006) Flies without centrioles. Cell 125:1375–1386. https://doi.org/10.1016/j.cell.2006.05.025

11. Bauer M, Cubizolles F, Schmidt A, Nigg EA (2016) Quantitative analysis of human centrosome architecture by targeted proteomics and fluorescence imaging. EMBO J 35:2152–2166. https://doi.org/10.15252/embj.201694462
12. Becker R, Leone M, Engel FB (2020) Microtubule Organization in Striated Muscle Cells. Cells 9:1395. https://doi.org/10.3390/cells9061395
13. Benvenuto G, Leone S, Astoricchio E, Bormke S, Jasek S, D'Aniello E, Kittelmann M, McDonald K, Hartenstein V, Baena V, Escrivà H, Bertrand S, Schierwater B, Burkhardt P, Ruiz-Trillo I, Jékely G, Ullrich-Lüter J, Lüter C, D'Aniello S, Arnone MI, Ferraro F (2024) Evolution of the ribbon-like organization of the Golgi apparatus in animal cells. Cell Rep 43: 113791. https://doi.org/10.1016/j.celrep.2024.113791
14. Bisel B, Wang Y, Wei JH, Xiang Y, Tang D, Miron-Mendoza M, Yoshimura S, Nakamura N, Seemann J (2008) ERK regulates Golgi and centrosome orientation towards the leading edge through GRASP65. J Cell Biol 182:837–843. https://doi.org/10.1083/jcb.200805045
15. Blanco-Ameijeiras J, Lozano-Fernández P, Martí E (2022) Centrosome maturation - in tune with the cell cycle. J Cell Sci 135:jcs259395. https://doi.org/10.1242/jcs.259395
16. Bornens M (2002) Centrosome composition and microtubule anchoring mechanisms. Curr Opin Cell Biol 14:25–34. https://doi.org/10.1016/S0955-0674(01)00290-3
17. Bornens M (2021) Centrosome organization and functions. Curr Opin Struct Biol 66:199–206. https://doi.org/10.1016/j.sbi.2020.11.002
18. Bracey KM, Gu G, Kaverina I (2022) Microtubules in pancreatic β cells: convoluted roadways toward precision. Front Cell Dev Biol 10:915206. https://doi.org/10.3389/fcell.2022.915206
19. Bracey KM, Ho KH, Yampolsky D, Gu G, Kaverina I, Holmes WR (2020) Microtubules regulate localization and availability of insulin granules in pancreatic Beta cells. Biophys J 118:193–206. https://doi.org/10.1016/j.bpj.2019.10.031
20. Brilot AF, Lyon AS, Zelter A, Viswanath S, Maxwell A, MacCoss MJ, Muller EG, Sali A, Davis TN, Agard DA (2012) CM1-driven assembly and activation of yeast γ-tubulin small complex underlies microtubule nucleation. elife 10:e65168. https://doi.org/10.7554/eLife.65168
21. Brito C, Serna M, Guerra P, Llorca O, Surrey T (2024) Transition of human gamma-tubulin ring complex into a closed *conformation* during microtubule nucleation. Science 383:870–876. https://doi.org/10.1126/science.adk6160
22. Brodu V, Baffet AD, Le Droguen PM, Casanova J, Guichet A (2010) A developmentally regulated two-step process generates a noncentrosomal microtubule network in Drosophila tracheal cells. Dev Cell 18:790–801. https://doi.org/10.1016/j.devcel.2010.03.015
23. Brouhard GJ, Stear JH, Noetzel TL, Al-Bassam J, Kinoshita K, Harrison SC, Howard J, Hyman AA (2008) XMAP215 is a processive microtubule polymerase. Cell 132:79–88. https://doi.org/10.1016/j.cell.2007.11.043
24. Brownhill K, Wood L, Allan V (2009) Molecular motors and the Golgi complex: staying put and moving through. Semin Cell Dev Biol 20:784–792. https://doi.org/10.1016/j.semcdb.2009.03.019
25. Bugnard E, Zaal KJ, Ralston E (2005) Reorganization of microtubule nucleation during muscle differentiation. Cell Motil Cytoskeleton 60:1–13. https://doi.org/10.1002/cm.20042
26. Burkhardt JK, Echeverri CJ, Nilsson T, Vallee RB (1997) Overexpression of the dynamitin (p50) subunit of the dynactin complex disrupts dynein-dependent maintenance of membrane organelle distribution. J Cell Biol 139:469–484. https://doi.org/10.1083/jcb.139.2.469
27. Chabin-Brion K, Marceiller J, Perez F, Settegrana C, Drechou A, Durand G, Pous C (2001) The Golgi complex is a microtubule-organizing organelle. Mol Biol Cell 12:2047–2060. https://doi.org/10.1091/mbc.12.7.2047
28. Chen F, Wu J, Iwanski MK, Jurriens D, Sandron A, Pasolli M, Puma G, Kromhout JZ, Yang C, Nijenhuis W, Kapitein LC, Berger F, Akhmanova A (2022) Self-assembly of pericentriolar material in interphase cells lacking centrioles. elife 11:e77892. https://doi.org/10.7554/eLife.77892

29. Choi YK, Liu P, Sze SK, Dai C, Qi RZ (2010) CDK5RAP2 stimulates microtubule nucleation by the gamma-tubulin ring complex. J Cell Biol 191:1089–1095. https://doi.org/10.1083/jcb.201007030
30. Cole NB, Sciaky N, Marotta A, Song J, Lippincott-Schwartz J (1996) Golgi dispersal during microtubule disruption: regeneration of Golgi stacks at peripheral endoplasmic reticulum exit sites. Mol Biol Cell 7:631–650. https://doi.org/10.1091/mbc.7.4.631
31. Conduit PT, Wainman A, Raff JW (2015) Centrosome function and assembly in animal cells. Nat Rev Mol Cell Biol 16:611–624. https://doi.org/10.1038/nrm4062
32. Consolati T, Locke J, Roostalu J, Chen ZA, Gannon J, Asthana J, Limn WM, Martino F, Cvetkovic MA, Rappsilber J, Costa A, Surrey T (2020) Microtubule nucleation properties of single human γTuRCs explained by their Cryo-EM structure. Dev Cell 53:603–617. https://doi.org/10.1016/j.devcel.2020.04.019
33. Corda D, Barretta ML, Cervigni RI, Colanzi A (2012) Golgi complex fragmentation in G2/M transition: an organelle-based cell-cycle checkpoint. IUBMB Life 64:661–670. https://doi.org/10.1002/iub.1054
34. Corthésy-Theulaz I, Pauloin A, Pfeffer SR (1992) Cytoplasmic dynein participates in the centrosomal localization of the Golgi complex. J Cell Biol 118:1333–1345. https://doi.org/10.1083/jcb.118.6.1333
35. Dendooven T, Yatskevich S, Burt A, Chen ZA, Bellini D, Rappsilber J, Kilmartin JV, Barford D (2024) Structure of the native γ-tubulin ring complex capping spindle microtubules. Nat Struct Mol Biol 31:1134–1144. https://doi.org/10.1038/s41594-024-01281-y
36. Desai A, Mitchinson TJ (1997) Microtubule polymerization dynamics. Annu Rev Cell Dev Biol 13:83–117. https://doi.org/10.1146/annurev.cellbio.13.1.83
37. Efimov A, Kharitonov A, Efimova N, Loncarek J, Miller PM, Andreyeva N, Gleeson P, Galjart N, Maia AR, McLeod IX, Yates JR, Maiato H, Khodjakov A, Akhmanova A, Kaverina I (2007) Asymmetric CLASP-dependent nucleation of noncentrosomal microtubules at the trans-Golgi network. Dev Cell 12:917–930. https://doi.org/10.1016/j.devcel.2007.04.002
38. Egea G, Rios RM (2008) The role of the cytoskeleton in the structure and function of the Golgi apparatus. In: Mironov AA, Pavelka M (eds) The Golgi apparatus. Springer, Vienna. https://doi.org/10.1007/978-3-211-76310-0_17
39. Egea G, Serra-Peinado C, Gavilan MP, Rios RM (2015) Cytoskeleton and Golgi-apparatus interactions: a two-way road of function and structure. Cell Health Cytoskelet 2015(7):37–54. https://doi.org/10.2147/CHC.S57108
40. Emr S, Glick BS, Linstedt AD, Lippincott-Schwartz J, Luini A, Malhotra V, Marsh BJ, Nakano A, Pfeffer SR, Rabouille C, Rothman JE, Warren G, Wieland FT (2009) Journeys through the Golgi—taking stock in a new era. J Cell Biol 187:449–453. https://doi.org/10.1083/jcb.200909011
41. Espigat-Georger A, Dyachuk V, Chemin C, Emorine L, Merdes A (2016) Nuclear alignment in myotubes requires centrosome proteins recruited by nesprin-1. J Cell Sci 129:4227–4237. https://doi.org/10.1242/jcs.191767
42. Fant X, Srsen V, Espigat-Georger A, Merdes A (2009) Nuclei of non-muscle cells bind centrosome proteins upon fusion with differentiating myoblasts. PLoS One 4:e8303. https://doi.org/10.1371/journal.pone.0008303
43. Farache D, Emorine L, Haren L, Merdes A (2018) Assembly and regulation of γ-tubulin complexes. Open Biol 8:170266. https://doi.org/10.1098/rsob.170266
44. Feiguin F, Ferreira A, Kosik KS, Caceres A (1994) Kinesin-mediated organelle translocation revealed by specific cellular manipulations. J Cell Biol 127:1021–1039
45. Feldman JL, Priess JR (2012) A role for the centrosome and PAR-3 in the hand-off of MTOC function during epithelial polarization. Curr Biol 22:575–582. https://doi.org/10.1016/j.cub.2012
46. Flor-Parra I, Iglesias-Romero AB, Chang F (2018) The XMAP215 ortholog Alp14 promotes microtubule nucleation in fission yeast. Curr Biol 28:1681–1691.e4. https://doi.org/10.1016/j.cub.2018.04.008

47. Fry AM, Sampson J, Shak C, Shackleton S (2017) Recent advances in pericentriolar material organization: ordered layers and scaffolding gels. F1000Res 6:1622. https://doi.org/10.12688/f1000research.11652.1
48. Frye K, Renda F, Fomicheva M, Zhu X, Gong L, Khodjakov A, Kaverina I (2020) Cell cycle-dependent dynamics of the Golgi-centrosome Association in motile cells. Cells 9:1069. https://doi.org/10.3390/cells9051069
49. Fu MM, McAlear TS, Nguyen H, Oses-Prieto JA, Valenzuela A, Shi RD, Perrino JJ, Huang TT, Burlingame AL, Bechstedt S, Barres BA (2019) The Golgi outpost protein TPPP nucleates microtubules and is critical for myelination. Cell 179:132–146.e14. https://doi.org/10.1016/j.cell.2019.08.025
50. Gaietta GM, Giepmans BN, Deerinck TJ, Smith WB, Ngan L, Llopis J, Adams SR, Tsien RY, Ellisman MH (2006) Golgi twins in late mitosis revealed by genetically encoded tags for live cell imaging and correlated electron microscopy. Proc Natl Acad Sci USA 103:17777–17782. https://doi.org/10.1073/pnas.0608509103
51. Gao Q, Vermeulen BJA, Würtz M, Shin H, Erdogdu D, Zheng A, Hofer FW, Neuner A, Pfeffer S, Schiebel E (2024) The structure of the γ-TuRC at the microtubule minus end - not just one solution. BioEssays 46:e2400117. https://doi.org/10.1002/bies.202400117
52. Gavilan MP, Gandolfo P, Balestra FR, Arias F, Bornens M, Rios RM (2018) The dual role of the centrosome in organizing the microtubule network in interphase. EMBO Rep 19:e45942. https://doi.org/10.15252/embr.201845942
53. Gimpel P, Lee YL, Sobota RM, Calvi A, Koullourou V, Patel R, Mamchaoui K, Nédélec F, Shackleton S, Schmoranzer J, Burke B, Cadot B, Gomes ER (2017) Nesprin-1α-dependent microtubule nucleation from the nuclear envelope via Akap450 is necessary for nuclear positioning in muscle cells. Curr Biol 27:2999–3009.e9. https://doi.org/10.1016/j.cub.2017.08.031
54. Gunzelmann J, Rüthnick D, Lin TC, Zhang W, Neuner A, Jäkle U, Schiebel E (2018) The microtubule polymerase Stu2 promotes oligomerization of the γ-TuSC for cytoplasmic microtubule nucleation. elife 7:e39932. https://doi.org/10.7554/eLife.39932
55. Guo H, Wei JH, Zhang Y, Seemann J (2021) Importin α phosphorylation promotes TPX2 activation by GM130 to control astral microtubules and spindle orientation. J Cell Sci 134:jcs258356. https://doi.org/10.1242/jcs.258356
56. Hao H, Niu J, Xue B, Su QP, Liu M, Yang J, Qin J, Zhao S, Wu C, Sun Y (2020) Golgi-associated microtubules are fast cargo tracks and required for persistent cell migration. EMBO Rep 21:e48385. https://doi.org/10.15252/embr.201948385
57. Harada A, Takei Y, Kanai Y, Tanaka Y, Nonaka S, Hirokawa N (1998) Golgi vesiculation and lysosome dispersion in cells lacking cytoplasmic dynein. J Cell Biol 141:51–59. https://doi.org/10.1083/jcb.141.1.51
58. Hirokawa N (1998) Kinesin and dynein superfamily proteins and the mechanism of organelle transport. Science:279–519–526. https://doi.org/10.1126/science.279.5350.519
59. Ho KH, Yang X, Osipovich AB, Cabrera O, Hayashi ML, Magnuson MA, Gu G, Kaverina I (2020) Glucose regulates microtubule disassembly and the dose of insulin secretion via tau phosphorylation. Diabetes 69:1936–1947. https://doi.org/10.2337/db19-1186
60. Ho WC, Allan VJ, Van Meer G, Berger EG, Kreis TE (1989) Reclustering of scattered Golgi elements occurs along microtubules. Eur J Cell Biol 48:250–263
61. Horton AC, Ehlers MD (2003) Dual modes of endoplasmic reticulum-to-Golgi transport in dendrites revealed by live-cell imaging. J Neurosci 23:6188–6199. https://doi.org/10.1523/JNEUROSCI.23-15-06188.2003
62. Horton AC, Rácz B, Monson EE, Lin AL, Weinberg RJ, Ehlers MD (2005) Polarized secretory trafficking directs cargo for asymmetric dendrite growth and morphogenesis. Neuron 48:757–771. https://doi.org/10.1016/j.neuron.2005.11.005
63. Hu R, Zhu X, Yuan M, Ho KH, Kaverina I, Gu G (2021) Microtubules and Gαo-signaling modulate the preferential secretion of young insulin secretory granules in islet β cells via

independent pathways. PLoS One 16:e0241939. https://doi.org/10.1371/journal.pone.0241939
64. Huang S, Wang Y (2017) Golgi structure formation, function, and post-translational modifications in mammalian cells. F1000Res 6:2050. https://doi.org/10.12688/f1000research.11900.1
65. Hudish LI, Reusch JE, Sussel L (2019) β cell dysfunction during progression of metabolic syndrome to type 2 diabetes. J Clin Invest 129:4001–4008. https://doi.org/10.1172/JCI129188
66. Hurtado L, Caballero C, Gavilan MP, Cardenas J, Bornens M, Rios RM (2011) Disconnecting the Golgi ribbon from the centrosome prevents directional cell migration and ciliogenesis. J Cell Biol 193:917–933. https://doi.org/10.1083/jcb.201011014
67. Ide K, Muko M, Hayashi K (2021) The Golgi apparatus is the main microtubule-organizing center in differentiating skeletal muscle cells. Histochem Cell Biol 156:273–281. https://doi.org/10.1007/s00418-021-01999-6
68. Janke C, Magiera MM (2020) The tubulin code and its role in controlling microtubule properties and functions. Nat Rev Mol Cell Biol 21:307–326. https://doi.org/10.1038/s41580-020-0214-3
69. Kelliher MT, Yue Y, Ng A, Kamiyama D, Huang B, Verhey KJ, Wildonger J (2018) Autoinhibition of kinesin-1 is essential to the dendrite-specific localization of Golgi outposts. J Cell Biol 217:2531–2547. https://doi.org/10.1083/jcb.201708096
70. Képès F, Rambourg A, Satiat-Jeunemaître B (2005) Morphodynamics of the secretory pathway. Int Rev Cytol 242:55–120. https://doi.org/10.1016/S0074-7696(04)42002-6
71. Kollman JM, Greenberg CH, Li S, Moritz M, Zelter A, Fong KK, Fernandez JJ, Sali A, Kilmartin J, Davis TN, Agard DA (2015) Ring closure activates yeast γTuRC for species-specific microtubule nucleation. Nat Struct Mol Biol 22:132–137. https://doi.org/10.1038/nsmb.2953
72. Kollman JM, Merdes A, Mourey L, Agard DA (2011) Microtubule nucleation by γ-tubulin complexes. Nat Rev Mol Cell Biol 12:709–721. https://doi.org/10.1038/nrm3209
73. Kollman JM, Polka JK, Zelter A, Davis TN, Agard DA (2010) Microtubule nucleating gamma-TuSC assembles structures with 13-fold microtubule-like symmetry. Nature 466:879–882. https://doi.org/10.1038/nature09207
74. Kraus J, Alfaro-Aco R, Gouveia B, Petry S (2023) Microtubule nucleation for spindle assembly: one molecule at a time. Trends Biochem Sci 48:761–775. https://doi.org/10.1016/j.tibs.2023.06.004
75. Ladinsky MS, Mastronarde DN, McIntosh JR, Howell KE, Staehelin LA (1999) Golgi structure in three dimensions: functional insights from the normal rat kidney cell. J Cell Biol 144:1135–1149. https://doi.org/10.1083/jcb.144.6.1135
76. Lawo S, Hasegan M, Gupta GD, Pelletier L (2012) Subdiffraction imaging of centrosomes reveals higher-order organizational features of pericentriolar material. Nat Cell Biol 14:1148–1158. https://doi.org/10.1038/ncb2591
77. Lawrence EJ, Zanic M, Rice LM (2020) CLASPs at a glance. J Cell Sci 133:jcs243097. https://doi.org/10.1242/jcs.243097
78. Lechler T, Fuchs E (2007) Desmoplakin: an unexpected regulator of microtubule organization in the epidermis. J Cell Biol 176:147–154. https://doi.org/10.1083/jcb.200609109
79. Lin TC, Neuner A, Schiebel E (2015) Targeting of γ-tubulin complexes to microtubule organizing centers: conservation and divergence. Trends Cell Biol 25:296–307. https://doi.org/10.1016/j.tcb.2014.12.002
80. Liu P, Zupa E, Neuner A, Bohler A, Loerke J, Flemming D, Ruppert T, Rudack T, Peter C, Spahn C, Gruss OJ, Pfeffer S, Schiebel E (2020) Insights into the assembly and activation of the microtubule nucleator γ-TuRC. Nature 578:467–471. https://doi.org/10.1038/s41586-019-1896-6
81. Lowe M, Gonatas NK, Warren G (2000) The mitotic phosphorylation cycle of the cis-Golgi matrix protein GM130. J Cell Biol 149:341–356. https://doi.org/10.1083/jcb.149.2.341

82. Lucas L, Cooper TA (2023) Insights into cell-specific functions of microtubules in skeletal muscle development and homeostasis. Int J Mol Sci 24:2903. https://doi.org/10.3390/ijms24032903. PMC3409831
83. Lüders J (2021) Nucleating microtubules in neurons: challenges and solutions. Dev Neurobiol 81:273–283. https://doi.org/10.1002/dneu.22751
84. Maia AR, Garcia Z, Kabeche L, Barisic M, Maffini S, Macedo-Ribeiro S, Cheeseman IM, Compton DA, Kaverina I, Maiato H (2012) Cdk1 and Plk1 mediate a CLASP2 phospho-switch that stabilizes kinetochore-microtubule attachments. J Cell Biol 199:285–301. https://doi.org/10.1083/jcb.201203091
85. Maia AR, Zhu X, Miller P, Gu G, Maiato H, Kaverina I (2013) Modulation of Golgi-associated microtubule nucleation throughout the cell cycle. Cytoskeleton 70:32–43. https://doi.org/10.1002/cm.21079
86. Mammri L, Conduit PT (2024) Cryo-EM structures of γ-TuRC reveal molecular insights into microtubule nucleation. Nat Struct Mol Biol 31:1004–1006. https://doi.org/10.1038/s41594-024-01345-z
87. Marks D, Larkin J, McNiven M (1994) Association of kinesin with the Golgi apparatus in rat hepatocytes. J Cell Sci 107:2417–2426. https://doi.org/10.1242/jcs.107.9.2417
88. Marsh BJ, Mastronarde DN, Buttle KF, Howell KE, McIntosh JR (2001) Organellar relationships in the Golgi region of the pancreatic beta cell line, HIT-T15, visualized by high resolution electron tomography. Proc Natl Acad Sci USA 98:2399–2406. https://doi.org/10.1073/pnas.051631998
89. Marsh BJ, Volkmann N, McIntosh JR, Howell KE (2004) Direct continuities between cisternae at different levels of the Golgi complex in glucose-stimulated mouse islet beta cells. Proc Natl Acad Sci USA 101:5565–5570. https://doi.org/10.1073/pnas.0401242101
90. Martin M, Veloso A, Wu J, Katrukha EA, Akhmanova A (2018) Control of endothelial cell polarity and sprouting angiogenesis by non-centrosomal microtubules. elife 7:e33864. https://doi.org/10.7554/eLife.33864
91. Mascanzoni F, Iannitti R, Colanzi A (2022) Functional coordination among the Golgi complex, the centrosome and the microtubule cytoskeleton during the cell cycle. Cells 11:354. https://doi.org/10.3390/cells11030354
92. Matsuoka R, Miki M, Mizuno S, Ito Y, Yamada C, Suzuki A (2022) MTCL2 promotes asymmetric microtubule organization by crosslinking microtubules on the Golgi membrane. J Cell Sci 135:jcs259374. https://doi.org/10.1242/jcs.259374
93. McManus CT, Travis SM, Jeffrey PD, Zhang R, Petry S (2024) Mechanism of how the universal module XMAP215 γ-TuRC nucleates microtubules. bioRxiv [Preprint]. 3: 2024.06.03.597159. https://doi.org/10.1101/2024.06.03.597159
94. Millarte V, Farhan H (2012) The Golgi in cell migration: regulation by signal transduction and its implications for cancer cell metastasis. ScientificWorldJournal 2012:498278. https://doi.org/10.1100/2012/498278
95. Miller PM, Folkmann AW, Maia AR, Efimova N, Efimov A, Kaverina I (2009) Golgi-derived CLASP-dependent microtubules control Golgi organization and polarized trafficking in motile cells. Nat Cell Biol 11:1069–1080. https://doi.org/10.1038/ncb1920
96. Mironov AA, Sesorova IS, Seliverstova EV, Beznoussenko GV (2017) Different Golgi ultrastructure across species and tissues: implications under functional and pathological conditions, and an attempt at classification. Tissue Cell 49:186–201. https://doi.org/10.1016/j.tice.2016.12.002
97. Mohan AG, Calenic B, Ghiurau NA, Duncea-Borca RM, Constantinescu AE, Constantinescu I (2023) The Golgi apparatus: a voyage through time, structure, function and implication in neurodegenerative disorders. Cells 12:1972. https://doi.org/10.3390/cells12151972
98. Moreno-Mateos MA, Espina ÁG, Torres B, Gámez del Estal MM, Romero-Franco A, Ríos RM, Pintor-Toro JA (2011) PTTG1/securin modulates microtubule nucleation and cell migration. Mol Biol Cell 22:4302–4311. https://doi.org/10.1091/mbc.E10-10-0838

99. Moudjou M, Bordes N, Paintrand M, Bornens M (1996) gamma-Tubulin in mammalian cells: the centrosomal and the cytosolic forms. J Cell Sci 109:875–887. https://doi.org/10.1242/jcs.109.4.875
100. Mukherjee A, Brooks PS, Bernard F, Guichet A, Conduit PT (2020) Microtubules originate asymmetrically at the somatic golgi and are guided via Kinesin2 to maintain polarity within neurons. elife 9:e58943. https://doi.org/10.7554/eLife.58943
101. Müller A, Schmidt D, Xu CS, Pang S, D'Costa JV, Kretschmar S, Münster C, Kurth T, Jug F, Weigert M, Hess HF, Solimena M (2021) 3D FIB-SEM reconstruction of microtubule-organelle interaction in whole primary mouse β cells. J Cell Biol 220:e202010039. https://doi.org/10.1083/jcb.202010039
102. Nakano A, Luini A (2010) Passage through the Golgi. Curr Opin Cell Biol 22:471–478. https://doi.org/10.1016/j.ceb.2010.05.003
103. Nessa A, Rahman SA, Hussain K (2016) Hyperinsulinemic Hypoglycemia - the molecular mechanisms. Front Endocrinol 7:29. https://doi.org/10.3389/fendo.2016.00029
104. Nguyen MM, McCracken CJ, Milner ES, Goetschius DJ, Weiner AT, Long MK, Michael NL, Munro S, Rolls MM (2014) Γ-tubulin controls neuronal microtubule polarity independently of Golgi outposts. Mol Biol Cell 25:2039–2050. https://doi.org/10.1091/mbc.E13-09-0515
105. Nguyen MM, Stone MC, Rolls MM (2011) Microtubules are organized independently of the centrosome in Drosophila neurons. Neural Dev 6:38. https://doi.org/10.1186/1749-8104-6-38
106. O'Rourke BP, Gomez-Ferreria MA, Berk RH, Hackl AM, Nicholas MP, O'Rourke SC, Pelletier L, Sharp DJ (2014) Cep192 controls the balance of centrosome and non-centrosomal microtubules during interphase. PLoS One 9:e101001. https://doi.org/10.1371/journal.pone.0101001
107. Oddoux S, Zaal KJ, Tate V, Kenea A, Nandkeolyar SA, Reid E, Liu W, Ralston E (2013) Microtubules that form the stationary lattice of muscle fibers are dynamic and nucleated at Golgi elements. J Cell Biol 203:205–213. https://doi.org/10.1083/jcb.201304063
108. Ori-McKenney KM, Jan LY, Jan YN (2012) Golgi outposts shape dendrite morphology by functioning as sites of acentrosomal microtubule nucleation in neurons. Neuron 76:921–930. https://doi.org/10.1016/j.neuron.2012.10.008
109. Paz J, Lüders J (2018) Microtubule-Organizing Centers: towards a minimal parts list. Trends Cell Biol 28:176–187. https://doi.org/10.1016/j.tcb.2017.10.005
110. Presley JF, Cole NB, Schroer TA, Hirschberg K, Zaal KJ, Lippincott-Schwartz J (1997) ER-to-Golgi transport visualized in living cells. Nature 389:81–85. https://doi.org/10.1038/38891
111. Prins KW, Humston JL, Mehta A, Tate V, Ralston E, Ervasti JM (2009) Dystrophin is a microtubule-associated protein. J Cell Biol 186:363–369. https://doi.org/10.1083/jcb.200905048
112. Quintyne NJ, Gill SR, Eckley DM, Crego CL, Compton DA, Schroer TA (1999) Dynactin is required for microtubule anchoring at centrosomes. J Cell Biol 147:321–334. https://doi.org/10.1083/jcb.147.2.321
113. Rai D, Song Y, Hua S, Stecker K, Monster JL, Yin V, Stucchi R, Xu Y, Zhang Y, Chen F, Katrukha EA, Altelaar M, Heck AJR, Wieczorek M, Jiang K, Akhmanova A (2024) CAMSAPs and nucleation-promoting factors control microtubule release from γ-TuRC. Nat Cell Biol 26:404–420. https://doi.org/10.1038/s41556-024-01366-2
114. Rale MJ, Romer B, Mahon BP, Travis SM, Petry S (2022) The conserved centrosomin motif, γTuNA, forms a dimer that directly activates microtubule nucleation by the γ-tubulin ring complex (γTuRC). elife 11:e80053. https://doi.org/10.7554/eLife.80053
115. Ralston E (1993) Changes in architecture of the Golgi complex and other subcellular organelles during myogenesis. J Cell Biol 120:399–409. https://doi.org/10.1083/jcb.120.2.399
116. Ralston E, Lu Z, Ploug T (1999) The organization of the Golgi complex and microtubules in skeletal muscle is fiber type-dependent. J Neurosci 19:10694–10705. https://doi.org/10.1523/JNEUROSCI.19-24-10694.1999

117. Reed NA, Cai D, Blasius TL, Jih GT, Meyhofer E, Gaertig J, Verhey KJ (2006) Microtubule acetylation promotes kinesin-1 binding and transport. Curr Biol 16:2166–2172. https://doi.org/10.1016/j.cub.2006.09.014
118. Rios RM (2014) The centrosome-Golgi apparatus nexus. Philos Trans R Soc Lond Ser B Biol Sci 369:20130462. https://doi.org/10.1098/rstb.2013.0462
119. Rios RM, Bornens M (2003) The Golgi apparatus at the cell centre. Curr Opin Cell Biol 15: 60–66. https://doi.org/10.1016/s0955-0674(02)00013-3
120. Rivero S, Cardenas J, Bornens M, Rios RM (2009) Microtubule nucleation at the cis-side of the Golgi apparatus requires AKAP450 and GM130. EMBO J 28:1016–1028. https://doi.org/10.1038/emboj.2009.47
121. Rolls MM, Jegla TJ (2015) Neuronal polarity: an evolutionary perspective. J Exp Biol 218: 572–580. https://doi.org/10.1242/jeb.112359
122. Rong Y, Yang W, Hao H, Wang W, Lin S, Shi P, Huang Y, Li B, Sun Y, Liu Z, Wu C (2021) The Golgi microtubules regulate single cell durotaxis. EMBO Rep 22:e51094. https://doi.org/10.15252/embr.202051094
123. Roostalu J, Surrey T (2017) Microtubule nucleation: beyond the template. Nat Rev Mol Cell Biol 18:702–710. https://doi.org/10.1038/nrm.2017.75
124. Rorsman P, Renström E (2003) Insulin granule dynamics in pancreatic beta cells. Diabetologia 46:1029–1045. https://doi.org/10.1007/s00125-003-1153-1
125. Roubin R, Acquaviva C, Chevrier V, Sedjaï F, Zyss D, Birnbaum D, Rosnet O (2013) Myomegalin is necessary for the formation of centrosomal and Golgi-derived microtubules. Biol Open 2:238–250. https://doi.org/10.1242/bio.20123392
126. Sallee MD, Feldman JL (2021) Microtubule organization across cell types and states. Curr Biol 31:R506–R511. https://doi.org/10.1016/j.cub.2021.01.042
127. Sanchez AD, Feldman JL (2017) Microtubule-organizing centers: from the centrosome to non-centrosomal sites. Curr Opin Cell Biol 44:93–101. https://doi.org/10.1016/j.ceb.2016.09.003
128. Sánchez-Huertas C, Freixo F, Viais R, Lacasa C, Soriano E, Lüders J (2016) Non-centrosomal nucleation mediated by augmin organizes microtubules in post-mitotic neurons and controls axonal microtubule polarity. Nat Commun 7:12187. https://doi.org/10.1038/ncomms12187
129. Sanders AAWM, Chang K, Zhu X, Thoppil RJ, Holmes WR, Kaverina I (2017) Nonrandom γ-TuNA-dependent spatial pattern of microtubule nucleation at the Golgi. Mol Biol Cell 28: 3181–3192. https://doi.org/10.1091/mbc.E17-06-0425
130. Saraste J, Prydz K (2019) A new look at the functional organization of the Golgi Ribbon. Front Cell Dev Biol 7:171. https://doi.org/10.3389/fcell.2019.00171
131. Sato Y, Hayashi K, Amano Y, Takahashi M, Yonemura S, Hayashi I, Hirose H, Ohno S, Suzuki A (2014) MTCL1 crosslinks and stabilizes non-centrosomal microtubules on the Golgi membrane. Nat Commun 5:5266. https://doi.org/10.1038/ncomms6266
132. Sawin KE, Lourenço PCC, Snaith HA (2004) Microtubule nucleation at non-spindle pole body microtubule-organizing centers requires fission yeast centrosomin-related protein mod20p. Curr Biol 14:763–775. https://doi.org/10.1016/j.cub.2004.03.042
133. Schroer TA (2004) Dynactin. Annu Rev Cell Dev Biol 20:759–779. https://doi.org/10.1146/annurev.cellbio.20.012103.094623
134. Skoufias DA, Burgess TL, Wilson L (1990) Spatial and temporal colocalization of the Golgi apparatus and microtubules rich in detyrosinated tubulin. J Cell Biol 111:1929–1937. https://doi.org/10.1083/jcb.111.5.1929
135. Stiess M, Maghelli N, Kapitein LC, Gomis-Rüth S, Wilsch-Bräuninger M, Hoogenraad CC, Tolić-Nørrelykke IM, Bradke F (2010) Axon extension occurs independently of centrosomal microtubule nucleation. Science 327:704–707. https://doi.org/10.1126/science.1182179
136. Stone MC, Roegiers F, Rolls MM (2008) Microtubules have opposite orientation in axons and dendrites of Drosophila neurons. Mol Biol Cell 19:4122–4129. https://doi.org/10.1091/mbc.e07-10-1079

137. Sweeney HL, Holzbaur ELF (2018) Motor Proteins. Cold Spring Harb Perspect Biol 10: a021931. https://doi.org/10.1101/cshperspect.a021931
138. Tas RP, Chazeau A, Cloin BMC, Lambers MLA, Hoogenraad CC, Kapitein LC (2017) Differentiation between oppositely oriented microtubules controls polarized neuronal transport. Neuron 96:1264–1271.e5. https://doi.org/10.1016/j.neuron.2017.11.018
139. Tassin AM, Maro B, Bornens M (1985a) Fate of microtubule-organizing centers during myogenesis in vitro. J Cell Biol 100:35–46. https://doi.org/10.1083/jcb.100.1.35
140. Tassin AM, Paintrand M, Berger EG, Bornens M (1985b) The Golgi apparatus remains associated with microtubule organizing centers during myogenesis. J Cell Biol 101:630–638. https://doi.org/10.1083/jcb.101.2.630
141. Teixidó-Travesa N, Roig J, Lüders J (2012) The where, when and how of microtubule nucleation-one ring to rule them all. J Cell Sci 125:4445–4456. https://doi.org/10.1242/jcs.106971
142. Thawani A, Kadzik RS, Petry S (2018) XMAP215 is a microtubule nucleation factor that functions synergistically with the γ-tubulin ring complex. Nat Cell Biol 20:575–585. https://doi.org/10.1038/s41556-018-0091-6
143. Thawani A, Petry S (2021) Molecular insight into how γ-TuRC makes microtubules. J Cell Sci 134:jcs245464. https://doi.org/10.1242/jcs.245464
144. Thawani A, Rale MJ, Coudray N, Bhabha G, Stone HA, Shaevitz JW, Petry S (2020) The transition state and regulation of γ-TuRC-mediated microtubule nucleation revealed by single molecule microscopy. elife 9:e54253. https://doi.org/10.7554/eLife.54253
145. Thyberg J, Moskalewski S (1993) Relationship between the Golgi complex and microtubules enriched in detyrosinated or acetylated alpha-tubulin: studies on cells recovering from nocodazole and cells in the terminal phase of cytokinesis. Cell Tissue Res 273:457–466. https://doi.org/10.1007/BF00333700
146. Thyberg J, Moskalewski S (1999) Role of microtubules in the Organization of the Golgi Complex. Exp Cell Res 246:263–279. https://doi.org/10.1016/s0014-4827(85)80032-x
147. Tormanen K, Ton C, Waring BM, Wang K, Sütterlin C (2019) Function of Golgi-centrosome proximity in RPE-1 cells. PLoS One 14:e0215215. https://doi.org/10.1371/journal.pone.0215215
148. Tovey CA, Conduit PT (2018) Microtubule nucleation by γ-tubulin complexes and beyond. Essays Biochem 62:765–780. https://doi.org/10.1042/EBC20180028
149. Tovey CA, Tsuji C, Egerton A, Bernard F, Guichet A, de la Roche M, Conduit PT (2021) Autoinhibition of Cnn binding to γ-TuRCs prevents ectopic microtubule nucleation and cell division defects. J Cell Biol 220:e202010020. https://doi.org/10.1083/jcb.202010020
150. Trogden KP, Lee J, Bracey KM, Ho KH, McKinney H, Zhu X, Arpag G, Folland TG, Osipovich AB, Magnuson MA, Zanic M, Gu G, Holmes WR, Kaverina I (2021) Microtubules regulate pancreatic β-cell heterogeneity via spatiotemporal control of insulin secretion hot spots. elife 10:e59912. https://doi.org/10.7554/eLife.59912
151. Trogden KP, Zhu X, Lee JS, Wright CVE, Gu G, Kaverina I (2019) Regulation of glucose-dependent Golgi-derived microtubules by cAMP/EPAC2 promotes secretory vesicle biogenesis in pancreatic β cells. Curr Biol 29:2339–2350.e5. https://doi.org/10.1016/j.cub.2019.06.032
152. Valenzuela A, Meservey L, Nguyen H, Fu M (2020) Golgi outposts nucleate microtubules in cells with specialized shapes. Trends Cell Biol 30:792–804. https://doi.org/10.1016/j.tcb.2020.07.004
153. van Beuningen SF, Hoogenraad CC (2016) Neuronal polarity: remodeling microtubule organization. Curr Opin Neurobiol 39:1–7. https://doi.org/10.1016/j.conb.2016.02.003
154. Varadi A, Tsuboi T, Johnson-Cadwell LI, Allan VJ, Rutter GA (2003) Kinesin I and cytoplasmic dynein orchestrate glucose-stimulated insulin-containing vesicle movements in clonal MIN6 beta-cells. Biochem Biophys Res Commun 311:272–282. https://doi.org/10.1016/j.bbrc.2003.09.208

155. Vasquez-Limeta A, Loncarek J (2021) Human centrosome organization and function in interphase and mitosis. Semin Cell Dev Biol 117:30–41. https://doi.org/10.1016/j.semcdb.2021.03.020
156. Vergarajauregui S, Becker R, Steffen U, Sharkova M, Esser T, Petzold J, Billing F, Kapiloff MS, Schett G, Thievessen I, Engel FB (2020) AKAP6 orchestrates the nuclear envelope microtubule-organizing center by linking golgi and nucleus via AKAP9. elife 9:e61669. https://doi.org/10.7554/eLife.61669
157. Vermeulen BJA, Bohler A, Gao Q, Neuner A, Župa E, Chu Z, Würtz M, Jäkle U, Gruss OJ, Pfeffer S, Schiebel E (2024) gamma-TuRC asymmetry induces local protofilament mismatch at the RanGTP-stimulated microtubule minus end. EMBO J 43:2062–2085. https://doi.org/10.1038/s44318-024-00087-4
158. Vineethakumari C, Lüders J (2022) Microtubule anchoring: attaching dynamic polymers to cellular structures. Front Cell Dev Biol 10:867870. https://doi.org/10.3389/fcell.2022.867870
159. Vinogradova T, Paul R, Grimaldi AD, Loncarek J, Miller PM, Yampolsky D, Magidson V, Khodjakov A, Mogilner A, Kaverina I (2012) Concerted effort of centrosomal and Golgi-derived microtubules is required for proper Golgi complex assembly but not for maintenance. Mol Biol Cell 23:820–833. https://doi.org/10.1091/mbc.E11-06-0550
160. Wade RH (2009) On and around microtubules: an overview. Mol Biotechnol 43:177–191. https://doi.org/10.2119/molecular%20medicine-2006-00038
161. Wang Z, Wu T, Shi L, Zhang L, Zheng W, Qu JY, Niu R, Qi RZ (2010) Conserved motif of CDK5RAP2 mediates its localization to centrosomes and the Golgi complex. J Biol Chem 285:22658–22665. https://doi.org/10.1074/jbc.M110.105965
162. Wang Z, Zhang C, Qi RZ (2014) A newly identified myomegalin isoform functions in Golgi microtubule organization and ER-Golgi transport. J Cell Sci 127:4904–4917. https://doi.org/10.1242/jcs.155408
163. Wei JH, Seemann J (2017) Golgi ribbon disassembly during mitosis, differentiation and disease progression. Curr Opin Cell Biol 47:43–51. https://doi.org/10.1016/j.ceb.2017.03.008
164. Wei JH, Zhang ZC, Wynn RM, Seemann J (2015) GM130 regulates Golgi-derived spindle assembly by activating TPX2 and capturing microtubules. Cell 162:287–299. https://doi.org/10.1016/j.cell.2015.06.014
165. Widlund PO, Stear JH, Pozniakovsky A, Zanic M, Reber S, Brouhard GJ, Hyman AA, Howard J (2011) XMAP215 polymerase activity is built by combining multiple tubulin-binding TOG domains and a basic lattice-binding region. Proc Natl Acad Sci USA 108:2741–2746. https://doi.org/10.1073/pnas.1016498108
166. Wieczorek M, Urnavicius L, Ti S, Kelly R, Molloy KR, Brian T, Chait BT, Kapoor TM (2020) Asymmetric molecular architecture of the human γ-Tubulin Ring complex. Cell 180:165–175. https://doi.org/10.1016/j.cell.2019.12.007
167. Wieczorek M, Ti SC, Urnavicius L, Molloy KR, Aher A, Chait BT, Kapoor TM (2021) Biochemical reconstitutions reveal principles of human γ-TuRC assembly and function. J Cell Biol 220:e202009146. https://doi.org/10.1083/jcb.202009146
168. Wong YL, Anzola JV, Davis RL, Yoon M, Motamedi A, Kroll A, Seo CP, Hsia JE, Kim SK, Mitchell JW, Mitchell BJ, Desai A, Gahman TC, Shiau AK, Oegema K (2015) Reversible centriole depletion with an inhibitor of Polo-like kinase 4. Science 348:1155–1160. https://doi.org/10.1126/science.aaa5111
169. Woodruff JB, Wueseke O, Hyman AA (2014) Pericentriolar material structure and dynamics. Philos Trans R Soc Lond Ser B Biol Sci 369:20130459. https://doi.org/10.1098/rstb.2013.0459
170. Wu J, de Heus C, Liu Q, Bouchet BP, Noordstra I, Jiang K, Hua S, Martin M, Yang C, Grigoriev I, Katrukha EA, Altelaar AFM, Hoogenraad CC, Qi RZ, Klumperman J, Akhmanova A (2016) Molecular pathway of Microtubule Organization at the Golgi Apparatus. Dev Cell 39(1):44–60. https://doi.org/10.1016/j.devcel.2016.08.009
171. Wu J, Akhmanova A (2017) Microtubule-Organizing Centers. Annu Rev Cell Dev Biol 33:51–75. https://doi.org/10.1146/annurev-cellbio-100616-060615

172. Würtz M, Zupa E, Atorino ES, Neuner A, Böhler A, Rahadian AS, Vermeulen BJA, Tonon G, Eustermann S, Schiebel E, Pfeffer S (2022) Modular assembly of the principal microtubule nucleator γ-TuRC. Nat Commun 13:473. https://doi.org/10.1038/s41467-022-28079-0
173. Yadav S, Linstedt AD (2011) Golgi positioning. Cold Spring Harb Perspect Biol 3:a000521. https://doi.org/10.1101/cshperspect.a005322
174. Yalgin C, Ebrahimi S, Delandre C, Yoong LF, Akimoto S, Tran H, Amikura R, Spokony R, Torben-Nielsen B, White KP, Moore AW (2015) Centrosomin represses dendrite branching by orienting microtubule nucleation. Nat Neurosci 18:1437–1445. https://doi.org/10.1038/nn.4099
175. Yang S, Franco KC, Li G, Lin J, Li XD, Qi RZ (2023) Autoinhibitory mechanism controls binding of centrosomin motif 1 to γ-tubulin ring complex. J Cell Biol 222:e202007101. https://doi.org/10.1083/jcb.202007101
176. Yang SZ, Wildonger J (2020) Golgi outposts locally regulate microtubule orientation in neurons but are not required for the overall polarity of the dendritic cytoskeleton. Genetics 215:435–447. https://doi.org/10.1534/genetics.119.302979
177. Yau KW, Schätzle P, Tortosa E, Pagès S, Holtmaat A, Kapitein LC, Hoogenraad CC (2016) Dendrites in vitro and in vivo contain microtubules of opposite polarity and axon formation correlates with uniform plus-end-out microtubule orientation. J Neurosci 36:1071–1085. https://doi.org/10.1523/JNEUROSCI.2430-15.2016
178. Yau KW, van Beuningen SF, Cunha-Ferreira I, Cloin BM, van Battum EY, Will L, Schätzle P, Tas RP, van Krugten J, Katrukha EA, Jiang K, Wulf PS, Mikhaylova M, Harterink M, Pasterkamp RJ, Akhmanova A, Kapitein LC, Hoogenraad CC (2014) Microtubule minus-end binding protein CAMSAP2 controls axon specification and dendrite development. Neuron 82:1058–1073. https://doi.org/10.1016/j.neuron.2014.04.019
179. Ye B, Zhang Y, Song W, Younger SH, Jan LY, Jan YN (2007) Growing dendrites and axons differ in their reliance on the secretory pathway. Cell 130:717–729. https://doi.org/10.1016/j.cell.2007.06.032
180. Zebrowski DC, Vergarajauregui S, Wu CC, Piatkowski T, Becker R, Leone M, Hirth S, Ricciardi F, Falk N, Giessl A, Just S, Braun T, Weidinger G, Engel FB (2015) Developmental alterations in centrosome integrity contribute to the post-mitotic state of mammalian cardiomyocytes. elife 4:e05563. https://doi.org/10.7554/eLife.05563
181. Zhang J, Megraw TL (2007) Proper recruitment of gamma-tubulin and D-TACC/Msps to embryonic Drosophila centrosomes requires Centrosomin Motif 1. Mol Biol Cell 18:4037–4049. https://doi.org/10.1091/mbc.e07-05-0474
182. Zheng Y, Wildonger J, Ye B, Zhang Y, Kita A, Younger SH, Zimmerman S, Jan LY, Jan YN (2008) Dynein is required for polarized dendritic transport and uniform microtubule orientation in axons. Nat Cell Biol 10:1172–1180. https://doi.org/10.1038/ncb1777
183. Zhou W, Chang J, Wang X, Savelieff MG, Zhao Y, Ke S, Ye B (2014) GM130 is required for compartmental organization of dendritic golgi outposts. Curr Biol 24:1227–1233. https://doi.org/10.1016/j.cub.2014.04.008
184. Zhu X, Hu R, Brissova M, Stein RW, Powers AC, Gu G, Kaverina I (2015) Microtubules negatively regulate insulin secretion in pancreatic β cells. Dev Cell 34:656–668. https://doi.org/10.1016/j.devcel.2015.08.020
185. Zimmermann F, Serna M, Ezquerra A, Fernandez-Leiro R, Llorca O, Luders J (2020) Assembly of the asymmetric human γ-tubulin ring complex by RUVBL1-RUVBL2 AAA ATPase. Sci Adv 6:eabe0894. https://doi.org/10.1126/sciadv.abe0894
186. Zobaroğlu-Özer P, Bora-Akoğlu G (2024) Split but merge: Golgi fragmentation in physiological and pathological conditions. Mol Biol Rep 51:214. https://doi.org/10.1007/s11033-023-09153-2

187. Zupa E, Liu P, Würtz M, Schiebel E, Pfeffer S (2021) The structure of the γ-TuRC: a 25-years-old molecular puzzle. Curr Opin Struct Biol 66:15–21. https://doi.org/10.1016/j.sbi.2020.08.008
188. Zupa E, Zheng A, Neuner A, Würtz M, Liu P, Böhler A, Schiebel E, Pfeffer S (2020) The cryo-EM structure of a γ-TuSC elucidates architecture and regulation of minimal microtubule nucleation systems. Nat Commun 11:5705. https://doi.org/10.1038/s41467-020-19456-8

Chapter 13
STING Innate Immunity Signalling from the Golgi

Tomohiko Taguchi

Abstract The cGAS-STING pathway has recently emerged as a critical driver of inflammation in a variety of settings, such as double-stranded DNA (dsDNA) virus infection, cellular stress, and tissue damage. The pathway senses microbial and host-derived dsDNA in the cytosol, and triggers the production of the type I interferons through the activation of TBK1 (kinase) and IRF3 (transcription factor). The detailed mechanistic understanding of the pathway has enabled the development of pharmacological agents for the treatment of chronic inflammation and cancer. STING is a transmembrane protein that localizes at the endoplasmic reticulum (ER). Once bound to cGAMP, a second messenger that is generated by dsDNA-bound cGAS, STING translocates from the ER to the Golgi. STING then activates TBK1 and IRF3 at the trans-Golgi network (TGN). Dysregulated Golgi-to-ER traffic underlies the pathogenesis of COPA syndrome, a monogenic autoinflammatory disease caused by missense mutations of coatomer protein complex subunit α (COP-α). In this chapter, I focus on emerging issues regarding the regulation of STING activation by membrane traffic between the ER and the Golgi, and also on the molecular mechanism underlying the specific activation of TBK1/IRF3 at the TGN.

Keywords Innate immune signalling · The trans-Golgi network · STING · TBK1 · IRF3 · Protein palmitoylation · Cholesterol

Main Part

STING Signalling Pathway

The innate immune system is an evolutionary ancient system to combat invading pathogens, such as viruses and bacteria. It serves as the first line of host defense, detecting the pathogens through pattern recognition receptors (PRRs) and activating

T. Taguchi (✉)
Graduate School of Life Sciences, Tohoku University, Sendai, Japan
e-mail: tomohiko.taguchi.b8@tohoku.ac.jp

A. Nakano, J. Saraste (eds.), *The Golgi Network, Volume II*, Subcellular Biochemistry 111, https://doi.org/10.1007/978-3-032-16833-7_13

intracellular signalling cascades that trigger the expression of proinflammatory cytokines, type I interferons, and other antiviral proteins that all facilitate the elimination of pathogens and infected cells [48, 63]. PRRs include Toll-like receptors [4], RIG-I-like receptors [69], nucleotide-binding domain and leucine-rich repeat-containing receptors [68], and C-type lectin receptors [12]. Furthermore, stimulator of interferon genes (STING) [16] (also known as MITA [72], ERIS [59], MPYS [20], or TMEM173) has been shown to be essential for the innate immune response against dsDNA viruses.

STING is an ER-localized transmembrane protein [16]. After binding to cyclic GMP-AMP (cGAMP [66]), which is produced by cGAMP synthase (cGAS) [57] in the presence of cytosolic dsDNA, STING can activate TBK1 kinase and the transcription factor interferon regulatory factor 3 (IRF3). Phosphorylated IRF3 by TBK1 dimerizes and translocates to the nucleus to induce transcription of genes that encode type I interferons such as interferon-β (IFNβ). STING also induces proinflammatory response via NF-κB by the activation of TBK1 and IKKε [1, 2].

While the STING-mediated dsDNA-sensing signalling pathway is critical for successful protection against infections of dsDNA viruses, dysregulated STING activity leads to the excessive production of inflammatory mediators with detrimental effects on surrounding cells and tissues [44]. Recent studies have revealed the critical role of STING in the pathogenesis of multiple autoinflammatory diseases (Aicardi-Goutières syndrome, STING-associated vasculopathy with onset in infancy (SAVI), COPA syndrome, etc.) and neurodegenerative diseases (Parkinson's disease, amyotrophic lateral sclerosis, multiple sclerosis, etc.) [60]. These findings underscore the critical roles of the STING pathway in human pathophysiology.

STING Activation by Exocytic Membrane Traffic to the Golgi

After STING binds cGAMP, STING translocates from the ER to the Golgi [17, 50]. The interference with the ER-to-Golgi traffic by a fungal metabolite brefeldin A (BFA) or expression of *Shigella* effector IpaJ, suppressed the phosphorylation of IRF3 and induction of IFNβ, suggesting that the ER-to-Golgi traffic is essential for the STING signalling [8, 27, 45].

The ER-to-Golgi traffic is facilitated by the coat protein complex-II (COP-II), a complex of highly conserved proteins that are responsible for generating COP-II membrane vesicles that bud from the ER [3]. The small GTPase Sar1, inner coat proteins Sec23/Sec24, and outer coat proteins Sec13/Sec31 are five cytosolic components of the COP-II complex. Knockdown of Sar1A and Sar1B, two mammalian Sar1 paralogs, suppressed the STING translocation from the ER to the Golgi and the phosphorylation of IRF3, suggesting that the translocation of STING from the ER with COP-II vesicles is required for the STING signalling [47]. The critical role of the COP-II-mediated transport of STING in the STING signalling is also demonstrated by knockdown of Sec13, Sec23, Sec24, or Sec31 [13, 49, 58].

Several COP-II-associated proteins are involved in the STING signalling. YIPF5, a mammalian ortholog of yeast Yip1p that regulates COP-II vesicle biogenesis, is essential for cytosolic DNA-induced translocation of STING from the ER and the induction of IFNβ [49]. Yeast Emp24p and Erv25p recruit specific cargo molecules into COP-II vesicles. TMED2, a mammalian ortholog of yeast Emp24p, is essential for herpes simplex virus (HSV)-induced translocation of STING from the ER and the STING signalling [58]. TMED10, a mammalian ortholog of yeast Erv25p, also participates in the STING signalling [58].

CxORF56, or STING ER exit protein (STEEP), facilitates the STING trafficking from the ER [71]. After cGAMP stimulation, STEEP stimulates phosphatidylinositol-3-phosphate [PtdIns(3)P] synthesis in the ER membrane and induces membrane curvature at the ER. The induction of the membrane curvature may promote the formation of COP-II vesicles. The rhomboid family are evolutionary conserved intramembrane proteases. The enzymatically inactive member iRhom2 in mammals localizes at the ER and is required for HSV-induced translocation of STING from the ER and the STING signalling [38].

STING becomes concentrated at non-canonical ER exit site upon cGAMP binding. Proteomic approach identifies ACBD3 protein (a Golgi-localized protein) as a component that concentrates near STING ER exit sites [42]. Depletion of ACBD3 suppresses the translocation of STING from the ER and the STING signalling [42].

How does the binding of cGAMP to STING facilitate the STING translocation with COP-II vesicles? One clue is provided by the cryo-electron microscopy structure of cGAMP-bound STING [52]. The structures show a 180° rotation of the ligand-binding domain of STING relative to the transmembrane domain after cGAMP binding. The conformational change of STING induced by cGAMP binding will expose and/or generate the new binding interface(s) with the COP-II components or the proteins essential for the STING translocation as described above.

Ca^{2+} sensor stromal interaction molecule 1 (STIM1) is an ER-localized protein. STIM1 deficiency leads to a low-level induction of IFNβ and a partial STING translocation without cGAMP stimulation [55]. STIM1 binds STING under unstimulated conditions, and its interaction with STING is reduced upon cGAMP binding. Thus, STIM1 may function as a tether to retain cGAMP-free STING at the ER.

STING Palmitoylation

The fact that the translocation of STING from the ER is essential for the activation of the STING signalling indicates that there should be a mechanism by which TBK1/IRF3 is activated at the post-ER compartments. Indeed, high-resolution fluorescence microscopy reveals that TBK1 becomes associated with the Golgi, in particular, at the TGN after STING stimulation [22]. In agreement with this, phosphorylated TBK1, a hallmark of the STING activation, is found specifically at the TGN [45].

A study reveals that palmitoylation of STING, which occurs at the Golgi, is essential for TBK1/IRF3 activation [45]. Treatment of cells with a pan-palmitoylation inhibitor 2-bromopalmitate (2-BP) inhibits the type I interferon response without interfering with the STING translocation from the ER to the Golgi. Mutation of two membrane-proximal Cys residues (Cys88/91) of STING largely abolishes its palmitoylation, and the palmitoylation-deficient STING variant (STING Cys88/91Ser) loses the ability to induce the STING-dependent signalling. The essential contribution of palmitoylation of Cys88/91 to STING signalling is further corroborated by the reports in which STING inhibitors are identified: C-176, C-178, H-151, nitro-unsaturated fatty acids, and 4-hydroxynonenal, which all covalently binds to Cys88 and/or Cys91, suppress STING palmitoylation and STING signalling [14, 15, 19] .

Protein palmitoylation has been implicated in the clustering of plasma membrane (PM) proteins into lipid rafts, specific lipid microdomains that consists of cholesterol and sphingomyelin (SM) [34]. Cholesterol and SM that is synthesized by a Golgi-localized SM synthase SMS1, may form lipid rafts at the TGN besides the PM [9]. Interference with the integrity of lipid rafts at the Golgi, by addition of C6-ceramide, inhibits STING signalling without affecting the translocation of STING from the ER to the Golgi [45]. In SMS1-knockout cells, STING-dependent phosphorylation of TBK1 and IRF3 is severely suppressed [10]. The essential role of the Golgi lipids in STING signalling is further supported by a cell-free assay to assess whether endogenous STING in microsomes can be phosphorylated by recombinant TBK1 [61]. Treating microsomes with sphingomyelinase or methyl-β-cyclodextrin, an agent to extract cholesterol from membranes, suppresses the phosphorylation of STING by TBK1 [61].

In sum, two Golgi-associated factors, *i.e.*, protein palmitoylation and the lipid rafts, are required for the STING signalling. Both may cooperate to promote STING clustering at the TGN, which then facilitates the activation of TBK1 and IRF3 (Fig. 13.1).

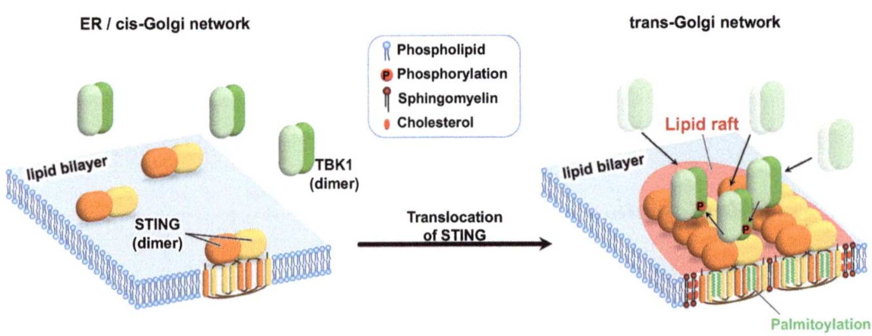

Fig. 13.1 The recruitment of TBK1 requires the clustering of STING. At the ER and the cis-Golgi network, STING exists as a dimer. At the TGN, STING becomes clustered, which is facilitated by STING palmitoylation and the Golgi raft lipids (cholesterol and sphingomyelin). The clustered STING at the TGN acts as a scaffold on which TBK1 is autophosphorylated and activated

STING Cluster Revealed by PALM

A recent study with photoactivated localization microscopy (PALM) provides the direct evidence of the formation of signalling-competent STING homocluster at the TGN [23]. A fluorescent protein mEos4b is attached to the *N*-terminus of STING, and mEos4b-STING is imaged by live-cell PALM. After 60 min stimulation with DMXAA (a membrane-permeable STING agonist), mEos4b-STING reaches the TGN and starts to exhibit numerous punctate structures, in which >20 mEos4b-STING molecules are present on average. In contrast, without STING agonist, the number of mEos4b-STING molecules per cluster is estimated to be <10. These results indicate the stimulation-dependent STING cluster formation at the TGN.

As described in the previous section, two Golgi-associated factors (protein palmitoylation and the lipid rafts) are essential for the STING signalling. Interfering with STING palmitoylation with 2-BP or H-151 reduces the number of mEos4b-STING molecules per cluster. Interfering with the lipid rafts with C6-ceramide also reduces the number of mEos4b-STING molecules per cluster. Therefore, live-cell PALM shows that STING palmitoylation and raft lipids are required for the STING clustering at the TGN.

Simultaneous dual-color live-cell PALM of STING and single-molecule imaging of TBK1 show that single molecules of TBK are recruited to the STING cluster. The lifetime of the colocalization durations of single-molecules of TBK1 with STING clusters is determined as around 150 ms in DMXAA-stimulated cells. Treatment of cells with H-151 significantly reduces it to 74 ms. These results suggest that the recruitment of TBK1 to STING is facilitated by a palmitoylation-dependent STING cluster formation.

STING-Associated Vasculopathy with Onset in Infancy

STING-associated vasculopathy with onset in infancy (SAVI) is a disorder involving inflammation throughout the body, especially in the skin, blood vessels, and lung. A number of STING variants (H72N, F153V, V147L, N154S, V155M, G158A, G166E, C206Y, G207E, R281Q/W, and R284G/S) have been identified in SAVI patients [18, 24–26, 32, 33, 36, 41, 51]. These disease-causative variants constitutively activate the STING-downstream signalling without cGAMP [18, 36]. Among them, three variants (V147L, N154S, and V155M) have been extensively characterized. These three do not localize at the ER, instead localize at the perinuclear compartments that include the Golgi [45]. Interference with the ER-to-Golgi traffic with BFA abolishes TBK1 binding to these variants. The interferon response triggered by these variants is inhibited by 2-BP or an introduction of Cys88/91Ser mutation [45]. These results suggest that the SAVI variants, as wild-type (WT) STING, require the ER exit and palmitoylation for their activity. The 3D modeling of the *C*-terminal cytosolic domain of STING predicts that V155M mutation increases the stability of the STING dimer [18]. The 3D structure of SAVI variants may be similar to that of cGAMP-bound WT STING, allowing SAVI variants to exit the ER without cGAMP binding.

SAVI mouse models have been developed, harboring common SAVI mutations, such as N153S and V154M, which correspond to the human N154S and V155M, respectively [43, 64]. These SAVI mutant mice express IFN and inflammatory gene signature, and have shorter lifespans than their WT littermate controls. Crosses of heterozygous SAVI mice do not yield homozygous SAVI mice as of embryonic day 14, indicating that homozygous SAVI embryos are not viable and that WT allele would function dominantly over SAVI alleles in terms of viability. The observation in mice matches well with the fact that all the SAVI patients, except SAVI patient (R281W), are heterozygous in STING alleles, namely with one WT allele and one SAVI allele.

WT STING and SAVI variants can form heterocomplex [53]. The heterocomplex localizes primarily at the ER and fails to reach the TGN. These results suggest that WT STING can suppress the activity of SAVI variants by forcing them to localize to the ER through the heterocomplex formation. The dormant WT/SAVI heterocomplex formation may underlie the dominance of STING WT allele over SAVI alleles in the STING-triggered inflammatory response.

COPA Disease

The COPA syndrome is an autoinflammatory and autoimmune disorder characterized by interstitial lung disease and high expression of type I interferon-stimulated genes [65]. The disease is caused by heterozygous mutations of the COPA gene that encodes COP-α protein of the COP-I complex. The COP-I complex mediates the retrograde membrane traffic from the Golgi to the ER through the formation of COP-I vesicles. All of the mutations reported so far (K230N, R233H, E241K, V242G, or D243G) lie in the *N*-terminal WD40 domain of COP-α, which is implicated in the recognition of cargo proteins [21, 65].

Five groups recently provide the evidence that COPA syndrome is caused by constitutive activation of STING [7, 21, 31, 46, 56]. In the presence of the disease-causative COP-α variants, STING is mislocalized to the Golgi, most likely due to the impaired COP-I transport. The forced Golgi localization of STING results in palmitoylation-dependent activation of STING at the TGN [46]. The mislocalization of STING to the Golgi occurs even in cGAS-knockout cells, indicating that STING constantly leaks out to the Golgi without its binding to cGAMP and is retrieved back to the ER with COP-I transport [46].

COP-α binds C-terminal di-lysine motifs (KKXX and KXKXX) of its cargo proteins. Since STING does not possess these motifs, COP-α may bind indirectly to STING through adapter proteins. By mass spectrometric analysis of STING-binding proteins at steady-state conditions, 18 proteins with these motifs are identified [46]. Knockdown of Surf4, not that of the other 17 proteins, relocates STING

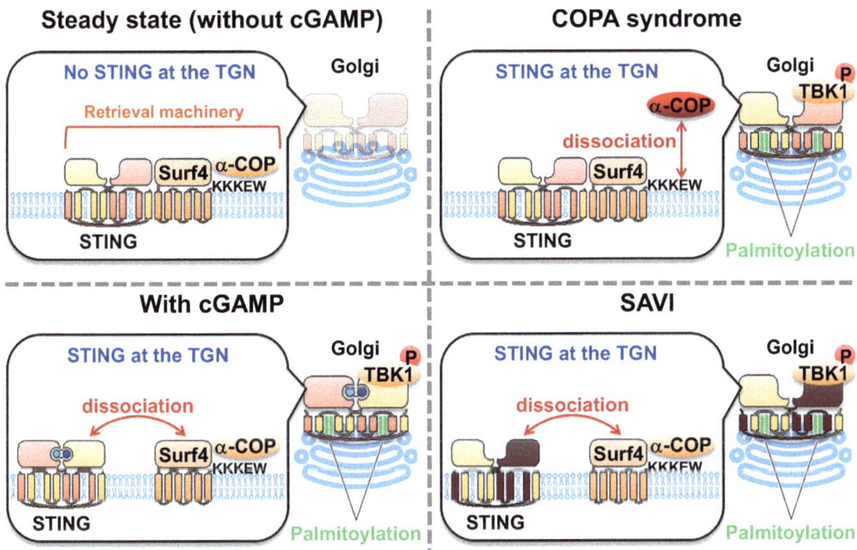

Fig. 13.2 A model of STING inactivation by retrograde membrane traffic from the Golgi to the ER. At steady state, STING constantly leaks out from the ER and is retrieved back from the Golgi to the ER with the COP-I-mediated transport. Thus, the predominant ER localization of STING is maintained, preventing STING traffic to the TGN. In contrast, in other settings, the association of STING with the retrieval machinery decreases, allowing STING to reach the TGN at which STING activates TBK1. This model does not exclude the possibility that anterograde trafficking of STING is accelerated by cGAMP binding and SAVI mutations

from the ER to the Golgi and results in the emergence of phosphorylated TBK1 [46]. STING/Surf4/COP-α complex is disrupted in the presence of the disease-causative COP-α variant [46]. These results indicate that Surf4 functions as the adaptor protein that links STING and COP-α, facilitating the retrograde transport of STING with COP-I vesicles (Fig. 13.2).

STING-Dependent LC3 Lipidation on the Golgi Membrane

During the trafficking through the Golgi where STING activates the downstream signalling cascade, STING also induces the lipidation of microtubule-associated protein 1 light chain 3 (LC3) [50]. In the context of canonical autophagy, LC3 is conjugated to the polar headgroup of phosphatidylethanolamine in double-membrane autophagosomes where it is essential for cargo loading and autophagosome maturation. In contrast, LC3-positive membranes that is induced by STING stimulation are single-membranes that cluster at the perinuclear region [11]. LC3-lipidation onto single-membrane vesicles is generally dubbed as

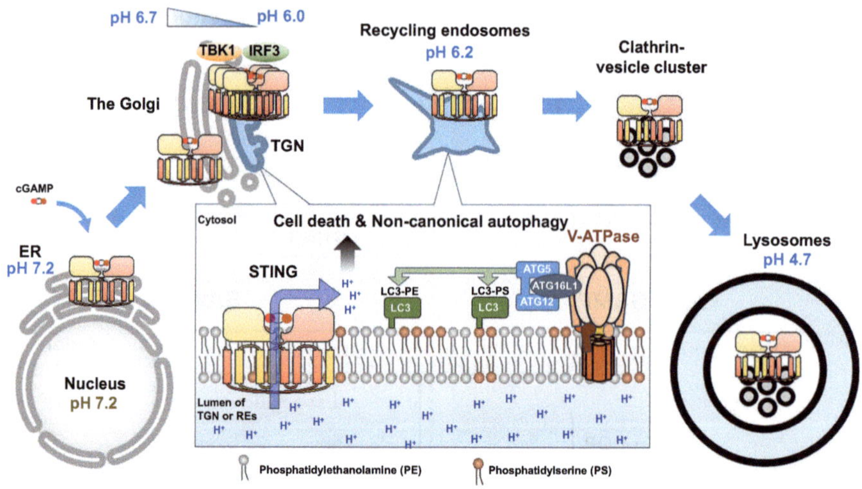

Fig. 13.3 STING triggers non-canonical autophagy through its proton channel activity. STING translocates to the Golgi upon binding its ligand cGAMP. At the TGN, STING clusters and activates innate immune signaling through activation of TBK1/IRF3. Subsequently, STING translocates to recycling endosomes, and is finally degraded in lysosomes through microautophagy. The luminal pH at the TGN and recycling endosomes is mildly acidic, resulting in proton efflux to the cytosol through the pore in the STING dimer. De-acidification of these organelles activates the v-ATPase/ATG16L1 axis to conjugate phosphatidylethanolamine (PE) and phosphatidylserine (PS) with LC3

"conjugation of ATG8 (LC3) to single membranes" (CASM). The molecular mechanism by which STING can drive CASM has not been clear.

Two recent studies independently solve this question [35, 67]. These studies reveal that STING is a proton channel and that STING activation de-acidifies the lumen of the Golgi and perinuclear endosomes using a Golgi pH sensor or an endosomal pH sensor, respectively. A chemical agonist of STING, compound 53 (C53) [37], previously shown to bind to the STING transmembrane domain, inhibits STING-mediated proton efflux *in vitro*, STING-dependent organelle de-acidification, and CASM. Thus, CASM depends on proton efflux through the STING pore (Fig. 13.3).

On its route to lysosomes from the ER [28, 30], STING encounters a gradual acidification of the luminal pH of organelles along the secretory pathway, *e.g.*, pH 7.2 (the ER), 6.7 (cis-Golgi), 6.0 (TGN), and 6.2 (recycling endosomes (REs)) [5, 62]. The study details the target organelles of CASM, and shows that TGN and REs are the primary targets of STING-induced LC3 lipidation [67]. Given that the luminal pH of TGN and REs is mildly acidic, it is quite reasonable that STING-induced CASM occurs exclusively at these organelles, but not at ER and cis-Golgi where the luminal pH is neutral [29].

Conclusion and Future Perspectives (Acknowledgements)

In the last 15 years, substantial progress has been made in the molecular mechanism underlying the innate immune response against cytosolic dsDNA. Besides the role in innate immune response against DNA viruses, the cGAS-STING pathway is now recognized to play a critical role in the pathogenesis of various autoinflammatory and neurodegenerative diseases, such as Parkinson's disease [54] and amyotrophic lateral sclerosis [40, 70]. Self-dsDNA from various sources including genomic DNA and mitochondrial DNA can aberrantly accumulate in the cytosol in certain settings, triggering inflammation through the activation of the cGAS-STING pathway.

The recent studies have also disclosed that the STING signalling is strictly regulated by intracellular membrane traffic [60]. The Golgi, or more specifically speaking the TGN, is now revealed to be the organelle where STING activates TBK1 and IRF3. The finding that STING needs palmitoylation for its activity paves the way to develop STING-specific antagonists [14, 15]. The precise understanding of the molecular mechanism underlying STING exit from the ER or retrograde retrieval of STING from the Golgi may contribute to the development of STING-specific agonist [6, 39].

Compliance with Ethical Standards T.T. has no competing/conflict of interest.

References

1. Abe T, Barber GN (2014) Cytosolic-DNA-mediated, STING-dependent proinflammatory gene induction necessitates canonical NF-κB activation through TBK1. J Virol 88:5328–5341
2. Balka KR, Louis C, Saunders TL, Smith AM, Calleja DJ, D'Silva DB, Moghaddas F, Tailler M, Lawlor KE, Zhan Y, Burns CJ, Wicks IP, Miner JJ, Kile BT, Masters SL, De Nardo D (2020) TBK1 and IKKε act redundantly to mediate STING-induced NF-κB responses in myeloid cells. Cell Rep 31:107492
3. Barlowe C, Helenius A (2016) Cargo capture and bulk flow in the early secretory pathway. Annu Rev Cell Dev Biol 32:197–222
4. Blasius AL, Beutler B (2010) Intracellular toll-like receptors. Immunity 32:305–315
5. Casey JR, Grinstein S, Orlowski J (2010) Sensors and regulators of intracellular pH. Nat Rev Mol Cell Biol 11:50–61
6. Chin EN, Sulpizio A, Lairson LL (2023) Targeting STING to promote antitumor immunity. Trends Cell Biol 33:189–203
7. Deng Z, Chong Z, Law CS, Mukai K, Ho FO, Martinu T, Backes BJ, Eckalbar WL, Taguchi T, Shum AK (2020) A defect in COPI-mediated transport of STING causes immune dysregulation in COPA syndrome. J Exp Med 217:e20201045
8. Dobbs N, Burnaevskiy N, Chen D, Gonugunta VK, Alto NM, Yan N (2015) STING activation by translocation from the ER Is associated with infection and autoinflammatory disease. Cell Host Microbe 18:157–168
9. Duran JM, Campelo F, van Galen J, Sachsenheimer T, Sot J, Egorov MV, Rentero C, Enrich C, Polishchuk RS, Goñi FM, Brügger B, Wieland F, Malhotra V (2012) Sphingomyelin organization is required for vesicle biogenesis at the Golgi complex. EMBO J 31:4535–4546

10. Fang R, Jiang Q, Jia X, Jiang Z (2023) ARMH3-mediated recruitment of PI4KB directs Golgi-to-endosome trafficking and activation of the antiviral effector STING. Immunity 56:500–515.e6
11. Fischer TD, Wang C, Padman BS, Lazarou M, Youle RJ (2020) STING induces LC3B lipidation onto single-membrane vesicles via the V-ATPase and ATG16L1-WD40 domain. J Cell Biol 219:e202009128
12. Geijtenbeek TB, Gringhuis SI (2009) Signalling through C-type lectin receptors: shaping immune responses. Nat Rev Immunol 9:465–479
13. Gui X, Yang H, Li T, Tan X, Shi P, Li M, Du F, Chen ZJ (2019) Autophagy induction via STING trafficking is a primordial function of the cGAS pathway. Nature 567:262–266
14. Haag SM, Gulen MF, Reymond L, Gibelin A, Abrami L, Decout A, Heymann M, van der Goot FG, Turcatti G, Behrendt R, Ablasser A (2018) Targeting STING with covalent small-molecule inhibitors. Nature 559:269–273
15. Hansen AL, Buchan GJ, Rühl M, Mukai K, Salvatore SR, Ogawa E, Andersen SD, Iversen MB, Thielke AL, Gunderstofte C, Motwani M, Møller CT, Jakobsen AS, Fitzgerald KA, Roos J, Lin R, Maier TJ, Goldbach-Mansky R, Miner CA, Qian W, Miner JJ, Rigby RE, Rehwinkel J, Jakobsen MR, Arai H, Taguchi T, Schopfer FJ, Olagnier D, Holm CK (2018) Nitro-fatty acids are formed in response to virus infection and are potent inhibitors of STING palmitoylation and signaling. Proc Natl Acad Sci U S A 115:E7768–E7775
16. Ishikawa H, Barber GN (2008) STING is an endoplasmic reticulum adaptor that facilitates innate immune signalling. Nature 455:674–678
17. Ishikawa H, Ma Z, Barber GN (2009) STING regulates intracellular DNA-mediated, type I interferon-dependent innate immunity. Nature 461:788–792
18. Jeremiah N, Neven B, Gentili M, Callebaut I, Maschalidi S, Stolzenberg MC, Goudin N, Frémond ML, Nitschke P, Molina TJ, Blanche S, Picard C, Rice GI, Crow YJ, Manel N, Fischer A, Bader-Meunier B, Rieux-Laucat F (2014) Inherited STING-activating mutation underlies a familial inflammatory syndrome with lupus-like manifestations. J Clin Invest 124:5516–5520
19. Jia M, Qin D, Zhao C, Chai L, Yu Z, Wang W, Tong L, Lv L, Wang Y, Rehwinkel J, Yu J, Zhao W (2020) Redox homeostasis maintained by GPX4 facilitates STING activation. Nat Immunol 21:727–735
20. Jin L, Waterman PM, Jonscher KR, Short CM, Reisdorph NA, Cambier JC (2008) MPYS, a novel membrane tetraspanner, is associated with major histocompatibility complex class II and mediates transduction of apoptotic signals. Mol Cell Biol 28:5014–5026
21. Kato T, Yamamoto M, Honda Y, Orimo T, Sasaki I, Murakami K, Hemmi H, Fukuda-Ohta Y, Isono K, Takayama S, Nakamura H, Otsuki Y, Miyamoto T, Takita J, Yasumi T, Nishikomori R, Matsubayashi T, Izawa K, Kaisho T (2021) Augmentation of stimulator of interferon genes-induced type I interferon production in COPA syndrome. Arthritis Rheumatol 73:2105–2115
22. Kemmoku H, Kuchitsu Y, Mukai K, Taguchi T (2022) Specific association of TBK1 with the trans-Golgi network following STING stimulation. Cell Struct Funct 47:19–30
23. Kemmoku H, Takahashi K, Mukai K, Mori T, Hirosawa KM, Kiku F, Uchida Y, Kuchitsu Y, Nishioka Y, Sawa M, Kishimoto T, Tanaka K, Yokota Y, Arai H, Suzuki KGN, Taguchi T (2024) Single-molecule localization microscopy reveals STING clustering at the trans-Golgi network through palmitoylation-dependent accumulation of cholesterol. Nat Commun 15:220
24. Keskitalo S, Haapaniemi E, Einarsdottir E, Rajamäki K, Heikkilä H, Ilander M, Pöyhönen M, Morgunova E, Hokynar K, Lagström S, Kivirikko S, Mustjoki S, Eklund K, Saarela J, Kere J, Seppänen MRJ, Ranki A, Hannula-Jouppi K, Varjosalo M (2019) Novel TMEM173 mutation and the role of disease modifying alleles. Front Immunol 10:2770
25. König N, Fiehn C, Wolf C, Schuster M, Cura Costa E, Tüngler V, Alvarez HA, Chara O, Engel K, Goldbach-Mansky R, Günther C, Lee-Kirsch MA (2017) Familial chilblain lupus due to a gain-of-function mutation in STING. Ann Rheum Dis 76:468–472

26. Konno H, Chinn IK, Hong D, Orange JS, Lupski JR, Mendoza A, Pedroza LA, Barber GN (2018) Pro-inflammation associated with a gain-of-function mutation (R284S) in the innate immune sensor STING. Cell Rep 23:1112–1123
27. Konno H, Konno K, Barber GN (2013) Cyclic dinucleotides trigger ULK1 (ATG1) phosphorylation of STING to prevent sustained innate immune signaling. Cell 155:688–698
28. Kuchitsu Y, Mukai K, Uematsu R, Takaada Y, Shinojima A, Shindo R, Shoji T, Hamano S, Ogawa E, Sato R, Miyake K, Kato A, Kawaguchi Y, Nishitani-Isa M, Izawa K, Nishikomori R, Yasumi T, Suzuki T, Dohmae N, Uemura T, Barber GN, Arai H, Waguri S, Taguchi T (2023) STING signalling is terminated through ESCRT-dependent microautophagy of vesicles originating from recycling endosomes. Nat Cell Biol 25:453–466
29. Kuchitsu Y, Taguchi T (2024a) STINGing organelle surface with acid. EMBO Rep 25:1708–1710
30. Kuchitsu Y, Taguchi T (2024b) Lysosomal microautophagy: an emerging dimension in mammalian autophagy. Trends Cell Biol 34:606–616
31. Lepelley A, Martin-Niclós MJ, Le Bihan M et al (2020) Mutations in COPA lead to abnormal trafficking of STING to the Golgi and interferon signaling. J Exp Med 217:e20200600
32. Lin B, Berard R, Al Rasheed A, Aladba B, Kranzusch PJ, Henderlight M, Grom A, Kahle D, Torreggiani S, Aue AG, Mitchell J, de Jesus AA, Schulert GS, Goldbach-Mansky R (2020) A novel STING1 variant causes a recessive form of STING-associated vasculopathy with onset in infancy (SAVI). J Allergy Clin Immunol 146:1204–1208.e6
33. Lin B, Torreggiani S, Kahle D, Rumsey DG, Wright BL, Montes-Cano MA, Silveira LF, Alehashemi S, Mitchell J, Aue AG, Ji Z, Jin T, de Jesus AA, Goldbach-Mansky R (2021) Case report: novel SAVI-causing variants in STING1 expand the clinical disease spectrum and suggest a refined model of STING activation. Front Immunol 12:636225
34. Linder ME, Deschenes RJ (2007) Palmitoylation: policing protein stability and traffic. Nat Rev Mol Cell Biol 8:74–84
35. Liu B, Carlson RJ, Pires IS, Gentili M, Feng E, Hellier Q, Schwartz MA, Blainey PC, Irvine DJ, Hacohen N (2023) Human STING is a proton channel. Science 381:508–514
36. Liu Y, Jesus AA, Marrero B et al (2014) Activated STING in a vascular and pulmonary syndrome. N Engl J Med 371:507–518
37. Lu D, Shang G, Li J, Lu Y, Bai XC, Zhang X (2022) Activation of STING by targeting a pocket in the transmembrane domain. Nature 604:557–562
38. Luo WW, Li S, Li C, Lian H, Yang Q, Zhong B, Shu HB (2016) iRhom2 is essential for innate immunity to DNA viruses by mediating trafficking and stability of the adaptor STING. Nat Immunol 17:1057–1066
39. Matsumoto K, Ni S, Arai H, Toyama T, Saito Y, Suzuki T, Dohmae N, Mukai K, Taguchi T (2023) A non-nucleotide agonist that binds covalently to cysteine residues of STING. Cell Struct Funct 48:59–70
40. McCauley ME, O'Rourke JG, Yáñez A, Markman JL, Ho R, Wang X, Chen S, Lall D, Jin M, Muhammad AKMG, Bell S, Landeros J, Valencia V, Harms M, Arditi M, Jefferies C, Baloh RH (2020) C9orf72 in myeloid cells suppresses STING-induced inflammation. Nature 585:96–101
41. Melki I, Rose Y, Uggenti C, Van Eyck L, Frémond ML, Kitabayashi N, Rice GI, Jenkinson EM, Boulai A, Jeremiah N, Gattorno M, Volpi S, Sacco O, Terheggen-Lagro SWJ, Tiddens HAWM, Meyts I, Morren MA, De Haes P, Wouters C, Legius E, Corveleyn A, Rieux-Laucat F, Bodemer C, Callebaut I, Rodero MP, Crow YJ (2017) Disease-associated mutations identify a novel region in human STING necessary for the control of type I interferon signaling. J Allergy Clin Immunol 140:543–552.e5
42. Motani K, Saito-Tarashima N, Nishino K, Yamauchi S, Minakawa N, Kosako H (2022) The Golgi-resident protein ACBD3 concentrates STING at ER-Golgi contact sites to drive export from the ER. Cell Rep 41:111868
43. Motwani M, Pawaria S, Bernier J, Moses S, Henry K, Fang T, Burkly L, Marshak-Rothstein A, Fitzgerald KA (2019a) Hierarchy of clinical manifestations in SAVI N153S and V154M mouse models. Proc Natl Acad Sci U S A 116:7941–7950

44. Motwani M, Pesiridis S, Fitzgerald KA (2019b) DNA sensing by the cGAS-STING pathway in health and disease. Nat Rev Genet 20:657–674
45. Mukai K, Konno H, Akiba T, Uemura T, Waguri S, Kobayashi T, Barber GN, Arai H, Taguchi T (2016) Activation of STING requires palmitoylation at the Golgi. Nat Commun 7:11932
46. Mukai K, Ogawa E, Uematsu R, Kuchitsu Y, Kiku F, Uemura T, Waguri S, Suzuki T, Dohmae N, Arai H, Shum AK, Taguchi T (2021) Homeostatic regulation of STING by retrograde membrane traffic to the ER. Nat Commun 12:61
47. Ogawa E, Mukai K, Saito K, Arai H, Taguchi T (2018) The binding of TBK1 to STING requires exocytic membrane traffic from the ER. Biochem Biophys Res Commun 503:138–145
48. Palm NW, Medzhitov R (2009) Pattern recognition receptors and control of adaptive immunity. Immunol Rev 227:221–233
49. Ran Y, Xiong MG, Xu ZS, Luo WW, Wang SY, Wang YY (2019) YIPF5 Is essential for innate immunity to DNA virus and facilitates COPII-dependent STING trafficking. J Immunol 203: 1560–1570
50. Saitoh T, Fujita N, Hayashi T, Takahara K, Satoh T, Lee H, Matsunaga K, Kageyama S, Omori H, Noda T, Yamamoto N, Kawai T, Ishii K, Takeuchi O, Yoshimori T, Akira S (2009) Atg9a controls dsDNA-driven dynamic translocation of STING and the innate immune response. Proc Natl Acad Sci U S A 106:20842–20846
51. Saldanha RG, Balka KR, Davidson S, Wainstein BK, Wong M, Macintosh R, Loo CKC, Weber MA, Kamath V, CIRCA, AADRY, Moghaddas F, De Nardo D, Gray PE, Masters SL (2018) A mutation outside the dimerization domain causing atypical STING-associated vasculopathy with onset in infancy. Front Immunol 9:1535
52. Shang G, Zhang C, Chen ZJ, Bai XC, Zhang X (2019) Cryo-EM structures of STING reveal its mechanism of activation by cyclic GMP-AMP. Nature 567:389–393
53. Shindo R, Kuchitsu Y, Mukai K, Taguchi T (2022) The activity of disease-causative STING variants can be suppressed by wild-type STING through heterocomplex formation. Front Cell Dev Biol 10:1037999
54. Sliter DA, Martinez J, Hao L, Chen X, Sun N, Fischer TD, Burman JL, Li Y, Zhang Z, Narendra DP, Cai H, Borsche M, Klein C, Youle RJ (2018) Parkin and PINK1 mitigate STING-induced inflammation. Nature 561:258–262
55. Srikanth S, Woo JS, Wu B, El-Sherbiny YM, Leung J, Chupradit K, Rice L, Seo GJ, Calmettes G, Ramakrishna C, Cantin E, An DS, Sun R, Wu TT, Jung JU, Savic S, Gwack Y (2019) The Ca^{2+} sensor STIM1 regulates the type I interferon response by retaining the signaling adaptor STING at the endoplasmic reticulum. Nat Immunol 20:152–162
56. Steiner A, Hrovat-Schaale K, Prigione I, Yu CH, Laohamonthonkul P, Harapas CR, Low RRJ, De Nardo D, Dagley LF, Mlodzianoski MJ, Rogers KL, Zillinger T, Hartmann G, Gantier MP, Gattorno M, Geyer M, Volpi S, Davidson S, Masters SL (2022) Deficiency in coatomer complex I causes aberrant activation of STING signalling. Nat Commun 13:2321
57. Sun L, Wu J, Du F, Chen X, Chen ZJ (2013) Cyclic GMP-AMP synthase is a cytosolic DNA sensor that activates the type I interferon pathway. Science 339:786–791
58. Sun MS, Zhang J, Jiang LQ, Pan YX, Tan JY, Yu F, Guo L, Yin L, Shen C, Shu HB, Liu Y (2018) TMED2 potentiates cellular IFN responses to DNA viruses by reinforcing MITA dimerization and facilitating its trafficking. Cell Rep 25:3086–3098.e3
59. Sun W, Li Y, Chen L, Chen H, You F, Zhou X, Zhou Y, Zhai Z, Chen D, Jiang Z (2009) ERIS, an endoplasmic reticulum IFN stimulator, activates innate immune signaling through dimerization. Proc Natl Acad Sci U S A 106:8653–8658
60. Taguchi T (2023) Membrane traffic governs the STING inflammatory signalling. J Biochem 174:483–490
61. Takahashi K, Niki T, Ogawa E, Fumika K, Nishioka Y, Sawa M, Arai H, Mukai K, Taguchi T (2021) A cell-free assay implicates a role of sphingomyelin and cholesterol in STING phosphorylation. Sci Rep 11:11996
62. Takahashi S, Kagami Y, Hanaoka K, Terai T, Komatsu T, Ueno T, Uchiyama M, Koyama-Honda I, Mizushima N, Taguchi T, Arai H, Nagano T, Urano Y (2018) Development of a series

of practical fluorescent chemical tools to measure pH values in living samples. J Am Chem Soc 140:5925–5933
63. Takeuchi O, Akira S (2010) Pattern recognition receptors and inflammation. Cell 140:805–820
64. Warner JD, Irizarry-Caro RA, Bennion BG, Ai TL, Smith AM, Miner CA, Sakai T, Gonugunta VK, Wu J, Platt DJ, Yan N, Miner JJ (2017) STING-associated vasculopathy develops independently of IRF3 in mice. J Exp Med 214:3279–3292
65. Watkin LB, Jessen B, Wiszniewski W et al (2015) COPA mutations impair ER-Golgi transport and cause hereditary autoimmune-mediated lung disease and arthritis. Nat Genet 47:654–660
66. Wu J, Sun L, Chen X, Du F, Shi H, Chen C, Chen ZJ (2013) Cyclic GMP-AMP is an endogenous second messenger in innate immune signaling by cytosolic DNA. Science 339: 826–830
67. Xun J, Zhang Z, Lv B, Lu D, Yang H, Shang G, Tan JX (2024) A conserved ion channel function of STING mediates noncanonical autophagy and cell death. EMBO Rep 25:544–569
68. Ye Z, Ting JP (2008) NLR, the nucleotide-binding domain leucine-rich repeat containing gene family. Curr Opin Immunol 20:3–9
69. Yoneyama M, Kikuchi M, Natsukawa T, Shinobu N, Imaizumi T, Miyagishi M, Taira K, Akira S, Fujita T (2004) The RNA helicase RIG-I has an essential function in double-stranded RNA-induced innate antiviral responses. Nat Immunol 5:730–737
70. Yu CH, Davidson S, Harapas CR, Hilton JB, Mlodzianoski MJ, Laohamonthonkul P, Louis C, Low RRJ, Moecking J, De Nardo D, Balka KR, Calleja DJ, Moghaddas F, Ni E, McLean CA, Samson AL, Tyebji S, Tonkin CJ, Bye CR, Turner BJ, Pepin G, Gantier MP, Rogers KL, McArthur K, Crouch PJ, Masters SL (2020) TDP-43 triggers mitochondrial DNA release via mPTP to activate cGAS/STING in ALS. Cell 183:636–649.e18
71. Zhang BC, Nandakumar R, Reinert LS et al (2020) STEEP mediates STING ER exit and activation of signaling. Nat Immunol 21:868–879
72. Zhong B, Yang Y, Li S, Wang YY, Li Y, Diao F, Lei C, He X, Zhang L, Tien P, Shu HB (2008) The adaptor protein MITA links virus-sensing receptors to IRF3 transcription factor activation. Immunity 29:538–550

Chapter 14
Golgi Membrane-Associated Degradation (GOMED), a New Intracellular Proteolysis Mechanism

Shigeomi Shimizu, Shinya Honda, Satoru Torii, Satoko Arakawa, Masatsune Tsujioka, Hirofumi Yamaguchi, and Hajime Tajima Sakurai

Abstract The Golgi apparatus is a crucial organelle that is involved in various cellular processes, including cellular secretion, and is primarily responsible for the modification, sorting, and transport of proteins and lipids. However, recently, we discovered a novel Golgi function, namely, Golgi membrane-associated degradation (GOMED). GOMED is a cellular function in which the *trans*-Golgi membrane undergoes deformation and becomes spherical, breaking down proteins and lipids that have been internalized. This process is conserved from yeast to mammalian cells. GOMED primarily plays a role in the quality control of hormones, cytokines, and cell membrane receptors that pass through the Golgi apparatus, and is crucial for maintaining the normal function of pancreatic β cell, neurons, and intestinal epithelial cells. Therefore, abnormalities in GOMED can lead to neurodegenerative diseases and inflammatory bowel diseases.

Keywords GOMED · Autophagy · Golgi apparatus · ULK1 · Wipi3

S. Shimizu (✉) · S. Honda · S. Torii · S. Arakawa · M. Tsujioka · H. Yamaguchi
Department of Pathological Cell Biology, Advanced Research Initiative, Institute of Science Tokyo, Chiyoda-ku, Tokyo, Japan
e-mail: shimizu.pcb@mri.tmd.ac.jp

H. T. Sakurai
Department of Pathological Cell Biology, Advanced Research Initiative, Institute of Science Tokyo, Chiyoda-ku, Tokyo, Japan

Department of Biochemistry and Molecular Biology, Graduate School of Science, University of Hyogo, Harima Science Garden City, Hyogo, Japan

A. Nakano, J. Saraste (eds.), *The Golgi Network, Volume II*, Subcellular Biochemistry 111, https://doi.org/10.1007/978-3-032-16833-7_14

Intracellular Protein Degradation System

Protein degradation mechanism within cells is crucial for managing the quality of cellular proteins and for maintaining cellular homeostasis. To date, the primary focus has been on analyzing the two main pathways of protein degradation, namely, the ubiquitin-proteasome system and autophagy [1].

The proteasome is a large protein complex found in the cytoplasm and nucleus [2]. When ubiquitin, a small protein, binds to a target protein, it acts as a marker that directs the protein to the proteasome, where it is degraded using the energy of ATP. This pathway plays a particularly important role in the degradation of short-lived and abnormal proteins.

Proteins that are synthesized in the endoplasmic reticulum (ER) must be properly folded, and misfolded proteins are harmful to the cell and must be rapidly degraded in the ER. This protein degradation process is called ER-associated degradation (ERAD), and is a key mechanism of the ER stress response [3, 4]. The proteasome also plays a role in the final stage of ERAD. Specifically, when misfolded proteins are produced in the ER, chaperone proteins and folding sensor proteins recognize them. The misfolded proteins are then retrotranslocated across the ER membrane into the cytoplasm, where they undergo ubiquitination and are subsequently degraded by the proteasome.

Autophagy is similarly a cellular function that degrades cytoplasmic proteins and plays an important role in cellular homeostasis and stress responses [5, 6]. Although ubiquitination often acts as a marker for degradation, in this case, the proteins are not degraded by the proteasome. Instead, they are surrounded by a double membrane originating from the ER membrane, forming a structure called the autophagosome. This autophagosome then fuses with a lysosome, and the various proteins, lipids and nucleic acids contained within the double membrane structure are degraded.

These two protein-degradation pathways are conserved across species, from yeast to mammalian cells, and are essential for the efficient processing of unnecessary proteins and damaged proteins within the cell, thereby maintaining normal cellular functions. Furthermore, when these functions are impaired, they can lead to the onset of various diseases. In this way, intracellular protein degradation mechanisms are indispensable for keeping cells healthy and for maintaining an organism's homeostasis. On the other hand, we have identified a novel protein degradation system that utilizes the Golgi membrane, which is distinct from these pathways, and we will outline its details in this paper.

Functions of the Golgi Apparatus

The Golgi apparatus is an organelle that is involved in crucial cellular processes, such as cellular secretion, and is primarily responsible for the modification, sorting, and transport of proteins and lipids. Specifically, the Golgi apparatus performs

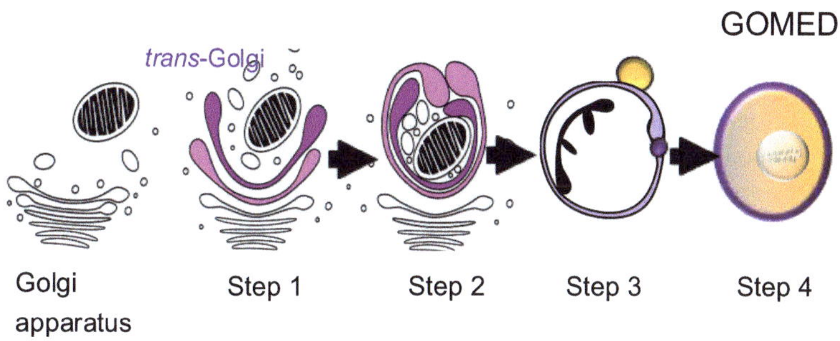

Fig. 14.1 Schematic diagram of GOMED. GOMED proceeds as follows: Step 1, *trans*-Golgi membrane curvature; Step 2, deepening of the curvature and divergence from the medial-Golgi; Step 3, generation of an autophagosome-like structure; Step 4, generation of an autolysosome-like structure by fusion with a lysosome

various modifications on proteins that are synthesized in the ER, including glycosylation, phosphorylation, and sulfation. These modifications are essential for the functional maturation of proteins and their transport to their designated destinations. Moreover, the modified proteins and lipids are sorted within the Golgi apparatus and transported to their appropriate destinations, such as the cell membrane, lysosomes, or secretory vesicles. This process involves specific transport proteins, such as clathrin and COPI/COPII coat proteins, which regulate the formation of transport vesicles and their targeting to their correct locations [7, 8].

The Golgi apparatus is composed of multiple membrane-bound, sac-like structures. In both plant and animal cells, these flattened sac-like structures are stacked to form *cis*, medial, and *trans* compartments, creating an arrangement that facilitates the sequential addition of sugar chains (glycosylation), and the transport of proteins. This stacked structure is well-suited for the stepwise processing and transport of molecules within a cell.

On the other hand, while conducting electron microscopy (EM) observations of mammalian cells subjected to DNA-damaging stress, we identified a substantially deformed Golgi apparatus. Specifically, we observed a Golgi apparatus with increased curvature in its *trans*-side cisternae, which eventually fused at both ends to form a spherical structure, enclosing proteins and other materials inside. Furthermore, we observed lysosomes fusing with the spherical Golgi membranes and breaking down the enclosed contents. Observation of this morphologically unique Golgi apparatus structure led to our discovery of a novel protein degradation mechanism [9] (Fig. 14.1).

Morphologically, the structures formed by the Golgi membrane enclosing the cytoplasm resemble autophagosomes formed from the ER membrane, and the structures formed by their subsequent fusion with lysosomes closely resemble autolysosomes. Therefore, we initially referred to this process as "alternative autophagy", "Atg5/7-independent autophagy", or "Rab9-dependent autophagy", but we subsequently found that it was a protein degradation mechanism distinct from autophagy (as discussed later). Moreover, as this is a cellular function that utilizes the Golgi membrane, we now frequently refer to it as Golgi membrane-associated degradation (GOMED) [10]. These findings indicate that the Golgi apparatus has two distinct forms and functions. The first is the Golgi apparatus described in textbooks, with its stacked cisternae, which functions in protein transport. The second is the newly identified form of the Golgi apparatus, in which the *trans*-Golgi membrane detaches from the medial Golgi, forms a spherical structure, and performs protein degradation.

Similarities Between Autophagy and GOMED

As previously reported, characteristic intracellular structures of autophagy, such as isolation membranes, autophagosomes, and autolysosomes, have also been observed in GOMED, showing similarities in the morphology of the two mechanisms (Table 14.1). On the other hand, differences include the following: (1) whereas the

Table 14.1 Comparison of autophagy and GOMED

		Autophagy	GOMED
Phylogenetically conservation	Common	yeast to mammals	
Morphology	Resemble	autophagosome-like structures autolysosome-like structures	
Membrane source	Different	ER membrane	Golgi membrane
Upstream molecules	Common	Ulk1, Beclin1	
Downstream molecules	Different	Atg5, Wipi1/2	Wipi3/4, Rab9
Degrading substrate	Totally Different	P62, LC3	Hormone (Insulin granules)
Inducing stimuli	Partly Common (DNA damage)	Starvation, mTor signal	Golgi stress

autophagic membrane is derived from the ER, the membrane involved in GOMED is derived from the Golgi apparatus [9–12]; and (2) the types of key effector molecules involved differ between the two pathways (discussed later). Furthermore, although there are stimuli that can induce both autophagy and GOMED, there are specific stimuli that activate only one of the pathways. For example, rapamycin and nutrient starvation selectively activate autophagy, whereas Golgi stress selectively activates GOMED. Additionally, substrate molecules targeted for degradation also differ substantially. For instance, p62 and LC3, which are markers of autophagy, are degraded only through autophagy, and not by GOMED. Conversely, insulin granules are degraded by GOMED but not by autophagy (discussed later) (Table 14.1).

Because the substrates that are degraded differ greatly between these two pathways, they cannot compensate for each other's functions, and they are considered to be independent protein degradation systems.

GOMED in Yeast

As research on autophagy advanced rapidly starting with yeast genetics [13, 14], we first planned experiments using yeast to elucidate the molecules required for GOMED. Specifically, to avoid confusion with autophagy, we used autophagy-deficient yeast cells (Atg5-deficient yeast cells), and analyzed whether GOMED was induced when these cells were exposed to various stimuli. The presence or absence of GOMED was determined by morphological analysis using phase-contrast and EM, as well as protein degradation analysis of cytoplasmic proteins by Western blotting [10]. As a result, we found that the antifungal drug amphotericin B has GOMED-inducing activity. Typically, the Golgi apparatus in yeast (*Saccharomyces cerevisiae*) differs from that in plant and mammalian cells, as the sac-like Golgi cisternae in yeast are scattered throughout the cytoplasm. However, upon treatment with amphotericin B, we observed Golgi apparatus aggregation and multilayering (stack formation). At the periphery of these stacks, we observed curvature of the *trans*-Golgi membrane, which enclosed cytoplasmic components and mitochondria. Additionally, fusion of the GOMED membrane with the vacuoles was observed, and Western blotting demonstrated the degradation of Golgi proteins and cytoplasmic-localized proteins [9]. Moreover, in yeast cells lacking Grh1, a molecule involved in Golgi stack formation, or Gvp36, a molecule involved in Golgi membrane curvature, no autophagosome-like structures were formed. Immunoelectron microscopy confirmed the localization of Golgi proteins on the autophagosome membrane, confirming that the autophagosome-like structures originate from the Golgi membrane [9].

Molecular Mechanism of GOMED

Investigation of the molecular mechanism of GOMED in mammalian cells was started by analysis of its association with existing autophagy-related molecules. In the early stages of autophagy, several molecules such as the serine/threonine kinase Ulk1, Atg9, and Beclin 1 (a component of the phosphatidylinositol 3-phosphate kinase class III complex) are crucial. Following these early steps, the elongation of the isolation membrane requires molecules with lipid-transfer activity, such as Wipi1 or Wipi2. Additionally, two ubiquitin-like conjugation systems play essential roles in this process. One is the Atg5 system, centered around the covalent bonding of Atg5 and Atg12 (involving Atg5, Atg7, Atg10, Atg12, and Atg16). The Atg5-Atg12 complex preferentially localizes to the outer membrane of the isolation membrane, promoting its elongation, and dissociates from the membrane before or after autophagosome formation. The other is the LC3 system, which involves the covalent attachment of LC3 to phosphatidylethanolamine (PE), in a pathway involving Atg3, Atg4, Atg7, and Atg8. The LC3-PE complex binds to the isolation membrane and autophagosome membrane in an Atg5-Atg12 complex-dependent manner, contributing to autophagosome formation [5, 6, 15] (Fig. 14.2). To determine which of these autophagy-associated molecules are involved in GOMED, we analyzed Atg5-deficient mouse fibroblasts. Specifically, we created Atg5-deficient cells in which these molecules were knocked down, and observed whether GOMED structures

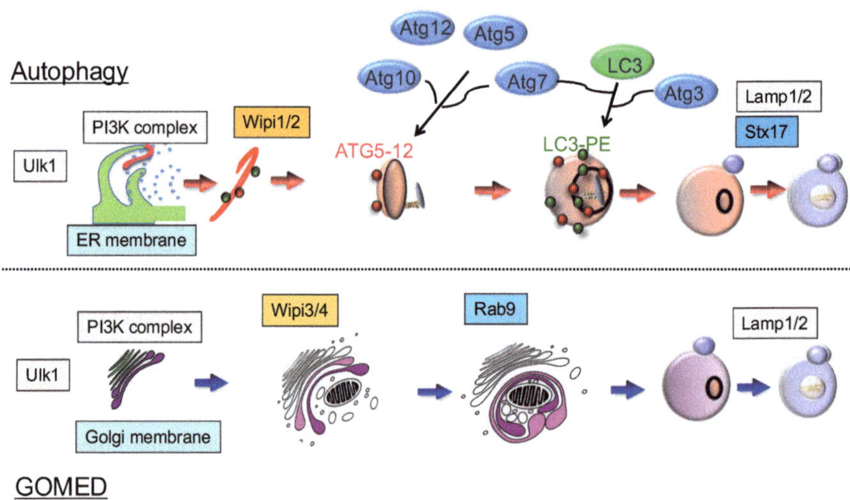

Fig. 14.2 Molecular mechanism of autophagy and GOMED. Both autophagy and GOMED start with Ulk1, and are induced via the PI3K complex. Autophagy originates from the ER membrane, and the phagophore membrane is elongated via Wipi1/2. Downstream of Atg5 and Atg7 are essential molecules, and LC3 binds to the phagophore membrane in an Atg5-dependent manner, contributing to autophagosome formation. In GOMED, autophagosome-like structure formation proceeds in a Wipi3/4- and Rab9-dependent manner using the Golgi membrane

formed after inducing GOMED through DNA-damage stress. The results demonstrated that molecules functioning relatively upstream in the autophagy pathway, such as Ulk1 and Beclin 1, also play important roles in GOMED. However, core autophagy molecules, such as Atg5, Atg7, Atg9, Atg12, Atg16, and LC3 were not involved in GOMED [9]. Additionally, through genetic analysis using yeast, we found that Hsv2 (the mammalian homolog of Wipi3 and Wipi4) is crucial for GOMED, helping to clarify the signaling mechanism involved [16, 17] (Fig. 14.2).

Ulk1

Ulk1 is the mammalian homolog of Atg1 in yeast, and functions as a key upstream effector in starvation-induced autophagy [18]. When mammalian cells are exposed to DNA damage, both autophagy and GOMED are induced. In contrast, in cells with double-knockout of Ulk1 and its compensatory homolog Ulk2, neither autophagy nor GOMED was induced. This clearly demonstrates that Ulk1 and Ulk2 are involved in both protein degradation mechanisms [9]. But how do Ulk1/Ulk2 molecules regulate such different cellular functions within the same cell?

Using cells subjected to DNA damage, we performed comprehensive mass spectrometric analysis of Ulk1 phosphorylation, leading to the following findings [19, 20]. (1) During both autophagy and GOMED, serine 637 on Ulk1 is dephosphorylated. (2) When a serine-to-aspartate substitution at this site was introduced to prevent dephosphorylation, neither protein degradation pathway was functional. (3) This dephosphorylation reaction is mediated by the protein phosphatase Mg^{2+}/Mn^{2+} dependent 1D (PPM1D), which is transcriptionally induced in a p53-dependent manner following DNA damage (Fig. 14.3). These findings were

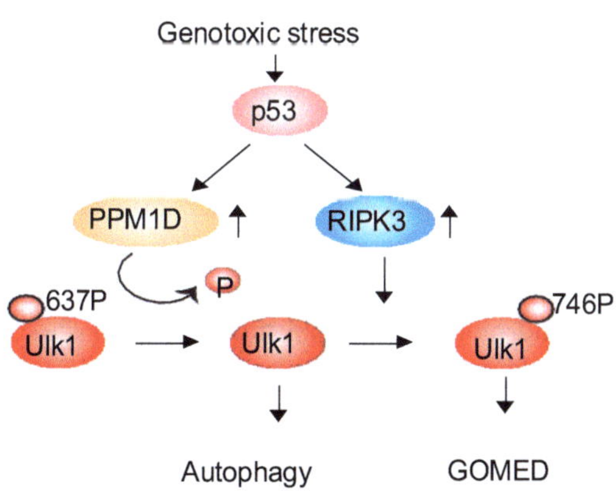

Fig. 14.3 Phosphorylation status of Ulk1 determines autophagy and GOMED. Schematic diagram of GOMED induction by genotoxic stress. Genotoxic stress induces the p53-dependent upregulation of PPM1D and RIPK3; PPM1D induces dephosphorylation of the 637th serine residue (637P) of Ulk1, which results in the induction of both autophagy and GOMED. Subsequent phosphorylation of the 746th serine residue (746P) of Ulk1 by RIPK3 activates GOMED, whereas the absence of this phosphorylation leads to autophagy

further confirmed in PPM1D-deficient cells, in which neither protein degradation system was induced [19].

Further mass spectrometry analysis demonstrated that during GOMED, after the dephosphorylation of serine 637, serine 746 on Ulk1 becomes phosphorylated. When a serine-to-alanine substitution was introduced in this position to prevent this phosphorylation, only GOMED was inhibited, whereas autophagy was still induced. This phosphorylation is carried out by the receptor interacting serine/threonine kinase 3 (RIPK3), which is also transcriptionally induced by p53 following DNA damage. These findings were corroborated in RIPK3-deficient cells, in which only GOMED was not induced in response to DNA damage [20]. Based on these results, we propose the following signaling mechanism for DNA damage-induced protein degradation. (1) p53-dependent transcription induces the expression of PPM1D and RIPK3. (2) PPM1D dephosphorylates serine 637 on Ulk1, resulting in the induction of both autophagy and GOMED. (3) Subsequently, the phosphorylation of serine 746 on Ulk1 activates GOMED, whereas the absence of this phosphorylation leads to normal autophagy being carried out. (4) When serine 746 is phosphorylated, Ulk1 does not form a complex with Atg13 and FIP200, as seen in typical autophagy. Instead, Ulk1 translocates to the Golgi apparatus, where it likely phosphorylates specific proteins, transmitting the signal downstream of GOMED (Fig. 14.3).

Interestingly, RIPK3 is also an essential molecule in necroptosis, which is a form of programmed cell death [21]. However, GOMED is not induced by necroptosis-inducing stimuli, such as TNF-α plus z-VAD (a caspase inhibitor), nor is necroptosis induced by DNA damage stimuli that trigger GOMED [20]. Additionally, other key necroptosis molecules, such as RIPK1 and mixed lineage kinase domain like pseudokinase (MLKL), were found not to play a role in GOMED [20]. These facts suggest that although RIPK3 is involved in both necroptosis and GOMED, the two systems operate independently without any interaction.

Wipi3

Because GOMED is conserved from yeast to mammals, we utilized yeast genetics techniques to identify genes necessary for GOMED, leading to the identification of Hsv2 [22]. Hsv2 is a protein belonging to the Proppin family [23, 24], which binds to phosphatidylinositol, and functions in lipid transfer. In Hsv2-deficient yeast cells, the elongation of the Golgi membrane does not occur, and GOMED is not induced [22], indicating that Hsv2 is an effector molecule of GOMED. For further investigation, we generated Atg5-deficient cells lacking Wipi3, the mammalian homolog of Hsv2, and subjected them to DNA-damage stimuli. As in yeast cells, elongation of the Golgi membrane did not occur in these cells; instead, the Golgi became abnormally swollen and rod-shaped, and GOMED was not induced. Additionally, neither autophagosome-like GOMED structures, nor isolation membrane-like GOMED structures were observed, suggesting that Wipi3 functions on the Golgi membrane before the formation of the isolation membrane [22].

Further analysis of Wipi3 localization demonstrated that, under normal conditions, Wipi3 is evenly distributed throughout the cytoplasm. However, after GOMED induction, a proportion of Wipi3 is translocated to the *trans*-Golgi membrane. Immunoelectron microscopy analysis of isolated GOMED structures from GOMED-induced cells showed that Wipi3 and GS15 (a *trans*-Golgi marker protein) signals were present on the convex surface of isolation membrane-like and autophagosome-like GOMED structures. This confirmed that Wipi3 translocates from the cytoplasm to the Golgi membrane, where it localizes to Golgi-derived isolation membranes, which subsequently form GOMED structures. Wipi3 is known to interact with phosphatidylinositol 3-phosphate (PI3P), and it has been suggested that two residues (R225 and R226) are essential for this interaction. Indeed, when mutant Wipi3 proteins (R225A or R226A) were expressed in cells and GOMED stimuli were applied, the mutant Wipi3 proteins remained localized throughout the cytoplasm, with no translocation to the Golgi, and GOMED was not induced. This demonstrated that the interaction between Wipi3 and PI3P is crucial for GOMED [22]. Wipi3 is a homolog of Wipi1 and Wipi2, which are involved in autophagy. The fact that one homolog regulates autophagy whereas the other regulates GOMED suggests that, despite their distinct functions, these pathways share a common evolutionary origin.

Rab9

Rab9 is a member of the Rab family of small GTP-binding proteins, which are key regulators of intracellular vesicle trafficking [25, 26]. Rab9 helps the retrograde transport of proteins and lipids between late endosomes and the *trans* Golgi network (TGN). Specifically, Rab9 cycles between an active, GTP-bound state and an inactive, GDP-bound state. When in the GTP-bound state, Rab9 interacts with various effector proteins, such as TIP47 and the retromer complex, mediating the docking and fusion processes of vesicles during transport. Abnormalities in Rab9 have been linked to the development of neurodegenerative diseases.

Considering that Rab9 is responsible for vesicle transport from late endosomes to the TGN, it was hypothesized that Rab9 is also involved in GOMED. In fact, the following findings support Rab9 as a key molecule in GOMED. (1) In Atg5-deficient mouse embryonic fibroblast (MEF) cells treated with etoposide, Rab9 was localized to autolysosome-like structures. (2) Knockdown of Rab9 reduced the induction of GOMED, indicating its crucial role in this process [9]. Importantly, Rab9 does not affect autophagy.

Detection of GOMED

In this section, we provide an overview of the detection methods for GOMED. Because the key structures involved in GOMED resemble autophagic structures, many of the assays that are presently used for GOMED detection are similar to those used for autophagy. Therefore, to facilitate the detection of GOMED, we often use Atg5-deficient or Atg7-deficient cells, in which autophagy cannot occur. Additionally, combining these tools with the detection of autophagy-specific markers, such as LC3, can help distinguish GOMED from autophagy and make GOMED detection more straightforward.

Analysis Using Electron Microscopy (EM)

As mentioned above, during GOMED, both autophagosome-like and autolysosome-like GOMED structures are formed. Although these structures can be relatively easily identified using EM, distinguishing them from similar structures formed by autophagy is crucial. One clue for their differentiation is the structure of the ER and the Golgi apparatus. In GOMED, many Golgi ribbons undergo ministack formation, whereas the structure of the ER remains mostly normal. In contrast, during autophagy, no such deformation of the Golgi apparatus is observed [9]. To accurately assess the dynamics of GOMED, it is essential to distinguish between phagophore-like GOMED structures (open structures) and autophagosome-like GOMED structures (closed structures) and measure them over time. When using conventional transmission electron microscopy (TEM), it is important to note that in phagophore-like GOMED structures, the electron density inside and outside the structure is the same, whereas in autophagosome-like GOMED structures, the density differs. Therefore, distinguishing between these two structures is crucial.

Analysis Using Lamp1 (or Lamp2)

Lamp1 (or Lamp2) is a protein localized on the lysosomal membrane. Lysosomes are small structures with a diameter ranging from 100 nm to 500 nm. However, when they fuse with autophagosome-like GOMED structures, they form larger autolysosome-like GOMED structures with diameters ranging from 300 nm to 1 μm. Therefore, by expressing Lamp1 (or Lamp2)-GFP in cells or by performing immunostaining for Lamp1 (or Lamp2), the formation of autolysosome-like GOMED structures can be analyzed based on the differences in the size of the fluorescent signals. Additionally, because the membrane is visualized by confocal microscopy, this technique often reveals ring-shaped fluorescence signals [9, 27]. Using the correlative light and electron microscopy (CLEM) method, autolysosome-like GOMED structures have been observed in areas corresponding to the Lamp1-GFP fluorescence signals. Moreover, when GFP is fused to the C-terminus of Lamp1, the GFP portion of Lamp1-GFP is located on the outer side of the autolysosome-like GOMED structures [9], meaning that the GFP signal is not diminished by the acidic environment or lysosomal enzymes within the autolysosome-like GOMED structures.

Analysis Using the mRFP-EGFP Tandem-Fluorescent Protein

When the mRFP-EGFP-LC3 tandem protein is expressed in cells, it is widely distributed throughout the cytoplasm. Upon autophagy induction, the LC3 portion responds by localizing to the phagophore membrane and is subsequently incorporated into autolysosomes. While localized on the phagophore membrane, the tandem protein exhibits yellow fluorescence. Once it is incorporated into autolysosomes, EGFP is quenched and degraded owing to the acidic environment and lysosomal enzymes, whereas mRFP is resistant to this degradation, and hence continues to emit red fluorescence [28, 29].

For the application of this autophagy assay, a version of the mRFP-EGFP tandem protein lacking LC3 was constructed [10, 30]. This protein is also widely expressed throughout the cytoplasm, showing yellow fluorescence. However, upon the induction of either autophagy or GOMED, EGFP is quenched and degraded in autolysosomes or autolysosome-like GOMED structures, causing the protein to emit red fluorescence (Fig. 14.4). In fact, the red punctate structures of mRFP-EGFP have been shown to colocalize with autolysosome-like GOMED structures using CLEM. Furthermore, GOMED can be quantitatively assessed using autophagy-deficient cells (such as Atg5-deficient cells).

Development of Fluorescence Loss in Photobleaching (FLIP)-Based Autophagy Detection Technology (FLAD)

To elucidate GOMED activity, it is crucial to quantitatively analyze the dynamics of two key steps; namely, the sequestration of intracellular substrate molecules into autophagosome-like GOMED structures, and their subsequent degradation in autolysosome-like GOMED structures. Recently, we developed a method to measure the dynamics from isolation membrane-like GOMED structures to autophagosome-like GOMED structures [31]. This method takes advantage of the fact that when autophagosome-like GOMED structures form from isolation membrane-like GOMED structures, some cytoplasmic molecules are nonselectively trapped inside. First, cells are engineered to express fluorescent proteins, such as EGFP or mRFP. By irradiating a specific part of the cell with light, the fluorescence

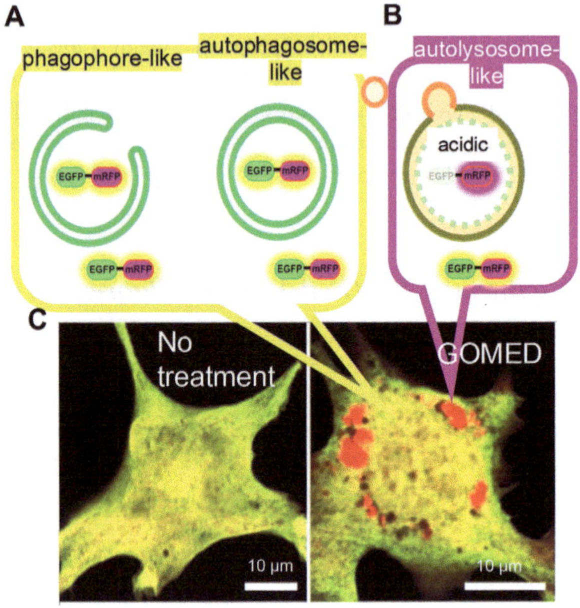

Fig. 14.4 Schematic diagram of the method of GOMED detection using mRFP-EGFP. (**a**) mRFP-EGFP shows yellow fluorescence even in phagophore-like structures and autophagosome-like structures. (**b**) In autolysosome-like structures, only EGFP is quenched and degraded in an acidic environment, and red fluorescence is observed. (**c**) Visualization of mRFP-EGFP-expressing $Atg5^{KO}$ MEFs. In untreated cells, yellow fluorescence from the mRFP-EGFP tandem protein was broadly observed throughout the cells. In GOMED-induced cells, red puncta were observed, because only EGFP is quenched and degraded. (Modified from Ref. [22])

in that region is initially lost. Subsequently, fluorescent proteins from the nonirradiated regions freely diffuse to the irradiated region. This causes a reduction in fluorescence in the nonirradiated region. However, if autophagosome-like GOMED structures are present in the nonirradiated region, the fluorescent proteins inside these closed structures cannot diffuse, and they remain visible as puncta. By quantifying these puncta, GOMED activity can be measured. We have named this method FLAD (Fig. 14.5). Whereas this method also detects autophagy, the coexpression of mRFP-LC3 enables the differentiation of autophagosomes and GOMED structures; i.e., mRFP-LC3-positive FLAD puncta correspond to autophagosomes, whereas mRFP-LC3-negative FLAD puncta correspond to autophagosome-like GOMED structures.

Small Fluorescent Probes

Many of the GOMED analysis methods described above rely on fluorescent proteins, which require gene introduction into cells. Therefore, in collaboration with Dojindo Laboratories, we developed three types of fluorescent probes; DAPGreen,

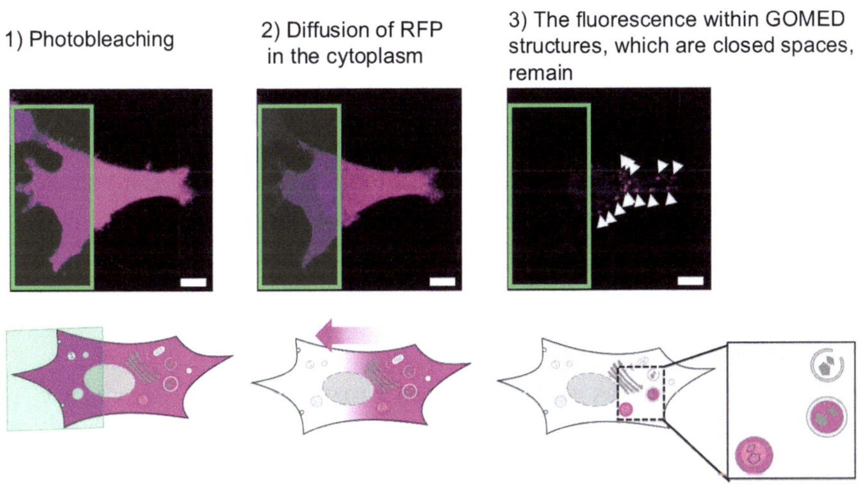

Fig. 14.5 Schematic model of the FLAD assay. When photobleaching is performed on the boxed regions (ROI) of RFP-expressing cells (1), RFP fluorescence outside the ROI is reduced owing to free diffusion of the RFP molecules (2). On the other hand, RFP sequestered in autophagosome-like structures cannot diffuse, and is detected as puncta (arrows) that retain their fluorescence intensity

DAPRed, and DALGreen [32, 33], which accumulate during autophagy and GOMED. These probes fluoresce when incorporated into membranes in a hydrophobic environment, and specifically recognize membranes by detecting membrane thickness and the presence of PI3P, which enables them to selectively label autophagy and GOMED membranes. We conducted a detailed analysis of which stages of autophagy and GOMED these fluorescent probes could label, and found the following [33]. DAPGreen labels all structures from the early stages of both autophagy and GOMED with green fluorescence, whereas DAPRed labels structures at the intermediate stages (just before the formation of autophagosomes or autophagosome-like structures) with red fluorescence. Additionally, DALGreen specifically labels autolysosomes and autolysosome-like GOMED structures of both autophagy and GOMED with green fluorescence. For actual analysis, a combination of DAPGreen and DAPRed, or DAPRed and DALGreen is used; the former can measure the dynamics of early GOMED/autophagy structure formation, whereas the latter can measure the transition from autophagosomes to autolysosomes or autophagosome-like to autolysosome-like structures. These probes can also be used in combination with LC3, enabling the distinction between autophagy and GOMED.

Analysis Using Signaling Molecules

A method to directly evaluate GOMED without the need to consider autophagy involves analyzing the dynamics of signaling molecules. Specifically, the phosphorylation of Ulk1 can be used for this purpose. As mentioned in the section on Ulk1, when GOMED is induced, serine 747 of human Ulk1 (serine 746 in mice) becomes phosphorylated, and Ulk1 translocates to the Golgi apparatus [20]. This reaction can be visualized by immunostaining with an antibody that specifically recognizes phosphorylated $Ulk1^{S746}$ (which can also detect the phosphorylation of human Ulk1). As this phosphorylation is not involved in autophagy, it enables the specific detection of GOMED. Additionally, Wipi3 also relocates from the cytoplasm to the Golgi apparatus during GOMED, and this can be tracked to detect GOMED.

GOMED Substrates and Biological Functions

The key difference between autophagy and GOMED lies in their degradation substrates. The well-known autophagy substrate p62 is not selectively degraded by GOMED. Conversely, autophagy does not participate in the degradation of substrates involved in various physiological phenomena associated with GOMED, as will be discussed below. In other words, there are currently no known cases in which the two pathways compensate for each other's functions. Because GOMED utilizes the Golgi membrane, molecules transported through the Golgi are expected to become its substrates. Additionally, when there is stress on this transport pathway, GOMED is likely induced to degrade proteins that accumulate in the Golgi. In fact, amphotericin B1, which induces GOMED in yeast, is known to act on ergosterol, and disrupts the protein transport system from the Golgi to the cell membrane, as well as secretion outside of the cell [10]. Besides amphotericin B, GOMED induction has been observed when molecules involved in transport from the Golgi to the cell membrane or outside of the cell are defective, or when the drug 1,3-cyclohexanebis(methylamine), which inhibits extracellular transport, is administered [10].

Substrate molecules degraded under these conditions include artificial proteins, such as VSVG-GFP, which are transported through the Golgi, and endogenous integrins. This suggests that the primary function of GOMED is to degrade proteins that accumulate in the Golgi owing to a block in transport from the Golgi to the cell membrane or outside the cell.

A representative function of GOMED is the regulation of insulin secretion. Insulin is synthesized in insulin-secreting cells (β cells) of the pancreas, and secreted via the Golgi apparatus. However, when glucose levels around the cells become low (i.e., blood sugar levels decrease), insulin secretion is suppressed, which then prevents further glucose reduction. At this time, GOMED is induced in β cells to inhibit the accumulation of insulin in the Golgi apparatus. This phenomenon has been observed when the culture medium of β-cell lines such as MIN6 cells, or isolated mouse pancreatic islet cells, is switched from high to low glucose [10]. Notably, autophagy is not involved in insulin degradation during hypoglycemia.

Removal of Mitochondria from Red Blood Cells by GOMED

GOMED degrades not only proteins transported via the Golgi apparatus, but also organelles, such as mitochondria. A representative example is the removal of mitochondria during the final stages of red blood cell differentiation. When red blood cells undergo final differentiation, nuclear elimination occurs first, followed by the removal of mitochondria and other organelles within 24 h (Fig. 14.6). This mitochondrial removal mechanism has been suggested to involve autophagosome-like structures. When we observed reticulocytes by EM, we found mitochondria that were surrounded by double membranes, forming autophagosome-like structures, as well as autolysosome-like structures showing partial mitochondrial degradation [9]. Additionally, from the high-magnification images of these structures, the origin of the membranes appeared to be the Golgi apparatus. Even in Atg5-deficient mice, autophagosome-like and autolysosome-like structures surrounding mitochondria were observed at similar levels. Furthermore, when we analyzed the abundance of mitochondria in fetal liver (the site of erythropoiesis) and in blood, mitochondrial removal in Atg5-deficient mice was induced to nearly the same extent as that in wild-type mice, with most red blood cells lacking mitochondria [9, 34]. This suggests that Atg5-dependent autophagy is not involved in the mitochondrial removal of red blood cells. On the other hand, in Ulk1-deficient mice (Ulk1 is involved in both autophagy and GOMED), a large number of mitochondria was found to remain in the red blood cells [34, 35]. To quantitatively evaluate the remaining mitochondria in red blood cells, we performed triple staining with the erythroid differentiation markers Ter119 and CD71, and MitoTracker Deep Red for mitochondria, followed by flow cytometry. Whereas red blood cells from wild-type mice and those from Atg5- or Atg7-deficient mice lacked mitochondria, Ulk1-deficient and Ulk1/Atg5 double-deficient red blood cells retained their mitochondria [34]. These analyses demonstrated that mitochondrial removal during the final stages of red blood cell differentiation occurs by Ulk1-dependent GOMED, rather than by Atg5-dependent autophagy.

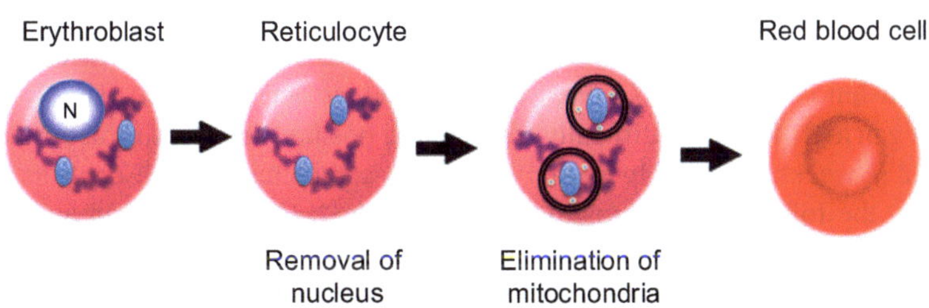

Fig. 14.6 Involvement of GOMED in mitophagy during erythrocyte maturation. Schematic diagram of red-blood-cell maturation. During erythrocyte maturation, erythroblasts lose their nuclei to become reticulocytes, and then the reticulocytes are transformed into erythrocytes by the elimination of mitochondria. GOMED is involved in the latter process.

GOMED-mediated mitochondrial removal has also been reported to occur during cellular dedifferentiation. When induced pluripotent stem cells are generated from normal cells, mitochondria are reduced, and this process was shown to be mediated by GOMED [36].

Cardiac Protection by GOMED

Mitochondrial dysfunction has been reported in various cardiac pathologies, such as ischemia-reperfusion injury, diabetic cardiomyopathy, pressure overload-induced cardiac hypertrophy, and heart failure [37–39]. The mitochondrial dysfunction involved in cardiac pathologies affects ATP production and calcium metabolism. On the other hand, these damaged mitochondria have been observed to be removed by autophagy, particularly by Pink1/Parkin-dependent autophagy [40]. In addition, there are various reports of mitochondrial removal via GOMED. For example, after 3–7 days of pressure overload caused by transverse aortic constriction, damaged mitochondria are removed, but autophagy does not occur during this period. Instead, the mitochondria are thought to be degraded through GOMED [41]. Furthermore, the following results have been observed under conditions of energy stress. (1) In heart-specific Atg7-deficient mice, mitochondria are degraded by GOMED, confirming that GOMED plays a role in mitochondrial removal under these conditions. (2) In cardiomyocyte-specific Ulk1-knockout mice, autophagy occurs, but mitochondrial removal does not. These results indicate that Ulk1-dependent GOMED plays an essential role in mitochondrial removal during energy stress [41]. Additionally, under energy stress, AMP-activated protein kinase (AMPK) phosphorylation is activated, which promotes the phosphorylation of Ulk1. Downstream of this, phosphorylation of serine 179 on Rab9 occurs, and this Rab9-dependent pathway is responsible for mitochondrial removal [41].

GOMED-Mediated Maintenance of Neuronal Cells

To understand the role of GOMED in the brain, we generated mice with a brain-specific deletion of the *Wipi3* gene, which is a key molecule in GOMED [24]. These mice began to demonstrate difficulty maintaining posture, and walking abnormalities at about 7 weeks of age. Upon analysis of the cerebellum, which controls motor function, the following abnormalities were observed: (1) degeneration and loss of Purkinje cells in the cerebellum; (2) abnormal Golgi apparatus morphology in neurons before degeneration; (3) failure of GOMED induction; and (4) increased glial fibrillary acidic protein, and the accumulation of lamellar bodies. Further analysis demonstrated the accumulation of iron and the iron transport protein ceruloplasmin in cerebellar neurons. Ceruloplasmin is primarily known as a copper transport protein, but also plays a role in iron metabolism. Whereas ceruloplasmin produced in the liver functions in the blood, neuronal ceruloplasmin is transported to the cell membrane via the Golgi apparatus, where it converts ferrous iron (Fe^{2+}) to ferric iron (Fe^{3+}). Ferric iron is then exported outside the cell via ferroportin. Based on lines of evidence that ceruloplasmin is a GOMED substrate, it is believed that in Wipi3-deficient mice, the dysfunction of GOMED prevents ceruloplasmin degradation [24], leading to its abnormal accumulation in the cytoplasm. This results in the production of excess ferric iron, and ultimately leads to iron deposition and cerebellar degeneration.

Similar neuronal dysfunction has been observed in the brains of mice lacking *Atg7* [42, 43], an essential autophagy gene. However, the internal structure of neurons of Wipi3-deficient mice was distinctly different, with no abnormalities in the Golgi apparatus, but structural abnormalities in the ER. In addition, there was no deposition of ceruloplasmin or iron, but ubiquitinated proteins were found to be accumulated in the cells without being degraded [24]. This indicates that autophagy and GOMED contribute to neuronal maintenance through different mechanisms. Indeed, when we generated mice with deletions in both *Wipi3* and *Atg7* genes, they died within 28 days after birth, and showed much more severe neurodegeneration than single-gene-knockout mice [24]. This demonstrates that autophagy and GOMED maintain neurons through distinct mechanisms.

In humans, neurodegeneration with brain iron accumulation, specifically the subtype known as static encephalopathy of childhood with neurodegeneration in adulthood (SENDA), is characterized by early-onset intellectual disability, and rapidly progressing extrapyramidal symptoms and dementia in adulthood [44]. SENDA is caused by mutations in the *Wipi4* gene, which is a homolog of Wipi3. However, whereas Wipi4-deficient mice do not demonstrate brain iron deposition or clear symptoms, Wipi3-deficient mice do, suggesting that they could be used as a model of SENDA.

GOMED and Inflammatory Bowel Disease (IBD)

IBD, which includes Crohn's disease and ulcerative colitis, is characterized by abnormal inflammation and immune responses in the intestines [45, 46]. Genome-wide association studies have identified a link between the autophagy-associated molecule Atg16 and IBD, with numerous follow-up reports supporting this association. On the other hand, GOMED has been linked to intestinal inflammation via the molecule tripartite motif containing 31 (TRIM31) [47]. TRIM31 is a ubiquitin ligase localized to mitochondria and lysosomes in the intestinal tract. In Crohn's disease patients, TRIM31 levels are reduced by 45% compared with healthy individuals. TRIM31 has been found to promote the formation of autophagy-like structures independently of Atg5/7 or LC3, suggesting that it is associated with GOMED-related proteins. These findings indicate that TRIM31-mediated GOMED may contribute to the prevention of IBD.

GOMED and Infectious Diseases

In the context of infectious diseases, GOMED induced by TRIM31 has been found to limit the growth of *Shigella* in intestinal epithelial cells, and to facilitate the clearance of intracellular *Shigella* from these cells. On the other hand, cytomegalovirus infection has been shown to significantly downregulate the expression of TRIM31, thereby impairing GOMED function [47]. Additionally, it has been reported that during group A streptococcal infection, Rab9A and Rab23 work together to carry out GOMED, playing a role in bacterial clearance [48]. These findings demonstrate that GOMED functions protectively in several infectious diseases by aiding in pathogen elimination.

Conclusion

In this chapter, we provided an overview of the history, molecular mechanisms, physiological functions, and established detection methods of GOMED. However, the complete picture of GOMED still remains unclear. In particular, many aspects of the mechanism of GOMED induction have not been elucidated. Further identification of GOMED substrates is also necessary. Most importantly, the physiological functions of GOMED and its association with diseases remain largely unknown, and future research in these areas is highly anticipated.

Funding This study was supported by JSPS KAKENHI Grant-in-Aid for Scientific Research (S) (23H05480), (A) (20H00467), (C) (20 K07353, 24 K09441), and by MEXT KAKENHI Grant-in-Aid for Scientific Research on Innovative Areas (22H04639, 23H04773, 24H01887). This study was also supported by AMED-CREST (JP23gm1410012) and by AMED under Grant Number JP21wm0525028 and JP24ama221130. This study was also supported by the foundation of Kinoshita Memorial Enterprise and Hyogo Science and Technology Association.

Conflicts of Interest The authors declare that they have no conflicts of interest associated with this manuscript.

References

1. Pohl C, Dikic I (2019) Cellular quality control by the ubiquitin-proteasome system and autophagy. Science 366:818–822
2. Tanaka K, Mizushima T, Saeki Y (2012) The proteasome: molecular machinery and pathophysiological roles. Biol Chem 393:217–234
3. Ninagawa S, George G, Mori K (2021) Mechanisms of productive folding and endoplasmic reticulum-associated degradation of glycoproteins and non-glycoproteins. Biochim Biophys Acta Gen Subj 1865:129812
4. Ruggiano A, Foresti O, Carvalho P (2014) ER-associated degradation: protein quality control and beyond. J Cell Biol 204:869–879

5. Morishita H, Mizushima N (2019) Diverse cellular roles of autophagy. Annu Rev Cell Dev Biol 35:453–475
6. Xie Z, Klionsky DJ (2007) Autophagosome formation: core machinery and adaptations. Nature 9:1102–1109
7. Farquhar MG, Palade GE (1998) The Golgi apparatus: 100 years of progress and controversy. Trends Cell Biol 8:2–10
8. Lowe M, Barr FA (2007) Inheritance and biogenesis of organelles in the secretory pathway. Nat Rev Mol Cell Biol 8:429–439
9. Nishida Y, Arakawa S, Fujitani K et al (2009) Discovery of Atg5/Atg7-independent alternative macroautophagy. Nature 461:654–658
10. Yamaguchi H, Arakawa S, Kanaseki T et al (2016) Golgi membrane-associated degradation pathway in yeast and mammals. EMBO J 35:1991–2007
11. Tooze SA, Yoshimori T (2010) The origin of the autophagosomal membrane. Nat Cell Biol 12: 831–835
12. Rubinsztein DC, Shpilka T, Elazar Z (2012) Mechanisms of autophagosome biogenesis. Curr Biol 22:R29–R34
13. Takeshige K, Baba M, Tsuboi S et al (1992) Autophagy in yeast demonstrated with proteinase-deficient mutants and conditions for its induction. J Cell Biol 119:301–311
14. Tsukada M, Ohsumi Y (1993) Isolation and characterization of autophagy-defective mutants of *Saccharomyces cervisiae*. FEBS Let 333:169–174
15. Debnath J, Gammoh N, Ryan KM (2023) Autophagy and autophagy-related pathways in cancer. Nat Rev Mol Cell Biol 24:560–575
16. Noguchi S, Shimizu S (2022) Molecular mechanisms and biological roles of GOMED. FEBS J 289:7213–7220
17. Feng H, Wang N, Zhang N et al (2022) Alternative autophagy: mechanisms and roles in different diseases. Cell Commun Signal 20:43
18. Rong Z, Zheng Z, Chen J et al (2022) Function and regulation of ULK1: from physiology to pathology. Gene 840:146772
19. Torii S, Yoshida T, Arakawa S et al (2016) Identification of PPM1D as an essential Ulk1 phosphatase for genotoxic stress-induced autophagy. EMBO Rep 17:1552–1564
20. Torii S, Yamaguchi H, Nakanishi A et al (2020) Identification of a phosphorylation site on Ulk1 required for genotoxic stress-induced alternative autophagy. Nat Commun 11:1754
21. Zhou Y, Xiang Y, Liu S et al (2024) RIPK3 signaling and its role in regulated cell death and diseases. Cell Death Discov 10:200
22. Yamaguchi H, Honda S, Torii S et al (2020) Wipi3 is essential for alternative autophagy and its loss causes neurodegeneration. Nat Commun 11:5311
23. Baskaran S, Ragusa MJ, Boura E et al (2012) Two-site recognition of phosphatidylinositol 3-phosphate by PROPPINs in autophagy. Mol Cell 47:339–348
24. Scacioc A, Schmidt C, Hofmann T et al (2017) Structure based biophysical characterization of the PROPPIN Atg18 shows Atg18 oligomerization upon membrane binding. Sci Rep 7:14008
25. Ganley IG, Carroll K, Bittova L et al (2004) Rab9 GTPase regulates late endosome size and requires effector interaction for its stability. Mol Biol Cell 15:5420–5430
26. Dong B, Kakihara K, Otani T et al (2013) Rab9 and retromer regulate retrograde trafficking of luminal protein required for epithelial tube length control. Nat Commun 4:1358
27. Nagata M, Arakawa S, Yamaguchi H et al (2018) Dram1 regulates DNA damage-induced alternative autophagy. Cell Stress 2:55–65
28. Kimura S, Noda T, Yoshimori T (2007) Dissection of the autophagosome maturation process by a novel reporter protein, tandem fluorescent-tagged LC3. Autophagy 3:452–460
29. Kaizuka T, Morishita H, Hama Y et al (2016) An Autophagic flux probe that releases an internal control. Mol Cell 64:835–849
30. Sakurai HT, Arakawa S, Yamaguchi H et al (2023) An overview of Golgi Membrane-Associated Degradation(GOMED) and its detection methods. Cells 12:2817

31. Sakurai HT, Arakawa S, Noguchi S et al (2022) FLIP-based autophagy-detecting technique reveals closed autophagic compartments. Sci Rep 12:22452
32. Iwashita H, Sakurai HT, Nagahora N et al (2018) Small fluorescent molecules for monitoring autophagic flux. FEBS Lett 592:559–567
33. Sakurai HT, Iwashita H, Arakawa S et al (2023) Development of small fluorescent probes for the analysis of autophagy kinetics. iScience 26:107218
34. Honda S, Arakawa S, Nishida Y et al (2014) Ulk1-mediated Atg5-independent macroautophagy mediates elimination of mitochondria from embryonic reticulocytes. Nat Commun 5:4004
35. Kundu M, Lindsten T, Yang CY (2008) Ulk1 plays a critical role in the autophagic clearance of mitochondria and ribosomes during reticulocyte maturation. Blood 112:1493–1502
36. Ma T, Li J, Xu Y et al (2015) Atg5-independent autophagy regulates mitochondrial clearance and is essential for iPSC reprogramming. Nat Cell Biol 17:1379–1387
37. Tahrir FG, Langford D, Amini S et al (2019) Mitochondrial quality control in cardiac cells: mechanisms and role in cardiac cell injury and disease. J Cell Physiol 234:8122–8133
38. Misra MK, Sarwat M, Bhakuni P et al (2009) Oxidative stress and ischemic myocardial syndromes. J Exp Clin Res 15:Ra209-219
39. Gollmer J, Zirlik A, Bugger H (2020) Mitochondrial mechanisms in diabetic cardiomyopathy. Diabetes Metab J 44:33–53
40. Koyano F, Okatsu K, Kosako H et al (2014) Ubiquitin is phosphorylated by PINK1 to activate parkin. Nature 510:162–166
41. Saito T, Nah J, Oka S et al (2019) An alternative mitophagy pathway mediated by Rab9 protects the heart against ischemia. J Clin Invest 129:802–819
42. Komatsu M, Waguri S, Chiba T et al (2006) Loss of autophagy in the central nervous system causes neurodegeneration in mice. Nature 441:880–884
43. Hara T, Nakamura K, Matsui M et al (2006) Suppression of basal autophagy in neural cells causes neurodegenerative disease in mice. Nature 441:885–889
44. Saitsu H, Nishimura T, Muramatsu K et al (2013) De novo mutations in the autophagy gene WDR45 cause static encephalopathy of childhood with neurodegeneration in adulthood. Nat Genet 45:445–449
45. Xavier RJ, Podolsky DK (2007) Unravelling the pathogenesis of inflammatory bowel disease. Nature 448:427–434
46. Larabi A, Barnich N, Nguyen HTT (2020) New insights into the interplay between autophagy, gut microbiota and inflammatory responses in IBD. Autophagy 16:38–51
47. Ra EA, Lee TA, Kim SW et al (2016) TRIM31 promotes Atg5/Atg7-independent autophagy in intestinal cells. Nat Commun 7:11726
48. Toh H, Nozawa T, Minowa-Nozawa A et al (2020) Group A Streptococcus modulates RAB1- and PIK3C3 complex-dependent autophagy. Autophagy 16:334–346

Chapter 15
Golgi Drivers of Cancer

Kyle A. Starost, Jagadeeswara R. Bommi, Prasenjit Mitra, and Seth J. Field

Abstract Unbiased analyses of cancer genomes, transcriptomes, and proteomes have led to the identification of numerous genes that, when perturbed, contribute to the formation of cancer in humans. These unbiased analyses, by identifying drivers of cancer that localize to and function at the Golgi, have illuminated a previously unappreciated role for the Golgi in driving oncogenic phenotypes. Here we review some of the cancer drivers that act at the Golgi. We highlight the evidence that they are functionally important in cancer, and the evidence that they localize to and function at the Golgi. We also discuss functional interactions between these proteins, and review some of the proposed mechanisms by which they promote oncogenic phenotypes.

Keywords Golgi · Cancer · Transformation · Oncogene · Phosphatidylinositol 4-phosphate · GOLPH3 · MYO18A · HRAS

K. A. Starost
Medical Scientist Training Program, Case Western Reserve University School of Medicine, Cleveland, OH, USA

Department of Medicine, Institute for Transformative Molecular Medicine, Case Western Reserve University School of Medicine, Cleveland, OH, USA

J. R. Bommi · P. Mitra
Department of Medicine, Institute for Transformative Molecular Medicine, Case Western Reserve University School of Medicine, Cleveland, OH, USA

S. J. Field (✉)
Department of Medicine, Institute for Transformative Molecular Medicine, Case Western Reserve University School of Medicine, Cleveland, OH, USA

Harrington Discovery Institute, University Hospitals Cleveland Medical Center, Cleveland, OH, USA
e-mail: seth.field@case.edu

A. Nakano, J. Saraste (eds.), *The Golgi Network, Volume II*, Subcellular Biochemistry 111, https://doi.org/10.1007/978-3-032-16833-7_15

Introduction

The discovery that perturbation of individual genes, and the function of the proteins that they express, can cause oncogenic transformation revolutionized our understanding of cancer over half a century ago [18, 185, 200]. A gold rush on the identification of oncogenes and their function implicated nearly every aspect of cell biology and cellular organelle in the pathogenesis of cancer [68, 69]. However, for most of this history, little was known about any possible role for the Golgi in oncogenic transformation.

Over the last 15 years, unbiased analyses of cancer genomes, transcriptomes, and proteomes have revealed roles for many Golgi proteins (and the genes that encode them) in human cancers. Increasingly sophisticated analysis of human cancers has also honed our definition of a cancer driver. The ability of a gene to cause oncogenic phenotypes in cell culture is not sufficient evidence to implicate a gene in human cancer. For the purposes of this review, we will focus our attention on genes and proteins for which there is stronger evidence for their role in human cancers. We will highlight some of the key findings (summarized in Table 15.1) that these genes and proteins function to promote human cancers, and the evidence that these proteins localize to, and function at the Golgi (more detailed discussion of mechanisms of Golgi function can be found elsewhere in this volume). Finally, where applicable we will discuss how some of these proteins are thought to function together, and mechanisms by which they may drive cancer. We note that for many of these proteins there is a sizable and diverse literature suggesting a large number of potential mechanisms by which they may drive cancer. Here we focus on mechanisms that involve the Golgi.

Many cancer drivers that reside and function at the plasma membrane pass through the Golgi. This includes proteins such as receptor tyrosine kinases, G protein coupled receptors, and integrin receptors. While many of these are bona fide cancer drivers, since their localization to the Golgi is incidental and not clearly related to their role in cancer, we do not discuss them in this review.

Golgi Cancer Drivers

GOLPH3

GOLPH3 (Golgi phosphoprotein 3, also known as GPP34, GMx33, and MIDAS) is widely expressed in all cell types. It is an effector of phosphatidylinositol 4-phosphate (PI(4)P), a phospholipid found in high concentrations in the cytosolic leaflet of the Golgi membrane [46, 58].

Table 15.1 Summary of key cancer drivers, detailing their specific Golgi locations, associated cancer types, and the proportion of cancers where the driver gene is amplified or mutated

Gene	Localization	Cancers	% Up reg or mutated	References
B4GALNT1	Transmembrane trans-Golgi	Sarcomas	11.37%	[214]
		Glioblastoma	9.80%	[214]
		Lung	5.12%	[214]
		Skin	2.25%	[214]
		Oral		[85]
FUT8	Transmembrane trans-Golgi	Hepatocellular Carcinoma	6%	[134]
		Lung	5.12%	[29, 101]
		Skin	3.83%	[1]
		Breast	0.83%	[80]
		Endometrial		[113]
GALNT7	Transmembrane trans-Golgi	Endometrial	6.62%	[108]
		Lung	4.31%	[108]
		Skin	4.28%	[108]
		Stomach	4.09%	[108]
		Prostate	1.42%	[108, 158]
GOLIM4	Transmembrane cis-Golgi	Lung	32–42%	[171]
		Esophagus	24%	[171]
		Head and neck	18%	[9, 171]
		Endometrial	3.78%	[176]
GOLPH3	Cytosolic face trans-Golgi	Lung	15.7–56%	[39, 67, 159, 196, 222]
		Ovarian	38%	[116, 159]
		Breast	16.5–32%	[39, 67, 159, 174, 217]
		Prostate	37%	[79, 159]
		Pancreatic	33%	[159, 220]
		Melanoma	32%	[159]
		Colorectal	24%	[159]
		Liver	29%	[78, 103, 159]
HRAS	Cytosolic face pan-Golgi	All RAS mutant cancers	7%	[142]
		Head and neck	3.3–9%	[89, 126, 142]
		Endocrine	2.2–7.0%	[142]
		Genito-urinary	0.2–7%	[142]
		Pheochromocytoma	7%	[165]
		Gynaecological	0.2–2.5%	[142]
		Skin	1.20%	[142]
		GI cancers	0–0.8%	[142]
		Haematological	0–0.8%	[142]
		Breast	0.60%	[142]

(continued)

Table 15.1 (continued)

Gene	Localization	Cancers	% Up reg or mutated	References
		Lung	0.2–0.6%	[142]
		CNS	0.1–0.5%	[142]
MMD	Transmembrane cis-Golgi	Ovarian		[138]
		Renal		[138]
		Lung		[169]
MYO18A	Cytosolic face trans-Golgi	Prostate	21%	[117]
		Breast	11%	[151]
PI4KB	Cytosolic face pan-Golgi	Breast	13–27%	[125, 170, 177]
		Lung	14%	[170]
		Ovarian	12%	[170]
PITPNC1	Cytosolic face trans-Golgi	Breast	46%	[66]
		Prostate	20%	[13]
		Melanoma	32%	[66]
		Colorectal	24%	[66, 173]
		Gastric		[172]
PRKD3	Cytosolic face trans-Golgi	Oral	42%	[32]
		Breast	40%	[21, 107]
		Prostate		[107]
ST6GAL1	Transmembrane trans-Golgi	Lung	30–40%	[47]
		Hepatocellular carcinoma	35%	[128]
		Ovarian	16–25%	[47, 204]
		Prostate		[157]
		Thyroid		[63]

See the text for more details

Cancer

GOLPH3 was the first example of a protein that plays a role in normal Golgi function to be identified as a cancer driver. A genome-wide analysis of human cancers found high frequency amplification of GOLPH3 in many solid tumors, including 56% of lung cancers, 38% of ovarian cancers, 32% of breast cancers, 33% of pancreatic cancers, 37% of prostate cancers, 32% of melanomas, and 24% of colon cancers [159]. The Cancer Genome Atlas (TCGA) reports rates of amplification of GOLPH3 pathway genes that are somewhat lower, although still significant, e.g., in 15.7% of non-small cell lung cancers and 16.5% of invasive breast carcinomas [39, 56, 67]. Hyperactivation of GOLPH3 is further seen due to increases in PI(4)P levels, protein overexpression, or activation by post-translational modification, beyond the rate of gene amplification [20, 52, 92]. GOLPH3 is a confirmed oncogene that cooperates with H-RAS-G12V or B-RAF-V600E to transform primary fibroblasts, lung cells, and melanocytes in culture and in xenograft mouse models [159]. Likewise, overexpression of GOLPH3 drives transformation in mammary and glioblastoma cell lines [150, 177, 217, 221, 224, 225]. Frequent overexpression of GOLPH3 and correlation of high expression with poor patient prognosis have been reported for non-small cell lung cancer [112, 162, 196, 222], breast cancer [217], prostate cancer [79, 99, 219], pancreatic ductal adenocarcinoma [220], hepatocellular carcinoma [78, 96], gastric cancer [77, 104, 136], renal cell carcinoma [211], glioblastoma [100, 224], esophageal squamous carcinoma [194], and ovarian carcinoma [115, 116]. In cell culture and mouse models of various cancers, interference with GOLPH3 abolishes cancer phenotypes [52, 66, 93, 159, 170, 210]. Furthermore, interference with GOLPH3 prevents tumor growth and metastasis in orthotopic xenograft and allogeneic lung cancer models [170]. Finally, heterozygous knockout of GOLPH3 prevents gastric cancer in an autochthonous genetically engineered mouse model (MUC6 homozygous knockout) [5].

Golgi

Proteomics of purified Golgi fractions discovered GOLPH3 (then referred to as GPP34 or GMx33) [12, 206]. Its localization to the Golgi was further validated by immunofluorescence [12, 206]. Subsequent studies found that GOLPH3 localizes to the Golgi via binding to PI(4)P [46]. Perturbation of GOLPH3 expression alters the morphology of the Golgi [46, 132, 209]. GOLPH3 has been found to be required for exit of vesicles from the Golgi for trafficking to the plasma membrane as part of the secretory pathway for a variety of cargoes [17, 46, 132, 145]. In addition, genetic interference with GOLPH3 has also been found to result in mislocalization of many Golgi glycosyltransferases [2, 137, 147, 149, 152, 155, 183, 203].

MYO18A

MYO18A (Myosin XVIIIA, also known as MYSPDZ, TIAF1, and SPR210) is an unconventional myosin that is widely expressed in all tissues. Uniquely, MYO18A contains a PDZ domain usually found in signaling proteins. It is also extensively phosphorylated [24].

Cancer

Sanchez-Garcia et al. developed a novel algorithm to mine genomic data from human breast cancers to identify cancer driver genes [151]. This unbiased analysis was validated by the identification of many previously known drivers of breast cancer (including ERBB2, CCND1, and MYC), and through demonstration that many of the newly identified genes were able to drive oncogenic transformation of cultured mammary epithelial cell lines. The screen identified MYO18A as a driver of breast cancer in humans and subsequent data has shown that overexpression of MYO18A is able to drive oncogenic features in cells in culture [25, 52, 117, 210].

Moreover, overexpression of MYO18A has been found to correlate with poor prognosis in acute myelogenous leukemia (AML) [3]. Fusions of MYO18A with other known cancer drivers have been observed, including MYO18A-FGFR1 resulting in the 8p11 myeloproliferative syndrome [190], MYO18A-PDGFRB resulting in eosinophilia-associated myeloproliferative neoplasms [160, 191], and a single case of MYO18A-MLL resulting in AML [184].

Golgi

Immunofluorescence found endogenous MYO18A to localize to the Golgi [46, 124]. Perturbation of MYO18A expression altered the morphology of the Golgi in cells in culture and in mice [46, 132, 145, 209, 210]. MYO18A has been found to be required for the exit of vesicles from the Golgi for trafficking to the plasma membrane as part of the secretory pathway for a variety of cargoes [17, 24–26, 46, 132, 145].

PI4KB

PI4KB (Phosphatidylinositol 4-kinase beta, also known as PI4KIIIbeta and PI4K92) is a Golgi-localized lipid kinase that phosphorylates phosphatidylinositol (PI) to generate PI(4)P [50, 59, 129, 205].

Cancer

Copy number analysis demonstrated frequent amplification of PI4KB, along with overexpression of the protein, in breast cancer [42, 125, 164, 199]. A previous study had found a requirement for PI4KB for growth and survival of MDA-MB-231 breast cancer cells in culture [35]. Furthermore, overexpression of PI4KB led to increased growth factor signaling in BT549 breast cancer cells [125].

A screen of genomic sequencing data from human lung, breast, and ovarian adenocarcinomas revealed frequent amplification of a region of chromosome 1q, containing PI4KB, with associated increased expression of the PI4KB transcript [170]. In both xenograft and syngeneic orthotopic mouse models of lung cancer, overexpression of PI4KB was found to synergize with KRAS-G12D to promote the development of lung cancers. Conversely, genetic interference of PI4KB expression in mice resulted in smaller tumors with decreased metastases. Pharmacologic inhibition of PI4KB function preferentially killed lung cancer cells as evidenced by increased apoptosis and decreased cell proliferation, colony formation, and cell migration and invasion [161, 170].

Golgi

PI4KB, the mammalian ortholog of yeast PIK1, is one of two distinct Golgi-resident PI 4-kinases that produce PI(4)P at the Golgi, the other being PI4K2A. FLAG-PI4KB expressed in COS cells was observed in the region of the Golgi [129]. Immunofluorescence of endogenous PI4KB in HeLa cells and NRK cells, and GFP-tagged PI4KB in MDCK cells revealed co-localization with Golgi markers [59, 202, 205]. PI(4)P is highly enriched in the cytosolic leaflet of the trans-Golgi and is required for normal Golgi morphology and function [8, 10, 46, 58, 106, 189, 198]. In model organisms, orthologs of PI4KB and PI4K2A both contribute to Golgi PI(4)P, with overlapping but distinct functions [22, 23, 168, 202]. Genetic interference with PI4KB in mammalian cells results in defects in Golgi to plasma membrane trafficking [16, 62, 170, 223].

PITPNC1

PITPNC1 (Phosphatidylinositol transfer protein cytoplasmic 1, also known as RDGBB and RDGB-beta) encodes a PI-transfer protein, a family of proteins that facilitate non-vesicular transfer of PI between membrane compartments [7, 38, 61, 181].

Cancer

Amplification of the PITPNC1 gene was observed in unbiased analysis of the genomes of 3131 cancers, including a large fraction of breast, melanoma, and colon cancers [14]. Overexpression of PITPNC1 was observed to correlate with metastatic progression in these cancers with increases in expression levels correlating with worse staging in cancer patient cohorts [66, 139]. Overexpression of PITPNC1 was sufficient to produce oncogenic transformation in cell culture and in murine xenograft models of breast cancer. It was also found to promote angiogenesis in metastatic breast cancer models through endothelial recruitment. Conversely, genetic interference with expression of PITPNC1 significantly impaired endothelial recruitment and metastasis in murine xenograft models of breast, colon, and melanoma cancers [66, 139]. Similarly, PITPNC1 overexpression was observed in lung and pancreatic ductal adenocarcinomas [51]. Genetic interference with expression of PITPNC1 impaired growth of cancer cell lines in culture and tumor growth in xenograft models of lung adenocarcinoma and pancreatic ductal adenocarcinoma.

Golgi

Overexpressed FLAG-PITPNC1 in HEK293 cells was observed diffusely throughout the cytoplasm [55]. However, later experiments in MDA-MB-231 breast cancer and PC12 rat pheochromocytoma cells observed GFP-tagged PITPNC1 localized throughout the cytoplasm with intense perinuclear staining that co-localized with the trans-Golgi marker p230 [66, 167]. Genetic interference with PITPNC1 resulted in compaction of the Golgi ribbon and reduced Golgi to plasma membrane trafficking as assessed by imaging a PI(4)P reporter or by measurement of secretion of MMP1, and several other proteins [66].

MMD

MMD (Monocyte to macrophage differentiation associated, also known as PAQR11 and MMA) is a 7-pass transmembrane protein (unrelated to 7-pass G protein coupled receptors) that physically interacts with ACSL4 and MBOAT7, enzymes involved in synthesis of PI [138].

Cancer

A screen of 185 specimens of surgically resected lung cancers found expression of five genes (MMD, DUSP6, STAT1, ERBB3, LCK) that significantly correlated with both patient relapse and overall survival. This result was validated in an independent cohort of 86 patients with non-small cell lung cancer [30]. Subsequent analysis of human lung adenocarcinoma cohorts in TCGA identified MMD expression to correlate with poor prognosis [169]. In cell culture assays, knockdown of MMD reduces lung adenocarcinoma cell invasion and migration. In syngeneic mouse models of lung cancer, genetic interference with MMD expression resulted in reduced metastasis [169].

Golgi

Immunofluorescence of both tagged and endogenous MMD showed localization to the Golgi [84, 169]. Furthermore, knockdown of MMD resulted in altered Golgi morphology and reduction in trafficking to the plasma membrane of a vesicular stomatitis virus G glycoprotein reporter [169].

GOLIM4

GOLIM4 (Golgi integral membrane protein 4, also known as GPP130, GIMPC, and GOLPH4) is a type II transmembrane protein [102, 130].

Cancer

A screen of TCGA cancer genomic data found GOLIM4 within a chromosome 3q amplicon in squamous cell carcinoma of the lung, esophagus, and head and neck. Overexpression of GOLIM4 transcript was observed to be increased in lung cancers and further increased in metastases [171]. Unbiased proteomic screening discovered elevated expression of GOLIM4 in endometrial cancers [176]. Genetic interference with GOLIM4 expression in lung cancer cell lines resulted in impaired tumor growth, metastatic activity, and longer survival duration in mice [171]. Conversely, overexpression of GOLIM4 synergized with KRAS-G12D and p53-R172H to drive lung adenocarcinoma in a syngeneic mouse model [171].

Golgi

Immunofluorescence microscopy revealed GOLIM4 co-localization with the Golgi marker Giantin in Caco-2, HeLa, Vero, and COS cells [102]. Further analysis by immunoelectron microscopy and cell fractionation revealed GOLIM4 associated with the cis-Golgi, but not the trans-Golgi [102]. Roles identified for GOLIM4 include tethering endosomes to the Golgi and modulating intracellular manganese homeostasis. For instance, genetic interference with GOLIM4 impaired endosome-to-Golgi retrieval of bypass pathway proteins GP73 and TGN46 [130]. Addition of exogenous manganese resulted in GOLIM4 aggregation and translocation away from the Golgi [127, 187]. Intriguingly, addition of manganese also attenuated proliferation and pro-metastatic processes of cancers harboring GOLIM4 amplification [171].

PRKD3

PRKD3 (Protein kinase D3, also known as PKD3 and PKC-NU) is a member of a family of serine/threonine protein kinases consisting of PRKD1, PRKD2, and PRKD3. They belong to the calcium/calmodulin-dependent kinase superfamily. Although there is literature about potential roles for all three in cancer, the data are most clear for PRKD3 [49, 64].

Cancer

Immunohistochemistry observed elevated levels of PRKD3 in prostate cancer specimens compared to normal prostate tissue [31]. In a panel of benign and malignant prostate cancer cell lines, PRKD3 expression was found to be upregulated in the more malignant lines such as DU145, PC3, and TRAMP-C2 cells. In aggressive prostate cancer cell lines, knockdown of PRKD3 significantly reduced cell proliferation, migration, and invasive behavior in cell culture [31, 226]. Furthermore, in the less malignant prostate cancer cell line LNCaP, overexpression of PRKD3 resulted in enhanced proliferation in culture [31].

Increased expression of PRKD3 was also observed in patient biopsies of triple negative breast cancer compared to normal tissue. Furthermore, cancer transcriptomics databases revealed that triple negative breast cancer biopsies showed significantly increased expression of PRKD3 compared to triple positive breast cancer samples [21]. This study found that PRKD3 is overexpressed in breast cancers of the estrogen receptor-negative luminal A, luminal B, and triple negative subtypes [21]. Genetic interference with PRKD3 expression impaired oncogenic phenotypes (cell proliferation, migration, and invasion) in breast cancer cell lines in culture and in xenograft mouse models of breast cancer [21]. Likewise, a small molecule inhibitor of PRKD protein kinase activity, CRT0066101, impaired tumor growth, invasion, and metastasis in a xenograft model of breast cancer [21]. The same small molecule inhibitor was also found to impair tumor growth in an orthotopic xenograft model of pancreatic ductal adenocarcinoma as well as cell culture and xenograft models of bladder cancer [70, 97].

Golgi

PRKD3 appears to have a localization that is cell type specific or perhaps context specific as it has been observed in multiple subcellular locations. Overexpression of tagged-PRKD3 in MDCK, HeLa, HEK293, and HCC 1806 cells showed diffuse cytoplasmic expression, some nuclear staining, and a stronger perinuclear signal [111, 143, 213]. Overexpression of a kinase-inactive PRKD3 mutant produced a very strong signal localized to the Golgi, potentially indicating a default localization when not active [213]. However, overexpression of tagged PRKD3 as well as observation of endogenous PRKD3 expression in PC3, PANC-1, MDCK, and COS-7 cells showed diffuse cytoplasmic and robust nuclear staining [31, 146] suggesting a differential localization based on cell type or experimental approach.

FUT8

FUT8 (Fucosyltransferase 8, also known as CDGF) is the only enzyme that catalyzes alpha-1,6-fucosylation in mammals, transferring fucose from GDP-fucose to the innermost N-acetyl glucosamine (GlcNAc) of an N-linked glycan [197].

Cancer

FUT8 expression is increased in several malignant cancers, including hepatocellular carcinoma [218], epithelial ovarian cancer [113], triple-negative breast cancer [80], melanoma [1], and non-small cell lung cancers [29]. Expression correlates with tumor metastasis, disease recurrence, and poor survival. Silencing of FUT8 suppressed the invasive ability of highly metastatic breast cancer cells [182] and its pharmacological inhibition by small molecule inhibitors prolonged the survival of mice with colorectal cancer pulmonary metastases [195]. A xenograft model of melanoma metastasis revealed a significant reduction of lung tumor burden upon FUT8 silencing. Likewise, silencing of FUT8 reduced the growth of established melanoma metastatic lesions in liver, kidney, and brain [1]. FUT8 also plays a role in the immune evasion mechanism of triple-negative breast cancer cells through aberrant N-glycosylation of B7 homolog 3 protein (B7-H3), an immune checkpoint protein involved in cell migration, proliferation, invasion, and angiogenesis [80].

Golgi

FUT8 is a type II membrane protein comprised of an N-terminal cytoplasmic region, a transmembrane region, a stem region, a coiled-coil domain, a catalytic domain, and an SH3 domain [178]. FUT8 co localized with GM130, demonstrating its residence in the Golgi [180]. The stem region, coiled-coil, SH3, and catalytic domains of FUT8 localized in the Golgi lumen [178]. Mutations in the stem region compromised the residence of FUT8 in the Golgi implicating the involvement of the domain in Golgi localization [179]. Like many other Golgi resident glycosyltransferases, FUT8 is also secreted, the mechanism and significance of which is yet unknown [88, 178].

ST6GAL1

ST6GAL1 (ST6 beta-galactoside alpha-2,6 sialyltransferase 1, also known as SIAT1) is a type II transmembrane protein that catalyzes the addition of alpha-2,6-linked sialic acids to the terminal galactose of N-glycosylated proteins. Both ST6GAL1 and ST6GAL2 catalyze the transfer of sialic acid residues in alpha-2,6 linkages to N-glycosylated proteins. However, while ST6GAL1 has a ubiquitous expression, ST6GAL2 displays a limited expression in the brain [95].

Cancer

Elevated surface sialylation, a consequence of increased expression of Golgi sialyltransferases like ST6GAL1, is found in pancreatic [15], cervical [192], breast [65], and prostate cancer [157]. ST6GAL1 overexpression was found to correlate with worse prognosis in ovarian cancer but was independent of prognosis in colon cancer [47, 188, 204]. Overexpression of ST6GAL1 in pancreatic cancer cell lines increased oncogenic behavior in culture and in xenograft mouse models, while knockdown of ST6GAL1 reduced tumor cell growth [156]. The expression of ST6GAL1 is upregulated in cancer by gene amplification, transcriptional activation, altered miRNA regulation, and post-translational modifications [43, 53, 193]. The human ST6GAL1 gene is present in the 3q26 amplicon which is frequently amplified in epithelial cancers, although that amplicon harbors other well-known cancer drivers such as PIK3CA, SOX2, ECT2, and PRKC1 [47, 53]. Interestingly, the 3Q26 amplicon also includes another Golgi cancer driver described in this review, GOLIM4.

Golgi

ST6GAL1 has been reported to be present in the Golgi as well as in exosomes [65]. In pancreatic ductal carcinoma, ST6GAL1 co-localized with the Golgi marker GM130 [15]. In ductal-like malignant pancreatic cancer, ST6GAL1 immunofluorescence was observed to be perinuclear, consistent with Golgi localization [47]. Chen et al., reported two ST6GAL1 isoforms, ST_{tyr} and ST_{cys}, the latter being prone to oligomerization and retention at the Golgi [27, 28].

B4GALNT1

B4GALNT1 (beta-1,4-N-acetyl-galactosaminyltransferase 1, also known as a GM2/GD2 synthase, and GALNACT) is a type II transmembrane protein that functions in the synthesis of GM2 and GD2 gangliosides by catalyzing the transfer of N-acetyl galactosamine to GM3 and GD3 by a beta-1,4 linkage [76, 212, 215].

Cancer

Overexpression of B4GALNT1 is reported in malignant melanoma and neuroblastoma [148], gastrointestinal cancers [216], oral squamous cell carcinoma [85], lung adenocarcinoma [83] and hepatocellular carcinoma [105]. In hepatocellular carcinoma, elevated B4GALNT1 expression correlated with reduced survival [175]. In hepatocellular carcinoma cell lines, overexpression of B4GALNT1 enhanced, while knockdown suppressed, oncogenic phenotypes in culture, and in a xenograft mouse model [175]. B4GALNT1 was also implicated in cancer stem cell propagation in lung adenocarcinoma [45]. In melanoma and neuroblastoma, overexpression of B4GALNT1 (with consequently increased levels of GM2 and GD2 gangliosides) promoted angiogenesis and anchorage-independent growth [215]. An anti-GD2 monoclonal antibody, dinutuximab, has been approved by the FDA for the treatment of high-risk neuroblastoma in pediatric patients [131], highlighting B4GALNT1 as a potential drug target.

Golgi

Subcellular fractionation of rat liver found B4GALNT1 in Golgi-enriched membranes [140]. Subsequent immunocytochemistry in CHO cells observed the co-localization of MYC-tagged B4GALNT1 with M6PR, a trans-Golgi marker [57]. Shishido et al., reported two isoforms of the enzyme, M1 and M2, with different subcellular localization. M1 had a retrograde transfer signal motif at the cytoplasmic tail and localized to the endoplasmic reticulum. The M2 isoform was seen predominantly localized to the Golgi. FLAG-tagged M2 co-localized with the Golgi marker GM130 [163]. The M1/M2 heteromer displayed sustained retention at the Golgi, the significance of which is incompletely understood. Nonetheless, the co-existence of several B4GALNT1 isoforms in different compartments of the secretory pathway has been conjectured to maintain the overall enzyme stability and glycosphingolipid homeostasis [163].

GALNT7

GALNT7 (N-acetylgalactosaminyltransferase 7, also known as GalNAcT7) exclusively glycosylates proteins that already possess N-acetyl galactosamine (GalNAc) moieties added by other GALNT-isozymes [44]. GALNT7 catalyzes the transfer of GalNAc onto the recipient glycoproteins. GALNT7 has been found to significantly affect the height of the glycocalyx at the cell surface [123].

Cancer

The TCGA database revealed overexpression of GALNT7 in 33 different types of cancers. Significantly elevated expression of the enzyme is reported in lung adenocarcinoma, breast cancer, esophageal cancer, stomach cancer, renal papillary carcinoma, prostate adenocarcinoma, endometrial carcinoma, lung squamous cell carcinoma, thyroid cancer, and cholangiocarcinoma [108]. However, GALNT7 expression was diminished in several other cancers, including colorectal and renal cell carcinoma [108]. Free GALNT7 was detectable in the blood and urine of patients with prostate cancer [158]. In cell culture models of prostate cancer, knockdown of GALNT7 inhibited cell proliferation and colony formation in cell culture. Knockdown of GALNT7 also suppressed the growth of CWR22RV1 prostate cancer cells in xenografts in mice. Conversely, overexpression of GALNT7 enhanced the growth of PC3 prostate cancer cells in xenografts in mice [158].

Golgi

GALNT7 glycosylates secretory proteins at the Golgi enroute to the cell surface. A notable protein that undergoes O-glycosylation by GALNT7 is Tn antigen [158] which is associated with tumorigenic processes and metastasis in several cancers [86]. GFP-tagged GALNT7, when expressed in HFF-1 fibroblasts, co-localized with the Golgi marker Giantin [11]. Similarly, Welch et al., also reported the co-localization of endogenous GALNT7 with GM130 in wild type U2OS cells [203]. A chemical biology approach revealed GALNT7-mediated glycosylation of several proteins at the cell surface and also along the secretory pathway, leading to the upregulation of Tn antigen in prostate cancer cells [158]. However, the precise mechanism by which GALNT7 upregulates Tn antigen is presently unknown.

HRAS

HRAS (Harvey rat sarcoma viral oncogene homolog, also known as C-HA-RAS) is a RAS-family small GTPase [121].

Cancer

HRAS was one of the earliest identified oncogenes, discovered based on the recognition that it had been captured, in activated form, by an oncogenic virus [71, 90, 201]. The literature on the role of HRAS in cancer is vast, and we cannot do justice to it within the confines of this review. Instead, we note that activating mutations in HRAS occur in a wide range of cancers. Compared to its KRAS and NRAS counterparts, HRAS is mutated at a relatively lower frequency and is mutated primarily in squamous epithelial carcinomas in various tissues including skin, lung, and head and neck cancers [41, 94, 98, 144]. Oncogenic mutations in HRAS include G12V, G12D, and Q61L. All impair the GTPase activity of HRAS, maintaining the GTP-locked, constitutively signaling conformation of the protein.

Apart from these well-known missense mutations, HRAS expression is remarkably up-regulated in gastric cancers, with elevated levels correlating with poor prognosis and enhanced aggressiveness [208]. Overexpression of wild-type protein was found to drive oncogenic properties in cell culture models, in chick chorioallantoic membrane assays, and in mouse models of gastric cancer. Down-regulation or silencing of HRAS notably reversed oncogenic characteristics [208].

Golgi

A significant portion of HRAS is found localized to the Golgi [34]. Through immunofluorescence and immunoelectron microscopy imaging, tagged and endogenous HRAS were found to localize to the Golgi in MDCK and COS-1 cells [33, 34, 114]. Both wild type and palmitoylated HRAS were present throughout the Golgi stacks [114]. We note that the canonical functions of RAS family proteins involves signaling downstream of receptor tyrosine kinases, including coupling signaling to the MEK-ERK and PI3K-AKT pathways, which are thought to occur at the plasma membrane [40, 90, 120, 141]. Nevertheless, several research teams have shown that Golgi-localized RAS activates signaling. For example, RAS reporters detected activity at the Golgi [19, 33, 75, 110]. RAS tethered to the Golgi was capable of driving signaling and oncogenic transformation [6, 74, 75].

Complexes, Pathways, and Potential Mechanisms

MMD-PITPNC1-PI4KB-GOLPH3-MYO18A

GOLPH3 is an effector of PI(4)P (Fig. 15.1) and an oncogenic driver that works with MYO18A to influence secretory trafficking and Golgi structure [46, 118, 132, 145, 166, 209, 210]. It drives cancer by activating growth factor signaling [147, 159], promoting survival following DNA damage [52], and driving cell migration [210]. MMD interacts with the lipid biosynthetic enzymes ACSL4 and MBOAT7. Genetic interference with MMD results in impaired synthesis of PI [138]. PI is transferred to Golgi membranes by PI transfer proteins, including PITPNC1 [36, 37, 66, 186]. At the Golgi, the PI-4-kinase, PI4KB, phosphorylates PI to produce PI(4)P [50, 59, 129, 205]. GOLPH3 binds tightly and specifically to PI(4)P, which recruits GOLPH3 to the Golgi [46, 132]. In addition, GOLPH3 binds directly to MYO18A [46, 132, 145, 166] functioning together in Golgi secretory trafficking [17, 46, 132, 145], promoting survival following DNA damage [52], and driving cell migration [177, 210]. The evidence suggests that this pathway drives oncogenic phenotypes at least in part through enhanced secretion of growth factors that promote cell proliferation and matrix metalloproteases that promote invasion into normal tissue [66, 169–171].

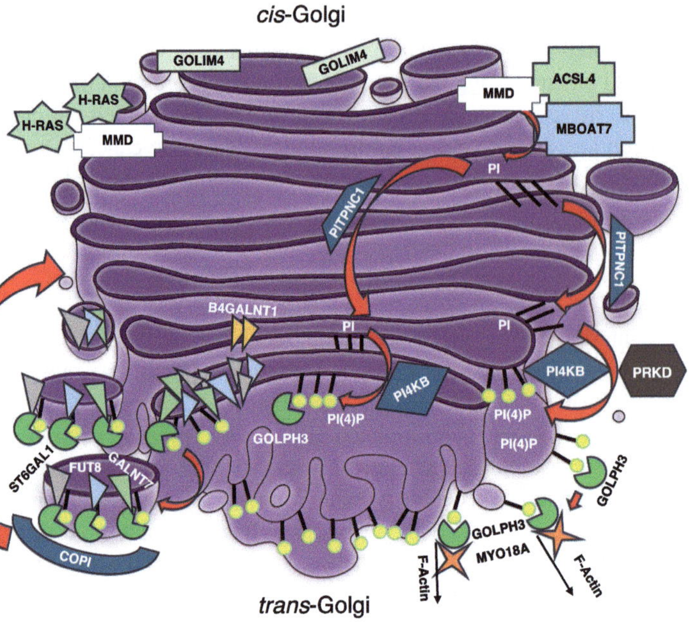

Fig. 15.1 Diagram of the Golgi depicting Golgi cancer drivers discussed in this review and the putative pathways they influence. PI(4)P appears to be a significant hub for Golgi oncogenic signaling. PI(4)P is downstream of MMD, PITPNC1, and PI4KB, and upstream of GOLPH3/MYO18A and GOLPH3/glycosyltransferases

Genetic data strongly support an oncogenic pathway involving PITPNC1-PI4KB-GOLPH3-MYO18A. Genetic interference with PITPNC1 or PI4KB results in a decrease in Golgi PI(4)P levels, and dissociation of GOLPH3 from the Golgi [66, 170]. Interference with PITPNC1 or PI4KB both result in compaction of the Golgi and a decrease in secretion, similar to the effect of genetic interference with GOLPH3 or MYO18A [46, 66, 132, 170]. Furthermore, the ability of PITPNC1 or PI4KB to drive oncogenic phenotypes in cell culture and in xenograft mouse models is abolished upon genetic interference with GOLPH3 [66, 170]. Finally, the ability of GOLPH3 to drive oncogenic phenotypes is dependent on MYO18A [25, 26, 52, 92, 119, 210]. Although it has not been tested directly, given MMD's role in PI synthesis, an attractive hypothesis is that it functions upstream of PITPNC1, PI4KB, GOLPH3, and MYO18A. Intriguingly, PI4KB has been identified as a substrate of PRKD [72]. Phosphorylation of PI4KB by PRKD may potentiate its PI-kinase activity, resulting in a rise in Golgi PI(4)P levels [72]. Furthermore, PRKD also phosphorylates PI(4)P effectors, specifically the non-vesicular lipid transport proteins OSBP and CERT [4, 54, 133, 135]. Therefore, PRKD may drive the activity of the PI4KB-GOLPH3-MYO18A oncogenic pathway.

GOLPH3-Glycosyltransferases

Knockout or knockdown of GOLPH3 (and/or its paralog GOLPH3L) results in mislocalization of many glycosyltransferases from the Golgi to the plasma membrane [82, 147, 152, 203]. Biochemical evidence supports the possibility of a physical interaction between GOLPH3 and the cytosolic tails of these glycosyltransferases, bridging to COPI vesicles, and serving to recycle these glycosyltransferases back to the Golgi (Fig. 15.1). Among the glycosyltransferases that are found to interact with, and depend on, GOLPH3 for their localization to the Golgi are the cancer drivers ST6GAL1, GALNT7, and FUT8. B4GALNT1, however, was not found associated with GOLPH3, nor dependent upon it for Golgi localization [203]. Glycosylation can regulate the function and stability of proteins [48, 81, 91, 207. Presumably, there may be targets of these glycosyltransferases that, upon increased glycosylation, are capable of enhancing oncogenic phenotypes.

One possible class of targets of glycosylation with relevance to cancer are the receptor tyrosine kinases, which undergo extensive glycosylation. Increased alpha-2,3 sialylation of the MET and RON receptor tyrosine kinases has been found to confer more invasive phenotypes in gastric carcinoma [60, 122]. The relationship between glycosylation and activity of receptor tyrosine kinases is likely complex, with examples of both increased and decreased signaling in response to increased glycosylation [73, 109]. Gangliosides also undergo glycosylation at the Golgi prior to trafficking to the plasma membrane. These have been reported to have a variety of effects on cells that may modulate cancer progression, including interaction with receptor tyrosine kinases [87, 153, 154]. Consistent with a role downstream of GOLPH3 in oncogenic transformation, knockdown of the ganglioside glycosylation enzyme B4GALT5 impaired the ability of GOLPH3 overexpression to drive oncogenic phenotypes in HeLa cells [147].

MMD-HRAS

There is some evidence that MMD plays a role in HRAS signaling through the retention of HRAS at the Golgi (Fig. 15.1). In HeLa cells, overexpression of MMD resulted in enhanced localization of HRAS to the Golgi and enhanced growth factor-induced phosphorylation of ERK. Conversely, depletion of MMD resulted in a reduction of HRAS at the Golgi and impaired ERK phosphorylation [84].

Conclusion

Our current picture of the role of the Golgi in driving cancer (Fig. 15.1) suggests that PI(4)P serves as an important signaling hub. Future efforts to further interrogate the PI(4)P signaling pathway and its role in cancer seem likely to help illuminate the role of the Golgi in cancer. More generally, as our understanding of cancer mechanisms expands, additional features of the Golgi are likely to be revealed to be important. This is an exciting area that may offer new targets for cancer therapeutics that are orthogonal to current treatment approaches.

Acknowledgements We thank the many labs that have contributed to our understanding of Golgi signaling pathways that contribute to oncogenesis and apologize for any unintentional omissions. We thank Caroline Jiang for her creative insights and artistic contributions for the Figure. We thank Natalia Aladyshkina and Anqi (Angela) Qiu for their comments on the manuscript.

Author Contribution K.A.S. and S.J.F. wrote the majority of the paper, with additional writing by J.R.B. and P.M. K.A.S. and S.J.F. produced the Figure.

Conflict of Interest The authors declare no conflicts of interest.

Funding We acknowledge support for K.A.S. in part by the National Institutes of Health under awards T32 GM007250, TL1 TR002549, and T32 GM152319 and support to S.J.F. by the National Institutes of Health National Cancer Institute (R01 CA254397), the Burroughs Wellcome Fund, the V Foundation for Cancer Research (V2006-003B), and the Harrington Discovery Institute of University Hospitals.

References

1. Agrawal P, Fontanals-Cirera B, Sokolova E et al (2017) A systems biology approach identifies FUT8 as a driver of melanoma metastasis. Cancer Cell 31:804–819.e7. https://doi.org/10.1016/j.ccell.2017.05.007
2. Ali MF, Chachadi VB, Petrosyan A, Cheng P-W (2012) Golgi phosphoprotein 3 determines cell binding properties under dynamic flow by controlling Golgi localization of core 2 N-acetylglucosaminyltransferase 1. J Biol Chem 287:39564–39577. https://doi.org/10.1074/jbc.M112.346528
3. Ali YO, Radwan SM, Saeed AM, El-Mesallamy HO (2023) Golgi signalling proteins GOLPH3, MYO18A, PITPNC1 and RAB1B: implications in prognosis and survival outcomes of AML patients. Biomarkers 28:387–395. https://doi.org/10.1080/1354750X.2023.2191166
4. Amako Y, Syed GH, Siddiqui A (2011) Protein kinase D negatively regulates hepatitis C virus secretion through phosphorylation of oxysterol-binding protein and ceramide transfer protein. J Biol Chem 286:11265–11274. https://doi.org/10.1074/jbc.M110.182097
5. Arai J, Hayakawa Y, Tateno H et al (2024) Impaired glycosylation of gastric mucins drives gastric tumorigenesis and serves as a novel therapeutic target. Gastroenterology 167:505–521.e19. https://doi.org/10.1053/j.gastro.2024.03.037
6. Aran V, Prior IA (2013) Compartmentalized Ras signaling differentially contributes to phenotypic outputs. Cell Signal 25:1748–1753. https://doi.org/10.1016/j.cellsig.2013.05.004

7. Ashlin TG, Blunsom NJ, Cockcroft S (2021) Courier service for phosphatidylinositol: PITPs deliver on demand. Biochim Biophys Acta BBA – Mol Cell Biol Lipids 1866:158985. https://doi.org/10.1016/j.bbalip.2021.158985
8. Audhya A, Foti M, Emr SD (2000) Distinct roles for the yeast phosphatidylinositol 4-kinases, Stt4p and Pik1p, in secretion, cell growth, and organelle membrane dynamics. Mol Biol Cell 11:2673–2689. https://doi.org/10.1091/mbc.11.8.2673
9. Bai Y, Cui X, Gao D et al (2018) Golgi integral membrane protein 4 manipulates cellular proliferation, apoptosis, and cell cycle in human head and neck cancer. Biosci Rep 38: BSR20180454. https://doi.org/10.1042/BSR20180454
10. Balla A, Tuymetova G, Tsiomenko A et al (2005) A plasma membrane pool of phosphatidylinositol 4-phosphate is generated by phosphatidylinositol 4-kinase type-III alpha: studies with the PH domains of the oxysterol binding protein and FAPP1. Mol Biol Cell 16:1282–1295. https://doi.org/10.1091/mbc.e04-07-0578
11. Becker JL, Tran DT, Tabak LA (2018) Members of the GalNAc-T family of enzymes utilize distinct Golgi localization mechanisms. Glycobiology 28:841–848. https://doi.org/10.1093/glycob/cwy071
12. Bell AW, Ward MA, Blackstock WP et al (2001) Proteomics characterization of abundant Golgi membrane proteins. J Biol Chem 276:5152–5165. https://doi.org/10.1074/jbc.M006143200
13. Beltran H, Prandi D, Mosquera JM et al (2016) Divergent clonal evolution of castration-resistant neuroendocrine prostate cancer. Nat Med 22:298–305. https://doi.org/10.1038/nm.4045
14. Beroukhim R, Mermel CH, Porter D et al (2010) The landscape of somatic copy-number alteration across human cancers. Nature 463:899–905. https://doi.org/10.1038/nature08822
15. Bhalerao N, Chakraborty A, Marciel MP et al (2023) ST6GAL1 sialyltransferase promotes acinar to ductal metaplasia and pancreatic cancer progression. JCI Insight 8. https://doi.org/10.1172/jci.insight.161563
16. Bilodeau P, Jacobsen D, Law-Vinh D, Lee JM (2020) Phosphatidylinositol 4-kinase III beta regulates cell shape, migration, and focal adhesion number. Mol Biol Cell 31:1904–1916. https://doi.org/10.1091/mbc.E19-11-0600
17. Bishé B, Syed GH, Field SJ, Siddiqui A (2012) Role of phosphatidylinositol 4-phosphate (PI4P) and its binding protein GOLPH3 in hepatitis C virus secretion. J Biol Chem 287: 27637–27647. https://doi.org/10.1074/jbc.M112.346569
18. Bishop JM (1991) Molecular themes in oncogenesis. Cell 64:235–248. https://doi.org/10.1016/0092-8674(91)90636-D
19. Bivona TG, Pérez de Castro I, Ahearn IM et al (2003) Phospholipase Cγ activates Ras on the Golgi apparatus by means of RasGRP1. Nature 424:694–698. https://doi.org/10.1038/nature01806
20. Blagoveshchenskaya A, Cheong FY, Rohde HM et al (2008) Integration of Golgi trafficking and growth factor signaling by the lipid phosphatase SAC1. J Cell Biol 180:803–812. https://doi.org/10.1083/jcb.200708109
21. Borges S, Perez EA, Thompson EA et al (2015) Effective targeting of estrogen receptor–negative breast cancers with the protein kinase D inhibitor CRT0066101. Mol Cancer Ther 14: 1306–1316. https://doi.org/10.1158/1535-7163.MCT-14-0945
22. Brill JA, Hime GR, Scharer-Schuksz M, Fuller MT (2000) A phospholipid kinase regulates actin organization and intercellular bridge formation during germline cytokinesis. Development 127:3855–3864. https://doi.org/10.1242/dev.127.17.3855
23. Burgess J, Del Bel LM, Ma C-IJ et al (2012) Type II phosphatidylinositol 4-kinase regulates trafficking of secretory granule proteins in Drosophila. Development 139:3040–3050. https://doi.org/10.1242/dev.077644
24. Buschman MD, Field SJ (2018) MYO18A: an unusual myosin. Adv Biol Regul 67:84–92. https://doi.org/10.1016/j.jbior.2017.09.005

25. Buschman MD, Rahajeng J, Field SJ (2015) GOLPH3 links the Golgi, DNA damage, and cancer. Cancer Res 75:624–627. https://doi.org/10.1158/0008-5472.CAN-14-3081
26. Buschman MD, Xing M, Field SJ (2015) The GOLPH3 pathway regulates Golgi shape and function and is activated by DNA damage. Front Neurosci 9:362. https://doi.org/10.3389/fnins.2015.00362
27. Chen C, Colley KJ (2000) Minimal structural and glycosylation requirements for ST6Gal I activity and trafficking. Glycobiology 10:531–583. https://doi.org/10.1093/glycob/10.5.531
28. Chen C, Ma J, Lazic A et al (2000) Formation of insoluble oligomers correlates with ST6Gal I stable localization in the Golgi. J Biol Chem 275:13819–13826. https://doi.org/10.1074/jbc.275.18.13819
29. Chen C-Y, Jan Y-H, Juan Y-H et al (2013) Fucosyltransferase 8 as a functional regulator of nonsmall cell lung cancer. Proc Natl Acad Sci 110:630–635. https://doi.org/10.1073/pnas.1220425110
30. Chen H-Y, Yu S-L, Chen C-H et al (2007) A five-gene signature and clinical outcome in non–small-cell lung cancer. N Engl J Med 356:11–20. https://doi.org/10.1056/NEJMoa060096
31. Chen J, Deng F, Singh SV, Wang QJ (2008) Protein kinase D3 (PKD3) contributes to prostate cancer cell growth and survival through a PKCε/PKD3 pathway downstream of Akt and ERK 1/2. Cancer Res 68:3844–3853. https://doi.org/10.1158/0008-5472.CAN-07-5156
32. Chen Z, Huang Q, Xu W et al (2020) PRKD3 promotes malignant progression of OSCC by downregulating KLF16 expression. Eur Rev Med Pharmacol Sci 24:12709–12716. https://doi.org/10.26355/eurrev_202012_24169
33. Chiu VK, Bivona T, Hach A et al (2002) Ras signalling on the endoplasmic reticulum and the Golgi. Nat Cell Biol 4:343–350. https://doi.org/10.1038/ncb783
34. Choy E, Chiu VK, Silletti J et al (1999) Endomembrane trafficking of Ras: the CAAX motif targets proteins to the ER and Golgi. Cell 98:69–80. https://doi.org/10.1016/S0092-8674(00)80607-8
35. Chu KME, Minogue S, Hsuan JJ, Waugh MG (2010) Differential effects of the phosphatidylinositol 4-kinases, PI4KIIα and PI4KIIIβ, on Akt activation and apoptosis. Cell Death Dis 1:e106. https://doi.org/10.1038/cddis.2010.84
36. Cockcroft S (2012) The diverse functions of phosphatidylinositol transfer proteins. In: Falasca M (ed) Phosphoinositides and disease. Springer Netherlands, Dordrecht, pp 185–208
37. Cockcroft S (2001) Phosphatidylinositol transfer proteins couple lipid transport to phosphoinositide synthesis. Semin Cell Dev Biol 12:183–191. https://doi.org/10.1006/scdb.2000.0235
38. Cockcroft S, Carvou N (2007) Biochemical and biological functions of class I phosphatidylinositol transfer proteins. Biochim Biophys Acta BBA – Mol Cell Biol Lipids 1771:677–691. https://doi.org/10.1016/j.bbalip.2007.03.009
39. Collisson EA, Campbell JD, Brooks AN et al (2014) Comprehensive molecular profiling of lung adenocarcinoma. Nature 511:543–550. https://doi.org/10.1038/nature13385
40. Cox AD, Der CJ (2010) Ras history: the saga continues. Small GTPases 1:2–27. https://doi.org/10.4161/sgtp.1.1.12178
41. Cox AD, Fesik SW, Kimmelman AC et al (2014) Drugging the undruggable RAS: mission possible? Nat Rev Drug Discov 13:828–851. https://doi.org/10.1038/nrd4389
42. Curtis C, Shah SP, Chin S-F et al (2012) The genomic and transcriptomic architecture of 2,000 breast tumours reveals novel subgroups. Nature 486:346–352. https://doi.org/10.1038/nature10983
43. Davidson MA, Shanks EJ (2017) 3q26-29 Amplification in head and neck squamous cell carcinoma: a review of established and prospective oncogenes. FEBS J 284:2705–2731. https://doi.org/10.1111/febs.14061
44. de las Rivas M, Lira-Navarrete E, Gerken TA, Hurtado-Guerrero R (2019) Polypeptide GalNAc-Ts: from redundancy to specificity. Curr Opin Struct Biol 56:87–96. https://doi.org/10.1016/j.sbi.2018.12.007

45. De Vitis C, Corleone G, Salvati V et al (2019) B4GALT1 is a new candidate to maintain the stemness of lung cancer stem cells. J Clin Med 8:1928. https://doi.org/10.3390/jcm8111928
46. Dippold HC, Ng MM, Farber-Katz SE et al (2009) GOLPH3 bridges phosphatidylinositol-4-phosphate and actomyosin to stretch and shape the Golgi to promote budding. Cell 139:337–351. https://doi.org/10.1016/j.cell.2009.07.052
47. Dorsett KA, Marciel MP, Hwang J et al (2021) Regulation of ST6GAL1 sialyltransferase expression in cancer cells. Glycobiology 31:530–539. https://doi.org/10.1093/glycob/cwaa110
48. Duchesne L, Tissot B, Rudd TR et al (2006) N-glycosylation of fibroblast growth factor receptor 1 regulates ligand and heparan sulfate co-receptor binding. J Biol Chem 281:27178–27189. https://doi.org/10.1074/jbc.M601248200
49. Durand N, Borges S, Storz P (2015) Functional and therapeutic significance of protein kinase D enzymes in invasive breast cancer. Cell Mol Life Sci 72:4369–4382. https://doi.org/10.1007/s00018-015-2011-2
50. Endemann G, Dunn SN, Cantley LC (1987) Bovine brain contains two types of phosphatidylinositol kinase. Biochemistry 26:6845–6852. https://doi.org/10.1021/bi00395a039
51. Entrialgo Cadierno R, Cueto Ureña C, Welch C et al (2023) The phospholipid transporter PITPNC1 links KRAS to MYC to prevent autophagy in lung and pancreatic cancer. Mol Cancer 22:86. https://doi.org/10.1186/s12943-023-01788-w
52. Farber-Katz SE, Dippold HC, Buschman MD et al (2014) DNA damage triggers Golgi dispersal via DNA-PK and GOLPH3. Cell 156:413–427. https://doi.org/10.1016/j.cell.2013.12.023
53. Fields AP, Justilien V, Murray NR (2016) The chromosome 3q26 OncCassette: a multigenic driver of human cancer. Adv Biol Regul 60:47–63. https://doi.org/10.1016/j.jbior.2015.10.009
54. Fugmann T, Hausser A, Schöffler P et al (2007) Regulation of secretory transport by protein kinase D-mediated phosphorylation of the ceramide transfer protein. J Cell Biol 178:15–22. https://doi.org/10.1083/jcb.200612017
55. Fullwood Y, dos Santos M, Hsuan JJ (1999) Cloning and characterization of a novel human phosphatidylinositol transfer protein, rdgBβ. J Biol Chem 274:31553–31558. https://doi.org/10.1074/jbc.274.44.31553
56. Gao J, Aksoy BA, Dogrusoz U et al (2013) Integrative analysis of complex cancer genomics and clinical profiles using the cBioPortal. Sci Signal 6:pl1–pl1. https://doi.org/10.1126/scisignal.2004088
57. Giraudo CG, Rosales Fritz VM, Maccioni HJ (1999) GA2/GM2/GD2 synthase localizes to the trans-golgi network of CHO-K1 cells. Biochem J 342:633–640
58. Godi A, Campli AD, Konstantakopoulos A et al (2004) FAPPs control Golgi-to-cell-surface membrane traffic by binding to ARF and PtdIns(4)P. Nat Cell Biol 6:393–404. https://doi.org/10.1038/ncb1119
59. Godi A, Pertile P, Meyers R et al (1999) ARF mediates recruitment of PtdIns-4-OH kinase-β and stimulates synthesis of PtdIns(4,5)P2 on the Golgi complex. Nat Cell Biol 1:280–287. https://doi.org/10.1038/12993
60. Gomes C, Osório H, Pinto MT et al (2013) Expression of ST3GAL4 leads to SLex expression and induces c-met activation and an invasive phenotype in gastric carcinoma cells. PLoS One 8:e66737. https://doi.org/10.1371/journal.pone.0066737
61. Grabon A, Bankaitis VA, McDermott MI (2019) The interface between phosphatidylinositol transfer protein function and phosphoinositide signaling in higher eukaryotes. J Lipid Res 60:242–268. https://doi.org/10.1194/jlr.R089730
62. Gromada J, Bark C, Smidt K et al (2005) Neuronal calcium sensor-1 potentiates glucose-dependent exocytosis in pancreatic β cells through activation of phosphatidylinositol 4-kinase β. Proc Natl Acad Sci 102:10303–10308. https://doi.org/10.1073/pnas.0504487102

63. Gunjača I, Benzon B, Pleić N et al (2023) Role of ST6GAL1 in thyroid cancers: insights from tissue analysis and genomic datasets. Int J Mol Sci 24:16334. https://doi.org/10.3390/ijms242216334
64. Gutiérrez-Galindo E, Yilmaz ZH, Hausser A (2023) Membrane trafficking in breast cancer progression: protein kinase D comes into play. Front Cell Dev Biol 11. https://doi.org/10.3389/fcell.2023.1173387
65. Hait NC, Maiti A, Wu R et al (2022) Extracellular sialyltransferase st6gal1 in breast tumor cell growth and invasiveness. Cancer Gene Ther 29:1662–1675. https://doi.org/10.1038/s41417-022-00485-y
66. Halberg N, Sengelaub CA, Navrazhina K et al (2016) PITPNC1 recruits RAB1B to the Golgi network to drive malignant secretion. Cancer Cell 29:339–353. https://doi.org/10.1016/j.ccell.2016.02.013
67. Hammerman PS, Lawrence MS, Voet D et al (2012) Comprehensive genomic characterization of squamous cell lung cancers. Nature 489:519–525. https://doi.org/10.1038/nature11404
68. Hanahan D (2022) Hallmarks of cancer: new dimensions. Cancer Discov 12:31–46. https://doi.org/10.1158/2159-8290.CD-21-1059
69. Hanahan D, Weinberg RA (2011) Hallmarks of cancer: the next generation. Cell 144:646–674. https://doi.org/10.1016/j.cell.2011.02.013
70. Harikumar KB, Kunnumakkara AB, Ochi N et al (2010) A novel small-molecule inhibitor of protein kinase D blocks pancreatic cancer growth in vitro and in vivo. Mol Cancer Ther 9:1136–1146. https://doi.org/10.1158/1535-7163.MCT-09-1145
71. Harvey JJ (1964) An unidentified virus which causes the rapid production of tumours in mice. Nature 204:1104–1105. https://doi.org/10.1038/2041104b0
72. Hausser A, Storz P, Märtens S et al (2005) Protein kinase D regulates vesicular transport by phosphorylating and activating phosphatidylinositol-4 kinase IIIβ at the Golgi complex. Nat Cell Biol 7:880–886. https://doi.org/10.1038/ncb1289
73. Head BP, Patel HH, Insel PA (2014) Interaction of membrane/lipid rafts with the cytoskeleton: impact on signaling and function: membrane/lipid rafts, mediators of cytoskeletal arrangement and cell signaling. Biochim Biophys Acta BBA – Biomembr 1838:532–545. https://doi.org/10.1016/j.bbamem.2013.07.018
74. Hernandez-Valladares M, Prior IA (2015) Comparative proteomic analysis of compartmentalised Ras signalling. Sci Rep 5:17307. https://doi.org/10.1038/srep17307
75. Herrero A, Casar B, Colón-Bolea P et al (2016) Defined spatiotemporal features of RAS-ERK signals dictate cell fate in MCF-7 mammary epithelial cells. Mol Biol Cell 27:1958–1968. https://doi.org/10.1091/mbc.e15-02-0118
76. Hidari KI-PJ, Ichikawa S, Furukawa K et al (1994) β1-4N-acetylgalactosaminyltransferase can synthesize both asialoglycosphingolipid GM2 and glycosphingolipid GM2in vitro and in vivo: isolation and characterization of a β1-4N-acetylgalactosaminyltransferase cDNA clone from rat ascites hepatoma cell line AH7974F. Biochem J 303:957–965. https://doi.org/10.1042/bj3030957
77. Hu B-S, Hu H, Zhu C-Y et al (2013) Overexpression of GOLPH3 is associated with poor clinical outcome in gastric cancer. Tumour Biol 34:515–520. https://doi.org/10.1007/s13277-012-0576-z
78. Hu G-S, Li Y-Q, Yang Y-M et al (2014) High expression of Golgi phosphoprotein-3 is associated with poor survival in patients with hepatocellular carcinoma. Tumour Biol 35:8625–8632. https://doi.org/10.1007/s13277-014-2105-8
79. Hua X, Yu L, Pan W et al (2012) Increased expression of Golgi phosphoprotein-3 is associated with tumor aggressiveness and poor prognosis of prostate cancer. Diagn Pathol 7:127. https://doi.org/10.1186/1746-1596-7-127
80. Huang Y, Zhang H-L, Li Z-L et al (2021) FUT8-mediated aberrant N-glycosylation of B7H3 suppresses the immune response in triple-negative breast cancer. Nat Commun 12:2672. https://doi.org/10.1038/s41467-021-22618-x

81. Ip CKM, Ng PKS, Jeong KJ et al (2018) Neomorphic PDGFRA extracellular domain driver mutations are resistant to PDGFRA targeted therapies. Nat Commun 9:4583. https://doi.org/10.1038/s41467-018-06949-w
82. Isaji T, Im S, Gu W et al (2014) An oncogenic protein Golgi phosphoprotein 3 up-regulates cell migration via sialylation. J Biol Chem 289:20694–20705. https://doi.org/10.1074/jbc.M113.542688
83. Jiang T, Wu H, Lin M et al (2021) B4GALNT1 promotes progression and metastasis in lung adenocarcinoma through JNK/c-Jun/Slug pathway. Carcinogenesis 42:621–630. https://doi.org/10.1093/carcin/bgaa141
84. Jin T, Ding Q, Huang H et al (2012) PAQR10 and PAQR11 mediate Ras signaling in the Golgi apparatus. Cell Res 22:661–676. https://doi.org/10.1038/cr.2011.161
85. Jing S, Deng Z, Liang L, Liang J (2020) B4GALNT1 enhances cell proliferation and growth in oral squamous cell carcinoma via p38 and JNK MAPK pathway. Transl Cancer Res 9. https://doi.org/10.21037/tcr.2020.03.73
86. Ju T, Aryal RP, Kudelka MR et al (2014) The Cosmc connection to the Tn antigen in cancer. Cancer Biomark 14:63–81. https://doi.org/10.3233/CBM-130375
87. Julien S, Bobowski M, Steenackers A et al (2013) How do gangliosides regulate RTKs signaling? Cells 2:751–767. https://doi.org/10.3390/cells2040751
88. Kamio K, Yoshida T, Gao C et al (2012) α1,6 Fucosyltransferase (Fut8) is implicated in vulnerability to elastase-induced emphysema in mice and a possible non-invasive predictive marker for disease progression and exacerbations in chronic obstructive pulmonary disease (COPD). Biochem Biophys Res Commun 424:112–117. https://doi.org/10.1016/j.bbrc.2012.06.081
89. Kandoth C, McLellan MD, Vandin F et al (2013) Mutational landscape and significance across 12 major cancer types. Nature 502:333–339. https://doi.org/10.1038/nature12634
90. Karnoub AE, Weinberg RA (2008) Ras oncogenes: split personalities. Nat Rev Mol Cell Biol 9:517–531. https://doi.org/10.1038/nrm2438
91. Kaszuba K, Grzybek M, Orłowski A et al (2015) N-Glycosylation as determinant of epidermal growth factor receptor conformation in membranes. Proc Natl Acad Sci 112:4334–4339. https://doi.org/10.1073/pnas.1503262112
92. Kuna RS, Field SJ (2019) GOLPH3: a Golgi phosphatidylinositol(4)phosphate effector that directs vesicle trafficking and drives cancer. J Lipid Res 60:269–275. https://doi.org/10.1194/jlr.R088328
93. Kunigou O, Nagao H, Kawabata N et al (2011) Role of GOLPH3 and GOLPH3L in the proliferation of human rhabdomyosarcoma. Oncol Rep 26:1337–1342. https://doi.org/10.3892/or.2011.1413
94. Lawrence MS, Sougnez C, Lichtenstein L et al (2015) Comprehensive genomic characterization of head and neck squamous cell carcinomas. Nature 517:576–582. https://doi.org/10.1038/nature14129
95. Lehoux S, Groux-Degroote S, Cazet A et al (2010) Transcriptional regulation of the human ST6GAL2 gene in cerebral cortex and neuronal cells. Glycoconj J 27:99–114. https://doi.org/10.1007/s10719-009-9260-y
96. Li Q, Ma Y, Xu W (2015) High GOLPH3 expression is associated with poor prognosis and invasion of hepatocellular carcinoma. Mol Med Rep 11:4315–4320. https://doi.org/10.3892/mmr.2015.3267
97. Li QQ, Hsu I, Sanford T et al (2018) Protein kinase D inhibitor CRT0066101 suppresses bladder cancer growth in vitro and xenografts via blockade of the cell cycle at G2/M. Cell Mol Life Sci 75:939–963. https://doi.org/10.1007/s00018-017-2681-z
98. Li S, Balmain A, Counter CM (2018) A model for RAS mutation patterns in cancers: finding the sweet spot. Nat Rev Cancer 18:767–777. https://doi.org/10.1038/s41568-018-0076-6
99. Li W, Guo F, Gu M et al (2015) Increased expression of GOLPH3 is associated with the proliferation of prostate cancer. J Cancer 6:420–429. https://doi.org/10.7150/jca.11228

100. Li X-Y, Liu W, Chen S-F et al (2011) Expression of the Golgi phosphoprotein-3 gene in human gliomas: a pilot study. J Neuro-Oncol 105:159–163. https://doi.org/10.1007/s11060-011-0573-x
101. Liang Y, Wang T, Gao R et al (2022) Fucosyltransferase 8 is overexpressed and influences clinical outcomes in lung adenocarcinoma patients. Pathol Oncol Res 28:1610116. https://doi.org/10.3389/pore.2022.1610116
102. Linstedt AD, Mehta A, Suhan J et al (1997) Sequence and overexpression of GPP130/GIMPc: evidence for saturable pH-sensitive targeting of a type II early Golgi membrane protein. Mol Biol Cell 8:1073–1087. https://doi.org/10.1091/mbc.8.6.1073
103. Liu H, Wang X, Feng B et al (2018) Golgi phosphoprotein 3 (GOLPH3) promotes hepatocellular carcinoma progression by activating mTOR signaling pathway. BMC Cancer 18:661. https://doi.org/10.1186/s12885-018-4458-7
104. Liu J, Wei H, Lai L et al (2019) Golgi phosphoprotein-3 promotes invasiveness of gastric cancer cells through the mTOR signalling pathway. Clin Invest Med 42:E38–E47. https://doi.org/10.25011/cim.v42i2.32815
105. Liu W, Chen Y, Yang J et al (2023) B4GALNT1 promotes carcinogenesis by regulating epithelial–mesenchymal transition in hepatocellular carcinoma based on pan-cancer analysis. J Gene Med 25:e3552. https://doi.org/10.1002/jgm.3552
106. Liu Y, Boukhelifa M, Tribble E et al (2008) The Sac1 phosphoinositide phosphatase regulates Golgi membrane morphology and mitotic spindle organization in mammals. Mol Biol Cell 19: 3080–3096. https://doi.org/10.1091/mbc.e07-12-1290
107. Liu Y, Song H, Zhou Y et al (2021) The oncogenic role of protein kinase D3 in cancer. J Cancer 12:735–739. https://doi.org/10.7150/jca.50899
108. Liu Y, Sun Y, Xiao M et al (2024) Comprehensive pan-cancer analysis reveals the versatile role of GALNT7 in epigenetic alterations and immune modulation in cancer. Heliyon 10. https://doi.org/10.1016/j.heliyon.2024.e31515
109. Liu Y-C, Yen H-Y, Chen C-Y et al (2011) Sialylation and fucosylation of epidermal growth factor receptor suppress its dimerization and activation in lung cancer cells. Proc Natl Acad Sci 108:11332–11337. https://doi.org/10.1073/pnas.1107385108
110. Lorentzen A, Kinkhabwala A, Rocks O et al (2010) Regulation of Ras localization by acylation enables a mode of intracellular signal propagation. Sci Signal 3:ra68. https://doi.org/10.1126/scisignal.20001370
111. Lu G, Chen J, Espinoza LA et al (2007) Protein kinase D 3 is localized in vesicular structures and interacts with vesicle-associated membrane protein 2. Cell Signal 19:867–879. https://doi.org/10.1016/j.cellsig.2006.10.012
112. Lu M, Tian Y, Yue W-M et al (2014) GOLPH3, a good prognostic indicator in early-stage NSCLC related to tumor angiogenesis. Asian Pac J Cancer Prev 15:5793–5798. https://doi.org/10.7314/APJCP.2014.15.14.5793
113. Lv X, Song J, Xue K et al (2019) Core fucosylation of copper transporter 1 plays a crucial role in cisplatin-resistance of epithelial ovarian cancer by regulating drug uptake. Mol Carcinog 58: 794–807. https://doi.org/10.1002/mc.22971
114. Lynch SJ, Snitkin H, Gumper I et al (2015) The differential palmitoylation states of N-Ras and H-Ras determine their distinct Golgi subcompartment localizations. J Cell Physiol 230:610–619. https://doi.org/10.1002/jcp.24779
115. Ma Y, Ren Y, Zhang X et al (2014) High GOLPH3 expression is associated with a more aggressive behavior of epithelial ovarian carcinoma. Virchows Arch 464:443–452. https://doi.org/10.1007/s00428-014-1536-3
116. Ma Y, Wang X, Wu Y et al (2014) Overexpression of GOLPH3 protein is associated with worse prognosis in patients with epithelial ovarian cancer. Tumour Biol 35:11845–11849. https://doi.org/10.1007/s13277-014-2411-1
117. Makowska KA, Hughes RE, White KJ et al (2015) Specific myosins control actin organization, cell morphology, and migration in prostate cancer cells. Cell Rep 13:2118–2125. https://doi.org/10.1016/j.celrep.2015.11.012

118. Makowski SL, Kuna RS, Field SJ (2020) Induction of membrane curvature by proteins involved in Golgi trafficking. Adv Biol Regul 75:100661. https://doi.org/10.1016/j.jbior.2019.100661
119. Makowski SL, Tran TT, Field SJ (2017) Emerging themes of regulation at the Golgi. Curr Opin Cell Biol 45:17–23. https://doi.org/10.1016/j.ceb.2017.01.004
120. Malumbres M, Barbacid M (2003) RAS oncogenes: the first 30 years. Nat Rev Cancer 3:459–465. https://doi.org/10.1038/nrc1097
121. McCormick F (2022) Chapter One – A brief history of RAS and the RAS initiative. In: O'Bryan JP, Piazza GA (eds) Advances in cancer research. Academic, pp 1–27
122. Mereiter S, Magalhães A, Adamczyk B et al (2016) Glycomic analysis of gastric carcinoma cells discloses glycans as modulators of RON receptor tyrosine kinase activation in cancer. Biochim Biophys Acta BBA – Gen Subj 1860:1795–1808. https://doi.org/10.1016/j.bbagen.2015.12.016
123. Möckl L, Pedram K, Roy AR et al (2019) Quantitative super-resolution microscopy of the mammalian glycocalyx. Dev Cell 50:57–72.e6. https://doi.org/10.1016/j.devcel.2019.04.035
124. Mori K, Matsuda K, Furusawa T et al (2005) Subcellular localization and dynamics of MysPDZ (Myo18A) in live mammalian cells. Biochem Biophys Res Commun 326:491–498. https://doi.org/10.1016/j.bbrc.2004.11.058
125. Morrow AA, Alipour MA, Bridges D et al (2014) The lipid kinase PI4KIIIβ is highly expressed in breast tumors and activates Akt in cooperation with Rab11a. Mol Cancer Res 12:1492–1508. https://doi.org/10.1158/1541-7786.MCR-13-0604
126. Mountzios G, Rampias T, Psyrri A (2014) The mutational spectrum of squamous-cell carcinoma of the head and neck: targetable genetic events and clinical impact. Ann Oncol 25:1889–1900. https://doi.org/10.1093/annonc/mdu143
127. Mukhopadhyay S, Bachert C, Smith DR, Linstedt AD (2010) Manganese-induced trafficking and turnover of the cis-Golgi glycoprotein GPP130. Mol Biol Cell 21:1282–1292. https://doi.org/10.1091/mbc.e09-11-0985
128. Myojin Y, Kodama T, Maesaka K et al (2021) ST6GAL1 is a novel serum biomarker for lenvatinib-susceptible FGF19-driven hepatocellular carcinoma. Clin Cancer Res 27:1150–1161. https://doi.org/10.1158/1078-0432.CCR-20-3382
129. Nakagawa T, Goto K, Kondo H (1996) Cloning, expression, and localization of 230-kDa phosphatidylinositol 4-kinase. J Biol Chem 271:12088–12094. https://doi.org/10.1074/jbc.271.20.12088
130. Natarajan R, Linstedt AD (2004) A cycling cis-Golgi protein mediates endosome-to-Golgi traffic. Mol Biol Cell 15:4798–4806. https://doi.org/10.1091/mbc.e04-05-0366
131. Nazha B, Inal C, Owonikoko TK (2020) Disialoganglioside GD2 expression in solid tumors and role as a target for cancer therapy. Front Oncol 10. https://doi.org/10.3389/fonc.2020.01000
132. Ng MM, Dippold HC, Buschman MD et al (2013) GOLPH3L antagonizes GOLPH3 to determine Golgi morphology. Mol Biol Cell 24:796–808. https://doi.org/10.1091/mbc.e12-07-0525
133. Nhek S, Ngo M, Yang X et al (2010) Regulation of oxysterol-binding protein Golgi localization through protein kinase D-mediated phosphorylation. Mol Biol Cell 21:2327–2337. https://doi.org/10.1091/mbc.e10-02-0090
134. Norton PA, Mehta AS (2019) Expression of genes that control core fucosylation in hepatocellular carcinoma: systematic review. World J Gastroenterol 25:2947–2960. https://doi.org/10.3748/wjg.v25.i23.2947
135. Olayioye MA, Hausser A (2012) Integration of non-vesicular and vesicular transport processes at the Golgi complex by the PKD-CERT network. Biochim Biophys Acta 1821:1096–1103. https://doi.org/10.1016/j.bbalip.2011.12.005
136. Peng J, Fang Y, Tao Y et al (2014) Mechanisms of GOLPH3 associated with the progression of gastric cancer: a preliminary study. PLoS One 9:e107362. https://doi.org/10.1371/journal.pone.0107362

137. Pereira NA, Pu HX, Goh H, Song Z (2014) Golgi phosphoprotein 3 mediates the Golgi localization and function of protein O-linked mannose β-1,2-N-acetlyglucosaminyltransferase 1. J Biol Chem 289:14762–14770. https://doi.org/10.1074/jbc.M114.548305
138. Phadnis VV, Snider J, Varadharajan V et al (2023) MMD collaborates with ACSL4 and MBOAT7 to promote polyunsaturated phosphatidylinositol remodeling and susceptibility to ferroptosis. Cell Rep 42:113023. https://doi.org/10.1016/j.celrep.2023.113023
139. Png KJ, Halberg N, Yoshida M, Tavazoie SF (2012) A microRNA regulon that mediates endothelial recruitment and metastasis by cancer cells. Nature 481:190–194. https://doi.org/10.1038/nature10661
140. Pohlentz G, Klein D, Schwarzmann G et al (1988) Both GA2, GM2, and GD2 synthases and GM1b, GD1a, and GT1b synthases are single enzymes in Golgi vesicles from rat liver. Proc Natl Acad Sci 85:7044–7048. https://doi.org/10.1073/pnas.85.19.7044
141. Prior IA, Hancock JF (2012) Ras trafficking, localization and compartmentalized signalling. Semin Cell Dev Biol 23:145–153. https://doi.org/10.1016/j.semcdb.2011.09.002
142. Prior IA, Hood FE, Hartley JL (2020) The frequency of Ras mutations in cancer. Cancer Res 80:2969–2974. https://doi.org/10.1158/0008-5472.CAN-19-3682
143. Pusapati GV, Krndija D, Armacki M et al (2010) Role of the second cysteine-rich domain and Pro275 in protein kinase D2 interaction with ADP-ribosylation factor 1, trans-Golgi network recruitment, and protein transport. Mol Biol Cell 21:1011–1022. https://doi.org/10.1091/mbc.e09-09-0814
144. Pylayeva-Gupta Y, Grabocka E, Bar-Sagi D (2011) RAS oncogenes: weaving a tumorigenic web. Nat Rev Cancer 11:761–774. https://doi.org/10.1038/nrc3106
145. Rahajeng J, Kuna RS, Makowski SL et al (2019) Efficient Golgi forward trafficking requires GOLPH3-driven, PI4P-dependent membrane curvature. Dev Cell 50:573–585.e5. https://doi.org/10.1016/j.devcel.2019.05.038
146. Rey O, Yuan J, Young SH, Rozengurt E (2003) Protein kinase Cν/protein kinase D3 nuclear localization, catalytic activation, and intracellular redistribution in response to G protein-coupled receptor agonists. J Biol Chem 278:23773–23785. https://doi.org/10.1074/jbc.M300226200
147. Rizzo R, Russo D, Kurokawa K et al (2021) Golgi maturation-dependent glycoenzyme recycling controls glycosphingolipid biosynthesis and cell growth via GOLPH3. EMBO J 40:e107238. https://doi.org/10.15252/embj.2020107238
148. Ruan S, Lloyd KO (1992) Glycosylation pathways in the biosynthesis of gangliosides in melanoma and neuroblastoma cells: relative glycosyltransferase levels determine ganglioside patterns. Cancer Res 52:5725–5731
149. Ruggiero FM, Martínez-Koteski N, Fidelio GD et al (2022) Golgi phosphoprotein 3 regulates the physical association of glycolipid glycosyltransferases. Int J Mol Sci 23:10354. https://doi.org/10.3390/ijms231810354
150. Salem AF, Whitaker-Menezes D, Lin Z et al (2012) Two-compartment tumor metabolism: autophagy in the tumor microenvironment and oxidative mitochondrial metabolism (OXPHOS) in cancer cells. Cell Cycle 11:2545–2559. https://doi.org/10.4161/cc.20920
151. Sanchez-Garcia F, Villagrasa P, Matsui J et al (2014) Integration of genomic data enables selective discovery of breast cancer drivers. Cell 159:1461–1475. https://doi.org/10.1016/j.cell.2014.10.048
152. Sardana R, Highland CM, Straight BE et al (2021) Golgi membrane protein Erd1 Is essential for recycling a subset of Golgi glycosyltransferases. eLife 10:e70774. https://doi.org/10.7554/eLife.70774
153. Sasaki N, Toyoda M, Ishiwata T (2021) Gangliosides as signaling regulators in cancer. Int J Mol Sci 22:5076. https://doi.org/10.3390/ijms22105076
154. Schengrund C-L (2023) The Ying and Yang of ganglioside function in cancer. Cancer 15:5362. https://doi.org/10.3390/cancers15225362
155. Schmitz KR, Liu J, Li S et al (2008) Golgi localization of glycosyltransferases requires a Vps74p oligomer. Dev Cell 14:523–534. https://doi.org/10.1016/j.devcel.2008.02.016

156. Schultz MJ, Holdbrooks AT, Chakraborty A et al (2016) The tumor-associated glycosyltransferase ST6Gal-I regulates stem cell transcription factors and confers a cancer stem cell phenotype. Cancer Res 76:3978–3988. https://doi.org/10.1158/0008-5472.CAN-15-2834
157. Scott E, Archer Goode E, Garnham R et al (2023) ST6GAL1-mediated aberrant sialylation promotes prostate cancer progression. J Pathol 261:71–84. https://doi.org/10.1002/path.6152
158. Scott E, Hodgson K, Calle B et al (2023) Upregulation of GALNT7 in prostate cancer modifies O-glycosylation and promotes tumour growth. Oncogene 42:926–937. https://doi.org/10.1038/s41388-023-02604-x
159. Scott KL, Kabbarah O, Liang M-C et al (2009) GOLPH3 modulates mTOR signalling and rapamycin sensitivity in cancer. Nature 459:1085–1090. https://doi.org/10.1038/nature08109
160. Sheng G, Zeng Z, Pan J et al (2017) Multiple MYO18A-PDGFRB fusion transcripts in a myeloproliferative neoplasm patient with t(5;17)(q32;q11). Mol Cytogenet 10:4. https://doi.org/10.1186/s13039-017-0306-8
161. Shi L, Tan X, Liu X et al (2021) Addiction to Golgi-resident PI4P synthesis in chromosome 1q21.3–amplified lung adenocarcinoma cells. Proc Natl Acad Sci 118:e2023537118. https://doi.org/10.1073/pnas.2023537118
162. Shi W, Feng W, Wang J et al (2019) Clinicopathologic features and prognostic implications of Golgi phosphoprotein 3 in non small cell lung cancer: a meta-analysis. J Cancer 10:5754–5763. https://doi.org/10.7150/jca.30067
163. Shishido F, Uemura S, Kashimura M, Inokuchi J (2017) Identification of a new B4GalNAcT1 (GM2/GD2/GA2 synthase) isoform, and regulation of enzyme stability and intracellular transport by arginine-based motif. Biochim Biophys Acta BBA – Biomembr 1859:2001–2011. https://doi.org/10.1016/j.bbamem.2017.07.006
164. Silva GO, He X, Parker JS et al (2015) Cross-species DNA copy number analyses identifies multiple 1q21-q23 subtype-specific driver genes for breast cancer. Breast Cancer Res Treat 152:347–356. https://doi.org/10.1007/s10549-015-3476-2
165. Stenman A, Welander J, Gustavsson I et al (2016) Mutation prevalence and associated expression patterns in pheochromocytoma. Genes Chromosomes Cancer 55:452–459. https://doi.org/10.1002/gcc.22347
166. Taft MH, Behrmann E, Munske-Weidemann L-C et al (2013) Functional characterization of human myosin-18A and its interaction with F-actin and GOLPH3. J Biol Chem 288:30029–30041. https://doi.org/10.1074/jbc.M113.497180
167. Takano N, Owada Y, Suzuki R et al (2003) Cloning and characterization of a novel variant (mM-rdgBβ1) of mouse M-rdgBs, mammalian homologs of Drosophila retinal degeneration B gene proteins, and its mRNA localization in mouse brain in comparison with other M-rdgBs. J Neurochem 84:829–839. https://doi.org/10.1046/j.1471-4159.2003.01591.x
168. Tan J, Brill JA (2014) Cinderella story: PI4P goes from precursor to key signaling molecule. Crit Rev Biochem Mol Biol 49:33–58. https://doi.org/10.3109/10409238.2013.853024
169. Tan X, Banerjee P, Guo H-F et al (2017) Epithelial-to-mesenchymal transition drives a pro-metastatic Golgi compaction process through scaffolding protein PAQR11. J Clin Invest 127:117–131. https://doi.org/10.1172/JCI88736
170. Tan X, Banerjee P, Pham EA et al (2020) PI4KIIIβ is a therapeutic target in chromosome 1q–amplified lung adenocarcinoma. Sci Transl Med 12:eaax3772. https://doi.org/10.1126/scitranslmed.aax3772
171. Tan X, Wang S, Xiao G-Y et al (2024) Chromosomal 3q amplicon encodes essential regulators of secretory vesicles that drive secretory addiction in cancer. J Clin Invest 134. https://doi.org/10.1172/JCI176355
172. Tan Y, Lin K, Zhao Y et al (2018) Adipocytes fuel gastric cancer omental metastasis *via* PITPNC1-mediated fatty acid metabolic reprogramming. Theranostics 8:5452–5468. https://doi.org/10.7150/thno.28219

173. Tan Y, Shao R, Li J et al (2020) PITPNC1 fuels radioresistance of rectal cancer by inhibiting reactive oxygen species production. Ann Transl Med 8:126. https://doi.org/10.21037/atm.2020.02.37
174. Tang S, Pan H, Wei W et al (2017) GOLPH3: a novel biomarker that correlates with poor survival and resistance to chemotherapy in breast cancer. Oncotarget 8:105155–105169. https://doi.org/10.18632/oncotarget.21927
175. Tang Y, Xu Z, Xu F et al (2023) B4GALNT1 promotes hepatocellular carcinoma stemness and progression via integrin α2β1-mediated FAK and AKT activation. JHEP Rep 5:100903. https://doi.org/10.1016/j.jhepr.2023.100903
176. Taylor AH, Konje JC, Ayakannu T (2023) Identification of potentially novel molecular targets of endometrial cancer using a non-biased proteomic approach. Cancer 15:4665. https://doi.org/10.3390/cancers15184665
177. Tokuda E, Itoh T, Hasegawa J et al (2014) Phosphatidylinositol 4-phosphate in the Golgi apparatus regulates cell–cell adhesion and invasive cell migration in human breast cancer. Cancer Res 74:3054–3066. https://doi.org/10.1158/0008-5472.CAN-13-2441
178. Tomida S, Nagae M, Kizuka Y (2024) Distinctive domains and activity regulation of core fucosylation enzyme FUT8. Biochim Biophys Acta BBA – Gen Subj 1868:130561. https://doi.org/10.1016/j.bbagen.2024.130561
179. Tomida S, Nagae M, Kizuka Y (2022) The stem region of α1,6-fucosyltransferase FUT8 is required for multimer formation but not catalytic activity. J Biol Chem 298:102676. https://doi.org/10.1016/j.jbc.2022.102676
180. Tomida S, Takata M, Hirata T et al (2020) The SH3 domain in the fucosyltransferase FUT8 controls FUT8 activity and localization and is essential for core fucosylation. J Biol Chem 295:7992–8004. https://doi.org/10.1074/jbc.RA120.013079
181. Trivedi D, Padinjat R (2007) RdgB proteins: functions in lipid homeostasis and signal transduction. Biochim Biophys Acta BBA – Mol Cell Biol Lipids 1771:692–699. https://doi.org/10.1016/j.bbalip.2007.04.014
182. Tu C-F, Wu M-Y, Lin Y-C et al (2017) FUT8 promotes breast cancer cell invasiveness by remodeling TGF-β receptor core fucosylation. Breast Cancer Res 19:111. https://doi.org/10.1186/s13058-017-0904-8
183. Tu L, Tai WCS, Chen L, Banfield DK (2008) Signal-mediated dynamic retention of glycosyltransferases in the Golgi. Science 321:404–407. https://doi.org/10.1126/science.1159411
184. Ussowicz M, Jaśkowiec A, Meyer C et al (2012) A three-way translocation of MLL, MLLT11, and the novel reciprocal partner gene MYO18A in a child with acute myeloid leukemia. Cancer Genet 205:261–265. https://doi.org/10.1016/j.cancergen.2012.02.006
185. Varmus HE (1990) Retroviruses and oncogenes I. Biosci Rep 10:413–430. https://doi.org/10.1007/BF01152288
186. Venditti R, Masone MC, Wilson C, De Matteis MA (2016) PI(4)P homeostasis: who controls the controllers? Adv Biol Regul 60:105–114. https://doi.org/10.1016/j.jbior.2015.09.007
187. Venkat S, Linstedt AD (2017) Manganese-induced trafficking and turnover of GPP130 is mediated by sortilin. Mol Biol Cell 28:2569–2578. https://doi.org/10.1091/mbc.e17-05-0326
188. Venturi G, Gomes Ferreira I, Pucci M et al (2019) Impact of sialyltransferase ST6GAL1 overexpression on different colon cancer cell types. Glycobiology 29:684–695. https://doi.org/10.1093/glycob/cwz053
189. Walch-Solimena C, Novick P (1999) The yeast phosphatidylinositol-4-OH kinase Pik1 regulates secretion at the Golgi. Nat Cell Biol 1:523–525. https://doi.org/10.1038/70319
190. Walz C, Chase A, Schoch C et al (2005) The t(8;17)(p11;q23) in the 8p11 myeloproliferative syndrome fuses MYO18A to FGFR1. Leukemia 19:1005–1009. https://doi.org/10.1038/sj.leu.2403712
191. Walz C, Haferlach C, Hänel A et al (2009) Identification of a MYO18A-PDGFRB fusion gene in an eosinophilia-associated atypical myeloproliferative neoplasm with a t(5;17)(q33-34;q11.2). Genes Chromosomes Cancer 48:179–183. https://doi.org/10.1002/gcc.20629

192. Wang J, Liu G, Liu M et al (2021) High-risk HPV16 E6 activates the cGMP/PKG pathway through glycosyltransferase ST6GAL1 in cervical cancer cells. Front Oncol 11. https://doi.org/10.3389/fonc.2021.716246
193. Wang J, Qian J, Hoeksema MD et al (2013) Integrative genomics analysis identifies candidate drivers at 3q26-29 amplicon in squamous cell carcinoma of the lung. Clin Cancer Res 19: 5580–5590. https://doi.org/10.1158/1078-0432.CCR-13-0594
194. Wang J-H, Chen X-T, Wen Z-S et al (2012) High expression of GOLPH3 in esophageal squamous cell carcinoma correlates with poor prognosis. PLoS One 7:e45622. https://doi.org/10.1371/journal.pone.0045622
195. Wang M, Zhang Z, Chen M et al (2023) FDW028, a novel FUT8 inhibitor, impels lysosomal proteolysis of B7-H3 via chaperone-mediated autophagy pathway and exhibits potent efficacy against metastatic colorectal cancer. Cell Death Dis 14:1–13. https://doi.org/10.1038/s41419-023-06027-0
196. Wang R, Ke Z, Wang F et al (2015) GOLPH3 overexpression is closely correlated with poor prognosis in human non-small cell lung cancer and mediates its metastasis through upregulating MMP-2 and MMP-9. Cell Physiol Biochem 35:969–982. https://doi.org/10.1159/000369753
197. Wang X, Gu J, Ihara H et al (2006) Core fucosylation regulates epidermal growth factor receptor-mediated intracellular signaling. J Biol Chem 281:2572–2577. https://doi.org/10.1074/jbc.M510893200
198. Wang YJ, Wang J, Sun HQ et al (2003) Phosphatidylinositol 4 phosphate regulates targeting of Clathrin adaptor AP-1 complexes to the Golgi. Cell 114:299–310. https://doi.org/10.1016/S0092-8674(03)00603-2
199. Waugh MG (2014) Amplification of chromosome 1q genes encoding the phosphoinositide signalling enzymes *PI4KB, AKT3, PIP5K1A* and *PI3KC2B* in breast cancer. J Cancer 5:790–796. https://doi.org/10.7150/jca.9794
200. Weinberg RA (2014) Coming full circle—from endless complexity to simplicity and back again. Cell 157:267–271. https://doi.org/10.1016/j.cell.2014.03.004
201. Weiss RA (2020) A perspective on the early days of RAS research. Cancer Metastasis Rev 39: 1023–1028. https://doi.org/10.1007/s10555-020-09919-1
202. Weixel KM, Blumental-Perry A, Watkins SC et al (2005) Distinct Golgi populations of phosphatidylinositol 4-phosphate regulated by phosphatidylinositol 4-kinases. J Biol Chem 280:10501–10508. https://doi.org/10.1074/jbc.M414304200
203. Welch LG, Peak-Chew S-Y, Begum F et al (2021) GOLPH3 and GOLPH3L are broad-spectrum COPI adaptors for sorting into intra-Golgi transport vesicles. J Cell Biol 220: e202106115. https://doi.org/10.1083/jcb.202106115
204. Wichert B, Milde-Langosch K, Galatenko V et al (2018) Prognostic role of the sialyltransferase ST6GAL1 in ovarian cancer. Glycobiology 28:898–903. https://doi.org/10.1093/glycob/cwy065
205. Wong K, Meyers R, Cantley LC (1997) Subcellular locations of phosphatidylinositol 4-kinase isoforms. J Biol Chem 272:13236–13241. https://doi.org/10.1074/jbc.272.20.13236
206. Wu CC, Taylor RS, Lane DR et al (2000) GMx33: a novel family of trans-Golgi proteins identified by proteomics. Traffic 1:963–975. https://doi.org/10.1111/j.1600-0854.2000.11206.x
207. Wu L, Cheng Y, Geng D et al (2022) O-GlcNAcylation regulates epidermal growth factor receptor intracellular trafficking and signaling. Proc Natl Acad Sci 119:e2107453119. https://doi.org/10.1073/pnas.2107453119
208. Wu XY, Liu WT, Wu ZF et al (2016) Identification of HRAS as cancer-promoting gene in gastric carcinoma cell aggressiveness. Am J Cancer Res 6:1935–1948
209. Xie Z, Hur SK, Zhao L et al (2018) A Golgi lipid signaling pathway controls apical Golgi distribution and cell polarity during neurogenesis. Dev Cell 44:725–740.e4. https://doi.org/10.1016/j.devcel.2018.02.025

210. Xing M, Peterman MC, Davis RL et al (2016) GOLPH3 drives cell migration by promoting Golgi reorientation and directional trafficking to the leading edge. Mol Biol Cell 27:3828–3840. https://doi.org/10.1091/mbc.E16-01-0005
211. Xue Y, Wu G, Liao Y et al (2014) GOLPH3 is a novel marker of poor prognosis and a potential therapeutic target in human renal cell carcinoma. Br J Cancer 110:2250–2260. https://doi.org/10.1038/bjc.2014.124
212. Yamashiro S, Haraguchi M, Furukawa K et al (1995) Substrate specificity of β1,4-N-acetylgalactosaminyltransferase in vitro and in cDNA-transfected cells: GM2/GD2 synthase efficiently generates ASIALO-GM2 in certain cells. J Biol Chem 270:6149–6155. https://doi.org/10.1074/jbc.270.11.6149
213. Yeaman C, Ayala MI, Wright JR et al (2004) Protein kinase D regulates basolateral membrane protein exit from trans-Golgi network. Nat Cell Biol 6:106–112. https://doi.org/10.1038/ncb1090
214. Yi H, Lin Y, Li Y et al (2022) Pan-cancer analysis of B4GALNT1 as a potential prognostic and immunological biomarker. J Immunol Res 2022:4355890. https://doi.org/10.1155/2022/4355890
215. Yoshida H, Koodie L, Jacobsen K et al (2020) B4GALNT1 induces angiogenesis, anchorage independence growth and motility, and promotes tumorigenesis in melanoma by induction of ganglioside GM2/GD2. Sci Rep 10:1199. https://doi.org/10.1038/s41598-019-57130-2
216. Yuyama Y, Dohi T, Morita H et al (1995) Enhanced expression of GM2/GD2 synthase mRNA in human gastrointestinal cancer. Cancer 75:1273–1280. https://doi.org/10.1002/1097-0142(19950315)75:6<1273::AID-CNCR2820750609>3.0.CO;2-O
217. Zeng Z, Lin H, Zhao X et al (2012) Overexpression of GOLPH3 promotes proliferation and tumorigenicity in breast cancer via suppression of the FOXO1 transcription factor. Clin Cancer Res 18:4059–4069. https://doi.org/10.1158/1078-0432.CCR-11-3156
218. Zhang C, Wu Q, Huang H et al (2020) Caveolin-1 upregulates Fut8 expression by activating the Wnt/β-catenin pathway to enhance HCC cell proliferative and invasive ability. Cell Biol Int 44:2202–2212. https://doi.org/10.1002/cbin.11426
219. Zhang L, Guo F, Gao X, Wu Y (2015) Golgi phosphoprotein 3 expression predicts poor prognosis in patients with prostate cancer undergoing radical prostatectomy. Mol Med Rep 12: 1298–1304. https://doi.org/10.3892/mmr.2015.3455
220. Zhang L-J, Wang K-B, Liu L et al (2014) Overexpression of GOLPH3 is associated with poor prognosis and clinical progression in pancreatic ductal adenocarcinoma. BMC Cancer 14:571. https://doi.org/10.1186/1471-2407-14-571
221. Zhang X, Ding Z, Mo J et al (2015) GOLPH3 promotes glioblastoma cell migration and invasion via the mTOR-YB1 pathway in vitro. Mol Carcinog 54:1252–1263. https://doi.org/10.1002/mc.22197
222. Zhang Y, Ma M, Han B (2014) GOLPH3 high expression predicts poor prognosis in patients with resected non-small cell lung cancer: an immunohistochemical analysis. Tumour Biol 35: 10833–10839. https://doi.org/10.1007/s13277-014-2357-3
223. Zheng Q, Bobich JA, Vidugiriene J et al (2005) Neuronal calcium sensor-1 facilitates neuronal exocytosis through phosphatidylinositol 4-kinase. J Neurochem 92:442–451. https://doi.org/10.1111/j.1471-4159.2004.02897.x
224. Zhou X, Xue P, Yang M et al (2014) Protein kinase D2 promotes the proliferation of glioma cells by regulating Golgi phosphoprotein 3. Cancer Lett 355:121–129. https://doi.org/10.1016/j.canlet.2014.09.008
225. Zhou X, Zhan W, Bian W et al (2013) GOLPH3 regulates the migration and invasion of glioma cells though RhoA. Biochem Biophys Res Commun 433:338–344. https://doi.org/10.1016/j.bbrc.2013.03.003
226. Zou Z, Zeng F, Xu W et al (2012) PKD2 and PKD3 promote prostate cancer cell invasion by modulating NF-κB- and HDAC1-mediated expression and activation of uPA. J Cell Sci 125: 4800–4811. https://doi.org/10.1242/jcs.106542

Chapter 16
Golgi Retention and Oncogenic Signaling of KIT Tyrosine Kinase in Gastrointestinal Stromal Tumor

Yuuki Obata and Toshirou Nishida

Abstract KIT, a receptor protein tyrosine kinase (RTK), is localized in the plasma membrane and binds stem cell factor. A cancer-causing permanently active mutant of KIT (KIT^{mut}) accumulates aberrantly in intracellular compartments. Oncogenic RTK mutants other than KIT^{mut} also exhibit characteristic organelle localization, unlike their wild-type counterparts. However, the precise subcellular localization of RTK^{mut} and its biological significance in oncogenic signaling remain unclear. Recent studies have shown that in gastrointestinal stromal tumors (GISTs), KIT^{mut} is inappropriately retained in the Golgi/*trans*-Golgi network (TGN) area in a manner dependent on its tyrosine kinase activity. It is trapped in the biosynthetic secretory pathway in which normal RTK trafficking occurs. Importantly, KIT^{mut} is activated and exerts oncogenic signaling predominantly in the Golgi/TGN of GIST cells. Recently, key players in the Golgi retention of KIT^{mut} have been reported, but they are unlikely to be involved in the intracellular retention of other RTKs, suggesting that the retention of individual RTKs is regulated by different molecular mechanisms. Here, we review the present knowledge on KIT retention in the Golgi/TGN region in GISTs, molecular mechanisms underlying this retention, and the aberrant localization of other RTK mutants. We also discuss strategies for the medical development of RTK inhibition by blocking intracellular trafficking.

Keywords RTK · GIST · Golgi/TGN · Organelle · KIT · PDGFRA

Y. Obata (✉)
Laboratory of Intracellular Traffic & Oncology, National Cancer Center Research Institute, Tokyo, Japan
e-mail: yuobata@ncc.go.jp

T. Nishida
National Cancer Center Hospital, Tokyo, Japan
e-mail: tnishida@ncc.go.jp

A. Nakano, J. Saraste (eds.), *The Golgi Network, Volume II*, Subcellular Biochemistry 111, https://doi.org/10.1007/978-3-032-16833-7_16

Introduction

Receptor Protein-Tyrosine Kinases

In normal mammalian cells, receptor protein-tyrosine kinases (RTKs), such as epidermal growth factor receptor (EGFR), stem cell factor receptor (SCFR: KIT), platelet-derived growth factor receptor A/B (PDGFRA/B), FMS-like tyrosine kinase 3 (FLT3), and hepatocyte growth factor receptor (HGFR: MET), localize in the plasma membrane (PM) to bind to their ligands [1, 2]. They are type I transmembrane glycosylated proteins with an amino-terminal extracellular domain (ECD), single transmembrane domain, and carboxy-terminal intracytoplasmic portion encompassing the tyrosine kinase domain (TKD) (Fig. 16.1). Basically, upon ligand binding at the PM, RTKs are homo- or hetero-dimerized, resulting in enhanced tyrosine kinase activity. Specific RTK, i.e., insulin-like growth factor 1 receptor (IGF-1R), forms oligomers in an inactive state in the absence of an activating ligand [3]. Activated RTKs autophosphorylate and/or transphosphorylate specific tyrosine residues in the cytoplasm, leading to changes in receptor conformation. In addition to the autophosphorylation, RTKs can directly tyrosine-phosphorylate and activate downstream targets, such as phospholipase C-γ (PLCγ) [4, 5]. Docking proteins

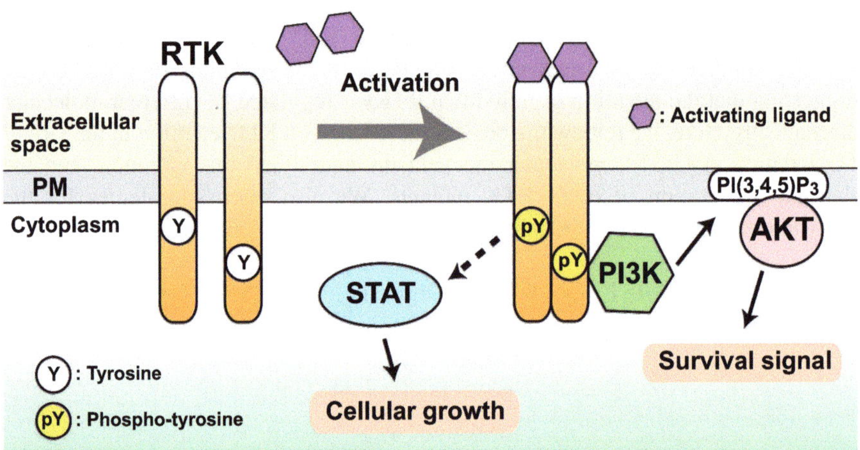

Fig. 16.1 Growth signaling via receptor protein-tyrosine kinases. In the absence of an activating ligand, the receptor protein-tyrosine kinase (RTK) in the plasma membrane (PM) exists in an inactive state. Upon ligand binding, RTK forms a homo−/heterooligomer and subsequently causes autophosphorylation of specific tyrosine residues. Src homology 2 (SH2) domain-containing proteins such as phosphatidylinositol 3-kinase (PI3K) and growth factor receptor-bound protein 2 (GRB2) are recruited to tyrosine-phosphorylated sites, resulting in the activation of AKT, extracellular signal-regulated kinase (ERK), and signal transducers and activators of transcription (STATs). This activation is closely associated with cell proliferation, survival, migration, and differentiation. Y, tyrosine residue; pY phospho-tyrosine residue; PI(3,4,5)P_3, phosphatidylinositol 3,4,5-triphosphate

bearing an Src homology 2 (SH2) domain that bind to a phosphotyrosine motif are recruited to the cytoplasmic domain of RTK, which serves as a scaffold for downstream signal propagation. The complex generally activates AKT, extracellular signal-regulated kinase (ERK), and signal transducers and activators of transcription (STATs), which play key roles in the regulation of gene expression, cytoskeletal rearrangement, and metabolism, leading to cell growth, survival, migration, differentiation, and cell cycle control. The termination of RTK signaling is tightly regulated by multiple mechanisms. After active RTKs are no longer required, they are inactivated by protein tyrosine phosphatases (PTPs) or receptor-mediated endocytosis, followed by their degradation inside lysosomes or proteasomes [6–10].

Intracellular Trafficking of RTKs

A summary of the intracellular trafficking of normal RTKs is shown in Fig. 16.2. RTKs are newly synthesized as a precursor in the rough endoplasmic reticulum (referred to henceforth as ER) [11, 12]. Their amino-terminal signal peptide is co-translationally removed, and ECD is partially glycosylated at the asparagine (N) residues inside the ER [13, 14]. The immature N-glycosylated form is known as the high mannose (HM) form. HM RTKs are incorporated into coat protein complex II-coated (COPII-coated) carriers and subsequently transported toward the *cis* face of the Golgi apparatus in a manner similar to that of other type I transmembrane proteins. During the transition from the *cis*-Golgi to *trans*-Golgi cisternae, RTKs undergo complex glycosylation (CG) to mature and reach the *trans*-Golgi network (TGN) [15]. It is well known that these glycosylation processes are important for the correct intracellular localization and regulation of RTK function. The ECDs of specific RTKs, such as MET and IGF-1R, are cleaved by furin convertase [16, 17]. The cleavage results in forming a heterodimeric receptor composed of an α subunit connected through a disulfide bond to a transmembrane β subunit. Mature RTKs are incorporated into TGN-derived carriers and transferred to the PM. This continuous pathway from the ER–Golgi system to the PM is called the secretory pathway. Under normal conditions, they remain in an inactive form during biosynthetic trafficking. Moreover, basal RTK activity is suppressed by PTPs [18]. When ligands exist in the extracellular space at an appropriate concentration, RTKs are activated and subsequently undergo receptor-mediated endocytosis, resulting in their localization to the endosomes. RTKs are subsequently sorted from endosomes to the PM for recycling or to lysosomes/proteasomes for degradation [19]. After activation, highly ubiquitinated RTKs are incorporated into proteolytic compartments. It should be noted that a subset of RTKs, such as EGFR and MET, can execute signaling on endosomes after internalization [20–22]. Recent evidence suggests that aberrant activation of RTK through genetic abnormalities results in the impairment of normal trafficking of the receptor. Constitutively active RTK mutants found in cancer cells seem to use organelles as signaling platforms.

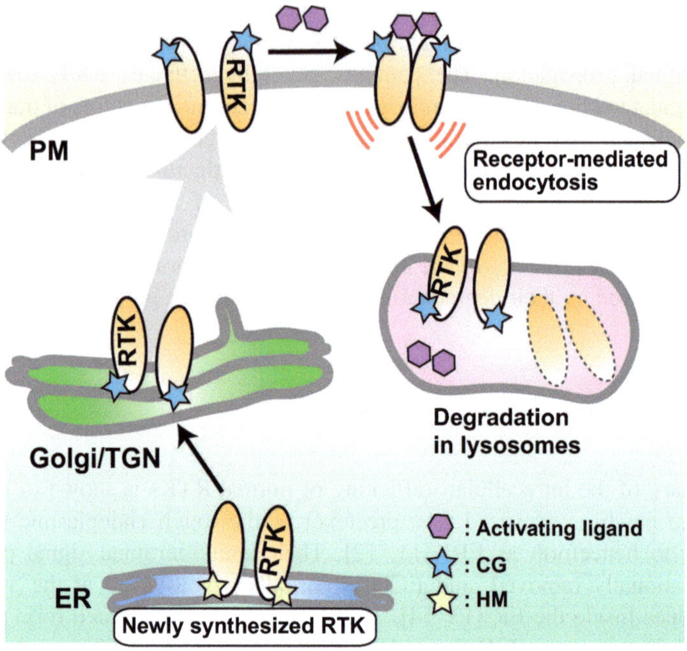

Fig. 16.2 Intracellular trafficking of normal RTK. Newly synthesized RTK in the endoplasmic reticulum (ER) exists in an immature glycosylated (high mannose, HM, yellow stars) form and subsequently moves to the Golgi apparatus. During the transition to the *trans*-Golgi network (TGN), RTK undergoes complex glycosylation (CG, blue stars) and is exported toward the plasma membrane (PM). Normal RTKs remain in an inactive state during biosynthetic secretory trafficking. Upon stimulation with ligand binding at the PM, RTK activates downstream cascades and is, in turn, internalized through receptor-mediated endocytosis. If required, the internalized RTK is recycled back to the PM; otherwise, it is incorporated into lysosomes or proteasomes for degradation

Growth Signaling by RTKs in Cancer Cells

Because RTK activity plays a key role in the proliferation and migration of normal cells, it is not surprising that its permanent activation could lead host cells to autonomous growth, resulting in the development of cancer and metastasis. Of the 55 members of the RTK family, more than half have mutated counterparts that are estimated to play pivotal roles in human malignancies [1]. In various cancer cells, constitutively active RTKs appear in a genetic abnormality-dependent manner as follows: **(i)** point mutations and in-frame insertions or deletions that prevent the formation of an inactive conformation and maintain the activated state, **(ii)** gene amplification resulting in overexpression, and **(iii)** gene fusion with a specific partner (Fig. 16.3) [12, 23]. Notably, the majority of oncogenic fusion RTK proteins, such as EML4-ALK and CCDC6-RET, lack entirely the ECD and the transmembrane domain due to in-frame gene fusion of the *TKD* with a cytosolic partner gene;

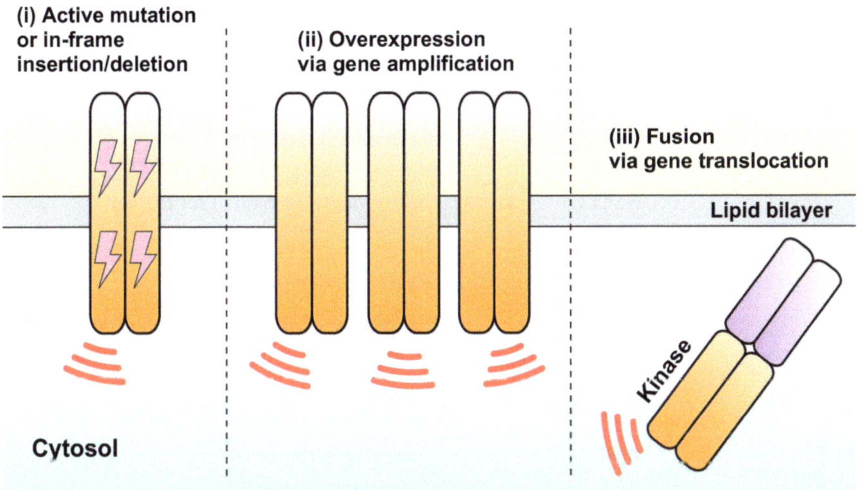

Fig. 16.3 Ligand-independent activation of RTK through genetic alterations. A constitutively active mutation or specific in-frame deletion/insertion (**i**), overexpression via gene amplification (**ii**), and fusion of the tyrosine kinase domain with a specific partner (**iii**) result in ligand-independent oligomerization. Note that the majority of oncogenic RTK fusion proteins lack the transmembrane domain, hence they are unable to enter the secretory pathway. In this chapter, we focus on (**i**) and (**ii**) oncogenic RTKs, that is, RTKs bearing a transmembrane domain that can migrate into the secretory pathway

thus, they are unable to enter the secretory pathway [24]. A membraneless organelle formed by the oncogenic fusion of RTK via a liquid–liquid phase separation mechanism serves as a signal platform [25]. Hence, in this chapter, we focus on RTK mutants harboring the transmembrane domain (hereafter referred to as RTK^{mut}), which can be recruited into the secretory pathway.

Representative examples of RTK^{mut} are EGFR^{mut} in non-small cell lung cancer (NSCLC), FLT3^{mut} in acute myelogenous leukemia (AML), and KIT^{mut} in gastrointestinal stromal tumors (GISTs) [26–28]. Because most cancer cells that have a permanently active RTK^{mut} are generally addicted to the mutant tyrosine phosphorylation signaling for their proliferation, small molecule tyrosine kinase inhibitors (TKIs) so called "molecular targeted drugs" have been developed for the treatment of malignant cancers [23, 29]. *In vitro* studies have demonstrated that the EGFR TKIs gefitinib and erlotinib, the FLT3 TKI quizartinib, and the KIT TKI imatinib markedly suppress the proliferation of RTK-addicted cancer cells by inhibiting tyrosine phosphorylation signaling [30–32]. Indeed, these TKIs significantly improved the prognosis of patients with advanced RTK-mutated NSCLCs, AMLs, and GISTs [33–36]. This is likely to be the case for other oncogenic tyrosine kinases (Table 16.1).

Several studies have shown that RTK localization is affected by permanently active mutations. RTK^{mut} seems to stop in the pathway on which normal RTK traffics [11, 12]. Inappropriate localizations vary from cancer type to cancer type

Table 16.1 Representative oncogenic tyrosine kinases and molecular targeted drugs

Oncogenic kinase mutant	Cancer	Targeted drug (TKI), Trade name
EGFR	NSCLC	Gefitinib (Iressa®)
	NSCLC	Erlotinib (Tarceva®)
	NSCLC	Afatinib (Giotrif®) for $EGFR^{T790M}$
	NSCLC	Osimertinib (Tagrisso®) for $EGFR^{T790M}$
KIT	GIST	Imatinib (Gleevec®)
	GIST	Sunitinib (Sutent®) for KIT^{Ex13}
	GIST	Regorafenib (Stivarga®) for KIT^{Ex17}
	GIST	Ripretinib (Qinlock®) for KIT^{Ex17}
	MCL	Midostaurin (Rydapt®) for KIT^{Ex17}
PDGFRA	GIST	Imatinib (Gleevec®)
	GIST	Avapritinib (Ayvakit®) for $PDGFRA^{Ex18}$
FLT3	AML	Midostaurin (Rydapt®)
	AML	Quizartinib (Vanflyta®)
	AML	Gilteritinib (Xospata®)
MET	NSCLC	Tepotinib (Tepmetko®)
	NSCLC	Capmatinib (Tabrecta®) for MET^{Ex14}
ALK	NSCLC	Crizotinib (Xalkori®)
	NSCLC	Alectinib (Alecensa®) for $EML4\text{-}ALK^{L1196M\ etc.}$
	NSCLC	Lorlatinib (Lorbrena®) for $EML4\text{-}ALK^{L1196M/G1202R\ etc.}$
RET	NSCLC	Pralsetinib (Gavreto®)
	Medullary thyroid cancer	Vandetanib (Caprelsa®)
ABL	CML	Imatinib (Gleevec®)
	CML	Ponatinib (Iclusig®) for $BCR\text{-}ABL^{T315I}$

NSCLC non-small cell lung cancer, *GIST* gastrointestinal stromal tumor, *MCL* mast cell leukemia, *AML* acute myelogenous leukemia, *CML* chronic myelogenous leukemia, *Ex* exon

and are classified into approximately three patterns: ER, Golgi/TGN, and endosome-lysosome. Although the characteristic locations of RTK^{mut} evoke important mechanisms for oncogenic signaling, the precise significance of localization in signaling and the causes for abnormal localization remain unknown.

KIT and PDGFRA in GIST Development

KIT Mutation as a Major Cause of GISTs

GISTs are the most common mesenchymal tumors of the gastrointestinal tract and account for approximately 20% of all sarcomas [37]. In 1998, gain-of-function mutations in the *KIT* gene were found in both solitary and familial GISTs [38, 39]. The majority of mutations are observed in *KIT* exon 11, which encodes the cytoplasmic juxtamembrane (JM) region and plays a critical role in the formation of an inhibitory conformation via intramolecular interactions [40–42]. Furthermore, progenitors of pacemaker cells, which control peristalsis in the gastrointestinal tract (called interstitial cells of Cajal) [38, 43, 44], were suggested to be the cellular origin of GISTs. *PDGFRA* mutations are frequently observed in GIST without *KIT* mutations [45, 46]. Currently, 80% and 10% of GISTs are causally associated with mutations in *KIT* and *PDGFRA*, respectively. In the case of *PDGFRA*-mutated GISTs, *TKD* mutations were more frequent than JM mutations. These two structurally related RTKs are classified into the type III RTK subfamily, which includes FLT3 and colony-stimulating factor 1 receptor (CSF-1R). The remaining 10% of patients with GISTs have no *RTK* mutation and develop tumors in an RTK-independent manner [47–49]. The driver mutations in GISTs are summarized in Fig. 16.4 and Table 16.2.

GISTs have been known to be highly resistant to radiotherapy and conventional chemotherapies, such as DNA-damaging agents. Therefore, no effective treatment methods occurred for recurrent cases or inoperable patients until the development of TKIs. Imatinib was originally developed as a molecular targeted drug for the treatment of chronic myelogenous leukemia (CML). It inhibits Abelson tyrosine kinase fused with breakpoint cluster region protein (BCR-ABL oncoprotein) by tightly binding to the ATP-binding pocket in the ABL TKD [50, 51]. Then, *in vitro* experiments have demonstrated that the targeted drug also blocks the enzymatic activation of both KIT and PDGFRA/B [52, 53]. Indeed, imatinib markedly and significantly extended the overall survival of patients with advanced GISTs.

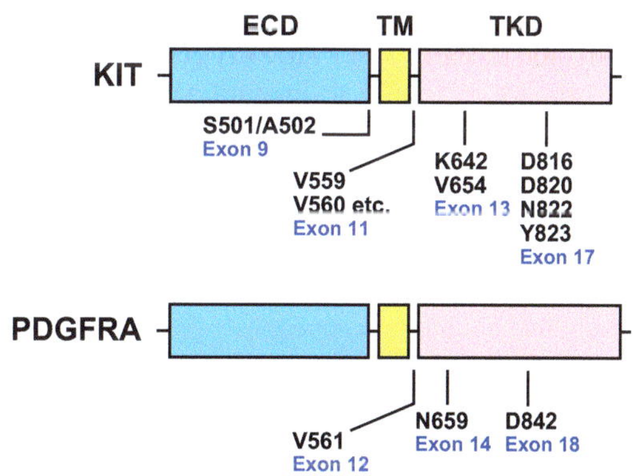

Fig. 16.4 Schematic structures and constitutively active mutations of KIT and PDGFRA. Schematic representation of KIT and PDGFRA showing the extracellular domain (ECD, blue), transmembrane domain (TM, yellow), and tyrosine kinase domain (TKD, pink). The positions of amino acids that could be substituted for other amino acids conferring constitutive activation of KIT or PDGFRA in GISTs are presented

Table 16.2 Driver mutations in GIST

Driver mutation	Estimated frequency (%)	Imatinib sensitivity
KIT Mutation		
Exon 9 or exon 8 (ECD)	5–10	Less sensitive
Exon 11 (JM)	60–70	Sensitive
Exon 13 (TKD)	<1	Resistant (sensitive in some patients)
Exon 17 (TKD)	1	Resistant
PDGFRA Mutation		
Exon 12 (JM)	<1	Sensitive
Exon 14 (TKD)	<1	Probably sensitive
Exon 18 (TKD)	10–15	Resistant
Wild-type KIT and *PDGFRA*		
NF1 inactive mutation	1–2	Resistant
BRAF mutation	<1	Resistant
SDHA, SDHB, SDHC, or *SDHB* inactive mutation	~3	Resistant

ECD extracellular domain, *JM* juxtamembrane, *TKD* tyrosine kinase domain, *NF1* neurofibromin 1, *SDH* succinate dehydrogenase complex flavoprotein subunit

To date, it is known that two–three years after the administration of imatinib, the targeted drug becomes ineffective for advanced GISTs, which is called acquired resistance [54, 55]. In the majority of cases, an imatinib-resistant mutation in the *KIT* gene appears [56]. Secondary mutations in exon 13 or exon 17, encoding *KIT TKD* conferred imatinib resistance to the primary mutants (Table 16.2). Most secondary mutations were found in the kinase domain of the same allele as the primary mutation (*cis*-position) [57]. So far, second to fourth line TKIs (sunitinib, regorafenib, and ripretinib) have been developed for the treatment of imatinib-resistant GISTs [47, 58, 59]. A few patients with GISTs originally had a TKD mutation in the *KIT* or *PDGFRA* gene (e.g., KIT^{D816V} and $PDGFRA^{D842V}$), resulting in primary resistance to imatinib. With the use of TKIs, the prognosis of patients with advanced GIST has improved significantly. Although the major causes of GISTs are well-characterized, how and where these RTK^{mut} initiate uncontrolled growth signaling remain unknown. We treated GISTs without fully understanding the characteristic features of RTK^{mut}.

RTK^{mut} Initiates the Growth Signaling on the Golgi/TGN in GIST Cells

Aberrant Subcellular Localization of KIT and PDGFRA

Before elucidating the precise subcellular localization of KIT^{mut} in GIST cells, several groups showed that mutant localization was clearly different from that of wild-type KIT (KIT-wt) in multiple cell lines other than GIST cells. In mast cell leukemia (MCL) [60], endogenous KIT^{mut} is mainly localized in endosomal–lysosomal compartments, where it uses as a signal platform for AKT activation (Fig. 16.5a), unlike wild-type counterpart that is found at the PM in normal mast cells [61–63]. Moreover, it activates STAT5 in the ER soon after synthesis. The endosomal localization of KIT^{mut} is also seen in AML cells [64]. In MCL and AML cells, mature KIT^{mut} normally moves to the PM along the secretory pathway, immediately undergoes endocytosis in a manner dependent on its kinase activity, and then accumulates on the endosome–lysosome. Other groups have shown that in KIT^{mut}-transfected NIH3T3, A375, or CHO cells, the mutant accumulates in an immaturely glycosylated form in the perinuclear region, probably in the ER or *cis*-Golgi [65–67]. On the basis of these reports, in identical cell lines, differences in the mutation site in the *KIT* gene (in other words, imatinib sensitivity) are unlikely to affect mutant localization, suggesting that the cellular environment determines the localization of KIT^{mut}. On immunohistochemistry staining, both KIT^{mut} and PDGFRA^{mut} were found in the paranuclear area in paraffin-embedded GIST tissue specimens [68–72], however, it was unclear what organelle these RTK^{mut} are retained in in GISTs.

Matured KIT^{mut} Mislocalizes to the Golgi/TGN Area Not Only in GIST Cell Lines but Also in Clinical Samples

GIST cell lines have been established [73–76], and competent antibodies against KIT, phospho-KIT, and organelle markers for immunofluorescence assays are currently available. Immunofluorescence assays on GIST cells performed by several groups showed that endogenous KIT^{mut} was found in the perinuclear area together with Golgi markers but not with markers for the ER, endosomes, or lysosomes (Fig. 16.5b, c) [77, 78]. Importantly, phospho-KIT Y703, Y721, Y730, and Y936 were predominantly located in the Golgi/TGN (Fig. 16.5c, lower panels) [79]. Immunoblot analyses suggested that most KIT proteins were autophosphorylated in a mature glycosylated form. In three-dimensional super-resolution confocal microscopic analysis [80, 81], phospho-KIT was significantly distributed in the TGN rather than on the *cis–medial* side of the Golgi apparatus [82], consistent with the observation that the majority of KIT is a mature glycosylated form in clinical samples as well as in GIST cell lines [66, 77, 83]. In paraffin-embedded GIST tissue specimens, KIT^{mut} accumulates in the perinuclear region together with a Golgi marker, whereas in specimens without KIT mutations, KIT-wt is predominantly observed in the PM [78]. Note that the localization of KIT^{mut} in the Golgi/TGN was observed, irrespective of the sensitivity of the mutant to imatinib. In addition, PDGFRA^{mut} is also found in the Golgi area of clinical samples. Thus, the localizations of KIT^{mut} and PDGFRA^{mut} were clearly different from those of their wild-type

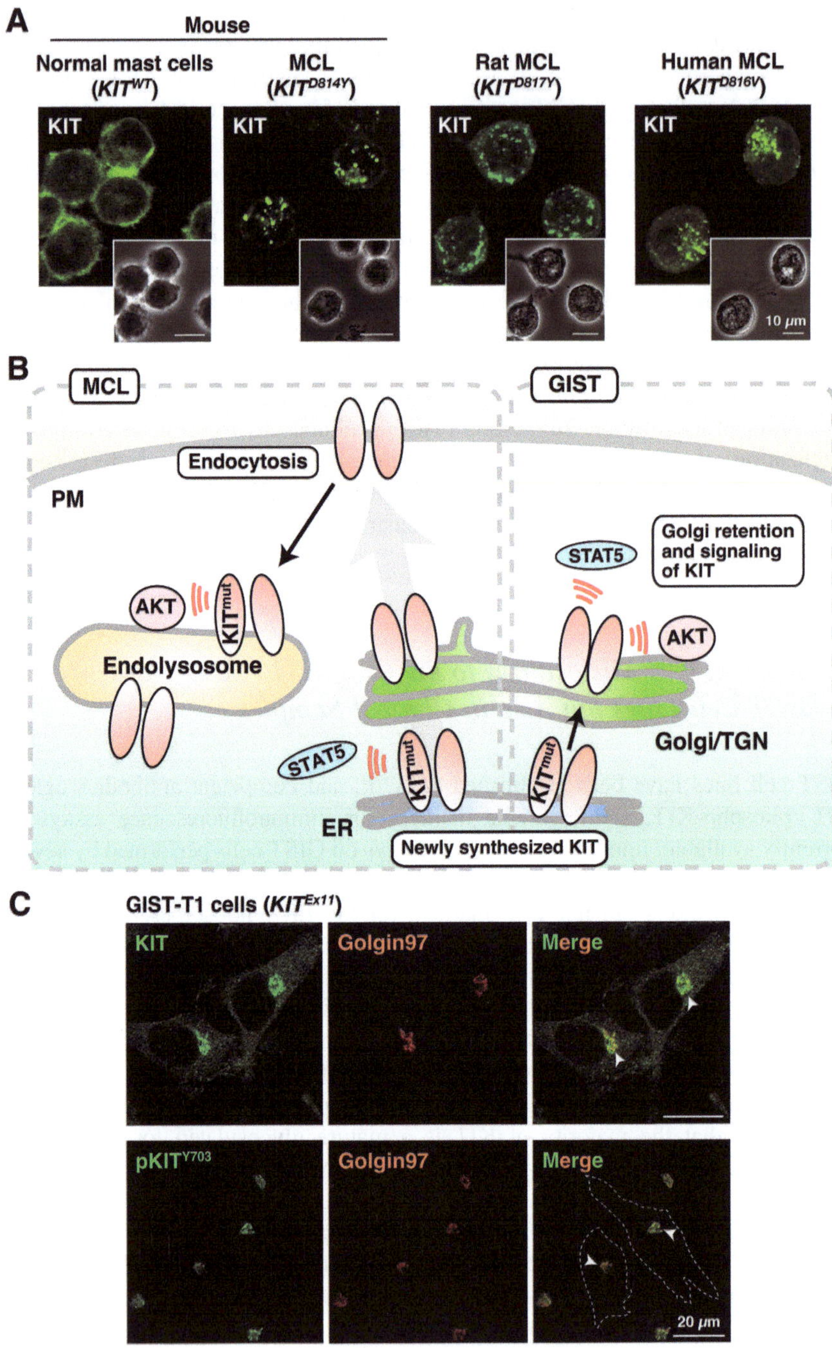

Fig. 16.5 Intracellular trafficking of KIT^{mut} in MCL and GIST. (**a**) Mouse, rat, and human mast cell lines are immunostained with anti-KIT antibody. Phase contrast images are shown. Bars, 10 μm. In normal mast cells, most wild-type KIT (KIT^{WT}) is at the PM, whereas in mast cell

counterparts. In summary, KIT^{mut} and probably $PDGFRA^{mut}$ are retained and activated in the Golgi/TGN region in mature glycosylated forms in GIST.

Figure 16.5b illustrates Golgi/TGN retention of KIT^{mut} during biosynthetic secretory trafficking in GIST cells. Brefeldin A (BFA), an inhibitor of protein transport from the ER to the Golgi apparatus [84–86], abolishes Golgi-localization of KIT^{mut}, whereas endocytosis inhibition does not affect the Golgi localization of KIT^{mut}. These data indicated that KIT^{mut} stops in the Golgi/TGN area during biosynthetic secretory trafficking. Interestingly, it is released from the Golgi/TGN region toward the PM under kinase activity inhibition conditions with TKIs, such as imatinib, midostaurin, and sunitinib [64, 87]. In HEK293 cells transfected with $PDGFRA^{mut}$, Golgi-localized $PDGFRA^{mut}$ disappears in the presence of TKI, whereas the PM levels of the mutant markedly increased [88], similar to the data obtained from KIT^{mut} in GIST cells. TKIs have two effects on RTK^{mut} localization by inhibiting mutant kinase activity: they move RTK^{mut} that is retained in the secretory compartment toward the PM and block kinase activity-dependent endocytosis, resulting in the trafficking of RTK^{mut} like that of wild-type RTK. Collectively, after biosynthesis in the ER, these RTK^{mut} are retained in the Golgi/TGN area in a manner dependent on its tyrosine kinase activity in GIST cells.

Oncogenic RTK Signaling Occurs on the Golgi/TGN Region in GIST Cells

What is the biological and pathogenic significance of mislocalized RTK^{mut} activation in oncogenic signaling in GIST cells? Previous studies have suggested that the Golgi/TGN contains various components for signal transduction, such as substrates for phosphorylation, protein kinases, and lipid kinases, including Src tyrosine kinases and phosphatidylinositol 4-kinase (PI4K) [89–92]. Together with the fact

Fig. 16.5 (continued) leukemia (MCL) cells, KIT^{mut} is mainly on endolysosomes. (**b**) (Left, MCL) Soon after synthesis, immature KIT^{mut} in the endoplasmic reticulum (ER) activates STAT5. It then migrates to the PM along the secretory pathway. After KIT^{mut} reaches the PM, it autonomously moves to endolysosomes through endocytosis in a kinase activity-dependent manner. KIT activates AKT specifically on endosome-lysosome compartments. (Right, GIST) Newly synthesized KIT^{mut} migrates normally from the ER to the Golgi apparatus, then undergoes complex glycosylation as normal. After reacting the Golgi/*trans*-Golgi network (TGN), KIT^{mut} is able to activate downstream molecules, such as AKT and STAT5. KIT^{mut} activation in the wrong subcellular compartment prevents its export from Golgi/TGN toward the plasma membrane (PM). In other words, KIT^{mut} stops in the Golgi/TGN area in a manner that is dependent on its tyrosine kinase activity. (Modified from Figure 7 in *Oncogene*, 2017, 36, 3661–3672, CC-BY 4.0). (**c**) GIST-T1 cells are immunostained for KIT (green), phospho-KIT Y703 ($pKIT^{Y703}$, green), and golgin97 (TGN marker, red). Both KIT and $pKIT^{Y703}$ are found in the perinuclear region together with golgin97. Similar results were obtained from other GIST cell lines. Note that localization of KIT^{mut} in the Golgi/TGN region is found irrespective of sensitivity of the mutant to imatinib. Scale bars, 20 μm. Arrowheads indicate the Golgi/TGN area. Dashed lines indicate cell borders

that KIT^{mut} is mainly activated in the Golgi/TGN in GISTs, it is possible that this organelle serves as a signaling platform for RTK^{mut}. Intriguingly, blocking ER export with BFA or 2-methycoprophilinamide (M-COPA) through ADP-ribosylation factor (ARF) inhibition [93–95] markedly suppressed the activation of KIT^{mut}, resulting in the inactivation of ERK, AKT, and STAT5 [79, 96]. In contrast, monensin and 30 N12, inhibitors of Golgi export [96, 97], had no effect on KIT activation or signaling. Therefore, KIT^{mut} initiated oncogenic signaling in the Golgi/TGN in GISTs (Fig. 16.5b). The PTP inhibitor, sodium orthovanadate [98], restores tyrosine phosphorylation of ER-retained KIT^{mut} [79], indicating that KIT in the ER is maintained in its dephosphorylated form by PTPs [99]. In other words, KIT^{mut} may avoid association with these negative regulators in the Golgi/TGN. Is the normal maturation of KIT^{mut} glycosylation required for mutant signaling? Swainsonine inhibits α-mannosidase II at the *medial*-Golgi, altering protein glycosylation [100]. This compound did not affect KIT localization or growth signaling, indicating that normal complex glycosylation is not essential for KIT^{mut} signaling in the Golgi/TGN. Moreover, cells transfected with PDGFRA^{mut} did not activate ERK or AKT in the presence of BFA [88, 101]. Taken together, these results suggest that oncogenic RTK signaling occurs in the Golgi/TGN of GIST cells.

KIT^{mut} has also been suggested to participate in various signal transduction cascades as an activator. It activates the RHOA–RHO kinase pathway [102] and the VAV1–RAC1 pathway [103] and tyrosine-phosphorylates Src-like adaptor protein (SLAP) and XK-related protein 5 (XKR5) [104, 105]. There is great interest in further investigating whether these are activated in the Golgi/TGN of GISTs.

Mechanism of Golgi Retention of KIT^{mut}

What factors cause the localization and/or retention of KIT^{mut} in the Golgi/TGN in GIST cells? Recently, multiple retention mechanisms have been reported, both simultaneously and independently. In this section, we review the mechanisms of Golgi retention of KIT^{mut} in GIST cells, which are shown in our paper and another group's reports.

Golgi Retention of KIT^{mut} Through a PKD2-Dependent Manner

Because KIT^{mut} stops in the Golgi/TGN area during biosynthetic secretory trafficking in a manner dependent on its tyrosine kinase activity, it was hypothesized that its downstream molecules play a role in abnormal retention. However, AKT, ERK, or STAT5 do not play a role in KIT^{mut} retention. We performed a mini-screening in GIST cells using compounds previously collected in our laboratory. Among these chemicals, only CRT0066101 markedly moved KIT^{mut} from the Golgi/TGN region toward the PM [82]. CRT0066101 was developed as a PKD inhibitor of protein kinase D (PKD) [106]. All mammalian PKDs consist of structurally related three proteins: PKD1 (PKCμ), PKD2, and PKD3 (PKCν), which play a role in the formation of transport carriers at the TGN through their serine/threonine kinase activity [107–112]. Recent studies provide evidence that PKD activity is intimately involved in the development of tumors [113–116]. We speculated that excess PKD activity, presumably dependent on KIT^{mut}, disturbs the normal trafficking of KIT from the Golgi/TGN region to the PM. Notably, PKD2, but not PKD1 or PKD3, participated in KIT^{mut} retention because siRNA-mediated knockdown of PKD2 selectively phenocopied the effect of CRT0066101 treatment. The reason why PKD1 and PKD3 are not involved in KIT retention remains unclear. It is important to note that PKD2 was significantly inactivated by KIT inhibitor treatment, indicating that it is downstream of KIT^{mut}. In support of this hypothesis, KIT^{mut} interacted with PKD2 in a manner dependent on its tyrosine kinase activity. Phospho-PKD2 S876, a sign of activation [117], is mainly found in the Golgi/TGN. The mediator between KIT^{mut} and PKD2 is considered to be PLCγ2. KIT^{mut} recruits and activates PLCγ2, resulting in the production of diacylglycerol (DAG) [4, 118], which has a critical role in PKD activation. Therefore, KIT is retained in the Golgi/TGN area in a PLCγ2–PKD2–dependent manner in GIST cells (Fig. 16.6).

In our view emerged from literature surveys [119–124], it was estimated that the PLCγ2–PKD2–PI4KIIIβ pathway and its effectors, i.e. Golgi-localized, γ-ear-containing, ARF-binding family of protein 1 (GGA1) and an adaptor protein-1 (AP1) component γ-adaptin [125, 126], are inappropriately recruited, leading to KIT retention in the Golgi/TGN (Fig. 16.6, left). Indeed, the loss of function of these factors induces the release of KIT^{mut} from the Golgi/TGN region. PKD2 activates PI4KIIIβ, then phosphatidylinositol 4-phosphate (PI4P) is generated. Interestingly, imatinib and CRT0066101 significantly inhibited PI4P production in the Golgi/TGN. These suggest that the abnormal and/or excess production of PI4P that is generated by the PLCγ2–PKD2–PI4KIIIβ cascade arrests KIT^{mut} in the Golgi/TGN area through recruiting GGA1 and AP1. GGA1 interacts with AP1 in a PKD2 activity-dependent manner. Both imatinib-sensitive and -resistant KIT^{mut} are retained in a manner dependent on the activation of the cascade.

Inappropriate activation/recruitment may perturb the trafficking of KIT^{mut} from the Golgi/TGN region toward the PM. Importantly, releasing KIT^{mut} from the Golgi/TGN through pharmacological inhibition of PKD2 or PI4KIIIβ suppresses KIT signaling. After release from the Golgi/TGN, KIT^{mut} moves to lysosomes via the PM and rapidly undergoes degradation (Fig. 16.6, right). These findings indicate that the release of RTKs from their signaling platforms could be a strategy for inhibiting oncogenic tyrosine phosphorylation. In preclinical investigations in tumor-bearing mice, CRT0066101 and a PI4KIIIβ inhibitor have been confirmed that they could be orally administrated and show an anti-cancer effect [127–131]; thus, investigating these compounds in animal models for GISTs will drive progress in the development of enhanced cancer treatments. KIT-wt and KIT^{mut} trafficking in leukemia cells appeared to be unaffected by PKD2 loss-of-function. Although it is later shown that $FLT3^{mut}$ is retained in the Golgi/TGN area [132], its retention is PKD2-independent.

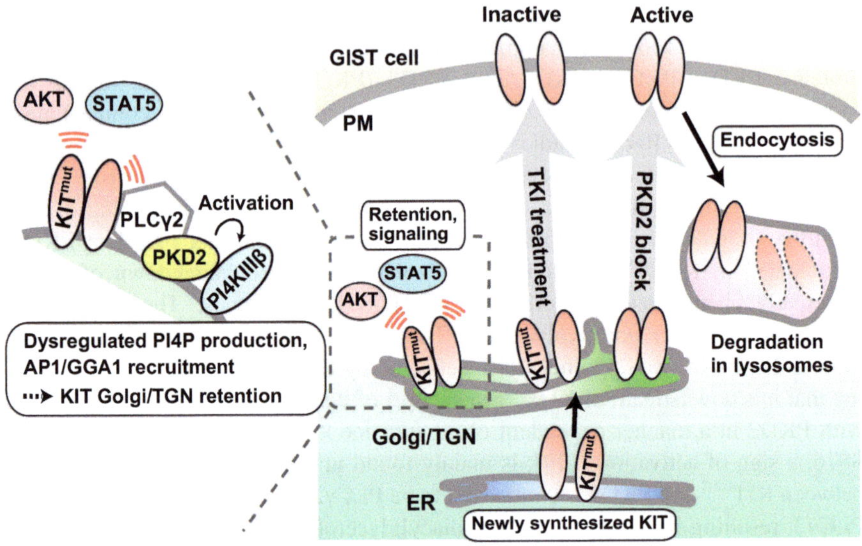

Fig. 16.6 Molecular mechanism of Golgi retention of KIT^{mut} in GIST cells. In GIST, KIT^{mut} is retained in the Golgi/*trans*-Golgi network (TGN) during its biosynthetic secretory transport. It activates the PLCγ2-PKD2-PI4KIIIβ cascade on the Golgi/TGN to generate phosphatidylinositol 4-phosphate (PI4P). Subsequently, PI4P recruits adaptor protein-1 (AP1) and Golgi-localized, γ-ear-containing, ARF-binding family of protein 1 (GGA1). AP1 and GGA1 adaptors and the complex trap KIT^{mut} in the Golgi/TGN region (left side). Suppression of this pathway with a PKD inhibitor, PKD2 knockdown, or tyrosine kinase inhibitors (TKIs) releases KIT^{mut} from the Golgi/TGN toward the plasma membrane (PM). TKI blocks tyrosine kinase activity-dependent endocytosis, resulting in the retention of KIT^{mut} in the PM (central gray arrow). KIT activity was retained under PKD2-inhibited conditions, and the mutant underwent endocytosis and was rapidly degraded in the lysosomes (right gray arrow). PLCγ2, phospholipase γ-2; PKD, protein kinase D; PI4KIIIβ, phosphatidylinositol 4-kinase IIIβ; ER, endoplasmic reticulum. (Modified from Figure 7 in *Cell Reports*, 2023, 42, 113035, CC-BY)

Therefore, PKD2 may selectively play a role in Golgi/TGN retention of KIT^{mut} in GIST cells.

Involvement of PKCθ and BLZF1 in the Golgi Localization of KIT^{mut} in GIST Cells

In addition to the above mechanisms, other molecules that may play a role in KIT^{mut} retention have been reported. Kim and colleagues showed that GISTs express protein kinase C-θ (PKCθ) to a greater extent than other KIT-positive cell lines from MCL, AML, and colon cancers [77], and several groups reported that expression levels of PKCθ are correlated with those of KIT in human GIST tissues from clinical samples as well as GIST cell lines [133–135]. KIT^{mut} requires the kinase

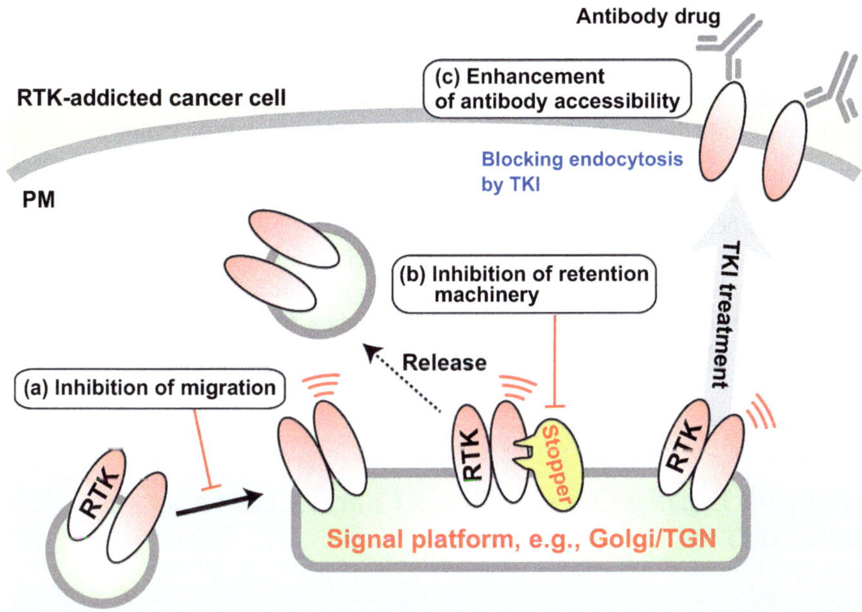

Fig. 16.7 Implications for cancer therapy devised from comprehension of the mechanism of RTK localization. In RTK-addicted cancer cells, the signalling platform for each RTK^{mut} appears to be confined to a specific compartment. Therefore, the blockade of RTK migration to the signal platform suppresses tyrosine phosphorylation signaling (**a**, i.e., M-COPA, a blocker of ER export toward the Golgi apparatus, suppresses FLT3^{mut} signaling in leukemia cells [132]). (**b**) If the cause of RTK retention is uncovered, it may be possible to establish a strategy for inhibiting growth signaling via the release of RTK from the platform (e.g., CRT0066101, a PKD inhibitor, releases KIT^{mut} from the Golgi/TGN of GIST cells, resulting in suppression of KIT signaling [82]). (**c**) As RTK^{mut} is retained in a manner dependent on its kinase activity, tyrosine kinase inhibitors (TKIs) move the mutant from the signal platform to the plasma membrane (PM). After the TKI bound RTK^{mut} reaches the PM, the targeted drug immobilizes the mutant in the PM by blocking kinase activity-dependent endocytosis, resulting in an increase in the PM levels of RTK^{mut}. Hence, the effects of RTK-targeting antibody drugs can be improved by combining them with TKIs [194, 195]

activity of PKCθ for its stability and localization in the Golgi region in both imatinib-sensitive and -resistant GIST cells. Furthermore, basic leucine zipper nuclear factor 1 (BLZF-1) may play a role in KIT^{mut} Golgi retention [96]. It is confirmed that both PKCθ and BLZF-1 associate with KIT^{mut}. Because the role of PKCθ, BLZF-1, and PKD2-related molecules in KIT^{mut} retention may be mutually inclusive, it is attractive to examine whether PKCθ and BLZF-1 are involved in the PLCγ2–PKD2–PI4KIIIβ cascade in the Golgi/TGN area. In addition, it is of substantial interest to determine whether these molecules are responsible for the Golgi retention of PDGFRA^{mut}.

Although the requirement of the PLCγ2–PKD2–PI4KIIIβ cascade and PKCθ/BLZF-1 for Golgi retention of KIT^{mut} is clear, the detailed mechanism of the mutant receptor trapped in the Golgi/TGN area is yet to be proven. Recent studies have shown that in the yeast *Saccharomyces cerevisiae*, vesicles budding from the TGN are recycled back to earlier Golgi cisternae in an AP1-dependent manner [136–138]. In this context, AP1 that is recruited by PLCγ2–PKD2–PI4KIIIβ activation might play a role in the intra-Golgi recycling of KIT^{mut}. By contrast, KIT^{mut} may be stably located in the Golgi/TGN membrane with the interactors including PKCθ and BLZF-1. Another possibility is that AP1–GGA1-dependent membrane carriers containing KIT^{mut} rapidly and repeatedly bounce in and out of the TGN membrane. Further studies are needed to understand how KIT^{mut} hijacks the physiological machinery. More advanced tools, including live imaging in conjunction with fluorescent protein tagging technology and synchronization methods of biosynthetic secretory trafficking, are essential for understanding local RTK dynamics in the Golgi/TGN, which will open new fields not only for cancer biology but also for basic cell biology.

Aberrant Localization of RTKs in Cancer Cells Other than GIST

In our view, emerging from literature surveys and our observations, the phenomenon that subcellular localization of the mutant is different from that of the wild-type can be seen in RTKs other than KIT and PDGFRA; however, the organelle where they localize varies from one cancer type to another. Based on previous studies, RTK mislocalization has been roughly classified into three types: ER, Golgi/TGN, and endosome/lysosome (Table 16.3).

Localization and Signaling in the Golgi/TGN Region

FLT3 FLT3 internal tandem duplication mutant (FLT3-ITD) that has an in-frame insertion into the JM domain is found in approximately 30% of AML patients [139–141]. FLT3-ITD expression is associated with poor prognosis of patients with AML. It is well-studied that FLT3-ITD is retained in intracellular compartments and initiates growth signaling during trafficking from the ER en route to the PM. Although the fluorescence images of the appearance of FLT3-ITD localization may be dependent on the procedures of staining, image acquisition, and whether endogenous level or not, the mutant is certainly retained in the ER and/or the Golgi/TGN region during biosynthetic trafficking [132, 142–146]. The PM levels of FLT3-ITD were significantly lower than those of FLT3-wt. FLT3-ITD can activate STAT5 at the ER [142, 147–149]. Golgi-retained FLT3-ITD initiates AKT and ERK activation [132]. Similar to GIST KIT^{mut}/$PDGFRA^{mut}$, FLT3-ITD is retained in secretory compartments in a manner dependent on its tyrosine kinase activity. Golgi retention of FLT3-ITD, however, is PKD2-independent [82], suggesting that

Table 16.3 Summary of subcellular localization of RTK in cancer cells

RTK	Cancer cells	Subcellular localization
KIT	GIST	Golgi/TGN
	MCL	Endosome–lysosome
	AML[a]	Endosome–lysosome
PDGFRA	GIST	Golgi/TGN
	HGG	Golgi/TGN
FLT3	AML	ER, Golgi/TGN
MET	HCC	Golgi/TGN
IGF-1R	Breast cancer	Golgi/TGN (after ligand stimulation)
FGFR3	MM	ER, Golgi/TGN
TRKA[b]	NB	ER–Golgi?
ALK	NB	ER, Golgi/TGN
RET	MEN 2B	ER
EGFR	NSCLC	Endosome

GIST gastrointestinal stromal tumor, *MCL* mast cell leukemia, *AML* acute myelogenous leukemia, *HGG* high-grade glioma, *HCC* hepatocellular carcinoma, *MM* multiple myeloma, *NB* neuroblastoma, *MEN 2B* multiple endocrine neoplasia type 2B, *NSCLC* non-small cell lung cancer, *ER* endoplasmic reticulum, *TGN trans*-Golgi network

Basically, wild-type RTKs localize to the plasma membrane in normal cells

[a]KIT^{mut} is mainly found in endosome–lysosome compartments, but it activates downstream molecules on the Golgi/TGN

[b]Farina et al. [196, 197]

the intracellular retention of individual RTKs is regulated by different molecular mechanisms. It is of particular interest to determine the mechanism underlying FLT3-ITD retention in AML. In addition, a mutation at D835 in the FLT3 TKD that confers constitutive activity to FLT3 was found in approximately 10% of patients with AML [141, 150]. It would be interesting to examine whether the FLT3 TKD mutant also causes growth signaling during biosynthetic secretory trafficking.

MET Ligand-independent activation is frequently found in various cancers, such as hepatocellular carcinoma (HCC), NSCLC, lung squamous adenocarcinoma, and gastric cancer [151, 152]. Frazier et al. showed that, in HCC cells, overexpressed MET is localized and autophosphorylated in the Golgi region. They also provided evidence that tyrosine kinase activity was required for MET localization to the Golgi region. In this compartment, it associates with HER3 RTK, resulting in the formation of a heterodimer and/or heterooligomer, and the complexes may induce growth signaling [153, 154]. Besides HCC, investigation of other *MET*-amplified cancers, such as lung/gastric cancers, is needed. Furthermore, the pathogenic role of constitutively active MET mutants and a MET exon 14 skipping mutant was recently reported [155, 156]. There is great interest in understanding the spatial organization of signaling.

IGF-1R IGF-1R is involved in the migration of several cancers [157]. Unlike KIT^{mut} in GISTs, IGF-1R localizes to the Golgi area in a retrograde trafficking-dependent manner [158, 159]. Upon IGF-1 stimulation at the PM, IGF-1R moves to the Golgi area, where its activation remains, inducing the phosphorylation of ERK and SH2 domain-containing (SHC) in breast cancer cells. Phosphorylation of IGF-1R, ERK, and SHC in the Golgi correlates with cell migration.

FGFR3 In multiple myeloma (MM) cells, $FGFR3^{mut}$ accumulates in the paranuclear region, presumably in the ER–Golgi region [160, 161]. However, the significance of the $FGFR3^{mut}$ accumulation in MM remains unclear. The possibility that it initiates growth signaling from secretory compartments is high [162, 163] because pan-phospho-tyrosine immunostaining demonstrates that tyrosine-phosphorylated proteins are abundantly found in the Golgi region in $FGFR3^{mut}$-transfected cells [164]. $FGFR3^{mut}$ is also found in approximately 80% of patients with achondroplasia [165]. As FGFR3 signaling is targeted for disease treatment [166], spatio-temporal analyses of $FGFR3^{mut}$ signaling are important to understand the pathological role of the mutant and the effect of targeted therapy.

PDGFRA Ip et al. found a Y288C substitution in PDGFRA ECD from glioma, NSCLC, colorectal cancer, and endometrial cancer, but not in GIST. This newly created cysteine residue contributes to PDGFRA dimerization through an intermolecular disulfide bridge, resulting in ligand-independent activation. Their immunofluorescence, fractionation, and immunoblot analyses have shown that $PDGFRA^{Y288C}$ is trapped in the ER–Golgi region, where it can initiate downstream activation [101]. Furthermore, Golgi localization of $PDGFRA^{D842V}$ in high-grade glioma (HGG) was reported [167].

Retention in the ER

ALK, LTK, and RET Permanently active *ALK* mutations are associated with poor prognosis in pediatric neuroblastomas (NB) [168–170]. The *ALK* gene lacking exons 2 and 3, which encodes a constitutively active form, is overexpressed in the pediatric neuroblastoma cell line NB-1 [171]. The deletion mutant was autophosphorylated and showed a unique localization. ALK^{mut} is predominantly seen in the ER region, unlike other RTKs described above. In addition, a deletion mutation was found in the ER region of cells transfected with ALK lacking the ECD domain [172]. In an NB cell line harboring ALK^{F1174L}, ALK was mainly distributed in the secretory compartments [173]. Leukocyte tyrosine kinase (LTK) is structurally associated with ALK. LTK is mainly localized in the ER and plays a key role in regulating its secretion [174]. Similar to ALK and LTK, RET^{mut}, which is associated with the development of multiple endocrine neoplasia type 2B (MEN 2B), is activated in the ER [175, 176]. Currently, it is unknown whether they play an oncogenic role in the ER.

EGFR and KIT in Endosomal Compartments

Under normal conditions, EGFR-wt is localized in the PM. After ligand stimulation, it undergoes endocytosis and is distributed in endosomes [22, 177]. On the other hand, once $EGFR^{mut}$ reaches the PM in cancer cells, it may spontaneously traffic to endosomes, thus the mutant is found in the compartments in the absence of EGF [178–180]. There is a high possibility that $EGFR^{mut}$ can induce growth signaling in endosomes because EGF-bound EGFR-wt activates downstream molecules. Simultaneous inhibition of ARF1, ARF4, and ARF5 with BFA-, M-COPA-, or siRNA-mediated knockdown in NSCLC cells blocked protein export from the ER, resulting in $EGFR^{mut}$ inactivation [94]. $EGFR^{mut}$ is not activated in ER. Similar to EGFR, KIT^{mut} in MCL autonomously accumulates in the endosome–lysosome compartments in a hypo-ubiquitinated state [61, 63]. It is of substantial interest to note that the trafficking of KIT^{mut} is clearly different between MCL and GIST as follows: KIT^{mut} in MCL activates STAT5 in the ER soon after synthesis, normally traffics to the PM, undergoes endocytosis, and accumulates in endosome–lysosome compartments, where it specifically activates AKT (Fig. 16.5b, left). In MCL, KIT^{mut} is unable to activate PKD2 via PLCγ2, resulting in the normal export from the Golgi/TGN. Further studies are needed to elucidate the cause of the difference in the trafficking of KIT^{mut} between MCL and GIST.

Implications of aberrant localization of RTK^{mut} in cancer cells are summarized as below.

(i) An organelle-based platform for tyrosine phosphorylation signaling is required for RTK^{mut}-addicted cancer cell proliferation.
(ii) RTK^{mut} is retained in specific organelles in a manner dependent on its kinase activity during the intracellular trafficking pathway in which normal RTK traffic occurs.
(iii) On the aberrant signaling platform, RTK^{mut} seems to escape negative regulators that induce dephosphorylation or degradation. Therefore, blockade of RTK^{mut} migration into the signal platform suppresses growth signaling.
(iv) Retention mechanisms exist for individual RTK^{mut} in each type of cancer. RTK^{mut} may hijack the physiological machinery for retention in an organelle.

It is estimated that constitutively active RTK^{mut} signaling occurs when the above four conditions are fulfilled.

Perspectives and Implications for Cancer Therapy

Because the signal platform for each RTK^{mut} seems to require a specific compartment to produce proliferative signaling, understanding this phenomenon may contribute to the establishment of a new strategy for the inhibition of tyrosine phosphorylation signaling. It should be emphasized that these could suppress growth signals not only by TKI-sensitive RTK^{mut} but also by TKI-resistant RTK^{mut}. Moreover, these studies provided a proof-of-concept (POC) or proof-of-mechanism (POM) for the development of targeted drugs (Fig. 16.7).

Control of RTK^{mut} Localization to the Signal Platform The blockade of RTK^{mut} migration to its specific signaling platform seems attractive (Fig. 16.7a) [181–183]. A previous study showed that the HMG-CoA reductase inhibitor fluvastatin, which is a known medicine for hypercholesterolemia, blocks FLT3-ITD trafficking from the ER by suppressing N-linked glycosylation, resulting in the inactivation of AKT and ERK [146]. An *in vivo* study demonstrated that fluvastatin markedly decreased the number of FLT3-ITD AML cells in mice, leading to a significant extension in the survival period. As described above, M-COPA that could be used for mice suppresses growth signaling of GIST KIT^{mut} and AML FLT3-ITD via blocking their migration to the Golgi apparatus [79, 132, 184]. Conversely, the release of RTK^{mut} may be a good strategy for blocking tyrosine phosphorylation signaling (Fig. 16.7b). PKD2 inhibition by CRT0066101 releases KIT^{mut} from the Golgi/TGN, resulting in the suppression of mutant signaling in GISTs [82]. Therefore, understanding the retention mechanisms of RTK^{mut} will help to develop new strategies for growth signaling.

Destabilization of KIT^{mut} Before and/or After Reaching the Golgi/TGN Inhibitors of heat shock protein 90 kDa (HSP90), such as 17-AAG and pimitespib (TAS-116), are known to decrease the protein levels of RTKs, which are client proteins of HSP90 [185–189]. HSP90 stabilizes target proteins by serving as a molecular chaperone; thus, RTKs are recruited to proteasomes for degradation in the presence of HSP90 inhibitors. In HSP90 inhibitor-treated GIST cells, autophosphorylated KIT^{mut} in the Golgi/TGN is diminished, resulting in the suppression of KIT signaling, suggesting that these inhibitors destabilize KIT^{mut} before and/or after the mutant reaches the Golgi/TGN [87, 190]. In clinical trials and animal experiments, pimitespib significantly suppressed GIST growth; therefore, it has been approved for treating patients with TKI-resistant metastatic GIST [191]. Several studies have reported that EGFR, FLT3, MET, and PDGFRA are also clients of HSP90. Thus, pimitespib could be effective for the treatment of RTK-addicted cancers by decreasing RTK levels in the intracellular compartments. Research on RTK localization would contribute to providing POC/POM for developing novel molecular targeted drugs.

Understanding the Effect of Combination Therapy of TKI With Antibody Anti-RTK antibodies have been developed as targeted drugs to block signal transduction and induce antibody-dependent cellular cytotoxicity (ADCC) [192]. It has been estimated that the therapeutic effects of antibody drugs correlate with the levels of cell-surface antigens. As mentioned above, TKI treatment restores the localization and trafficking of KIT^{mut} to that of normal KIT, indicating that PM localization of KIT^{mut} is enhanced by TKI treatment [61, 64, 65, 193]. It is likely to be the case also in other oncogenic RTKs, e.g., FLT3-ITD, EGFR^{mut}, and PDGFRA^{mut} [88, 132, 180]. Several groups have attempted to investigate whether

TKIs improve the effect of antibody drugs or T cells expressing chimeric antigen receptors (CAR) by enhancing PM levels of targeted RTK [194, 195]. For example, in preclinical studies in AML-bearing mice, ADCC using an anti-FLT3 antibody was enhanced by increasing the PM levels of FLT3-ITD through TKI treatment (Fig. 16.7c). Combining TKIs with immunotherapy may improve the prognosis of cancer patients.

Conclusions and Future Perspectives

Unlike KIT-wt in normal cells, KIT^{mut} is retained in the Golgi/TGN region during secretory trafficking in its mature glycosylated form in GIST. The tyrosine kinase activity is critical for cell retention. $PDGFRA^{mut}$ also stopped in a tyrosine kinase-dependent manner. Both imatinib-sensitive and imatinib-resistant KIT^{mut} cause growth signaling in the Golgi/TGN. The PLCγ2–PKD2–PI4KIIIβ pathway is activated by KIT^{mut}, resulting in Golgi retention of KIT^{mut}. PKCθ and BNZF-1 are also involved in the retention. They may have exerted this phenomenon independently or in a mutually inclusive manner. There is a high possibility that the Golgi retention of individual RTKs is regulated by different molecular mechanisms. Together with the fact that the subcellular localization of RTK mutants cannot be understood uniformly among cancer types, the mechanisms of intracellular RTK retention are far more complex than initially thought. If the causes of Golgi retention and trafficking machinery to the Golgi/TGN for each RTK^{mut} are uncovered, novel strategies for inhibiting RTK^{mut} might be established. To achieve this, a careful analysis of individual RTK and each type of cancer is required.

Acknowledgements This work was supported by a grant-in-aid for Scientific Research from the Japan Society for the Promotion of Science, Japan, a research grant from the Research Foundation for Pharmaceutical Sciences, Japan, the Japan Foundation for Applied Enzymology, Japan, the Life Science Foundation of JAPAN, and the Foundation for Promotion of Cancer Research, Japan, the Mochida Memorial Foundation for Medical and Pharmaceutical Research, Japan, the Sumitomo Foundation, Japan, and the Princess Takamatsu Cancer Research Fund, Japan.

Compliance with Ethical Standards The authors declare no conflict of interest. This work does not contain any studies with human or animal subjects performed by any of the authors.

References

1. Blume-Jensen P, Hunter T (2001) Oncogenic kinase signalling. Nature 411:355–365. https://doi.org/10.1038/35077225
2. Lemmon MA, Schlessinger J (2010) Cell signaling by receptor tyrosine kinases. Cell 141: 1117–1134. https://doi.org/10.1016/j.cell.2010.06.011
3. Ward CW, Lawrence MC, Streltsov VA et al (2007) The insulin and EGF receptor structures: new insights into ligand-induced receptor activation. Trends Biochem Sci 32:129–137. https://doi.org/10.1016/j.tibs.2007.01.001

4. Carpenter G, Qs J (1999) Phospholipase C-gamma as a signal-transducing element. Exp Cell Res 253:15–24. https://doi.org/10.1006/excr.1999.4671
5. Owusu Obeng E, Rusciano I, Marvi MV et al (2020) Phosphoinositide-dependent signaling in cancer: a focus on phospholipase C isozymes. Int J Mol Sci 21:2581. https://doi.org/10.3390/ijms21072581
6. Acconcia F, Sigismund S, Polo S (2009) Ubiquitin in trafficking: the network at work. Exp Cell Res 315:1610–1618. https://doi.org/10.1016/j.yexcr.2008.10.014
7. Hendriks WJ, Pulido R (2013) Protein tyrosine phosphatase variants in human hereditary disorders and disease susceptibilities. Biochim Biophys Acta 1832:1673–1696. https://doi.org/10.1016/j.bbadis.2013.05.022
8. Julien SG, Dubé N, Hardy S et al (2011) Inside the human cancer tyrosine phosphatome. Nat Rev Cancer 11:35–49. https://doi.org/10.1038/nrc2980
9. Kreitman M, Noronha A, Yarden Y (2018) Irreversible modifications of receptor tyrosine kinases. FEBS Lett 592:2199–2212. https://doi.org/10.1002/1873-3468.13095
10. Miaczynska M (2013) Effects of membrane trafficking on signaling by receptor tyrosine kinases. Cold Spring Harb Perspect Biol 5:a009035. https://doi.org/10.1101/cshperspect.a009035
11. Schmidt-Arras D, Böhmer FD (2020) Mislocalisation of activated receptor Tyrosine Kinases - challenges for cancer therapy. Trends Mol Med 26:833–847. https://doi.org/10.1016/j.molmed.2020.06.002
12. Toffalini F, Demoulin JB (2010) New insights into the mechanisms of hematopoietic cell transformation by activated receptor tyrosine kinases. Blood 116:2429–2437. https://doi.org/10.1182/blood-2010-04-279752
13. Helenius A, Aebi M (2004) Roles of N-linked glycans in the endoplasmic reticulum. Annu Rev Biochem 73:1019–1049. https://doi.org/10.1146/annurev.biochem.73.011303.073752
14. Stanley P (2011) Golgi glycosylation. Cold Spring Harb Perspect Biol 3:a005199. https://doi.org/10.1101/cshperspect.a005199
15. Slieker LJ, Martensen TM, Lane MD (1986) Synthesis of epidermal growth factor receptor in human A431 cells. Glycosylation-dependent acquisition of ligand binding activity occurs post-translationally in the endoplasmic reticulum. J Biol Chem 261:15233–15241. https://doi.org/10.1016/S0021-9258(18)66858-9
16. Khatib AM, Siegfried G, Prat A et al (2001) Inhibition of proprotein convertases is associated with loss of growth and tumorigenicity of HT-29 human colon carcinoma cells: importance of insulin-like growth factor-1 (IGF-1) receptor processing in IGF-1-mediated functions. J Biol Chem 276:30686–30693. https://doi.org/10.1074/jbc.M101725200
17. Komada M, Hatsuzawa K, Shibamoto S et al (1993) Proteolytic processing of the hepatocyte growth factor/scatter factor receptor by furin. FEBS Lett 328:25–29. https://doi.org/10.1016/0014-5793(93)80958-w
18. Östman A, Hellberg C, Böhmer FD (2006) Protein-tyrosine phosphatases and cancer. Nat Rev Cancer 6:307–320. https://doi.org/10.1038/nrc1837
19. Goh LK, Sorkin A (2013) Endocytosis of receptor tyrosine kinases. Cold Spring Harb Perspect Biol 5:a017459. https://doi.org/10.1101/cshperspect.a017459
20. Joffre C, Barrow R, Ménard L et al (2011) A direct role for Met endocytosis in tumorigenesis. Nat Cell Biol 13:827–837. https://doi.org/10.1038/ncb2257
21. Ménard L, Parker PJ, Kermorgant S (2014) Receptor tyrosine kinase c-Met controls the cytoskeleton from different endosomes via different pathways. Nat Commun 5:3907. https://doi.org/10.1038/ncomms4907
22. Sigismund S, Argenzio E, Tosoni D et al (2008) Clathrin-mediated internalization is essential for sustained EGFR signaling but dispensable for degradation. Dev Cell 15:209–219. https://doi.org/10.1016/j.devcel.2008.06.012
23. Du Z, Lovly CM (2018) Mechanisms of receptor tyrosine kinase activation in cancer. Mol Cancer 17:58. https://doi.org/10.1186/s12943-018-0782-4

24. Nelson KN, Peiris MN, Meyer AN et al (2017) Receptor Tyrosine Kinases: translocation partners in hematopoietic disorders. Trends Mol Med 23:59–79. https://doi.org/10.1016/j.molmed.2016.11.002
25. Tulpule A, Guan J, Neel DS et al (2021) Kinase-mediated RAS signaling via membraneless cytoplasmic protein granules. Cell 184:2649–2664.e18. https://doi.org/10.1016/j.cell.2021.03.031
26. Daver N, Schlenk RF, Russell NH et al (2019) Targeting FLT3 mutations in AML: review of current knowledge and evidence. Leukemia 33:299–312. https://doi.org/10.1038/s41375-018-0357-9
27. Kitamura Y, Hirota S, Nishida T (2003) Gastrointestinal stromal tumors (GIST): a model for molecule-based diagnosis and treatment of solid tumors. Cancer Sci 94:315–320. https://doi.org/10.1111/j.1349-7006.2003.tb01439.x
28. Linardou H, Dahabreh IJ, Bafaloukos D et al (2009) Somatic EGFR mutations and efficacy of tyrosine kinase inhibitors in NSCLC. Nat Rev Clin Oncol 6:352–366. https://doi.org/10.1038/nrclinonc.2009.62
29. Saraon P, Pathmanathan S, Snider J et al (2021) Receptor tyrosine kinases and cancer: oncogenic mechanisms and therapeutic approaches. Oncogene 40:4079–4093. https://doi.org/10.1038/s41388-021-01841-2
30. Arao T, Fukumoto H, Takeda M et al (2004) Small in-frame deletion in the epidermal growth factor receptor as a target for ZD6474. Cancer Res 64:9101–9104. https://doi.org/10.1158/0008-5472.CAN-04-2360
31. Bauer S, Duensing A, Demetri GD et al (2007) KIT oncogenic signaling mechanisms in imatinib-resistant gastrointestinal stromal tumor: PI3-kinase/AKT is a crucial survival pathway. Oncogene 26:7560–7568. https://doi.org/10.1038/sj.onc.1210558
32. Pratz KW, Sato T, Murphy KM et al (2010) FLT3-mutant allelic burden and clinical status are predictive of response to FLT3 inhibitors in AML. Blood 115:1425–1432. https://doi.org/10.1182/blood-2009-09-242859
33. Chang A, Parikh P, Thongprasert S et al (2006) Gefitinib (IRESSA) in patients of Asian origin with refractory advanced non-small cell lung cancer: subset analysis from the ISEL study. J Thorac Oncol 1:847–855. https://doi.org/10.1016/S1556-0864(15)30415-9
34. De Giorgi U, Verweij J (2005) Imatinib and gastrointestinal stromal tumors: where do we go from here? Mol Cancer Ther 4:495–501. https://doi.org/10.1158/1535-7163.MCT-04-0302
35. Erba HP, Montesinos P, Kim HJ et al (2023) Quizartinib plus chemotherapy in newly diagnosed patients with FLT3-internal-tandem-duplication-positive acute myeloid leukaemia (QuANTUM-First): a randomised, double-blind, placebo-controlled, phase 3 trial. Lancet 401:1571–1583. https://doi.org/10.1016/S0140-6736(23)00464-6
36. Joensuu H, Roberts PJ, Sarlomo-Rikala M et al (2001) Effect of the tyrosine kinase inhibitor STI571 in a patient with a metastatic gastrointestinal stromal tumor. N Engl J Med 344:1052–1056. https://doi.org/10.1056/NEJM200104053441404
37. Dematteo RP, Heinrich MC, El Rifai WM et al (2002) Clinical management of gastrointestinal stromal tumors: before and after STI-571. Hum Pathol 33:466–477. https://doi.org/10.1053/hupa.2002.124122
38. Hirota S, Isozaki K, Moriyama Y et al (1998) Gain-of-function mutations of c-kit in human gastrointestinal stromal tumors. Science 279:577–580. https://doi.org/10.1126/science.279.5350.577
39. Nishida T, Hirota S, Taniguchi M et al (1998) Familial gastrointestinal stromal tumours with germline mutation of the KIT gene. Nat Genet 19:323–324. https://doi.org/10.1038/1209
40. Chan PM, Ilangumaran S, La Rose J et al (2003) Autoinhibition of the kit receptor tyrosine kinase by the cytosolic juxtamembrane region. Mol Cell Biol 23:3067–3078. https://doi.org/10.1128/MCB.23.9.3067-3078.2003
41. Lennartsson J, Rönnstrand L (2012) Stem cell factor receptor/c-Kit: from basic science to clinical implications. Physiol Rev 92:1619–1649. https://doi.org/10.1152/physrev.00046.2011

42. Roskoski R Jr (2005) Structure and regulation of Kit protein-tyrosine kinase--the stem cell factor receptor. Biochem Biophys Res Commun 338:1307–1315. https://doi.org/10.1016/j.bbrc.2005.09.150
43. Maeda H, Yamagata A, Nishikawa S et al (1992) Requirement of c-kit for development of intestinal pacemaker system. Development 116:369–375. https://doi.org/10.1242/dev.116.2.369
44. Thomsen L, Robinson TL, Lee JC et al (1998) Interstitial cells of Cajal generate a rhythmic pacemaker current. Nat Med 4:848–851. https://doi.org/10.1038/nm0798-848
45. Heinrich MC, Corless CL, Duensing A et al (2003) PDGFRA activating mutations in gastrointestinal stromal tumors. Science 299:708–710. https://doi.org/10.1126/science.1079666
46. Hirota S, Ohashi A, Nishida T et al (2003) Gain-of-function mutations of platelet-derived growth factor receptor alpha gene in gastrointestinal stromal tumors. Gastroenterology 125:660–667. https://doi.org/10.1016/s0016-5085(03)01046-1
47. Blay JY, Kang YK, Nishida T et al (2021) Gastrointestinal stromal tumours. Nat Rev Dis Primers 7:22. https://doi.org/10.1038/s41572-021-00254-5
48. Corless CL, Barnett CM, Heinrich MC (2011) Gastrointestinal stromal tumours: origin and molecular oncology. Nat Rev Cancer 11:865–878. https://doi.org/10.1038/nrc3143
49. Klug LR, Khosroyani HM, Kent JD et al (2022) New treatment strategies for advanced-stage gastrointestinal stromal tumours. Nat Rev Clin Oncol 19:328–341. https://doi.org/10.1038/s41571-022-00606-4
50. Druker BJ, Tamura S, Buchdunger E et al (1996) Effects of a selective inhibitor of the Abl tyrosine kinase on the growth of Bcr-Abl positive cells. Nat Med 2:561–566. https://doi.org/10.1038/nm0596-561
51. Druker BJ, Talpaz M, Resta DJ et al (2001) Efficacy and safety of a specific inhibitor of the BCR-ABL tyrosine kinase in chronic myeloid leukemia. N Engl J Med 344:1031–1037. https://doi.org/10.1056/NEJM200104053441401
52. Heinrich MC, Griffith DJ, Druker BJ et al (2000) Inhibition of c-kit receptor tyrosine kinase activity by STI 571, a selective tyrosine kinase inhibitor. Blood 96:925–932. https://doi.org/10.1182/blood.V96.3.925
53. McGary EC, Weber K, Mills L et al (2002) Inhibition of platelet-derived growth factor-mediated proliferation of osteosarcoma cells by the novel tyrosine kinase inhibitor STI571. Clin Cancer Res 8:3584–3591
54. Lasota J, Miettinen M (2008) Clinical significance of oncogenic KIT and PDGFRA mutations in gastrointestinal stromal tumours. Histopathology 53:245–266. https://doi.org/10.1111/j.1365-2559.2008.02977.x
55. Nishida T, Yoshinaga S, Takahashi T et al (2021) Recent progress and challenges in the diagnosis and treatment of gastrointestinal stromal tumors. Cancers (Basel) 13:3158. https://doi.org/10.3390/cancers13133158
56. Rubin BP, Duensing A (2006) Mechanisms of resistance to small molecule kinase inhibition in the treatment of solid tumors. Lab Investig 86:981–986. https://doi.org/10.1038/labinvest.3700466
57. Nishida T, Kanda T, Nishitani A et al (2008) Secondary mutations in the kinase domain of the KIT gene are predominant in imatinib-resistant gastrointestinal stromal tumor. Cancer Sci 99:799–804. https://doi.org/10.1111/j.1349-7006.2008.00727.x
58. Demetri GD, Reichardt P, Kang YK et al (2013) Efficacy and safety of regorafenib for advanced gastrointestinal stromal tumours after failure of imatinib and sunitinib (GRID): an international, multicentre, randomised, placebo-controlled, phase 3 trial. Lancet 381:295–302. https://doi.org/10.1016/S0140-6736(12)61857-1
59. Smith BD, Kaufman MD, Lu WP et al (2019) Ripretinib (DCC-2618) is a switch control kinase inhibitor of a broad spectrum of oncogenic and drug-resistant KIT and PDGFRA variants. Cancer Cell 35:738–751.e9. https://doi.org/10.1016/j.ccell.2019.04.006

60. Boissan M, Feger F, Guillosson JJ et al (2000) c-Kit and c-kit mutations in mastocytosis and other hematological diseases. J Leukoc Biol 67:135–148. https://doi.org/10.1002/jlb.67.2.135
61. Obata Y, Toyoshima S, Wakamatsu E et al (2014) Oncogenic Kit signals on endolysosomes and endoplasmic reticulum are essential for neoplastic mast cell proliferation. Nat Commun 5: 5715. https://doi.org/10.1038/ncomms6715
62. Hara Y, Obata Y, Horikawa K et al (2017) M-COPA suppresses endolysosomal Kit-Akt oncogenic signalling through inhibiting the secretory pathway in neoplastic mast cells. PLoS One 12:e0175514. https://doi.org/10.1371/journal.pone.0175514
63. Rai S, Tanaka H, Suzuki M et al (2020) Chlorpromazine eliminates acute myeloid leukemia cells by perturbing subcellular localization of FLT3-ITD and KIT-D816V. Nat Commun 11: 4147. https://doi.org/10.1038/s41467-020-17666-8
64. Obata Y, Hara Y, Shiina I et al (2019) N822K- or V560G-mutated KIT activation preferentially occurs in lipid rafts of the Golgi apparatus in leukemia cells. Cell Commun Signal 17: 114. https://doi.org/10.1186/s12964-019-0426-3
65. Bougherara H, Subra F, Crépin R et al (2009) The aberrant localization of oncogenic kit tyrosine kinase receptor mutants is reversed on specific inhibitory treatment. Mol Cancer Res 7:1525–1533. https://doi.org/10.1158/1541-7786.MCR-09-0138
66. Tabone-Eglinger S, Subra F, El Sayadi H et al (2008) KIT mutations induce intracellular retention and activation of an immature form of the KIT protein in gastrointestinal stromal tumors. Clin Cancer Res 14:2285–2294. https://doi.org/10.1158/1078-0432.CCR-07-4102
67. Xiang Z, Kreisel F, Cain J et al (2007) Neoplasia driven by mutant c-KIT is mediated by intracellular, not plasma membrane, receptor signaling. Mol Cell Biol 27:267–282. https://doi.org/10.1128/MCB.01153-06
68. Agaimy A, Otto C, Braun A et al (2013) Value of epithelioid morphology and PDGFRA immunostaining pattern for prediction of PDGFRA mutated genotype in gastrointestinal stromal tumors (GISTs). Int J Clin Exp Pathol 6:1839–1846
69. Emile JF, Stock N, Corless CL et al (2009) Dotlike or Golgi-like KIT and PDGFRA Staining in GISTs. Am J Surg Pathol 33:157–158. https://doi.org/10.1097/PAS.0b013e3181883765
70. González-Cámpora R, Delgado MD, Amate AH et al (2011) Old and new immunohistochemical markers for the diagnosis of gastrointestinal stromal tumors. Anal Quant Cytol Histol 33: 1–11
71. Jaramillo S, Ríos-Moreno MJ, Hernández A et al (2012) Gastrointestinal stromal tumors (GISTs): role of CD 117 and PDGFRA Golgi-like staining pattern in the recognition of mutational status. Rev Esp Enferm Dig 104:128–133. https://doi.org/10.4321/s1130-01082012000300005
72. Peterson MR, Piao Z, Weidner N et al (2006) Strong PDGFRA positivity is seen in GISTs but not in other intra-abdominal mesenchymal tumors: Immunohistochemical and mutational analyses. Appl Immunohistochem Mol Morphol 14:390–396. https://doi.org/10.1097/01.pai.0000203038.33414.a3
73. Rossi S, Ou W, Tang D et al (2006) Gastrointestinal stromal tumours overexpress fatty acid synthase. J Pathol 209:369–375. https://doi.org/10.1002/path.1983
74. Taguchi T, Sonobe H, Toyonaga S et al (2002) Conventional and molecular cytogenetic characterization of a new human cell line, GIST-T1, established from gastrointestinal stromal tumor. Lab Investig 82:663–665. https://doi.org/10.1038/labinvest.3780461
75. Takahashi T, Elzawahry A, Mimaki S et al (2017) Genomic and transcriptomic analysis of imatinib resistance in gastrointestinal stromal tumors. Genes Chromosomes Cancer 56:303–313. https://doi.org/10.1002/gcc.22438
76. Tuveson DA, Willis NA, Jacks T et al (2001) STI571 inactivation of the gastrointestinal stromal tumor c-KIT oncoprotein: biological and clinical implications. Oncogene 20:5054–5058. https://doi.org/10.1038/sj.onc.1204704
77. Kim WK, Yun S, Park CK et al (2017) Sustained mutant KIT activation in the Golgi complex is mediated by PKC-θ in gastrointestinal stromal tumors. Clin Cancer Res 23:845–856. https://doi.org/10.1158/1078-0432.CCR-16-0521

78. Obata Y, Horikawa K, Takahashi T et al (2017) Oncogenic signaling by Kit tyrosine kinase occurs selectively on the Golgi apparatus in gastrointestinal stromal tumors. Oncogene 36: 3661–3672. https://doi.org/10.1038/onc.2016.519
79. Obata Y, Horikawa K, Shiina I et al (2018) Oncogenic Kit signalling on the Golgi is suppressed by blocking secretory trafficking with M-COPA in gastrointestinal stromal tumours. Cancer Lett 415:1–10. https://doi.org/10.1016/j.canlet.2017.11.032
80. Kurokawa K, Osakada H, Kojidani T et al (2019) Visualization of secretory cargo transport within the Golgi apparatus. J Cell Biol 218:1602–1618. https://doi.org/10.1083/jcb.201807194
81. Kurokawa K, Nakano A (2020) Live-cell imaging by Super-resolution Confocal Live Imaging Microscopy (SCLIM): simultaneous three-color and four-dimensional live cell imaging with high space and time resolution. Bio Protoc 10:e3732. https://doi.org/10.21769/BioProtoc.3732
82. Obata Y, Kurokawa K, Tojima T et al (2023) Golgi retention and oncogenic KIT signaling via PLCγ2-PKD2-PI4KIIIβ activation in gastrointestinal stromal tumor cells. Cell Rep 42:113035. https://doi.org/10.1016/j.celrep.2023.113035
83. Duensing A, Medeiros F, McConarty B et al (2004a) Mechanisms of oncogenic KIT signal transduction in primary gastrointestinal stromal tumors (GISTs). Oncogene 23:3999–4006. https://doi.org/10.1038/sj.onc.1207525
84. Dascher C, Balch WE (1994) Dominant inhibitory mutants of ARF1 block endoplasmic reticulum to Golgi transport and trigger disassembly of the Golgi apparatus. J Biol Chem 269:1437–1448. https://doi.org/10.1016/S0021-9258(17)42277-0
85. Hayashi T, Takatsuki A, Tamura G (1974) The action mechanism of brefeldin A. I. Growth recovery of Candida albicans by lipids from the action of brefeldin A. J Antibiot (Tokyo) 27: 65–72. https://doi.org/10.7164/antibiotics.27.65
86. Lippincott-Schwartz J, Yuan LC, Bonifacino JS et al (1989) Rapid redistribution of Golgi proteins into the ER in cells treated with brefeldin A: evidence for membrane cycling from Golgi to ER. Cell 56:801–813. https://doi.org/10.1016/0092-8674(89)90685-5
87. Teranishi R, Takahashi T, Obata Y et al (2023) Combination of pimitespib (TAS-116) with sunitinib is an effective therapy for imatinib-resistant gastrointestinal stromal tumors. Int J Cancer 152:2580–2593. https://doi.org/10.1002/ijc.34461
88. Bahlawane C, Eulenfeld R, Wiesinger MY et al (2015) Constitutive activation of oncogenic PDGFRα-mutant proteins occurring in GIST patients induces receptor mislocalisation and alters PDGFRα signalling characteristics. Cell Commun Signal 13:21. https://doi.org/10.1186/s12964-015-0096-8
89. Del Giudice S, De Luca V, Parizadeh S et al (2022) Endogenous and exogenous regulatory signaling in the secretory pathway: role of Golgi signaling molecules in cancer. Front Cell Dev Biol 10:833663. https://doi.org/10.3389/fcell.2022.833663
90. Mayinger P (2011) Signaling at the Golgi. Cold Spring Harb Perspect Biol 3:a005314. https://doi.org/10.1101/cshperspect.a005314
91. Pulvirenti T, Giannotta M, Capestrano M et al (2008) A traffic-activated Golgi-based signalling circuit coordinates the secretory pathway. Nat Cell Biol 10:912–922. https://doi.org/10.1038/ncb1751
92. Weixel KM, Blumental-Perry A, Watkins SC et al (2005) Distinct Golgi populations of phosphatidylinositol 4-phosphate regulated by phosphatidylinositol 4-kinases. J Biol Chem 280:10501–10508. https://doi.org/10.1074/jbc.M414304200
93. Klausner RD, Donaldson JG, Lippincott-Schwartz J (1992) Brefeldin A: insights into the control of membrane traffic and organelle structure. J Cell Biol 116:1071–1080. https://doi.org/10.1083/jcb.116.5.1071
94. Natsume M, Niwa M, Ichikawa S et al (2024) Brefeldin A and M-COPA block the export of RTKs from the endoplasmic reticulum via simultaneous inactivation of ARF1, ARF4, and ARF5. J Biol Chem 300:107327. https://doi.org/10.1016/j.jbc.2024.107327

95. Shiina I, Umezaki Y, Ohashi Y et al (2013) Total synthesis of AMF-26, an antitumor agent for inhibition of the Golgi system, targeting ADP-ribosylation factor 1. J Med Chem 56:150–159. https://doi.org/10.1021/jm301695c
96. Kwon Y, Kim J, Cho SY et al (2023) Identification of novel pathogenic roles of BLZF1/ATF6 in tumorigenesis of gastrointestinal stromal tumor showing Golgi-localized mutant KIT. Cell Death Differ 30:2309–2321. https://doi.org/10.1038/s41418-023-01220-2
97. Griffiths G, Quinn P, Warren G (1983) Dissection of the Golgi complex. I. Monensin inhibits the transport of viral membrane proteins from medial to trans Golgi cisternae in baby hamster kidney cells infected with Semliki Forest virus. J Cell Biol 96:835–850. https://doi.org/10.1083/jcb.96.3.835
98. Yonemoto W, Filson AJ, Queral-Lustig AE et al (1987) Detection of phosphotyrosine-containing proteins in polyomavirus middle tumor antigen-transformed cells after treatment with a phosphotyrosine phosphatase inhibitor. Mol Cell Biol 7:905–913. https://doi.org/10.1128/mcb.7.2.905-913.1987
99. Stuible M, Tremblay ML (2010) In control at the ER: PTP1B and the down-regulation of RTKs by dephosphorylation and endocytosis. Trends Cell Biol 20:672–679. https://doi.org/10.1016/j.tcb.2010.08.013
100. Goss PE, Baker MA, Carver JP et al (1995) Inhibitors of carbohydrate processing: a new class of anticancer agents. Clin Cancer Res 1:935–944
101. Ip CKM, Ng PKS, Jeong KJ et al (2018) Neomorphic PDGFRA extracellular domain driver mutations are resistant to PDGFRA targeted therapies. Nat Commun 9:4583. https://doi.org/10.1038/s41467-018-06949-w
102. Mali RS, Ramdas B, Ma P et al (2011) Rho kinase regulates the survival and transformation of cells bearing oncogenic forms of KIT, FLT3, and BCR-ABL. Cancer Cell 20:357–369. https://doi.org/10.1016/j.ccr.2011.07.016
103. Martin H, Mali RS, Ma P et al (2013) Pak and Rac GTPases promote oncogenic KIT-induced neoplasms. J Clin Invest 123:4449–4463. https://doi.org/10.1172/JCI67509
104. Kazi JU, Agarwal S, Sun J et al (2014) Src-like-adaptor protein (SLAP) differentially regulates normal and oncogenic c-Kit signaling. J Cell Sci 127:653–662. https://doi.org/10.1242/jcs.140590
105. Sun J, Thingholm T, Højrup P et al (2018) XK-related protein 5 (XKR5) is a novel negative regulator of KIT/D816V-mediated transformation. Oncogene 7:48. https://doi.org/10.1038/s41389-018-0057-3
106. Lv D, Chen H, Feng Y et al (2021) Small-molecule inhibitor targeting protein Kinase D: a potential therapeutic strategy. Front Oncol 11:680221. https://doi.org/10.3389/fonc.2021.680221
107. Baron CL, Malhotra V (2002) Role of diacylglycerol in PKD recruitment to the TGN and protein transport to the plasma membrane. Science 295:325–328. https://doi.org/10.1126/science.1066759
108. Liljedahl M, Maeda Y, Colanzi A et al (2001) Protein kinase D regulates the fission of cell surface destined transport carriers from the trans-Golgi network. Cell 104:409–420. https://doi.org/10.1016/s0092-8674(01)00228-8
109. Malhotra V, Campelo F (2011) PKD regulates membrane fission to generate TGN to cell surface transport carriers. Cold Spring Harb Perspect Biol 3:a005280. https://doi.org/10.1101/cshperspect.a005280
110. Van Lint J, Rykx A, Maeda Y et al (2002) Protein kinase D: an intracellular traffic regulator on the move. Trends Cell Biol 12:193–200. https://doi.org/10.1016/s0962-8924(02)02262-6
111. Wakana Y, Campelo F (2021) The PKD-dependent biogenesis of TGN-to-plasma membrane transport carriers. Cells 10:1618. https://doi.org/10.3390/cells10071618
112. Wille C, Köhler C, Armacki M et al (2014) Protein kinase D2 induces invasion of pancreatic cancer cells by regulating matrix metalloproteinases. Mol Biol Cell 25:324–336. https://doi.org/10.1091/mbc.E13-06-0334

113. Armacki M, Joodi G, Nimmagadda SC et al (2014) A novel splice variant of calcium and integrin-binding protein 1 mediates protein kinase D2-stimulated tumour growth by regulating angiogenesis. Oncogene 33:1167–1180. https://doi.org/10.1038/onc.2013.43
114. Cilleros-Rodríguez D, Toledo-Lobo MV, Martínez-Martínez D et al (2024) Protein kinase D activity is a risk biomarker in prostate cancer that drives cell invasion by a Snail/ERK dependent mechanism. Biochim Biophys Acta Mol basis Dis 1870:166851. https://doi.org/10.1016/j.bbadis.2023.166851
115. Harikumar KB, Kunnumakkara AB, Ochi N et al (2010) A novel small-molecule inhibitor of protein kinase D blocks pancreatic cancer growth in vitro and in vivo. Mol Cancer Ther 9: 1136–1346. https://doi.org/10.1158/1535-7163.MCT-09-1145
116. Roy A, Ye J, Deng F, Wang QJ (2017) Protein kinase D signaling in cancer: a friend or foe? Biochim Biophys Acta Rev Cancer 1868:283–294. https://doi.org/10.1016/j.bbcan.2017.05.008
117. Azoitei N, Cobbaut M, Becher A et al (2018) Protein kinase D2: a versatile player in cancer biology. Oncogene 37:1263–1278. https://doi.org/10.1038/s41388-017-0052-8
118. Cooke M, Kazanietz MG (2022) Overarching roles of diacylglycerol signaling in cancer development and antitumor immunity. Sci Signal 15:eabo0264. https://doi.org/10.1126/scisignal.abo0264
119. Bankaitis VA, Garcia-Mata R, Mousley CJ (2012) Golgi membrane dynamics and lipid metabolism. Curr Biol 22:R414–R424. https://doi.org/10.1016/j.cub.2012.03.004
120. Graham TR, Burd CG (2011) Coordination of Golgi functions by phosphatidylinositol 4-kinases. Trends Cell Biol 21:113–121. https://doi.org/10.1016/j.tcb.2010.10.002
121. Hausser A, Storz P, Märtens S et al (2005) Protein kinase D regulates vesicular transport by phosphorylating and activating phosphatidylinositol-4 kinase IIIbeta at the Golgi complex. Nat Cell Biol 7:880–886. https://doi.org/10.1038/ncb1289
122. Roussel É, Lippé R (2018) Cellular protein Kinase D modulators play a role during multiple steps of Herpes Simplex Virus 1 Egress. J Virol 92:e01486–e01418. https://doi.org/10.1128/JVI.01486-18
123. Wang YJ, Wang J, Sun HQ et al (2003) Phosphatidylinositol 4 phosphate regulates targeting of clathrin adaptor AP-1 complexes to the Golgi. Cell 114:299–310. https://doi.org/10.1016/s0092-8674(03)00603-2
124. Wang J, Sun HQ, Macia E et al (2007) PI4P promotes the recruitment of the GGA adaptor proteins to the trans-Golgi network and regulates their recognition of the ubiquitin sorting signal. Mol Biol Cell 18:2646–2655. https://doi.org/10.1091/mbc.e06-10-0897
125. Doray B, Ghosh P, Griffith J et al (2002) Cooperation of GGAs and AP-1 in packaging MPRs at the trans-Golgi network. Science 297:1700–1703. https://doi.org/10.1126/science.1075327
126. Hirst J, Borner GH, Antrobus R et al (2012) Distinct and overlapping roles for AP-1 and GGAs revealed by the "knocksideways" system. Curr Biol 22:1711–1716. https://doi.org/10.1016/j.cub.2012.07.012
127. Borges S, Perez EA, Thompson EA et al (2015) Effective targeting of Estrogen receptor-negative breast cancers with the protein Kinase D inhibitor CRT0066101. Mol Cancer Ther 14:1306–1316. https://doi.org/10.1158/1535-7163.MCT-14-0945
128. Boura E, Nencka R (2015) Phosphatidylinositol 4-kinases: function, structure, and inhibition. Exp Cell Res 337:136–145. https://doi.org/10.1016/j.yexcr.2015.03.028
129. Shi L, Tan X, Liu X et al (2021) Addiction to Golgi-resident PI4P synthesis in chromosome 1q21.3-amplified lung adenocarcinoma cells. Proc Natl Acad Sci USA 118:e2023537118. https://doi.org/10.1073/pnas.2023537118
130. Tan X, Banerjee P, Pham EA et al (2020) PI4KIIIβ is a therapeutic target in chromosome 1q-amplified lung adenocarcinoma. Sci Transl Med 12:eaax3772. https://doi.org/10.1126/scitranslmed.aax3772
131. Wei N, Chu E, Wipf P et al (2014) Protein kinase d as a potential chemotherapeutic target for colorectal cancer. Mol Cancer Ther 13:1130–1141. https://doi.org/10.1158/1535-7163.MCT-13-0880

132. Yamawaki K, Shiina I, Murata T et al (2021) FLT3-ITD transduces autonomous growth signals during its biosynthetic trafficking in acute myelogenous leukemia cells. Sci Rep 11: 22678. https://doi.org/10.1038/s41598-021-02221-2
133. Blay P, Astudillo A, Buesa JM et al (2004) Protein kinase C theta is highly expressed in gastrointestinal stromal tumors but not in other mesenchymal neoplasias. Clin Cancer Res 10: 4089–4095. https://doi.org/10.1158/1078-0432.CCR-04-0630
134. Duensing A, Joseph NE, Medeiros F et al (2004b) Protein Kinase C theta (PKCtheta) expression and constitutive activation in gastrointestinal stromal tumors (GISTs). Cancer Res 64:5127–5131. https://doi.org/10.1158/0008-5472.CAN-04-0559
135. Ou WB, Zhu MJ, Demetri GD et al (2008) Protein kinase C-theta regulates KIT expression and proliferation in gastrointestinal stromal tumors. Oncogene 27:5624–5634. https://doi.org/10.1038/onc.2008.177
136. Casler JC, Papanikou E, Barrero JJ et al (2019) Maturation-driven transport and AP-1-dependent recycling of a secretory cargo in the Golgi. J Cell Biol 218:1582–1601. https://doi.org/10.1083/jcb.201807195
137. Casler JC, Glick BS (2020) A microscopy-based kinetic analysis of yeast vacuolar protein sorting. elife 9:e56844. https://doi.org/10.7554/eLife.56844
138. Casler JC, Johnson N, Krahn AH et al (2022) Clathrin adaptors mediate two sequential pathways of intra-Golgi recycling. J Cell Biol 221:e202103199. https://doi.org/10.1083/jcb.202103199
139. Kazi JU, Rönnstrand L (2019) FMS-like Tyrosine Kinase 3/FLT3: from basic science to clinical implications. Physiol Rev 99:1433–1466. https://doi.org/10.1152/physrev.00029.2018
140. Kiyoi H, Ohno R, Ueda R et al (2002) Mechanism of constitutive activation of FLT3 with internal tandem duplication in the juxtamembrane domain. Oncogene 21:2555–2563. https://doi.org/10.1038/sj.onc.1205332
141. Meshinchi S, Appelbaum FR (2009) Structural and functional alterations of FLT3 in acute myeloid leukemia. Clin Cancer Res 15:4263–4269. https://doi.org/10.1158/1078-0432.CCR-08-1123
142. Choudhary C, Olsen JV, Brandts C et al (2009) Mislocalized activation of oncogenic RTKs switches downstream signaling outcomes. Mol Cell 36:326–339. https://doi.org/10.1016/j.molcel.2009.09.019
143. Koch S, Jacobi A, Ryser M et al (2008) Abnormal localization and accumulation of FLT3-ITD, a mutant receptor tyrosine kinase involved in leukemogenesis. Cells Tissues Organs 188: 225–235. https://doi.org/10.1159/000118788
144. Köthe S, Müller JP, Böhmer SA et al (2013) Features of Ras activation by a mislocalized oncogenic tyrosine kinase: FLT3 ITD signals through K-Ras at the plasma membrane of acute myeloid leukemia cells. J Cell Sci 126:4746–4755. https://doi.org/10.1242/jcs.131789
145. Lv K, Ren JG, Han X et al (2021) Depalmitoylation rewires FLT3-ITD signaling and exacerbates leukemia progression. Blood 138:2244–2255. https://doi.org/10.1182/blood.2021011582
146. Williams AB, Li L, Nguyen B et al (2012) Fluvastatin inhibits FLT3 glycosylation in human and murine cells and prolongs survival of mice with FLT3/ITD leukemia. Blood 120:3069–3079. https://doi.org/10.1182/blood-2012-01-403493
147. Moloney JN, Stanicka J, Cotter TG (2017) Subcellular localization of the FLT3-ITD oncogene plays a significant role in the production of NOX- and p22phox-derived reactive oxygen species in acute myeloid leukemia. Leuk Res 52:34–42. https://doi.org/10.1016/j.leukres.2016.11.006
148. Schmidt-Arras D, Böhmer SA, Koch S et al (2009) Anchoring of FLT3 in the endoplasmic reticulum alters signaling quality. Blood 113:3568–3576. https://doi.org/10.1182/blood-2007-10-121426

149. Takahashi S (2019) Mutations of FLT3 receptor affect its surface glycosylation, intracellular localization, and downstream signaling. Leuk Res Rep 13:100187. https://doi.org/10.1016/j.lrr.2019.100187
150. Yamamoto Y, Kiyoi H, Nakano Y et al (2001) Activating mutation of D835 within the activation loop of FLT3 in human hematologic malignancies. Blood 97:2434–2439. https://doi.org/10.1182/blood.v97.8.2434
151. Fernandes M, Jamme P, Cortot AB et al (2021) When the MET receptor kicks in to resist targeted therapies. Oncogene 40:4061–4078. https://doi.org/10.1038/s41388-021-01835-0
152. Guo R, Luo J, Chang J et al (2020) MET-dependent solid tumours - molecular diagnosis and targeted therapy. Nat Rev Clin Oncol 17:569–587. https://doi.org/10.1038/s41571-020-0377-z
153. Frazier NM, Brand T, Gordan JD et al (2019) Overexpression-mediated activation of MET in the Golgi promotes HER3/ERBB3 phosphorylation. Oncogene 38:1936–1950. https://doi.org/10.1038/s41388-018-0537-0
154. Stern YE, Al-Ghabkari A, Monast A et al (2022) Met-HER3 crosstalk supports proliferation via MPZL3 in MET-amplified cancer cells. Cell Mol Life Sci 79:178. https://doi.org/10.1007/s00018-022-04149-w
155. Cerqua M, Botti O, Arigoni M et al (2022) METΔ14 promotes a ligand-dependent, AKT-driven invasive growth. Life Sci Alliance 5:e202201409. https://doi.org/10.26508/lsa.202201409
156. Cortot AB, Kherrouche Z, Descarpentries C et al (2017) Exon 14 deleted MET receptor as a new biomarker and target in cancers. J Natl Cancer Inst 109:djw262. https://doi.org/10.1093/jnci/djw262
157. Pollak M (2012) The insulin and insulin-like growth factor receptor family in neoplasia: an update. Nat Rev Cancer 12:159–169. https://doi.org/10.1038/nrc3215
158. Rieger L, O'Shea S, Godsmark G et al (2020) IGF-1 receptor activity in the Golgi of migratory cancer cells depends on adhesion-dependent phosphorylation of Tyr1250 and Tyr1251. Sci Signal 13:eaba3176. https://doi.org/10.1126/scisignal.aba3176
159. Rieger L, O'Connor R (2021) Controlled signaling-insulin-like growth factor receptor endocytosis and presence at intracellular compartments. Front Endocrinol (Lausanne) 11:620013. https://doi.org/10.3389/fendo.2020.620013
160. Bonaventure J, Horne WC, Baron R (2007) The localization of FGFR3 mutations causing thanatophoric dysplasia type I differentially affects phosphorylation, processing and ubiquitylation of the receptor. FEBS J 274:3078–3093. https://doi.org/10.1111/j.1742-4658.2007.05835.x
161. Ronchetti D, Greco A, Compasso S et al (2001) Deregulated FGFR3 mutants in multiple myeloma cell lines with t(4;14): comparative analysis of Y373C, K650E and the novel G384D mutations. Oncogene 20:3553–3562. https://doi.org/10.1038/sj.onc.1204465
162. Lievens PM, Mutinelli C, Baynes D et al (2004) The kinase activity of fibroblast growth factor receptor 3 with activation loop mutations affects receptor trafficking and signaling. J Biol Chem 279:43254–43260. https://doi.org/10.1074/jbc.M405247200
163. Lievens PM, Roncador A, Liboi E (2006) K644E/M FGFR3 mutants activate Erk1/2 from the endoplasmic reticulum through FRS2 alpha and PLC gamma-independent pathways. J Mol Biol 357:783–792. https://doi.org/10.1016/j.jmb.2006.01.058
164. Gibbs L, Legeai-Mallet L (2007) FGFR3 intracellular mutations induce tyrosine phosphorylation in the Golgi and defective glycosylation. Biochim Biophys Acta 1773:502–512. https://doi.org/10.1016/j.bbamcr.2006.12.010
165. Ozono K, Kubota T, Michigami T (2024) Promising horizons in achondroplasia along with the development of new drugs. Endocr J 71:643–650. https://doi.org/10.1507/endocrj.EJ24-0109
166. Tavormina PL, Bellus GA, Webster MK et al (1999) A novel skeletal dysplasia with developmental delay and acanthosis nigricans is caused by a Lys650Met mutation in the fibroblast growth factor receptor 3 gene. Am J Hum Genet 64:722–731. https://doi.org/10.1086/302275

167. Mayr L, Neyazi S, Schwark K et al (2025) Effective targeting of PDGFRA-altered high-grade glioma with avapritinib. Cancer Cell 43:740–756.e8. https://doi.org/10.1016/j.ccell.2025.02.018
168. Carpenter EL, Mossé YP (2012) Targeting ALK in neuroblastoma--preclinical and clinical advancements. Nat Rev Clin Oncol 9:391–399. https://doi.org/10.1038/nrclinonc.2012.72
169. Chen Y, Takita J, Choi YL et al (2008) Oncogenic mutations of ALK kinase in neuroblastoma. Nature 455:971–974. https://doi.org/10.1038/nature07399
170. George RE, Sanda T, Hanna M et al (2008) Activating mutations in ALK provide a therapeutic target in neuroblastoma. Nature 455:975–978. https://doi.org/10.1038/nature07397
171. Okubo J, Takita J, Chen Y et al (2012) Aberrant activation of ALK kinase by a novel truncated form ALK protein in neuroblastoma. Oncogene 31:4667–4676. https://doi.org/10.1038/onc.2011.616
172. Fukuhara S, Nomoto J, Kim SW et al (2018) Partial deletion of the ALK gene in ALK-positive anaplastic large cell lymphoma. Hematol Oncol 36:150–158. https://doi.org/10.1002/hon.2455
173. Mazot P, Cazes A, Boutterin MC et al (2011) The constitutive activity of the ALK mutated at positions F1174 or R1275 impairs receptor trafficking. Oncogene 30:2017–2025. https://doi.org/10.1038/onc.2010.595
174. Centonze FG, Reiterer V, Nalbach K et al (2019) LTK is an ER-resident receptor tyrosine kinase that regulates secretion. J Cell Biol 218:2470–2480. https://doi.org/10.1083/jcb.201903068
175. Krampitz GW, Norton JA (2014) RET gene mutations (genotype and phenotype) of multiple endocrine neoplasia type 2 and familial medullary thyroid carcinoma. Cancer 120:1920–1931. https://doi.org/10.1002/cncr.28661
176. Runeberg-Roos P, Virtanen H, Saarma M (2007) RET(MEN 2B) is active in the endoplasmic reticulum before reaching the cell surface. Oncogene 26:7909–7915. https://doi.org/10.1038/sj.onc.1210591
177. Sigismund S, Confalonieri S, Ciliberto A et al (2012) Endocytosis and signaling: cell logistics shape the eukaryotic cell plan. Physiol Rev 92:273–366. https://doi.org/10.1152/physrev.00005.2011
178. Chung BM, Raja SM, Clubb RJ et al (2009) Aberrant trafficking of NSCLC-associated EGFR mutants through the endocytic recycling pathway promotes interaction with Src. BMC Cell Biol 10:84. https://doi.org/10.1186/1471-2121-10-84
179. Ménard L, Floc'h N, Martin MJ et al (2018) Reactivation of mutant-EGFR degradation through Clathrin inhibition overcomes resistance to EGFR Tyrosine Kinase inhibitors. Cancer Res 78:3267–3279. https://doi.org/10.1158/0008-5472.CAN-17-2195
180. Watanuki Z, Kosai H, Osanai N et al (2014) Synergistic cytotoxicity of afatinib and cetuximab against EGFR T790M involves Rab11-dependent EGFR recycling. Biochem Biophys Res Commun 455:269–276. https://doi.org/10.1016/j.bbrc.2014.11.003
181. Lee ZY, Lee WH, Lim JS et al (2024) Golgi apparatus targeted therapy in cancer: are we there yet? Life Sci 352:122868. https://doi.org/10.1016/j.lfs.2024.122868
182. Martins M, Vieira J, Pereira-Leite C et al (2023) The Golgi apparatus as an anticancer therapeutic target. Biology (Basel) 13:1. https://doi.org/10.3390/biology13010001
183. Prieto-Dominguez N, Parnell C, Teng Y (2019) Drugging the small GTPase pathways in cancer treatment: promises and challenges. Cells 8:255. https://doi.org/10.3390/cells8030255
184. Watari K, Nakamura M, Fukunaga Y et al (2012) The antitumor effect of a novel angiogenesis inhibitor (an octahydronaphthalene derivative) targeting both VEGF receptor and NF-κB pathway. Int J Cancer 131:310–321. https://doi.org/10.1002/ijc.26356
185. Bauer S, Yu LK, Demetri GD et al (2006) Heat shock protein 90 inhibition in imatinib-resistant gastrointestinal stromal tumor. Cancer Res 66:9153–9161. https://doi.org/10.1158/0008-5472.CAN-06-0165

186. Garcia-Carbonero R, Carnero A, Paz-Ares L (2013) Inhibition of HSP90 molecular chaperones: moving into the clinic. Lancet Oncol 14:e358–e369. https://doi.org/10.1016/S1470-2045(13)70169-4
187. Miyajima N, Tsutsumi S, Sourbier C et al (2013) The HSP90 inhibitor ganetespib synergizes with the MET kinase inhibitor crizotinib in both crizotinib-sensitive and -resistant MET-driven tumor models. Cancer Res 73:7022–7033. https://doi.org/10.1158/0008-5472.CAN-13-1156
188. Ohkubo S, Kodama Y, Muraoka H et al (2015) TAS-116, a highly selective inhibitor of heat shock protein 90α and β, demonstrates potent antitumor activity and minimal ocular toxicity in preclinical models. Mol Cancer Ther 14:14–22. https://doi.org/10.1158/1535-7163.MCT-14-0219
189. Smyth T, Van Looy T, Curry JE et al (2012) The HSP90 inhibitor, AT13387, is effective against imatinib-sensitive and -resistant gastrointestinal stromal tumor models. Mol Cancer Ther 11:1799–1808. https://doi.org/10.1158/1535-7163.MCT-11-1046
190. Saito Y, Takahashi T, Obata Y et al (2020) TAS-116 inhibits oncogenic KIT signalling on the Golgi in both imatinib-naïve and imatinib-resistant gastrointestinal stromal tumours. Br J Cancer 122:658–667. https://doi.org/10.1038/s41416-019-0688-y
191. Doi T, Yamamoto N, Ohkubo S (2024) Pimitespib for the treatment of advanced gastrointestinal stromal tumors and other tumors. Future Oncol 20:507–519. https://doi.org/10.2217/fon-2022-1172
192. Pinto S, Pahl J, Schottelius A, Carter PJ, Koch J (2022) Reimagining antibody-dependent cellular cytotoxicity in cancer: the potential of natural killer cell engagers. Trends Immunol 43: 932–946. https://doi.org/10.1016/j.it.2022.09.007
193. Bougherara H, Georgin-Lavialle S, Damaj G et al (2013) Relocalization of KIT D816V to cell surface after dasatinib treatment: potential clinical implications. Clin Lymphoma Myeloma Leuk 13:62–69. https://doi.org/10.1016/j.clml.2012.08.004
194. Jetani H, Garcia-Cadenas I, Nerreter T et al (2018) CAR T-cells targeting FLT3 have potent activity against FLT3-ITD+ AML and act synergistically with the FLT3-inhibitor crenolanib. Leukemia 32:1168–1179. https://doi.org/10.1038/s41375-018-0009-0
195. Reiter K, Polzer H, Krupka C et al (2018) Tyrosine kinase inhibition increases the cell surface localization of FLT3-ITD and enhances FLT3-directed immunotherapy of acute myeloid leukemia. Leukemia 32:313–322. https://doi.org/10.1038/leu.2017.257
196. Farina AR, Cappabianca L, Ruggeri P et al (2015) Retrograde TrkAIII transport from ERGIC to ER: a re-localisation mechanism for oncogenic activity. Oncotarget 6:35636–35651. https://doi.org/10.18632/oncotarget.5802
197. Farina AR, Cappabianca L, Ruggeri P et al (2018) The oncogenic neurotrophin receptor tropomyosin-related kinase variant, TrkAIII. J Exp Clin Cancer Res 37:119. https://doi.org/10.1186/s13046-018-0786-3

Chapter 17
The Role of the Golgi Apparatus in Neurodegeneration

Sina Shadfar, Sara Assar Kashani, Shashi Gautam, Zeinab Takalloo, Fabiha Farzana, Sonam Parakh, and Julie D. Atkin

Abstract The Golgi apparatus has important, well characterised functions in the trafficking, processing, and post-translational modification of proteins and lipids. However, roles in other cellular processes are increasingly reported, including autophagy, apoptosis, DNA repair, and cytoskeletal (microtubules and actin) function. The Golgi therefore serves as a regulatory hub for multiple signalling pathways that maintain essential cellular activities. The Golgi normally consists of flattened stacks of membrane (cisternae), but during normal physiology and pathological conditions it 'fragments', resulting in altered morphology and distribution. This is well described as an early pathological feature of many neurodegenerative diseases, including Alzheimer's (AD), Parkinson's (PD), Huntington's (HD) and prion diseases, and amyotrophic lateral sclerosis (ALS). These age-related conditions are characterised by the death of neurons: highly specialised, unique cells that form the foundation of the nervous system. Interestingly, many Golgi-related functions are also dysregulated in these diseases. However, this has received relatively little attention compared to other pathogenic mechanisms. The Golgi apparatus in neurons shares features common to other eukaryotic cells but it also has unique properties, such as the presence of distinctive assemblies: Golgi outposts and satellites, which remain poorly characterised. Here we discuss the increasing evidence describing dysfunction and fragmentation of the Golgi apparatus and its possible role in the pathogenesis of neurodegenerative diseases.

Keywords Neurodegenerative diseases · Golgi apparatus · Neurons · Golgi outposts · Neurodegeneration · Vesicular transport

---

Shashi Gautam, Zeinab Takalloo, Fabiha Farzana and Sonam Parakh contributed equally.

S. Shadfar · S. Assar Kashani · S. Gautam · Z. Takalloo · F. Farzana · S. Parakh · J. D. Atkin (✉)
Motor Neuron Disease Research Centre, Macquarie Medical School, Faculty of Medicine, Health and Human Sciences, Macquarie University, Sydney, NSW, Australia
e-mail: julie.atkin@mq.edu.au

A. Nakano, J. Saraste (eds.), *The Golgi Network, Volume II*, Subcellular Biochemistry 111, https://doi.org/10.1007/978-3-032-16833-7_17

Introduction

The Golgi apparatus is a highly dynamic organelle present in most eukaryotic cells, including neurons. It was first identified by the Italian neuroscientist Camillo Golgi in 1898 in neurons of the cerebellum. However, the Golgi apparatus in neurons remains poorly characterised compared to other cell types.

The classical functions of the Golgi apparatus are well characterised. It plays a crucial role in the processing and sorting of proteins and lipids into membrane-bound vesicles before they are transported to their final destinations. The Golgi is also the major cellular site for post-translational modification of proteins and lipids including glycosylation [1], sulfation [2], and proteolytic cleavage [3]. However, it is increasingly linked to many other cellular activities and signalling mechanisms including autophagy, apoptosis, DNA repair, and the cytoskeleton (both microtubules and actin). The Golgi resides centrally within the cell, at the intersection of the secretory, endocytic and lysosomal transport pathways and it is recognised to be a central hub for regulating multiple cellular events.

The Golgi Apparatus

Detailed structural and functional descriptions of the Golgi apparatus are covered in other chapters in this volume and thus will not be covered in detail. A concise summary of those aspects relevant to neurodegenerative diseases is provided below.

Golgi Structure and Morphology

The Golgi apparatus consists of flattened stacks of membrane or 'cisternae', and it is usually located within the perinuclear region of the cell, near the centrosome. Like other organelles it has distinct zones that regulate specific functions [4] due to differences in protein composition of the cisterna. Its three separate functional compartments are the *cis*-Golgi - the closest region to the ER, the *medial*-Golgi, where proteins are modified, sorted and packaged for transportation, and the *trans*-Golgi network (TGN), closest to the plasma membrane.

Specific types of vesicles also associate and either fuse with or bud from the Golgi apparatus. Newly synthesised protein cargo destined for secretion or transmembrane constituents are transported from the endoplasmic reticulum (ER) to the Golgi in the anterograde direction via vesicles coated in coatomer protein complex II (COPII). This produces the ER-Golgi intermediate compartment (ERGIC) which eventually fuses with the *cis*-Golgi [5]. Vesicles transporting proteins from the *cis*-Golgi back to the ER in the retrograde direction are coated with a different coatomer complex,

COPI. In addition, clathrin-coated vesicles are produced by the reticular TGN and targeted to endosomes or the cell surface [5].

The morphology of the Golgi is complex and highly dynamic, and it can quickly disassemble and reassemble during normal physiological events including mitosis, cell migration, polarization, differentiation, and cell death. Structural impairments of the Golgi morphology, resulting in its dispersal or fragmentation, are well described in many pathological conditions, including cancer, congenital disorders, and infectious and neurodegenerative diseases.

The Central Role of the Golgi in Signalling Pathways

Rab GTPases are critical regulators of vesicle trafficking and membrane dynamics. Rab1 and Rab2 are localized at the *cis* Golgi where they mediate ER-to-Golgi and intra-Golgi transport, and Rab8 localizes to the TGN/recycling endosomal system. Disruption of their functions leads to protein mislocalization and Golgi fragmentation. A wide array of signalling proteins are present on the membranes of the Golgi [5], including G-proteins, kinases, phosphatases and phosphoinositide kinases [5]. The Golgi also serves as a calcium reservoir critical for cellular signalling. Hence, not surprisingly, the Golgi is implicated in a wide range of cellular pathways, both related to, and distinct from, its trafficking functions [6].

The Golgi is a critical component of the unfolded protein response (UPR), which is activated in response to the accumulation of misfolded proteins in the ER [7]. It is also an important non-centrosomal microtubule organising centre (MTOC), from which microtubules are produced by a process known as nucleation [8]. Golgi-localised proteins Golgi Reassembly and Stacking Protein of 65 kDa (GRASP65) and Golgi Coiled-Coil Protein of 88 kDa (GCC88) also mediate the polymerisation of actin, another cytoskeletal component [9]. The Golgi is also implicated in DNA repair in the cytoplasmic, rather than nuclear, response to DNA damage [10]. It also regulates autophagy, an important mechanism by which damaged or redundant proteins/organelles are degraded by the lysosome. The Golgi apparatus can in turn be selectively degraded as autophagic cargo by 'Golgiphagy' [11].

There is evidence that the Golgi functions in pathways that regulate cell growth, proliferation, metabolism and/or survival, which involve mechanistic target of rapamycin (mTOR), phosphoinositide 3-kinase (PI3K)/AKT and transforming growth factor-beta (TGF-β) signalling [12]. The Golgi is also involved the nuclear factor kappa-light-chain-enhancer of activated B cells (NF-κB) response [13] which mediates immune responses, inflammation, and cell survival, as well as in Wnt signalling, which is critical for embryonic development, cell proliferation, and differentiation [13].

The Golgi Stress Response

The Golgi stress response is emerging as an important and normal cellular mechanism akin to the UPR and mitochondria stress responses. However, in comparison it is poorly characterised [14]. The Golgi apparatus is regulated in accordance with cellular demands, but under certain conditions, its capacity is insufficient, which triggers the Golgi stress response. This mechanism is activated in response to disruption of the function or structure of the Golgi and it aims to relieve the stress and restore homeostasis to the Golgi. However, if the stress cannot be resolved, the Golgi fragments and the cell undergoes apoptosis [14, 15]. Hence, the Golgi also acts as a sensor that regulates the entry of the cell into apoptosis. Several mechanisms can activate the Golgi stress response, including inhibition of Golgi-resident enzymes or transporters, alterations in signalling pathways impacting Golgi function, disruption of Golgi structure or trafficking, overproduction of Golgi cargo proteins, or changes in lipid composition. The Golgi stress response also contributes to maintaining redox homeostasis.

When the Golgi apparatus is stressed, cells activate several specific signalling pathways, involving transcription factor EB3 (TFE3), heat shock protein (HSP)-47 and c-AMP response element-binding protein 3 (CREB4-ARF4). TFE3 is a basic helix–loop–helix type transcription factor and a major regulator of the Golgi stress response. In normal cells, TFE3 is highly phosphorylated and retained in the cytoplasm. However, when glycosylation of proteins is impaired in the Golgi, TFE3 becomes dephosphorylated, and it translocates to the nucleus and binds to the Golgi apparatus stress response element (GASE). The latter is present in the promoters of specific genes activated during Golgi stress that are associated with Golgi structure and function. The TFE3 pathway is associated with the *trans*-Golgi network rather than the *cis*-Golgi [16].

HSP47 is a member of the heat shock protein family of chaperones. During stress to the Golgi, the levels of HSP47 increase. The Golgi fragments when expression of HSP47 is inhibited by siRNA [17], implying that HSP47 protects cells from Golgi stress-induced apoptosis. However, it remains unclear how a chaperone localized in the ER can protect cells from Golgi stress [16].

CREB3/Luman is a basic leucine zipper-containing ER transcription factor while ARF4 is a GTPase localized in the Golgi membrane that regulates Golgi-to-ER COPI-mediated transport. During Golgi stress, CREB3 is activated, resulting in release of its cytoplasmic domain from the ER membrane to the nucleus. This upregulates transcription of ARF4, resulting in Golgi disruption and apoptosis. Both the CREB4-ARF4 and HSP-47 pathways are closely associated with the ER, implying that these two mechanisms are associated with the *cis*-Golgi rather than the *trans*-Golgi [16].

The Structure and Function of Neurons

Neurons are unique and highly specialized cells that form the fundamental units of the central and peripheral nervous systems (CNS and PNS) [18, 19]. They mediate many of the complex activities mediated by the brain and spinal cord including thought, emotion, movement, motor co-ordination and sensory processing [18]. These functions are facilitated by the formation of vast, intricate neural networks through synapses; specialized junctions where neurons communicate with each other (or muscles) electrically and chemically [18]. Each neuron has the capacity to connect to thousands of neigbours [18, 19].

The cell body (soma) of a neuron contains the nucleus and other organelles present in most cell types. However, there are two structures unique to neurons that facilitate their specialised functions: dendrites and axons [18]. These two assemblies are essential for the extensive information processing required for complex neuronal functions. Dendrites are branch-like structures that receive electrical impulses (action potentials) from other neurons at synapses and transmit them to the cell body, whereas axons transmit electrical impulses from the cell body to the synapse. A typical multi-polar neuron contains only one axon but many dendritic trees [18]. Some neurons are myelinated, whereby their axons are covered with a myelin sheath, which insulates the axon and speeds up electrical transmission [18]. This is important because axons can be very long in some types of neurons, such as motor neurons where axons can be up to several cm in length. Transporting proteins and RNA along the axon involves a specialised process driven by motor proteins along microtubules - 'axonal transport' [18, 19].

While most other cell types are regularly renewed, neurons are post-mitotic, thus they do not divide and are not replaced throughout an organism's lifespan [20]. This longevity is coupled with a limited capacity for repair and regeneration, which renders them particularly vulnerable to degeneration [21, 22]. Neurons are also large cells that consume a lot of energy to maintain their resting potential and support synaptic activity [20]. Thus, they rely heavily on a constant supply of oxygen and glucose and hence they are also highly susceptible to metabolic dysfunction. Whilst other cell types can dilute their waste materials by cell division, neurons cannot perform this activity [20, 23]. Hence, they are prone to the accumulation of waste products, including misfolded proteins [23]. This is the major pathological hallmark of neurodegenerative diseases.

Non-neuronal Cells in the Nervous System

The nervous system is comprised of more cells than neurons only. Glial cells provide support to neurons and perform many important homeostatic and signalling functions. Glial cells are much more abundant than neurons and they include astrocytes, microglia, ependymal cells and oligodendrocytes in the CNS, and Schwann cells, satellite cells, and enteric cells in the PNS. Unlike neurons, glial cells undergo mitosis.

Astrocytes are large, star-shaped cells that function in synaptogenesis, synapse removal during embryonic development, and maintenance of the blood–brain barrier (BBB). Microglia are the immune cells of the CNS, converting from a resting or surveillant (ramified) state to an amoeboid phagocytic state following inflammation. Oligodendrocytes form the myelin sheaths of axons in the CNS and Schwann cells perform an equivalent function in the PNS.

The Golgi Apparatus in Neurons and Glial Cells

Neurons have a vast surface area due to specialized extensions in both dendrites and axons—dendritic spines and axonal growth cones, respectively. Remarkably, up to 99% of a neuron's cytoplasm is contained within these structures, emphasizing their crucial role in cellular function and communication. Synapses are densely packed with proteins, reflecting the highly specialized demands of neurons for neurotransmitter release, receptor recycling, and synaptic plasticity. Synaptic terminals in particular require substantial protein synthesis to support these functions. The Golgi apparatus plays a key role in regulating the synthesis, modification, and transport of these essential synaptic proteins, presenting a challenge in efficiently managing their supply. However, it is now well established that neurons synthesize proteins locally in both axons and dendrites, without involving the cell body (soma). This enables neurons to more rapidly respond to stimuli without relying solely on the cell body for protein production. Axons and dendrites contain mRNA, ribosomes and translation machinery. In dendrites, local protein synthesis is crucial for synaptic plasticity, learning, and memory, while in axons, it facilitates growth cone guidance during development, and regeneration after injury. Besides synaptic activity, the Golgi is also implicated in neuron-specific activities such as neurogenesis, migration, maturation of postmitotic neurons, and myelination.

There are several unusual features of the neuronal Golgi that reflect the specialized structure and function of these cells. These include the unique, but poorly characterised, mini-Golgi organelles present in both neurons and glia: Golgi outposts and satellites. Neurons also contain Nissl bodies - specialised regions of the rough ER - that function closely with the Golgi to ensure efficient processing and transport of proteins required for neuronal function. The Golgi apparatus in neurons is thus uniquely adapted to meet the complex and highly regulated demands of these cells (Fig. 17.1). [24].

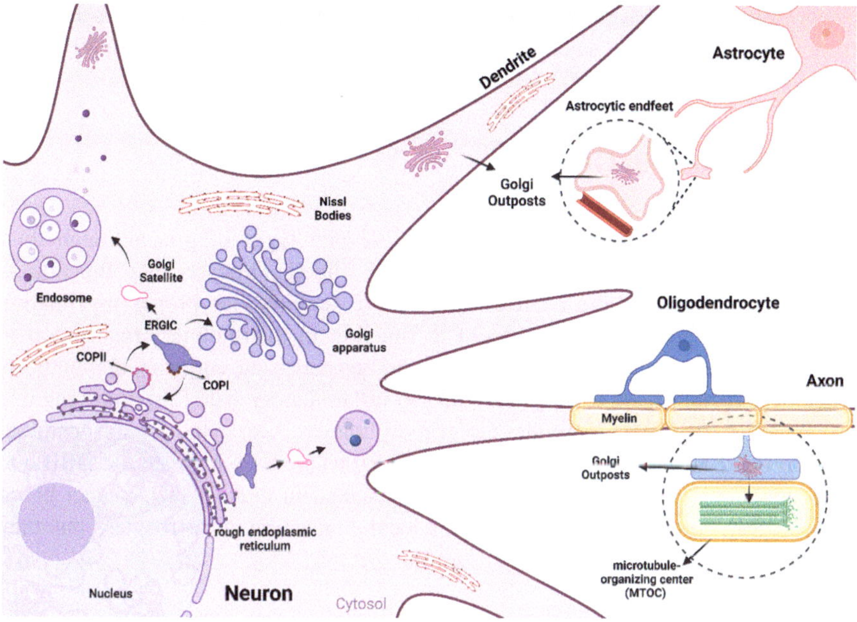

Fig. 17.1 Schematic figure illustrating the properties of the Golgi apparatus in neurons and glial cells. COPI and COPII vesicles mediate bidirectional transport between the ER and the Golgi. Golgi satellites function as local secretory compartments between the ER-Golgi intermediate compartment (ERGIC) and endosomes. Golgi outposts, present in dendrites and glia, contribute to local protein processing and microtubule organization. In oligodendrocytes, Golgi outposts support myelination, while in astrocytes, they are found in endfeet near blood vessels

Golgi Outposts and Satellites

Golgi outposts and satellites are separate structures that are not connected to the somatic Golgi although they share its functions and display similar molecular markers. They are much smaller than the cell body Golgi and have been identified in many types of neurons, including those affected in neurodegenerative diseases such as the hippocampus, cortex and cerebellum [25].

Golgi outposts are relatively abundant in dendrites but their presence and/or functionality in axons has been debated [26]. They are not present in all dendrites, and are estimated to exist in 5–20% of neurons [26], where they are restricted to the main dendritic branches. Golgi outposts are known to express membrane proteins and they modify locally synthesised glycoproteins in neurons [26]. The motility of Golgi outposts and the outgrowth of dendrites appear to be inter-related, implying that the Golgi regulates the morphology of dendrites [26].

Golgi satellites are usually much smaller than Golgi outposts and they function as an intermediate local secretory compartment between the ERGIC and endosomes. They are also implicated in mediating local protein glycosylation. In neurons post-Golgi vesicles have also been described. They are small vesicles that transport from Golgi outposts and satellites to dendritic or axonal membranes [26, 27].

Golgi Outposts in Glia

Golgi outposts have also been detected in several types of glia although their functions have not been well characterised. In oligodendrocytes they function as MTOCs distinct from the centrosome and are required for proper myelination [28]. Golgi outposts are also thought to function in microtubule organisation in microglia. Microglia in the normal ramified homeostatic state contain Golgi outposts whereas they are absent in the amoeboid pro-inflammatory state [29], implying that inflammation regulates microglial Golgi morphology. Astrocytes contain 'endfeet' that surround all blood vessels within the CNS, where they regulate the BBB and other functions. Golgi outposts have been detected in endfeet that contact blood vessels [30], implying they function in local translation in astrocytes, but they remain poorly characterised [25].

Golgi Fragmentation in Neurons

Golgi fragmentation occurs when its characteristic ribbon-like structure is disrupted into multiple mini-stacks. This process can be either reversible or irreversible and it is triggered by disruptions in various cellular mechanisms. These include imbalances in vesicle trafficking, particularly between anterograde and retrograde transport at the Golgi, as well as cytoskeletal instability caused by microtubule destabilization, which compromises Golgi ribbon integrity. Additionally, elevated oxidative stress damages Golgi membranes, accelerating fragmentation and linking this process to mitochondrial dysfunction. Despite fragmentation, some Golgi functions, such as membrane transport (except for that involving large cargo) and protein/lipid glycosylation, remain intact. This is expected because Golgi fragmentation occurs reversibly under normal physiological conditions, such as during ER-Golgi transport inhibition, neuronal death, differentiation, and polarization [31]. This suggests that reversible Golgi fragmentation may serve as a compensatory mechanism for rapid protein transport under cellular stress rather than a pathological response. However, Golgi fragmentation can severely impact other functions of the Golgi, including initiating the stress response, apoptosis, cell migration cell polarity, and entry into mitosis [31].

The Golgi apparatus in neurons is often more fragmented and dispersed than in other cell types, especially in axons and dendrites [26]. This fragmentation appears to be a normal physiological response, because it occurs reversibly during electrical stimulation of neurons [32]. In contrast, irreversible Golgi fragmentation is described during pathological conditions and in response to environmental stress.

This results in permanent structural breakdown of the Golgi, which disrupts its normal functions and leads to impaired protein sorting, post-translational modification and trafficking, cellular stress and neuronal death [33]. Hyperexcitability or excessive stimulation of neurons also leads to irreversible Golgi fragmentation [32]. Fragmentation of the neuronal Golgi apparatus has been consistently reported as an early, preclinical event in neurodegenerative diseases.

Neurodegenerative Diseases

Neurodegenerative diseases are a wide range of conditions characterised by the progressive loss of neurons [21]. These include dementias such as Alzheimer's disease (AD) and frontotemporal dementia (FTD), movement disorders such as Parkinson's disease (PD) and Huntington's disease (HD), diseases affecting motor neurons such as amyotrophic lateral sclerosis (ALS), and prion diseases such as Creutzfeldt-Jakob disease (CJD). Most cases of neurodegenerative diseases arise sporadically although there is also a significant genetic component in most diseases, where mutations in specific genes are present in a small proportion of cases. However, some conditions, such as polyglutamine diseases, are entirely genetic.

The main risk factor for neurodegenerative diseases is increasing age, and due to the global rapidly aging population, the incidence of these conditions is predicted to significantly increase in the coming decades [22, 34]. By 2050, the global population of AD patients alone is predicted to exceed 150 million individuals [35]. Due to the limited treatment options for neurodegenerative diseases, they are a significant burden on society. The alarming rise in cases in the future also poses a significant economic burden, straining healthcare systems and caregiving resources worldwide, representing a major public health crisis of the 21st century [36].

The common pathological hallmark of neurodegenerative diseases is the presence of misfolded protein aggregates, either extracellular or intracellular, localised within both neurons and glia. Whilst the symptoms of neurodegenerative diseases can vary widely, as well as the location of neurons affected and the protein that misfolds in each condition, they share many common pathological processes at the cellular and molecular level, including Golgi fragmentation. However, this has generally received relatively little attention compared to other pathological processes. Below we discuss neurodegenerative diseases and the evidence describing how Golgi disruption relates to pathophysiology in these conditions.

The Golgi Apparatus in Neurodegeneration

As the Golgi is integral to maintaining cellular homeostasis, it is not surprising that dysregulation of the Golgi is implicated in the pathophysiology of neurodegenerative conditions [37]. Whilst Golgi fragmentation is a widely described pathological event, there are other ways that dysfunction to the Golgi is implicated in these conditions. Many cellular activities and signalling mechanisms linked to the Golgi, including autophagy, apoptosis, DNA repair, vesicular trafficking and cytoskeletal function (both microtubules and actin), are dysregulated in neurodegenerative diseases. Interestingly, >40% of genes associated with the Golgi are also implicated in diseases affecting the CNS or PNS, highlighting the importance of the Golgi in neuronal functions [24].

Alzheimer's Disease (AD)

AD is the leading cause of dementia worldwide [38] and it is caused by degeneration of the hippocampus and entorhinal cortex in the brain [39]. The symptoms of AD range from memory loss, cognitive impairment (aphasia and agnosia), hallucinations, and delusions, to behavioural disturbances [40, 41]. Two key pathological features involving misfolded proteins are characteristic of AD: amyloid plaques and neurofibrillary tangles (NFTs). Amyloid plaques are extracellular deposits of β-amyloid (Aβ) protein, primarily composed of Aβ40 or Aβ42 [42] produced from amyloid precursor protein (APP). In contrast, NFTs consist of twisted strands of hyperphosphorylated tau protein that accumulate and aggregate within neurons [43]. There are over 40 genes linked to familial forms of AD.

Golgi morphological alterations are present in postmortem neurons of AD patients [44, 45] and in tissue culture and mouse models [46]. It is among the earliest phenotypic changes in human induced pluripotent stem cell (iPSC) and mouse AD models, suggesting that protein processing and post-translational modifications are disrupted [47]. Notably, Golgi fragmentation has been observed before the formation of Aβ plaques and tau aggregation in a transgenic mouse model (APPswe/PS1ΔE9), suggesting it is an early event in disease progression [46, 48].

The mechanisms underlying Golgi fragmentation in AD remain poorly understood but are likely multifactorial. Golgi fragmentation is induced by Aβ accumulation in cultured neurons and other cells, which is reversible upon removal of Aβ [46]. Disruption of the microtubule network due to tau aggregation and NFT formation also induces Golgi fragmentation [49]. Tau may also disrupt vesicle trafficking and thus induce Golgi fragmentation by inhibiting the binding of motor proteins, kinesin and dynein, to microtubules [50]. Another possibility is that it acts through modulation of Golgi structural proteins [51].

Neuronal atrophy is a key indicator of declining cellular metabolism, a factor extensively studied in relation to AD [46]. One highly sensitive marker for metabolic alterations in neurons is change in the size of the Golgi apparatus, which reflects disruptions in protein processing and function [46]. The size of the Golgi apparatus in the cholinergic nucleus in the forebrain gradually increases after the age of 50 years [52]. This enlargement may occur earlier and more prominently in AD, suggesting a potential link between Golgi dysfunction and disease progression [52]. However, contradictory findings were obtained in a more recent study that reported a significant decrease in Golgi size in AD neurons compared to healthy controls, which correlated with tau hyperphosphorylation [33].

Cellular regions associated with the Golgi are also implicated in AD. Aβ can be produced within endosomes [53] and the TGN. APP and the enzyme beta-site amyloid precursor protein cleaving enzyme 1 (BACE1), which cleaves APP, are processed within the Golgi apparatus [54]. This limits APP cleavage by BACE1 and reduces Aβ production [55, 56]. However, when transport of BACE1 within the Golgi is disrupted, APP cleavage increases, leading to higher Aβ levels. Elevated Aβ can raise intracellular calcium levels, activating calpain and triggering excessive activation of cyclin-dependent kinase 5 (CDK5) [46]. This phosphorylates GRASP65 and disrupts the interaction between Golgi Matrix protein of 130 kDa (GM130) and p115, inducing Golgi stress [46]. Golgi stress also increases vesicle budding and disrupts protein glycosylation [57] leading to enhanced Aβ deposition by accelerating APP transport and BACE1 accumulation [46, 58]. Overall, the connection between Golgi stress and Aβ plaque formation suggests a complex interplay that requires further investigation.

Under physiological conditions, tau plays an essential role in microtubule binding and stabilization, thereby orchestrating vesicle formation, membrane trafficking, and protein transport [59–61]. However, Golgi stress compromises normal tau functions and disrupts microtubule structure [62, 63], leading to alterations in Golgi positioning and vesicular transport [59, 64, 65]. These disruptions impair post-translational modification, processing, and sorting of proteins, thereby contributing to the pathological cascade in AD [33]. However, the interplay between Golgi stress and tau hyperphosphorylation remains under debate. Some studies suggest that tau hyperphosphorylation induces Golgi stress prior to NFT formation [49]. Conversely, other research indicates that Golgi stress in AD instead promotes tau hyperphosphorylation by activating CDK5 and extracellular signal-regulated kinases, leading to Golgin-84 depletion, NFT formation, and microtubule destabilization [64]. Furthermore, Golgi stress has been linked to a transient increase in GSK-3β activity, which enhances tau phosphorylation, disrupts microtubules, and compromises neuronal function [66]. Collectively, these findings suggest that Golgi stress may serve as a key early event in tau hyperphosphorylation in AD. However, further studies are necessary to elucidate the precise regulatory mechanisms underlying this relationship.

Parkinson's Disease

Parkinson's Disease (PD) primarily affects the motor system, resulting in bradykinesia, resting tremor, rigidity, and postural instability [67]. However significant non-motor symptoms are also present: cognitive decline, sleep disturbances, gastrointestinal dysfunction, and depression, which often precede the motor symptoms [68]. PD is characterized by the loss of dopaminergic neurons in the substantia nigra pars compacta region of the brain [69, 70], and the accumulation of misfolded α-synuclein protein into intracellular inclusions known as Lewy bodies [69]. There are over 20 genes linked to familial forms of PD, and several encode proteins associated with the Golgi apparatus.

Golgi fragmentation is also well described in PD. Postmortem analyses of dopaminergic neurons with Lewy bodies in the substantia nigra pars compacta has revealed fragmented Golgi structures in PD patients [71]. Notably, neurons containing early-stage Lewy bodies exhibit a significantly higher prevalence of fragmented Golgi structures (19%) compared to neurons with mature Lewy bodies (5%) or without inclusions (3%) [72]. This suggests that Golgi fragmentation occurs early in disease progression and may be driven by the cytotoxic effects of α-synuclein aggregates before they mature into Lewy bodies. In cellular PD models, Golgi fragmentation is a reproducible feature that occurs in response to oxidative stress, cytoskeletal disruption [73] and prior to α-synuclein aggregation and neurodegeneration, also indicating that it is an early pathological event [37].

Under physiological conditions, α-synuclein associates with Golgi membranes, where it interacts with lipid bilayers and the vesicular trafficking machinery. However, in PD, aggregation of α-synuclein impairs ER-to-Golgi trafficking, leading to fragmentation of the Golgi ribbon [37]. Overexpression of Rab1 or Rab8 can restore Golgi morphology in PD models, highlighting the importance of maintaining vesicle trafficking in the integrity of the Golgi [74]. Additionally, aggregated α-synuclein destabilizes microtubules, disrupting the cytoskeletal support required for maintaining Golgi structure and vesicle transport [75].

Leucine-Rich Repeat Kinase 2 (LRRK2) is one of the most frequently mutated genes in both familial and sporadic PD. It encodes a protein that plays a critical role in vesicle trafficking and the function of the Golgi. LRRK2 maintains stationary Golgi outposts and prevents their movement, as well as restricts dendritic branching [26]. LRRK2 interacts with ArfGAP1 at Golgi membranes and a PD mutation in LRRK2, G2019S, enhances its kinase activity, leading to hyperphosphorylation of Rab GTPases, which disrupts vesicle trafficking at the Golgi [76], inducing Golgi fragmentation.

Rab29 plays an important role in vesicle trafficking in the lysosomal system, including autophagy and synaptic function, and it is also mutated in familial forms of PD. Rab29 is specifically localized to the TGN, and it acts as a master regulator of LRRK2 [77]. Dysregulation of the Rab29-LRRK2 interaction leads to LRKK2 recruitment to the Golgi apparatus, and enhancement of its kinase activity. Expression of pathogenic LRRK2 mutants such as G2019S and R1441G lead to Golgi fragmentation and impaired vesicle trafficking [78]. These disruptions are linked to endolysosomal dysfunction, a process that is relevant to PD pathogenesis [78]. PD-associated Rab29 mutations result in impairment of autophagy and trafficking [79]. Mutations in another Rab, Rab39B, are also present in PD. Rab39B is highly expressed in neurons, and it is localized to the Golgi apparatus, specifically at the junction of the *cis*-Golgi and ER. Mutations in Rab39B in PD lead to alterations in Golgi morphology, decreased vesicle trafficking, synaptic dysfunction and α-synuclein aggregation [80].

Mutations in vacuolar protein sorting 35 (VPS35), a core component of the retromer complex, are also present in PD. VPS35 is essential for retrograde transport from endosomes to the TGN [81]. PD mutations disrupt Golgi associated trafficking by impairing the retrieval of critical proteins such as LAMP-2a and cathepsin D [82] and inducing α-synuclein aggregation [83]. These mutations also enhance LRRK2 kinase activity at the TGN by promoting hyperphosphorylation of Rab proteins. Dysfunctional VPS35-LRRK2 interactions contribute to protein mislocalization [84].

In addition to structural changes, functional impairments of the Golgi apparatus have been observed in PD. Disruptions in ER-to-Golgi and intra-Golgi transport pathways are observed in cellular and animal PD models [85]. Defective glycosylation patterns in PD brains, including changes in sialylation and fucosylation, likely due to Golgi dysfunction, also alter protein processing and aggregation [86]. In cellular [87] and animal [88] PD models, disruptions in Golgi calcium homeostasis exacerbate ER stress and mitochondrial dysfunction, impairing neuronal vulnerability [89].

MPTP (1-Methyl-4-phenyl-1,2,3,6-tetrahydropyridine) is a toxin used to selectively damage dopaminergic neurons in the substantia nigra to produce animal models of PD. Cyclin-dependent kinase 5 (CDK5) regulates the structure, function and fragmentation of the Golgi [90] and in MPTP-induced mouse models, CDK5 hyperactivation exacerbates dopaminergic neuron loss in the substantia nigra pars compacta. In contrast, inhibition of CDK5 protects against neurodegeneration [91].

ALS

Amyotrophic lateral sclerosis (ALS), also known as motor neuron disease (MND), is characterized by the progressive degeneration of motor neurons in the spinal cord, brainstem and brain [92]. It leads to gradual muscle weakness, atrophy and paralysis, typically resulting in death within 3 to 5 years after diagnosis due to failure of respiratory muscles [93]. The major pathological hallmark of ALS is the presence of pathological forms of Tar DNA binding protein-43 (TDP-43) in affected tissues. This involves the misfolding, truncation and hyperphosphorylation of TDP-43 into aggregates that mis-localise from the nucleus to the cytoplasm in almost all cases (>97%). More than 40 genes are now associated with ALS, including several Golgi-associated proteins such as optineurin and vesicle-associated membrane protein-associated protein B (VAPB). Several Golgi-associated pathways are associated with pathophysiology in ALS including autophagy and dysfunctional ER-Golgi trafficking.

Hexanucleotide repeat expansions (GGGGCC) in a non-coding region of the chromosome open reading frame 72 (C9orf72) gene cause most familial ALS cases (40–50%) [94], followed by mutations in superoxide dismutase 1 (SOD1, 20%) and TDP-43 (4–5%) [95]. The C9orf72 mutations lead to pathological RNA foci and dipeptide repeat proteins (DPRs) produced by repeat-associated translation of the repeat expansion. The C9orf72 protein functions in endosomal-lysosomal dynamics, regulating autophagosome formation and vesicular trafficking [96].

Golgi fragmentation is a well described pathological feature [37, 97] that was identified first in ALS among all neurodegenerative diseases. It has been detected in motor neurons of the spinal cord, brain and brainstem of sporadic and familial ALS patients [98–100]. Importantly, it is associated with the major pathological hallmark in sporadic ALS patients: mislocalisation of TDP-43 in the cytoplasm [98]. Golgi fragmentation is also present in the widely used transgenic mice model expressing mutant SOD1^{G93A} where it precedes the formation of SOD1 inclusions, neuromuscular denervation and apoptosis [101]. It has also been detected in TDP-43 rodent models: rats expressing TDP-43^{M337V} [102]. Similarly, in neuronal cells, expression of C9orf72 DPRs (polyGA, polyGP and polyGR), or ALS-associated mutant forms of SOD1 (both extra- and intracellular), TDP-43, Fused in Sarcoma (FUS) and VAPB, as well as knockdown of VAPB induce Golgi fragmentation [103–107]. Mutations in other ALS-associated genes, including ubiquilin 2 (UBQLN2) [108] and cyclin F [109] also induce Golgi fragmentation in neuronal cell lines or human cortical neurons. Together this implies that Golgi fragmentation is an upstream, pathogenic event common to most ALS cases rather than a secondary consequence of neurodegeneration.

Disruption of microtubule-associated proteins crucial for maintaining Golgi structure, including tubulin and stathmin 1/2, are implicated as triggers for Golgi dysfunction in ALS. Mutant SOD1 reduces acetylated tubulin levels [106] and increased levels of stathmin-1, which destabilises microtubules, are present in the SOD1^{G93A} mouse model [110]. Golgi fragmentation in transgenic SOD1^{G93A} mice motor neurons, and in another SOD1 model, mutant SOD1^{G85R}, is triggered by progressive up-regulation of stathmins 1 and 2, leading to impaired polymerization of Golgi-derived microtubules and loss of the COPI coat [111]. Knockdown or cytoplasmic mislocalisation of TDP-43 decreases stathmin-2 levels in iPSC-derived motor neurons and spinal cords [112]. Importantly, mislocalised TDP-43 induces dysregulated RNA splicing and cryptic exon inclusion in stathmin-2 pre-mRNA, which is implicated as an important pathogenic mechanism in ALS/FTD [113]. ALS-associated mutations in dynactin, which mediates dynein-mediated retrograde transport [114], and the alpha isoform of tubulin, also induce Golgi dysfunction [115]. However, Golgi fragmentation is not present in transgenic mice with motor neuronopathy overexpressing the heavy subunit of human neurofilament (NF-H), suggesting that microtubule dysfunction alone does not trigger Golgi dysfunction [116].

ER-Golgi trafficking is impaired by multiple ALS-associated mutations, including C9orf72 DPRs polyGA, polyGP and polyGR [103], and mutant forms of SOD1, TDP-43, FUS UBQLN2 and cyclin F [105, 106, 108, 109]. Whilst mutant TDP-43 and FUS specifically disrupt ER-to-ERGIC transport, mutant SOD1 dysregulates ERGIC-to-Golgi transport [106]. In contrast, ALS-associated mutations in optineurin inhibit trafficking from the Golgi to the plasma membrane [97]. Golgi dysfunction has also been linked to aberrant lipid metabolism and glycosylation patterns of gangliosides GM1, GM3 and Glycosyltransferase 8 domain containing

1 (GLT8D1) in ALS patients and $SOD1^{G93A}$ transgenic mice [117]. Together these findings imply that Golgi dysfunction is important in pathophysiology in ALS.

Huntington's and Other Polyglutamine Diseases

Polyglutamine (polyQ) diseases are inherited neurodegenerative diseases caused by expansion of trinucleotide CAG repeat sequences (encoding glutamine, Q) within the protein-coding regions of specific genes. This results in an abnormally long polyQ tract in the corresponding proteins. These conditions include Huntington's disease (HD), spinocerebellar ataxias (SCA) and Machado-Joseph disease (MJD). The symptoms of polyQ diseases vary depending on the condition, the affected brain region, and the length of the CAG repeat expansion. Unlike AD, PD and ALS, where most cases arise sporadically, polyQ conditions are entirely genetic in origin.

Huntington's disease (HD) is characterized by disturbances in cognition, behavior, neuropsychiatry, and the presence of choreiform (involuntary, rapid, irregular) movements due to degeneration of neurons in the corpus striatum [118]. HD is caused by autosomal dominant CAG repeats in the huntingtin gene (*HTT*) on chromosome 4 [118–120]. Normally, there are 10 to 35 CAG repeats in *HTT* whereas in HD, this is expanded to more than 40 repeats [120]. Huntingtin is a large scaffolding protein that is associated with endocytic and exocytic vesicles to facilitate protein trafficking at multiple stages of the secretory pathway, including those involving the Golgi apparatus. It has multiple binding partners and coordinates trafficking along both microtubule and actin components of the cytoskeleton.

Huntingtin mediates both ER-Golgi and post-Golgi trafficking of proteins [121, 122], as well as vesicle fusion at the plasma membrane. Huntingtin also plays a role in selective autophagy by interacting with optineurin at the Golgi apparatus via myosin VI [123–125]. Optineurin is a Rab8 effector that mediates the post-Golgi transport of proteins [126] and huntingtin facilitates localization of the optineurin/Rab8 complex at the Golgi [127]. Mutant huntingtin displays reduced localization in the Golgi and formation of the optineurin/Rab8 complex is impaired, disrupting transport at the TGN [127].

Golgi fragmentation has not yet been detected in HD patient tissues, although it has been identified in disease models. Overexpression of mutant huntington in cellular models leads to alteration of Golgi morphology and formation of more clathrin-coated vesicles [127]. The Golgi-specific protein acyl-CoA-binding domain-containing 3 (ACBD3) induces toxicity in HD by formation of a complex with striatal-specific protein Rhes and mutant huntingtin [128]. ACBD3 levels are increased in the striatum of HD patients, in brains of HD mouse models and striatal cell lines expressing polyQ, and are further elevated by induction of Golgi stress [128]. Oxidative stress, an imbalance between pro-oxidant and antioxidant pathways, is a feature of all neurodegenerative diseases, including HD. In cellular models of HD, oxidative stress induces Golgi stress by reducing the levels of antioxidant transporter sodium-ascorbate co-transporter 2 (SVCT2) [129]. Similarly, depletion of cysteine is implicated in HD but this can be reversed by induction of the Golgi stress response [128].

Spinocerebellar ataxias (SCAs) are a diverse group of up to 40 polyQ disorders. Spinocerebellar ataxia type 2 (SCA2) is an autosomal dominant disease that results from expansion of the polyQ repeat in ataxin-2. This protein is predominantly localized in the ER and Golgi apparatus and widely expressed in the human brain. SCA2 primarily affects Purkinje neurons and select brainstem neurons, resulting in progressive ataxia and premature death [130, 131]. In cell culture and animal models expressing mutant ataxin-2, the presence of the expanded CAG repeat alters its Golgi localization, lead to Golgi dispersion, and disrupts the normal morphology and function of the Golgi complex [132].

Machado-Joseph disease, or spinocerebellar ataxia type 3 is characterised by lack of balance and coordination. Expression of the pathogenic 78 repeat MJD protein MJDtr-78Q results in loss of Golgi outposts and downregulation of COPII and secretion-related genes in *Drosophila* neurons, resulting in dendritic pathology [133].

Prion Diseases

Transmissible spongiform encephalopathies (TSEs) or prion diseases are a group of rare but rapidly progressing neurodegenerative disorders associated with misfolding of cellular prion protein (PrP^C). They can arise either sporadically, genetically, or through acquired transmission. Creutzfeldt-Jakob disease (CJD) is the most common sporadic prion disease (85% of cases) which is characterized by psychological symptoms, loss of physical co-ordination, muscle spasms and memory loss. Genetic forms result from mutations in the gene encoding PrP^C, *PRNP,* which include fatal familial insomnia. This condition is characterized by progressive insomnia, autonomic dysfunction and hypertension, and defects in cognition and endocrine dysfunction. Acquired prion diseases include Kuru, which resulted from human-to-human transmission through ritualistic endocannibalism, as well as iatrogenic forms of CJD caused by medical exposure, and zoonotic transmission from infected animals, such as bovine spongiform encephalopathy (BSE), leading to variant CJD in humans [134]. PrP^C is normally found at the cell surface in cholesterol-rich lipid raft membranes, but in prion diseases, it misfolds and aggregates to form the pathological form, β-sheet-rich PrP^{Sc} [135, 136]. PrP is a glycosylphosphatidylinositol (GPI)‐anchored membrane protein that follows the conventional ER–Golgi secretory pathway, being synthesized in the ER and processed in the Golgi apparatus, where it undergoes maturation through post translational modifications such as N-glycosylation [134].

Fragmentation of the Golgi in affected neurons has been described in five patients with CJD [137]. In familial TSEs, mutant PrP^C is thought to undergo spontaneous misfolding in the ER and/or Golgi apparatus which disrupts vesicular trafficking [138, 139]. Similarly, another study found that both recombinant DsRed– PrP^C–GFP fusion protein and endogenous PrP^C exhibit abnormal accumulation in the Golgi apparatus [140]. Infection of transgenic-PrP-EGFP/*Prn-p*$^{+/+}$ mice with scrapie (PrP^{Sc} obtained from sheep) resulted in the formation of aggregates within the neuronal Golgi apparatus of neurons [141]. Similarly, the fatal familial insomnia PrP mutant D178N/Met-129 aggregates prominently in the Golgi apparatus of neuronal cell lines [142]. In transgenic mice models, the D177N/M128 mutant PrP also accumulates in the Golgi, which becomes enlarged [143]. Trafficking from the Golgi apparatus to the plasma membrane is also disrupted in PrP^{Sc}-infected neuronal cells [136] and in mice prior to the onset of symptoms [144] implying that impairment of Golgi trafficking is linked to pathophysiology of prion diseases.

Rare Neurodegenerative Diseases

Hereditary spastic paraplegia (HSP) is a rare neurodegenerative disorder characterized by axonal degeneration, primarily affecting the corticospinal tracts and dorsal spinal cord. Clinically, HSP manifests as bilateral lower limb spasticity, weakness, hyperreflexia, extensor plantar responses, and gait abnormalities of varying severity [145]. Over 50 genes are linked to HSP and a subset of these encode Golgi-related proteins [146]. These include genes encoding atlastin-1 [147], acetyl-CoA transporter [148], maspardin [149], and DDHD domain containing proteins (DDHD), each associated with different HSP subtypes [145].

Atlastin-1, encoded by the *ATL1* gene, is a *cis*-Golgi-localized protein essential for Golgi morphogenesis. Mutations in atlastin-1 are the most common cause of hereditary spastic paraplegia (HSP) [147, 150]. Disruption of the *ATL1* gene leads to Golgi abnormalities, resulting in structural defects in *Drosophila* neurons [151] and impaired Golgi development [152]. Another gene mutated in HSP, *SLC33A1*, encodes acetyl-CoA transporter which facilitates movement of acetyl-CoA from the ER to Golgi [153]. Mutations in *SLC33A1* disrupt ganglioside and glycoprotein processing in zebrafish models [154]. Similarly, maspardin, encoded by the *ACP33* gene, is a resident protein of endosomes/TGN. It interacts with Rab7, suggesting a role in protein transport and sorting [155, 156]. Pathogenic variants in *ACP33* have been linked to Mast syndrome, a complex form of HSP associated with cognitive decline and dementia [149]. Mutations in genes encoding phospholipases DDHD1 and DDHD2, linked to HSP type 28 and HSP type 54 respectively, further support a role of Golgi dysfunction in HSP [157–160] because they are essential for maintaining Golgi-ER membrane dynamics [161, 162]. DDHD1 is critical for preserving ER-Golgi integrity [163], while DDHD2 regulates membrane transport from the Golgi to the plasma membrane [164]. To date, around 35 distinct mutations in the *DDHD2* gene have been identified in individuals diagnosed with HSP type 54 [165, 166]. Morphological changes in the Golgi and compromised vesicle trafficking have also been observed in cellular models expressing mutant DDHD2 [164, 167].

Progressive cerebello-cerebral atrophy (PCCA), also known as pontocerebellar hypoplasia (PCH), is a group of rare neurodegenerative disorders primarily affecting the pons and cerebellum [168]. More than 15 subtypes of PCH have been identified, which are associated with at least 25 different genes [169]. Autosomal recessive PCCA is caused by mutations in *VPS53*, which encodes a key component of the Golgi-associated retrograde protein (GARP) complex. This protein has crucial role in trafficking from recycling endosomes to the late Golgi complex [170]. Mutations in the *VPS53* gene result in a severe neurodevelopmental disorder characterized by profound intellectual disability, progressive microcephaly, spasticity, and early-onset epilepsy [171, 172]. Loss of VPS53 function disrupts proper localization and recycling of late Golgi membrane proteins, leading to neurodegeneration and the clinical manifestations of one type of PCH, PCH2E, in yeast and animal models [170, 173].

Warburg Micro syndrome (WARBM) is a severe, inherited neurological disorder characterized by profound intellectual disability, microcephaly, and other neurological manifestations [174, 175]. The genetic basis of WARM is heterogeneous, but four known subtypes are linked to mutations in genes encoding Golgi-associated proteins *RAB18, RAB3GAP1, RAB3GAP2,* and *TBC1D20* [176, 177]. Mutations in *RAB3GAP1*, the most frequent cause of WARBM, affect protein localization within the ER, Golgi apparatus, and the ERGIC [178, 179]. *RAB18* is implicated in the regulation of secretory granule transport, ER-Golgi trafficking, and lipid metabolism [180–182]. TBC1D20 is a GAP for Rab1, which mediates traffic between the ER and the Golgi complex.

Spinal Muscular Atrophy (SMA) is another entirely genetic disorder caused by degeneration of motor neurons in the spinal cord, leading to progressive muscle weakness, atrophy, and impaired movement. It is a leading cause of infant mortality [183, 184] and is caused by mutations in the survival motor neuron (*SMN*) gene, leading to loss of SMN protein. There are three types of SMA: Type 1; severe or Werdnig-Hoffmann Disease, Type 2: intermediate [185] and Type 3 (mild or Kugelberg-Welander Disease. There is limited evidence for Golgi fragmentation in SMA. However, loss of SMN protein leads to abnormal Golgi morphology in neuronal cell lines and fibroblasts derived from Type I and Type II SMA patients [186]. SMN protein interacts with the coatomer subunit α (Cop-α) of the COPI complex that regulates vesicle trafficking at the ER-Golgi interface and between Golgi compartments [187, 188]. Knockdown of α-COP in cultured neurons disrupts SMN localization, leading to its accumulation in the TGN [187].

Conclusion and Future Perspectives

The Golgi apparatus is central to maintaining the viability of neurons through its functions in trafficking, protein processing, and lipid metabolism, amongst many other processes. In neurons the Golgi has several distinct features that distinguish it from that found in other cell types. These include Golgi outposts and other Golgi structures, the ability to manage extensive trafficking requirements for synaptic function and long-distance transport along long axons, and local protein transport in neurites, that is independent of the cell body.

Fragmentation of the Golgi apparatus is detected more commonly in neurons than other cell types. Reversible fragmentation is thought to be a normal physiological response to the unique electrical activity of neurons. However, irreversible Golgi fragmentation is a widely described pathological feature in neurodegeneration, although for some of these conditions there is limited evidence from patient tissues. Nevertheless, findings obtained from cellular and rodent disease models suggest that Golgi dysfunction is an early event in the pathogenesis of many neurodegenerative diseases, involving both morphological and functional abnormalities. This implies it is upstream in the pathogenic cascade of events. However, the role of the Golgi in the pathophysiology of these conditions has been poorly studied in comparison to other disease mechanisms. Defects in the Golgi are also implicated through other more indirect mechanisms, including impaired autophagy, protein and lipid processing, increased cellular stress, disrupted vesicular transport involving the Golgi, cytoskeletal abnormalities, amongst other processes.

Figure 17.2 summarises our current understanding of Golgi dysfunction in four of the most common neurodegenerative diseases: AD, PD, ALS and HD. In AD, it contributes to key pathological processes: Aβ accumulation, tau hyperphosphorylation, and neuronal degeneration [33]. In PD, several Golgi genes are mutated in familial forms, while functional impairment of the Golgi exacerbates cellular stress pathways including oxidative and ER stress, mitochondrial dysfunction, and lysosomal deficits. In ALS, Golgi dysfunction driven by major ALS-associated proteins: TDP-43, SOD1, C9orf72, and FUS disrupts protein trafficking, impairing axonal transport, inducing ER stress and leading to neurodegeneration. The huntingtin protein normally functions in vesicle transport involving the Golgi. In HD, expression of the huntingtin protein containing abnormal expansion of the CAG repeat sequence, leads to impairment of trafficking and other Golgi-related defects. Similarly in prion diseases and several of the rarer neurodegenerative conditions, Golgi dysfunction is also implicated in pathophysiology. Together these findings imply that the impairments in the structure and function of the Golgi apparatus are important disease-related events in neurodegenerative conditions. They also imply that novel potential therapeutic strategies targeting the Golgi apparatus may be beneficial for neurodegenerative diseases in the future. Golgi dysfunction thus represents an important but somewhat neglected focus for research and therapeutic development for these conditions.

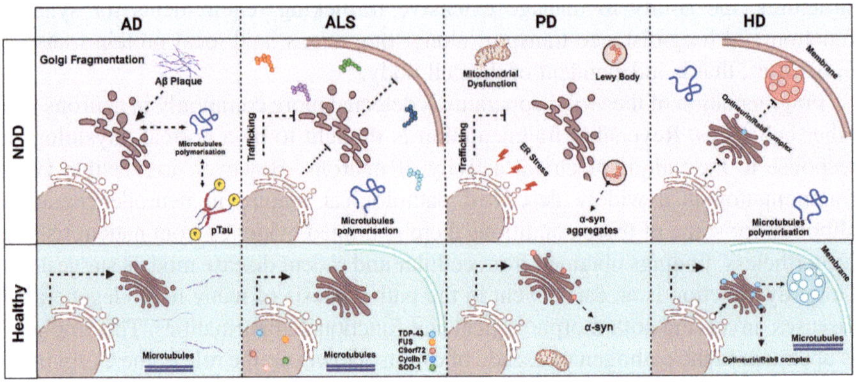

Fig. 17.2 Golgi Dysfunction in neurodegenerative diseases (NDD). Golgi dysfunction is implicated in multiple neurodegenerative diseases, disrupting protein trafficking, cytoskeletal dynamics, and contributing to disease pathology. Four of the most common conditions are illustrated. In Alzheimer's disease (AD), β-amyloid (Aβ) and phosphorylated tau (pTau) disrupt Golgi structure, impair vesicular trafficking, and destabilize the microtubule network, affecting axonal transport. In Parkinson's disease (PD), α-synuclein aggregates within Lewy bodies, lead to Golgi fragmentation and inhibition of Rab1-mediated ER-to-Golgi trafficking. Mitochondrial dysfunction, a hallmark of PD, also exacerbates cellular stress, impairing Golgi function. In ALS, mutant SOD1, TDP-43, C9orf72, FUS, and cyclin F disrupt Golgi integrity, impair vesicle transport, and induce Golgi fragmentation, while also affecting microtubule stability and actin dynamics. In HD, mutant huntingtin (mHTT) disrupts Golgi-associated vesicular trafficking, leading to protein mislocalization and cellular toxicity, alongside alterations in microtubule-based transport and cytoskeletal organization

References

1. Stanley P (2011) Golgi glycosylation. Cold Spring Harb Perspect Biol 3(4)
2. Baeuerle PA, Huttner WB (1987) Tyrosine sulfation is a trans-Golgi-specific protein modification. J Cell Biol 105(6 Pt 1):2655–2664
3. Xu H, Shields D (1993) Prohormone processing in the trans-Golgi network: endoproteolytic cleavage of prosomatostatin and formation of nascent secretory vesicles in permeabilized cells. J Cell Biol 122(6):1169–1184
4. Sasaki K, Yoshida H (2019) Golgi stress response and organelle zones. FEBS Lett 593(17):2330–2340
5. Appenzeller-Herzog C, Hauri HP (2006) The ER-Golgi intermediate compartment (ERGIC): in search of its identity and function. J Cell Sci 119(Pt 11):2173–2183
6. Makowski SL, Tran TTT, Field SJ (2017) Emerging themes of regulation at the Golgi. Curr Opin Cell Biol 45:17–23
7. Bravo R et al (2013) Endoplasmic reticulum and the unfolded protein response: dynamics and metabolic integration. Int Rev Cell Mol Biol 301:215–290
8. Zhu X, Kaverina I (2013) Golgi as an MTOC: making microtubules for its own good. Histochem Cell Biol 140(3):361–367
9. Kulkarni-Gosavi P, Makhoul C, Gleeson PA (2019) Form and function of the Golgi apparatus: scaffolds, cytoskeleton and signalling. FEBS Lett 593(17):2289–2305
10. Farber-Katz SE et al (2014) DNA damage triggers Golgi dispersal via DNA-PK and GOLPH3. Cell 156(3):413–427

11. Chen Y et al (2024) Golgiphagy: a novel selective autophagy to the fore. Cell Biosci 14(1):130
12. Kaeser-Pebernard S et al (2022) mTORC1 controls Golgi architecture and vesicle secretion by phosphorylation of SCYL1. Nat Commun 13(1):4685
13. Carrà G et al (2022) Shedding light on NF-κB functions in cellular organelles. Front Cell Dev Biol 10:841646
14. Machamer CE (2015) The Golgi complex in stress and death. Front Neurosci 9:421
15. Nakagomi S et al (2008) A Golgi fragmentation pathway in neurodegeneration. Neurobiol Dis 29(2):221–231
16. Sasaki K, Yoshida H (2015) Organelle autoregulation-stress responses in the ER, Golgi, mitochondria and lysosome. J Biochem 157(4):185–195
17. Miyata S et al (2013) The endoplasmic reticulum-resident chaperone heat shock protein 47 protects the Golgi apparatus from the effects of O-glycosylation inhibition. PLoS One 8(7):e69732
18. Ludwig PE, Reddy V, Varacallo MA (2025) Neuroanatomy, neurons. In: StatPearls. StatPearls publishing. Copyright © 2025. StatPearls Publishing LLC, Treasure Island (FL)
19. Noback CR et al (eds) (2005) Neurons and associated cells, in the human nervous system: structure and function. Humana Press, Totowa, pp 11–39
20. Harris JJ, Jolivet R, Attwell D (2012) Synaptic energy use and supply. Neuron 75(5):762–777
21. Shadfar S, Brocardo M, Atkin JD (2022) The complex mechanisms by which neurons die following DNA damage in neurodegenerative diseases. Int J Mol Sci 23(5)
22. Shadfar S et al (2023) Redox dysregulation as a driver for DNA damage and its relationship to neurodegenerative diseases. Transl Neurodegener 12(1):18
23. Kaur J et al (2021) Waste clearance in the brain. Front Neuroanat 15:665803
24. Zappa F, Failli M, De Matteis MA (2018) The Golgi complex in disease and therapy. Curr Opin Cell Biol 50:102–116
25. Valenzuela A et al (2020) Golgi outposts nucleate microtubules in cells with specialized shapes. Trends Cell Biol 30(10):792–804
26. Chen M et al (2023) The organization and function of the Golgi apparatus in dendrite development and neurological disorders. Genes Dis 10(6):2425–2442
27. Kemal S et al (2022) ER and Golgi trafficking in axons, dendrites, and glial processes. Curr Opin Cell Biol 78:102119
28. Fu MM et al (2019) The Golgi outpost protein TPPP nucleates microtubules and is critical for myelination. Cell 179(1):132–146.e14
29. Weigel M, Wang L, Fu MM (2021) Microtubule organization and dynamics in oligodendrocytes, astrocytes, and microglia. Dev Neurobiol 81(3):310–320
30. Boulay AC et al (2017) Translation in astrocyte distal processes sets molecular heterogeneity at the gliovascular interface. Cell Discov 3:17005
31. Zobaroğlu-Özer P, Bora-Akoğlu G (2024) Split but merge: Golgi fragmentation in physiological and pathological conditions. Mol Biol Rep 51(1):214
32. Thayer DA, Jan YN, Jan LY (2013) Increased neuronal activity fragments the Golgi complex. Proc Natl Acad Sci USA 110(4):1482–1487
33. Antón-Fernández A et al (2017) Morphometric alterations of Golgi apparatus in Alzheimer's disease are related to tau hyperphosphorylation. Neurobiol Dis 97(Pt A):11–23
34. Jagaraj CJ et al (2024) Molecular hallmarks of ageing in amyotrophic lateral sclerosis. Cell Mol Life Sci 81(1):111
35. Peng Y et al (2024) True or false? Alzheimer's disease is type 3 diabetes: evidences from bench to bedside. Ageing Res Rev 99:102383
36. Ramezani M et al (2023) Ketone bodies mediate alterations in brain energy metabolism and biomarkers of Alzheimer's disease. Front Neurosci 17:1297984
37. Martínez-Menárguez J et al (2019) Golgi fragmentation in neurodegenerative diseases: is there a common cause? Cells 8(7)
38. Abyadeh M et al (2023) Apolipoprotein ε in brain and retinal neurodegenerative diseases. Aging Dis 14(4):1311–1330

39. Perl DP (2010) Neuropathology of Alzheimer's disease. Mt Sinai J Med 77(1):32–42
40. Peng Y et al (2023) Current and future therapeutic strategies for Alzheimer's disease: an overview of drug development bottlenecks. Front Aging Neurosci 15:1206572
41. (2022) 2022 Alzheimer's disease facts and figures. Alzheimers Dement 18(4):700–789
42. Abyadeh M et al (2024) Amyloid-beta and tau protein beyond Alzheimer's disease. Neural Regen Res 19(6):1262–1276
43. DeTure MA, Dickson DW (2019) The neuropathological diagnosis of Alzheimer's disease. Mol Neurodegener 14(1):32
44. Huse JT et al (2002) Beta-secretase processing in the trans-Golgi network preferentially generates truncated amyloid species that accumulate in Alzheimer's disease brain. J Biol Chem 277(18):16278–16284
45. Gonatas NK, Gonatas JO, Stieber A (1998) The involvement of the Golgi apparatus in the pathogenesis of amyotrophic lateral sclerosis, Alzheimer's disease, and ricin intoxication. Histochem Cell Biol 109(5–6):591–600
46. Joshi G et al (2014) Aβ-induced Golgi fragmentation in Alzheimer's disease enhances Aβ production. Proc Natl Acad Sci USA 111(13):E1230–E1239
47. Haukedal H et al (2023) Golgi fragmentation – one of the earliest organelle phenotypes in Alzheimer's disease neurons. Front Neurosci 17:1120086
48. Roh JH et al (2012) Disruption of the sleep-wake cycle and diurnal fluctuation of β-amyloid in mice with Alzheimer's disease pathology. Sci Transl Med 4(150):150ra122
49. Liazoghli D et al (2005) Fragmentation of the Golgi apparatus induced by the overexpression of wild-type and mutant human tau forms in neurons. Am J Pathol 166(5):1499–1514
50. Seitz A et al (2002) Single-molecule investigation of the interference between kinesin, tau and MAP2c. EMBO J 21(18):4896–4905
51. Wang Y, Seemann J (2011) Golgi biogenesis. Cold Spring Harb Perspect Biol 3(10):a005330
52. Ishunina TA, Swaab DF (2003) Increased neuronal metabolic activity and estrogen receptors in the vertical limb of the diagonal band of Broca in Alzheimer's disease: relation to sex and aging. Exp Neurol 183(1):159–172
53. Thinakaran G, Koo EH (2008) Amyloid precursor protein trafficking, processing, and function*. J Biol Chem 283(44):29615–29619
54. Toh WH et al (2017) Amyloid precursor protein traffics from the Golgi directly to early endosomes in an Arl5b- and AP4-dependent pathway. Traffic 18(3):159–175
55. Tan JZA et al (2020) Distinct anterograde trafficking pathways of BACE1 and amyloid precursor protein from the TGN and the regulation of amyloid-β production. Mol Biol Cell 31(1):27–44
56. Tan JZA, Gleeson PA (2019) The trans-Golgi network is a major site for α-secretase processing of amyloid precursor protein in primary neurons. J Biol Chem 294(5):1618–1631
57. Xiang Y et al (2013) Regulation of protein glycosylation and sorting by the Golgi matrix proteins GRASP55/65. Nat Commun 4:1659
58. Joshi G, Wang Y (2015) Golgi defects enhance APP amyloidogenic processing in Alzheimer's disease. BioEssays 37(3):240–247
59. Fokin AI et al (2014) Interaction of early secretory pathway and Golgi membranes with microtubules and microtubule motors. Biochemistry (Mosc) 79(9):879–893
60. Hancock WO (2014) Bidirectional cargo transport: moving beyond tug of war. Nat Rev Mol Cell Biol 15(9):615–628
61. Gong CX, Alonso Adel C (2013) Tau and beyond for Alzheimer's disease. J Alzheimer's Dis 37(3):465–466
62. Bloom GS (2014) Amyloid-β and tau: the trigger and bullet in Alzheimer disease pathogenesis. JAMA Neurol 71(4):505–508
63. Morishima-Kawashima M, Ihara Y (2002) Alzheimer's disease: beta-amyloid protein and tau. J Neurosci Res 70(3):392–401

64. Jiang Q et al (2014) Golgin-84-associated Golgi fragmentation triggers tau hyperphosphorylation by activation of cyclin-dependent kinase-5 and extracellular signal-regulated kinase. Neurobiol Aging 35(6):1352–1363
65. Baloyannis SJ (2014) Golgi apparatus and protein trafficking in Alzheimer's disease. J Alzheimer's Dis 42(Suppl 3):S153–S162
66. Elyaman W, Yardin C, Hugon J (2002) Involvement of glycogen synthase kinase-3beta and tau phosphorylation in neuronal Golgi disassembly. J Neurochem 81(4):870–880
67. Jung YY et al (2017) Enhanced dopaminergic neurotoxicity mediated by MPTP in IL-32β transgenic mice. Neurochem Int 102:79–88
68. Varesi A et al (2022) The interplay between gut microbiota and Parkinson's disease: implications on diagnosis and treatment. Int J Mol Sci 23(20)
69. Shadfar S et al (2024) Correction: Methanolic extract of Boswellia serrata gum protects the Nigral dopaminergic neurons from rotenone-induced neurotoxicity. Mol Neurobiol 62(4): 5352–5353
70. Shadfar S et al (2018) Neuroprotective effects of antidepressants via upregulation of neurotrophic factors in the MPTP model of Parkinson's disease. Mol Neurobiol 55(1): 554–566
71. Henderson MX, Trojanowski JQ, Lee VM (2019) α Synuclein pathology in Parkinson's disease and related α-synucleinopathies. Neurosci Lett 709:134316
72. Fujita Y et al (2006) Fragmentation of Golgi apparatus of nigral neurons with alpha-synuclein-positive inclusions in patients with Parkinson's disease. Acta Neuropathol 112(3):261–265
73. Wei Y et al (2023) The function of Golgi apparatus in LRRK2-associated Parkinson's disease. Front Mol Neurosci 16:1097633
74. Rendón WO et al (2013) Golgi fragmentation is Rab and SNARE dependent in cellular models of Parkinson's disease. Histochem Cell Biol 139(5):671–684
75. Mazzetti S et al (2024) Linking acetylated α-tubulin redistribution to α-Synuclein pathology in brain of Parkinson's disease patients. NPJ Parkinsons Dis 10(1):2
76. Stormo AED et al (2022) The E3 ligase TRIM1 ubiquitinates LRRK2 and controls its localization, degradation, and toxicity. J Cell Biol 221(4)
77. Komori T, Kuwahara T (2023) An update on the interplay between LRRK2, Rab GTPases and Parkinson's disease. Biomolecules 13(11)
78. Purlyte E et al (2018) Rab29 activation of the Parkinson's disease-associated LRRK2 kinase. EMBO J 37(1):1–18
79. Boecker CA et al (2021) Increased LRRK2 kinase activity alters neuronal autophagy by disrupting the axonal transport of autophagosomes. Curr Biol 31(10):2140–2154.e6
80. Wilson GR et al (2014) Mutations in RAB39B cause X-linked intellectual disability and early-onset Parkinson disease with α-synuclein pathology. Am J Hum Genet 95(6):729–735
81. Rowlands J, Moore DJ (1899) VPS35 and retromer dysfunction in Parkinson's disease. Philos Trans R Soc Lond Ser B Biol Sci 2024(379):20220384
82. Tang FL et al (2015) VPS35 in dopamine neurons is required for endosome-to-Golgi retrieval of Lamp2a, a receptor of chaperone-mediated autophagy that is critical for α-Synuclein degradation and prevention of pathogenesis of Parkinson's disease. J Neurosci 35(29): 10613–10628
83. Williams ET, Chen X, Moore DJ (2017) VPS35, the Retromer complex and Parkinson's disease. J Parkinsons Dis 7(2):219–233
84. Chen X et al (2019) Parkinson's disease-linked D620N VPS35 knockin mice manifest tau neuropathology and dopaminergic neurodegeneration. Proc Natl Acad Sci USA 116(12): 5765–5774
85. Singh PK, Muqit MMK (2020) Parkinson's: a disease of aberrant vesicle trafficking. Annu Rev Cell Dev Biol 36:237–264
86. Rebelo AL et al (2024) Changes in tissue protein N-glycosylation and associated molecular signature occur in the human Parkinsonian brain in a region-specific manner. PNAS Nexus 3(1):pgad439

87. Guardia-Laguarta C et al (2015) Novel subcellular localization for α-synuclein: possible functional consequences. Front Neuroanat 9:17
88. Celardo I et al (2016) Mitofusin-mediated ER stress triggers neurodegeneration in pink1/parkin models of Parkinson's disease. Cell Death Dis 7(6):e2271
89. Schapira AH (2013) Calcium dysregulation in Parkinson's disease. Brain 136(Pt 7): 2015–2016
90. Sun KH et al (2008) Novel genetic tools reveal Cdk5's major role in Golgi fragmentation in Alzheimer's disease. Mol Biol Cell 19(7):3052–3069
91. Cheung ZH, Ip NY (2012) Cdk5: a multifaceted kinase in neurodegenerative diseases. Trends Cell Biol 22(3):169–175
92. Feldman EL et al (2025) Transforming amyotrophic lateral sclerosis into a liveable disease. Lancet Neurol 24(2):100–101
93. Ragagnin AMG et al (2019) Motor neuron susceptibility in ALS/FTD. Front Neurosci 13:532
94. Brown CA et al (2021) Estimated prevalence and incidence of amyotrophic lateral sclerosis and SOD1 and C9orf72 genetic variants. Neuroepidemiology 55(5):342–353
95. Jo M et al (2020) The role of TDP-43 propagation in neurodegenerative diseases: integrating insights from clinical and experimental studies. Exp Mol Med 52(10):1652–1662
96. Amick J, Ferguson SM (2017) C9orf72: at the intersection of lysosome cell biology and neurodegenerative disease. Traffic 18(5):267–276
97. Sundaramoorthy V, Sultana JM, Atkin JD (2015) Golgi fragmentation in amyotrophic lateral sclerosis, an overview of possible triggers and consequences. Front Neurosci 9:400
98. Fujita Y et al (2008) Anterior horn cells with abnormal TDP-43 immunoreactivities show fragmentation of the Golgi apparatus in ALS. J Neurol Sci 269(1–2):30–34
99. Gonatas N et al (1992) Fragmentation of the Golgi apparatus of motor neurons in amyotrophic lateral sclerosis. Am J Pathol 140(3):731
100. Ito H et al (2011) Clinicopathologic study on an ALS family with a heterozygous E478G optineurin mutation. Acta Neuropathol 122:223–229
101. van Dis V et al (2014) Golgi fragmentation precedes neuromuscular denervation and is associated with endosome abnormalities in SOD1-ALS mouse motor neurons. Acta Neuropathol Commun 2:1–15
102. Tong J et al (2012) XBP 1 depletion precedes ubiquitin aggregation and Golgi fragmentation in TDP-43 transgenic rats. J Neurochem 123(3):406–416
103. Sultana J et al (2024) C9orf72-associated dipeptide repeat expansions perturb ER-Golgi vesicular trafficking, inducing Golgi fragmentation and ER stress, in ALS/FTD. Mol Neurobiol:1–21
104. Sundaramoorthy V et al (2013) Extracellular wildtype and mutant SOD1 induces ER–Golgi pathology characteristic of amyotrophic lateral sclerosis in neuronal cells. Cell Mol Life Sci 70:4181–4195
105. Parakh S et al (2021) Protein disulphide isomerase (PDI) is protective against amyotrophic lateral sclerosis (ALS)-related mutant Fused in Sarcoma (FUS) in in vitro models. Sci Rep 11(1):17557
106. Soo KY et al (2015) Rab1-dependent ER–Golgi transport dysfunction is a common pathogenic mechanism in SOD1, TDP-43 and FUS-associated ALS. Acta Neuropathol 130:679–697
107. Teuling E et al (2007) Motor neuron disease-associated mutant vesicle-associated membrane protein-associated protein (VAP) B recruits wild-type VAPs into endoplasmic reticulum-derived tubular aggregates. J Neurosci 27(36):9801–9815
108. Halloran M et al (2020) Amyotrophic lateral sclerosis-linked UBQLN2 mutants inhibit endoplasmic reticulum to Golgi transport, leading to Golgi fragmentation and ER stress. Cell Mol Life Sci 77:3859–3873
109. Ragagnin AM et al (2023) ALS/FTD-associated mutation in cyclin F inhibits ER-Golgi trafficking, inducing ER stress, ERAD and Golgi fragmentation. Sci Rep 13(1):20467

110. Ferraiuolo L et al (2007) Microarray analysis of the cellular pathways involved in the adaptation to and progression of motor neuron injury in the SOD1 G93A mouse model of familial ALS. J Neurosci 27(34):9201–9219
111. Bellouze S et al (2016) Stathmin 1/2-triggered microtubule loss mediates Golgi fragmentation in mutant SOD1 motor neurons. Mol Neurodegener 11(1):43
112. Klim JR et al (2019) ALS-implicated protein TDP-43 sustains levels of STMN2, a mediator of motor neuron growth and repair. Nat Neurosci 22(2):167–179
113. Mehta PR et al (2023) The era of cryptic exons: implications for ALS-FTD. Mol Neurodegener 18(1):16
114. Hafezparast M et al (2003) Mutations in dynein link motor neuron degeneration to defects in retrograde transport. Science 300(5620):808–812
115. Bellouze S et al (2014) Golgi fragmentation in pmn mice is due to a defective ARF1/TBCE cross-talk that coordinates COPI vesicle formation and tubulin polymerization. Hum Mol Genet 23(22):5961–5975
116. Stieber A et al (2000) The neuronal Golgi apparatus is fragmented in transgenic mice expressing a mutant human SOD1, but not in mice expressing the human NF-H gene. J Neurol Sci 173(1):63–72
117. Dodge JC et al (2015) Glycosphingolipids are modulators of disease pathogenesis in amyotrophic lateral sclerosis. Proc Natl Acad Sci 112(26):8100–8105
118. Tabrizi SJ et al (2020) Huntington disease: new insights into molecular pathogenesis and therapeutic opportunities. Nat Rev Neurol 16(10):529–546
119. McColgan P, Tabrizi SJ (2018) Huntington's disease: a clinical review. Eur J Neurol 25(1): 24–34
120. Medina A et al (2022) Prevalence and incidence of Huntington's disease: an updated systematic review and meta-analysis. Mov Disord 37(12):2327–2335
121. del Toro D et al (2006) Mutant huntingtin impairs the post-Golgi trafficking of brain-derived neurotrophic factor but not its Val66Met polymorphism. J Neurosci 26(49):12748–12757
122. Brandstaetter H, Kruppa AJ, Buss F (2014) Huntingtin is required for ER-to-Golgi transport and for secretory vesicle fusion at the plasma membrane. Dis Model Mech 7(12):1335–1340
123. Hattula K, Peränen J (2000) FIP-2, a coiled-coil protein, links Huntingtin to Rab8 and modulates cellular morphogenesis. Curr Biol 10(24):1603–1606
124. Sahlender DA et al (2005) Optineurin links myosin VI to the Golgi complex and is involved in Golgi organization and exocytosis. J Cell Biol 169(2):285–295
125. Bond LM et al (2011) Myosin VI and its binding partner optineurin are involved in secretory vesicle fusion at the plasma membrane. Mol Biol Cell 22(1):54–65
126. Homma Y, Hiragi S, Fukuda M (2021) Rab family of small GTPases: an updated view on their regulation and functions. FEBS J 288(1):36–55
127. del Toro D et al (2009) Mutant huntingtin impairs post-Golgi trafficking to lysosomes by delocalizing optineurin/Rab8 complex from the Golgi apparatus. Mol Biol Cell 20(5): 1478–1492
128. Sbodio JI et al (2013) Golgi protein ACBD3 mediates neurotoxicity associated with Huntington's disease. Cell Rep 4(5):890–897
129. Covarrubias-Pinto A et al (2021) Impaired intracellular trafficking of sodium-dependent vitamin C transporter 2 contributes to the redox imbalance in Huntington's disease. J Neurosci Res 99(1):223–235
130. Sanpei K et al (1996) Identification of the spinocerebellar ataxia type 2 gene using a direct identification of repeat expansion and cloning technique, DIRECT. Nat Genet 14(3):277–284
131. Pulst SM et al (1996) Moderate expansion of a normally biallelic trinucleotide repeat in spinocerebellar ataxia type 2. Nat Genet 14(3):269–276
132. Ng H, Pulst S-M, Huynh DP (2007) Ataxin-2 mediated cell death is dependent on domains downstream of the polyQ repeat. Exp Neurol 208(2):207–215
133. Chung CG et al (2017) Golgi outpost synthesis impaired by toxic Polyglutamine proteins contributes to dendritic pathology in neurons. Cell Rep 20(2):356–369

134. Zerr I et al (2024) Creutzfeldt-Jakob disease and other prion diseases. Nat Rev Dis Primers 10(1):14
135. Halliday M, Radford H, Mallucci GR (2014) Prions: generation and spread versus neurotoxicity. J Biol Chem 289(29):19862–19868
136. Uchiyama K et al (2013) Prions disturb post-Golgi trafficking of membrane proteins. Nat Commun 4:1846
137. Sakurai A et al (2000) Fragmentation of the Golgi apparatus of the ballooned neurons in patients with corticobasal degeneration and Creutzfeldt-Jakob disease. Acta Neuropathol 100(3):270–274
138. Caughey B et al (2009) Getting a grip on prions: oligomers, amyloids, and pathological membrane interactions. Annu Rev Biochem 78:177–204
139. Chiesa R, Harris DA (2001) Prion diseases: what is the neurotoxic molecule? Neurobiol Dis 8(5):743–763
140. Castle AR, Westaway D (2025) Prion protein Endoproteolysis: cleavage sites, mechanisms and connections to prion disease. J Neurochem 169(1):e16310
141. Barmada SJ, Harris DA (2005) Visualization of prion infection in transgenic mice expressing green fluorescent protein-tagged prion protein. J Neurosci 25(24):5824–5832
142. Massignan T et al (2010) Mutant prion protein expression is associated with an alteration of the Rab GDP dissociation inhibitor alpha (GDI)/Rab11 pathway. Mol Cell Proteomics 9(4): 611–622
143. Bouybayoune I et al (2015) Transgenic fatal familial insomnia mice indicate prion infectivity-independent mechanisms of pathogenesis and phenotypic expression of disease. PLoS Pathog 11(4):e1004796
144. Stewart RS et al (2005) Neurodegenerative illness in transgenic mice expressing a transmembrane form of the prion protein. J Neurosci 25(13):3469–3477
145. Fink JK (2013) Hereditary spastic paraplegia: clinico-pathologic features and emerging molecular mechanisms. Acta Neuropathol 126(3):307–328
146. Lo Giudice T et al (2014) Hereditary spastic paraplegia: clinical-genetic characteristics and evolving molecular mechanisms. Exp Neurol 261:518–539
147. Kelly CM et al (2022) A novel insertion mutation in atlastin 1 is associated with spastic quadriplegia, increased membrane tethering, and aberrant conformational switching. J Biol Chem 298(1):101438
148. Akefe IO et al (2024) The DDHD2-STXBP1 interaction mediates long-term memory via generation of saturated free fatty acids. EMBO J 43(4):533–567
149. Simpson MA et al (2003) Maspardin is mutated in mast syndrome, a complicated form of hereditary spastic paraplegia associated with dementia. Am J Hum Genet 73(5):1147–1156
150. Byrnes LJ, Sondermann H (2011) Structural basis for the nucleotide-dependent dimerization of the large G protein atlastin-1/SPG3A. Proc Natl Acad Sci 108(6):2216–2221
151. Montagna A et al (2020) In vivo analysis of CRISPR/Cas9 induced Atlastin pathological mutations in Drosophila. Front Neurosci 14:547746
152. Bonnon C et al (2010) Selective export of human GPI-anchored proteins from the endoplasmic reticulum. J Cell Sci 123(Pt 10):1705–1715
153. Peng Y et al (2018) Increased transport of acetyl-CoA into the endoplasmic reticulum causes a progeria-like phenotype. Aging Cell 17(5):e12820
154. Liu P et al (2017) S113R mutation in SLC33A1 leads to neurodegeneration and augmented BMP signaling in a mouse model. Dis Model Mech 10(1):53–62
155. Hanna MC, Blackstone C (2009) Interaction of the SPG21 protein ACP33/maspardin with the aldehyde dehydrogenase ALDH16A1. Neurogenetics 10(3):217–228
156. McCray BA, Skordalakes E, Taylor JP (2010) Disease mutations in Rab7 result in unregulated nucleotide exchange and inappropriate activation. Hum Mol Genet 19(6):1033–1047
157. Miura S et al (2016) A novel frameshift mutation of DDHD1 in a Japanese patient with autosomal recessive spastic paraplegia. Eur J Med Genet 59(8):413–416

158. Mignarri A et al (2016) Mitochondrial dysfunction in hereditary spastic paraparesis with mutations in DDHD1/SPG28. J Neurol Sci 362:287–291
159. Jia F et al (2024) DDHD2, whose mutations cause spastic paraplegia type 54, enhances lipophagy via engaging ATG8 family proteins. Cell Death Differ 31(3):348–359
160. Gonzalez M et al (2013) Mutations in phospholipase DDHD2 cause autosomal recessive hereditary spastic paraplegia (SPG54). Eur J Hum Genet 21(11):1214–1218
161. Inoue H et al (2012) Roles of SAM and DDHD domains in mammalian intracellular phospholipase A1 KIAA0725p. Biochim Biophys Acta 1823(4):930–939
162. Sato S et al (2010) Golgi-localized KIAA0725p regulates membrane trafficking from the Golgi apparatus to the plasma membrane in mammalian cells. FEBS Lett 584(21):4389–4395
163. Kunduri G et al (2014) Phosphatidic acid phospholipase A1 mediates ER-Golgi transit of a family of G protein-coupled receptors. J Cell Biol 206(1):79–95
164. Sato, S.-i., et al., Golgi-localized KIAA0725p regulates membrane trafficking from the Golgi apparatus to the plasma membrane in mammalian cells. FEBS Lett, 2010. 584(21): p. 4389–4395
165. Schiavoni S et al (2020) Paediatric-onset hereditary spastic paraplegias: a retrospective cohort study. Dev Med Child Neurol 62(9):1068–1074
166. Nicita F et al (2019) Defining the clinical-genetic and neuroradiological features in SPG54: description of eight additional cases and nine novel DDHD2 variants. J Neurol 266(11): 2657–2664
167. Morikawa RK et al (2009) Intracellular phospholipase A1gamma (iPLA1gamma) is a novel factor involved in coat protein complex I- and Rab6-independent retrograde transport between the endoplasmic reticulum and the Golgi complex. J Biol Chem 284(39):26620–26630
168. Namavar Y et al (2011) Clinical, neuroradiological and genetic findings in pontocerebellar hypoplasia. Brain 134(Pt 1):143–156
169. Barth PG (1993) Pontocerebellar hypoplasias. An overview of a group of inherited neurodegenerative disorders with fetal onset. Brain Dev 15(6):411–422
170. Conibear E, Stevens TH (2000) Vps52p, Vps53p, and Vps54p form a novel multisubunit complex required for protein sorting at the yeast late Golgi. Mol Biol Cell 11(1):305–323
171. Feinstein M et al (2014) VPS53 mutations cause progressive cerebello-cerebral atrophy type 2 (PCCA2). J Med Genet 51(5):303–308
172. van Dijk T et al (2018) What's new in pontocerebellar hypoplasia? An update on genes and subtypes. Orphanet J Rare Dis 13(1):92
173. Fröhlich F et al (2015) The GARP complex is required for cellular sphingolipid homeostasis. elife:4
174. Warburg M et al (1993) Autosomal Recesssive microcephaly, microcornea, congenital cataract, mental retardation, optic atrophy, and Hypogenitalism: Micro syndrome. Am J Dis Child 147(12):1309–1312
175. Handley MT et al (2013) Mutation spectrum in RAB3GAP1, RAB3GAP2, and RAB18 and genotype–phenotype correlations in Warburg Micro syndrome and Martsolf syndrome. Hum Mutat 34(5):686–696
176. Kerkeni N et al (2022) Novel RAB3GAP1 mutation in the first Tunisian family with Warburg Micro syndrome. J Clin Neurol 18(2):214–222
177. Wu Q et al (2016) RAB18, a protein associated with Warburg Micro syndrome, controls neuronal migration in the developing cerebral cortex. Mol Brain 9(1):19
178. Borck G et al (2011) A homozygous RAB3GAP2 mutation causes Warburg Micro syndrome. Hum Genet 129(1):45–50
179. Aligianis IA et al (2005) Mutations of the catalytic subunit of RAB3GAP cause Warburg Micro syndrome. Nat Genet 37(3):221–223
180. Dejgaard SY et al (2008) Rab18 and Rab43 have key roles in ER-Golgi trafficking. J Cell Sci 121(Pt 16):2768–2781
181. Zappa F, Venditti R, De Matteis MA (2017) TRAPPing Rab18 in lipid droplets. EMBO J 36(4):394–396

182. Gerondopoulos A et al (2014) Rab18 and a Rab18 GEF complex are required for normal ER structure. J Cell Biol 205(5):707–720
183. D'Amico A et al (2011) Spinal muscular atrophy. Orphanet J Rare Dis 6(1):71
184. Laufersweiler-Plass C et al (2003) Behavioural problems in children and adolescents with spinal muscular atrophy and their siblings. Dev Med Child Neurol 45(1):44–49
185. Lefebvre S et al (1995) Identification and characterization of a spinal muscular atrophy-determining gene. Cell 80(1):155–165
186. Custer SK et al (2019) Abnormal Golgi morphology and decreased COPI function in cells with low levels of SMN. Brain Res 1706:135–146
187. Ting CH et al (2012) The spinal muscular atrophy disease protein SMN is linked to the Golgi network. PLoS One 7(12):e51826
188. Peter CJ et al (2011) The COPI vesicle complex binds and moves with survival motor neuron within axons. Hum Mol Genet 20(9):1701–1711

Chapter 18
The Dynamic Role of the Golgi Apparatus in Neuronal Function and Alzheimer's Disease

Lou Fourriere and Paul A. Gleeson

Abstract The Golgi apparatus is a highly dynamic organelle and central to the regulation of a variety of cell processes. Here we review the relationship between the morphology and function of this organelle, especially to in relation to neuronal development and maintenance. The morphology of the Golgi organelle in neurons is unique as it is present as a Golgi ribbon in the cell body and also as dendritic Golgi elements, known as Golgi outposts, in the neuronal extensions. Many neurodegenerative diseases are associated with the loss of the Golgi ribbon structure and the appearance of dispersed and fragmented Golgi units in the cell body/soma. Here we summarise a key role of the Golgi to limit the proteolytic processing of amyloid precursor protein (APP) by secretases and protect healthy cells, and the perturbations of Golgi morphology arising from increased APP processing and amyloid β production in Alzheimer's disease, the most common of the neurodegenerative disease. The potential consequence of changes in the Golgi architecture and increase of amyloid β production on neuronal function is also considered.

Keywords Golgi apparatus · Mini-stacks · Golgi outposts · Amyloid precursor protein (APP) · Golgins · Amyloid β · Membrane trafficking · Alzheimer's disease (AD)

Abbreviations

Aβ	amyloid β peptide
BACE1	beta site APP-cleaving enzyme
APP	amyloid precursor protein
APPswe	Swedish APP mutant
EM	electron microscopy

L. Fourriere · P. A. Gleeson (✉)
The Department of Biochemistry and Pharmacology, Bio21 Molecular Science and Biotechnology Institute, The University of Melbourne, Parkville, VIC, Australia
e-mail: pgleeson@unimelb.edu.au

A. Nakano, J. Saraste (eds.), *The Golgi Network, Volume II*, Subcellular Biochemistry 111, https://doi.org/10.1007/978-3-032-16833-7_18

EOAD	early onset Alzheimer's disease
TGN	*trans*-Golgi network
ER	endoplasmic reticulum

Introduction

Over the past two decades, cell biologists have recognised that intracellular organelles represent a highly integrated system and that many cellular pathways are interconnected [24]. The Golgi apparatus is now recognised to play roles beyond the classical functions of protein sorting and glycosylation. Although somewhat overlooked for many years it should not have been a surprise, given the central location of the Golgi apparatus in cells, that the Golgi is a significant contributor to the regulation of a range of cellular processes. Knowledge of these processes has continued to expand and includes cell polarization [47], directed migration [59], stress [80, 82], mitosis [12], metabolism [1], pro-inflammatory responses [8], autophagy [48] and calcium store regulation [104]. While the architecture of the Golgi is complex, it is clear that the precise morphology of the Golgi provides a window into the regulation of a number of these cell processes, a topic we recently reviewed [53]. The Golgi architecture is sculptured by interactions of the Golgi matrix proteins and membrane-associated golgins with the cytoskeleton [46, 74]. A major discovery has been the appreciation of a strong relationship between signalling pathways and Golgi morphology, as a genome wide kinome and phosphatome screen identified >100 signalling genes which influenced the morphology of the Golgi, either inducing fragmentation or compaction [9] indicating an extensive network of pathways which are wired to Golgi membranes.

The relationship between the structure and function of the Golgi is especially relevant to understand neuronal development, and the establishment and maintenance of neuronal networks required for cognitive processes. Alterations in the normal morphology of the neuronal somatic Golgi is associated with many neurological diseases [71]; for many years it was unclear whether perturbations in Golgi morphology were a secondary downstream consequence to the primary pathways driving disease; however, direct evidence has emerged for a number of neurological diseases demonstrating that the loss of integrity of the Golgi structure is an early event in the progression of the disease and, moreover, contributes directly to disease pathogenesis [41].

In this Chapter, we discuss the regulation of Golgi morphology, the dynamics of the Golgi architecture in neuronal development and maintenance, and its perturbation in neurodegenerative diseases. Alzheimer's disease is the most common neurodegenerative disease and is associated with the production of amyloid β (Aβ). Here we discuss the key role of the Golgi apparatus in regulating physiologically levels of Aβ in health, to ensure that the production of the toxic amyloid peptides is below threshold levels, and in disease where there is dysfunctional production of high levels of toxic products from the processing of amyloid precursor protein (APP). In addition, the impact of the toxic amyloid products on the integrity of the Golgi and membrane trafficking will also be discussed.

Golgi Morphology

The structure of the Golgi apparatus differs between species and the difference between invertebrates and vertebrates is particularly striking. Invertebrates, as well as protists and plants have multiple Golgi elements, usually mini-stacks, scattered independently throughout the cytoplasm of the cell [26, 61]. The individual mini-stacks are functional units and are comprised of *cis*-Golgi cisternae which receive cargo from the ER, medial-cisternae and *trans*-cisternae where cargoes are exported along the secretory pathways to various cell destination. The Golgi, in most vertebrate cells, has a ribbon-like structure during interphase and is typically positioned in close proximity to the centrosome, the microtubule organising centre (MTOC), and nucleus during interphase. The ribbon architecture is formed by lateral fusion of individual Golgi mini-stacks [17, 76] [61]. There is the capacity to both generate the higher order ribbon structure from individual mini-stacks and to dissociate the ribbon into individual mini-stacks via specific machinery which regulates these processes [79].

A few specialised mammalian cells do not have a Golgi ribbon structure but rather only mini-stacks, for example gastric parietal cells [29] and differentiated myoblasts [52], reflecting the important relationship between the specialised requirements of the organisation of particular cell types and function. Neurons are unique in possessing both a Golgi ribbon, in the cell body (known as soma) of the neuron, and distinct and independent mini-stacks in dendritic neurites, some distance away from the cell body [44, 101, 106, 107]. This arrangement in neurons highlights specific contributions of the Golgi to neuronal function, as discussed below. An important distinction is that the *Drosophila* (fruit fly) neuron, as an invertebrate, lacks a Golgi ribbon and is composed exclusively of Golgi ministacks [45]. Differences in the organisation of the dendritic Golgi have been observed between rat and human neurons [107], highlighting an important consideration in the choice of organism in studying human neurological diseases.

Structure of the Neuronal Golgi

The architecture of the neuron is fundamental for synaptic transmission between neurons. Neurons are highly polarised cells consisting of the somato-dendritic domain, responsible for receiving signals, and axonal compartment, responsible for transmitting signals. The surface area of neurons is greater than most other cells by many thousand-fold [34]. Transport of newly synthesised membrane proteins from the endoplasmic reticulum (ER) and Golgi in the central cell body of the neuron to their functional destination at remote sites on the neurite extensions represents a challenge given the physical distance involved. In particular, delivery of newly synthesized cargo to the synapse needs to be highly efficient to maintain neuronal function, and long-distance transport of newly synthesised cargo from the soma could be problematic for a rapid delivery. Secretory organelles identified in dendrites, including ER and Golgi structures [22, 33, 67], could overcome this long-distance transport challenge. Dendritic Golgi, also called Golgi outposts [33], are able to support local protein synthesis, glycosylation, and trafficking distinct from the conventional anterograde transport pathways of the cell body as revealed by recent studies. In addition, smaller Golgi structures, referred to as Golgi satellites, have also been described in rodent systems [57]. Hence the organization of the Golgi as both a ribbon structure located centrally and Golgi mini-stacks as outposts in dendrites represents a unique adaption of the neuronal secretory pathway tailored for the landscape of a neuron.

Dynamic Golgi Structures: Ribbons and Mini-stacks

As mentioned above, Golgi matrix proteins are essential to maintain the structural integrity of the ribbon and some non-classical Golgi functions are dependent on a dynamic ribbon architecture, for example a Golgi checkpoint mechanism to regulate entry into mitosis [14]. Mice with a disrupted Golgi ribbon architecture, due to the lack of a key matrix Golgi protein, GM130, die neonatally ([51]. However, the Golgi ribbon architecture is not static but rather highly dynamic and can undergo a rapid remodelling of the Golgi ribbon into distinct Golgi elements (and vice versa) in numerous physiological conditions [73, 114]. For example, the Golgi is extensively fragmented during mitosis and subsequently a compact ribbon is reformed in interphase [13, 62]. Disruption of the Golgi is also associated with directed secretion [47, 114]. Hence, the pathways which mediate the unlinking of the stacks of the Golgi ribbon are coupled to normal physiological processes. Other cell processes associated with the dispersal of the Golgi ribbon include stress responses, DNA repair, apoptosis and autophagy [26, 54, 55, 59, 80, 118]. Of note, many signalling pathways have been identified as the driver of the fragmentation of the Golgi ribbon for cellular processes (for review see Table 1 [53].

To understand the relationship between the Golgi ribbon and ministacks it is important to appreciate the machinery regulating this process. The formation of the Golgi ribbon is regulated by both microtubules and actin filaments [78, 91]. A number of Golgi membrane-associated proteins interact with the cytoskeleton; these Golgi proteins act as molecular scaffolds as they interact with multiple binding partners, including signalling components [53]. In particular, ~12 Golgi-localised molecular scaffolds have been identified which interact with either the microtubule and/or the actin cytoskeleton, and also in some cases molecular motors, and these interactions influence and regulate Golgi morphology [46]. Golgi scaffolds include the golgins, large coiled-coil membrane tethers which also interact with small G proteins and regulate membrane trafficking, and the GRASPs. For recent reviews on the molecular regulation of Golgi morphology by scaffold molecules see [46, 79].

Microtubules can regulate the location of the Golgi via the centrosome and, in addition, the *cis* and *trans* Golgi membranes can nucleate microtubules and stabilise the Golgi structure by interaction with Golgi matrix proteins and golgins. The current model implicates the recruitment of the γ-tubulin ring complex (γ-TuRC) and other factors to the *cis*-Golgi membrane to nucleate microtubules and CLASPs and golgins on the *trans*-Golgi network (TGN) co-operate to polymerise and stabilize the microtubules [78]. In addition, many Golgi scaffolds can interact with the actin cytoskeleton on either the *cis* face of the Golgi or the TGN. For example, our lab has recently identified the TGN membrane tether, called GCC88, which links Golgi membranes to actin filaments. GCC88 is a peripheral membrane protein, located at the TGN, and *via* an interaction with the long isomer of ITSN-1, a guanine nucleotide exchange factor for Cdc42, mediates an actin-dependent pathway to regulate Golgi structure [54]. Stable cell lines overexpressing GCC88 lack a Golgi ribbon. Our work [27] made the unexpected discovery that the organization of the Golgi as a ribbon is essential for regulating the mTOR complex 1 (mTORC1) pathway and autophagy. mTORC1 is a major signalling pathway of eukaryotic cells regulating metabolism and proliferation and is a negative regulator of autophagy [112]. By manipulating the levels of the TGN golgin, GCC88, we demonstrated that the Golgi ribbon is a site for localization and activation of mTOR, a process dependent on the ribbon structure [27]. These findings are important as various diseases, including cancer and neurodegenerative diseases, are associated with the loss of the Golgi ribbon [71, 90].

Fragmentation of the Golgi Ribbon

Often the loss of the Golgi ribbon is referred to as fragmentation, however, the pathways that regulate the ribbon morphology are dynamic and the balance between the ribbon morphology and ministacks and other Golgi elements may be very fluid. It is important to distinguish between the different physiological and dynamic states of the Golgi morphology and the presence of aberrant Golgi structures which may be associated with disease and with defects in membrane trafficking and signalling.

Golgi components with cytoskeletal connections are wired into different pathways which regulate the integrity and dynamics of the Golgi ribbon. At least four different Golgi morphologies have been identified following loss or perturbation of the Golgi ribbon as defined by ultrastructural electron microscopic (EM) studies [55]. The precise Golgi structures differ depending on the pathways involved and include: (1) Conversion of the Golgi ribbon to intact Golgi mini-stacks; (2) loss of both Golgi ribbon and integrity of Golgi stacks, for example swollen cisternae; (3) dispersal of one Golgi compartment i.e. dispersal of the TGN associated with the activation of NLRP3 inflammasome; and (4) conversion of Golgi ribbon to

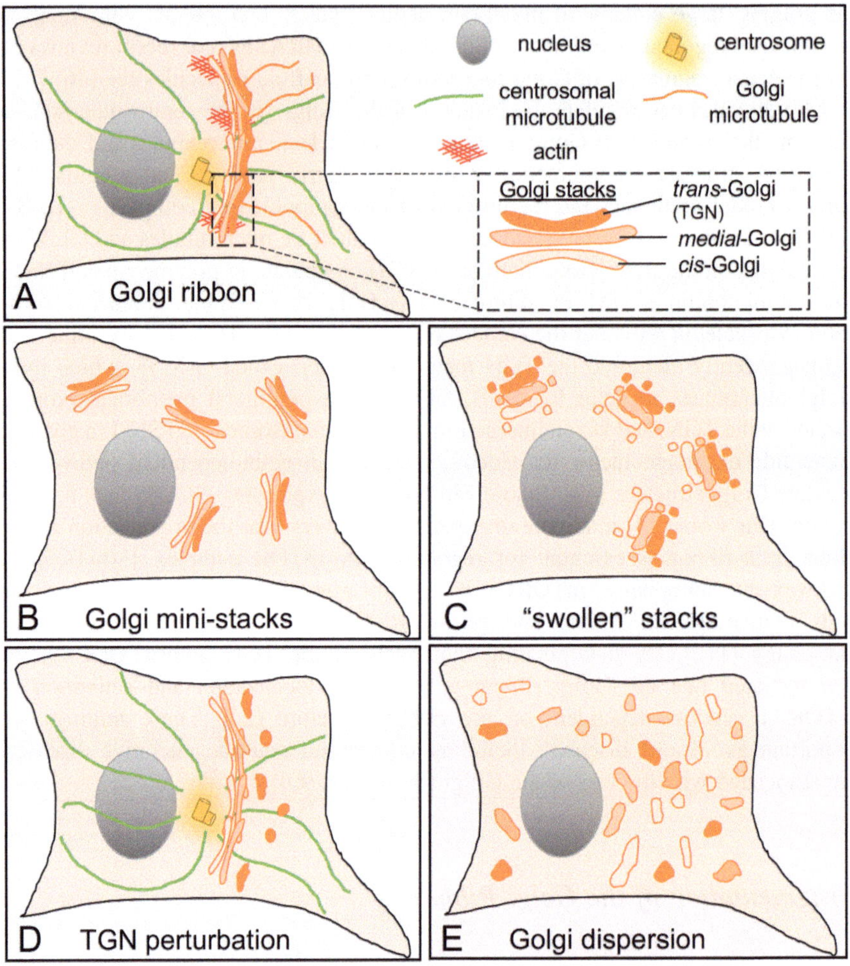

Fig. 18.1 Model depicting different Golgi morphologies following fragmentation of the Golgi ribbon. (**a**) Intact Golgi ribbon structure. (**b–e**) Morphology of the Golgi structure following loss of the integrity of the Golgi ribbon under different conditions, based on optical and EM data. (**b**) Golgi mini-stacks dispersed throughout the cytoplasm, (**c**) loss of the integrity of the dispersed Golgi mini-stacks with shortened cisternae, swelling of the cisternae and increase in the Golgi associated tubules, (**d**) dispersal of one compartment of the Golgi stack, depicting dissociation of the TGN with the remainder of the Golgi stack within a ribbon structure; (**e**) loss of the ribbon and Golgi stacks with the Golgi membranes dispersed as tubules and vesicles

tubulovesicular elements (Fig. 18.1). The functional outcome associated with the conversion of the Golgi ribbon into these different morphologies differs according to the associated cell process. Overall, there is a complex network of signalling pathways, Golgi morphologies and functional activities and the molecular details underlying these relationships are not well understood in many cases.

The Neuronal Golgi

An important issue currently under investigation is the relationship between the somatic Golgi ribbon and the Golgi outposts in dendrites. Golgi outposts were initially identified in rodent and *Drosophila* neurons [33, 35] [115] and more recently have also been identified in human iPSC-derived neurons [107], human brain organoids [107] and human brain sections [96]. Rodent outposts are restricted to the proximal region of large dendrites often at dendrite branch points [33] [44, 70] whereas in primary human neurons the Golgi outposts are located throughout the dendrites [107]. In rat hippocampal neurons, about 20% of the neurons contain one Golgi outpost in one of the dendrites; in addition to Golgi outposts, rat hippocampal primary neurons also contain smaller Golgi satellite elements widely distributed in dendrites [57] [28]. Golgi satellites have also been reported in peripheral axons [15]. It is notable that Golgi outposts in human neurons are more abundant than that reported in rodent neurons as Golgi outposts were observed in the dendrites of >70% of iPSC derived human neurons. Long-term live cell imaging using an IncuCyte and a ceramide Golgi dye identified very dynamic Golgi structures in human primary neurons, with bidirectional movement from the soma to the dendrites [107, 108]. These dendritic Golgi structures represent intact Golgi stacks based on *cis*- and *trans*-Golgi markers staining observed by immunofluorescence.

Studies on the function of Golgi outposts come from model systems, particularly *Drosophila* and rodents, and outposts have been shown to be required for dendrite maintenance [6, 35, 36]. A number of different membrane proteins have been shown to transit through the dendritic Golgi outposts [44]. Of note is that different subtypes of glutamate receptors show differences in whether they traffic through the conventional somatic Golgi or the dendritic Golgi outposts. NMDA receptors pass through the dendrite Golgi outpost whereas, following a block in Golgi exit, AMPA receptors accumulated in the somatic Golgi [38] and these different trafficking routes may reflect fine tuning of the synaptic receptors by independent regulation of the trafficking of glutamate receptors with different activities at the synapse. NMDA receptors modulate the strength of the synaptic transmission and the local delivery of either newly synthesised or recycling NMDA receptors may be important in the fine-tuning process.

Dynamics of the Golgi During Neurogenesis

The Golgi apparatus plays a fundamental role in the establishment of cell polarity and cell migration during neurogenesis. Cell migration is essential for the assembly of functional networks. During the migration phase of neurogenesis in rodents, the entire Golgi moves towards the migration front to support polarised trafficking, a process regulated by the re-location of the MTOC [16, 42, 103]. A defect in Golgi dendritic relocation during neurogenesis is linked to neurodegenerative diseases

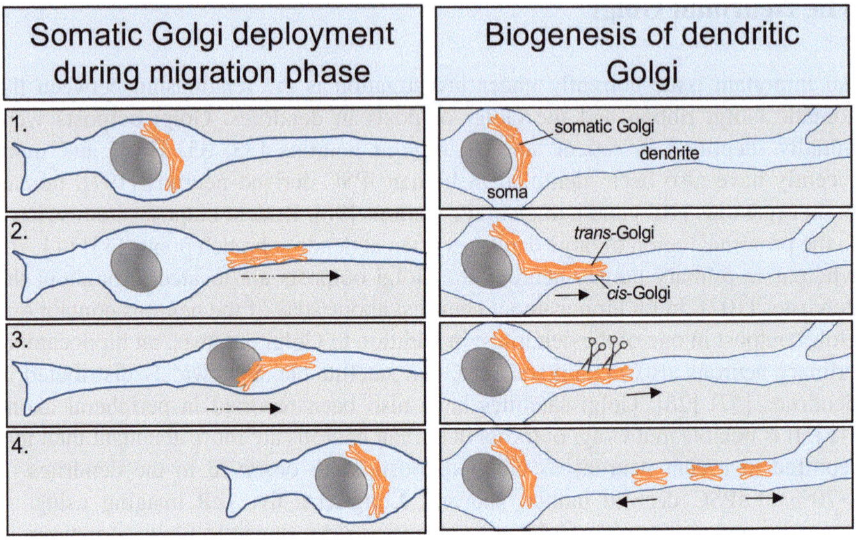

Fig. 18.2 Neurogenesis and the behaviour of the dendritic Golgi in primary human neurons. Model showing the organisation and dynamics of the Golgi apparatus during differentiation of neural precursor cells to human iPSC derived neurons. Left panels: The intact Golgi ribbon is transiently translocated from the cell body to the major neurite during the migration phase of neurogenesis. The cell body then follows and the Golgi ribbon returned to the cell body. Right panels: The biogenesis of dendritic Golgi outposts from the somatic Golgi ribbon during neuronal maturation. The dendritic Golgi outposts arise from the somatic Golgi and are highly dynamic and display bilateral movement along the dendrites

[5]. A similar process has been observed early in neuronal differentiation of human iPSC-derived neurons; live imaging identified the transient translocation of the entire somatic Golgi into a neurite extension of the cell which faced the direction of migration [107]. Following cell migration, the Golgi was returned to a perinuclear location in the soma (Fig. 18.2). In the later stage of differentiation, Golgi ministacks containing *cis* and *trans*-cisternae are observed in dendrites and these Golgi outposts are very dynamic with bidirectional transport between the proximal and distal regions of dendrites [107]. Live imaging over an extended time period revealed that Golgi outposts in human iPSC-derived neurons are present in a large number of dendrites; these outposts are functional and can receive and retain resident Golgi glycosyltransferses [107].

Biogenesis of the Golgi Outposts

The formation of Golgi outposts in rodent neurons is regulated by the RhoA signalling pathway [70] which is linked with the actin cytoskeleton. RhoA/ROCK signalling pathway has also been shown to be important in the biogenesis of Golgi outposts in human iPSC-derived neurons [107]. The outposts originate from membranes of the somatic Golgi rather than *de novo* outgrowth from the dendritic ER and are generated in an actin and myosin II dependent process. With the use of ROCK and actin inhibitors we have proposed a two-step process for the generation of Golgi outposts in human neurons, firstly an actin-dependent conversion of the compact ribbon into an extended elongated ribbon structure which protrudes into the proximal dendrite and, secondly, ROCK signalling regulated fission of the extended Golgi structure into the mini-stacks typical of the outpost [107] (Fig. 18.2). Hence, signalling pathways and actin remodelling play a role in releasing the mini-stacks from the somatic Golgi ribbon into the dendrites. There is currently very limited information available on whether the dendritic Golgi is compromised in neurological diseases; this is an important question, particularly as changes in the morphology of the somatic Golgi could have consequences for the generation of the Golgi outposts. Notably, one report indicates that the dendritic Golgi might be affected in disease. Indeed, elongated Golgi in the dendrites of patient iPSC-derived neurons with a lysosomal disorder were observed [49], suggesting a block in the fission step to form the individual Golgi mini-stacks in the neuronal dendrites.

Alzheimer's Disease and the Golgi

The major hallmarks of Alzheimer's disease (AD) are extracellular amyloid-β plaques, intracellular tangles and inflammation [98, 111]. Amyloid plaques in the brain consist of aggregated Aβ peptides derived from sequential proteolytic processing of amyloid precursor protein (APP) by an enzyme family of secretases. It is likely that the toxic Aβ peptides seed the onset of the disease whereas the neurofibrillary tangles (hyperphosphorylated tau protein) and inflammation contribute to driving the progression of the disease. The initial processing of APP by the β-secretase BACE1 is regulated by intracellular membrane sorting events and represents the rate-limiting step in Aβ production in healthy neurons [94, 102]. Several AD susceptibility genes are regulators of membrane trafficking [77, 111] and there is now strong evidence that AD arises from defective membrane trafficking of the secretases or APP in the secretory and/or endocytic pathways [66, 94, 99, 105]. BACE1 requires an acidic environment for its optimal activity [83] and the early endosomes and TGN present compatible locations based on the pH of these compartments and the convergence of APP and BACE1 [94]. Work from various laboratories have demonstrated that both the secretory and endocytic pathways are sites for APP processing by BACE1 (Fig. 18.3) (for reviews see [20, 109]).

Observations over 25 years have reported perturbations in the neuronal Golgi ribbon architecture as a hallmark of familial and sporadic AD in humans [25, 41, 71, 97]; "fragmentation" of the Golgi is observed in a diverse range of AD models [32] and directly contributes to the pathology of the disease [56]. For example, Golgi fragmentation is associated with enhanced processing and Aβ production by familial APP mutant human iPSC-derived neurons [32]. Golgi fragmentation was also

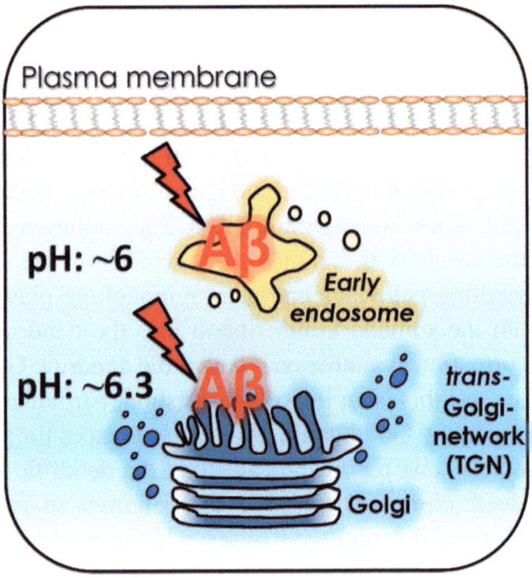

Fig. 18.3 Intracellular locations of Aβ production. The major sites of convergence of APP and BACE1 which are compatible with APP processing and Aβ production are highlighted

triggered in a transgenic mouse model of AD expressing familial mutations [40]. Strikingly, inhibition of the GRASP-mediated pathway of Golgi fragmentation restored a compact Golgi in this transgenic animal model and protected from neuronal disease progression [39, 40].

Role of the Golgi in APP Processing and Production of Aβ Peptides

Perturbation of the trafficking of BACE1 and/or APP into or from the TGN, in either non-neuronal or neuronal cells, can lead to an increase in APP processing and intracellular Aβ production in the Golgi and more generally the secretory pathway [7, 95, 100]. For example, redirecting the recycling of BACE1 from the recycling endosomes to the TGN, by modification of the cytoplasmic tail, enhances APP processing [10]. On the other hand, retaining APP in the TGN, by knocking-down the sorting adaptor AP-4, increases processing of APP and Aβ production and secretion [7, 100]. These studies demonstrate that secretase processing of APP and Aβ production can occur in the Golgi when the trafficking pathways of BACE1 and/or APP are perturbed; however, these are artificial conditions and a key question that follows from these findings is whether APP processing in the TGN occurs in disease conditions.

APP processing and Aβ production in the endosomal pathway has attracted considerable attention (see recent reviews [85, 92]). The majority of AD (95%) cases are late-onset where symptoms occur at or after age 65. Genetic risk factors have been identified associated with late onset AD and a number of these disease risk factors have a role in endosomal sorting, including retromer components, SORL1, BIN1 and PDL3 [37, 43, 111]. Numerous studies in cell and animal models confirm the importance of endosomes in the generation of Aβ associated with these late-onset AD risk alleles. The early endosome is also the major site for the limited processing of wild-type APP in neurons [110]. APP is delivered to the early endosomes either from the plasma membrane [65, 72] or directly from the TGN by an AP-4-dependent pathway [7, 100].

A role for the secretory pathway in APP processing has also been explored in both non-neuronal and neuronal cells. Given that APP processing can occur along the secretory pathway if trafficking is perturbed, the question arises how APP is protected from undue processing along the secretory pathway in healthy neurons. Recent work has shown there are multiple cellular mechanisms to partition APP and BACE1 along the secretory pathway. These two membrane proteins are sorted into distinct transport pathways from the TGN via the interaction of sorting motifs on their cytoplasmic tails with different adaptor complexes [95]. Hence the segregation into different subdomains of the TGN and transport pathways limits the opportunity for BACE1 cleavage of APP. We recently investigated whether APP and BACE1 were transported through the Golgi stack together or were partitioned into distinct membrane subdomains. Using super-resolution microscopy, only minimum co-localisation of APP and BACE1 was detected in the *cis*-Golgi and throughout the Golgi stack, hence APP and BACE1 are segregated as they transit through the Golgi [21]. Live imaging to track both cargoes simultaneously demonstrated that APP and BACE1 were segregated on exit from the ER hence their partitioning occurs very early in the secretory pathway [21]. The segregation of these two membrane proteins is likely to be due to partitioning into different membrane subdomains. A question that arises from this finding is the impact of familial APP mutations on the segregation and trafficking, given that many of the familiar mutations are found within the transmembrane domain.

Enhanced Processing and Aβ Production of Familial APP Mutants in the Golgi

Familial mutations leading to early onset AD (EOAD) have increased processing of APP which results in elevated levels of Aβ [87, 94]. The familial mutations of APP are dominant-acting missense mutations either flanking or within the transmembrane domain. More than 100 mutations have been identified to date [3]. The APP Swedish mutation (APP_{SWE}), named as it was first identified in two Swedish families, is the most well-described familial pathogenic AD mutation. APP_{SWE} has a double mutation (K595N and M596L) adjacent to the ß-cleavage site [11], is associated with increased levels of Aß [11], and is preferentially localised in the Golgi fraction of cell homogenates [113]. The Swedish mutation enhances the rate of cleavage by BACE1

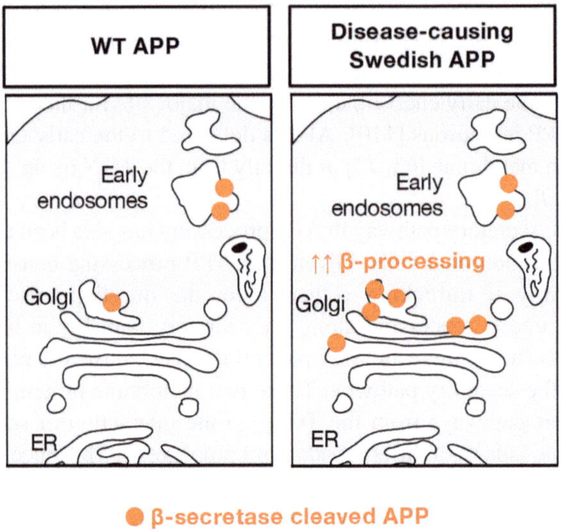

Fig. 18.4 Intracellular sites of amyloidogenic processing of wild-type APP and pathogenic familial Swedish APP mutant differ. Wild-type APP and disease-related Swedish APP mutant differ in their major intracellular sites of secretase processing as well as the level of processing and Aβ production. Only low levels of secretase processing of wild-type APP are observed and the majority takes place in the early endosomes, whereas the disease-causing Swedish APP mutant is associated with high levels of secretase processing and Aβ production within the Golgi apparatus. Modified from ([110], CC-BY 4.0)

[50]. The Golgi is fragmented in the hippocampal tissue of mice expressing APP_{SWE} [40].

Early studies showed that pathogenic Swedish APP mutant was processed in the secretory pathway [30] and biochemical evidence suggested that the Aβ was generated predominantly in the Golgi [113]. Our recent studies using the RUSH system have confirmed that the Golgi represents the major site of APPswe processing [110] (Fig. 18.4). In addition to the impact of the APP_{SWE} mutations on the changes in the rate of APP_{SWE} cleavage by BACE1 [50], our lab has shown that APP_{SWE} is transported through the Golgi at a slower rate compared with wild-type APP and results in elevated APP processing and Aß production [107]. Interestingly, the inhibition of BACE1 activity restored the anterograde trafficking of APPswe to that of APPwt [110], hence the generation of Aß suggests a potential feedback loop which extends the residency time of APP_{SWE} in the Golgi. We have shown that the Golgi is dispersed in HeLa cells expressing APP_{SWE}, but not in cells expressing wild-type APP, and that this phenotype is likely to be dependent on Aß production [107]. The Golgi in human iPSC-derived human neurons from APP_{SWE} patients has also been reported to be fragmented [32] and the addition of exogenous Aβ to neurons also results in Golgi fragmentation [40]. These findings provide strong evidence for an Aβ mediated perturbation of Golgi morphology. The basis for the difference in transport rates between wild-type APP and APP_{SWE} through the Golgi is unknown; one possibility is a differential partitioning between lipid subdomains with different trafficking kinetics. Regardless, the regulation of trafficking in the Golgi is an important consideration in APP processing and intracellular Aβ production.

Using the RUSH system in conjunction with immunoblotting to detect the processed APP products, we have shown that there are different intracellular locations for the preferential secretase cleavage of wild-type APP and familial mutants [110]. These studies demonstrate the importance of the Golgi in both regulating APP processing to sub-toxic levels in heathy conditions and a contribution to elevated Aβ levels in neurons bearing familial AD bearing APP mutations.

Impact of Fragmented Golgi on Aβ Production, Neuronal Function and Disease

Perturbation of the normal Golgi ribbon architecture in Alzheimer's disease has been observed in patients' iPSC-derived neurons [32] and clinical samples [89] and in cell and animal models of the disease [39].

In most of the studies, the "fragmented" Golgi morphology has been characterised by immunofluorescence. In addition, a few studies have analysed the ultrastructural morphology of the Golgi by EM. EM showed disrupted and swollen Golgi cisternae of the hypothalamus of human Alzheimer's brains [4]. EM analysis of hippocampal neurons from transgenic mice expressing familial APPswe and presenilin 1, showed swollen Golgi cisternae and a reduction in number and length of the cisternae compared with wild type mice [40]. Primary neurons treated with exogenous Aβ also showed similar ultrastructural changes of the Golgi morphology [40].

Another approach involving the disruption of the Golgi ribbon by targeting golgins has also been shown to directly result in neuronal degeneration. Selective knock-out of one of the regulators of the Golgi ribbon architecture, namely GM130, in Purkinje neurons, results in Golgi fragmentation, neuronal degeneration and dendritic atrophy in these mice [51]. Quantitative EM revealed an alteration in the location of the Golgi in the soma, and a major reduction in Golgi cisternae length and the number of cisternae per stack. More extensive EM studies are now required to determine the precise morphology of the Golgi architecture associated with different familial AD mutations and risk factors in AD in the soma and dendrites. Knowledge of the Golgi ultrastructure in disease is relevant as there are a variety of different Golgi morphologies (see Fig. 18.1), including distinct Golgi mini-stacks or loss of the integrity of the ministacks (as in the case of GM130 loss), and the precise morphology is relevant to an appreciation of the functional effect of these morphological changes).

Hyperphosphorylated tau is also a characteristic hallmark of AD which probably occurs downstream of Aβ production resulting in toxic tau aggregates and neurofibrillary tangles [60, 64, 117]. Given the functional importance of tau protein in stabilising microtubules, and the role of microtubule organising centres on the organisation of Golgi membranes [93], the generation of tau bundles may also influence the dynamics and architecture of the Golgi.

There are many potential consequences of loss of the Golgi ribbon in neurons. The following lists a number of possibilities based on various cell and animal model systems.

1. Alteration of the Golgi ribbon is likely to disrupt organisation of membrane subdomains [20], especially the organisation of lipid disordered and lipid ordered domains. Notably, APP and the secretases have been shown to partition to different membrane subdomains which is important for regulating secretase cleavage of APP [19, 75, 84] and disruption of these membrane subdomains may result in enhanced physical contact between APP and the secretases promoting an excessive production of Aβ.
2. Disruption of the ribbon architecture can affect the organization of the glycosylation enzymes within cisternae. Using super-resolution microscopy, small domains or units of glycosylation enzymes have recently been discovered [31]; these zones are dynamic and essential for the regulation of glycan synthesis. Loss of the ribbon has been shown to impact on the functional activity of these highly organised zones of glycosyltransferases [31].
3. Golgi fragmentation may affect the efficiency of anterograde trafficking of membrane proteins and lipids from the Golgi to their final destination, including synaptic receptors at the dendritic spines.
4. The Golgi ribbon plays a fundamental role in the biogenesis of dendritic Golgi [70, 107] and the alteration of Golgi morphology in the soma may impede the biogenesis of dendritic Golgi outposts and thereby affect cargo trafficking in dendrites and maintenance of functional synapses.
5. The loss of the Golgi ribbon has been shown to affect glycosylation [69] which can influence the levels of glycoproteins at the cell surface [116]. Neurons are particularly sensitive to changes in glycosylation such as polysialylation [81]. Polysialylated neural cell adhesion molecules are considered to play an important role in the regulation of higher cognitive functions and neurological diseases, and the levels of these polysialylated molecules have been reported to be altered in AD [23, 58]. Golgi fragmentation can result in changes in glycosylation due to mis-localisation of Golgi glycosyltransferases, as an indirect consequence of the location of coat adaptors and gatekeepers required for recycling of Golgi glycosyltransferases [18].
6. Golgi fragmentation is known to increase autophagy and autophagic processes are known to contribute to AD [27, 63].
7. The loss of Golgi ribbon can affect membrane flux between the TGN and endosomal compartment and disturb the homeostasis of the endosomal-lysosomal pathway and lysosomal function, as shown for knock downs of TGN components [2, 61].

8. The Golgi communicates with other organelles, including the ER, endosomes and possibly lysosomes, via membrane contacts sites [68, 88]. Dispersal of the Golgi ribbon may affect the function of these contact sites and of relevance is that dysfunctional membrane contact sites has been linked to neurological diseases [68].
9. The Golgi has intrinsic stress responses [82] and morphological changes in the Golgi architecture may enhance these stress responses and promote neuronal apoptosis.
10. The Golgi is a reserve of calcium which is regulated by actin pathways [104]. Fragmentation of the Golgi ribbon may influence Golgi calcium reserves and calcium dependent neuronal signalling pathways [56].
11. Inflammation is a hallmark of AD and the prolonged dispersal of the TGN, associated with chronic activation of NLRP3 inflammasome [8], may result in a block in sorting and export functions of the *trans*-Golgi.

Collectively these pathways can be affected by the loss of the Golgi ribbon and can contribute significantly to neuronal loss of function, neurological defects and progression of disease.

Future Studies

Given the unique features of the neuronal Golgi, and the role of the dendritic Golgi in delivering newly synthesized and/or recycling cargo to the dendritic spines for synaptic function, it will be important to determine the impact of alterations of the somatic Golgi structure on the secretory pathways mediated by the somatic Golgi and the dendritic Golgi outposts in dendrites. In addition, transport between the TGN/Golgi and endocytic system is critical for maintenance of endosomal-lysosomal homeostasis [61] and fragmentation of the Golgi ribbon may indirectly affect the endocytic pathway and the lysosomes. Conversely, the perturbation of the early endosomes associated with the late onset AD risk alleles may indirectly affect the TGN/Golgi by a block in the recycling of essential regulators of Golgi trafficking, back to the TGN. Furthermore, the Golgi-endocytic system is important for the recycling of trafficking machinery, which are essential for maintaining the correct location of glycosyltransferases throughout the Golgi stack [18]. Therefore, it is important to analyse the impact of Golgi fragmentation in AD on changes in glycosylation, particularly of glycoproteins at the synapses, and the consequences to neuronal signalling.

The challenge now is to define networks which regulate and are regulated by changes in the Golgi morphology and identify functional disruptions in the cell pathways arising from sustained Golgi fragmentation. Cryo-EM tomography has enormous potential to provide a deep understanding of the *in-situ* ultrastructural changes in the early phase of the disease, and coupled with multiOmics, to define the temporal sequence of the pathways perturbed by enhanced Aβ production and to identify the targets responsible for impaired transport and neuronal dysfunction.

There is now the ability to expand the use of cultured monolayers of human iPSC-derived neurons to 3D organoids and dissect the role of Golgi fragmentation on the organisation of neuronal networks within the complex cell environment which mimics the brain. Moreover, findings from such studies would have considerable potential to be integrated into the emerging AI field of organoid learning and memory [86] to bridge defects in Golgi function with neuronal cognition.

Finally, the identification of pathways responsible for the initiation of neurotoxicity has considerable potential to be exploited for development of neuronal cell-based assays for identification of novel druggable targets and biomarkers for treatment and detection of AD.

Acknowledgments This work was supported by funding from the National Health and Medical Research Council of Australia (2013384).

Conflict of Interest The authors have no conflict of interest to declare.

References

1. Abdel Rahman AM, Ryczko M, Nakano M, Pawling J, Rodrigues T, Johswich A, Taniguchi N, Dennis JW (2015) Golgi N-glycan branching N-acetylglucosaminyltransferases I, V and VI promote nutrient uptake and metabolism. Glycobiology 25:225–240
2. Akaaboune SR, Wang Y (2024) Golgi defect as a major contributor to lysosomal dysfunction. Front Cell Dev Biol 12:1386149
3. Alzforum (2024) Alzforum: Alzheimer's disease mutations [Online]. Available: https://www.alzforum.org/mutations [Accessed]
4. Baloyannis SJ, Mavroudis I, Mitilineos D, Baloyannis IS, Costa VG (2015) The hypothalamus in Alzheimer's disease: a Golgi and electron microscope study. Am J Alzheimers Dis Other Dement 30:478–487
5. Barnat M, Capizzi M, Aparicio E, Boluda S, Wennagel D, Kacher R, Kassem R, Lenoir S, Agasse F, Braz BY, Liu JP, Ighil J, Tessier A, Zeitlin SO, Duyckaerts C, Dommergues M, Durr A, Humbert S (2020) Huntington's disease alters human neurodevelopment. Science 369:787–793
6. Bowen AB, Bourke AM, Hiester BG, Hanus C, Kennedy MJ (2017) Golgi-independent secretory trafficking through recycling endosomes in neuronal dendrites and spines. elife 6
7. Burgos PV, Mardones GA, Rojas AL, Dasilva LL, Prabhu Y, Hurley JH, Bonifacino JS (2010) Sorting of the Alzheimer's disease amyloid precursor protein mediated by the AP-4 complex. Dev Cell 18:425–436
8. Chen J, Chen ZJ (2018) PtdIns4P on dispersed trans-Golgi network mediates NLRP3 inflammasome activation. Nature 564:71–76
9. Chia J, Goh G, Racine V, Ng S, Kumar P, Bard F (2012) RNAi screening reveals a large signaling network controlling the Golgi apparatus in human cells. Mol Syst Biol 8:629
10. Chia CP, Toh W, Sharples R, Gasnereau IJ, Hill AF, Gleeson PA (2013) Intracellular itinerary of internalised β-secretase, BACE1, and its potential impact on β-amyloid peptide biogenesis. Traffic 14:997–1013

11. Citron M, Oltersdorf T, Haass C, Mcconlogue L, Hung AY, Seubert P, Vigo-Pelfrey C, Lieberburg I, Selkoe DJ (1992) Mutation of the beta-amyloid precursor protein in familial Alzheimer's disease increases beta-protein production. Nature 360:672–674
12. Colanzi A, Corda D (2007) Mitosis controls the Golgi and the Golgi controls mitosis. Curr Opin Cell Biol 19:386–393
13. Colanzi A, Suetterlin C, Malhotra V (2003) Cell-cycle-specific Golgi fragmentation: how and why? Curr Opin Cell Biol 15:462–467
14. Corda D, Barretta ML, Cervigni RI, Colanzi A (2012) Golgi complex fragmentation in G2/M transition: an organelle-based cell-cycle checkpoint. IUBMB Life 64:661–670
15. Cornejo VH, Gonzalez C, Campos M, Vargas-Saturno L, Juricic MLA, Miserey-Lenkei S, Pertusa M, Madrid R, Couve A (2020) Non-conventional axonal organelles control TRPM8 ion channel trafficking and peripheral cold sensing. Cell Rep 30:4505–4517 e5
16. de Anda FC, Pollarolo G, da Silva JS, Camoletto PG, Feiguin F, Dotti CG (2005) Centrosome localization determines neuronal polarity. Nature 436:704–708
17. de Matteis MA, Mironov AA, Beznoussenko GV (2008) The Golgi ribbon and the function of the golgins. Springer-Verlag/Wein, New York
18. D'souza Z, Sumya FT, Khakurel A, Lupashin V (2021) Getting sugar coating right! The role of the Golgi trafficking machinery in glycosylation. Cells 10(12):3275
19. Ehehalt R, Keller P, Haass C, Thiele C, Simons K (2003) Amyloidogenic processing of the Alzheimer beta-amyloid precursor protein depends on lipid rafts. J Cell Biol 160:113–123
20. Fourriere L, Gleeson PA (2021) Amyloid beta production along the neuronal secretory pathway: dangerous liaisons in the Golgi? Traffic 22:319–327
21. Fourriere L, Cho EH, Gleeson PA (2022) Segregation of the membrane cargoes, BACE1 and amyloid precursor protein (APP) throughout the Golgi apparatus. Traffic 23:158–173
22. Gardiol A, Racca C, Triller A (1999) Dendritic and postsynaptic protein synthetic machinery. J Neurosci 19:168–179
23. Gascon E, Vutskits L, Kiss JZ (2007) Polysialic acid-neural cell adhesion molecule in brain plasticity: from synapses to integration of new neurons. Brain Res Rev 56:101–118
24. Gleeson PA (2023) Interorganellar communication – overview. In: Bradshaw RA, Gerald WH, Stahl PD (eds) Encyclopedia of cell biology, 2nd edn. Elsevier, Oxford
25. Gonatas NK, Stieber A, Gonatas JO (2006) Fragmentation of the Golgi apparatus in neurodegenerative diseases and cell death. J Neurol Sci 246:21–30
26. Gosavi P, Gleeson PA (2017) The function of the Golgi ribbon structure- an enduring mytery unfolds! BioEssays 39(11)
27. Gosavi P, Houghton FJ, Mcmillan PJ, Hanssen E, Gleeson PA (2018) The Golgi ribbon in mammalian cells negatively regulates autophagy by modulating mTOR activity. J Cell Sci 131:jcs211987
28. Govind AP, Jeyifous O, Russell TA, Yi Z, Weigel AV, Ramaprasad A, Newell L, Ramos W, Valbuena FM, Casler JC, Yan JZ, Glick BS, Swanson GT, Lippincott-Schwartz J, Green WN (2021) Activity-dependent Golgi satellite formation in dendrites reshapes the neuronal surface glycoproteome. elife 10:e68910
29. Gunn P, Gliddon BL, Londrigan SL, Lew AM, van Driel IR, Gleeson PA (2011) The Golgi apparatus in the endomembrane-rich gastric parietal cells exist as functional stable mini-stacks dispersed throughout the cytoplasm. Biol Cell 103:559–572
30. Haass C, Lemere CA, Capell A, Citron M, Seubert P, Schenk D, Lannfelt L, Selkoe DJ (1995) The Swedish mutation causes early-onset Alzheimer's disease by beta-secretase cleavage within the secretory pathway. Nat Med 1:1291–1296
31. Harada A, Kunii M, Kurokawa K, Sumi T, Kanda S, Zhang Y, Nadanaka S, Hirosawa KM, Tokunaga K, Tojima T, Taniguchi M, Moriwaki K, Yoshimura SI, Yamamoto-Hino M, Goto S, Katagiri T, Kume S, Hayashi-Nishino M, Nakano M, Miyoshi E, Suzuki KGN, Kitagawa H, Nakano A (2024) Dynamic movement of the Golgi unit and its glycosylation enzyme zones. Nat Commun 15:4514

32. Haukedal H, Corsi GI, Gadekar VP, Doncheva NT, Kedia S, de Haan N, Chandrasekaran A, Jensen P, Schionning P, Vallin S, Marlet FR, Poon A, Pires C, Agha FK, Wandall HH, Cirera S, Simonsen AH, Nielsen TT, Nielsen JE, Hyttel P, Muddashetty R, Aldana BI, Gorodkin J, Nair D, Meyer M, Larsen MR, Freude K (2023) Golgi fragmentation – one of the earliest organelle phenotypes in Alzheimer's disease neurons. Front Neurosci 17:1120086
33. Horton AC, Ehlers MD (2003) Dual modes of endoplasmic reticulum-to-Golgi transport in dendrites revealed by live-cell imaging. J Neurosci 23:6188–6199
34. Horton AC, Ehlers MD (2004) Secretory trafficking in neuronal dendrites. Nat Cell Biol 6: 585–591
35. Horton AC, Racz B, Monson EE, Lin AL, Weinberg RJ, Ehlers MD (2005) Polarized secretory trafficking directs cargo for asymmetric dendrite growth and morphogenesis. Neuron 48:757–771
36. Jan YN, Jan LY (2010) Branching out: mechanisms of dendritic arborization. Nat Rev Neurosci 11:316–328
37. Jansen IE, Savage JE, Watanabe K, Bryois J, Williams DM, Steinberg S, Sealock J, Karlsson IK, Hagg S, Athanasiu L, Voyle N, Proitsi P, Witoelar A, Stringer S, Aarsland D, Almdahl IS, Andersen F, Bergh S, Bettella F, Bjornsson S, Braekhus A, Brathen G, de Leeuw C, Desikan RS, Djurovic S, Dumitrescu L, Fladby T, Hohman TJ, Jonsson PV, Kiddle SJ, Rongve A, Saltvedt I, Sando SB, Selbaek G, Shoai M, Skene NG, Snaedal J, Stordal E, Ulstein ID, Wang Y, White LR, Hardy J, Hjerling-Leffler J, Sullivan PF, van der Flier WM, Dobson R, Davis LK, Stefansson H, Stefansson K, Pedersen NL, Ripke S, Andreassen OA, Posthuma D (2019) Genome-wide meta-analysis identifies new loci and functional pathways influencing Alzheimer's disease risk. Nat Genet 51:404–413
38. Jeyifous O, Waites CL, Specht CG, Fujisawa S, Schubert M, Lin EI, Marshall J, Aoki C, de Silva T, Montgomery JM, Garner CC, Green WN (2009) SAP97 and CASK mediate sorting of NMDA receptors through a previously unknown secretory pathway. Nat Neurosci 12:1011–1019
39. Joshi G, Wang Y (2015) Golgi defects enhance APP amyloidogenic processing in Alzheimer's disease. BioEssays 37:240–247
40. Joshi G, Chi Y, Huang Z, Wang Y (2014) Abeta-induced Golgi fragmentation in Alzheimer's disease enhances Abeta production. Proc Natl Acad Sci USA 111:E1230–E1239
41. Joshi G, Bekier ME 2nd, Wang Y (2015) Golgi fragmentation in Alzheimer's disease. Front Neurosci 9:340
42. Kaneko N, Sawada M, Sawamoto K (2017) Mechanisms of neuronal migration in the adult brain. J Neurochem 141:835–847
43. Karch CM, Goate AM (2015) Alzheimer's disease risk genes and mechanisms of disease pathogenesis. Biol Psychiatry 77:43–51
44. Kennedy MJ, Hanus C (2019) Architecture and dynamics of the neuronal secretory network. Annu Rev Cell Dev Biol 35:543–566
45. Kondylis V, Rabouille C (2009) The Golgi apparatus: lessons from Drosophila. FEBS Lett 583:3827–3838
46. Kulkarni-Gosavi P, Makhoul C, Gleeson PA (2019) Form and function of the Golgi apparatus: scaffolds, cytoskeleton and signalling. FEBS Lett 593:2289–2305
47. Kupfer A, Dennert G, Singer SJ (1983) Polarization of the Golgi apparatus and the microtubule-organizing center within cloned natural killer cells bound to their targets. Proc Natl Acad Sci USA 80:7224–7228
48. Lamb CA, Yoshimori T, Tooze SA (2013) The autophagosome: origins unknown, biogenesis complex. Nat Rev Mol Cell Biol 14:759–774
49. Lemonnier T, Blanchard S, Toli D, Roy E, Bigou S, Froissart R, Rouvet I, Vitry S, Heard JM, Bohl D (2011) Modeling neuronal defects associated with a lysosomal disorder using patient-derived induced pluripotent stem cells. Hum Mol Genet 20:3653–3666

50. Lin X, Koelsch G, Wu S, Downs D, Dashti A, Tang J (2000) Human aspartic protease memapsin 2 cleaves the beta-secretase site of beta-amyloid precursor protein. Proc Natl Acad Sci USA 97:1456–1460
51. Liu C, Mei M, Li Q, Roboti P, Pang Q, Ying Z, Gao F, Lowe M, Bao S (2017) Loss of the golgin GM130 causes Golgi disruption, Purkinje neuron loss, and ataxia in mice. Proc Natl Acad Sci USA 114:346–351
52. Lu Z, Joseph D, Bugnard E, Zaal KJ, Ralston E (2001) Golgi complex reorganization during muscle differentiation: visualization in living cells and mechanism. Mol Biol Cell 12:795–808
53. Makhoul C, Gosavi P, Gleeson PA (2018) The Golgi architecture and cell sensing. Biochem Soc Trans 46:1063–1072
54. Makhoul C, Gosavi P, Duffield R, Delbridge B, Williamson NA, Gleeson PA (2019a) Intersectin-1 interacts with the golgin GCC88 to couple the actin network and Golgi architecture. Mol Biol Cell 30:370–386
55. Makhoul C, Gosavi P, Gleeson PA (2019b) Golgi dynamics: the morphology of the mammalian Golgi apparatus in health and disease. Front Cell Dev Biol 7:112
56. Martinez-Menarguez JA, Tomas M, Martinez-Martinez N, Martinez-Alonso E (2019) Golgi fragmentation in neurodegenerative diseases: is there a common cause? Cells 8(7):748
57. Mikhaylova M, Bera S, Kobler O, Frischknecht R, Kreutz MR (2016) A dendritic Golgi satellite between ERGIC and retromer. Cell Rep 14:189–199
58. Mikkonen M, Soininen H, Tapiola T, Alafuzoff I, Miettinen R (1999) Hippocampal plasticity in Alzheimer's disease: changes in highly polysialylated NCAM immunoreactivity in the hippocampal formation. Eur J Neurosci 11:1754–1764
59. Millarte V, Farhan H (2012) The Golgi in cell migration: regulation by signal transduction and its implications for cancer cell metastasis. ScientificWorldJournal 2012:498278
60. Muratore CR, Rice HC, Srikanth P, Callahan DG, Shin T, Benjamin LN, Walsh DM, Selkoe DJ, Young-Pearse TL (2014) The familial Alzheimer's disease APPV717I mutation alters APP processing and Tau expression in iPSC-derived neurons. Hum Mol Genet 23:3523–3536
61. Nakano A (2022) The Golgi apparatus and its next-door neighbors. Front Cell Dev Biol 10: 884360
62. Nelson WJ (2000) W(h)ither the Golgi during mitosis? J Cell Biol 149:243–248
63. Nixon RA, Rubinsztein DC (2024) Mechanisms of autophagy-lysosome dysfunction in neurodegenerative diseases. Nat Rev Mol Cell Biol 25(11):926–946
64. Nizynski B, Dzwolak W, Nieznanski K (2017) Amyloidogenesis of Tau protein. Protein Sci 26:2126–2150
65. Park H, Hundley FV, Yu Q, Overmyer KA, Brademan DR, Serrano L, Paulo JA, Paoli JC, Swarup S, Coon JJ, Gygi SP, Wade Harper J (2022) Spatial snapshots of amyloid precursor protein intramembrane processing via early endosome proteomics. Nat Commun 13:6112
66. Peric A, Annaert W (2015) Early etiology of Alzheimer's disease: tipping the balance toward autophagy or endosomal dysfunction? Acta Neuropathol 129:363–381
67. Pierce JP, van Leyen K, Mccarthy JB (2000) Translocation machinery for synthesis of integral membrane and secretory proteins in dendritic spines. Nat Neurosci 3:311–313
68. Prinz WA, Toulmay A, Balla T (2020) The functional universe of membrane contact sites. Nat Rev Mol Cell Biol 21:7–24
69. Puthenveedu MA, Bachert C, Puri S, Lanni F, Linstedt AD (2006) GM130 and GRASP65-dependent lateral cisternal fusion allows uniform Golgi-enzyme distribution. Nat Cell Biol 8: 238–248
70. Quassollo G, Wojnacki J, Salas DA, Gastaldi L, Marzolo MP, Conde C, Bisbal M, Couve A, Caceres A (2015) A RhoA signaling pathway regulates dendritic Golgi outpost formation. Curr Biol 25:971–982
71. Rabouille C, Haase G (2015) Editorial: Golgi pathology in neurodegenerative diseases. Front Neurosci 9:489
72. Rajendran L, Annaert W (2012) Membrane trafficking pathways in Alzheimer's disease. Traffic 13:759–770

73. Ramirez IB, Lowe M (2009) Golgins and GRASPs: holding the Golgi together. Semin Cell Dev Biol 20:770–779
74. Ravichandran Y, Goud B, Manneville JB (2020) The Golgi apparatus and cell polarity: roles of the cytoskeleton, the Golgi matrix, and Golgi membranes. Curr Opin Cell Biol 62:104–113
75. Refolo LM, Pappolla MA, Lafrancois J, Malester B, Schmidt SD, Thomas-Bryant T, Tint GS, Wang R, Mercken M, Petanceska SS, Duff KE (2001) A cholesterol-lowering drug reduces beta-amyloid pathology in a transgenic mouse model of Alzheimer's disease. Neurobiol Dis 8: 890–899
76. Rios RM, Borens M (2003) The Golgi apparatus at the cell centre. Curr Opin Cell Biol 15:60–66
77. Rosenthal SL, Kamboh MI (2014) Late-onset Alzheimer's disease genes and the potentially implicated pathways. Curr Genet Med Rep 2:85–101
78. Sanders AA, Kaverina I (2015) Nucleation and dynamics of Golgi-derived microtubules. Front Neurosci 9:431
79. Saraste J, Prydz K (2019) A new look at the functional organization of the Golgi ribbon. Front Cell Dev Biol 7:171
80. Sasaki K, Yoshida H (2015) Organelle autoregulation-stress responses in the Er, Golgi, mitochondria and lysosome. J Biochem 157:185–195
81. Sato C, Kitajima K (2021) Polysialylation and disease. Mol Asp Med 79:100892
82. Serebrenik YV, Hellerschmied D, Toure M, Lopez-Giraldez F, Brookner D, Crews CM (2018) Targeted protein unfolding uncovers a Golgi-specific transcriptional stress response. Mol Biol Cell 29:1284–1298
83. Shimizu H, Tosaki A, Kaneko K, Hisano T, Sakurai T, Nukina N (2008) Crystal structure of an active form of BACE1, an enzyme responsible for amyloid beta protein production. Mol Cell Biol 28:3663–3671
84. Simons K, Ehehalt R (2002) Cholesterol, lipid rafts, and disease. J Clin Invest 110:597–603
85. Small SA, Petsko GA (2020) Endosomal recycling reconciles the Alzheimer's disease paradox. Sci Transl Med 12(572):eabb1717
86. Smirnova L, Morales Pantoja IE, Hartung T (2023) Organoid intelligence (OI) – the ultimate functionality of a brain microphysiological system. ALTEX 40:191–203
87. Song MS, Baker GB, Todd KG, Kar S (2011) Inhibition of beta-amyloid(1-42) internalization attenuates neuronal death by stabilizing the endosomal-lysosomal system in rat cortical cultured neurons. Neuroscience 178:181–188
88. Starling GP, Yip YY, Sanger A, Morton PE, Eden ER, Dodding MP (2016) Folliculin directs the formation of a Rab34-RILP complex to control the nutrient-dependent dynamic distribution of lysosomes. EMBO Rep 17:823–841
89. Stieber A, Mourelatos Z, Gonatas NK (1996) In Alzheimer's disease the Golgi apparatus of a population of neurons without neurofibrillary tangles is fragmented and atrophic. Am J Pathol 148:415–426
90. Sundaramoorthy V, Sultana JM, Atkin JD (2015) Golgi fragmentation in amyotrophic lateral sclerosis, an overview of possible triggers and consequences. Front Neurosci 9:400
91. Sutterlin C, Colanzi A (2010) The Golgi and the centrosome: building a functional partnership. J Cell Biol 188:621–628
92. Szabo MP, Mishra S, Knupp A, Young JE (2022) The role of Alzheimer's disease risk genes in endolysosomal pathways. Neurobiol Dis 162:105576
93. Takemura R, Okabe S, Umeyama T, Kanai Y, Cowan NJ, Hirokawa N (1992) Increased microtubule stability and alpha tubulin acetylation in cells transfected with microtubule-associated proteins MAP 1b, MAP 2 or tau. J Cell Sci 103(Pt 4):953–964
94. Tan JZA, Gleeson PA (2019) The role of membrane trafficking in the processing of amyloid precursor protein and production of amyloid peptides in Alzheimer's disease. Biochim Biophys Acta Biomembr 1861:697–712

95. Tan JZA, Fourriere L, Wang J, Perez F, Boncompain G, Gleeson PA (2020) Distinct anterograde trafficking pathways of BACE1 and amyloid precursor protein from the TGN and the regulation of amyloid-beta production. Mol Biol Cell 31:27–44
96. Taverna E, Mora-Bermudez F, Strzyz PJ, Florio M, Icha J, Haffner C, Norden C, Wilsch-Brauninger M, Huttner WB (2016) Non-canonical features of the Golgi apparatus in bipolar epithelial neural stem cells. Sci Rep 6:21206
97. Tcw J, Goate AM (2017) Genetics of beta-amyloid precursor protein in Alzheimer's disease. Cold Spring Harb Perspect Med 7(6):a024539
98. Therriault J, Zimmer ER, Benedet AL, Pascoal TA, Gauthier S, Rosa-Neto P (2022) Staging of Alzheimer's disease: past, present, and future perspectives. Trends Mol Med 28(9):726–741
99. Toh WH, Gleeson PA (2016) Dysregulation of intracellular trafficking and endosomal sorting in Alzheimer's disease: controversies and unanswered questions. Biochem J 473:1977–1993
100. Toh WH, Tan JZ, Zulkefli KL, Houghton FJ, Gleeson PA (2017) Amyloid precursor protein traffics from the Golgi directly to early endosomes in an Arl5b- and AP4-dependent pathway. Traffic 18:159–175
101. Valenzuela JI, Perez F (2015) Diversifying the secretory routes in neurons. Front Neurosci 9: 358
102. Vassar R, Bennett BD, Babu-Khan S, Kahn S, Mendiaz EA, Denis P, Teplow DB, Ross S, Amarante P, Loeloff R, Luo Y, Fisher S, Fuller J, Edenson S, Lile J, Jarosinski MA, Biere AL, Curran E, Burgess T, Louis JC, Collins F, Treanor J, Rogers G, Citron M (1999) Beta-secretase cleavage of Alzheimer's amyloid precursor protein by the transmembrane aspartic protease BACE. Science 286:735–741
103. Vinopal S, Dupraz S, Alfadil E, Pietralla T, Bendre S, Stiess M, Falk S, Camargo Ortega G, Maghelli N, Tolic IM, Smejkal J, Gotz M, Bradke F (2023) Centrosomal microtubule nucleation regulates radial migration of projection neurons independently of polarization in the developing brain. Neuron 111(1241–1263):e16
104. von Blume J, Alleaume AM, Cantero-Recasens G, Curwin A, Carreras-Sureda A, Zimmermann T, van Galen J, Wakana Y, Valverde MA, Malhotra V (2011) ADF/cofilin regulates secretory cargo sorting at the TGN via the Ca2+ ATPase SPCA1. Dev Cell 20:652–662
105. Wang X, Huang T, Bu G, Xu H (2014) Dysregulation of protein trafficking in neurodegeneration. Mol Neurodegener 9:31
106. Wang J, Fourriere L, Gleeson PA (2020) Local secretory trafficking pathways in neurons and the role of dendritic Golgi outposts in different cell models. Front Mol Neurosci 13:597391
107. Wang J, Daniszewski M, Hao MM, Hernandez D, Pebay A, Gleeson PA, Fourriere L (2023a) Organelle mapping in dendrites of human iPSC-derived neurons reveals dynamic functional dendritic Golgi structures. Cell Rep 42:112709
108. Wang J, Gleeson PA, Fourriere L (2023b) Long-term live cell imaging during differentiation of human iPSC-derived neurons. STAR Protoc 4:102699
109. Wang J, Fourriere L, Gleeson PA (2024a) Advances in the cell biology of the trafficking and processing of amyloid precursor protein: impact of familial Alzheimer's disease mutations. Biochem J 481(19):1297–1325
110. Wang J, Gleeson PA, Fourriere L (2024b) Spatial-temporal mapping reveals the Golgi as the major processing site for the pathogenic Swedish APP mutation: familial APP mutant shifts the major APP processing site. Traffic 25:e12932
111. Webers A, Heneka MT, Gleeson PA (2020) The role of innate immune responses and neuroinflammation in amyloid accumulation and progression of Alzheimer's disease. Immunol Cell Biol 98:28–41
112. Wullschleger S, Loewith R, Hall MN (2006) TOR signaling in growth and metabolism. Cell 124:471–484
113. Xu H, Sweeney D, Wang R, Thinakaran G, Lo AC, Sisodia SS, Greengard P, Gandy S (1997) Generation of Alzheimer beta-amyloid protein in the *trans*-Golgi network in the apparent absence of vesicle formation. Proc Natl Acad Sci USA 94:3748–3752

114. Yadav S, Puri S, Linstedt AD (2009) A primary role for Golgi positioning in directed secretion, cell polarity, and wound healing. Mol Biol Cell 20:1728–1736
115. Ye B, Zhang Y, Song W, Younger SH, Jan LY, Jan YN (2007) Growing dendrites and axons differ in their reliance on the secretory pathway. Cell 130:717–729
116. Zhang X (2021) Alterations of Golgi structural proteins and glycosylation defects in cancer. Front Cell Dev Biol 9:665289
117. Zhang Y, Chen H, Li R, Sterling K, Song W (2023) Amyloid beta-based therapy for Alzheimer's disease: challenges, successes and future. Signal Transduct Target Ther 8:248
118. Farhan H, Rabouille C (2011) Signalling to and from the secretory pathway. J Cell Sci 124:171–180

Chapter 19
Golgi Fragmentation as a Potential Link Between SARS-CoV-2 Infection and Alzheimer's Disease: Mechanisms and Implications for Neurodegeneration in Long COVID

Yanzhuang Wang and Sam Gandy

Abstract The COVID-19 pandemic has impacted millions of people worldwide, and recent studies have shown that SARS-CoV-2 infection can lead to an Alzheimer's-like neuropathological and biomarker phenotype, as well as clinical symptoms of "brain fog". This raises an intriguing question: "How and where might the molecular pathways underlying SARS-CoV-2 infection and Alzheimer's disease (AD) converge?" One common feature of both SARS-CoV-2 infection and AD is the alteration of the endomembrane system, particularly the fragmentation of the Golgi apparatus. In this review article, we summarize the existing literature on SARS-CoV-2 infection biology and speculate about the potential mechanisms linking Golgi defects, SARS-CoV-2 infection, and neurodegeneration.

Keywords Alzheimer's disease · Autophagy · Brain fog · COVID · Golgi · Glycolipids · Glycosylation · GRASP55 · GRASP65 · Heparan sulfate proteoglycans · Intracellular trafficking · Myelin · O-GlcNAcylation · Phosphorylation · SARS-CoV-2 infection · Secretion · Tau

---

Y. Wang (✉)
Institute of Neurological and Psychiatric Disorders, Shenzhen Bay Laboratory, Shenzhen, Guangdong, China
e-mail: yzwang@szbl.ac.cn

S. Gandy (✉)
Departments of Neurology and Psychiatry and the Mount Sinai Alzheimer's Disease Research Center, Icahn School of Medicine at Mount Sinai, New York, NY, USA

The James J Peters VA Medical Center, Bronx, NY, USA
e-mail: samuel.gandy@mssm.edu

A. Nakano, J. Saraste (eds.), *The Golgi Network, Volume II*, Subcellular Biochemistry 111, https://doi.org/10.1007/978-3-032-16833-7_19

Abbreviations

ACE2	Angiotensin converting enzyme-2
AD	Alzheimer's disease
APP	Amyloid precursor protein
COVID-19	Coronavirus disease 2019
DMVs	Double-membrane vesicles
ER	Endoplasmic reticulum
ERGIC	Endoplasmic reticulum-Golgi intermediate compartment
Gb3	Globotriaosylceramide
GM1	Monosialotetrahexosylganglioside
GRASPs	Golgi reassembly stacking proteins
HSPGs	Heparan sulfate proteoglycans
OGA	O-GlcNAcase
O-GlcNAc	O-linked N-acetylglucosamine
OGT	O-GlcNAc transferase
SARS-CoV-2	Severe acute respiratory syndrome coronavirus 2
T2DM	Type 2 diabetes mellitus
TMPRSS2	Transmembrane serine protease 2
TNFα	Tumor necrosis factor α
UVRAG	UV radiation resistance-associated gene

Introduction

The Golgi apparatus is a complex network of membranes that plays a critical role in protein sorting, modification, and transport within cells. Dysfunction or fragmentation of the Golgi apparatus has been linked to several human diseases, including neurodegenerative disorders such as Alzheimer's disease (AD) (Fig. 19.1) [1–4]. Interestingly, recent studies have shown that SARS-CoV-2 infection can also lead to Golgi fragmentation in affected cells (Fig. 19.2) [5–8]. In addition to causing respiratory illness, SARS-CoV-2 infection can affect the central nervous system, leading to a range of neurological symptoms including "brain fog", loss of smell or taste, and even more severe complications like encephalitis [9–12]. The post-acute phase of COVID-19 increases the risk for long-term cognitive, memory, movement, and mental disorders, which are symptoms commonly found in AD and AD-related dementias (AD/ADRD) [13].

SARS-CoV-2 can reach the central nerve system (CNS) through a variety of pathways. For example, the loss of smell, a common symptom of SAR-CoV-2 infection, suggests that the virus invades the olfactory tissue and can reach CNS through peripheral nerve terminals [14–16]. The cribriform plate, which is close to the olfactory bulb, provides access to the brain [17]. Studies of mice nasally infected with SARS-CoV-2 and MERS-CoV showed that viral particles can reach the

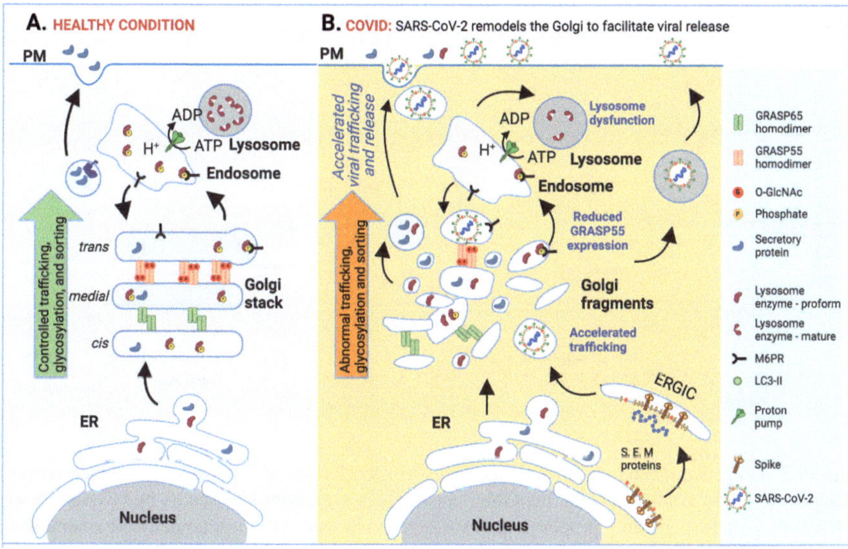

Fig. 19.1 SARS-CoV-2 virus remodels the Golgi structure by decreasing GRASP55 expression to accelerate viral trafficking and release. (**a**) In wild-type cells, the Golgi apparatus is organized into stacks by GRASP55 and GRASP65 proteins, ensuring proper trafficking, glycosylation, and sorting. (**b**) In SARS-CoV-2 infected cells, the virus causes Golgi fragmentation by downregulating GRASP55 expression, which may accelerate virus maturation and release [7]. Golgi fragmentation caused by SARS-CoV-2 infection may also impair the sorting and delivery of lysosomal enzymes to lysosomes, which can result in lysosomal dysfunction [62, 70, 136]

thalamus and brainstem [18]. The gastrointestinal tract is another point of viral entry through the brain-gut axis (vagus nerve) [17]. SARS-CoV-2 can also reach the CNS through the epithelium, which is connected to the peripheral or lymphatic circulation [19]. Viruses can further infect myeloid cells through which viral particles can disseminate to other tissues, including the CNS [17, 19–21]. Additionally, SARS-CoV-2 can infect cells through receptors other than angiotensin converting enzyme-2 (ACE2), the common viral entry receptor. This versatile entry pathway through ACE2-independent mechanisms allows the virus to invade many tissues, including neurons, providing multiple entry points into the brain [22–26].

There is growing evidence that SARS-CoV-2 infection may contribute to neurodegeneration [13, 27–31], although the precise mechanisms involved remain unclear. Given the link between Golgi fragmentation and both SARS-CoV-2 infection and AD, it is tempting to speculate that there may be common molecular pathways that contribute to neurodegeneration in these two contexts (Fig. 19.3) [32]. Both conditions involve type I membrane proteins such as the SARS-CoV-2 spike protein and the amyloid precursor protein (APP), whose trafficking, glycosylation, and processing are inevitably affected by Golgi fragmentation (Fig. 19.1). There is a possibility that SARS-CoV-2 infection-induced Golgi fragmentation could potentially accelerate APP trafficking and processing, initiating an AD-like

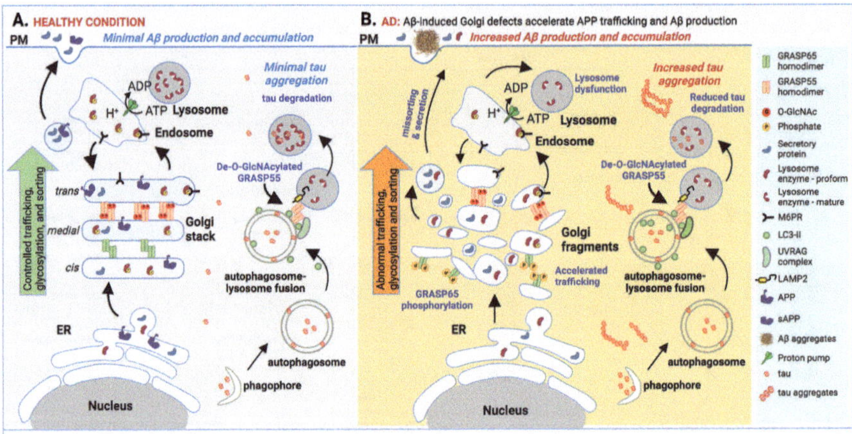

Fig. 19.2 Golgi fragmentation in Alzheimer's disease accelerates APP trafficking and increases Aβ production. (**a**) In wild-type cells, the Golgi apparatus is organized into stacks by GRASP55 and GRASP65 proteins, ensuring proper protein sorting. Lysosomal enzymes, tagged with mannose-6-phosphate (M6P) in the endoplasmic reticulum (ER), are recognized by the M6P receptor (M6PR), and transported to lysosomes via endosomes. Secretory proteins are transported through the Golgi to the plasma membrane (PM) for secretion. There is minimal amount of extracellular Aβ accumulation and intracellular tau aggregation [62, 67]. (**b**) In AD neurons, Aβ accumulation activates cdk5, leading to tau and GRASP65 phosphorylation, causing tau aggregation and Golgi fragmentation [1–3]. Golgi fragmentation accelerates APP trafficking and increases Aβ production, impacting the trafficking and processing of other neuronal proteins. It also causes lysosomal dysfunction, thereby reducing tau aggregate degradation and leading to its accumulation

process. In this chapter, we will discuss the potential mechanisms and implications of Golgi defects for neurodegeneration after long COVID, proposing Golgi fragmentation as a potential link between SARS-CoV-2 infection and AD. Understanding these pathways could shed light on the potential long-term neurological consequences of COVID-19 and provide insights into new therapeutic targets for AD and other neurodegenerative diseases.

SARS-CoV-2 Infection and Its Impact on the Secretory Pathway in Host Cells

SARS-CoV-2 is an enveloped, single-stranded, positive-sense RNA virus in the beta-coronavirus family. SARS-CoV-2 uses human ACE2 as the entry receptor. Upon interaction with ACE2, the spike (S) protein is activated by transmembrane serine protease 2 (TMPRSS2) at the cell surface, which induces viral fusion with host cell membranes. When TMPRSS2 activity is low, SARS-CoV-2 is endocytosed, and spike is cleaved by cathepsin L in lysosomes, leading to membrane fusion and release of viral RNA into host cell cytosol. After viral RNA release into

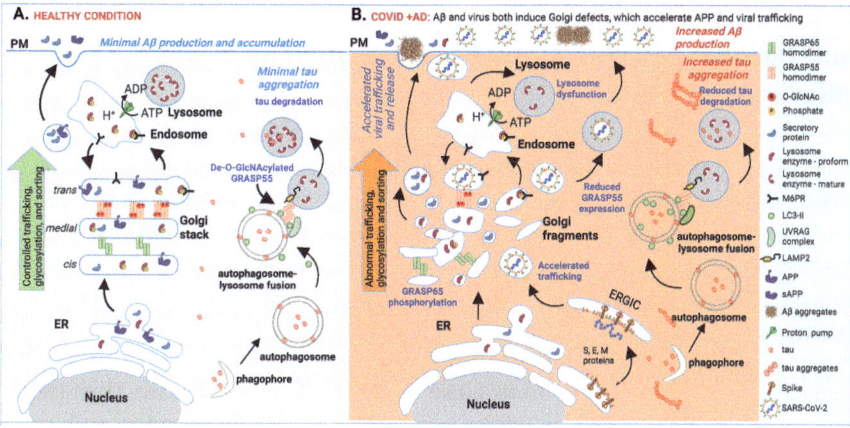

Fig. 19.3 Potential convergence of Aβ toxicity and SARS-CoV-2 at the Golgi. (**a**) In wild-type cells, the Golgi apparatus is organized into stacks by GRASP55 and GRASP65 proteins, ensuring proper trafficking, glycosylation, and sorting of secretory and lysosomal proteins. (**b**) In SARS-CoV-2 infected neurons, Aβ accumulation causes Golgi fragmentation by triggering cdk5-mediated GRASP65 phosphorylation [2], whereas the virus induces Golgi fragmentation by reducing GRASP55 expression [7], which may accelerate both APP and virus trafficking and release [7]. As a result, Aβ accumulation and viral infection provide additive effects on the Golgi, which subsequently increase the toxicity of Aβ and viral infection. Golgi fragmentation caused by SARS-CoV-2 infection and Aβ accumulation may also impair the trafficking, glycosylation, and sorting of secretory proteins and lysosomal enzymes, leading to neuronal dysfunction [62, 70, 136]

the host cell, the SARS-CoV-2 genome is translated into a total of 29 proteins [33]. In the host cell, viral proteins remodel the endoplasmic reticulum (ER) to form double-membrane vesicles (DMVs) that facilitate viral genomic RNA replication [34, 35].

Three of the four viral structural proteins [spike, envelope (E), and membrane (M)] are synthesized in the ER and transported to the ER-Golgi intermediate compartment (ERGIC) and *cis* Golgi, where virions are assembled and budded into the lumen of the ERGIC and Golgi (Fig. 19.1). In the Golgi, the spike protein is highly glycosylated [36, 37], a post-translational event that affects its folding, stability, interaction with ACE2, and susceptibility to vaccines. Mature viral particles are then transported by Golgi-derived vesicles to the plasma membrane (either directly [38] or via lysosomes [39]) and released into the extracellular space (Fig. 19.2). Compared to SARS-CoV, SARS-CoV-2 spike possesses a unique site that is cleaved by the furin protease in the *trans*-Golgi network (TGN), which facilitates viral entry.

Potential Mechanism of Golgi Structure Formation and Possible Effect of Golgi Fragmentation on SARS-CoV-2 Infection

The secretory pathway (including the Golgi apparatus) has been reported to be the site of key events in the infection cycle of multiple viruses including SARS-CoV-2 [5, 6, 40]. The list includes, but is not limited to: arteri- [41], bunyamwera [42, 43], cytomegalo- [44–46], flavi- (Dengue or Zika) [47], Hepatitis [48, 49], papilloma- [50, 51], polio- [52], and rhino viruses [53]. While SARS-CoV-2 is the most recent addition to the list, our (Y.W.) lab [7] and other labs [5, 6, 40] have recently provided evidence in support of the basic premise of our hypothesis that SARS-CoV-2 remodels the Golgi apparatus to facilitate viral trafficking and release. High resolution, spatiotemporal dynamics studies have confirmed that the secretory pathway is involved in the replication and assembly of SARS-CoV-2 virus [5–7, 40]. In addition, integrative imaging reveals that SARS-CoV-2 infection triggers Golgi structural defects [5], including Golgi ribbon unlinking, cisternal unstacking, and vesiculation, which we collectively refer to as Golgi fragmentation in this study, but the cause and consequence of Golgi fragmentation remain largely unknown.

The Golgi is a central hub for membrane trafficking, processing, and sorting. To perform these functions, the Golgi needs to form a multilayer stacked structure. While the exact mechanism of Golgi stack formation is not fully understood, one hypothesis is that the flat cisternal membranes of this organelle are held in place by a scaffold of Golgi structural proteins. The best characterized ones are the *G*olgi *rea*ssembly *s*tacking *p*roteins (GRASPs), GRASP55 and GRASP65 [54–60]. Our previous work demonstrated that GRASP55 and its homologue GRASP65 play essential roles in Golgi structure formation; they form *trans*-oligomers that "glue" adjacent Golgi cisternae together into stacks (Fig. 19.1) [57, 58, 60]. Depletion of GRASP proteins leads to Golgi cisternal unstacking.

Golgi structure disruption impairs its normal function [55, 61–64]. Inhibiting Golgi stacking by depleting GRASP55 and GRASP65 in cells accelerates protein trafficking, impairs accurate glycosylation and sorting, and increases the synthesis of heparan sulfate [54, 60, 65] that is known to enhance spike-ACE2 interaction [66]. A plausible explanation for these results is that stacking reduces the accessibility of coat proteins to the rims of the stacked Golgi membranes, which decreases the rate of vesicle budding and transport (Fig. 19.1) [67, 68]. In support of this hypothesis, unstacking increased the rate of COPI vesicle formation from Golgi membranes [61], suggesting a possibility that Golgi fragmentation may also facilitate SARS-CoV-2 viral assembly (Fig. 19.1). In addition, the spike-ACE2 interaction may be enhanced by heparan sulfate proteoglycans (HSPGs) [66], and interestingly, Golgi fragmentation induced by GRASP depletion increases HSPG synthesis [65]. Our preliminary evidence that SARS-CoV-2 infection induces Golgi fragmentation by reducing GRASP55 expression supports our formulation that SARS-CoV-2 infection-induced Golgi fragmentation facilitates viral trafficking and release (Fig. 19.1) [7].

Golgi Defects in Alzheimer's Disease

Despite the intensive efforts made in the last two decades, the causes of typical, sporadic, late-onset AD remain largely unknown. Two protein aggregates—the extracellular amyloid plaques formed by amyloid-beta (Aβ) peptides and the intracellular neurofibrillary tangles formed by the hyperphosphorylated microtubule-associated protein tau—are two of the most readily recognizable pathological hallmarks of AD, although many other factors also contribute to the development of the disease. In 2014, one of us (Y.W.) discovered that Aβ oligomer accumulation triggers Golgi fragmentation by activating cyclin-dependent protein kinase 5 (cdk5), which, in turn, phosphorylates GRASP65 [2]. GRASP65 phosphorylation in AD inhibits its oligomerization and causes Golgi fragmentation, which subsequently accelerates APP trafficking and processing, increasing Aβ production (Fig. 19.2). Conversely, rescue of the Golgi structure by the expression of a phosphorylation-deficient mutant of GRASP65 reduces Aβ production by increasing non-amyloidogenic α-cleavage of APP [1–3]. Given that cdk5 is a major kinase for tau phosphorylation, this observation links enhanced APP amyloidogenic cleavage with tau phosphorylation (Fig. 19.2). As expected, Golgi fragmentation is a common occurrence in brain samples of AD patients and can be triggered by neuronal overactivation [69].

Golgi Defects May Impact a Variety of Cellular Activities

Given the important role of the Golgi in membrane trafficking, long term Golgi defects may alter neuronal functions by impairing the trafficking, processing, and sorting of many neuronal proteins localized to the cell surface (including synapses) as well as multiple subcellular compartments and organelles. In addition to the accelerated protein trafficking described above, Golgi defects can cause missorting of lysosomal enzymes (Fig. 19.1) [62, 70, 71]. Golgi defects may also impact protein secretion. For example, it has been shown in a recent study [72] that treatment of naive hamsters with serum from SARS CoV2-infected hamsters causes downregulation of olfactory receptor expression in olfactory sensory neurons, indicating that SARS-CoV2 infection may trigger the secretion of an unknown molecule (s) into the serum. In addition, disruption of the Golgi structure by GRASP depletion also alters other cellular activities such as cell attachment, migration, and growth [73]. Therefore, we believe that molecular and cellular dysfunction related to Golgi fragmentation may extend beyond the Golgi *per se*, resulting in dysfunction of more distal compartments of the secretory pathway including autophagosomes and lysosomes, and impacting a variety of cellular activities.

One intriguing molecular pathway linking type 2 diabetes mellitus (T2DM) to AD is a nutrient-sensing pathway that involves the post-translational modification of nuclear and cytoplasmic proteins by O-linked N-acetylglucosamine (O-GlcNAc)

[74–77], a well-established protective stress response [78]. Inhibition of O-GlcNAcase (OGA) reduces tau aggregation and Aβ production [79–82]. To determine how the Golgi responds to energy deprivation, we made an unexpected discovery that GRASP55 is O-GlcNAcylated by the O-GlcNAc transferase (OGT) under growth conditions. Glucose deprivation reduces GRASP55 O-GlcNAcylation, allowing it to relocate from the Golgi to the interface between autophagosomes and lysosomes, where it functions as a membrane tether to facilitate autophagosome-lysosome fusion (Fig. 19.2) [83–86]. GRASP55 depletion causes autophagosome accumulation [84]. In addition, GRASP55 interacts with beclin 1 to facilitate the assembly and membrane association of the UV radiation-resistance-associated gene (UVRAG)-phosphoinositide 3-kinase (PI3K) complex (Fig. 19.2) [83], which is known to facilitate autophagosome maturation [87]. Therefore, GRASP55 plays a dual role in Golgi structure formation and autophagosome maturation [83–86].

Under stress condition, re-localization of GRASP55 from the Golgi to autophagosomes has been shown to be required for the unconventional secretion of cytosolic neural toxic proteins such as mutant huntingtin (Htt) and tau [88], whereas HSPGs are essential for tau uptake [89, 90], both of which are required for the spread of protein aggregates such as tau from diseased neurons to healthy neurons. HSPGs are synthesized in the Golgi; and Golgi structural defects impair glycosaminoglycan synthesis, sulfation, and secretion [65]. In summary, various AD pathologies, such as aberrant protein glycosylation and sorting, compromised lysosomal/autophagosomal degradation, heightened production of toxic Aβ peptide, and tau phosphorylation and aggregation, may be associated with Golgi malfunctions. As a result of Golgi fragmentation, proteotoxicity to neurons is amplified at numerous locations along the secretory pathway, as well as in the cytoplasm (Fig. 19.3).

How Might SARS-CoV-2 Infection Cause Neurodegeneration?

Although the underlying mechanism of viral involvement in neurodegeneration remains unknown, in principle, it can occur *via* two major alternative mechanisms [91]. First, SARS-CoV-2 viral particles and proteins may promote amyloid deposition and exacerbate Aβ neurotoxicity. In COVID-19, we found that SARS-CoV-2 infection induces Golgi fragmentation by reducing GRASP55 expression, which facilitates viral trafficking and release [7]. In AD, activated cdk5 phosphorylates and inactivates the other Golgi stacking protein GRASP65 to cause Golgi fragmentation, which in turn accelerates APP trafficking and increases Aβ production [2]. Therefore, SARS-CoV-2-induced Golgi fragmentation may accelerate APP trafficking and increase Aβ production (Fig. 19.3).

Second, SARS-CoV-2 viral proteins may modulate pathogenic protein aggregation and trigger prion-like propagation of neurotoxic species of misfolded proteins associated with AD/ADRD [92–94]. These neurotoxic, amyloid-forming, aggregation-prone proteins, including Aβ peptides, tau, α-synuclein (α-syn), TDP43, and their mutants, are collectively referred to as proteopathic proteins [95]. It has been shown that SARS-CoV-2 infection increases tau hyperphosphorylation and aggregation [96]. SARS-CoV-2 nucleocapsid (N) protein accelerates α-syn and TDP43 aggregation *in vitro* and causes cell death [97, 98]. Peptides from SARS-CoV-2 viral proteins can self-assemble in neurotoxic amyloid aggregates [99]. Thus, SARS-CoV-2 proteins may directly nucleate tau, α-syn, and TDP43 to form prion-like aggregates in the host cell. Alternatively, SARS-CoV-2 infection may facilitate proteotoxic aggregation by reducing their secretion. Research from our lab (Y.W.) and others demonstrate that GRASP55 controls autophagy-dependent unconventional secretion and aggregation of proteopathic proteins [88, 100, 101]. Consequently, SARS-CoV-2 induced GRASP55 down-regulation may increase the cellular level of proteopathic proteins and indirectly promote their aggregation (Fig. 19.3).

GRASP Proteins Can Protect Neurons from Various Neurodegenerative Protein Aggregates

We mentioned above the observation reported by Joshi et al. (2014) that expression of a phosphorylation-deficient mutants of GRASP55 and GRASP65 reverses the Golgi fragmentation that occurs in the hippocampal neurons of the PS1/APP mouse [1–3]. One of us (Y.W.) has also reported that GRASP55 expression can reduce the toxicity of a pathogenic mutant form of Htt protein in cultured striatal cells by increasing its secretion [88]. This raises a possibility that GRASPs *per se*, or GRASP-mimetic molecules or drugs, might be useful in the treatment or prevention of neurodegenerative or other diseases. Efforts to test this possibility in various cellular and animal models of human diseases are underway.

Could the Golgi Apparatus Be a Potential Subcellular Locus Where COVID19 and AD Converge?

The convergence of COVID-19 and AD in causing Golgi fragmentation suggests that the Golgi could be a potential subcellular location where the two diseases intersect [32]. Both the SARS-CoV-2 spike protein and APP are type I membrane proteins that are synthesized in the ER and transported through the secretory pathway. In the Golgi, they undergo N- and O-glycosylation modifications and are processed by proteases in the later stages of their trafficking. Golgi fragmentation associated with these diseases affects the trafficking and processing of both spike and APP. It is plausible that in COVID-19 patients, Golgi fragmentation caused by

SARS-CoV-2 infection could hasten APP trafficking and processing, thereby contributing to AD neuropathology (Fig. 19.3) [102].

SARS-CoV-2 infection can trigger a cytokine storm. Secretion of TNFα requires proper Golgi function, while processing and secretion of IL-1β requires Golgi fragmentation [103]. In COVID-19, SARS-CoV-2 infection-induced Golgi fragmentation may enhance the secretion of cytokines such as TNFα and IL-1β. Consistently, SARS-CoV-2 infects all neuronal types and microglia cells except GFAP-positive astrocytes [102]. It is also possible that SARS-CoV-2 infection-induced Golgi fragmentation may cause trafficking and glycosylation defects and result in important changes of protein, lipid, and glycan compositions at the cell surface. These changes, may, in turn, trigger activation of microglia and neuroinflammation. While speculative, these points support the possibility that Golgi fragmentation may play an important role in both COVID-19 and AD (Fig. 19.3).

Possible Links Between Golgi Defects and "Brain Fog"?

"Brain fog" is a general term used to describe cognitive impairment, perceived fatigue, cognitive difficulties, or more simply, slow thinking, which can occur under many different circumstances including SARS-CoV-2 infection. Although SARS-CoV-2 is primarily a respiratory illness, neuropsychiatric dysfunction represents a major cause of chronic disability ("long COVID") [104–107]. The basis for the neuropsychiatric dysfunction remains poorly understood. Immune/inflammatory factors may be released from the lungs of SARS-CoV2-infected patients, and these molecules may be transported via the circulation to the brain where they have been proposed to cause direct neuronal injury and/or an indirect neuronal injury by disrupting the vascular supply to neurons [108, 109]. It has also been reported that SARS-CoV-2 *per se* can be found in the brains of COVID-19 patients [110–113]. Regardless of whether brain injury is direct or indirect, integrative neuroimaging reveals that several months after SARS-CoV-2 infections, study participants had more gray matter loss and tissue abnormalities (mainly in the areas of the brain associated with smell) and more brain shrinkage than participants who had not been infected with the virus [114]. Similar observations have been made in the brain of SARS-CoV-2 infected non-human primates [115]. In the brain, olfactory epithelial support cells and stem cells express both ACE2 and TMPRSS2, as do cells in the nasal respiratory epithelium [111].

The exact number of COVID-19 survivors with brain fog symptoms is not known, but it has been reported to occur in about one-third or more of COVID-19 survivors in the context of substantial neurological and psychiatric morbidity in the 6 months following SARS-CoV-2 infection [102, 116–120]. A recent study reported that the omicron variant is much less likely than delta to cause long COVID, but still, almost 5% of people infected with the omicron variant experience fatigue, brain fog, headaches, or other health issues at least a month after getting infected

[121]. Regardless of these numbers, it is well accepted that millions of individuals surviving severe COVID-19 may be at increased risk of neuropsychiatric sequelae [105].

While the underlying mechanism of brain fog is currently unclear, one possible explanation is that SARS-CoV-2 infection causes a defect in myelination [122–124]. The Golgi is the cellular "factory" where sphingolipids and glycolipids are synthesized. Golgi fragmentation induced by GRASP depletion leads to a substantial elevation in the levels of monosialotetrahexosylganglioside (GM1) with concomitant reductions in the levels of globotriaosylceramide (Gb3) [55]. While both GM1 and Gb3 play important roles in myelination and function, GM1 has also been shown to enhance Aβ aggregation and toxicity [125, 126]. SARS-CoV-2 infection-induced Golgi fragmentation may alter sphingolipid and glycolipid synthesis and cause defects in myelination [122]. Consistent with this hypothesis is the observation that reduced levels of sphingosine (a major component of sphingomyelin) are highly associated with the development of symptomatic COVID-19 in the majority of SARS-CoV-2-infected patients compared to asymptomatic counterparts [123, 127].

A similar scenario has been described in the case of methotrexate toxicity [128]. In this situation, multiple brain cell types appear to fail because of exposure to this drug that is a mainstay of cancer chemotherapy [129], despite its association with "chemo brain", a neuropsychiatric syndrome. In our view, the triad of hypomyelination, Golgi fragmentation, and clinical "chemo brain" bears a striking similarity to the triad of hypomyelination, Golgi fragmentation, and clinical "brain fog" associated with SARS-CoV-2 infection. Yet another example of this "triad" may be emerging from a subset of First Responders exposed to inhaled silicates and organic neurotoxins following the 9/11 attacks at the World Trade Center (WTC) [130, 131]. Recent reports indicate that AD-related biomarkers are altered in some SARS-CoV-2 patients suffering from long COVID and brain fog [9, 102] as well as in some WTC responders suffering from a post-traumatic stress disorder/cognitive impairment syndrome [132, 133].

We speculate that Golgi fragmentation may be a common feature shared by neuropsychiatric illnesses associated with AD-like clinical and biomarker changes. In the case of typical AD, Golgi fragmentation may be initiated by the accumulation of oligomeric Aβ (Fig. 19.2) [1–3], while in SARS-CoV-2 infection or exposure to neurotoxins (methotrexate, inhaled silicates and organic neurotoxins), the Golgi fragmentation is due to immune/inflammatory or neurotoxic insults (Fig. 19.3). In all these situations, tissue and body fluid Aβ and phospho-tau are similarly affected. Given the increasing reliance on these biomarkers for confirming the diagnosis of AD and on defining patient populations for randomized clinical trials, another important implication of our speculation is that Golgi fragmentation may, in some cases (e.g., SARS-CoV-2 infection, neurotoxic exposure), give rise to false positive AD-like biomarker profiles in body fluids.

Concluding Remarks and Future Directions

In summary, understanding the mechanisms by which SARS-CoV-2 affects brain cells at the molecular and cellular levels is crucial, and the observation of Golgi fragmentation could provide a valuable starting point. Exploring Golgi abnormalities may enhance our comprehension of AD pathogenesis, AD biomarkers, and clinical dementia. These abnormalities encompass not only events associated with Aβ fibrils, which have shown no consistent correlation with cognitive status in the living human brain [134], but also those related to the accumulation of Aβ oligomers [2]. Dutch mutant APP transgenic mice accumulate Aβ oligomers but do not develop fibrils, yet they experience learning deficits similar to mice that accumulate both oligomeric and fibrillar Aβ [135]. To make sense of the molecular cell pathology of AD, we need to investigate how these fibrillar and oligomeric Aβ conformations induce flaws in protein trafficking and sorting in various brain cell types, which contribute to abnormalities in glycosylation, lysosomal function, and autophagy.

Contemplating the potential involvement of Golgi fragmentation in the mechanisms underlying neurodegenerative diseases, neurotoxic exposures, and SARS-CoV-2 infection may not be merely an academic pursuit. If the encouraging leads from cellular and animal models are confirmed, it is possible that manipulating GRASPs and other Golgi proteins therapeutically could present novel opportunities to treat a range of currently untreatable major human diseases.

Acknowledgments The authors would like to express their appreciation to Dr. Jianchao Zhang and other members of the Wang, Ehrlich-Gandy, and Tortorella labs, for their stimulating discussions and insightful suggestions. Y.W. was supported by NIH/NIGMS (R35GM130331), while S.G. is supported by RF1AG058469, RF1AG059319, U01AG046170, R01AG061894, P30 AG066514 (to Mary Sano), and the Cure Alzheimer's Fund.

COI Statement Y.W. is a founder of Camillo, Inc., a biotechnology company that focuses on the potential therapeutic use of GRASPs and other Golgi structural proteins in the treatment of neurodegenerative and other diseases. S.G. is a co-founder of Recuerdo Pharmaceuticals. He has served as a consultant in the past for J&J, Diagenic, and Pfizer, and he currently consults for Cognito Therapeutics, GLG Group, SVB Securities, Guidepoint, Third Bridge, MEDACORP, Altpep, Vigil Neurosciences, and Eisai. He has received research support in the past from Warner-Lambert, Pfizer, Baxter, and Avid.

References

1. Joshi G, Bekier ME 2nd, Wang Y (2015) Golgi fragmentation in Alzheimer's disease. Front Neurosci 9:340. https://doi.org/10.3389/fnins.2015.00340
2. Joshi G, Chi Y, Huang Z, Wang Y (2014) Abeta-induced Golgi fragmentation in Alzheimer's disease enhances Abeta production. Proc Natl Acad Sci USA 111(13):E1230–E1239. https://doi.org/10.1073/pnas.1320192111
3. Joshi G, Wang Y (2015) Golgi defects enhance APP amyloidogenic processing in Alzheimer's disease. Bioessays 37(3):240–247. https://doi.org/10.1002/bies.201400116

4. Gonatas NK, Gonatas JO, Stieber A (1998) The involvement of the Golgi apparatus in the pathogenesis of amyotrophic lateral sclerosis, Alzheimer's disease, and ricin intoxication. Histochem Cell Biol 109(5–6):591–600. https://doi.org/10.1007/s004180050257
5. Cortese M, Lee JY, Cerikan B, Neufeldt CJ, Oorschot VMJ, Kohrer S et al (2020) Integrative imaging reveals SARS-CoV-2-induced reshaping of subcellular morphologies. Cell Host Microbe 28(6):853–66 e5. https://doi.org/10.1016/j.chom.2020.11.003
6. Scherer KM, Mascheroni L, Carnell GW, Wunderlich LCS, Makarchuk S, Brockhoff M et al (2022) SARS-CoV-2 nucleocapsid protein adheres to replication organelles before viral assembly at the Golgi/ERGIC and lysosome-mediated egress. Sci Adv 8(1):eabl4895. https://doi.org/10.1126/sciadv.abl4895
7. Zhang J, Kennedy A, de Melo Jorge DM, Xing L, Reid W, Bui S et al (2025) SARS-CoV-2 remodels the Golgi apparatus to facilitate viral assembly and secretion. PLoS Pathog 21(6): e1013295. https://doi.org/10.1371/journal.ppat.1013295
8. Toader C, Eva L, Covache-Busuioc RA, Costin HP, Glavan LA, Corlatescu AD et al (2023) Unraveling the multifaceted role of the Golgi apparatus: insights into neuronal plasticity, development, neurogenesis, Alzheimer's disease, and SARS-CoV-2 interactions. Brain Sci 13(10). https://doi.org/10.3390/brainsci13101363
9. Reiken S, Sittenfeld L, Dridi H, Liu Y, Liu X, Marks AR (2022) Alzheimer's-like signaling in brains of COVID-19 patients. Alzheimers Dement. https://doi.org/10.1002/alz.12558
10. Rhodes CH, Priemer DS, Karlovich E, Perl DP, Goldman JE (2022) β-amyloid deposits in young COVID patients. SSRN Electronic Journal. https://doi.org/10.2139/SSRN.4003213
11. Wong AC, Devason AS, Umana IC, Cox TO, Dohnalova L, Litichevskiy L et al (2023) Serotonin reduction in post-acute sequelae of viral infection. Cell 186(22):4851–67 e20. https://doi.org/10.1016/j.cell.2023.09.013
12. Antar AAR, Yu T, Demko ZO, Hu C, Tornheim JA, Blair PW et al (2023) Long COVID brain fog and muscle pain are associated with longer time to clearance of SARS-CoV-2 RNA from the upper respiratory tract during acute infection. Front Immunol 14:1147549. https://doi.org/10.3389/fimmu.2023.1147549
13. Xu E, Xie Y, Al-Aly Z (2022) Long-term neurologic outcomes of COVID-19. Nat Med 28(11):2406–2415. https://doi.org/10.1038/s41591-022-02001-z
14. McGavern DB, Kang SS (2011) Illuminating viral infections in the nervous system. Nat Rev Immunol 11(5):318–329. https://doi.org/10.1038/nri2971
15. Lochhead JJ, Thorne RG (2012) Intranasal delivery of biologics to the central nervous system. Adv Drug Deliv Rev 64(7):614–628. https://doi.org/10.1016/j.addr.2011.11.002
16. Dahm T, Rudolph H, Schwerk C, Schroten H, Tenenbaum T (2016) Neuroinvasion and inflammation in viral central nervous system infections. Mediat Inflamm 2016:8562805. https://doi.org/10.1155/2016/8562805
17. Baig AM, Khaleeq A, Ali U, Syeda H (2020) Evidence of the COVID-19 virus targeting the CNS: tissue distribution, host-virus interaction, and proposed neurotropic mechanisms. ACS Chem Neurosci 11(7):995–998. https://doi.org/10.1021/acschemneuro.0c00122
18. Gandhi S, Srivastava AK, Ray U, Tripathi PP (2020) Is the collapse of the respiratory center in the brain responsible for respiratory breakdown in COVID-19 patients? ACS Chem Neurosci 11(10):1379–1381
19. Desforges M, Miletti TC, Gagnon M, Talbot PJ (2007) Activation of human monocytes after infection by human coronavirus 229E. Virus Res 130(1–2):228–240. https://doi.org/10.1016/j.virusres.2007.06.016
20. Stegelmeier AA, van Vloten JP, Mould RC, Klafuric EM, Minott JA, Wootton SK et al (2019) Myeloid cells during viral infections and inflammation. Viruses 11(2):168. https://doi.org/10.3390/v11020168
21. Rebholz H, Braun RJ, Ladage D, Knoll W, Kleber C, Hassel AW (2020) Loss of olfactory function-early indicator for Covid-19, other viral infections and neurodegenerative disorders. Front Neurol 11:569333. https://doi.org/10.3389/fneur.2020.569333

22. Chen J, Fan J, Chen Z, Zhang M, Peng H, Liu J et al (2021) Nonmuscle myosin heavy chain IIA facilitates SARS-CoV-2 infection in human pulmonary cells. Proc Natl Acad Sci USA 118(50):e2111011118. https://doi.org/10.1073/pnas.2111011118
23. Gao C, Zeng J, Jia N, Stavenhagen K, Matsumoto Y, Zhang H et al (2020) SARS-CoV-2 spike protein interacts with multiple innate immune receptors. bioRxiv: The Preprint Server for Biology. https://doi.org/10.1101/2020.07.29.227462
24. Amraei R, Yin W, Napoleon MA, Suder EL, Berrigan J, Zhao Q et al (2021) CD209L/L-SIGN and CD209/DC-SIGN act as receptors for SARS-CoV-2. ACS Cent Sci 7(7):1156–1165. https://doi.org/10.1021/acscentsci.0c01537
25. Gu Y, Cao J, Zhang X, Gao H, Wang Y, Wang J et al (2022) Author Correction: Receptome profiling identifies KREMEN1 and ASGR1 as alternative functional receptors of SARS-CoV-2. Cell Res 32(6):600. https://doi.org/10.1038/s41422-022-00654-6
26. Yang L, Kim TW, Han Y, Nair MS, Harschnitz O, Zhu J et al (2024) SARS-CoV-2 infection causes dopaminergic neuron senescence. Cell Stem Cell 31(2):196–211 e6. https://doi.org/10.1016/j.stem.2023.12.012
27. Dolatshahi M, Sabahi M, Aarabi MH (2021) Pathophysiological clues to how the emergent SARS-CoV-2 can potentially increase the susceptibility to neurodegeneration. Mol Neurobiol 58(5):2379–2394
28. Serrano-Castro PJ, Estivill-Torrus G, Cabezudo-Garcia P, Reyes-Bueno JA, Ciano Petersen N, Aguilar-Castillo MJ et al (2020) Impact of SARS-CoV-2 infection on neurodegenerative and neuropsychiatric diseases: a delayed pandemic? Neurologia (Engl Ed) 35(4):245–251. https://doi.org/10.1016/j.nrl.2020.04.002
29. Lippi A, Domingues R, Setz C, Outeiro TF, Krisko A (2020) SARS-CoV-2: at the crossroad between aging and neurodegeneration. Mov Disord 35(5):716
30. Idrees D, Kumar V (2021) SARS-CoV-2 spike protein interactions with amyloidogenic proteins: potential clues to neurodegeneration. Biochem Biophys Res Commun 554:94–98. https://doi.org/10.1016/j.bbrc.2021.03.100
31. Gomez-Pinedo U, Matias-Guiu J, Sanclemente-Alaman I, Moreno-Jimenez L, Montero-Escribano P, Matias-Guiu JA (2020) SARS-CoV-2 as a potential trigger of neurodegenerative diseases. Mov Disord 35(7):1104
32. Wang Y, Gandy S (2022) The Golgi apparatus: Site for convergence of COVID-19 brain fog and Alzheimer's disease? Mol Neurodegener 17(1):67. https://doi.org/10.1186/s13024-022-00568-2
33. Gordon DE, Jang GM, Bouhaddou M, Xu J, Obernier K, White KM et al (2020) A SARS-CoV-2 protein interaction map reveals targets for drug repurposing. Nature 583(7816): 459–468. https://doi.org/10.1038/s41586-020-2286-9
34. Oudshoorn D, Rijs K, Limpens R, Groen K, Koster AJ, Snijder EJ et al (2017) Expression and cleavage of middle east respiratory syndrome coronavirus nsp3–4 polyprotein induce the formation of double-membrane vesicles that mimic those associated with coronaviral rna replication. MBio 8(6). https://doi.org/10.1128/mBio.01658-17
35. Ricciardi S, Guarino AM, Giaquinto L, Polishchuk EV, Santoro M, Di Tullio G et al (2022) The role of NSP6 in the biogenesis of the SARS-CoV-2 replication organelle. Nature 606(7915):761–768. https://doi.org/10.1038/s41586-022-04835-6
36. Sanda M, Morrison L, Goldman R (2021) N- and O-glycosylation of the SARS-CoV-2 spike protein. Anal Chem 93(4):2003–2009. https://doi.org/10.1021/acs.analchem.0c03173
37. Tian W, Li D, Zhang N, Bai G, Yuan K, Xiao H et al (2021) O-glycosylation pattern of the SARS-CoV-2 spike protein reveals an "O-Follow-N" rule. Cell Res 31(10):1123–1125. https://doi.org/10.1038/s41422-021-00545-2
38. Eymieux S, Uzbekov R, Rouille Y, Blanchard E, Hourioux C, Dubuisson J et al (2021) Secretory vesicles are the principal means of SARS-CoV-2 egress. Cells 10(8). https://doi.org/10.3390/cells10082047

39. Ghosh S, Dellibovi-Ragheb TA, Kerviel A, Pak E, Qiu Q, Fisher M et al (2020) beta-coronaviruses use lysosomes for egress instead of the biosynthetic secretory pathway. Cell 183(6):1520–35 e14. https://doi.org/10.1016/j.cell.2020.10.039
40. Hackstadt T, Chiramel AI, Hoyt FH, Williamson BN, Dooley CA, Beare PA et al (2021) Disruption of the Golgi apparatus and contribution of the endoplasmic reticulum to the SARS-CoV-2 replication complex. Viruses 13(9). https://doi.org/10.3390/v13091798
41. Matczuk AK, Zhang M, Veit M, Ugorski M (2022) Expression of the heterotrimeric GP2/GP3/GP4 spike of an arterivirus in mammalian cells. Viruses 14(4). https://doi.org/10.3390/v14040749
42. Salanueva IJ, Novoa RR, Cabezas P, Lopez-Iglesias C, Carrascosa JL, Elliott RM et al (2003) Polymorphism and structural maturation of bunyamwera virus in Golgi and post-Golgi compartments. J Virol 77(2):1368–1381. https://doi.org/10.1128/jvi.77.2.1368-1381.2003
43. Ronnholm R (1992) Localization to the Golgi complex of Uukuniemi virus glycoproteins G1 and G2 expressed from cloned cDNAs. J Virol 66(7):4525–4531. https://doi.org/10.1128/JVI.66.7.4525-4531.1992
44. Lucin P, Jug Vucko N, Karleusa L, Mahmutefendic Lucin H, Blagojevic Zagorac G, Lisnic B et al (2020) Cytomegalovirus generates assembly compartment in the early phase of infection by perturbation of host-cell factors recruitment at the early endosome/endosomal recycling compartment/trans-Golgi interface. Front Cell Dev Biol 8:563607. https://doi.org/10.3389/fcell.2020.563607
45. Gardner TJ, Tortorella D (2016) Virion glycoprotein-mediated immune evasion by human cytomegalovirus: a sticky virus makes a slick getaway. Microbiol Mol Biol Rev 80(3):663–677. https://doi.org/10.1128/MMBR.00018-16
46. Jean Beltran PM, Cristea IM (2014) The life cycle and pathogenesis of human cytomegalovirus infection: lessons from proteomics. Expert Rev Proteomics 11(6):697–711. https://doi.org/10.1586/14789450.2014.971116
47. Viettri M, Zambrano JL, Rosales R, Caraballo GI, Gutierrez-Escolano AL, Ludert JE (2021) Flavivirus infections induce a Golgi stress response in vertebrate and mosquito cells. Sci Rep 11(1):23489. https://doi.org/10.1038/s41598-021-02929-1
48. Hansen MD, Johnsen IB, Stiberg KA, Sherstova T, Wakita T, Richard GM et al (2017) Hepatitis C virus triggers Golgi fragmentation and autophagy through the immunity-related GTPase M. Proc Natl Acad Sci USA 114(17):E3462–E3E71. https://doi.org/10.1073/pnas.1616683114
49. Dubois-Dalcq ME, Doller EW, Haspel MV, Holmes KV (1982) Cell tropism and expression of mouse hepatitis viruses (MHV) in mouse spinal cord cultures. Virology 119(2):317–331. https://doi.org/10.1016/0042-6822(82)90092-7
50. Young JM, Zine El Abidine A, Gomez-Martinez RA, Ozbun MA (2019) The known and potential intersections of Rab-GTPases in human papillomavirus infections. Front Cell Dev Biol 7:139. https://doi.org/10.3389/fcell.2019.00139
51. King K, Larsen BB, Gryseels S, Richet C, Kraberger S, Jackson R et al (2022) Coevolutionary analysis implicates toll-like receptor 9 in papillomavirus restriction. MBio 13(2):e0005422. https://doi.org/10.1128/mbio.00054-22
52. Sandoval IV, Carrasco L (1997) Poliovirus infection and expression of the poliovirus protein 2B provoke the disassembly of the Golgi complex, the organelle target for the antipoliovirus drug Ro-090179. J Virol 71(6):4679–4693. https://doi.org/10.1128/JVI.71.6.4679-4693.1997
53. Quiner CA, Jackson WT (2010) Fragmentation of the Golgi apparatus provides replication membranes for human rhinovirus 1A. Virology 407(2):185–195. https://doi.org/10.1016/j.virol.2010.08.012
54. Tang D, Yuan H, Wang Y (2010) The role of GRASP65 in Golgi cisternal stacking and cell cycle progression. Traffic 11(6):827–842. https://doi.org/10.1111/j.1600-0854.2010.01055.x
55. Bekier ME 2nd, Wang L, Li J, Huang H, Tang D, Zhang X et al (2017) Knockout of the Golgi stacking proteins GRASP55 and GRASP65 impairs Golgi structure and function. Mol Biol Cell 28(21):2833–2842. https://doi.org/10.1091/mbc.E17-02-0112

56. Barr FA, Puype M, Vandekerckhove J, Warren G (1997) GRASP65, a protein involved in the stacking of Golgi cisternae. Cell 91(2):253–262. https://doi.org/10.1016/s0092-8674(00)80407-9
57. Shorter J, Watson R, Giannakou ME, Clarke M, Warren G, Barr FA (1999) GRASP55, a second mammalian GRASP protein involved in the stacking of Golgi cisternae in a cell-free system. EMBO J 18(18):4949–4960. https://doi.org/10.1093/emboj/18.18.4949
58. Wang Y, Seemann J, Pypaert M, Shorter J, Warren G (2003) A direct role for GRASP65 as a mitotically regulated Golgi stacking factor. EMBO J 22(13):3279–3290. https://doi.org/10.1093/emboj/cdg317
59. Wang Y, Satoh A, Warren G (2005) Mapping the functional domains of the Golgi stacking factor GRASP65. J Biol Chem 280(6):4921–4928. https://doi.org/10.1074/jbc.M412407200
60. Xiang Y, Wang Y (2010) GRASP55 and GRASP65 play complementary and essential roles in Golgi cisternal stacking. J Cell Biol 188(2):237–251. https://doi.org/10.1083/jcb.200907132
61. Wang Y, Wei JH, Bisel B, Tang D, Seemann J (2008) Golgi cisternal unstacking stimulates COPI vesicle budding and protein transport. PLoS One 3(2):e1647. https://doi.org/10.1371/journal.pone.0001647
62. Xiang Y, Zhang X, Nix DB, Katoh T, Aoki K, Tiemeyer M et al (2013) Regulation of protein glycosylation and sorting by the Golgi matrix proteins GRASP55/65. Nat Commun 4:1659. https://doi.org/10.1038/ncomms2669
63. Lee I, Tiwari N, Dunlop MH, Graham M, Liu X, Rothman JE (2014) Membrane adhesion dictates Golgi stacking and cisternal morphology. Proc Natl Acad Sci USA 111(5):1849–1854. https://doi.org/10.1073/pnas.1323895111
64. Zhang X, Wang Y (2016) Glycosylation quality control by the Golgi structure. J Mol Biol 428(16):3183–3193. https://doi.org/10.1016/j.jmb.2016.02.030
65. Ahat E, Song Y, Xia K, Reid W, Li J, Bui S et al (2022) GRASP depletion-mediated Golgi fragmentation impairs glycosaminoglycan synthesis, sulfation, and secretion. Cell Mol Life Sci 79(4):199. https://doi.org/10.1007/s00018-022-04223-3
66. Clausen TM, Sandoval DR, Spliid CB, Pihl J, Perrett HR, Painter CD et al (2020) SARS-CoV-2 infection depends on cellular heparan sulfate and ACE2. Cell 183(4):1043–57 e15. https://doi.org/10.1016/j.cell.2020.09.033
67. Zhang X, Wang Y (2015) GRASPs in Golgi structure and function. Front Cell Dev Biol 3:84. https://doi.org/10.3389/fcell.2015.00084
68. Zhang X, Wang Y (2020) Nonredundant roles of GRASP55 and GRASP65 in the Golgi apparatus and beyond. Trends Biochem Sci 45(12):1065–1079. https://doi.org/10.1016/j.tibs.2020.08.001
69. Thayer DA, Jan YN, Jan LY (2013) Increased neuronal activity fragments the Golgi complex. Proc Natl Acad Sci USA 110(4):1482–1487. https://doi.org/10.1073/pnas.1220978110
70. Akaaboune SR, Wang Y (2024) Golgi defect as a major contributor to lysosomal dysfunction. Front Cell Dev Biol 12:1386149. https://doi.org/10.3389/fcell.2024.1386149
71. Akaaboune SR, Javed A, Bui S, Wierenga A, Wang Y (2025) GRASP55 regulates sorting and maturation of the lysosomal enzyme beta-hexosaminidase A. Mol Biol Cell:mbcE24100452. https://doi.org/10.1091/mbc.E24-10-0452
72. Zazhytska M, Kodra A, Hoagland DA, Frere J, Fullard JF, Shayya H et al (2022) Non-cell-autonomous disruption of nuclear architecture as a potential cause of COVID-19-induced anosmia. Cell 185(6):1052–64 e12. https://doi.org/10.1016/j.cell.2022.01.024
73. Ahat E, Xiang Y, Zhang X, Bekier ME 2nd, Wang Y (2019) GRASP depletion-mediated Golgi destruction decreases cell adhesion and migration via the reduction of alpha5beta1 integrin. Mol Biol Cell 30(6):766–777. https://doi.org/10.1091/mbc.E18-07-0462
74. Accardi G, Caruso C, Colonna-Romano G, Camarda C, Monastero R, Candore G (2012) Can Alzheimer disease be a form of type 3 diabetes? Rejuvenation Res 15(2):217–221. https://doi.org/10.1089/rej.2011.1289

75. Arnold SE, Arvanitakis Z, Macauley-Rambach SL, Koenig AM, Wang HY, Ahima RS et al (2018) Brain insulin resistance in type 2 diabetes and Alzheimer disease: concepts and conundrums. Nat Rev Neurol 14(3):168–181. https://doi.org/10.1038/nrneurol.2017.185
76. Franco KS, Bronson D (2005) Diabetes mellitus and Alzheimer disease. Arch Neurol 62(2): 330; author reply −1. https://doi.org/10.1001/archneur.62.2.330-a
77. Rasool M, Malik A, Qazi AM, Sheikh IA, Manan A, Shaheen S et al (2014) Current view from Alzheimer disease to type 2 diabetes mellitus. CNS Neurol Disord Drug Targets 13(3): 533–542. https://doi.org/10.2174/18715273113126660167
78. Zachara NE, O'Donnell N, Cheung WD, Mercer JJ, Marth JD, Hart GW (2004) Dynamic O-GlcNAc modification of nucleocytoplasmic proteins in response to stress. A survival response of mammalian cells. J Biol Chem 279(29):30133–30142. https://doi.org/10.1074/jbc.M403773200
79. Kim C, Nam DW, Park SY, Song H, Hong HS, Boo JH et al (2013) O-linked beta-N-acetylglucosaminidase inhibitor attenuates beta-amyloid plaque and rescues memory impairment. Neurobiol Aging 34(1):275–285. https://doi.org/10.1016/j.neurobiolaging.2012.03.001
80. Yuzwa SA, Shan X, Macauley MS, Clark T, Skorobogatko Y, Vosseller K et al (2012) Increasing O-GlcNAc slows neurodegeneration and stabilizes tau against aggregation. Nat Chem Biol 8(4):393–399. https://doi.org/10.1038/nchembio.797
81. Yuzwa SA, Cheung AH, Okon M, McIntosh LP, Vocadlo DJ (2014) O-GlcNAc modification of tau directly inhibits its aggregation without perturbing the conformational properties of tau monomers. J Mol Biol 426(8):1736–1752. https://doi.org/10.1016/j.jmb.2014.01.004
82. Yuzwa SA, Shan X, Jones BA, Zhao G, Woodward ML, Li X et al (2014) Pharmacological inhibition of O-GlcNAcase (OGA) prevents cognitive decline and amyloid plaque formation in bigenic tau/APP mutant mice. Mol Neurodegener 9:42. https://doi.org/10.1186/1750-1326-9-42
83. Zhang X, Wang L, Ireland SC, Ahat E, Li J, Bekier ME 2nd et al (2019) GORASP2/GRASP55 collaborates with the PtdIns3K UVRAG complex to facilitate autophagosome-lysosome fusion. Autophagy 15(10):1787–1800. https://doi.org/10.1080/15548627.2019.1596480
84. Zhang X, Wang L, Lak B, Li J, Jokitalo E, Wang Y (2018) GRASP55 senses glucose deprivation through O-GlcNAcylation to promote autophagosome-lysosome fusion. Dev Cell 45(2):245–61 e6. https://doi.org/10.1016/j.devcel.2018.03.023
85. Zhang X, Wang Y (2018) GRASP55 facilitates autophagosome maturation under glucose deprivation. Mol Cell Oncol 5(4):e1494948. https://doi.org/10.1080/23723556.2018.1494948
86. Zhang X, Wang Y (2018) The Golgi stacking protein GORASP2/GRASP55 serves as an energy sensor to promote autophagosome maturation under glucose starvation. Autophagy 14(9):1649–1651. https://doi.org/10.1080/15548627.2018.1491214
87. Liang C, Feng P, Ku B, Dotan I, Canaani D, Oh BH et al (2006) Autophagic and tumour suppressor activity of a novel Beclin1-binding protein UVRAG. Nat Cell Biol 8(7):688–699. https://doi.org/10.1038/ncb1426
88. Ahat E, Bui S, Zhang J, da Veiga Leprevost F, Sharkey L, Reid W et al (2022) GRASP55 regulates the unconventional secretion and aggregation of mutant huntingtin. J Biol Chem 298(8):102219. https://doi.org/10.1016/j.jbc.2022.102219
89. Holmes BB, DeVos SL, Kfoury N, Li M, Jacks R, Yanamandra K et al (2013) Heparan sulfate proteoglycans mediate internalization and propagation of specific proteopathic seeds. Proc Natl Acad Sci USA 110(33):E3138–E3147. https://doi.org/10.1073/pnas.1301440110
90. Zhao J, Zhu Y, Song X, Xiao Y, Su G, Liu X et al (2020) 3-O-sulfation of heparan sulfate enhances tau interaction and cellular uptake. Angew Chem Int Ed Engl 59(5):1818–1827. https://doi.org/10.1002/anie.201913029
91. Zhou Y, Xu J, Hou Y, Leverenz JB, Kallianpur A, Mehra R et al (2021) Network medicine links SARS-CoV-2/COVID-19 infection to brain microvascular injury and neuroinflammation in dementia-like cognitive impairment. Alzheimer's Res Ther 13(1):110. https://doi.org/10.1186/s13195-021-00850-3

92. Michiels E, Rousseau F, Schymkowitz J (2021) Mechanisms and therapeutic potential of interactions between human amyloids and viruses. Cell Mol Life Sci 78(6):2485–2501
93. Liu S, Hossinger A, Heumuller SE, Hornberger A, Buravlova O, Konstantoulea K et al (2021) Highly efficient intercellular spreading of protein misfolding mediated by viral ligand-receptor interactions. Nat Commun 12(1):5739. https://doi.org/10.1038/s41467-021-25855-2
94. Hara H, Chida J, Uchiyama K, Pasiana AD, Takahashi E, Kido H et al (2021) Neurotropic influenza A virus infection causes prion protein misfolding into infectious prions in neuroblastoma cells. Sci Rep 11(1):10109. https://doi.org/10.1038/s41598-021-89586-6
95. Asher DM, Belay E, Bigio E, Brandner S, Brubaker SA, Caughey B et al (2020) Risk of transmissibility from neurodegenerative disease-associated proteins: experimental knowns and unknowns. J Neuropathol Exp Neurol 79(11):1141–1146. https://doi.org/10.1093/jnen/nlaa109
96. Ramani A, Muller L, Ostermann PN, Gabriel E, Abida-Islam P, Muller-Schiffmann A et al (2020) SARS-CoV-2 targets neurons of 3D human brain organoids. EMBO J 39(20):e106230. https://doi.org/10.15252/embj.2020106230
97. Semerdzhiev SA, Fakhree MAA, Segers-Nolten I, Blum C, Claessens M (2022) Interactions between SARS-CoV-2 N-protein and alpha-synuclein accelerate amyloid formation. ACS Chem Neurosci 13(1):143–150. https://doi.org/10.1021/acschemneuro.1c00666
98. Li Y, Lu S, Gu J, Xia W, Zhang S, Zhang S et al (2022) SARS-CoV-2 impairs the disassembly of stress granules and promotes ALS-associated amyloid aggregation. Protein Cell 13(8): 602–614. https://doi.org/10.1007/s13238-022-00905-7
99. Charnley M, Islam S, Bindra GK, Engwirda J, Ratcliffe J, Zhou J et al (2022) Neurotoxic amyloidogenic peptides in the proteome of SARS-COV2: potential implications for neurological symptoms in COVID-19. Nat Commun 13(1):3387. https://doi.org/10.1038/s41467-022-30932-1
100. van Ziel AM, Largo-Barrientos P, Wolzak K, Verhage M, Scheper W (2019) Unconventional secretion factor GRASP55 is increased by pharmacological unfolded protein response inducers in neurons. Sci Rep 9(1):1567. https://doi.org/10.1038/s41598-018-38146-6
101. Beckmann ND, Lin WJ, Wang M, Cohain AT, Charney AW, Wang P et al (2020) Multiscale causal networks identify VGF as a key regulator of Alzheimer's disease. Nat Commun 11(1): 3942. https://doi.org/10.1038/s41467-020-17405-z
102. Shen WB, Logue J, Yang P, Baracco L, Elahi M, Reece EA et al (2022) SARS-CoV-2 invades cognitive centers of the brain and induces Alzheimer's-like neuropathology. bioRxiv: The Preprint Server for Biology. https://doi.org/10.1101/2022.01.31.478476
103. Chen J, Chen ZJ (2018) PtdIns4P on dispersed trans-Golgi network mediates NLRP3 inflammasome activation. Nature 564(7734):71–76. https://doi.org/10.1038/s41586-018-0761-3
104. Efstathiou V, Stefanou MI, Demetriou M, Siafakas N, Makris M, Tsivgoulis G et al (2022) Long COVID and neuropsychiatric manifestations (Review). Exp Ther Med 23(5):363. https://doi.org/10.3892/etm.2022.11290
105. Clift AK, Ranger TA, Patone M, Coupland CAC, Hatch R, Thomas K et al (2022) Neuropsychiatric ramifications of severe COVID-19 and other severe acute respiratory infections. JAMA Psychiatry 79(7):690–698. https://doi.org/10.1001/jamapsychiatry.2022.1067
106. Frontera JA, Simon NM (2022) Bridging knowledge gaps in the diagnosis and management of neuropsychiatric sequelae of COVID-19. JAMA Psychiatry. https://doi.org/10.1001/jamapsychiatry.2022.1616
107. Frere JJ, Serafini RA, Pryce KD, Zazhytska M, Oishi K, Golynker I et al (2022) SARS-CoV-2 infection in hamsters and humans results in lasting and unique systemic perturbations post recovery. Sci Transl Med:eabq3059. https://doi.org/10.1126/scitranslmed.abq3059
108. Francistiova L, Klepe A, Curley G, Gulya K, Dinnyes A, Filkor K (2021) Cellular and molecular effects of SARS-CoV-2 linking lung infection to the brain. Front Immunol 12: 730088. https://doi.org/10.3389/fimmu.2021.730088

109. Lee MH, Perl DP, Steiner J, Pasternack N, Li W, Maric D et al (2022) Neurovascular injury with complement activation and inflammation in COVID-19. Brain 145(7):2555–2568. https://doi.org/10.1093/brain/awac151
110. Natoli S, Oliveira V, Calabresi P, Maia LF, Pisani A (2020) Does SARS-Cov-2 invade the brain? Translational lessons from animal models. Eur J Neurol 27(9):1764–1773. https://doi.org/10.1111/ene.14277
111. Bilinska K, von Bartheld CS, Butowt R (2021) Expression of the ACE2 virus entry protein in the nervus terminalis reveals the potential for an alternative route to brain infection in COVID-19. Front Cell Neurosci 15:674123. https://doi.org/10.3389/fncel.2021.674123
112. Butowt R, von Bartheld CS (2022) The route of SARS-CoV-2 to brain infection: have we been barking up the wrong tree? Mol Neurodegener 17(1):20. https://doi.org/10.1186/s13024-022-00529-9
113. Meinhardt J, Radke J, Dittmayer C, Franz J, Thomas C, Mothes R et al (2021) Olfactory transmucosal SARS-CoV-2 invasion as a port of central nervous system entry in individuals with COVID-19. Nat Neurosci 24(2):168–175. https://doi.org/10.1038/s41593-020-00758-5
114. Douaud G, Lee S, Alfaro-Almagro F, Arthofer C, Wang C, McCarthy P et al (2022) SARS-CoV-2 is associated with changes in brain structure in UK Biobank. Nature 604(7907): 697–707. https://doi.org/10.1038/s41586-022-04569-5
115. Rutkai I, Mayer MG, Hellmers LM, Ning B, Huang Z, Monjure CJ et al (2022) Neuropathology and virus in brain of SARS-CoV-2 infected non-human primates. Nat Commun 13(1): 1745. https://doi.org/10.1038/s41467-022-29440-z
116. Rass V, Beer R, Schiefecker AJ, Lindner A, Kofler M, Ianosi BA et al (2022) Neurological outcomes 1 year after COVID-19 diagnosis: a prospective longitudinal cohort study. Eur J Neurol 29(6):1685–1696. https://doi.org/10.1111/ene.15307
117. Apple AC, Oddi A, Peluso MJ, Asken BM, Henrich TJ, Kelly JD et al (2022) Risk factors and abnormal cerebrospinal fluid associate with cognitive symptoms after mild COVID-19. Ann Clin Transl Neurol 9(2):221–226. https://doi.org/10.1002/acn3.51498
118. Al-Aly Z, Xie Y, Bowe B (2021) High-dimensional characterization of post-acute sequelae of COVID-19. Nature 594(7862):259–264. https://doi.org/10.1038/s41586-021-03553-9
119. Xie Y, Xu E, Al-Aly Z (2022) Risks of mental health outcomes in people with covid-19: cohort study. BMJ 376:e068993. https://doi.org/10.1136/bmj-2021-068993
120. Gurin L, Evangelist M, Laverty P, Hanley K, Corcoran J, Herbsman J et al (2022) Early neurorehabilitation and recovery from disorders of consciousness after severe COVID-19. Neurocrit Care 36(2):357–371. https://doi.org/10.1007/s12028-021-01359-1
121. Antonelli M, Capdevila Pujol J, Spector T, Ourselin S, Steves C (2022) Risk of long COVID associated with delta versus omicron variants of SARS-CoV-2. Lancet 399(10343): 2263–2264
122. Fernandez-Castaneda A, Lu P, Geraghty AC, Song E, Lee MH, Wood J et al (2022) Mild respiratory COVID can cause multi-lineage neural cell and myelin dysregulation. Cell 185(14):2452–68 e16. https://doi.org/10.1016/j.cell.2022.06.008
123. Kao J, Frankland PW (2022) COVID fog demystified. Cell 185(14):2391–2393. https://doi.org/10.1016/j.cell.2022.06.020
124. Ho CY, Salimian M, Hegert J, O'Brien J, Choi SG, Ames H et al (2022) Postmortem assessment of olfactory tissue degeneration and microvasculopathy in patients with COVID-19. JAMA Neurol 79(6):544–553. https://doi.org/10.1001/jamaneurol.2022.0154
125. Vahed M, Neya S, Matsuzaki K, Hoshino T (2018) Analysis of physicochemical interaction of Abeta(40) with a GM1 ganglioside-containing lipid membrane. J Phys Chem B 122(14): 3771–3781. https://doi.org/10.1021/acs.jpcb.8b00139
126. Yanagisawa K, Odaka A, Suzuki N, Ihara Y (1995) GM1 ganglioside-bound amyloid beta-protein (A beta): a possible form of preamyloid in Alzheimer's disease. Nat Med 1(10): 1062–1066. https://doi.org/10.1038/nm1095-1062
127. Janneh AH, Kassir MF, Dwyer CJ, Chakraborty P, Pierce JS, Flume PA et al (2021) Alterations of lipid metabolism provide serologic biomarkers for the detection of

asymptomatic versus symptomatic COVID-19 patients. Sci Rep 11(1):14232. https://doi.org/10.1038/s41598-021-93857-7
128. Wen J, Maxwell RR, Wolf AJ, Spira M, Gulinello ME, Cole PD (2018) Methotrexate causes persistent deficits in memory and executive function in a juvenile animal model. Neuropharmacology 139:76–84. https://doi.org/10.1016/j.neuropharm.2018.07.007
129. Kozminski P, Halik PK, Chesori R, Gniazdowska E (2020) Overview of dual-acting drug methotrexate in different neurological diseases, autoimmune pathologies and cancers. Int J Mol Sci 21(10). https://doi.org/10.3390/ijms21103483
130. Alper HE, Tuly RA, Seil K, Brite J (2020) Post-9/11 mental health comorbidity predicts self-reported confusion or memory loss in world trade center health registry enrollees. Int J Environ Res Public Health 17(19). https://doi.org/10.3390/ijerph17197330
131. Kuan PF, Clouston S, Yang X, Che C, Gandy S, Kotov R et al (2021) Single-cell transcriptomics analysis of mild cognitive impairment in World Trade Center disaster responders. Alzheimers Dement (Amst) 13(1):e12154. https://doi.org/10.1002/dad2.12154
132. Clouston SA, Kotov R, Pietrzak RH, Luft BJ, Gonzalez A, Richards M et al (2016) Cognitive impairment among World Trade Center responders: long-term implications of re-experiencing the 9/11 terrorist attacks. Alzheimers Dement (Amst) 4:67–75. https://doi.org/10.1016/j.dadm.2016.08.001
133. Clouston SAP, Hall CB, Kritikos M, Bennett DA, DeKosky S, Edwards J et al (2022) Cognitive impairment and World Trade Centre-related exposures. Nat Rev Neurol 18(2):103–116. https://doi.org/10.1038/s41582-021-00576-8
134. Ackley SF, Zimmerman SC, Brenowitz WD, Tchetgen Tchetgen EJ, Gold AL, Manly JJ et al (2021) Effect of reductions in amyloid levels on cognitive change in randomized trials: instrumental variable meta-analysis. BMJ 372:n156. https://doi.org/10.1136/bmj.n156
135. Gandy S, Simon AJ, Steele JW, Lublin AL, Lah JJ, Walker LC et al (2010) Days to criterion as an indicator of toxicity associated with human Alzheimer amyloid-beta oligomers. Ann Neurol 68(2):220–230. https://doi.org/10.1002/ana.22052
136. Zhang X, Wang Y (2020) Non-redundant roles of GRASP55 and GRASP65 in the Golgi apparatus and beyond. Trends Biochem Sci 45(12):1065–1079

MIX
Papier aus verantwortungsvollen Quellen
Paper from responsible sources
FSC® C105338

If you have any concerns about our products,
you can contact us on
ProductSafety@springernature.com

In case Publisher is established outside the EU,
the EU authorized representative is:
Springer Nature Customer Service Center GmbH
Europaplatz 3, 69115 Heidelberg, Germany

Printed by Libri Plureos GmbH
in Hamburg, Germany